R. Williams 1992

OXFORD STUDIES

IN

NUCLEAR PHYSICS

GENERAL EDITOR

P. E. HODGSON

THEORY OF THE NUCLEAR SHELL MODEL

R. D. LAWSON

Argonne National Laboratory

CLARENDON PRESS · OXFORD
1980

Oxford University Press, Walton Street, Oxford OX2 6DP

OXFORD LONDON GLASGOW
NEW YORK TORONTO MELBOURNE WELLINGTON
KUALA LUMPUR SINGAPORE JAKARTA HONG KONG TOKYO
DELHI BOMBAY CALCUTTA MADRAS KARACHI
NAIROBI DAR ES SALAAM CAPE TOWN

Published in the United States by Oxford University Press, New York

British Library Cataloguing in Publication Data

Lawson, R D
Theory of the nuclear shell model. -
(Oxford studies in nuclear physics).
1. Nuclear physics
I. Title II. Series
539.7′4 QC776 79-40480

ISBN 0-19-851516-2

Printed in the United States

PREFACE

This book developed out of lectures I presented at several different institutions over a number of years. I have not tried to be either elegant or complete in my presentation, but instead have tried to write a "how to do it" book for calculating those quantities that one frequently encounters in a Shell Model description of the atomic nucleus. At each stage when a new technique is introduced it is immediately followed by an example of its use and the calculated results are compared to experimental data.

Since the manuscript was completed, several new data compilations have appeared:

(1) The 1971 Wapstra-Gove binding energy and separation energy tables have been updated and a more recent reference is

A. H. Wapstra and K. Bos, *Atomic Data and Nuclear Data Tables,* **19,** 177, 1977.

(2) More recent compilations of the properties of nuclei with $A \leq 20$ will be found in

F. Ajzenberg-Selove, $A = 18$–20 Nuclei, *Nucl. Phys.* **A300,** 1, 1978.
$A = 5$–10 Nuclei, *Nucl. Phys.* **A320,** 1, 1979.

(3) The 1973 Endt and van der Leun tabulation of the properties of $A = 21$–44 nuclei has been replaced by

P. M. Endt and C. van der Leun, *Nucl. Phys.* **A310,** 1, 1978.
P. M. Endt, Spectroscopic Factors for Single Nucleon Transfer, *Atomic Data and Nuclear Data Tables,* **19,** 23, 1977.

(4) A more recent tabulation of the gamma-ray lifetimes in light nuclei is given by

P. M. Endt, *Atomic Data and Nuclear Data Tables,* **23,** 3, 1979.

(5) The beta decay lifetimes have been updated in a publication by

S. Raman, C. A. Hauser, T. A. Walkiewicz and I. S. Towner, *Atomic Data and Nuclear Data Tables,* **21,** 567, 1978.

If one wishes to know where the most recent compilation of data for a given A is to be found, one has only to consult the most recent issue of *Nuclear Data Sheets* (published by Academic Press) and on page (ii) and (iii) of each issue a list will be found.

During the past twenty years I have been affiliated with Argonne National Laboratory. During that time I have worked closely with and learned from a large number of physicists. In particular I would like to acknowledge my colleagues Stan Cohen, Dieter Kurath, Malcolm Macfarlane and the late John Soper for their contributions to my understanding of nuclear physics.

Several people have read various portions of the manuscript at various stages of its completion. I particularly would like to thank Soumya Chakravarti, David Gloeckner, Henk Hasper, Thijs Koeling, Arnold Müller-Arnke, Frank Serduke, William Teeters, Rolf Siemssen, and Sigmund van der Werf for their comments. The assistance of Ms. Dorothy Burdzinski, Nancy Williams, and Joke Wöhler in typing parts of the manuscript is gratefully acknowledged.

Finally I would like to thank my wife, Sally, for the encouragement she gave and the patience she showed during the final stages of the writing of this book.

CONTENTS

1. SINGLE CLOSED-SHELL NUCLEI 1

1. Coupling rules and ground rules 3

1.1. Allowable angular-momentum states for two particles 8

1.2. Allowable states of the configuration j^n 16

1.3. Many particles in different orbits 21

1.4. The shell-model Hamiltonian 25

2. Two particles outside a closed shell 30

2.1. $d_{\frac{5}{2}}s_{\frac{1}{2}}$ model of ${}^{18}_{8}O_{10}$ 31

2.2. Core-excited states 35

3. The configuration j^n 38

3.1. Parentage coefficients 38

3.2. The spectra of the proton $f_{\frac{7}{2}}$ nuclei 42

4. Several particles in different levels 49

4.1. Resumé of diagrammatic rules for recoupling 49

4.2. Counting and phase factors 52

4.3. Energy matrix elements 54

4.4. Spectrum of ${}^{19}_{8}O_{11}$ 63

5. Seniority 68

5.1. The seniority quantum number 68

5.2. How good is the seniority quantum number? 73

5.3. Seniority eigenfunctions 76

5.4. The Talmi binding-energy formula 79

5.5. Many single-particle levels 83

2. NEUTRON-PROTON PROBLEMS 87

1. The residual nucleon-nucleon interaction 87

1.1. Charge symmetry and charge independence 87

1.2. Isospin 91

1.3. How good is the isospin quantum number? 98

1.4. What about seniority? 105

2. Isospin/non-isospin methods of calculation 107
2.1. Some diagrammatic rules for recoupling including isospin 107
2.2. Isospin calculation for ${}^{35}_{17}Cl_{18}$ 109
2.3. Neutron-proton calculation for ${}^{35}_{17}Cl_{18}$ 112
2.4. ${}^{39}_{18}Ar_{21}$ spectrum using the neutron-proton formalism 115
2.5. ${}^{39}_{18}Ar_{21}$ using isospin 118

3. Weak-coupling model 122
3.1. Centre-of-gravity theorem 124
3.2. Negative-parity states in ${}^{19}_{9}F_{10}$ 126
3.3. Inelastic scattering; the positive parity septuplet in ${}^{209}_{83}Bi_{126}$ 128
3.4. More detailed calculations for ${}^{63}_{29}Cu_{34}$ 131
3.5. Failure of weak coupling through violation of the Pauli principle 135

4. Weak coupling including isospin 136
4.1. The Bansal–French formula 137
4.2. $d_{\frac{3}{2}}$ hole states near $A=40$ 142
5. Single-nucleon transfer 148
5.1. Spectroscopic factors for the configuration j^n 154
5.2. Spectroscopic factors when more than one level is involved 159
5.3. Sum rules 162

3. PARTICLES AND HOLES 169
1. Particle-particle and particle-hole spectra 169
1.1. The Pandya transformation 169
1.2. The ${}^{38}_{17}Cl_{21}$—${}^{40}_{19}K_{21}$ spectra 171
1.3. The ${}^{42}_{21}Sc_{21}$—${}^{48}_{21}Sc_{27}$ spectra 172
1.4. The ${}^{92}_{41}Nb_{51}$—${}^{96}_{41}Nb_{55}$ spectra 175
2. The particle-hole conjugation operator 176
2.1. Explicit form of the particle-hole conjugation operator 178
2.2. Hole-hole interaction energies 184
2.3. Transformation of single-particle operators 188
2.4. General particle-hole theorem 193

3. Particle-hole transformation including isospin 197
3.1. The yrast negative-parity states in $^{40}_{20}Ca_{20}$ 199

4. HARMONIC OSCILLATOR WAVE FUNCTIONS 203
1. Energy matrix elements calculated with oscillator functions 204
1.1. Central spin-dependent interaction 208
1.2. Tensor force 209
1.3. Two-body spin-orbit potential 211
2. The $0p$ shell 212
2.1. The $T=1$ interaction in the $0p$ shell 214
2.2. The $T=0$ interaction in the $0p$ shell 219
3. Spurious centre-of-mass motion 223
3.1. Elimination of spurious centre-of-mass states 228
3.2. Negative-parity states in $^{4}_{2}He_{2}$ 231
3.3. Amount of spurious state in a given wave function 238
4. Two-nucleon transfer 245
4.1. Wave functions for the 0^+ states in $^{18}_{8}O_{10}$ 257
4.2. (t, p) reaction on the calcium isotopes 262
4.3. Comparison of the (p, t) and $(p, {}^3He)$ reactions 264

5. ELECTROMAGNETIC PROPERTIES 266
1. Single-particle estimate for gamma decay 266
2. Effective operators 277
2.1. Unlike-particle mixing 277
2.2. Like-particle mixing 279
3. Matrix elements and phases 281
3.1. Gamma transitions between two-particle states; application to $^{18}_{8}O_{10}$ 283
4. $E1$ transitions 287
4.1. Isospin effects 288
4.2. Deformation effects 290
5. $M2$ transitions 293
6. Magnetic dipole properties 295
6.1. Magnetic moments 297
6.1.1. Single-particle estimate and the configuration j^n 297

6.1.2. Weak-coupling multiplet 300
6.1.3. The configuration $(\pi j)^n(\nu j)^{\pm n}$ 301
6.2. *M*1 transitions 303
6.2.1. The identical-nucleon configuration 303
6.2.2. Transitions within a weak-coupling multiplet 304
6.2.3. The signature-selection rule 306
6.2.4. Isospin effects 308
6.2.5. Analogue to anti-analogue decays 309
6.2.6. *l*-forbidden transitions 315
7. Electric quadrupole properties 316
7.1. Effective charges 317
7.2. Quadrupole moments 321
7.3. $\Delta T \neq 0$ transitions 323
7.4. Properties of the half-filled shell 324
7.5. Particle-hole multiplets 325
7.6. Signature rules 326
7.7. Weak-coupling multiplets 327
8. Collective states 328

6. QUASI-PARTICLES 337
1. BCS approximation 339
1.1. Variational calculation for the ground-state energy 341
1.2. The degenerate model; comparison of BCS and exact solutions 345
1.3. Lead isotopes with a pairing force 348
2. Quasi-particle calculations 351
2.1. Energy calculations with quasi-particles 353
2.2. *M*4 transitions in the lead nuclei 361
2.3. Pick-up and stripping on the tin isotopes 364
2.4. Beta decay of the indium isotopes 371
3. Some limitations of quasi-particle calculations 381

7. POOR MAN'S HARTREE-FOCK 387
1. Nilsson eigenfunctions 390
2. Projected wave functions 398

2.1. Wave functions for ${}^{7}_{3}Li_4$ 399
2.2. Wave function for the ${}^{6}_{3}Li_3$ ground state; band mixing 406
2.3. Intruder states 411
2.4. Beta decay involving the odd-A calcium isotopes 413

APPENDIX 1. CLEBSCH-GORDAN COEFFICIENTS 420
1. Angular-momentum operators 420
2. Single-particle eigenfunctions 422
3. Clebsch–Gordan coefficients 424
4. Useful formulae involving Clebsch–Gordan coefficients 426
5. Analytic expressions for Clebsch–Gordan coefficients 427

APPENDIX 2. REDUCED MATRIX ELEMENTS 430
1. The Wigner–Eckart theorem 430
2. $\langle Y_{l'}\| Y_\lambda \|Y_l\rangle$ 432
3. $\langle R_{l'} Y_{l'm'}| r_\mu |R_l Y_{lm}\rangle$ 432
4. $\langle R_{l'} Y_{l'm'}| \nabla_\mu |R_l Y_{lm}\rangle$ 433
5. $\langle \phi_{j'}\| \sigma \|\phi_j\rangle$ 433
6. $\langle \phi_{j'}\| \ell \|\phi_j\rangle$ 434
7. $\langle \phi_{j'}\| Y_{l''} \|\phi_j\rangle$ 434
8. $\langle \phi_{j'}\| [\sigma \times Y_{l''}]_L \|\phi_j\rangle$ 435
9. The delta-function potential 436

APPENDIX 3. QUASI-SPIN AND NUMBER DEPENDENCE OF MATRIX ELEMENTS 440
1. Relationship between quasi-spin and seniority 440
2. Number dependence of the fractional-parentage coefficients 442
3. Quasi-spin tensors of rank 0 and 1 444
3.1. Number dependence of the two-particle parentage coefficients 445
3.2. Number dependence of gamma-decay matrix elements 448
4. Two-body interaction in terms of quasi-spin 450
4.1. Number dependence of matrix elements off-diagonal in seniority 451
4.2. Seniority conservation with the delta-function interaction 452

4.3. Number dependence of matrix elements diagonal in seniority 455

APPENDIX 4. RACAH AND 9j COEFFICIENTS 459
1. Racah coefficients 459
2. 9j coefficients 469
3. Some formulae involving Racah and 9j coefficients 472
4. Analytic expressions for Racah coefficients 474

APPENDIX 5. FRACTIONAL PARENTAGE COEFFICIENTS 477
1. Single-particle parentage coefficients 477
2. Double-parentage coefficients 480
3. Particle-hole relationship 482
4. Isospin-parentage coefficients 484
5. Some useful formulae 487
6. Tables of one- and two- nucleon parentage coefficients 488

APPENDIX 6. RELATIVE CENTRE-OF-MASS TRANSFORMATION COEFFICIENTS 492
1. Transformation brackets when $n_1 = n_2 = 0$ 494
2. Transformation brackets for the general case 496
3. Symmetry properties of the transformation brackets 497

APPENDIX 7. THE ROTATION MATRIX 500
1. Coupling rules for D-functions 503
2. Explicit form for the D-functions 504
3. Integration of D-functions 505
4. The D-functions as angular momentum eigenfunctions 506

REFERENCES 509

AUTHOR INDEX 519

SUBJECT INDEX 524

1

SINGLE CLOSED-SHELL NUCLEI

The aim of this book is to provide a treatise on nuclear phenomenology intermediate between that of Mayer and Jensen (1955) and that of de Shalit and Talmi (1963). The former gives the experimental facts on which the shell model is based together with simple examples of how the ideas can be applied, and the latter provides an excellent compendium of formulae needed to carry out shell-model calculations but does not illustrate the concepts with specific examples. In this work we shall assume a knowledge of the experimental facts on which the shell model is based. Although we shall not try to be as complete in our derivations as de Shalit and Talmi, we shall derive a variety of equations using the diagrammatic methods of Macfarlane and French (1960), and at each stage where a new idea is introduced we shall use it to perform a simple calculation and compare the result with experiment. Once the concept has been illustrated, it is always possible to complicate the procedure by extending the configuration space. We shall not attempt the extension which is straight forward but tedious. Because we shall try to keep our calculations simple it will often turn out that agreement between theory and experiment is somewhat 'grey'. In general, the agreement can be improved by extending the configuration space.

In phenomenological shell-model calculations there are two approaches that are used.

(a) One assumes that the valence nucleons occupy a selected set of single-particle levels (called the model space). One diagonalizes the shell-model Hamiltonian on the assumption that the residual two-body interaction between the valence nucleons has a simple radial form, i.e. a delta function, a Yukawa, or a Gaussian. The matrix elements of the potential are generally evaluated using harmonic-oscillator eigenfunctions for the single-particle radial wave functions, and we shall discuss this evaluation in some detail in Appendix 2 and in Chapter 4. One can easily include central, tensor, and two-body spin-orbit interactions and, if one adjusts the various potential strengths and ranges to give a best fit to the experimental data (number of data points at least twice the number of parameters), one can usually fit spectra to within about 500 keV if the model space is adequate.

(b) A second approach is the one championed by Talmi and co-workers (Meshkov and Ufford 1956, Talmi and Unna 1960, Talmi 1962). One again assumes a simple model space, but now the residual two-body interaction is parametrized in terms of its matrix elements. This is a much more flexible description of the potential and, when the number of experimental data are of the order of twice the number of parameters, the spectra can be fitted to within about 150 keV if the model space is adequate.

In these two approaches no attempt is made to correlate the required residual interaction with the free nucleon–nucleon potential. On the other hand, in recent years calculations have been carried out with the purpose of deducing the appropriate matrix elements from the fundamental nucleon-nucleon interaction (Kuo and Brown 1966, 1968, Elliott *et al.* 1968, Barrett and Kirson 1973, Kuo 1974, Barrett 1975, Ellis and Osnes 1977). However, since the free-nucleon interaction has a strong repulsive core (Hamada and Johnston 1962, Reid 1968), it is difficult to compute these matrix elements and effective operators, and at the present time one is not sure of the convergence of the calculation. For this reason we shall have little to say about these so called 'first-principles' calculations.

It often happens that a very simple configuration assignment can lead to an understanding of the observed spectra of several nuclei, so that as far as binding energies and excitation energies are concerned the nucleons act *as if* they occupy the single-particle state j. Moreover, the 'pseudonium' calculations of Cohen *et al.* (1966), Lawson and Soper (1966), and Soper (1970) show that other nuclear properties may also be interpreted by assuming the valence nucleons occupy the single-particle state j even when one knows this is not the case. Consequently, one has a situation similar to that encountered with the nucleon and its associated meson cloud—a nucleon in the nucleus that has many properties in common with a bare nucleon in the single-particle state j may, in reality, be a 'clothed nucleon' or 'quasi-particle'. Clearly it is important to explore the make-up of these quasi-particles, since it is only by probing their structure that one will be able to understand nuclear properties. Typical experiments that are likely to shed light on this question are as follows:

(a) With a given configuration assignment some transitions will be predicted to be much stronger than others. It is important to check whether or not this is true experimentally.

(b) A certain configuration assignment may lead to a selection rule on gamma decay, beta decay etc. One should look carefully to see if this prediction is satisfied.
(c) Often within a given configuration various quantities will be 'geometrically' related; i.e. the ratio of two transition rates may be given simply by a Racah or Clebsch-Gordan coefficient. For such cases the structure of the transition-rate operator cancels out, and only the fact that it is a spherical-tensor operator of a certain rank is important.
(d) The first states of each spin and parity in a nucleus comprise a group of levels that are known as the yrast levels. It often happens that the properties of these yrast states can be easily understood, and it is only when one deals with the non-yrast levels that one learns about the detailed structure of the nucleons that act *as if* they had angular momentum j. Thus, even though it is more difficult to determine the properties of the non-yrast levels, a knowledge of their behaviour will often shed considerable light on the structure of the 'clothed nucleon'.

In general, the easiest nuclear properties to understand are the energies of states relative to an assumed closed shell. We shall therefore begin with a study of these properties for nuclei in which either the protons or the neutrons form an inert closed shell. In this case the properties of the low-lying states will be determined by one type of nucleon alone, and the nuclei are called single-closed-shell nuclei (s.c.s, nuclei). Before showing how the spectra of these nuclei can be calculated, we shall discuss some rules concerning the coupling of angular momenta and also set down the phase conventions and other ground rules that will be used in this book.

1. Coupling rules and ground rules

In its simplest form the shell model treats the nucleons in a nucleus as moving independently with the interaction between nucleon number n and all others approximated by an average potential $V(\underline{r}_n, \underline{\sigma}_n, \underline{\ell}_n)$ where $\underline{r}_n$ is the coordinate of the nth particle, $\underline{\ell}_n$ is the orbital angular-momentum operator

$$\underline{\ell}_n = \frac{1}{\hbar}(\underline{r}_n \times \underline{p}_n) \tag{1.1}$$

and σ_n is the Pauli spin operator with components

$$\sigma_x = \begin{pmatrix} 0 & 1 \\ 1 & 0 \end{pmatrix} \quad \sigma_y = \begin{pmatrix} 0 & -i \\ i & 0 \end{pmatrix} \quad \sigma_z = \begin{pmatrix} 1 & 0 \\ 0 & -1 \end{pmatrix}. \tag{1.2}$$

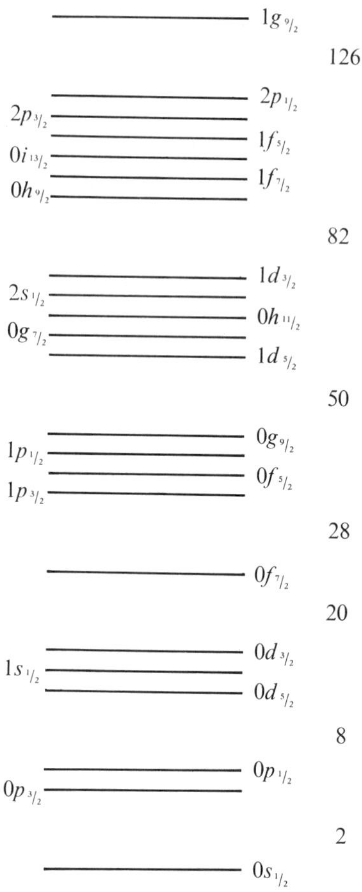

Fig. 1.1. Experimental single-particle level sequence. At the magic numbers 2, 8, 20, 28, 50, 82, and 126 an energy break occurs.

This average potential should be chosen to reproduce the shell breaks at N and $Z = 2$, 8, 20, 28, 50, 82, and 126 (the so-called magic numbers) and the experimental level sequence shown in Fig. 1.1. Since there is a large energy difference between states with the same orbital angular momentum but different total angular momentum, a strong one-body spin-orbit potential is required. Thus the Hamiltonian describing the independent-particle motion of the nth nucleon is taken to have the form

$$(H_0)_n = \frac{p_n^2}{2m} + V_1(r_n) + V_2(r_n)\underset{\sim}{\sigma}_n \cdot \underset{\sim}{\ell}_n \tag{1.3}$$

where $V_1(r_n)$ and $V_2(r_n)$ are functions of the scalar coordinate r_n and

$$p_n^2 = (p_x^2)_n + (p_y^2)_n + (p_z^2)_n.$$

Because of the strong spin-orbit splitting (it amounts to 5·08 MeV for the d states in $^{17}_{8}O_9$, Ajzenberg-Selove 1977) j–j coupled wave functions are appropriate for the description of the single-particle states. These eigenfunctions will be denoted by ϕ_{jm}, where j and m are the total angular momentum and z component, respectively. ϕ_{jm} corresponds to coupling the orbital angular momentum l and the spin angular momentum $s = \frac{1}{2}$ to give j and the *order of coupling used throughout this book will be* $l + s = j$: in other words

$$\phi_{jm} = R_j(r) \sum_{k\mu} (l\tfrac{1}{2}k\mu \mid jm) Y_{lk}(\theta, \phi) \chi_\mu \tag{1.4}$$

where $Y_{lk}(\theta, \phi)$ is the spherical harmonic

$$Y_{lk}(\theta, \phi) = \frac{(-1)^{l+k}}{(2l)!!} \left\{ \frac{(2l+1)(l-k)!}{4\pi(l+k)!} \right\}^{\frac{1}{2}} (\sin\theta)^k \, e^{ik\phi} \frac{d^{l+k}(\sin\theta)^{2l}}{\{d(\cos\theta)\}^{l+k}} \tag{1.5}$$

χ_μ are the spin eigenfunctions

$$\chi_{\frac{1}{2}} = \begin{pmatrix} 1 \\ 0 \end{pmatrix}$$
$$\chi_{-\frac{1}{2}} = \begin{pmatrix} 0 \\ 1 \end{pmatrix} \tag{1.6}$$

and $R_j(r)$ is the radial wave function. Note that the vector operators for orbital and total angular momentum will be denoted by $\underline{\ell}$ and $\underline{j}$ respectively and their eigenvalues by l and j.

The bracket $(l\frac{1}{2}k\mu \mid jm)$ is the Clebsch–Gordan coefficient (discussed in Appendix 1) with the phase convention of Condon and Shortley (1951). Obviously ϕ_{jm} should also have the label l associated with it. However, within a given shell the j value alone is sufficient to tell us the l value and consequently unless confusion arises this index will be suppressed.

The nucleus measures a few fermis (1 fm $= 10^{-13}$ cm) in radius and consequently $V_1(r)$ should be attractive and confine the radial extent of the nucleon motion to conform with this. A potential that gives the desired level sequence and is consistent with the nuclear size is the Woods–Saxon well (Woods and Saxon 1954, Ross *et al.* 1956)

$$V_1(r) = \frac{-V_0}{[1 + \exp\{(r-R)/a\}]} \tag{1.7}$$

where V_0 is the strength of the interaction, a measures the diffuseness of the well ($a \simeq 0{\cdot}65$ fm), $R = r_0 A^{\frac{1}{3}}$ is the nuclear radius which increases as the cube root of the number of nucleons A in the nucleus, and $r_0 \simeq 1{\cdot}2$ fm.

Since the total single-particle angular-momentum operator is

$$\underset{\sim}{j} = \underset{\sim}{\ell} + \tfrac{1}{2}\underset{\sim}{\sigma} \tag{1.8}$$

it is clear that

$$\begin{aligned}\underset{\sim}{\sigma}\,.\,\underset{\sim}{\ell}\phi_{jm} &= (j^2 - \ell^2 - \tfrac{1}{4}\sigma^2)\phi_{jm}\\ &= \{j(j+1) - l(l+1) - \tfrac{3}{4}\}\phi_{jm}\\ &= l\phi_{jm} \quad \text{if} \quad j = l + \tfrac{1}{2}\\ &= -(l+1)\phi_{jm} \quad \text{if} \quad j = l - \tfrac{1}{2}.\end{aligned}$$

Because the state with $j = l + \frac{1}{2}$ lies lower in energy, $V_2(r)$ is an attractive potential and frequently is taken to have the Thomas form (Thomas 1926)

$$V_2(r) = \frac{-\lambda\hbar^2}{4m^2c^2}\frac{1}{r}\frac{\mathrm{d}V_1(r)}{\mathrm{d}r}$$

where λ is a dimensionless constant and $V_1(r)$ is the central potential of equation (1.3). In order to fit the experimental level sequence $\lambda \simeq 40$.

We shall assume that the parameters characterizing the potential in equation (1.3) have been chosen so as to reproduce the observed energies of a single particle outside a closed shell. For example, in ${}^{17}_{8}O_9$ the energy of the $d_{\frac{5}{2}}$ state (the ground state of the nucleus) relative to the ${}^{16}_{8}O_8$ core is

$$\begin{aligned}\varepsilon_d &= \mathrm{BE}({}^{17}_{8}O_9) - \mathrm{BE}({}^{16}_{8}O_8)\\ &= -4{\cdot}143\ \mathrm{MeV}\end{aligned} \tag{1.9}$$

where $\mathrm{BE}(X)$ *is the negative of the total binding energy of the nucleus* (Wapstra and Gove 1971). Since the $s_{\frac{1}{2}}$ state lies at an excitation energy of 871 keV in this nucleus (Ajzenberg-Selove 1977)

$$\begin{aligned}\varepsilon_s &= -4{\cdot}143 + 0{\cdot}871\\ &= -3{\cdot}272\ \mathrm{MeV}.\end{aligned} \tag{1.10}$$

Thus in the next section, where we discuss the properties of the oxygen isotopes, we shall assume the eigenvalues of H_0 reproduce these values.

The eigenfunctions of the Woods–Saxon potential, $V_1(r)$ of equation (1.7), do not lend themselves to analytic calculations. Consequently, unless otherwise stated, we shall approximate these wave functions by harmonic-oscillator eigenfunctions which satisfy the Schrödinger equation

$$\left(\frac{p^2}{2m} + \tfrac{1}{2}m\omega^2 r^2\right)\phi_{jm} = (2n + l + \tfrac{3}{2})\hbar\omega\phi_{jm}$$

where ϕ_{jm} is given by equation (1.4) with

$$R_j(r) = R_{nl}(r)$$
$$= \left[\frac{2^{l-n+2}(2l+2n+1)!!\alpha^{2l+3}}{\sqrt{\pi} n!\{(2l+1)!!\}^2}\right]^{\frac{1}{2}}$$
$$\times (\exp - \tfrac{1}{2}\alpha^2 r^2) r^l \sum_{k=0}^{n} \frac{(-1)^k 2^k n!(2l+1)!!(\alpha^2 r^2)^k}{k!(n-k)!(2l+2k+1)!!}$$

with

$$\alpha^2 = m\omega/\hbar.$$

(See Chapter 4 and Appendix 6 for a discussion of the properties of the oscillator functions.) In most calculations $\hbar\omega$ will be taken to have the value

$$\hbar\omega = 41/A^{\frac{1}{3}} \tag{1.11}$$

and in this approximation $R_j(r)$ depends on l and not on j. Furthermore, *the radial wave function will be normalized so that it is positive for $r = 0$. The convention on the number of radial nodes n will be that $n = 0$ the first time a state with given l occurs.*

We shall frequently have to evaluate matrix elements of $r^\lambda Y_{\lambda\mu}(\theta, \phi)$ and when oscillator wave functions are used one can obtain an analytic expression for the radial integral (Nilsson 1955). When $(l + l' + \lambda)$ is even

$$\int R_{n'l'}(\alpha r) R_{nl}(\alpha r) r^{\lambda+2}\, dr = (-1)^{n-n'} \left\{\frac{n!n'!2^{n+n'-\lambda}}{(2n+2l+1)!!(2n'+2l'+1)!!}\right\}^{\frac{1}{2}}$$
$$\times \frac{1}{\alpha^\lambda}\left\{\frac{1+(-1)^{l+l'+\lambda}}{2}\right\}\{\tfrac{1}{2}(l'-l+\lambda)\}!\{\tfrac{1}{2}(l-l'+\lambda)\}!$$
$$\times \sum_q \frac{(l+l'+\lambda+2q+1)!!}{2^q q!(n-q)!(n'-q)!\{q+\frac{1}{2}(l-l'+\lambda)-n'\}!\{q+\frac{1}{2}(l'-l+\lambda)-n\}!} \tag{1.11a}$$

and the sum on q is over all integers for which the arguments of the factorials are non-negative numbers.

In any shell-model calculation one tries to calculate the energy relative to a closed shell rather than the total energy of the system. For a single nucleon outside a doubly magic core this energy, as we have discussed, is taken to be the eigenvalue of H_0. However, when there is more than one nucleon outside the core one must replace the simple shell model in which

$$H' = \sum_n (H_0)_n$$

by the interacting shell model

$$H = \sum_n (H_0)_n + \sum_{i<j} V_{ij} \tag{1.12}$$

where V_{ij} is the residual two-body interaction which exists in addition to the average shell-model potential. That such an added potential is necessary may be seen from the following experimental facts:

(a) According to the level sequence shown in Fig. 1.1 the lowest state of ${}^{18}_{8}O_{10}$ should correspond to two neutrons in the $0d_{\frac{5}{2}}$ single-particle state. If H' alone is appropriate for the description of this state the binding energy of ${}^{18}_{8}O_{10}$ relative to ${}^{16}_{8}O_8$ should be precisely $2\varepsilon_d$, where ε_d is given by equation (1.9). Experimentally (Wapstra and Gove 1971) the energy of the ground state relative to ${}^{16}_{8}O_8$ is

$$\tilde{E}_0 = \mathrm{BE}({}^{18}_{8}O_{10}) - \mathrm{BE}({}^{16}_{8}O_8)$$
$$= -12{\cdot}189 \text{ MeV}.$$

In other words, one needs an additional attraction of 3·903 MeV in excess of the single-particle energies.

(b) Again using ${}^{18}_{8}O_{10}$ as an example, two neutrons in the $0d_{\frac{5}{2}}$ orbit can couple their spins to $I = 0$, 2 and 4 (see section 1.1 of this Chapter where this will be proved). If H' alone is appropriate, all three of these states would be degenerate in energy. Experimentally (Ajzenberg-Selove 1972) this is not the case—the yrast 2^+ state in ${}^{18}_{8}O_{10}$ lies 1·98 MeV above the 0^+ ground state and the yrast 4^+ level is at 3·55 MeV.

Thus in order to describe the observed energies of low-lying nuclear states one must modify H' and include some residual two-body interaction V_{ij}. In section 1.4 of this Chapter we shall discuss the mathematical formulation of these ideas. However, before going into this we shall deduce the possible angular-momentum states that can arise when there is more than one nucleon outside an inert core.

1.1. Allowable angular-momentum states for two particles

As discussed in Appendix 1 the single-particle angular-momentum operator $\underset{\sim}{j}$ given in equation (1.8), obeys the commutation relationship

$$\underset{\sim}{j} \times \underset{\sim}{j} = i\underset{\sim}{j}. \tag{1.13}$$

ϕ_{jm} (equation (1.4)) is an eigenfunction of this operator and has the properties

$$\mathrm{j}^2\phi_{jm} = j(j+1)\phi_{jm}$$
$$\mathrm{j}_z\phi_{jm} = m\phi_{jm} \tag{1.14}$$

and

$$j_{\pm}\phi_{jm} = \surd\{(j \mp m)(j \pm m + 1\}\phi_{jm\pm 1} \tag{1.15}$$

where

$$\mathrm{j}_{\pm} = \mathrm{j}_x \pm i\mathrm{j}_y. \tag{1.16}$$

One may use these results together with the fact that the total angular-momentum operator $\underline{\mathrm{J}}$ for the nucleus is

$$\underline{\mathrm{J}} = \sum_{i=1}^{A} \underline{\mathrm{j}}_i \tag{1.17}$$

to deduce the allowable angular-momentum states of the many-body system.

To start with it is clear from the definition of $\underline{\mathrm{J}}$ that the total z component of angular momentum of the many-particle wave function

$$\phi_M = \phi_{j_1m_1}(1)\phi_{j_2m_2}(2)\ldots\phi_{j_nm_n}(n) \qquad \left(M = \sum_i m_i\right)$$

is

$$\begin{aligned}\mathrm{J}_z\phi_M &= \sum_i (\mathrm{j}_z)_i\phi_M \\ &= \sum_i m_i\phi_M \\ &= M\phi_M.\end{aligned}$$

From this result the following theorem may easily be proved.

Theorem 1. A many-particle wave function with z-component of angular momentum M which has the property that

$$\mathrm{J}_+\phi_M = 0$$

is an eigenfunction of the operator J^2 and describes a state with total angular momentum M.

Proof. From the definition of $\underline{\mathrm{J}}$, J_+, J_- and the commutation relationship (equation (1.13)) it follows that

$$\begin{aligned}\mathrm{J}^2\phi_M &= \tfrac{1}{2}(\mathrm{J}_+\mathrm{J}_- + \mathrm{J}_-\mathrm{J}_+)\phi_M + \mathrm{J}_z^2\phi_M \\ &= \tfrac{1}{2}\mathrm{J}_+\mathrm{J}_-\phi_M + M^2\phi_M.\end{aligned}$$

However,

$$\mathrm{J}_+\mathrm{J}_- = \mathrm{J}_-\mathrm{J}_+ + 2\mathrm{J}_z$$

so that

$$\mathrm{J}^2\phi_M = M(M+1)\phi_M.$$

This theorem may be used in conjunction with the Pauli principle to show that the angular momentum carried by a fully occupied single-particle orbit is zero. To see this we note that the Pauli principle requires that the wave function describing a neutron state or a proton state must be antisymmetric to the interchange of any two particles. Thus, for example, an allowable state of the two-neutron system with particles in the state (j_1m_1) and (j_2m_2) is

$$\phi_M = 2^{-\frac{1}{2}}\{\phi_{j_1m_1}(1)\phi_{j_2m_2}(2) - \phi_{j_1m_1}(2)\phi_{j_2m_2}(1)\}.$$

In general this condition can be expressed in terms of a Slater determinant (Slater 1929)

$$(n!)^{-\frac{1}{2}}\begin{vmatrix} \phi_{j_1m_1}(1)\phi_{j_2m_2}(1) & \dots & \phi_{j_nm_n}(1) \\ \phi_{j_1m_1}(2)\phi_{j_2m_2}(2) & \dots & \phi_{j_nm_n}(2) \\ \cdot \quad\quad \cdot & & \cdot \\ \cdot \quad\quad \cdot & & \cdot \\ \cdot \quad\quad \cdot & & \cdot \\ \phi_{j_1m_1}(n)\phi_{j_2m_2}(n) & \dots & \phi_{j_nm_n}(n) \end{vmatrix} = |\phi_{j_1m_1}\phi_{j_2m_2}\dots\phi_{j_nm_n}| \tag{1.18}$$

which, from the properties of determinants, is antisymmetric to the interchange of any two particles and vanishes when two particles occupy the same quantum state. Because of this condition the only allowable M state for the configuration $(j)^{2j+1}$ (i.e. the state in which $(2j+1)$ nucleons occupy the single-particle orbit j) is

$$\Phi_M = |\phi_{jj}\phi_{jj-1}\dots\phi_{j-j+1}\phi_{j-j}|$$

with

$$M = \sum_i m_i = 0.$$

J_+ operating on Φ_M leads to the result

$$\begin{aligned}\mathrm{J}_+\Phi_M = |(\mathrm{j}_+\phi_{jj})\phi_{jj-1}\dots\phi_{j-j+1}\phi_{j-j}| + |\phi_{jj}(\mathrm{j}_+\phi_{jj-1})\dots\phi_{j-j+1}\phi_{j-j}| \\ +\dots+|\phi_{jj}\phi_{jj-1}\dots\phi_{j-j+1}(\mathrm{j}_+\phi_{j-j})|.\end{aligned} \tag{1.19}$$

From equation (1.15) it is clear that $\mathrm{j}_+\phi_{jj}=0$. Furthermore, in the second term of this equation, $\mathrm{j}_+\phi_{jj-1}=\sqrt{(2j)}\phi_{jj}$ so that the j_+ operation produces a situation in which two particles are in the same quantum state (jj) and hence the Slater determinant vanishes. The same arguments apply for the remaining terms in equation (1.19) so that

$$\mathrm{J}_+\Phi_{M=0}=0$$

and by virtue of Theorem 1 Φ_M is an eigenfunction of J^2 and J_z with eigenvalue 0.

Because of this the total angular momentum of the nucleus is carried by particles moving outside the inert fully occupied orbits. The simplest situation is the two-particle case and we shall discuss this first. Two particles in the single-particle states j_1 and j_2 have maximum angular momentum

$$I_{\max} = j_1 + j_2.$$

This follows from the fact that the wave function with maximum M

$$\Phi_M = |\phi_{j_1 j_1} \phi_{j_2 j_2}|$$

has the property that $J_+\Phi_M = 0$.

By use of the lowering operator, J_-, one can produce a state with I unchanged but M decreased by one unit. According to equation (1.15)

$$\begin{aligned} J_-\Phi_{I=j_1+j_2, M=j_1+j_2} &= \sqrt{\{2(j_1+j_2)\}}\Phi_{I=j_1+j_2, M=j_1+j_2-1} \\ &= \sqrt{(2j_1)}\,|\phi_{j_1 j_1-1}\phi_{j_2 j_2}| + \sqrt{(2j_2)}\,|\phi_{j_1 j_1}\phi_{j_2 j_2-1}|. \end{aligned}$$

Thus

$$\Phi_{I=j_1+j_2, M=j_1+j_2-1} = \frac{1}{\sqrt{(j_1+j_2)}}\left(\sqrt{j_1}\,|\phi_{j_1 j_1-1}\phi_{j_2 j_2}| + \sqrt{j_2}\,|\phi_{j_1 j_1}\phi_{j_2 j_2-1}\right). \quad (1.20)$$

Proceeding in this way one may show that

$$\begin{aligned} \Phi_{I=j_1+j_2, M=j_1+j_2-2} &= \frac{1}{\sqrt{\{(j_1+j_2)(2j_1+2j_2-1)\}}} \\ &\quad \times \left(\sqrt{\{j_1(2j_1-1)\}}\,|\phi_{j_1 j_1-2}\phi_{j_2 j_2}| + 2\sqrt{(j_1 j_2)}|\,\phi_{j_1 j_1-1}\phi_{j_2 j_2-1}|\right. \\ &\quad \left. + \sqrt{\{j_2(2j_2-1)\}}\,|\phi_{j_1 j_1}\phi_{j_2 j_2-2}|\right) \end{aligned}$$

and so forth until we reach the state with $M = -(j_1+j_2)$.

As we go to lower values of M more terms appear in the wave function. In particular in equation (1.20) there are two Slater determinants entering into the make-up of the state. Therefore a second linear combination

$$\Psi_{M=j_1+j_2-1} = \alpha\,|\phi_{j_1 j_1-1}\phi_{j_2 j_2}| + \beta\,|\phi_{j_1 j_1}\phi_{j_2 j_2-1}| \quad (1.21)$$

can be formed and the coefficients adjusted so that $J_+\Psi_{M=j_1+j_2-1} = 0$. Since

$$\begin{aligned} J_+\Psi_{M=j_1+j_2-1} &= \alpha\,|(j_+\phi_{j_1 j_1-1})\phi_{j_2 j_2}| + \beta\,|\phi_{j_1 j_1}(j_+\phi_{j_2 j_2-1})| \\ &= \{\sqrt{(2j_1)}\alpha + \sqrt{(2j_2)}\beta\}\,|\phi_{j_1 j_1}\phi_{j_2 j_2}| \end{aligned}$$

the choice

$$\alpha = -\left(\frac{j_2}{j_1+j_2}\right)^{\frac{1}{2}}$$
$$\beta = \left(\frac{j_1}{j_1+j_2}\right)^{\frac{1}{2}} \tag{1.22}$$

ensures that the state given by equation (1.21) is normalized, orthogonal to $\Phi_{I=j_1+j_2, M=j_1+j_2-1}$, and has angular momentum $I = j_1 + j_2 - 1$.

In general we have the following theorem:

Theorem 2. In a given configuration $(j_1, j_2, \ldots j_k)$ with n particles let us suppose there are p distinct Slater determinants $\Phi_M(i)$ $(i = 1, \ldots, p)$ with the property $J_z\Phi_M(i) = M\Phi_M(i)$. If there are $(p+q)$ $(q > 0)$ linearly independent Slater determinants that have the property $J_z\Phi_{M-1}(i) = (M-1)\Phi_{M-1}(i)$, then q states with angular momentum $I = M-1$ can be constructed.

Proof. Apply the J_+ operator to the linear combination

$$\Phi = \alpha_1\Phi_{M-1}(1) + \alpha_2\Phi_{M-1}(2) + \ldots + \alpha_{p+q}\Phi_{M-1}(p+q).$$

Since the effect of J_+ is to increase M by one unit

$$J_+\Phi = f_1(\alpha_1, \ldots, \alpha_{p+q})\Phi_M(1) + \ldots + f_p(\alpha_1, \ldots, \alpha_{p+q})\Phi_M(p)$$

where $f_1(\alpha_1, \ldots, \alpha_{p+q})$ is a linear function of $\alpha_1, \alpha_2, \ldots, \alpha_{p+q}$. The requirement that $J_+\Phi = 0$ therefore leads to the p conditions

$$f_1(\alpha_1, \ldots, \alpha_{p+q}) = 0$$
$$\ldots$$
$$f_p(\alpha_1, \ldots, \alpha_{p+q}) = 0.$$

Thus there are p equations in $(p+q)$ unknowns and hence q solutions. Because of Theorem 1 each of these solutions has angular momentum $I = M-1$.

For the special case $n = 2$, Theorem 2 leads to the following result:

Theorem 3. For two particles in the states j_1 and j_2 $(j_1 \neq j_2)$ the allowable angular-momentum values are

$$I = j_1 + j_2,\, j_1 + j_2 - 1,\, j_1 + j_2 - 2, \ldots, |j_1 - j_2|.$$

Proof. For a given M we can construct Slater determinants

$$|\phi_{j_1 M-j_2}\phi_{j_2 j_2}|,\ |\phi_{j_1 M-j_2+1}\phi_{j_2 j_2-1}|, \ldots, |\phi_{j_1 j_1}\phi_{j_2 M-j_1}|.$$

It is apparent that if we continuously decrease M by one unit a single additional Slater determinant appears at each step provided that

$$M-j_2 \geq -j_1$$
$$M-j_1 \geq -j_2$$

i.e. provided that $M \geq |j_1-j_2|$. Thus from Theorems 1 and 2 it follows that the allowable angular-momentum states are those given by the statement of the Theorem.

One can check that this coupling rule is realized experimentally by looking, for example, at the nucleus ${}^{38}_{17}\mathrm{Cl}_{21}$. From Fig. 1.1 it can be seen that the first 16 protons in the nucleus fill the $0s_{\frac{1}{2}}$, $0p_{\frac{3}{2}}$, $0p_{\frac{1}{2}}$, $0d_{\frac{5}{2}}$, and $1s_{\frac{1}{2}}$ orbits and the 17th proton occupies the $0d_{\frac{3}{2}}$ single-particle state. For the neutrons the $0d_{\frac{3}{2}}$ orbit is also completely occupied and the final neutron is in the $0f_{\frac{7}{2}}$ level. Thus the entire angular momentum of the nucleus should be carried by the $0d_{\frac{3}{2}}$ proton and $0f_{\frac{7}{2}}$ neutron so that the low-lying states in ${}^{38}_{17}\mathrm{Cl}_{21}$ should have

$$I=5, 4, 3, \text{and } 2.$$

In Fig. 1.2 the experimentally observed levels (Endt and Van der Leun 1973) are shown and it is seen that the theoretical expectations are borne out.

The numbers generated by use of the J_- and J_+ operators in equations (1.20) and (1.21) are coefficients that ensure that the two angular momenta couple to $I=j_1+j_2$ and $I=j_1+j_2-1$, respectively, with z-

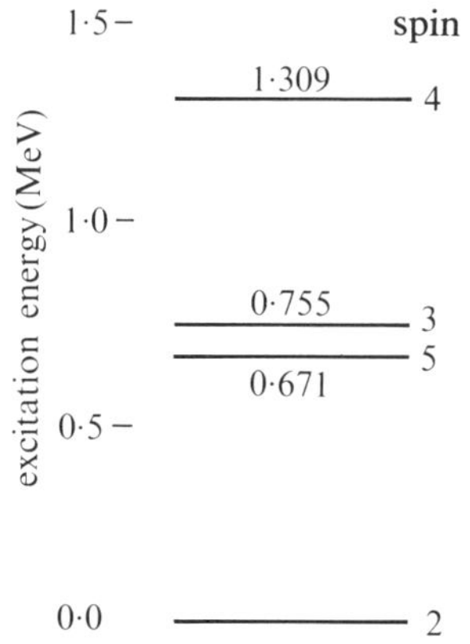

Fig. 1.2. The experimental states observed in ${}^{38}_{17}\mathrm{Cl}_{21}$ (Endt and Van der Leun 1973). The fact that the low-lying states have $I=2$, 3, 4, and 5 is consistent with the valence proton and neutron occupying the $0d_{\frac{3}{2}}$ and $0f_{\frac{7}{2}}$ orbits, respectively.

component $M = j_1 + j_2 - 1$. *These are known as Clebsch–Gordan coefficients.* In terms of these a state with angular momentum (I, M) compounded from j_1 and j_2 is written as

$$\psi_{IM}(1,2) = \sum_{m_1 m_2} (j_1 j_2 m_1 m_2 \mid IM) |\phi_{j_1 m_1} \phi_{j_2 m_2}| \tag{1.23}$$

where $(j_1 j_2 m_1 m_2 \mid IM)$ is the Clebsch–Gordan coefficient. The square of this coefficient gives the probability that in the state $\psi_{IM}(1,2)$ we shall find the configuration $|\phi_{j_1 m_1} \phi_{j_2 m_2}|$.

In Appendix 1 the properties of the Clebsch–Gordan coefficients are discussed and it is shown that

$$(j_1 j_2 m_1 m_2 \mid IM) = (-1)^{j_1 + j_2 - I} (j_2 j_1 m_2 m_1 \mid IM). \tag{1.24}$$

This property may be used to show that

Theorem 4. Two neutrons or two protons in the same single-particle orbit j (j half integral) can only couple their spins to even values of I; i.e.

$$I = 0, 2, 4, \ldots, (2j-1).$$

Proof. In terms of the Clebsch–Gordan coefficients the general wave function for two identical particles may be written as

$$\psi_{IM}(1,2) = \frac{B}{\sqrt{2}} \sum_{m_1 m_2} (j_1 j_2 m_1 m_2 \mid IM) \{\phi_{j_1 m_1}(1) \phi_{j_2 m_2}(2) - \phi_{j_1 m_1}(2) \phi_{j_2 m_2}(1)\}$$

where B is to be determined from the normalization condition. Thus

$$\begin{aligned}
\langle \psi_{IM}(1,2) \mid \psi_{IM}(1,2) \rangle &= B^2 \sum_{m_1 m_2} \sum_{m_1' m_2'} (j_1 j_2 m_1 m_2 \mid IM)(j_1 j_2 m_1' m_2' \mid IM) \\
&\quad \times \{\delta_{m_1 m_1'} \delta_{m_2 m_2'} - \delta_{j_1 j_2} \delta_{m_1 m_2'} \delta_{m_2 m_1'}\} \\
&= B^2 \{1 - \delta_{j_1 j_2} (-1)^{j_1 + j_2 - I}\} \sum_{m_1 m_2} (j_1 j_2 m_1 m_2 \mid IM)^2 \\
&= B^2 \{1 - \delta_{j_1 j_2} (-1)^{j_1 + j_2 - I}\}
\end{aligned}$$

where use has been made of the summation property given in equation (A1.23) in writing the last line of this equation. Consequently if $j_1 = j_2$ it is impossible to write a normalized antisymmetric wave function when I is odd, so that only even values of I are possible.

One can check that this expectation is borne out experimentally by looking at the spectrum of ${}^{50}_{22}Ti_{28}$. In this case one would expect ${}^{48}_{20}Ca_{28}$ to form an inert closed shell and as a consequence the low-lying states

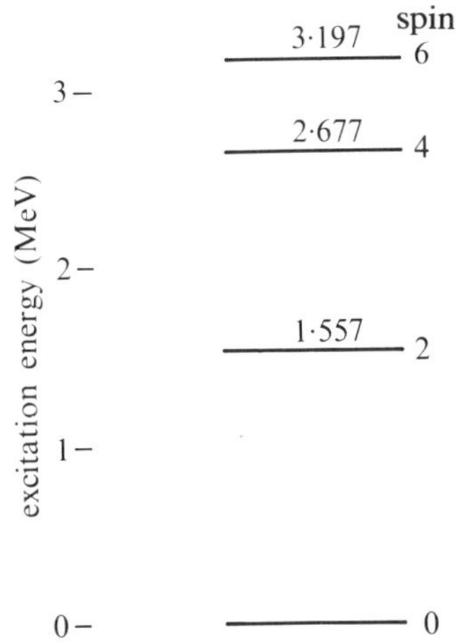

Fig. 1.3. The experimental spectrum of ${}^{50}_{22}Ti_{28}$ (Nomura *et al.* 1970). No odd-I states are seen below 3 MeV excitation energy, and this is consistent with a description of the nucleus in which the two valence protons move in a $0f_{\frac{7}{2}}$ orbit outside an inert ${}^{48}_{20}Ca_{28}$ core.

should be those corresponding to two protons in the $0f_{\frac{7}{2}}$ orbit. In Fig. 1.3 we present the experimental results of Nomura *et al.* (1970) which show that below 3 MeV excitation energy no odd-I states are seen. (In Fig. 2.1 we show that the odd angular-momentum states of $(0f_{\frac{7}{2}})^2$, $I=1, 3, 5$, and 7, are found below 2 MeV excitation energy when one has a neutron and a proton in the $0f_{\frac{7}{2}}$ orbit. These states arise because the neutron-proton wave function does not have to be antisymmetric.)

Even when we do not deal with single-particle states the Clebsch–Gordan coupling of equation (1.23) must be carried out if one wishes to have an angular-momentum eigenfunction. Since we shall usually want to do this, it is convenient to introduce the shorthand notation $[\Psi_{I_1}\times\Phi_{I_2}]_{IM}$ to stand for the Clebsch–Gordan summation involved in obtaining the state (IM) by adding I_1 to I_2. Thus

$$[\Psi_{I_1}\times\Phi_{I_2}]_{IM}=\sum_{M_1M_2}(I_1I_2M_1M_2\mid IM)\Psi_{I_1M_1}\Phi_{I_2M_2}. \qquad (1.25)$$

Furthermore, Theorem 3 tells us that

$$|I_1-I_2|\leq I\leq I_1+I_2$$

i.e. I_1, I_2, and I must form a triangle. Because of this it is convenient to picture this coupling as a triangle. Thus we shall often represent the state

$[\Psi_{I_1} \times \Phi_{I_2}]_{IM}$ as (Macfarlane and French 1960)

$$[\Psi_{I_1} \times \Phi_{I_2}]_{IM} =$$ 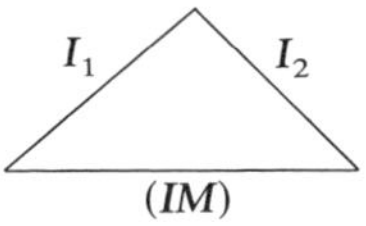

(1.26a)

or

$$[\Phi_{I_2} \times \Psi_{I_1}]_{IM} =$$ 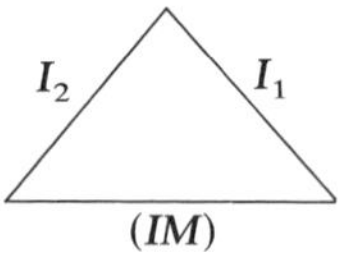

. (1.26b)

Moreover, because of equation (1.24) it follows that

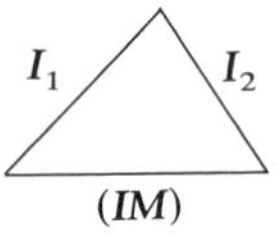

$$= (-1)^{I_1 + I_2 - I}$$ 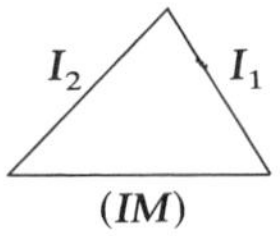

(1.27)

1.2. Allowable states of the configuration j^n

In this Section we deal with identical nucleons (either neutrons or protons) in the same single-particle orbit. There are several simple theorems that can be derived concerning the allowable angular-momentum states.

Theorem 5. The maximum possible angular momentum that can arise in the configuration j^n is

$$I_M = n\{j - (n-1)/2\}.$$

Proof. The state with maximum M is

$$\Phi_{M_{\max}} = |\phi_{jj}\phi_{jj-1} \dots \phi_{jj-n+2}\phi_{jj-n+1}|.$$

J_+ operating on this wave function gives zero and consequently from Theorem 1 it follows that $I_M = M_{\max}$. But

$$\begin{aligned} M_{\max} &= j + (j-1) + \dots + (j-n+2) + (j-n+1) \\ &= nj - \sum_{k=1}^{n-1} k \\ &= n\{j - (n-1)/2\}. \end{aligned}$$

Theorem 6. There is no state of the configuration j^n with $I = I_M - 1$.

Proof. The only possible way of obtaining the M value $(M_{\max} - 1)$ is

$$\Phi_{M_{\max}-1} = |\phi_{jj}\phi_{jj-1} \dots \phi_{jj-n+2}\phi_{jj-n}|.$$

Thus according to Theorem 2 no state with $I=I_M-1$ can be constructed. $\Phi_{M_{\max}-1}$ is precisely the wave function obtained by applying J_- to $\Phi_{M_{\max}}$ of the preceding theorem and consequently is the state $\Phi_{I=I_M,M=I_M-1}$.

Theorem 7. In the configuration j^n there is one state with

$$\begin{aligned} I &= I_M-2 \\ &= n\{j-(n-1)/2\}-2. \end{aligned}$$

Proof. There are two Slater determinants with $M=n\{j-(n-1)/2\}-2$, namely

$$\begin{aligned} \Phi(1) &= |\phi_{jj}\phi_{jj-1}\ldots\phi_{jj-n+3}\phi_{jj-n+2}\phi_{jj-n-1}| \\ \Phi(2) &= |\phi_{jj}\phi_{jj-1}\ldots\phi_{jj-n+3}\phi_{jj-n+1}\phi_{jj-n}|. \end{aligned}$$

By combining this result with Theorems 2 and 6 we see that a state with $I=I_M-2$ can be formed. (Note if $(I_M-2)<0$—as would be the case for $n=3$, $j=\frac{3}{2}$—the state cannot be produced since all I must be >0.)

To facilitate the enumeration and explicit construction of many-particle wave functions it is convenient to simplify the notation and introduce the language of second quantization. To do this we define the particle-creation operator $a^\dagger_{jm}$ which creates a particle in the state ϕ_{jm}, and examine what properties $a^\dagger_{jm}$ must possess so that

$$|\phi_{j_1m_1}\phi_{j_2m_2}\ldots\phi_{j_nm_n}| = a^\dagger_{j_1m_1}a^\dagger_{j_2m_2}\ldots a^\dagger_{j_nm_n}|0\rangle$$

where $|0\rangle$ is the vacuum which is either a state with no particles or a state with no nucleons in the single-particle orbits $(j_1, j_2, \ldots, j_n)$.

Because the Slater determinant changes sign when two nucleons are interchanged and is zero when there are two nucleons in the same state, it follows that the creation operator must satisfy the relationship

$$a^\dagger_{j_1m_1}a^\dagger_{j_2m_2}|\Phi\rangle = -a^\dagger_{j_2m_2}a^\dagger_{j_1m_1}|\Phi\rangle$$

where $|\Phi\rangle$ is any arbitrary state vector within the Hilbert space. Thus

$$\{a^\dagger_{jm}, a^\dagger_{j'm'}\} = a^\dagger_{jm}a^\dagger_{j'm'} + a^\dagger_{j'm'}a^\dagger_{jm} = 0. \tag{1.28}$$

If $a^\dagger_{jm}$ is to stand for the state ϕ_{jm} then the normalization condition imposes the requirement that

$$\langle 0|(a^\dagger_{jm})^\dagger a^\dagger_{jm}|0\rangle = \langle 0|a_{jm}a^\dagger_{jm}|0\rangle = 1$$

In other words a_{jm}, the Hermitian conjugate of $a_{jm}^\dagger$, is the operator that destroys a particle in the state (jm). Thus

$$a_{jm}a_{jm}^\dagger|0\rangle=|0\rangle. \tag{1.29}$$

By taking the Hermitian adjoint of equation (1.28) one sees that

$$\{a_{jm}, a_{j'm'}\}=a_{jm}a_{j'm'}+a_{j'm'}a_{jm}=0. \tag{1.30}$$

We next consider the effect of the operator $\{a_{jm}^\dagger, a_{j'm'}\}$ on any arbitrary state vector $|\Phi\rangle$

$$|\Phi\rangle=a_{k_1}^\dagger a_{k_2}^\dagger\ldots a_{k_n}^\dagger|0\rangle$$

where k_i stands for the dual index $(j_i m_i)$. If $(jm)\neq(j'm')$ it is clear that $\{a_{jm}^\dagger, a_{j'm'}\}|\Phi\rangle=0$ if either the state (jm) is occupied or the state $(j'm')$ is unoccupied in $|\Phi\rangle$. However, if (jm) is unoccupied and $(j'm')$ is occupied

$$\begin{aligned} a_{jm}^\dagger a_{j'm'}|\Phi\rangle &= a_{jm}^\dagger a_{j'm'}a_{k_1}^\dagger\ldots a_{k_q}^\dagger a_{j'm'}^\dagger\ldots a_{k_n}^\dagger|0\rangle \\ &=(-1)^q a_{jm}^\dagger a_{k_1}^\dagger\ldots a_{k_q}^\dagger\ldots a_{k_n}^\dagger|0\rangle \end{aligned}$$

where use has been made of equations (1.28) and (1.29). Alternatively

$$\begin{aligned} a_{j'm'}a_{jm}^\dagger|\Phi\rangle &= a_{j'm'}a_{jm}^\dagger a_{k_1}^\dagger\ldots a_{k_q}^\dagger a_{j'm'}^\dagger\ldots a_{k_n}^\dagger|0\rangle \\ &=(-1)^{q+1} a_{jm}^\dagger a_{k_1}^\dagger\ldots a_{k_q}^\dagger\ldots a_{k_n}^\dagger|0\rangle. \end{aligned}$$

Thus if $(jm)\neq(j'm')$, $\{a_{jm}^\dagger, a_{j'm'}\}$ vanishes when applied to any state vector.

For $(jm)=(j'm')$ the commutator has the value unity independent of whether the state (jm) is occupied or unoccupied in $|\Phi\rangle$. Consequently

$$\{a_{jm}^\dagger, a_{j'm'}\}=a_{jm}^\dagger a_{j'm'}+a_{j'm'}a_{jm}^\dagger=\delta_{jj'}\delta_{mm'}. \tag{1.31}$$

Provided that $a_{jm}^\dagger$ and a_{jm} satisfy equations (1.28)–(1.31), their products may be used to represent the Slater determinant of equation (1.18).

With the aid of this shorthand notation we shall now discuss how one enumerates the allowable states of the configuration j^n. The procedure is straightforward but tedious and consists of writing down all possible states $\Phi_M(i)$ $(M\geq 0)$

$$\Phi_M(i)=a_{jm_1}^\dagger a_{jm_2}^\dagger\ldots a_{jm_n}^\dagger|0\rangle$$

with

$$M=\sum_q m_q.$$

Once this is done we may use Theorem 2 which says that if there are $n_{M+1}\Phi_{M+1}$ and $n_M\Phi_M$, then $n=(n_M-n_{M+1})$ states with $I=M$ can be constructed.

As an example we consider the $(\frac{7}{2})^4$ configuration. Since in this case we always deal with the same j value we suppress this index and write

$$a^{\dagger}_{jm} = a^{\dagger}_{m}.$$

The simplest way is to proceed in odometer fashion and write

$M=8$

$\Phi_8(1) = a^{\dagger}_{\frac{7}{2}} a^{\dagger}_{\frac{5}{2}} a^{\dagger}_{\frac{3}{2}} a^{\dagger}_{\frac{1}{2}} |0\rangle$

$M=7$

$\Phi_7(1) = a^{\dagger}_{\frac{7}{2}} a^{\dagger}_{\frac{5}{2}} a^{\dagger}_{\frac{3}{2}} a^{\dagger}_{-\frac{1}{2}} |0\rangle$

$M=6$

$\Phi_6(1) = a^{\dagger}_{\frac{7}{2}} a^{\dagger}_{\frac{5}{2}} a^{\dagger}_{\frac{3}{2}} a^{\dagger}_{-\frac{3}{2}} |0\rangle$
$\Phi_6(2) = a^{\dagger}_{\frac{7}{2}} a^{\dagger}_{\frac{5}{2}} a^{\dagger}_{\frac{1}{2}} a^{\dagger}_{-\frac{1}{2}} |0\rangle$

$M=5$

$\Phi_5(1) = a^{\dagger}_{\frac{7}{2}} a^{\dagger}_{\frac{5}{2}} a^{\dagger}_{\frac{3}{2}} a^{\dagger}_{-\frac{5}{2}} |0\rangle$
$\Phi_5(2) = a^{\dagger}_{\frac{7}{2}} a^{\dagger}_{\frac{5}{2}} a^{\dagger}_{\frac{1}{2}} a^{\dagger}_{-\frac{3}{2}} |0\rangle$
$\Phi_5(3) = a^{\dagger}_{\frac{7}{2}} a^{\dagger}_{\frac{3}{2}} a^{\dagger}_{\frac{1}{2}} a^{\dagger}_{-\frac{1}{2}} |0\rangle$

$M=4$

$\Phi_4(1) = a^{\dagger}_{\frac{7}{2}} a^{\dagger}_{\frac{5}{2}} a^{\dagger}_{\frac{3}{2}} a^{\dagger}_{-\frac{7}{2}} |0\rangle$
$\Phi_4(2) = a^{\dagger}_{\frac{7}{2}} a^{\dagger}_{\frac{5}{2}} a^{\dagger}_{\frac{1}{2}} a^{\dagger}_{-\frac{5}{2}} |0\rangle$
$\Phi_4(3) = a^{\dagger}_{\frac{7}{2}} a^{\dagger}_{\frac{5}{2}} a^{\dagger}_{-\frac{1}{2}} a^{\dagger}_{-\frac{3}{2}} |0\rangle$
$\Phi_4(4) = a^{\dagger}_{\frac{7}{2}} a^{\dagger}_{\frac{3}{2}} a^{\dagger}_{\frac{1}{2}} a^{\dagger}_{-\frac{3}{2}} |0\rangle$
$\Phi_4(5) = a^{\dagger}_{\frac{5}{2}} a^{\dagger}_{\frac{3}{2}} a^{\dagger}_{\frac{1}{2}} a^{\dagger}_{-\frac{1}{2}} |0\rangle$

$M=3$

$\Phi_3(1) = a^{\dagger}_{\frac{7}{2}} a^{\dagger}_{\frac{5}{2}} a^{\dagger}_{\frac{1}{2}} a^{\dagger}_{-\frac{7}{2}} |0\rangle$
$\Phi_3(2) = a^{\dagger}_{\frac{7}{2}} a^{\dagger}_{\frac{5}{2}} a^{\dagger}_{-\frac{1}{2}} a^{\dagger}_{-\frac{5}{2}} |0\rangle$
$\Phi_3(3) = a^{\dagger}_{\frac{7}{2}} a^{\dagger}_{\frac{3}{2}} a^{\dagger}_{\frac{1}{2}} a^{\dagger}_{-\frac{5}{2}} |0\rangle$
$\Phi_3(4) = a^{\dagger}_{\frac{7}{2}} a^{\dagger}_{\frac{3}{2}} a^{\dagger}_{-\frac{1}{2}} a^{\dagger}_{-\frac{3}{2}} |0\rangle$
$\Phi_3(5) = a^{\dagger}_{\frac{5}{2}} a^{\dagger}_{\frac{3}{2}} a^{\dagger}_{\frac{1}{2}} a^{\dagger}_{-\frac{3}{2}} |0\rangle$

$M=2$

$\Phi_2(1) = a^{\dagger}_{\frac{7}{2}} a^{\dagger}_{\frac{5}{2}} a^{\dagger}_{-\frac{1}{2}} a^{\dagger}_{-\frac{7}{2}} |0\rangle$
$\Phi_2(2) = a^{\dagger}_{\frac{7}{2}} a^{\dagger}_{\frac{5}{2}} a^{\dagger}_{-\frac{3}{2}} a^{\dagger}_{-\frac{5}{2}} |0\rangle$
$\Phi_2(3) = a^{\dagger}_{\frac{7}{2}} a^{\dagger}_{\frac{3}{2}} a^{\dagger}_{\frac{1}{2}} a^{\dagger}_{-\frac{7}{2}} |0\rangle$
$\Phi_2(4) = a^{\dagger}_{\frac{7}{2}} a^{\dagger}_{\frac{3}{2}} a^{\dagger}_{-\frac{1}{2}} a^{\dagger}_{-\frac{5}{2}} |0\rangle$
$\Phi_2(5) = a^{\dagger}_{\frac{7}{2}} a^{\dagger}_{\frac{1}{2}} a^{\dagger}_{-\frac{1}{2}} a^{\dagger}_{-\frac{3}{2}} |0\rangle$
$\Phi_2(6) = a^{\dagger}_{\frac{5}{2}} a^{\dagger}_{\frac{3}{2}} a^{\dagger}_{\frac{1}{2}} a^{\dagger}_{-\frac{5}{2}} |0\rangle$
$\Phi_2(7) = a^{\dagger}_{\frac{5}{2}} a^{\dagger}_{\frac{3}{2}} a^{\dagger}_{-\frac{1}{2}} a^{\dagger}_{-\frac{3}{2}} |0\rangle$

$M=1$

$\Phi_1(1) = a^{\dagger}_{\frac{7}{2}} a^{\dagger}_{\frac{5}{2}} a^{\dagger}_{-\frac{3}{2}} a^{\dagger}_{-\frac{7}{2}} |0\rangle$
$\Phi_1(2) = a^{\dagger}_{\frac{7}{2}} a^{\dagger}_{\frac{3}{2}} a^{\dagger}_{-\frac{1}{2}} a^{\dagger}_{-\frac{7}{2}} |0\rangle$
$\Phi_1(3) = a^{\dagger}_{\frac{7}{2}} a^{\dagger}_{\frac{3}{2}} a^{\dagger}_{-\frac{3}{2}} a^{\dagger}_{-\frac{5}{2}} |0\rangle$
$\Phi_1(4) = a^{\dagger}_{\frac{7}{2}} a^{\dagger}_{\frac{1}{2}} a^{\dagger}_{-\frac{1}{2}} a^{\dagger}_{-\frac{5}{2}} |0\rangle$
$\Phi_1(5) = a^{\dagger}_{\frac{5}{2}} a^{\dagger}_{\frac{3}{2}} a^{\dagger}_{\frac{1}{2}} a^{\dagger}_{-\frac{7}{2}} |0\rangle$
$\Phi_1(6) = a^{\dagger}_{\frac{5}{2}} a^{\dagger}_{\frac{3}{2}} a^{\dagger}_{-\frac{1}{2}} a^{\dagger}_{-\frac{5}{2}} |0\rangle$
$\Phi_1(7) = a^{\dagger}_{\frac{5}{2}} a^{\dagger}_{\frac{1}{2}} a^{\dagger}_{-\frac{1}{2}} a^{\dagger}_{-\frac{3}{2}} |0\rangle$

$M=0$

$\Phi_0(1) = a^{\dagger}_{\frac{7}{2}} a^{\dagger}_{\frac{5}{2}} a^{\dagger}_{-\frac{5}{2}} a^{\dagger}_{-\frac{7}{2}} |0\rangle$
$\Phi_0(2) = a^{\dagger}_{\frac{7}{2}} a^{\dagger}_{\frac{3}{2}} a^{\dagger}_{-\frac{3}{2}} a^{\dagger}_{-\frac{7}{2}} |0\rangle$
$\Phi_0(3) = a^{\dagger}_{\frac{7}{2}} a^{\dagger}_{\frac{1}{2}} a^{\dagger}_{-\frac{1}{2}} a^{\dagger}_{-\frac{7}{2}} |0\rangle$
$\Phi_0(4) = a^{\dagger}_{\frac{7}{2}} a^{\dagger}_{\frac{1}{2}} a^{\dagger}_{-\frac{3}{2}} a^{\dagger}_{-\frac{5}{2}} |0\rangle$
$\Phi_0(5) = a^{\dagger}_{\frac{5}{2}} a^{\dagger}_{\frac{3}{2}} a^{\dagger}_{-\frac{1}{2}} a^{\dagger}_{-\frac{7}{2}} |0\rangle$
$\Phi_0(6) = a^{\dagger}_{\frac{5}{2}} a^{\dagger}_{\frac{3}{2}} a^{\dagger}_{-\frac{3}{2}} a^{\dagger}_{-\frac{5}{2}} |0\rangle$
$\Phi_0(7) = a^{\dagger}_{\frac{5}{2}} a^{\dagger}_{\frac{1}{2}} a^{\dagger}_{-\frac{1}{2}} a^{\dagger}_{-\frac{5}{2}} |0\rangle$
$\Phi_0(8) = a^{\dagger}_{\frac{3}{2}} a^{\dagger}_{\frac{1}{2}} a^{\dagger}_{-\frac{1}{2}} a^{\dagger}_{-\frac{3}{2}} |0\rangle$

Consequently from Theorem 2 it follows that the allowable states of $(\frac{7}{2})^4$ are

$$I = 0, 2^2, 4^2, 5, 6, \text{and } 8$$

where the superscripts give the number of orthogonal states with that value of I that can be constructed. Fortunately Bayman and Lande (1966) have given a complete enumeration of the possible states for all values of n provided $j \leq \frac{11}{2}$ and for $n \leq 5$ when $j = \frac{13}{2}$ and $\frac{15}{2}$ so that, in general, one does not have to resort to this type of calculation.

Finally, to find explicitly the wave function for the state with angular momentum I and $M = I$ one must take a linear combination of all the possible $\Phi_M(i)$ and arrange that J_+ acting on the combination gives zero. For example, in the $(\frac{7}{2})^4$ configuration the combination with $M = 4$ is

$$\begin{aligned}\chi_4 = \{&\alpha a^\dagger_{\frac{7}{2}} a^\dagger_{\frac{5}{2}} a^\dagger_{\frac{3}{2}} a^\dagger_{-\frac{7}{2}} + \beta a^\dagger_{\frac{7}{2}} a^\dagger_{\frac{5}{2}} a^\dagger_{\frac{1}{2}} a^\dagger_{-\frac{5}{2}} \\ &+ \gamma a^\dagger_{\frac{7}{2}} a^\dagger_{\frac{5}{2}} a^\dagger_{-\frac{1}{2}} a^\dagger_{-\frac{3}{2}} + \delta a^\dagger_{\frac{7}{2}} a^\dagger_{\frac{3}{2}} a_{\frac{1}{2}} a^\dagger_{-\frac{3}{2}} \\ &+ \varepsilon a^\dagger_{\frac{5}{2}} a^\dagger_{\frac{3}{2}} a^\dagger_{\frac{1}{2}} a^\dagger_{-\frac{1}{2}}\} \,|0\rangle.\end{aligned}$$

The operator

$$\mathrm{J}_+ = \sum_i (\mathrm{j}_+)_i$$

has the property that it destroys a particle in the state (jm) and recreates it in the state $(jm+1)$ and at the same time multiplies the wave function by $\sqrt{\{(j-m)(j+m+1)\}}$ (see equation (1.15)). Thus in terms of annihilation and creation operators the J_+-operation takes the form

$$\mathrm{J}_+ = \sum_{jm} \sqrt{\{(j-m)(j+m+1)\}}\, a^\dagger_{jm+1} a_{jm}. \tag{1.32}$$

With this in mind it is easy to see that $\mathrm{J}_+\chi_4 = 0$ provided that

$$\begin{aligned}(&\{\sqrt{7}\alpha + \sqrt{15}\beta\} a^\dagger_{\frac{7}{2}} a^\dagger_{\frac{5}{2}} a^\dagger_{\frac{3}{2}} a^\dagger_{-\frac{5}{2}} \\ &+ \{2\sqrt{3}\beta + 4\gamma + 2\sqrt{3}\delta\} a^\dagger_{\frac{7}{2}} a^\dagger_{\frac{5}{2}} a^\dagger_{\frac{1}{2}} a^\dagger_{-\frac{3}{2}} \\ &+ \{\sqrt{15}\delta + \sqrt{7}\varepsilon\} a^\dagger_{\frac{7}{2}} a^\dagger_{\frac{3}{2}} a^\dagger_{\frac{1}{2}} a^\dagger_{-\frac{1}{2}}) \,|0\rangle = 0.\end{aligned}$$

Since the various Φ_5 are linearly independent, the coefficient of each term must vanish. Consequently there are three conditions and five unknowns so that two solutions are possible. For the first we choose $\gamma = 0$, and in this case

$$\begin{aligned}\Phi_{44} = \frac{1}{\sqrt{44}} (&\sqrt{15}\, a^\dagger_{\frac{7}{2}} a^\dagger_{\frac{5}{2}} a^\dagger_{\frac{3}{2}} a^\dagger_{-\frac{7}{2}} - \sqrt{7}\, a^\dagger_{\frac{7}{2}} a^\dagger_{\frac{5}{2}} a^\dagger_{\frac{1}{2}} a^\dagger_{-\frac{5}{2}} \\ &+ \sqrt{7}\, a^\dagger_{\frac{7}{2}} a^\dagger_{\frac{3}{2}} a^\dagger_{\frac{1}{2}} a^\dagger_{-\frac{3}{2}} - \sqrt{15}\, a^\dagger_{\frac{5}{2}} a^\dagger_{\frac{3}{2}} a^\dagger_{\frac{1}{2}} a^\dagger_{-\frac{1}{2}}) \,|0\rangle.\end{aligned} \tag{1.33}$$

A state orthogonal to this and still satisfying the condition $\mathrm{J}_+\chi_4 = 0$ results if

$$\sqrt{15}\alpha - \sqrt{7}\beta + \sqrt{7}\delta - \sqrt{15}\varepsilon = 0$$

and for this choice

$$\Psi_{44}=\frac{1}{\surd 65}\{\surd 15 a^{\dagger}_{\frac{7}{2}}a^{\dagger}_{\frac{5}{2}}a^{\dagger}_{\frac{3}{2}}a^{\dagger}_{-\frac{7}{2}}-\surd 7 a^{\dagger}_{\frac{7}{2}}a^{\dagger}_{\frac{5}{2}}a^{\dagger}_{\frac{1}{2}}a^{\dagger}_{-\frac{5}{2}}$$
$$+\surd 21 a^{\dagger}_{\frac{7}{2}}a^{\dagger}_{\frac{5}{2}}a^{\dagger}_{-\frac{1}{2}}a^{\dagger}_{-\frac{3}{2}}-\surd 7 a^{\dagger}_{\frac{7}{2}}a^{\dagger}_{\frac{3}{2}}a^{\dagger}_{\frac{1}{2}}a^{\dagger}_{-\frac{3}{2}}$$
$$+\surd 15 a^{\dagger}_{\frac{5}{2}}a^{\dagger}_{\frac{3}{2}}a^{\dagger}_{\frac{1}{2}}a^{\dagger}_{-\frac{1}{2}}\}\,|0\rangle. \qquad (1.34)$$

The $I=4$ states with $M\neq 4$ can be obtained, if desired, by repeated application of J_- to Φ_{44} and Ψ_{44}.

The basic difference between these two wave functions is that Φ_{44} has the combination $a^{\dagger}_m a^{\dagger}_{-m}$ in all terms. In other words, if the single-particle state m is occupied so is the state $-m$. Thus this wave function has two paired particles and two unpaired nucleons. On the other hand, Ψ_{44} does not possess this property since the term $a_{\frac{7}{2}}{}^{\dagger}a_{\frac{5}{2}}{}^{\dagger}a_{-\frac{1}{2}}{}^{\dagger}a_{-\frac{3}{2}}{}^{\dagger}\,|0\rangle$ does not vanish and consequently Ψ_{44} is said to have four unpaired nucleons. The states described by equations (1.33) and (1.34) have seniority 2 and 4, respectively, where the seniority quantum number gives the number of unpaired particles in the state. This quantum number will be discussed in detail in section 5 of this Chapter.

1.3. Many particles in different orbits

Although the Pauli principle excludes certain states of the configuration j^n, it provides no limitation on the values of I that can arise when I_1 and I_2 are the angular momenta of the configurations $(j_1)^n$ and $(j_2)^m$, respectively. In this case all values that satisfy the triangle inequality can be realized:

$$|I_1-I_2|\leq I\leq I_1+I_2.$$

We now use this fact to show the number of ways a state of given I can be obtained when the model space has several different single-particle states.

To be specific we deduce the number of spin-2 states that occur in ${}^{62}_{28}Ni_{34}$ when we assume $N=Z=28$ is an inert closed shell and that the six valence nucleons are distributed in the $0f_{\frac{5}{2}}$, $1p_{\frac{3}{2}}$ and $1p_{\frac{1}{2}}$ orbits (see Fig. 1.1). Although there are several ways one can proceed, we choose the following.

Let I_1 be the angular momentum of n_1 nucleons in the $0f_{\frac{5}{2}}$ state and I_2 that of the n_2 particles in the $1p_{\frac{3}{2}}$ level. The remaining $n_3=6-n_1-n_2$ nucleons are in the $p_{\frac{1}{2}}$ orbit and have angular momentum I_3. We first couple I_1 and I_2 to I_{12} and then add I_3 to obtain the total spin I. The enumeration is done odometer style starting with the maximum number of particles in $0f_{\frac{5}{2}}$ and working down. With the exception of the $(0f_{\frac{5}{2}})^3$ configuration, the allowable values of I_1, I_2, and I_3 can be written down simply if one makes use of the fact that j^n and j^{-n} have the same allowable spin values. A calculation similar to that of the preceding Section shows that the I values

$$I_1 = \tfrac{3}{2}, \tfrac{5}{2}, \text{ and } \tfrac{9}{2}$$

are allowed for the configuration $(0f_{\frac{5}{2}})^3$.

All possible ways of making a spin-2 state are listed in Table 1.1 and in total there are 33. Thus with this model space a shell-model calculation to describe the structure of the low-lying spin-2 states in ${}^{62}_{28}Ni_{34}$ would involve the construction and diagonalization of a 33×33 matrix. Although not all of these 33 states will lie low in excitation energy, many of them will. Thus it is not surprising that Fanger *et al.* (1970) see at least five and perhaps eight spin-2 states below 3·5 MeV excitation energy in this nucleus.

One can express the couplings listed in Table 1.1 in terms of diagrams similar to those introduced at the end of section 1.1. For example, the first way of obtaining the spin-2 state is from the configuration

$$\Psi_{2M} = [(f_{\frac{5}{2}})^5_{\frac{5}{2}} \times p_{\frac{3}{2}}]_{2M}$$

where the notation $(f_{\frac{5}{2}})^5_{\frac{5}{2}}$ stands for the configuration in which five particles in the $f_{\frac{5}{2}}$ single-particle orbit couple their spins to $I_1 = \frac{5}{2}$. In general, the configuration $(j)^n_{IM}$ will be described by the diagram

$(j)^n_{IM} =$ 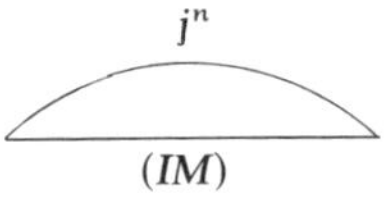 (1.35)

Thus in analogy with equation (1.26a)

$\Psi_{2M} = [(f_{\frac{5}{2}})^5_{\frac{5}{2}} \times p_{\frac{3}{2}}]_{2M} =$ 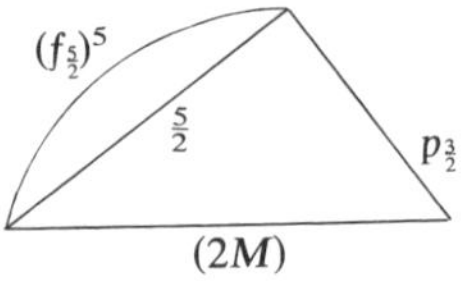

A more complicated diagram would result when one deals with the coupling (line 26 of Table 1.1)

$$\Phi_{2M} = [[(f_{\frac{5}{2}})^2_2 \times (p_{\frac{3}{2}})^3_{\frac{3}{2}}]_{\frac{5}{2}} \times p_{\frac{1}{2}}]_{2M}.$$

In this case the two $f_{\frac{5}{2}}$ nucleons couple to spin 2, the three $p_{\frac{3}{2}}$ particles to $\frac{3}{2}$, and these two angular momenta are compounded to give $I_{12} = \frac{5}{2}$. Finally the spin of $\frac{5}{2}$ is added to $\frac{1}{2}$ (the $p_{\frac{1}{2}}$ particle) to form the resultant angular

TABLE 1.1

Possible ways of making a spin-2 state by putting six neutrons into the $0f_{\frac{5}{2}}$, $1p_{\frac{3}{2}}$ and $1p_{\frac{1}{2}}$ single-particle orbits

$f_{\frac{5}{2}}$ level		$p_{\frac{3}{2}}$ level		Angular momentum of coupled	$p_{\frac{1}{2}}$ level	
Number of nucleons	Angular momentum	Number of nucleons	Angular momentum	$f_{\frac{5}{2}}$, $p_{\frac{3}{2}}$ system, I_{12}	Number of nucleons	Angular momentum
5	$\frac{5}{2}$	1	$\frac{3}{2}$	2		
5	$\frac{5}{2}$				1	$\frac{1}{2}$
4	0	2	2	2		
4	2	2	0	2		
4	2	2	2	2		
4	4	2	2	2		
4	0	1	$\frac{3}{2}$	$\frac{3}{2}$	1	$\frac{1}{2}$
4	2	1	$\frac{3}{2}$	$\frac{3}{2}$	1	$\frac{1}{2}$
4	2	1	$\frac{3}{2}$	$\frac{5}{2}$	1	$\frac{1}{2}$
4	4	1	$\frac{3}{2}$	$\frac{5}{2}$	1	$\frac{1}{2}$
4	2				2	0
3	$\frac{3}{2}$	3	$\frac{3}{2}$	2		
3	$\frac{5}{2}$	3	$\frac{3}{2}$	2		
3	$\frac{3}{2}$	2	0	$\frac{3}{2}$	1	$\frac{1}{2}$
3	$\frac{3}{2}$	2	2	$\frac{3}{2}$	1	$\frac{1}{2}$
3	$\frac{3}{2}$	2	2	$\frac{5}{2}$	1	$\frac{1}{2}$
3	$\frac{5}{2}$	2	0	$\frac{5}{2}$	1	$\frac{1}{2}$
3	$\frac{5}{2}$	2	2	$\frac{3}{2}$	1	$\frac{1}{2}$
3	$\frac{5}{2}$	2	2	$\frac{5}{2}$	1	$\frac{1}{2}$
3	$\frac{9}{2}$	2	2	$\frac{5}{2}$	1	$\frac{1}{2}$
3	$\frac{3}{2}$	1	$\frac{3}{2}$	2	2	0
3	$\frac{5}{2}$	1	$\frac{3}{2}$	2	2	0
2	2	4	0	2		
2	0	3	$\frac{3}{2}$	$\frac{3}{2}$	1	$\frac{1}{2}$
2	2	3	$\frac{3}{2}$	$\frac{3}{2}$	1	$\frac{1}{2}$
2	2	3	$\frac{3}{2}$	$\frac{5}{2}$	1	$\frac{1}{2}$
2	4	3	$\frac{3}{2}$	$\frac{5}{2}$	1	$\frac{1}{2}$
2	0	2	2	2	2	0
2	2	2	0	2	2	0
2	2	2	2	2	2	0
2	4	2	2	2	2	0
1	$\frac{5}{2}$	4	0	$\frac{5}{2}$	1	$\frac{1}{2}$
1	$\frac{5}{2}$	3	$\frac{3}{2}$	2	2	0

momentum of 2. In terms of diagrams this would be pictured as

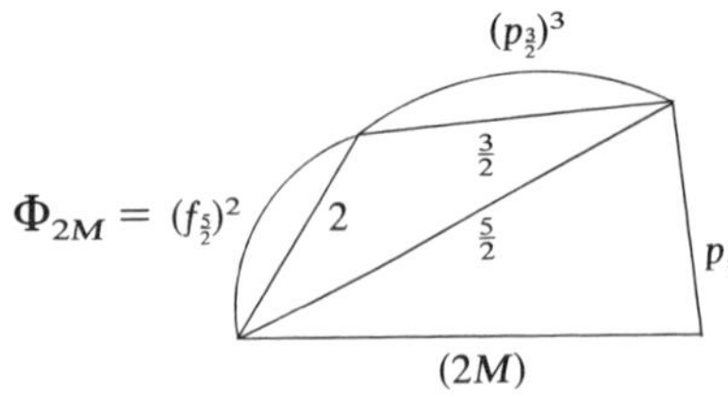

(1.36)

In deducing the possible couplings in Table 1.1 we have always added the spin of the $(f_{\frac{5}{2}})^n$ configuration to that of $(p_{\frac{3}{2}})^{n'}$ to form a resultant I_{12} and then added I_{12} and the spin of $(p_{\frac{1}{2}})^{n''}$ together to give the final spin of two. One could, of course, have numerated the possibilities in a different fashion. For example, we might have combined the spins associated with $(p_{\frac{3}{2}})^{n'}$ and $(p_{\frac{1}{2}})^{n''}$ to a resultant K and then added the angular momentum of the $(f_{\frac{5}{2}})^n$ configuration to K to form $I=2$. An example of a spin-2-state obtained in this way would be

$$\Phi'_{2M}=[(f_{\frac{5}{2}})^2_2\times[(p_{\frac{3}{2}})^3_{\frac{3}{2}}\times p_{\frac{1}{2}}]_2]_{2M}$$

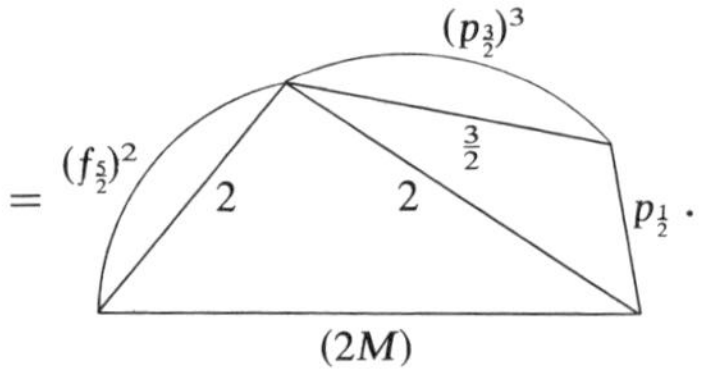

Clearly, as long as we consider all possibilities one representation is as good as another, and any state such as the one given above can be expressed as a linear combination of the configurations listed in Table 1.1. Moreover, since the states $(f_{\frac{5}{2}})^n_{I_1M_1}$, $(p_{\frac{3}{2}})^{n'}_{I_2M_2}$, and $(p_{\frac{1}{2}})^{n''}_{I_3M_3}$ are orthogonal

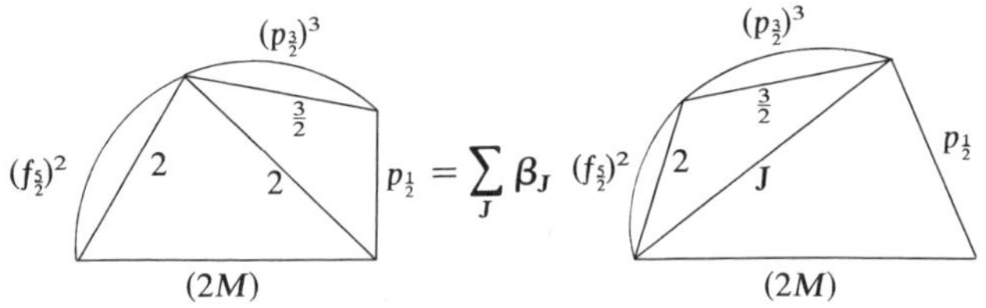

where β_J is given by the overlap integral

$$\beta_J=\langle[[(f_{\frac{5}{2}})^2_2\times(p_{\frac{3}{2}})^3_{\frac{3}{2}}]_J\times p_{\frac{1}{2}}]_{2M}\,|\,[(f_{\frac{5}{2}})^2_2\times[(p_{\frac{3}{2}})^3_{\frac{3}{2}}\times p_{\frac{1}{2}}]_2]_{2M}\rangle$$

$$=\sum_{M'm'm''M''}\;\sum_{\bar{M}'\bar{m}'\bar{m}''\bar{M}''}(2\tfrac{3}{2}M'm'\,|\,JM'')(J\tfrac{1}{2}M''m''\,|\,2M)(22\bar{M}'\bar{M}''\,|\,2M)$$

$$\times(\tfrac{3}{2}\tfrac{1}{2}\bar{m}'\bar{m}''\,|\,2\bar{M}'')\langle(f_{\frac{5}{2}})^2_{2M'}(p_{\frac{3}{2}})^3_{\frac{3}{2}m'}(p_{\frac{1}{2}})^1_{\frac{1}{2}m''}\,|\,(f_{\frac{5}{2}})^2_{2\bar{M}'}(p_{\frac{3}{2}})^3_{\frac{3}{2}\bar{m}'}(p_{\frac{1}{2}})^1_{\frac{1}{2}\bar{m}''}\rangle \qquad (1.37)$$

and β_J^2 is the probability that the state $[[(f_{\frac{5}{2}})^2_2\times(p_{\frac{3}{2}})^3_{\frac{3}{2}}]_J\times p_{\frac{1}{2}}]_{2M}$ will be found in the configuration $[(f_{\frac{5}{2}})^2_2\times[(p_{\frac{3}{2}})^3_{\frac{3}{2}}\times p_{\frac{1}{2}}]_2]_{2M}$. Since the $(j)^n_{IM}$ wave functions are all orthonormal it follows that $M'=\bar{M}'$, $m'=\bar{m}'$ and $m''=\bar{m}''$. Furthermore in Appendix 4 we show that this overlap is independent of M, the z component of the resultant angular momentum. Thus

$$\beta_J=\sum_{M'm'm''M''\bar{M}''}(2\tfrac{3}{2}M'm'\mid JM'')(J\tfrac{1}{2}M''m''\mid 2M)(22M'\bar{M}''\mid 2M)(\tfrac{3}{2}\tfrac{1}{2}m'm''\mid 2\bar{M}'')$$
$$=\sqrt{\{5(2J+1)\}}W(2\tfrac{3}{2}2\tfrac{1}{2};J2)$$

where $W(2\frac{3}{2}2\frac{1}{2};J2)$ is the Racah coefficient (Racah 1942).

From equation (1.37) it is clear that the detailed structure of the $(f_{\frac{5}{2}})^2_{2M'}$ $(p_{\frac{3}{2}})^3_{\frac{3}{2}m'}$ and $(p_{\frac{1}{2}})^1_{\frac{1}{2}m''}$ states is not important, and one could equally well have had any other orthonormal wave functions ψ_{2M}, $\zeta_{\frac{3}{2}m'}$, and $\chi_{\frac{1}{2}m''}$ involved in the recoupling. Thus in general one may write

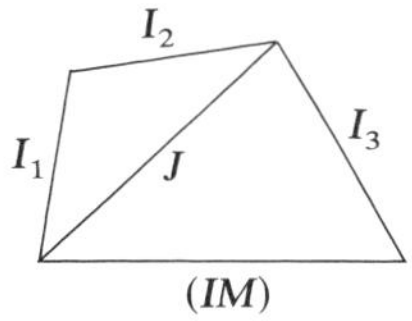

$$=\sum_k\sqrt{\{(2J+1)(2K+1)\}}W(I_1I_2II_3;JK)$$

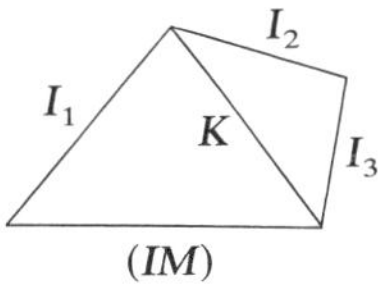

(1.38)

and

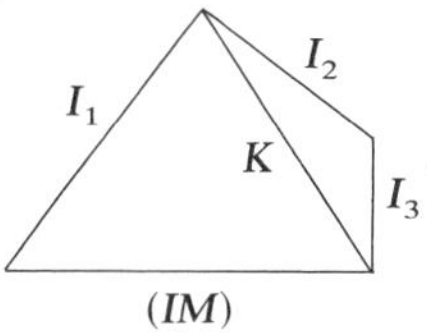

$$=\sum_J\sqrt{\{(2J+1)(2K+1)\}}\,W(I_1I_2II_3;JK)$$

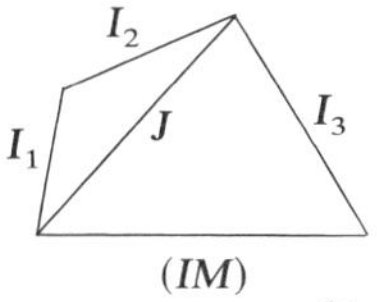

(1.39)

Recouplings such as these are frequently encountered in nuclear-structure calculations, and because of this we discuss in detail the properties of the Racah coefficients in Appendix 4.

1.4. *The shell-model Hamiltonian*

In shell-model theory the eigenvalues and eigenvectors used to describe nuclear states are obtained by diagonalizing the Hamiltonian, equation (1.12). Since the number of ways of obtaining a given I becomes excessively large when many single-particle orbits are involved, one must decide on a truncation procedure—in other words one must select carefully the appropriate set of single-particle levels in terms of which to diagonalize H. Since energy breaks occur at N and/or $Z=2$, 8, 20, 28, 50, 82, and 126, these numbers of neutrons and protons are usually characterized (at least to start with) as forming an inert core. In order to find the single-particle levels that the extra-core nucleons should occupy it is useful to consult the most recent issue of *Nuclear Data Sheets.** From the experimental spectrum of the nucleus with one nucleon outside the assumed closed shell one can determine which single-particle levels should be taken into account. The observed excitation energies together with the Wapstra–Gove (1971) binding-energy tables give the values of ε_j (the single-particle energies) to be used in the calculation. Once a model space has been decided upon the shell-model Hamiltonian including the residual two-body interaction is diagonalized using many-particle-basis states which are eigenfunctions of H_0.

Thus for n nucleons outside an inert core one proceeds (as outlined previously) to construct states of good angular momentum using the single particle ϕ_{jm} of equation (1.4). In addition to its spin, a nuclear state also has a parity Π defined by

$$\Pi = (-1)^{\Sigma l_i} \tag{1.40}$$

where the summation is over all occupied states. However, since $\sum l_i$ for a fully occupied level is even, the parity of a state is determined by the valence nucleons. It is designated by a + or − superscript to the right of the spin value.

In the next three Sections of this Chapter we shall discuss examples of shell-model calculations for single closed-shell nuclei (s.c.s. nuclei)—i.e. nuclei in which either the neutrons or protons form an inert closed shell. However, before doing this it will be convenient to give expressions for H in terms of the annihilation and creation operators introduced in section 1.2. We define $a^\dagger_{jm}$ and a_{jm} as the creation and destruction operators for particles in the state ϕ_{jm} which is an eigenfunction of H_0. Thus the operator

$$\mathrm{N}_j = \sum_m a^\dagger_{jm} a_{jm} \tag{1.41}$$

has the property that when it operates on any many-particle wave function constructed from ϕ_{jm} it gives the wave function back again multiplied by the number of particles in the state j. Thus, for example, when $\mathrm{N}_{j=\frac{7}{2}}$ operates on the state Φ_{44} of equation (1.33) it gives

$$\mathrm{N}_j \Phi_{44} = 4\Phi_{44}.$$

Consequently N_j is called the number operator for the state j, and if ε_j is the single-particle energy associated with this state one may write

$$H_0 = \sum_n (H_0)_n$$
$$= \sum_{jm} \varepsilon_j a^\dagger_{jm} a_{jm}.$$

* *Nuclear Data Sheets* is published by Academic Press at approximately monthly intervals and on page (ii) or (iii) of each issue references are given to the most recent compilations of the properties of nuclei.

The two-particle operator

$$A^{\dagger}_{JM}(j_1j_2)=(1+\delta_{j_1j_2})^{-\frac{1}{2}}\sum_{m_1m_2}(j_1j_2m_1m_2\mid JM)a^{\dagger}_{j_1m_1}a^{\dagger}_{j_2m_2} \tag{1.42}$$

creates a pair of nucleons in the single-particle states j_1 and j_2 with total angular momentum (JM). (For the special case $j_1=j_2$ only even values of J are allowed—Theorem 4, section 1.1.) The Hermitian adjoint operator

$$\begin{aligned}A_{JM}(j_3j_4)&=\{A^{\dagger}_{JM}(j_3j_4)\}^{\dagger}\\&=(1+\delta_{j_3j_4})^{-\frac{1}{2}}\sum_{m_3m_4}(j_3j_4m_3m_4\mid JM)a_{j_4m_4}a_{j_3m_3}\end{aligned} \tag{1.43}$$

is the one that destroys a pair of particles in the states j_3 and j_4 coupled to angular momentum (JM). Thus if only two-body residual interactions are considered the shell-model Hamiltonian of equation (1.12) may be written as

$$\begin{aligned}H&=\sum_n (H_0)_n+\sum_{i<j}V_{ij}\\&=\sum_{jm}\varepsilon_j a^{\dagger}_{jm}a_{jm}\\&\quad+\sum_{JM}\sum_{\{j_1j_2\}}\sum_{\{j_3j_4\}}E_J(j_1j_2;j_3j_4)A^{\dagger}_{JM}(j_1j_2)A_{JM}(j_3j_4)\end{aligned} \tag{1.44}$$

where the notation $\{j_ij_k\}$ means that in the summation $j_i\geq j_k$. The energies $E_J(j_1j_2;j_3j_4)$ are the matrix elements of the residual two-body interaction

$$E_J(j_1j_2;j_3j_4)=\langle\psi_{JM}(j_1j_2)|\,V_{12}\,|\psi_{JM}(j_3j_4)\rangle \tag{1.45}$$

where the two-particle wave functions $\psi_{JM}(j_1j_2)$ and $\psi_{JM}(j_3j_4)$ are the normalized antisymmetric states created by the $A^{\dagger}_{JM}(j_1j_2)$ of equation (1.42) and destroyed by $A_{JM}(j_3j_4)$ of equation (1.43).

It is sometimes convenient to get rid of the restriction $j_i\geq j_k$ in equation (1.44). This may be done by noting that, because of the anti-commutation relationship satisfied by $a^{\dagger}_{jm}$ (equation (1.28)) and the symmetry property (equation (A1.18)) of the Clebsch–Gordan coefficients,

$$A^{\dagger}_{JM}(j_1j_2)=-(-1)^{j_1+j_2-J}A^{\dagger}_{JM}(j_2j_1).$$

A similar relationship holds for $A_{JM}(j_3j_4)$ so that $E_J(j_1j_2;j_3j_4)$ has the following properties:

$$\begin{aligned}E_J(j_1j_2;j_3j_4)&=-(-1)^{j_1+j_2-J}E_J(j_2j_1;j_3j_4)\\&=-(-1)^{j_3+j_4-J}E_J(j_1j_2;j_4j_3)\\&=(-1)^{j_1+j_2+j_3+j_4}E_J(j_2j_1;j_4j_3)\\&=E_J(j_3j_4;j_1j_2)\end{aligned} \tag{1.46}$$

where the last of these properties arises because V is an Hermitian operator. By using these relationships it is possible to write

$$H = \sum_{jm} \varepsilon_j a^\dagger_{jm} a_{jm} + \tfrac{1}{4} \sum_{JM} \sum_{j_i} (1+\delta_{j_1 j_2})(1+\delta_{j_3 j_4}) E_J(j_1 j_2; j_3 j_4) A^\dagger_{JM}(j_1 j_2) A_{JM}(j_3 j_4) \quad (1.47)$$

where the sum on j_i is over j_1, j_2, j_3, j_4 with all restrictions removed.

Since $A^\dagger_{JM}(j_1 j_2)$ is an irreducible tensor operator of rank J (see Appendix 2, section 1), it follows that its modified adjoint operator, defined in general by equation (A2.9), is

$$\begin{aligned}\tilde{A}_{J-M} &= (-1)^{J-M}(A^\dagger_{JM})^\dagger \\ &= (-1)^{J-M} A_{JM}.\end{aligned}$$

This is also an irreducible tensor operator of rank J and because the Clebsch-Gordan coefficient $(JJM-M\,|\,00)$ given by equation (A1.21) is $(-1)^{J-M}/\sqrt{(2J+1)}$, an alternative way of writing H is

$$\begin{aligned}H &= \sum_{jm} \varepsilon_j a^\dagger_{jm} a_{jm} + \sum_J \sum_{\{j_1 j_2\}} \sum_{\{j_3 j_4\}} \sqrt{(2J+1)} E_J(j_1 j_2; j_3 j_4)[A^\dagger_J(j_1 j_2) \times \tilde{A}_J(j_3 j_4)]_{00} \\ &= \sum_{jm} \varepsilon_j a^\dagger_{jm} a_{jm} + \tfrac{1}{4} \sum_J \sum_{j_i} (1+\delta_{j_1 j_2})(1+\delta_{j_3 j_4}) \sqrt{(2J+1)} E_J(j_1 j_2; j_3 j_4)[A^\dagger_J(j_1 j_2) \times \tilde{A}_J(j_3 j_4)]_{00}.\end{aligned} \quad (1.48)$$

The extension of these ideas to the case where there are both neutrons and protons is straightforward and will be discussed in Chapter 2.

In addition to the j_+ operator in equation (1.32) and the number operator in equation (1.41) we shall frequently encounter other single-particle operators the matrix elements of which we may wish to know. For example, if one wants to calculate the quadrupole moment of the nucleus one must compute matrix elements of the operator $\sqrt{(16\pi/5)} \times \sum_i e_i r_i^2 Y_{20}(\theta_i, \phi_i)$ where e_i is the electric charge of the ith particle. One may easily show that this operator is an irreducible tensor of rank 2 (see Appendix 2, section 1); i.e. it satisfies the relationship

$$\begin{aligned}[\mathrm{J}_z, Q_{\lambda\mu}] &= \mu Q_{\lambda\mu} \\ [\mathrm{J}_\pm, Q_{\lambda\mu}] &= \sqrt{\{(\lambda \mp \mu)(\lambda \pm \mu + 1)\}} Q_{\lambda\mu\pm 1}\end{aligned} \quad (1.49)$$

with $\lambda = 2$ and $\mu = 0$. Matrix elements of operators satisfying equation (1.49) obey the Wigner–Eckart theorem proved in Appendix 2, section 1 (Eckart 1930, Wigner 1959) and have the property that

$$\langle \Psi_{I'M'} | \, Q_{\lambda\mu} \, | \Psi_{IM} \rangle = (I\lambda M\mu \mid I'M') \langle \Psi_{I'} || \, Q_\lambda \, || \Psi_I \rangle \tag{1.50}$$

where the reduced matrix element $\langle \Psi_{I'} || Q_\lambda || \Psi_I \rangle$ is independent of the z components of angular momentum. Thus all the m-dependence of the matrix element is contained in the Clebsch–Gordan coefficient $(I\lambda M\mu \mid I'M')$. Consequently any irreducible tensor operator of rank λ which is the sum of single-particle operators can be written as

$$\begin{aligned} Q_{\lambda\mu} &= \sum_i (Q_{\lambda\mu})_i \\ &= \sum_{jm} \sum_{j_1 m_1} \langle \phi_{j_1} || \, Q_\lambda \, || \phi_j \rangle (j\lambda m\mu \mid j_1 m_1) a^\dagger_{j_1 m_1} a_{jm} \end{aligned} \tag{1.51}$$

where $\langle \phi_{j_1} || Q_\lambda || \phi_j \rangle$ is the single-particle reduced matrix element. In Appendix 2 the commonly encountered single-particle reduced matrix elements are evaluated.

Equation (1.51) can also be written in a somewhat different way. As shown in Appendix 2 the modified Hermitian adjoint operator

$$\begin{aligned} \tilde{a}_{j-m} &= (-1)^{j-m} (a^\dagger_{jm})^\dagger \\ &= (-1)^{j-m} a_{jm} \end{aligned} \tag{1.52}$$

is an irreducible tensor operator of rank $(j, -m)$. When this result is combined with the symmetry properties given in equations (A1.18)–(A1.20) of the Clebsch–Gordan coefficients

$$\begin{aligned} (j\lambda m\mu \mid j_1 m_1) &= (-1)^{j-m} \left(\frac{2j_1+1}{2\lambda+1} \right)^{\frac{1}{2}} (j j_1 m - m_1 \mid \lambda - \mu) \\ &= (-1)^{j-m} \left(\frac{2j_1+1}{2\lambda+1} \right)^{\frac{1}{2}} (j_1 j m_1 - m \mid \lambda\mu) \end{aligned}$$

one sees that

$$\begin{aligned} Q_{\lambda\mu} &= \sum_i (Q_{\lambda\mu})_i \\ &= \sum_{jj_1} \left(\frac{2j_1+1}{2\lambda+1} \right)^{\frac{1}{2}} \langle \phi_{j_1} || \, Q_\lambda \, || \phi_j \rangle [a^\dagger_{j_1} \times \tilde{a}_j]_{\lambda\mu}. \end{aligned} \tag{1.53}$$

By use of the completeness relationship $\sum_{I'M'} |\Psi_{I'M'}\rangle\langle\Psi_{I'M'}| = 1$, the ortho-normality of the Clebsch-Gordan coefficients, equation (A1.23), and the definition of the reduced matrix element, equation (1.50), one sees that

$$\sum_{M\mu} (I\lambda M\mu \mid I'M')Q_{\lambda\mu}\Psi_{IM} = \langle\Psi_{I'}\| Q_\lambda \|\Psi_I\rangle\Psi_{I'M'}.$$

In terms of diagrams this relationship takes the form

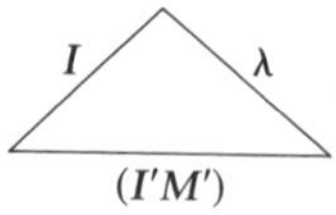

$$= \langle\Psi_{I'}\| Q_\lambda \|\Psi_I\rangle \underset{(I'M')}{\underline{\qquad\qquad}}\,. \tag{1.54}$$

2. Two particles outside a closed shell

To find the appropriate single-particle levels (model space) to be used in the description of the two-particle system, one must first look at the experimentally observed states for a single nucleon outside the assumed inert core. If there is only one low-lying state j an adequate model for the two-particle system will be one in which both nucleons are restricted to this orbit. Thus the wave function of the two-particle system would be

$$\psi_{IM}(jj) = 2^{-\frac{1}{2}} \sum_{m_1 m_2} (jjm_1 m_2 \mid IM) a^\dagger_{jm_1} a^\dagger_{jm_2} |0\rangle.$$

If ε_j is the energy of the single-particle state j relative to the closed core and $E_I(jj; jj)$ is the interaction energy when the two particles couple to angular momentum I, the energy of this state relative to the closed shell is

$$\begin{aligned} \langle\psi_{IM}(jj)| H |\psi_{IM}(jj)\rangle &= 2\varepsilon_j + \langle\psi_{IM}(jj)| V |\psi_{IM}(jj)\rangle. \\ &= 2\varepsilon_j + E_I(jj; jj) \end{aligned} \tag{1.55}$$

An example of a situation where such a model could be used is ${}^{50}_{22}\mathrm{Ti}_{28}$. In this case ${}^{48}_{20}\mathrm{Ca}_{28}$ is assumed to be the inert core. The only level in ${}^{49}_{21}\mathrm{Sc}_{28}$ below 2 MeV excitation energy is the $f_{\frac{7}{2}}$ ground state (Raman 1970). Consequently as long as we deal with states below about 2 MeV excitation energy an adequate model space for treating the two extracore protons should be one in which they are confined to the $0f_{\frac{7}{2}}$ level.

More usual is the case that several single-particle levels must be considered in the description of low-lying states. When this is so there will generally be more than one way to construct a state with given angular momentum. For simplicity assume there are two states denoted by $\psi_{IM}(j_1 j_2)$ and $\psi_{IM}(j_3 j_4)$. For this value of I the eigenfunctions of H will be linear combinations of these two wave functions satisfying the equation

$$H\{\alpha_1\psi_{IM}(j_1 j_2) + \alpha_2\psi_{IM}(j_3 j_4)\} = \tilde{E}\{\alpha_1\psi_{IM}(j_1 j_2) + \alpha_2\psi_{IM}(j_3 j_4)\}.$$

Since the ψ_{IM} are products of the eigenfunctions ϕ_{jm} of H_0 it follows that if we multiply this equation on the left by $\psi_{IM}(j_1j_2)$ and integrate over all space we obtain the result

$$\alpha_1\{\varepsilon_{j_1}+\varepsilon_{j_2}+E_I(j_1j_2;\, j_1j_2)\}+\alpha_2E_I(j_1j_2;\, j_3j_4)=\alpha_1\tilde{E}$$

where the $E_I(j_1j_2;\, j_3j_4)$ are given by equation (1.45). Alternatively, if we multiply by $\psi_{IM}(j_3j_4)$ we find

$$\alpha_1E_I(j_3j_4;\, j_1j_2)+\alpha_2\{\varepsilon_{j_3}+\varepsilon_{j_4}+E_I(j_3j_4;\, j_3j_4)\}=\alpha_2\tilde{E}.$$

These two equations can be written as

$$\sum_k\{\langle H\rangle_{ik}-\tilde{E}\delta_{ik}\}\alpha_k=0 \qquad (i=1 \text{ and } 2) \tag{1.56}$$

where

$$\langle H\rangle_{11}=\varepsilon_{j_1}+\varepsilon_{j_2}+E_I(j_1j_2;\, j_1j_2)$$
$$\langle H\rangle_{22}=\varepsilon_{j_3}+\varepsilon_{j_4}+E_I(j_3j_4;\, j_3j_4)$$
$$\langle H\rangle_{12}=\langle H\rangle_{21}=E_I(j_1j_2;\, j_3j_4).$$

(In writing the last of these equations we have made use of the fact that V is Hermitian.)

If

$$\left|\frac{E_I(j_1j_2;\, j_3j_4)}{\varepsilon_{j_1}+\varepsilon_{j_2}+E_I(j_1j_2;\, j_1j_2)-\varepsilon_{j_3}-\varepsilon_{j_4}-E_I(j_3j_4;\, j_3j_4)}\right|\ll 1$$

one can use perturbation theory to compute $\tilde{E}$. However, if this is not the case one must find $\tilde{E}$ from the condition that the determinant of the coefficients of α_k must vanish. In this case the eigenvalues and eigenfunctions are

$$\tilde{E}_\pm=\frac{(\langle H\rangle_{11}+\langle H\rangle_{22})\pm\sqrt{\{(\langle H\rangle_{11}-\langle H\rangle_{22})^2+4\langle H\rangle_{12}^2\}}}{2} \tag{1.57}$$

$$\psi_+=\{(\tilde{E}_+-\langle H\rangle_{11})^2+\langle H\rangle_{12}^2\}^{-\frac{1}{2}}(\langle H\rangle_{12}\psi_{IM}(j_1j_2)+(\tilde{E}_+-\langle H\rangle_{11})\psi_{IM}(j_3j_4)) \tag{1.58a}$$

$$\psi_-=\{(\tilde{E}_+-\langle H\rangle_{11})^2+\langle H\rangle_{12}^2\}^{-\frac{1}{2}}((\tilde{E}_+-\langle H\rangle_{11})\psi_{IM}(j_1j_2)-\langle H\rangle_{12}\psi_{IM}(j_3j_4)). \tag{1.58b}$$

2.1. $d_{\frac{5}{2}}s_{\frac{1}{2}}$ *model of* ${}^{18}_{8}O_{10}$

The simplest non-trivial shell-model calculation is the one involving two nucleons outside an inert core. To illustrate how this calculation is carried out we treat the nucleus ${}^{18}_{8}O_{10}$ on the assumption that ${}^{16}_{8}O_8$ is an inert closed shell. From the compilation of Ajzenberg-Selove (1977) we see that below 3 MeV excitation energy the only states observed in the 'single-particle nucleus' ${}^{17}_{8}O_9$ are the $d_{\frac{5}{2}}$ and $s_{\frac{1}{2}}$ levels whose energies, relative to the closed ${}^{16}_{8}O_8$ core, are given by equations (1.9) and (1.10). Consequently an appropriate model space for a study of the low-lying states of ${}^{18}_{8}O_{10}$ is one in which the two valence neutrons are restricted to these two orbits. The possible angular momenta are

$$\begin{aligned} &(d_{\frac{5}{2}})^2_I && I=0, 2, \text{and } 4\\ &(d_{\frac{5}{2}}s_{\frac{1}{2}})_I && I=2 \text{ and } 3\\ &(s_{\frac{1}{2}})^2_I && I=0. \end{aligned}$$

Since there is only one way to obtain $I=3$ and 4, the Hamiltonian matrix for these states is (1×1) and their energies are given by an expression similar to equation (1.55). However, to find the eigenvalues and eigenfunctions for the $I=0$ and 2 states one must construct and diagonalize a (2×2) matrix.

To calculate the spectrum of this nucleus we assume that the residual nucleon–nucleon interaction V_{ij} is the surface-delta potential (see Appendix 2, section 9). This interaction is the delta-function potential

$$V_{ij} = -4\pi V_0 \delta(\underline{r}_i - \underline{r}_j)$$

together with the restriction that the magnitude of the radial integrals involved

$$\bar{R} = \int R_{j_1}(r)R_{j_2}(r)R_{j_3}(r)R_{j_4}(r)r^2\,\mathrm{d}r$$

be set equal to a constant. As discussed in Appendix 2, section 4.9, this latter condition is approximately the result that would emerge if the $R_j(r)$ were eigenfunctions of the Woods–Saxon potential of equation (1.7). To make the sign of $\bar{R}$ the same as would be given if the matrix element was explicitly evaluated with wave functions that are positive at $r\to 0$ we must set

$$\bar{R} = (-1)^{n_1+n_2+n_3+n_4}R_0$$

where n_i is the number of radial nodes for the state j_i and R_0 is a positive number.

When we deal with two identical particles, the matrix elements of this potential are given by equation (A2.29) with $T=1$

$$\begin{aligned} E_I(j_1j_2; j_3j_4) = {} & (-1)^{j_1+j_3+l_2+l_4+n_1+n_2+n_3+n_4}\{1+(-1)^{l_1+l_2+l_3+l_4}\}\\ & \times\{1+(-1)^{l_3+l_4+I}\}\\ & \times\frac{V_0R_0}{4(2I+1)}\left\{\frac{(2j_1+1)(2j_2+1)(2j_3+1)(2j_4+1)}{(1+\delta_{j_1j_2})(1+\delta_{j_3j_4})}\right\}^{\frac{1}{2}}\\ & \times(j_1j_2\tfrac{1}{2}-\tfrac{1}{2}\mid I0)(j_3j_4\tfrac{1}{2}-\tfrac{1}{2}\mid I0). \end{aligned} \tag{1.59}$$

By use of the tables of Clebsch–Gordan coefficients (Rotenberg *et al.* 1959) one easily shows that

$$\langle H_{I=0}\rangle = \begin{pmatrix} 2\varepsilon_d - 3V_0R_0 & -\sqrt{3}V_0R_0 \\ -\sqrt{3}V_0R_0 & 2\varepsilon_s - V_0R_0 \end{pmatrix}$$

$$\langle H_{I=2}\rangle = \begin{pmatrix} 2\varepsilon_d - \frac{24}{35}V_0R_0 & \dfrac{-12\sqrt{7}}{35}V_0R_0 \\ \dfrac{-12\sqrt{7}}{35}V_0R_0 & \varepsilon_d + \varepsilon_s - \frac{6}{5}V_0R_0 \end{pmatrix}$$

$$\langle H_{I=3}\rangle = \varepsilon_d + \varepsilon_s$$

$$\langle H_{I=4}\rangle = 2\varepsilon_d - \tfrac{2}{7}V_0R_0.$$

One may fix the value of V_0R_0 by requiring that the observed excitation energy of one of the states be reproduced or by requiring that the experimental binding energy of ${}^{18}_{8}O_{10}$ relative to ${}^{16}_{8}O_8$ be obtained. We somewhat arbitrarily require the latter condition to be fulfilled—i.e. the lower eigenvalue of $\langle H_{I=0}\rangle$ should be

$$E_0 = \mathrm{BE}({}^{18}_{8}O_{10}) - \mathrm{BE}({}^{16}_{8}O_8)$$
$$= -12{\cdot}189\ \mathrm{MeV}.$$

For this to be true

$$V_0R_0 = 1{\cdot}057\ \mathrm{MeV}. \tag{1.60}$$

In Fig. 1.4 we show the spectrum that would result with this choice of V_0R_0 and use of the $d_{\frac{5}{2}}$ and $s_{\frac{1}{2}}$ single-particle energies given by equations (1.9) and (1.10). Clearly, with the exception of the 3·63 MeV 0^+ level, which we shall discuss in a moment, there is a one-to-one correspondence between the predicted and observed levels up to about 5 MeV excitation energy. The quality of agreement between theory and experiment can be somewhat improved by taking a finite-range spin-dependent residual two-body interaction. However, with the exception of the first excited 0^+ state, the main qualitative features are just those predicted by the surface-delta interaction.

In addition to obtaining the spectrum of the nucleus, we obtain wave functions for the various states. The eigenfunction corresponding to the ground state of ${}^{18}_{8}O_{10}$ is

$$\psi_{00} = 0{\cdot}929(d_{\frac{5}{2}})^2_{00} + 0{\cdot}371(s_{\frac{1}{2}})^2_{00}. \tag{1.61}$$

Thus the ground state is about 86% $(d_{\frac{5}{2}})^2_{00}$ and 14% $(s_{\frac{1}{2}})^2_{00}$.

In a similar manner the eigenfunction of the first excited 2^+ state is

$$\psi_{2M} = 0{\cdot}764(d_{\frac{5}{2}})^2_{2M} + 0{\cdot}645(d_{\frac{5}{2}}s_{\frac{1}{2}})_{2M} \tag{1.62}$$

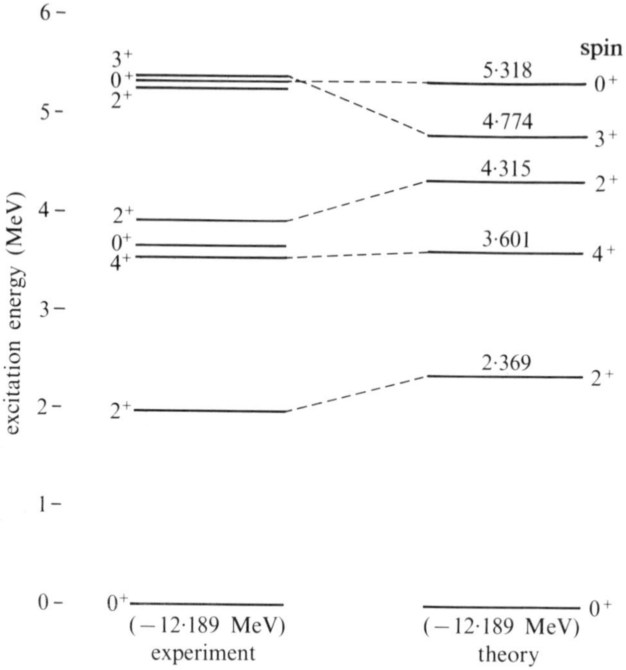

Fig. 1.4. The spectrum ${}^{18}_{8}O_{10}$. The ground-state energy is measured in MeV relative to ${}^{16}_{8}O_8$, but all other energies are excitation energies. The experimental spectrum was taken from the compilation of Ajzenberg–Selove (1972). The theoretical spectrum is based on the $(d_{\frac{5}{2}}, s_{\frac{1}{2}})$ model space with single-particle energies given by equations (1.9) and (1.10). The surface-delta interaction given by equation (1.59) with $V_0R_0 = 1{\cdot}057$ MeV was used for the residual two-body potential.

so that the 2^+ state is much more strongly mixed than the ground state. In calculating the energies of the system the only place where the phase factor $(-1)^{n_1+n_2+n_3+n_4}$ gave other than unity was for the matrix element $E_2(\frac{5}{2}, \frac{5}{2}; \frac{5}{2}, \frac{1}{2}) \equiv \langle H_{I=2}\rangle_{12}$. However, in the energy calculation $\langle H_{I=2}\rangle_{12}$ always comes in squared so that the energy eigenvalues do not depend on the sign of this matrix element. On the other hand, if one changes the sign of $\langle H_{I=2}\rangle_{12}$ the coefficient of $(d_{\frac{5}{2}}s_{\frac{1}{2}})_{2M}$ in equation (1.62) changes sign and consequently the contribution that would come from this term to say the $2^+ \rightarrow 0^+$ gamma decay would have its sign changed. (We shall discuss this further in Chapter 5, section 2.) It is for this reason that one must be careful and take the sign of $\bar{R}$ in equation (A2.29) to be consistent with that of the radial wave functions used in calculating transition rates or other properties.

Returning now to the excited 0^+ level, it turns out that if one uses oscillator wave functions and a finite-range potential the predicted excitation energy of the level with a structure that is predominantly $(s_{\frac{1}{2}})_0^2$ is about 3 MeV (Elliott and Flowers 1955). The reason for this is that the oscillator which has infinite walls confines the radial motion of a particle much more than does the Woods–Saxon potential. This is particularly true for small values of the orbital angular momentum (Inoue *et al.* 1964). For high l values the particle is naturally constrained by the centrifugal barrier $l(l+1)/r^2$. Thus oscillator wave functions badly overestimate the value of $E_0(\frac{1}{2}\frac{1}{2};\frac{1}{2}\frac{1}{2})$ and hence lead to the prediction that the $(s_{\frac{1}{2}})_0^2$ state lies low in excitation energy. On the other hand, we shall now show that a 0^+ level is predicted at an excitation energy of about 3·5 MeV when break-up of the ${}^{16}_{8}O_8$ core is taken into account. Once this is demonstrated it follows that the spectrum of ${}^{18}_{8}O_{10}$ up to 5 MeV excitation energy can be quite adequately explained when the residual nucleon–nucleon interaction is the surface delta potential (i.e. when the potential has short range and the matrix elements are calculated with single-particle radial wave functions that take into account the finite well depth of the shell-model potential).

2.2. *Core-excited states*

So far in describing the states of ${}^{18}_{8}O_{10}$ we have assumed ${}^{16}_{8}O_8$ is inert and that the valence nucleons are in the $(0d, 1s)$ shell. These states are shown schematically in Fig. 1.5 and correspond to the configuration $(0s)^4(0p)^{12}(0d, 1s)^2$. It was pointed out by Brown (1964) and Engeland (1965) that the configuration $(0s)^4(0p)^{10}(0d, 1s)^4$, in which two protons are excited out of the $0p$ shell up to the $(0d, 1s)$ shell, should lie at low-excitation energy in ${}^{18}_{8}O_{10}$. This configuration is also shown schematically in Fig. 1.5 and we now estimate where one might expect to find it in ${}^{18}_{8}O_{10}$.

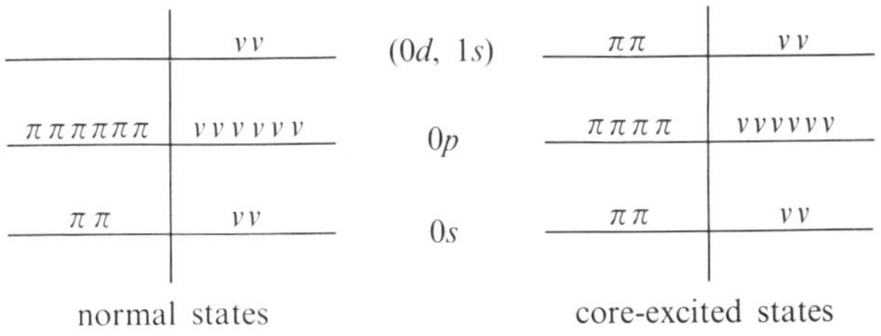

Fig. 1.5. Schematic diagram of the normal (model-space) states of ${}^{18}_{8}O_{10}$ and the core-excited 0^+ level corresponding to a four-particle two-hole configuration. π stands for protons and ν for neutrons.

If the spins of the four particles in the $(0d, 1s)$ shell are coupled so that the lowest possible energy state is realized, the nucleons will be in the configuration associated with the ${}^{20}_{10}Ne_{10}$ ground state. The energy of these particles relative to ${}^{16}_{8}O_8$ is therefore given by

$$E_{ds} = BE({}^{20}_{10}Ne_{10}) - BE({}^{16}_{8}O_8)$$
$$= -33{\cdot}027 \text{ MeV}$$

where BE(X) is the negative of the binding energy of the nucleus X and can be found in the tabulation of Wapstra and Gove (1971). To have the lowest possible energy, the particles in the $0s$ and $0p$ orbits will be in the configuration attributed to the ${}^{14}_{6}C_8$ ground state. The energy of these nucleons relative to ${}^{16}_{8}O_8$ is

$$E_{sp} = BE({}^{14}_{6}C_8) - BE({}^{16}_{8}O_8)$$
$$= 22{\cdot}335 \text{ MeV}.$$

Since there are four particles in the $(0d, 1s)$ shell and two holes in the $0p$ orbit, one must take into account the particle-hole interaction. If we define the average interaction between an (sd) particle and a $0p$ hole as E', the particle-hole interaction in this four-particle two-hole state will be $8E'$. Thus the energy of the four-particle two-hole state relative to ${}^{16}_{8}O_8$ is

$$\Delta E = E_{ds} + E_{sp} + 8E'$$
$$= (-10{\cdot}692 + 8E') \text{ MeV}. \qquad (1.63)$$

To estimate E' we make use of the fact that the first excited state in ${}^{19}_{9}F_{10}$ is a $\frac{1}{2}^-$ level corresponding to the excitation of a proton from $0p$ up to the $(0d, 1s)$ shell. The structure of the $\frac{1}{2}^+$ ground state and this $\frac{1}{2}^-$ level are shown in Fig. 1.6. Again, for lowest energy the four (sd) particles will be in the configuration associated with the ${}^{20}_{10}Ne_{10}$ ground state and their energy relative to ${}^{16}_{8}O_8$ is given by E_{ds}. For minimum energy, the $0s$ and $0p$ orbits should have exactly the structure of the ${}^{15}_{7}N_8$ ground state so

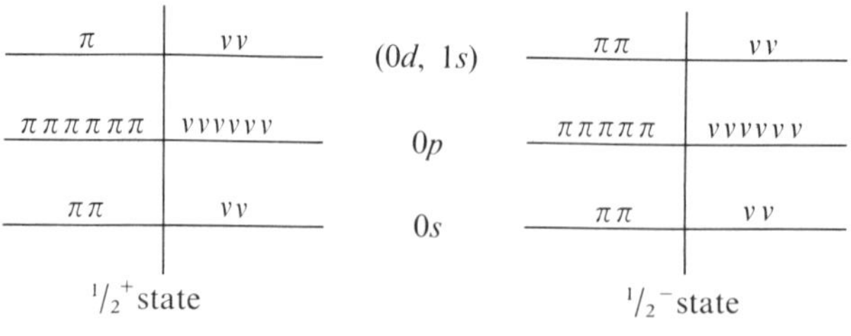

Fig. 1.6. Schematic diagram for the $\frac{1}{2}^+$ ground state of ${}^{19}_{9}F_{10}$ and the 110 keV $\frac{1}{2}^-$ level which is assumed to be a four-particle one-hole state.

that

$$E'_{sp} = \mathrm{BE}({}^{15}_{7}\mathrm{N}_8) - \mathrm{BE}({}^{16}_{8}\mathrm{O}_8)$$
$$= 12{\cdot}128 \text{ MeV}.$$

In this description of the $\frac{1}{2}^-$ state there is only one $0p$ hole so the particle-hole interaction is $4E'$. Consequently the energy of this state relative to ${}^{16}_{8}\mathrm{O}_8$ is

$$E_{\frac{1}{2}^-} = (-20{\cdot}899 + 4E') \text{ MeV}.$$

Since the energy of the ${}^{19}_{9}\mathrm{F}_{10}$ ground state relative to ${}^{16}_{8}\mathrm{O}_8$ is

$$E_{\frac{1}{2}^+} = \mathrm{BE}({}^{19}_{9}\mathrm{F}_{10}) - \mathrm{BE}({}^{16}_{8}\mathrm{O}_8)$$
$$= -20{\cdot}182 \text{ MeV}$$

it follows that if the $\frac{1}{2}^-$ level is to lie 110 keV above the $\frac{1}{2}^+$ level

$$4E' = 0{\cdot}827 \text{ MeV}.$$

With this value of E', ΔE given by equation (1.63) is

$$\Delta E = -9{\cdot}038 \text{ MeV}.$$

The energy of the ${}^{18}_{8}\mathrm{O}_{10}$ ground state relative to ${}^{16}_{8}\mathrm{O}_8$ was already found to be $-12{\cdot}189$ MeV so that the four-particle two-hole state shown in Fig. 1.5 is estimated to lie at an excitation energy of $(-9{\cdot}038 + 12{\cdot}189) = 3{\cdot}151$ MeV. Experimentally the first excited 0^+ state is seen at 3·63 MeV.

Thus the fact that the surface-delta interaction places the state that is dominantly $(s_{\frac{1}{2}})^2_0$ near 5 MeV (where a second excited 0^+ is seen) is in agreement with the expectation that the first excited 0^+ in ${}^{18}_{8}\mathrm{O}_{10}$ is a four-particle two-hole state where the four particles have the structure of the ${}^{20}_{10}\mathrm{Ne}_{10}$ ground state. If this picture is correct one would expect to see additional states corresponding to excited states of ${}^{20}_{10}\mathrm{Ne}_{10}$. For example, the analogue of the 1·63 MeV 2^+ state in ${}^{20}_{10}\mathrm{Ne}_{10}$ (Ajzenberg–Selove 1972) would be expected to occur about 1·63 MeV above the 3·63 MeV first-excited 0^+ level in ${}^{18}_{8}\mathrm{O}_{10}$—i.e. at about 5·26 MeV. This prediction is in excellent agreement with the experimental observation of a 2^+ level at 5·25 MeV excitation energy. The next excited state in ${}^{20}_{10}\mathrm{Ne}_{10}$ is the 4^+ level at 4·25 MeV and its analogue would be expected to lie at about $3{\cdot}63 + 4{\cdot}25 = 7{\cdot}88$ MeV. Although not shown in Fig. 1.4, a 4^+ level does exist in ${}^{18}_{8}\mathrm{O}_{10}$ at 7·11 MeV. An analysis of all the data on ${}^{18}_{8}\mathrm{O}_{10}$ including static moments, gamma-decay transition rates, and one- and two-nucleon transfer cross-sections shows that the wave functions of the 3·63 MeV 0^+, the 5·25 MeV 2^+, and the 7·11 MeV 4^+ states are dominated by a four-particle two-hole component (Lawson *et al.* 1976).

3. *The configuration j^n*

In section 1.2 we showed how one can enumerate the possible angular-momentum states for n particles in the single-particle state j. In this Section we shall show how to calculate the spectrum of a nucleus in which the valence nucleons are all in the same single-particle state. To do this we must first introduce the idea of parentage coefficients.

3.1. *Parentage coefficients*

To illustrate the concepts involved in fractional-parentage coefficients we consider the calculation of the matrix elements of $V=\sum_{i<j} V_{ij}$ for the three-particle configuration $(0f_{\frac{7}{2}})^3_{I=\frac{15}{2}}$. Since V is rotationally invariant it is an irreducible tensor operator of rank zero: i.e. it satisfies equation (1.49) with $\lambda=\mu=0$. Since the Clebsch–Gordan coefficient $(I0M0\mid I'M')=\delta_{II'}\delta_{MM'}$ it follows from the Wigner-Eckart theorem (equation (1.50)) that V has non-vanishing matrix elements only when $I=I'$ and moreover that these matrix elements are independent of M. Thus in our example it is sufficient to evaluate V in the $M=\frac{15}{2}$ state.

From Theorem 1, section 1.1 one sees that the single Slater determinant

$$\Psi_{\frac{15}{2}\frac{15}{2}}=|\phi_{\frac{7}{2}\frac{7}{2}}\phi_{\frac{7}{2}\frac{5}{2}}\phi_{\frac{7}{2}\frac{3}{2}}|$$

$$=6^{-\frac{1}{2}}\begin{vmatrix}\phi_{\frac{7}{2}\frac{7}{2}}(1)\phi_{\frac{7}{2}\frac{5}{2}}(1)\phi_{\frac{7}{2}\frac{3}{2}}(1)\\ \phi_{\frac{7}{2}\frac{7}{2}}(2)\phi_{\frac{7}{2}\frac{5}{2}}(2)\phi_{\frac{7}{2}\frac{3}{2}}(2)\\ \phi_{\frac{7}{2}\frac{7}{2}}(3)\phi_{\frac{7}{2}\frac{5}{2}}(3)\phi_{\frac{7}{2}\frac{3}{2}}(3)\end{vmatrix}$$

gives the wave function of the $I=\frac{15}{2}M=\frac{15}{2}$ state of $(0f_{\frac{7}{2}})^3$. In this case the matrix element of V becomes a sum of three terms

$$\langle\Psi_{\frac{15}{2}\frac{15}{2}}|\sum_{i<j} V_{ij}\,|\Psi_{\frac{15}{2}\frac{15}{2}}\rangle=\langle\Psi_{\frac{15}{2}\frac{15}{2}}|\,V_{12}+V_{13}+V_{23}\,|\Psi_{\frac{15}{2}\frac{15}{2}}\rangle.$$

To evaluate the first term, V_{12}, it is convenient to rewrite the wave function in a form in which particles 1 and 2 are explicitly separated from particle 3. This may be done by expanding the Slater determinant in terms of the minors of the last row. Thus

$$\Psi_{\frac{15}{2}\frac{15}{2}}=3^{-\frac{1}{2}}\{|\phi_{\frac{7}{2}\frac{5}{2}}\phi_{\frac{7}{2}\frac{3}{2}}|\,\phi_{\frac{7}{2}\frac{7}{2}}(3)-|\phi_{\frac{7}{2}\frac{7}{2}}\phi_{\frac{7}{2}\frac{3}{2}}|\,\phi_{\frac{7}{2}\frac{5}{2}}(3)+|\phi_{\frac{7}{2}\frac{7}{2}}\phi_{\frac{7}{2}\frac{5}{2}}|\,\phi_{\frac{7}{2}\frac{3}{2}}(3)\}.$$

One may make an angular-momentum decomposition of the two-particle determinant by using the result proved in Theorem 4, section 1.1, namely that for even J

$$\Phi_{JM}(1,2)=\tfrac{1}{2}\sum_{mm'}(\tfrac{7}{2}\tfrac{7}{2}mm'\mid JM)\{\phi_{\frac{7}{2}m}(1)\phi_{\frac{7}{2}m'}(2)-\phi_{\frac{7}{2}m}(2)\phi_{\frac{7}{2}m'}(1)\}$$

$$=2^{-\frac{1}{2}}\sum_{mm'}(\tfrac{7}{2}\tfrac{7}{2}mm'\mid JM)\,|\phi_{\frac{7}{2}m}\phi_{\frac{7}{2}m'}|$$

When use is made of the completeness relationship for the Clebsch–Gordan coefficients (equation (A1.24)) one finds that

$$|\phi_{\frac{7}{2}m}\phi_{\frac{7}{2}m'}|=\sqrt{2}\sum_{JM}(\tfrac{7}{2}\tfrac{7}{2}mm'\mid JM)\Phi_{JM}(1,2).$$

By using the tables of Clebsch–Gordan coefficients (Rotenberg *et al.* 1959) one can show that

$$\begin{aligned}\Psi_{\frac{15}{2}\frac{15}{2}}=3^{-\frac{1}{2}}[&-\sqrt{(\tfrac{15}{22})}\Phi_{44}(1,2)\phi_{\frac{7}{2}\frac{7}{2}}(3)\\&+\sqrt{(\tfrac{51}{22})}\{\sqrt{(\tfrac{22}{51})}\Phi_{66}(1,2)\phi_{\frac{7}{2}\frac{3}{2}}(3)\\&-\sqrt{(\tfrac{22}{51})}\Phi_{65}(1,2)\phi_{\frac{7}{2}\frac{5}{2}}(3)\\&+\sqrt{(\tfrac{7}{51})}\Phi_{64}(1,2)\phi_{\frac{7}{2}\frac{7}{2}}(3)\}].\end{aligned}$$

The coefficients $\sqrt{\tfrac{22}{51}}$, $-\sqrt{\tfrac{22}{51}}$, and $\sqrt{\tfrac{7}{51}}$ are precisely the Clebsch–Gordan coefficients that ensure that spin 6 and spin $\tfrac{7}{2}$ couple to $I=\tfrac{15}{2}$ $M=\tfrac{15}{2}$. Furthermore, the Clebsch–Gordan coefficient $(4\tfrac{7}{2}4\tfrac{7}{2}\mid\tfrac{15}{2}\tfrac{15}{2})\equiv 1$ so that

$$\begin{aligned}\Psi_{\frac{15}{2}\frac{15}{2}}=&\sqrt{(\tfrac{17}{22})}[\Phi_6(1,2)\times\phi_{\frac{7}{2}}(3)]_{\frac{15}{2}\frac{15}{2}}\\&-\sqrt{(\tfrac{5}{22})}[\Phi_4(1,2)\times\phi_{\frac{7}{2}}(3)]_{\frac{15}{2}\frac{15}{2}}.\end{aligned}\tag{1.65}$$

Because V_{12} is a scalar it cannot change the M quantum number of Φ_{IM}. In addition since V_{12} does not operate on particle 3 the m value of $\phi_{\frac{7}{2}m}(3)$ is unchanged. Thus

$$\begin{aligned}\langle\Psi_{\frac{15}{2}\frac{15}{2}}|\,V_{12}\,|\Psi_{\frac{15}{2}\frac{15}{2}}\rangle=&\tfrac{17}{22}\sum_{mM}(6\tfrac{7}{2}Mm\mid\tfrac{15}{2}\tfrac{15}{2})^2\langle\Phi_{6M}|\,V_{12}\,|\Phi_{6M}\rangle\\&+\tfrac{5}{22}\sum_{mM}(4\tfrac{7}{2}Mm\mid\tfrac{15}{2}\tfrac{15}{2})^2\langle\Phi_{4M}|\,V_{12}\,|\Phi_{4M}\rangle\\=&\tfrac{17}{22}E_6(\tfrac{7}{2}\tfrac{7}{2};\tfrac{7}{2}\tfrac{7}{2})+\tfrac{5}{22}E_4(\tfrac{7}{2}\tfrac{7}{2};\tfrac{7}{2}\tfrac{7}{2})\end{aligned}$$

and in writing the last line of this equation use has been made of the fact that the matrix element of V_{12} is independent of M and the normalization property of the Clebsch–Gordan coefficients (equation (A1.23)).

One can evaluate V_{13} and V_{23} in the same manner; namely expand the Slater determinant in terms of the minors of the second row and then the first row. Clearly the same coefficients will come in as above so that

$$\begin{aligned}\langle\Psi_{\frac{15}{2}\frac{15}{2}}|\,H_0+V_{12}+V_{13}+V_{23}\,|\Psi_{\frac{15}{2}\frac{15}{2}}\rangle=3\varepsilon_{\frac{7}{2}}&\\+3(\tfrac{17}{22}E_6(\tfrac{7}{2}\tfrac{7}{2};\tfrac{7}{2}\tfrac{7}{2})+\tfrac{5}{22}E_4(\tfrac{7}{2}\tfrac{7}{2};\tfrac{7}{2}\tfrac{7}{2})).&\end{aligned}\tag{1.66}$$

The numbers $\sqrt{\tfrac{17}{22}}$ and $-\sqrt{\tfrac{5}{22}}$ in equation (1.65) are known as the fractional-parentage coefficients and their square give the probability that in the antisymmetric three-particle $(0f_{\frac{7}{2}})^3_{I=\frac{15}{2}}$ state one will find the configuration $[\Phi_6(1,2)\times\phi_{\frac{7}{2}}(3)]_{\frac{15}{2}\frac{15}{2}}$ and $[\Phi_4(1,2)\phi_{\frac{7}{2}}(3)]_{\frac{15}{2}\frac{15}{2}}$ respectively.

One may simply extend these ideas to the n-particle case and write

$$\Psi_{IM}(1,\ldots,n)=\sum_{J\beta}\langle j^{n-1}J\beta, j|\}j^nI\alpha\rangle\,[\Phi_{J\beta}(1,\ldots,n-1)\times\phi_j(n)]_{IM} \quad (1.67)$$

where the fractional-parentage coefficients (c.f.p.'s) $\langle j^{n-1}J\beta, j|\}j^nI\alpha\rangle$ are chosen so that the wave function is antisymmetric to the interchange of any two particles (Racah 1943), and in the representation we have used they are real quantities. (In writing this equation we have included additional quantum numbers α and β in case the angular momentum alone is not sufficient to specify completely the state of the many-particle system; see equations (1.33) and (1.34) for an example of this.) In Appendix 5 we discuss the properties of these coefficients and show that they are independent of M (as implied by equation (1.67)) and that they are related to the reduced matrix element of the creation operator $a^{\dagger}_{jm}$,

$$\langle j^{n-1}J\beta, j|\}j^nI\alpha\rangle=\frac{(-1)^{n-1}\langle\Psi_{I\alpha}\|a^{\dagger}_j\|\Phi_{J\beta}\rangle}{\sqrt{n}}. \quad (1.68)$$

Fortunately one does not need to go through the type of calculation used in deducing equation (1.65) since an extensive tabulation of these coefficients has been given by Bayman and Lande (1966). The c.f.p.'s encountered in this book are given explicitly in Appendix 5.

These one-particle parentage coefficients are precisely the coefficients encountered when one wishes to calculate matrix elements of single-particle operators (see Chapter 5) or when one is evaluating $\sum_{i<j} V_{ij}$ for the three-particle system. One can also use these coefficients to evaluate V for the many-particle system if one is willing to go through a chain procedure. To see this we rewrite V for the n-particle system as a sum of interactions:

$$V_k=\sum_{\substack{i<j\\(i,j\neq k)}} V_{ij}$$

where V_k is the total residual interaction excluding particle k. Thus, for example, $V_1=\sum_{i<j=2}^{n} V_{ij}$ etc. If V is written as $\sum_{k=1}^{n} V_k$ one obviously overcounts i.e. the interaction V_{12} which occurs in each term except those with $k=1$ and 2 is counted $(n-2)$ times. Thus when this is taken into account one may write V as

$$\begin{aligned} V&=\frac{1}{n-2}\sum_{k=1}^{n} V_k\\ &=\frac{1}{n-2}\sum_{k=1}^{n}\sum_{\substack{i<j\\(i,j\neq k)}}^{n} V_{ij}. \end{aligned} \quad (1.69)$$

One can evaluate each term in this sum by expanding the Slater determinants characterizing the wave function $\Psi_{IM\alpha}(1, \ldots, n)$ in terms of the minors of the kth row. Consequently by analogy with equation (1.66) one sees that

$$\langle\Psi_{IM\alpha'}|\, H_0 + \sum_{i<j=1}^{n} V_{ij}\, |\Psi_{IM\alpha}\rangle = n\varepsilon_j\delta_{\alpha\alpha'} + \left(\frac{n}{n-2}\right) \sum_{J\beta\beta'} \langle j^{n-1}J\beta', j|\}j^n I\alpha'\rangle \langle j^{n-1}J\beta, j|\}j^n I\alpha\rangle \times \langle\Phi_{JM'\beta'}|\sum_{i<j=1}^{n-1} V_{ij}\, |\Phi_{JM'\beta}\rangle. \tag{1.70}$$

Consequently once the interaction energies in all the three particle states $(j)^3_I$ are known one can simple calculate $\langle V\rangle$ for the four-body system etc.

This procedure is tedious and a simpler method is to introduce the idea of double parentage coefficients; i.e. instead of treating only particle number n in a special manner one treats both n and $n-1$ as special. In this case we may write

$$\Psi_{IM\alpha}(1, \ldots, n) = \sum_{J\beta K} \langle j^{n-2}J\beta, j^2K|\}j^n I\alpha\rangle \times [\Phi_{J\beta}(1, \ldots, n-2) \times \psi_K(n-1, n)]_{IM} \tag{1.71}$$

where the double parentage coefficient, $\langle j^{n-2}J\beta, j^2K|\}j^n I\alpha\rangle$, must be chosen so that equation (1.71) is antisymmetric to the interchange of any two particles. In Appendix 5 we discuss some useful properties of these coefficients and show that they are independent of the z components of angular momentum and are given by the reduced matrix element

$$\langle j^{n-2}J\beta, j^2K|\}j^n I\alpha\rangle = \left\{\frac{2}{n(n-1)}\right\}^{\frac{1}{2}} \langle\Psi_{I\alpha}||A^{\dagger}_K(jj)||\Phi_{J\beta}\rangle \tag{1.72}$$

where $A^{\dagger}_{KM}(jj)$ is the two-particle creation operator of equation (1.42).

A tabulation of the double-parentage coefficients is given by Towner and Hardy (1969) for $j \leq \frac{9}{2}$. Their phase convention is the same as that of Bayman and Lande (1966). Thus in any calculation that simultaneously uses one- and two-nucleon parentage coefficients the coefficients tabulated by these two groups should be used. The double-parentage coefficients for the $(f_{\frac{7}{2}})^4$ configuration are given in Appendix 5, Table (A5.5).

Once the double-parentage coefficients are known the energy of the state $(j)^n_I$ can be written down. Since there are $n(n-1)/2$ pairs it follows that

$$\langle \Psi_{IM\alpha'} | H_0 + \sum_{i<j} V_{ij} | \Psi_{IM\alpha} \rangle = n\varepsilon_j \delta_{\alpha\alpha'} + \frac{n(n-1)}{2} \sum_{J\beta K} \langle j^{n-2} J\beta, j^2 K | \} j^n I\alpha' \rangle \langle j^{n-2} J\beta, j^2 K | \} j^n I\alpha \rangle \times E_K(jj; jj) \quad (1.73)$$

where $E_K(jj; jj)$ is the two-body matrix element of equation (1.45)

Finally the ideas embodied in equations (1.67) and (1.70) can be written in terms of diagrams. From equations (1.26) and (1.35) one sees that

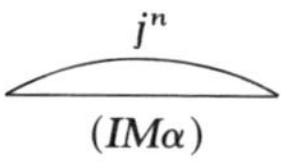

$$= \sum_{J\beta} \langle j^{n-1} j\beta, j \} j^n I\alpha \rangle \quad (1.74)$$

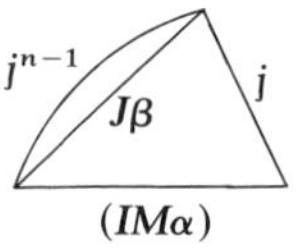

and

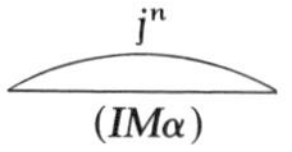

$$\sum_{J\beta K} \langle j^{n-2} J\beta, j^2 K | \} j^n I\alpha \rangle \quad (1.75)$$

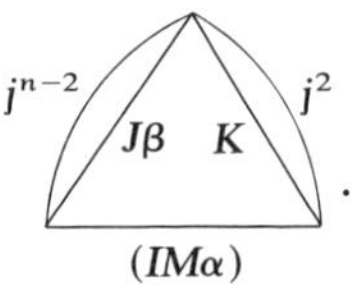

3.2. The spectra of the proton $f_{\frac{7}{2}}$ nuclei

In this Section we consider the spectra of nuclei with $N = 28$ and $Z = 21$–24. We assume that ${}^{48}_{20}\mathrm{Ca}_{28}$ forms an inert core and this is consistent with the experimental observation (Raman 1970) that the only state below 2 MeV excitation energy in ${}^{49}_{21}\mathrm{Sc}_{28}$ is the $f_{\frac{7}{2}}$ ground state. Thus a reasonable model for the low-lying states of ${}^{50}_{22}\mathrm{Ti}_{28}$, ${}^{51}_{23}\mathrm{V}_{28}$, and ${}^{52}_{24}\mathrm{Cr}_{28}$ is one in which the two-, three-, and four-valence protons are restricted to the $0f_{\frac{7}{2}}$ shell.

To use equations (1.70) and (1.73) to calculate the spectrum of the configuration $(f_{\frac{7}{2}})^n$ one must know the single-particle energy ε_f and the four two-body energies $E_J(\frac{7}{2}\frac{7}{2}; \frac{7}{2}\frac{7}{2})$ $(J = 0, 2, 4, 6)$ that characterize the residual nucleon–nucleon interaction within the $f_{\frac{7}{2}}$ configuration. ε_f may be obtained from the binding-energy tables of Wapstra and Gove (1971):

$$\begin{aligned} \varepsilon_f &= \mathrm{BE}({}^{49}_{21}\mathrm{Sc}_{28}) - \mathrm{BE}({}^{48}_{20}\mathrm{Ca}_{28}) \\ &= -9{\cdot}619 \text{ MeV}. \end{aligned} \quad (1.76)$$

Previously we calculated the two-body matrix elements $E_J(j_1 j_2; j_3 j_4)$ by making an explicit choice for the form of V_{ij}. However, instead of doing this one may take the values directly from the spectrum of the two-particle nucleus ${}^{50}_{22}\mathrm{Ti}_{28}$ and use these values to make predictions about the structure of ${}^{51}_{23}V_{28}$ and ${}^{52}_{24}Cr_{28}$. To do this we observe that if there were no mutual interaction between the two valence protons in the 0^+ ground state of ${}^{50}_{22}Ti_{28}$ the difference in the binding energy of ${}^{50}_{22}Ti_{28}$ and ${}^{48}_{20}Ca_{28}$ would be just $2\varepsilon_f$. If we define $E_0(\frac{7}{2}\frac{7}{2};\frac{7}{2}\frac{7}{2})=E_0$ as the additional interaction energy of the two nucleons when they couple to spin zero, then

$$\mathrm{BE}({}^{50}_{22}\mathrm{Ti}_{28})-\mathrm{BE}({}^{48}_{20}\mathrm{Ca}_{28})=2\varepsilon_f+E_0$$

and by use of the binding-energy tables one finds $E_0=-2{\cdot}552$ MeV. In Fig. 1.7 the experimental spectrum of ${}^{50}_{22}Ti_{28}$ is shown (Nomura *et al.* 1970). The 2^+, 4^+, and 6^+ states associated with the configuration $(f_{\frac{7}{2}})^2$ lie at excitation energies of 1·557, 2·677, and 3·197 MeV, respectively. Thus the experimental values for the residual two-body interaction matrix

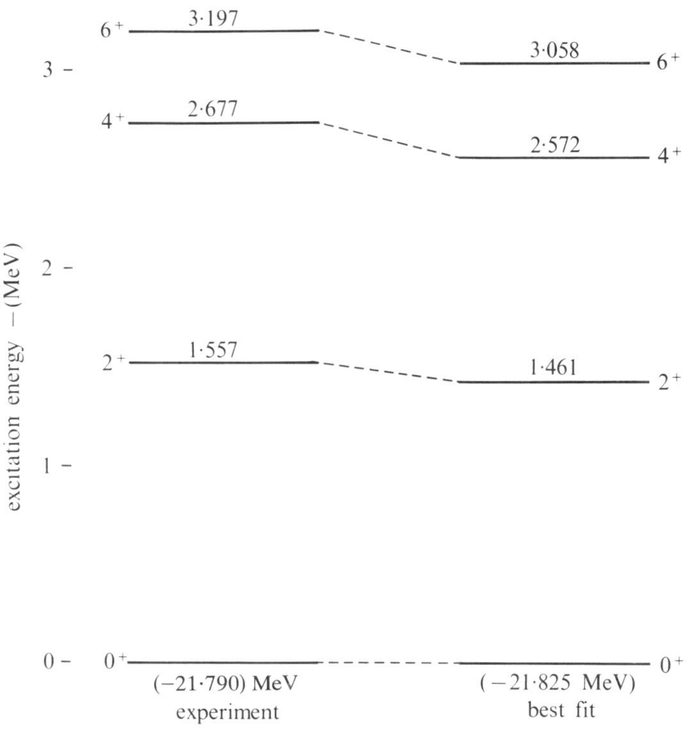

Fig. 1.7. Spectrum of ${}^{50}_{22}Ti_{28}$. The ground-state energy is measured relative to ${}^{48}_{20}Ca_{28}$ and is given at the bottom of each column in MeV. Other energies are excitation energies taken from the work of Nomura *et al.* (1970). The best-fit spectrum corresponds to the matrix elements of equation (1.83).

elements are

$$\begin{aligned}
E_0 &= -2{\cdot}552\ \text{MeV}\\
E_2 &= -2{\cdot}552+1{\cdot}557 = -0{\cdot}995\ \text{MeV}\\
E_4 &= -2{\cdot}552+2{\cdot}677 = 0{\cdot}125\ \text{MeV}\\
E_6 &= -2{\cdot}552+3{\cdot}197 = 0{\cdot}645\ \text{MeV}.
\end{aligned} \tag{1.77}$$

From equation (1.70) it follows that the energies of the $(f_{\frac{7}{2}})^3$ state are

$$\tilde{E}'_I = 3\varepsilon_f + 3\sum_J \langle j^2J, j|\}j^3I\rangle^2 E_J$$

with the fractional-parentage coefficients for the various I states given in Table A5.2.

If we define the six components of the column vector $\tilde{E}'$ to stand for the energies of the $I=\frac{3}{2}, \frac{5}{2}, \frac{7}{2}, \frac{9}{2}, \frac{11}{2}$, and $\frac{15}{2}$ states of ${}^{51}_{23}V_{28}$ and the five-component column vector E to denote the energies ε_f, E_0, E_2, E_4, and E_6 given by equations (1.76) and (1.77) one may write

$$\tilde{E}' = A'E \tag{1.78}$$

where A' is the (6×5) matrix

$$A' = \begin{pmatrix}
3 & 0 & \frac{9}{14} & \frac{33}{14} & 0\\
3 & 0 & \frac{11}{6} & \frac{2}{11} & \frac{65}{66}\\
3 & \frac{3}{4} & \frac{5}{12} & \frac{3}{4} & \frac{13}{12}\\
3 & 0 & \frac{13}{42} & \frac{150}{77} & \frac{49}{66}\\
3 & 0 & \frac{5}{6} & \frac{13}{22} & \frac{52}{33}\\
3 & 0 & 0 & \frac{15}{22} & \frac{51}{22}
\end{pmatrix}. \tag{1.79}$$

(As a check on the algebra involved in obtaining A' let $V_{ij}\to 1$. Thus all E_J become unity and the sum of the coefficients of E_0, E_2, E_4, and E_6 must be equal to $n(n-1)/2=3$, the number of $f_{\frac{7}{2}}$ pairs.)

When the values of ε_f and E_J given by equations (1.76) and (1.77) are used the theoretical spectrum for ${}^{51}_{23}V_{28}$ that results is shown in Fig. 1.8. The prediction for the binding energy relative to ${}^{48}_{20}Ca_{28}$ is in reasonable agreement with the experimental value deduced from the Wapstra–Gove (1971) tables and the excitation energies are in accordance with the experimental findings of Horoshko *et al.* (1970). In the region of 2.5 MeV excitation energy some states that cannot be fitted into the pure $f_{\frac{7}{2}}$ description of ${}^{51}_{23}V_{28}$ are seen. However, this is to be expected since above 2 MeV in ${}^{49}_{21}Sc_{28}$ one begins to see states other than the $f_{\frac{7}{2}}$ level.

The low-lying states of ${}^{52}_{24}Cr_{28}$ are those of the $(f_{\frac{7}{2}})^4$ configuration and as

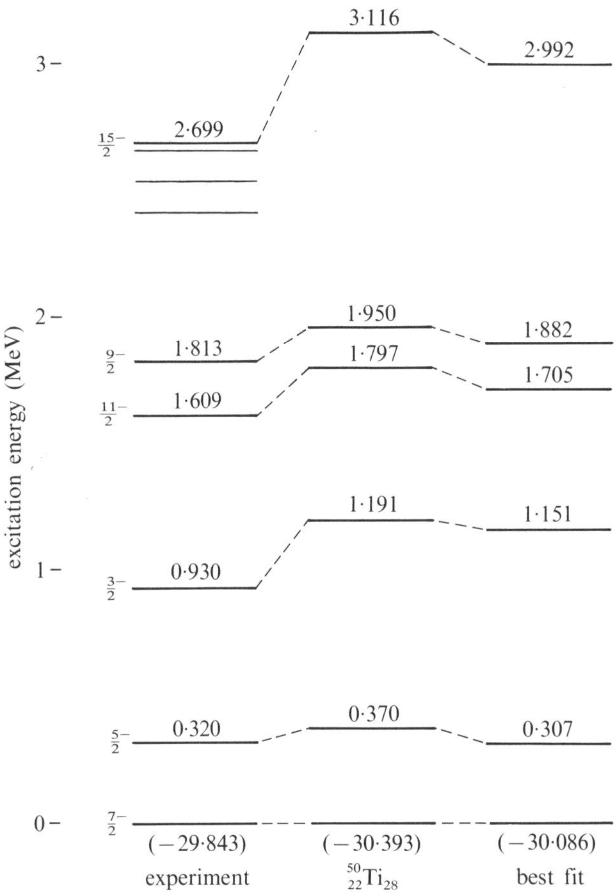

Fig. 1.8. Spectrum of $^{51}_{23}V_{28}$. The ground-state energy is measured relative to $^{48}_{20}Ca_{28}$ whereas the others are excitation energies in MeV taken from the work of Horoshko *et al.* (1970). The spectrum calculated by use of the $^{49}_{21}Sc_{28}$ and $^{50}_{22}Ti_{28}$ energies given in equations (1.76) and (1.77) is shown in the centre column, and the best-fit spectrum calculated by use of the matrix elements of equation (1.83) is shown on the right. All experimental states below 2·7 MeV are shown, and up to this energy three states that arise from outside the $f_{\frac{7}{2}}$ model space are observed. These are drawn with lighter lines.

shown in section 1.2 the possible spins are

$$I = 0, 2^2, 4^2, 5, 6, \text{and } 8$$

where the superscript indicates the number of states with that angular momentum. Thus for both $I = 2$ and $I = 4$ one must set up and diagonalize a (2×2) energy matrix. However, if the states of the configuration are classified according to the seniority quantum number v (for a discussion of seniority see section 5 of this Chapter) explicit use of equation (1.73) together with the double-parentage coefficients given in Appendix 5 (Table A5.5) shows that the off-diagonal matrix element between the two 2^+ and two 4^+ states vanishes for any two-body interaction (see also the discussion in section 4.1, Appendix 3).

If we denote by $\tilde{E}''$ the column vector whose eight components are the energies of the $I=0^+$, $2^+(v=2)$, $2^+(v=4)$, $4^+(v=2)$, $4^+(v=4)$, 5^+, 6^+, and 8^+ states in ${}^{52}_{24}Cr_{28}$, we can write

$$\tilde{E}''=A''E \tag{1.80}$$

where A'' is the (8×5) matrix

$$A''=\begin{pmatrix} 4 & \frac{3}{2} & \frac{5}{6} & \frac{3}{2} & \frac{13}{6} \\ 4 & \frac{1}{2} & \frac{11}{6} & \frac{3}{2} & \frac{13}{6} \\ 4 & 0 & 1 & \frac{42}{11} & \frac{13}{11} \\ 4 & \frac{1}{2} & \frac{5}{6} & \frac{5}{2} & \frac{13}{6} \\ 4 & 0 & \frac{7}{3} & 1 & \frac{8}{3} \\ 4 & 0 & \frac{8}{7} & \frac{192}{77} & \frac{182}{77} \\ 4 & \frac{1}{2} & \frac{5}{6} & \frac{3}{2} & \frac{19}{6} \\ 4 & 0 & \frac{10}{21} & \frac{129}{77} & \frac{127}{33} \end{pmatrix}. \tag{1.81}$$

Apart from the first column the coefficients in the matrix are determined by inserting the double-parentage coefficients (Table A5.5) into equation (1.73). A check on the algebra is again provided by the condition that the sum of the coefficients of E_0, E_2, E_4, and E_6 must be equal to the number of pairs which in this case is six.

When the values of ε_f and E_J given by equations (1.76) and (1.77) are used the predicted spectrum for the nucleus is shown in Fig. 1.9. Experimentally (Freedman *et al.* 1966) only the 2^+ $(v=4)$ and 8^+ states of the $(f_{\frac{7}{2}})^4$ configuration have so far escaped detection. As for the seniority-four 2^+, it is predicted at around 3 MeV and several 2^+ states have been seen in this region. Other predicted excitation energies are in quite good agreement with experiment; however, the theoretical binding energy relative to ${}^{48}_{20}Ca_{28}$ is off by more than 1 MeV.

One can obviously improve the overall agreement between theory and experiment by considering the five energies ε_f, E_0, E_2, E_4, and E_6 to be

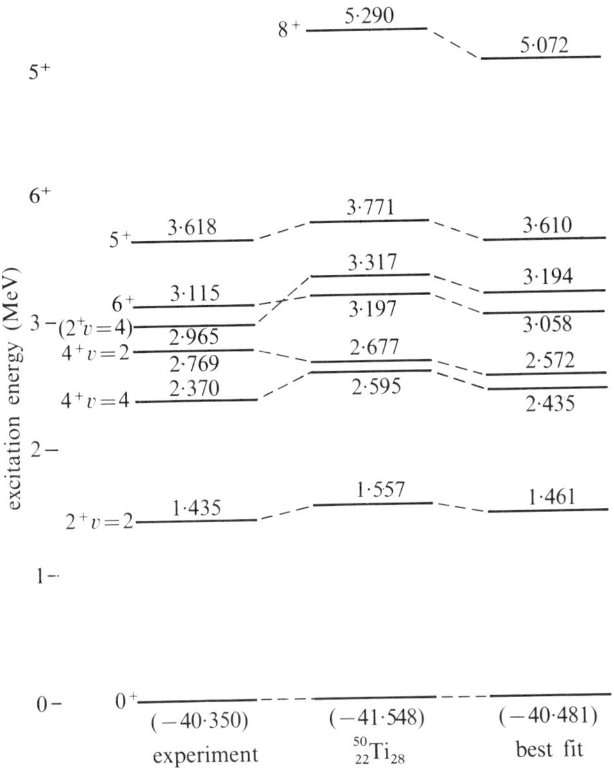

Fig. 1.9. Spectrum of $^{52}_{24}Cr_{28}$. The experimental data are those of Freedman *et al.* (1966). The identification of the $I=2^+$ state observed at 2·965 MeV as the seniority-four 2^+ state of the $(\pi f_{\frac{7}{2}})^4$ configuration is tentative. See the caption of Fig. 1.8 for additional details.

parameters and adjust them to give a best fit to all the data. In total there are seventeen pieces of experimental data—the binding energies of $^{49}_{21}Sc_{28}$, $^{50}_{22}Ti_{28}$, $^{51}_{23}V_{28}$, and $^{52}_{24}Cr_{28}$ relative to $^{48}_{20}Ca_{28}$; the energies of the 2^+, 4^+, and 6^+ states in $^{50}_{22}Ti_{28}$; the energies of the $\frac{3}{2}^-$, $\frac{5}{2}^-$, $\frac{9}{2}^-$, $\frac{11}{2}^-$, and $\frac{15}{2}^-$ levels in $^{51}_{23}V_{28}$; and the experimentally seen 2^+ $(v=2)$, 4^+ $(v=2$ and $4)$, 5^+, and 6^+ states in $^{52}_{24}Cr_{28}$. To make a best fit we define a 17-component column vector $\tilde{E}(\text{expt})$ that contains the experimental energies. By analogy with equations (1.78) and (1.80) one can express the theoretical energies as

$$\tilde{E}(\text{th}) = AE$$

where E is again the five-component column vector the components of which are ε_f, E_0, E_2, E_4, and E_6 and A is a (17×5) matrix. To find E that best reproduces the experimental energies we minimize

$$\chi^2 = \sum_{i=1}^{17} \{\tilde{E}_i(\text{expt}) - \tilde{E}_i(\text{th})\}^2$$

$$= \sum_{i=1}^{17} \{\tilde{E}_i(\text{expt}) - \sum_k A_{ik}E_k\}^2$$

as a function of E_m, the five components of the column vector E. For a best fit to the data

$$\frac{\partial \chi^2}{\partial E_m} \equiv 0.$$

Thus

$$\sum_{i=1}^{17} \{\tilde{E}_i(\text{expt}) - \sum_k A_{ik}E_k\}A_{im} \equiv 0 \tag{1.82}$$

for the five possible values of m. This provides five homogeneous equations for the five unknown E_m, and in matrix notation the single equation embodying these conditions is

$$A^T\tilde{E}(\text{expt}) = A^TAE$$

$$= \Gamma E$$

where A^T is the transpose of the matrix A and $\Gamma = A^TA$. Thus the value of E that best reproduces the data is

$$E = \Gamma^{-1}A^TE(\text{expt})$$

with Γ^{-1} the inverse of Γ.

When the 17 pieces of data described above are used, the best-fit values of the $0f_{\frac{7}{2}}$ proton matrix elements are

$$\begin{aligned} \varepsilon_f &= -9{\cdot}846 \text{ MeV} \\ E_0 &= -2{\cdot}133 \text{ MeV} \\ E_2 &= -0{\cdot}672 \text{ MeV} \\ E_4 &= 0{\cdot}439 \text{ MeV} \\ E_6 &= 0{\cdot}925 \text{ MeV} \end{aligned} \tag{1.83}$$

and the theoretical predictions for the spectra are shown in the column headed 'best-fit' in Figs. 1.7–1.9.

If we define the r.m.s. error in any one of the calculated level-energies as

$$\Delta E = \left\{\left[\sum_{i=1}^{17} \{\tilde{E}_i(\text{expt}) - \tilde{E}_i(th)\}^2\right] \Big/ 17\right\}^{\frac{1}{2}}$$

$\Delta E = 152$ keV when the energies of equation (1.83) are used. Thus by use of the $f_{\frac{7}{2}}$ model space one can reproduce quite satisfactorily the energies of the low-lying states in ${}^{50}_{22}Ti_{28}$, ${}^{51}_{23}V_{28}$ and ${}^{52}_{24}Cr_{28}$. The r.m.s. error found in this calculation is typical of the kind of fit that can be made to experimental data when the matrix elements of the residual two-body interaction are taken as parameters and a reasonable model space is used.

The correlation of data using the residual two-body matrix elements as parameters (and not calculated from a potential) was introduced into atomic spectroscopy by Bacher and Goudsmit (1934) and has been exploited with considerable success in nuclear physics by Talmi and his co-workers (Talmi and Unna 1960, Talmi 1962).

4. Several particles in different levels

To calculate the spectrum associated with a nucleus in which several (>2) particles occupy two or more levels it is often necessary to recouple angular momenta, and this leads to the Racah coefficients (Racah 1942) introduced in section 1.3. In this Section we discuss the counting and phase factors that must be taken into account in computing the Hamiltonian matrix and then show how the diagrammatic methods for describing wave functions can be used to compute these energy matrix elements. Finally we shall apply the ideas developed to calculate the spectrum of ${}^{19}_{8}O_{11}$.

4.1. Resumé of diagrammatic rules for recoupling

In section 1 of this Chapter we introduced the idea of expressing the angular-momentum coupling in terms of diagrams (Macfarlane and French 1960), and in this Section we collect these various rules together. As discussed in section 1.1, two angular momenta I_1 and I_2 can couple to a total spin I which satisfies the triangle inequality

$$|I_1 - I_2| \leq I \leq I_1 + I_2.$$

Because of this it is convenient to express the coupling in terms of a triangle

$[\Psi_{I_1} \times \Phi_{I_2}]_{IM} =$ 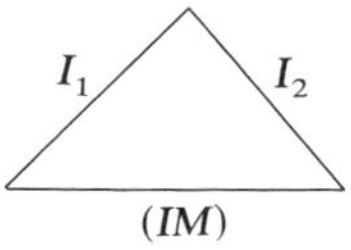

. (1.84)

Furthermore, because of the symmetry property of the Clebsch–Gordan coefficients given in equation (A1.18) it follows that

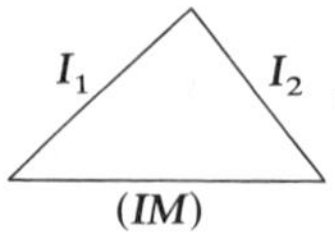

$$= (-1)^{I_1+I_2-I} \tag{1.85}$$
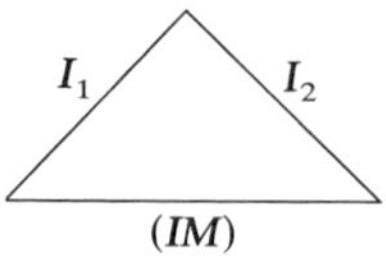

One may express the result of an operator $Q_{\lambda u}$ acting on the state $\psi_{I_iM_i}$ as

$$\begin{aligned} Q_{\lambda u}\psi_{I_iM_i} &= \sum_{I_fM_f} \langle \psi_{I_fM_f}|\, Q_{\lambda u}\, |\psi_{I_iM_i}\rangle \psi_{I_fM_f} \\ &= \sum_{I_fM_f} (I_i\lambda M_i\mu \mid I_fM_f)\langle \psi_{I_f}||Q_\lambda||\psi_{I_i}\rangle \psi_{I_fMf} \\ &= \sum_{I_fM_f} (I_i\lambda M_i\mu \mid I_fM_f)\langle \psi_{I_f}||Q_\lambda||\psi_{I_i}\rangle \underset{(I_fM_f)}{\underline{\hspace{2cm}}} \end{aligned} \tag{1.85a}$$

provided the $\psi_{I_fM_f}$ form a complete set of states. (In this equation $\langle \psi_{I_f}||Q_\lambda||\psi_{I_i}\rangle$ is the reduced matrix element defined by equation (1.50).) Alternatively, if $Q_{\lambda u}$ and $\psi_{I_iM_i}$ are coupled to a resultant angular momentum (I_fM_f) one obtains the result given by equation (1.54)

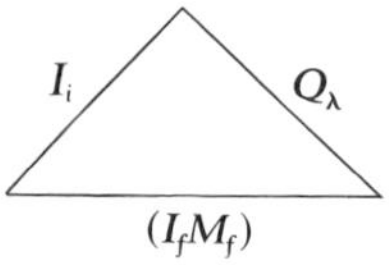

$$= \langle \psi_{I_f}||Q_\lambda||\psi_{I_i}\rangle \underset{(I_fM_f)}{\underline{\hspace{2cm}}}\,. \tag{1.85b}$$

In terms of diagrams, the fractional-parentage decompositions introduced in section 3.1 can be expressed as

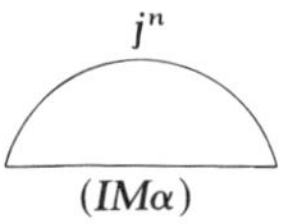

$$= \sum_{J\beta} \langle j^{n-1}J\beta, j|\}j^nI\alpha\rangle \tag{1.86}$$
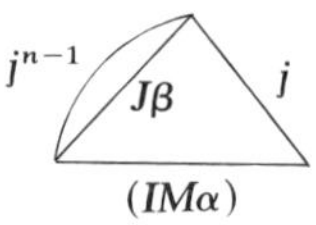

and

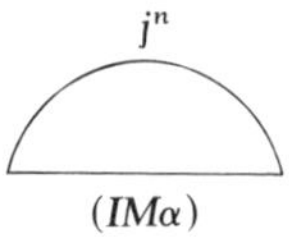

$$= \sum_{J\beta K} \langle j^{n-2}J\beta, j^2K|\}j^nI\alpha\rangle \tag{1.87}$$
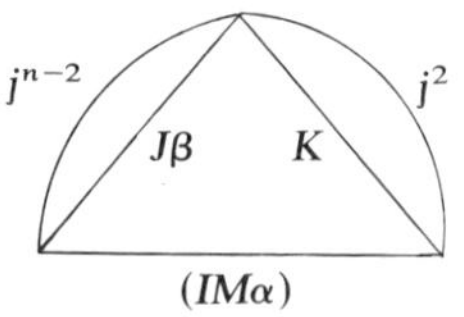

where 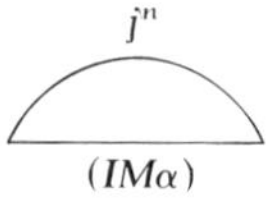

stands for the wave function in which n particles in the

single-particle state j couple to spin (IM) and α is any additional quantum number necessary to specify the state completely.

When three angular momenta are compounded the order of the coupling is not unique. Thus one may write

$$[[\Psi_{I_1}\times\Phi_{I_2}]_J\times\chi_{I_3}]_{IM} =$$ 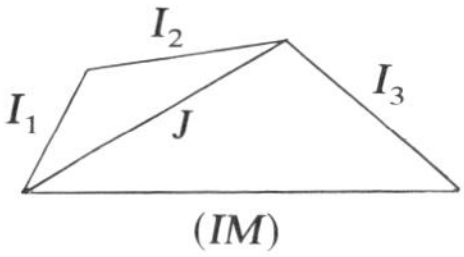

or

$$[\Psi_{I_1}\times[\Phi_{I_2}\times\chi_{I_3}]_K]_{IM} =$$ 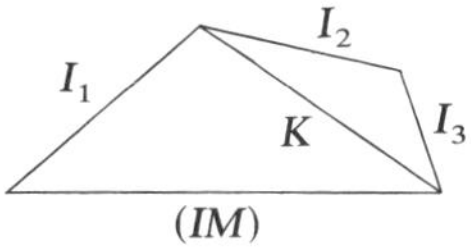

As shown in Appendix 4 and as discussed in section 1.3 these two ways of coupling are related and

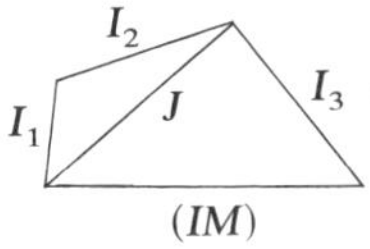

$$= \sum_K \surd\{(2J+1)(2K+1)\}W(I_1I_2II_3;\, JK)$$
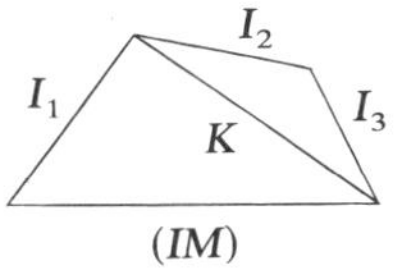

(1.88)

and

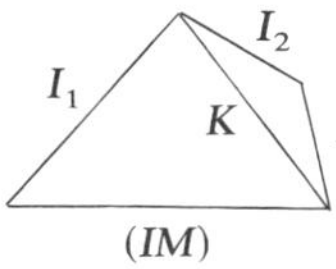

$$= \sum_J \surd\{(2J+1)(2K+1)\}W(I_1I_2II_3\colon JK)$$
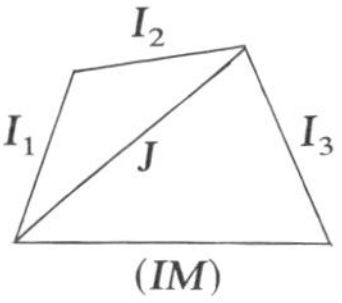

(1.89)

where $W(I_1I_2II_3;JK)$ is the Racah coefficient discussed in Appendix 4 and tabulated by Rotenberg *et al.* (1959).

Finally one also encounters situations in which four angular momenta are coupled to a resultant (IM). Again the coupling is not unique. One possibility is

$$[[\Psi_{I_1}\times\Phi_{I_2}]_{I_{12}}\times[\chi_{I_3}\times\zeta_{I_4}]_{I_{34}}]_{IM} =$$ 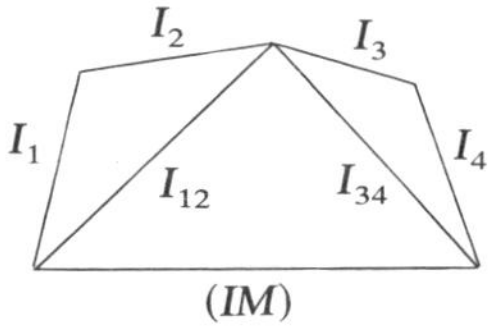

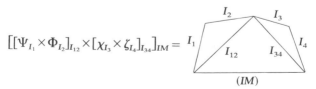

whereas a second is provided by

$$[[\Psi_{I_1}\times\chi_{I_3}]_{I_{13}}\times[\Phi_{I_2}\times\zeta_{I_4}]_{I_{24}}]_{IM} =$$

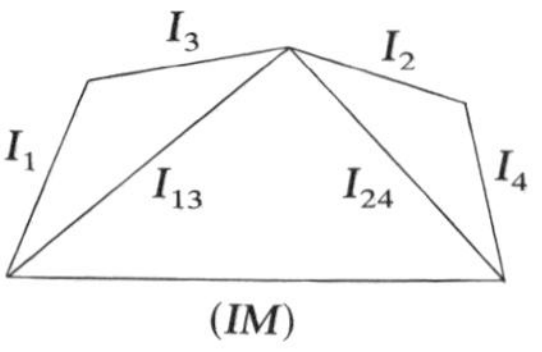

As might be expected these two coupling schemes are related and one may write

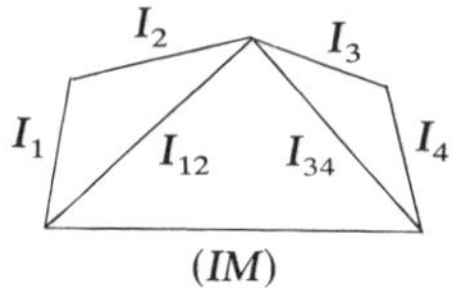

$$= \sum_{I_{13}} \sum_{I_{24}} \sqrt{\{(2I_{12}+1)(2I_{34}+1)(2I_{13}+1)(2I_{24}+1)\}}$$

$$\times \begin{Bmatrix} I_1 & I_2 & I_{12} \\ I_3 & I_4 & I_{34} \\ I_{13} & I_{24} & I \end{Bmatrix}$$

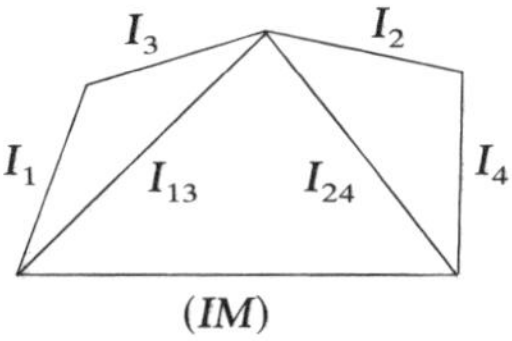

(1.90)

The symbol {} in equation (1.90) is known as the $9j$ coefficient and in Appendix 4 we discuss some of its important properties.

4.2. Counting and phase factors

In addition to the angular-momentum recoupling that occurs in nuclear-structure calculations, there are also some phase factors that arise because we deal with fermions and some counting factors that must be taken into account. In this Section we present two rules that will be of great help in ascertaining these factors properly.

If we write the shell-model Hamiltonian in terms of annihilation and creation operators, equation (1.48), it is clear that within the identical nucleon configuration one encounters matrix elements of $a^{\dagger}_{jm}$, $\tilde{a}_{jm}$, $A^{\dagger}_{KM'}(jj)$, and $\tilde{A}_{KM'}(jj)$. Consequently it is always factors of the form $\langle (j)^n_{IM}|\, a^{\dagger}_{jm}\, |(j)^{n-1}_{JM'}\rangle$, $\langle (j)^{n-1}_{JM'}|\, \tilde{a}_{jm}\, |(j)^n_{IM}\rangle$, $\langle (j)^n_{IM}|\, A^{\dagger}_{KM''}(jj)\, |(j)^{n-2}_{JM'}\rangle$, and $\langle (j)^{n-2}_{JM'}|\, \tilde{A}_{KM''}(jj)\, |(j)^n_{IM}\rangle$ that come into a calculation. As shown in Appendix 5 the reduced matrix elements of these quantities are related to the one- and two-nucleon c.f.p.'s by a phase and counting factor. According to equations (A5.6) and (A5.14)

$$\langle (j)^n_{I\alpha}|\, |a^{\dagger}_j|\, |(j)^{n-1}_{J\beta}\rangle = (-1)^{n-1}\sqrt{n}\langle j^{n-1}J\beta, j|\}j^n I\alpha\rangle$$

$$\langle (j)^n_{I\alpha}|\, |A^{\dagger}_K(jj)|\, |(j)^{n-2}_{J\beta}\rangle = \left\{\frac{n(n-1)}{2}\right\}^{\frac{1}{2}}\langle j^{n-2}J\beta, j^2K|\}j^n I\alpha\rangle$$

and from equation (A2.10)

$$\langle (j)^{n-1}_{J\beta} \| \tilde{a}_j \| (j)^n_{I\alpha}\rangle = (-1)^{n-1+J+j-I}\left(\frac{2I+1}{2J+1}\right)^{\frac{1}{2}} \langle j^{n-1}J\beta, j| \} j^n I\alpha\rangle$$

$$\langle (j)^{n-2}_{J\beta} \| \tilde{A}_K(jj) \| (j)^n_{I\alpha}\rangle = (-1)^{J+K-I}\left\{\frac{n(n-1)(2I+1)}{2(2J+1)}\right\}^{\frac{1}{2}}$$

$$\times \{j^{n-2}J\beta, j^2K| \} j^n I\alpha\rangle.$$

Although all angular-momentum factors are taken into account by the methods outlined in section 4.1 the counting and phase factors are not and consequently must be explicitly inserted. Thus we arrive at the rule

Rule 1: To obtain the proper counting factors in any shell-model calculation one must remember to multiply each single-nucleon parentage coefficient $\langle j^{n-1}J\beta, j| \} j^n I\alpha\rangle$ by $(-1)^{n-1}\sqrt{n}$ and each two-nucleon parentage coefficient $\langle j^{n-2}J\beta, j^2K| \} j^n I\alpha\rangle$ by $\sqrt{\{n(n-1)/2\}}$.

This is, of course, the origin of the factor $n(n-1)/2$ in equation (1.73) since in that case there were two double-parentage coefficients, each one involving a state with n particles. There is one case where one is likely to forget that a parentage coefficient has arisen since it is trivially equal to unity. This is for the configuration $(j)^2_I$. Thus although

$$\langle j^1 J = j, j| \} j^2 I\rangle = \delta_{I,\text{even}}$$

and

$$\langle j^0 J = 0, j^2 I| \} j^2 I\rangle = \delta_{I,\text{even}}$$

one must remember that a parentage coefficient was really there and one must multiply by $-\sqrt{2}$ when the one-nucleon c.f.p. is involved or 1 when it is the two-particle parentage coefficient that is encountered.

In addition to this rule there is one other that comes about because we deal with fermions. Let us denote by $A^\dagger_J(j, n)$ the operator that creates the state $(j)^n_J$ and by $A_J(j, n)$ the operator that destroys this state. Thus, for example, to calculate the matrix element of $a^\dagger_{j_1m_1}a_{j_2m_2}$ we have to evaluate

$$\begin{aligned}
&\langle (j_1)^{n'+1}_{J'_1}(j_2)^{n''-1}_{J'_2} |\, a^\dagger_{j_1m_1}a_{j_2m_2}\, |(j_1)^{n'}_{J_1}(j_2)^{n''}_{J_2}\rangle \\
&\quad = \langle 0|\, A_{J'_2}(j_2, n''-1)A_{J'_1}(j_1, n'+1)a^\dagger_{j_1m_1}a_{j_2m_2}A^\dagger_{J_1}(j_1, n')A^\dagger_{J_2}(j_2, n'')\,|0\rangle \\
&\quad = (-1)^{n'}\langle 0|\, A_{J'_2}(j_2, n''-1)a_{j_2m_2}A^\dagger_{J_2}(j_2, n'')\,|0\rangle \\
&\qquad \times \langle 0|\, A_{J'_1}(j_1, n'+1)a^\dagger_{j_1m_1}A^\dagger_{J_1}(j_1, n')\,|0\rangle \\
&\quad = (-1)^{n'}\langle (j_2)^{n''-1}_{J'_2} |\, a_{j_2m_2}\, |(j_2)^{n''}_{J_2}\rangle\langle (j_1)^{n'+1}_{J'_1} |\, a^\dagger_{j_1m_1}\, |(j_1)^{n'}_{J_1}\rangle
\end{aligned}$$

where the phase factor $(-1)^{n'}$ arises because each of the n' fermion creation operators in $A^{\dagger}_{J_1}(j_1, n')$ must be brought through $a_{j_2m_2}$ so that all operators involving $a^{\dagger}_{j_1m_1}$ and $a_{j_1m_1}$ stand together. Consequently the second rule that must be taken into account in the calculation of matrix elements is

Rule 2: If P denotes the total number of particle interchanges necessary to bring all the creation and destruction operators associated with the same single-particle states together then, because we deal with fermions, the matrix element must be multiplied by the phase factor $(-1)^P$.

4.3. Energy matrix elements

To illustrate the use of the recoupling rules we consider the calculation of the Hamiltonian matrix when the valence nucleons are restricted to two single-particle orbits j_1 and j_2. In this case the wave functions involved have the form

$$[(j_1)^{n_1}_{J_1\beta_1} \times (j_2)^{n_2}_{J_2\beta_2}]_{IM} =$$

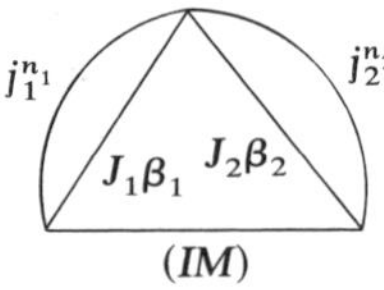

and because the Hamiltonian is a scalar the only matrix elements that must be calculated are

$$\langle H\rangle = \langle[(j_1)^{n'_1}_{J'_1\beta'_1} \times (j_2)^{n'_2}_{J'_2\beta'_2}]_{IM}|\, H\, |[(j_1)^{n_1}_{J_1\beta_1} \times (j_2)^{n_2}_{J_2\beta_2}]_{IM}\rangle. \tag{1.91}$$

Furthermore, because the shell-model Hamiltonian in equation (1.44) contains at most two-body interactions that conserve the number of particles,

$$n_1 + n_2 = n'_1 + n'_2$$

so that

$$\begin{aligned} n'_1 &= n_1, & n'_2 &= n_2 \\ n'_1 &= n_1 \pm 1, & n'_2 &= n_2 \mp 1 \\ n'_1 &= n_1 \pm 2, & n'_2 &= n_2 \mp 2. \end{aligned}$$

The residual two-body interaction may be divided into three parts:

(i) an interaction within the configuration j_1 alone;
(ii) an interaction within the configuration j_2 alone;
(iii) an interaction between the particles in j_1 and those in j_2.

In both (i) and (ii) V_{ij} acts between particles in the same single-particle orbit and may be calculated by using equation (1.73). Thus if we denote by $\langle H\rangle_{(i)+(ii)}$ the matrix element when the interaction between particles in levels j_1 and j_2 is turned off we arrive at the result

$$\langle[(j_1)^{n_1'}_{J_1'\beta_1'}\times(j_2)^{n_2'}_{J_2'\beta_2'}]_{IM}|\,H\,|[(j_1)^{n_1}_{J_1\beta_1}\times(j_2)^{n_2}_{J_2\beta_2}]_{IM}\rangle_{(i)+(ii)}$$

$$\begin{aligned}
&=\delta_{J_1J_1'}\delta_{J_2J_2'}\delta_{n_1n_1'}\delta_{n_2n_2'}\Big(\{n_1\varepsilon_{j_1}+n_2\varepsilon_{j_2}\}\delta_{\beta_1'\beta_1}\delta_{\beta_2'\beta_2}\\
&\quad+\delta_{\beta_2'\beta_2}\frac{n_1(n_1-1)}{2}\sum_{J\beta K}\langle j_1^{n_1-2}J\beta,\,j_1^2K|\}j_1^{n_1}J_1\beta_1\rangle\\
&\quad\times\langle j_1^{n_1-2}J\beta,\,j_1^2K|\}j_1^{n_1}J_1\beta_1'\rangle E_K(j_1j_1;\,j_1j_1)\\
&\quad+\delta_{\beta_1'\beta_1}\frac{n_2(n_2-1)}{2}\sum_{J\beta K}\langle j_2^{n_2-2}J\beta,\,j_2^2K|\}j_2^{n_2}J_2\beta_2\rangle\\
&\quad\times\langle j_2^{n_2-2}J\beta,\,j_2^2K|\}j_2^{n_2}J_2\beta_2'\rangle E_K(j_2j_2;\,j_2j_2)\Big).
\end{aligned}\tag{1.92}$$

According to equation (1.44) the effect of the potential V operating on the two-particle state $[\phi_{j_1}\times\phi_{j_2}]_{KM}$ is

$$V[\phi_{j_1}\times\phi_{j_2}]_{KM}=\sum_{\{j_3j_4\}}E_K(j_3j_4;\,j_1j_2)[\phi_{j_3}\times\phi_{j_4}]_{KM}$$

where $\{j_3j_4\}$ implies an ordered sum in which $j_3\geq j_4$. For the two-level case the diagrammatic form of this relationship is

V 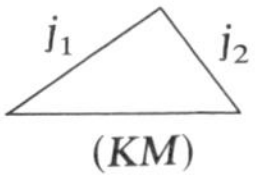

$=\sum_{\{j_3j_4\}}E_K(j_3j_4;\,j_1j_2)$ 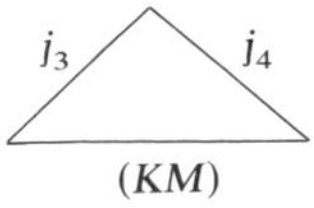

$=E_K(j_1j_1;\,j_1j_2)$ 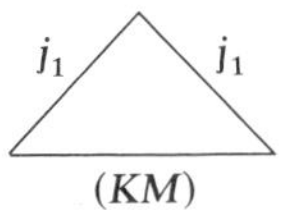

$+E_K(j_1j_2;\,j_1j_2)$ 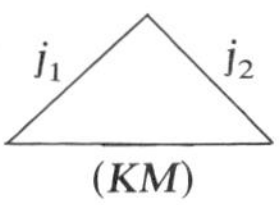

$+E_K(j_2j_2;\,j_1j_2)$ 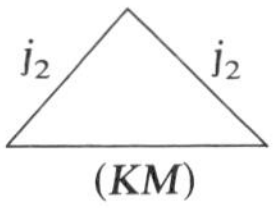

. (1.93)

Thus in order to calculate the contribution to $\langle H\rangle$ due to the interaction between particles in j_1 and j_2 it is convenient to get the two interacting nucleons into the same triangle. This involves a fractional-parentage decomposition of the state $[(j_1)^{n_1}_{J_1\beta_1}\times(j_2)^{n_2}_{J_2\beta_2}]_{IM}$ and by use of equation

(1.86) one finds

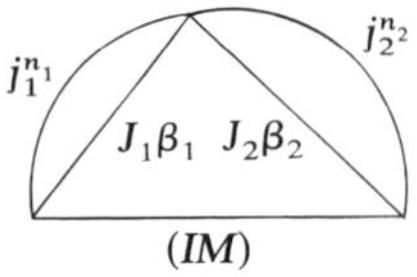

$$= \sum_{J_3\beta_3J_4\beta_4} \langle j_1^{n_1-1}J_3\beta_3, j_1| \} j_1^{n_1}J_1\beta_1\rangle$$

$$\times \langle j_2^{n_2-1}J_4\beta_4, j_2| \} j_2^{n_2}J_2\beta_2\rangle$$

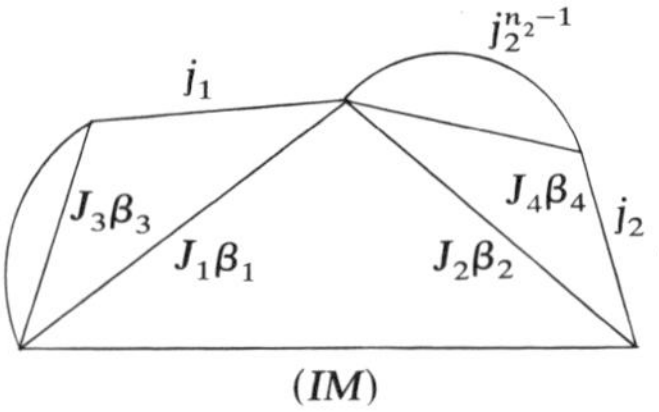

One can bring the 'split-off' particles into the same triangle by use of the $9j$ recoupling relationship in equation (1.90). However, it is more convenient to carry out the recoupling in a slightly different way. By use of equations (1.85) and (1.89) one can write

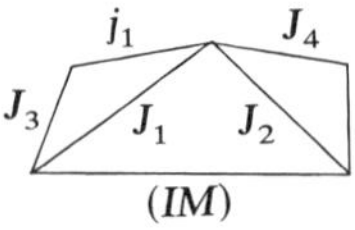

$$= (-1)^{j_2+J_4-J_2}$$

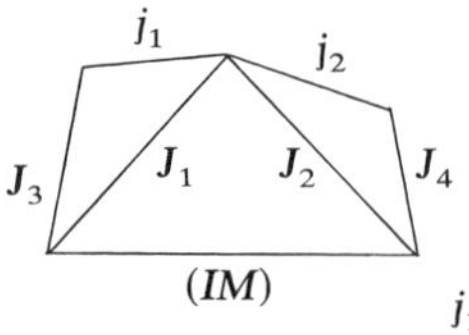

$$= (-1)^{j_2+J_4-J_2} \sum_L \surd\{(2J_2+1)(2L+1)\} \times W(J_1j_2IJ_4; LJ_2)$$

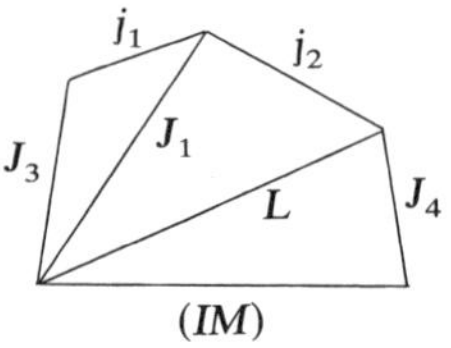

(1.94)

where the irrelevant quantum numbers β (as far as angular-momentum coupling is concerned) have been suppressed. A further recoupling characterized by equation (1.87) brings j_1 and j_2 into the same triangle. Thus

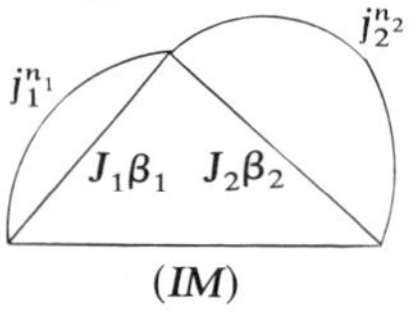

$$= \sum_{J_3\beta_3J_4\beta_4LK} (-1)^{j_2+J_4-J_2}\surd\{(2J_1+1)(2J_2+1)(2L+1)(2K+1)\}$$
$$\times \langle j_1^{n_1-1}J_3\beta_3, j_1| \} j_1^{n_1}J_1\beta_1\rangle \langle j_2^{n_2-1}J_4\beta_4, j_2| \} j_2^{n_2}J_2\beta_2\rangle$$
$$\times W(J_1j_2IJ_4; LJ_2)\, W(J_3j_1Lj_2; J_1K)$$

$$\times$$

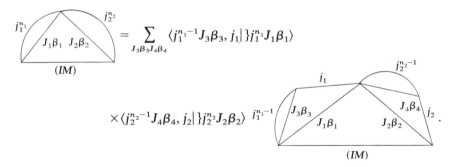

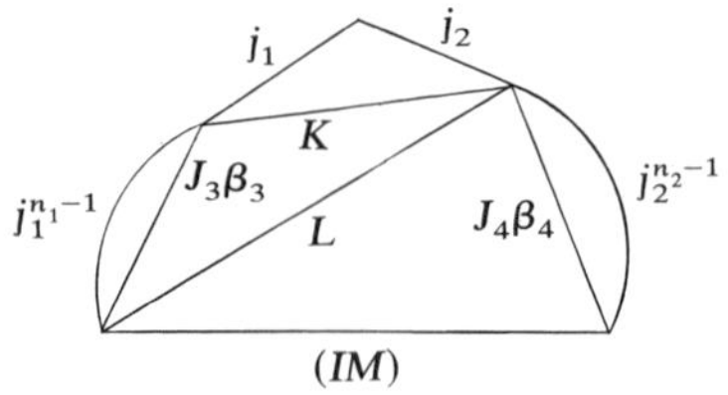

(1.95)

One can use this recoupled form of the wave function to calculate the contribution to $\langle H\rangle$ when $n_1'=n_1$, $n_2'=n_2$. In this case only the second term in equation (1.93) (the one that does not change the number of particles in either orbit) can contribute. The only effect of V when it operates on equation (1.95) is to reproduce identically each term in the summation on the right-hand side of the equation multiplied by $E_K(j_1j_2;j_1j_2)$. To complete the calculation of the matrix element one must make the wave function on the left-hand side of equation (1.91) have the same form as that arising from the operation of V on $[(j_1)^{n_1}_{J_1\beta_1}\times(j_2)^{n_2}_{J_2\beta_2}]_{IM}$; in other words a decomposition identical to that just carried out must be made for the configuration $[(j_1)^{n_1}_{J_1'\beta_1'}\times(j_2)^{n_2}_{J_2'\beta_2'}]_{IM}$. The overlap of the two resulting diagrams will be zero unless corresponding legs of the two multi-gons have the same quantum numbers so that

$$\begin{aligned}\langle[(j_1)^{n_1}_{J_1'\beta_1'}\times(j_2)^{n_2}_{J_2'\beta_2'}]_{IM}|\,H\,\|[(j_1)^{n_1}_{J_1\beta_1}\times(j_2)^{n_2}_{J_2\beta_2}]_{IM}\rangle&=\langle H\rangle_{(\mathrm{i})+(\mathrm{ii})}\\ &+(-1)^{J_2'-J_2}n_1n_2\sqrt{\{(2J_1+1)(2J_2+1)(2J_1'+1)(2J_2'+1)\}}\\ &\times\sum_{J_3\beta_3J_4\beta_4LK}(2L+1)(2K+1)\langle j_1^{n_1-1}J_3\beta_3,j_1|\}j_1^{n_1}J_1\beta_1\rangle\\ &\times\langle j_1^{n_1-1}J_3\beta_3,j_1|\}J_1'\beta_1'\rangle\langle j_2^{n_2-1}J_4\beta_4,j_2|\}j_2^{n_2}J_2\beta_2\rangle\\ &\times\langle j_2^{n_2-1}J_4\beta_4,j_2|\}j_2^{n_2}J_2'\beta_2'\rangle\,W(J_1j_2IJ_4;LJ_2)W(J_1'j_2IJ_4;LJ_2')\\ &\times W(J_3j_1Lj_2;J_1K)W(J_3j_1Lj_2;J_1'K)\,E_K(j_1j_2;j_1j_2)\end{aligned}\qquad(1.96)$$

where $\langle H\rangle_{(\mathrm{i})+(\mathrm{ii})}$ is given by equation (1.92). The factor n_1n_2 in the matrix element arises because of Rule 1 in section 4.1 which says that a factor $(-1)^{1-n}\sqrt{n}$ is to be associated with each c.f.p. for an n-particle state. As for a possible phase factor arising from Rule 2, note that the term in V that contributes has the form $a^\dagger_{j_1}a^\dagger_{j_2}a_{j_2}a_{j_1}$ and consequently to get all the operators involving j_1 together means one must bring them through the bilinear product $a^\dagger_{j_2}a_{j_2}$ and this leads to a plus sign. Consequently there is no additional phase factor arising from Rule 2.

We next consider the calculation of off-diagonal matrix elements. There are two possible contributions to the one in which $n_1'=n_1+1$, $n_2'=n_2-1$. The first possibility involves the destruction of one particle in j_1 and one in j_2 and the subsequent creation of both of them in the state j_1 (the first term in equation (1.93)). If we denote that part of V which mediates this by V_1 we have

$$V_1=\sum_{K'M'}E_{K'}(j_1j_1;j_1j_2)A^\dagger_{K'M'}(j_1j_1)A_{K'M'}(j_1j_2).$$

The second possibility is that V destroys two particles in the state j_2 and then creates one in j_1 and one in j_2. This part of V we denote by V_2:

$$V_2=\sum_{K'M'}E_{K'}(j_1j_2;j_2j_2)A^\dagger_{K'M'}(j_1j_2)A_{K'M'}(j_2j_2).$$

In order to make the wave function of the left-hand side of the matrix element in equation (1.91) look like the result of V_1 operating on equation (1.95) we must make a double-parentage decomposition of the state $(j_1)^{n_1+1}_{J'_1\beta'_1}$. Thus

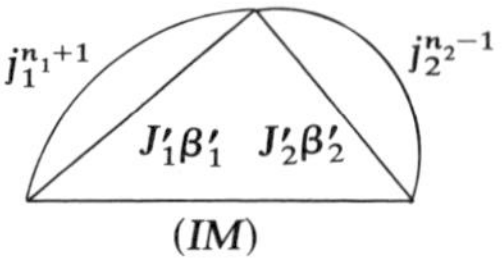

$$= \sum_{J_5\beta_5K'} \langle j_1^{n_1-1}J_5\beta_5, j_1^2K'|\}j_1^{n_1+1}J'_1\beta'_1\rangle$$ 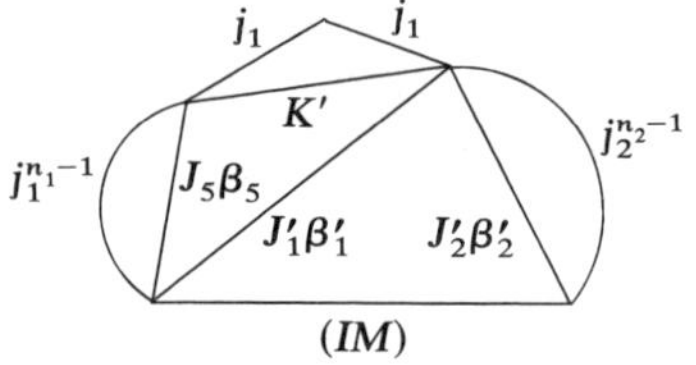 (1.97)

Since

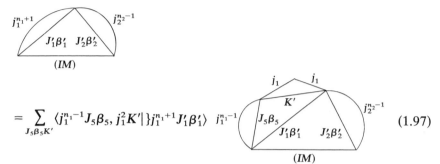

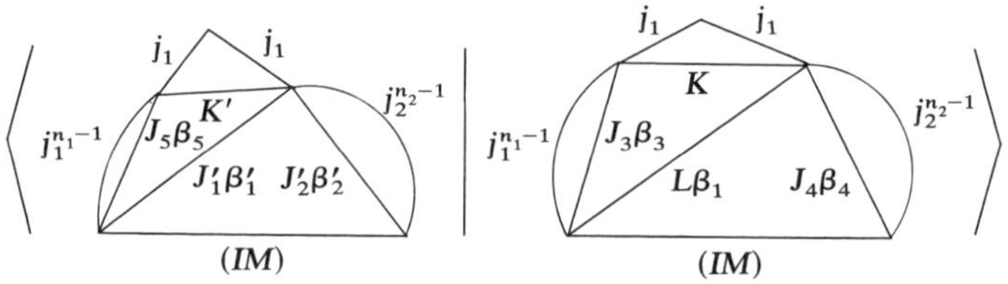

$$= \delta_{KK'}\delta_{J_3J_5}\delta_{\beta_3\beta_5}\delta_{LJ'_1}\delta_{J_4J'_2}\delta_{\beta_1\beta'_1}$$

it follows that

$$\begin{aligned}
\langle V_1\rangle &= \langle[(j_1)^{n_1+1}_{J'_1\beta'_1}\times(j_2)^{n_2-1}_{J'_2\beta'_2}]_{IM}|\ V_1\ |[(j_1)^{n_1}_{J_1\beta_1}\times(j_2)^{n_2}_{J_2\beta_2}]_{IM}\rangle \\
&= (-1)^{n_2+1+j_2+J'_2-J_2}\, n_1\sqrt{\{n_2(n_1+1)(2J_1+1)(2J_2+1)(2J'_1+1)/2\}} \\
&\quad \times \sum_{J_3\beta_3K}\sqrt{(2K+1)}\langle j_1^{n_1-1}J_3\beta_3, j_1|\}j_1^{n_1}J_1\beta_1\rangle \\
&\quad \times\langle j_2^{n_2-1}J'_2\beta'_2, j_2|\}j_2^{n_2}J_2\beta_2\rangle \\
&\quad \times\langle j_1^{n_1-1}J_3\beta_3, j_1^2K|\}j_1^{n_1+1}J'_1\beta'_1\rangle W(J_1j_2IJ'_2; J'_1J_2) \\
&\quad \times W(J_3j_1J'_1j_2; J_1K)E_K(j_1j_1; j_1j_2).
\end{aligned} \tag{1.98}$$

To see the origin of the counting- and number-dependent phase factor in this equation we note that Rule 1 leads to the multiplication factor $(-1)^{n_1+n_2}\, n_1\sqrt{\{n_2(n_1+1)/2\}}$. To determine the phase induced by Rule 2 we define $A_J^\dagger(j, n)$ as the n-particle operator that creates the state $(j)^n_J$ and $A_J(j, n)$ as the operator that destroys the state. In terms of these operators the above matrix element has the form

$$\langle 0|\, A_{J_2'}(j_2, n_2-1)A_{J_1'}(j_1, n_1+1)a^\dagger_{j_1}a^\dagger_{j_1}a_{j_2}a_{j_1}A^\dagger_{J_1}(j_1, n_1)A^\dagger_{J_2}(j_2, n_2)\,|0\rangle$$
$$=(-1)^{n_1+1}\langle 0|\, A_{J_1'}(j_1, n_1+1)a^\dagger_{j_1}a^\dagger_{j_1}a_{j_1}A^\dagger_{J_1}(j_1, n_1)\,|0\rangle$$
$$\times\langle 0|\, A_{J_2'}(j_2, n_2-1)a_{j_2}A^\dagger_{J_2}(j_2, n_2)\,|0\rangle.$$

Since $(-1)^{2n}=1$, it follows that the effect of these two rules is to multiply the matrix element by $(-1)^{n_2+1}n_1\sqrt{\{n_2(n_1+1)/2\}}$.

We now turn to the contribution arising from V_2. To calculate this it is clear that a double-parentage decomposition of the state $(j_2)^{n_2}_{J_2\beta_2}$ is called for:

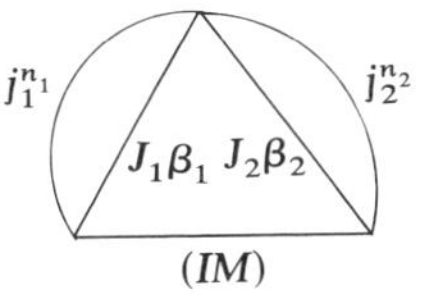

$$=\sum_{J_4\beta_4K}(-1)^{J_4+K-J_2}\langle j_2^{n_2-2}J_4\beta_4, j_2^2K|\}j_2^{n_2}J_2\beta_2\rangle$$

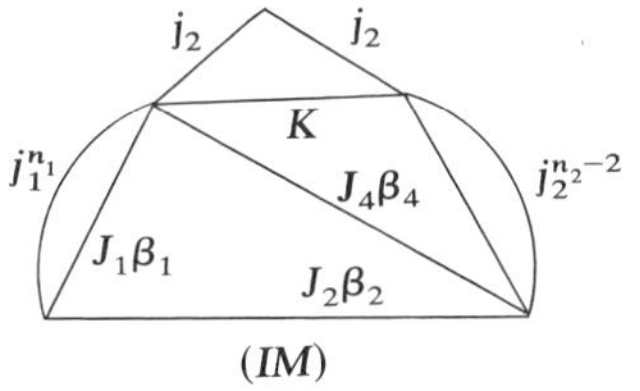

(1.99)

Thus

$$V_2\,|[(j_1)^{n_1}_{J_1\beta_1}\times(j_2)^{n_2}_{J_2\beta_2}]_{IM}\rangle$$
$$=\sum_{J_4\beta_4K}(-1)^{J_4+K-J_2}E_K(j_1j_2;\,j_2j_2)\langle j_2^{n_2-2}J_4\beta_4, j_2^2K|\}j_2^{n_2}J_2\beta_2\rangle$$

$\times$ 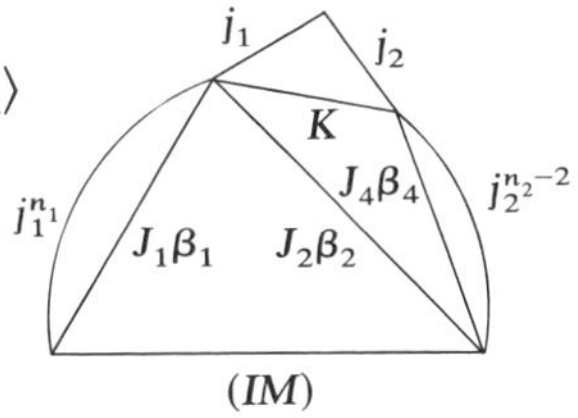

$$=\sum_{J_4\beta_4KL}(-1)^{J_4+K-J_2}\sqrt{\{(2K+1)(2L+1)\}}E_K(j_1j_2;\,j_2j_2)$$
$$\times\langle j_2^{n_2-2}J_4\beta_4, j_2^2K|\}j_2^{n_2}J_2\beta_2\rangle W(j_1j_2J_2J_4;\,KL)$$

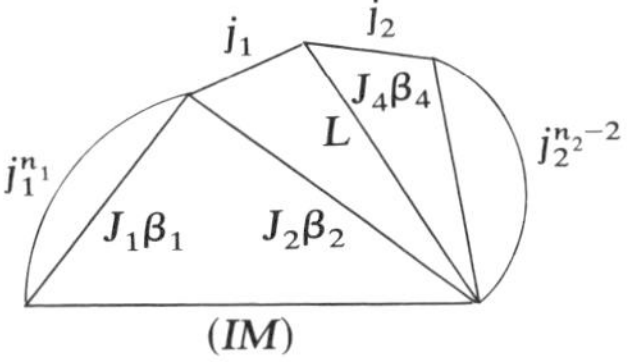

$$=\sum_{J_4\beta_4KLL'}(-1)^{J_4+K-J_2}\sqrt{\{(2K+1)(2L+1)(2J_2+1)(2L'+1)\}}$$
$$\times E_K(j_1j_2;\,j_2j_2)\langle j_2^{n_2-2}J_4\beta_4, j_2^2K|\}j_2^{n_2}J_2\beta_2\rangle$$
$$\times W(j_1j_2J_2J_4;\,KL)W(J_1j_1IL;\,L'J_2)$$

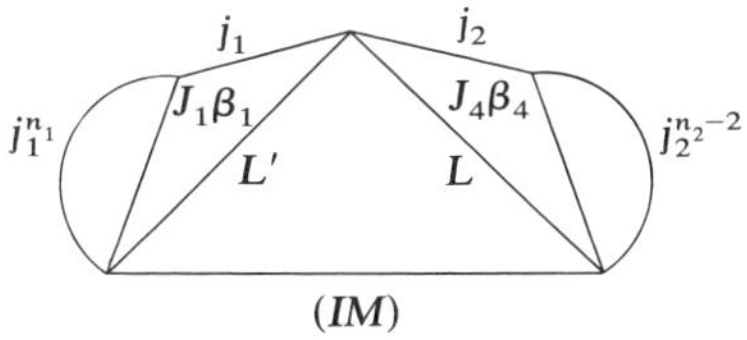

The above term clearly has the form that would arise when a fractional-parentage decomposition of the state $[(j_1)^{n_1+1}_{J'_1\beta'_1}\times(j_2)^{n_2-1}_{J'_2\beta'_2}]_{IM}$ is carried out. Thus

$$\begin{aligned}\langle V_2\rangle &= \langle[(j_1)^{n_1+1}_{J'_1\beta'_1}\times(j_2)^{n_2-1}_{J'_2\beta'_2}]_{IM}|\,V_2\,\|[(j_1)^{n_1}_{J_1\beta_1}\times(j_2)^{n_2}_{J_2\beta_2}]_{IM}\rangle\\ &=(-1)^{n_2}(n_2-1)\surd\{n_2(n_1+1)(2J'_1+1)(2J_2+1)(2J'_2+1)/2\}\\ &\quad\times\sum_{J_4\beta_4K}(-1)^{j_2+J_2-J'_2-K}\surd(2K+1)\langle j_1^{n_1}J_1\beta_1,j_1|\}j_1^{n_1+1}J'_1\beta'_1\rangle\\ &\quad\times\langle j_2^{n_2-2}J_4\beta_4,j_2|\}j_2^{n_2-1}J'_2\beta'_2\rangle\langle j_2^{n_2-2}J_4\beta_4,j_2^2K|\}j_2^{n_2}J_2\beta_2\rangle\\ &\quad\times W(j_1j_2J_2J_4;KJ'_2)W(J_1j_1IJ'_2;J'_1J_2)E_K(j_1j_2;j_2j_2).\qquad(1.100)\end{aligned}$$

In this equation Rule 1 gives the counting factor $(-1)^{n_1+n_2}(n_2-1)\times\surd\{n_2(n_1+1)/2\}$ whereas Rule 2 gives the phase factor $(-1)^{n_1}$ since $A^\dagger_{J_1}(j_1,n_1)$ must be brought through the product $a^\dagger_{j_2}a_{j_2}a_{j_2}$ in order that all operators involving the orbit j_1 stand together. Combining these results we see that

$$\langle[(j_1)^{n_1+1}_{J'_1\beta'_1}\times(j_2)^{n_2-1}_{J'_2\beta'_2}]_{IM}|\,H\,\|[(j_1)^{n_1}_{J_1\beta_1}\times(j_2)^{n_2}_{J_2\beta_2}]_{IM}\rangle=\langle V_1\rangle+\langle V_2\rangle\qquad(1.101)$$

where $\langle V_1\rangle$ and $\langle V_2\rangle$ are given by equations (1.98) and (1.100).

To find the off-diagonal matrix element when $n'_1=n_1-1$ and $n'_2=n_2+1$ one can proceed in exactly the same manner as used for equation (1.101). Alternatively, if one takes into account the anticommutation of $a^\dagger_{j_1m_1}$ and $a^\dagger_{j_2m_2}$ together with the symmetry property of the Clebsch–Gordan coefficients one sees that

$$[(j_1)^{n_1}_{J_1\beta_1}\times(j_2)^{n_2}_{J_2\beta_2}]_{IM}=(-1)^{J_1+J_2-I+n_1n_2}[(j_2)^{n_2}_{J_2\beta_2}\times(j_1)^{n_1}_{J_1\beta_1}]_{IM}$$

and

$$[(j_1)^{n_1-1}_{J'_1\beta'_1}\times(j_2)^{n_2+1}_{J'_2\beta'_2}]_{IM}=(-1)^{J'_1+J'_2-I+(n_1-1)(n_2+1)}[(j_2)^{n_2+1}_{J'_2\beta'_2}\times(j_1)^{n_1-1}_{J'_1\beta'_1}]_{IM}$$

from which it follows that

$$\begin{aligned}&\langle[(j_1)^{n_1-1}_{J'_1\beta'_1}\times(j_2)^{n_2+1}_{J'_2\beta'_2}]_{IM}|\,H\,\|[(j_1)^{n_1}_{J_1\beta_1}\times(j_2)^{n_2}_{J_2\beta_2}]_{IM}\rangle\\ &=(-1)^{J_1+J_2-J'_1-J'_2+n_1-n_2-1}\\ &\quad\times\langle[(j_2)^{n_2+1}_{J'_2\beta'_2}\times(j_1)^{n_1-1}_{J'_1\beta'_1}]_{IM}|\,H\,\|[(j_2)^{n_2}_{J_2\beta_2}\times(j_1)^{n_1}_{J_1\beta_1}]_{IM}\rangle.\qquad(1.102)\end{aligned}$$

(In writing the phase factor we have made use of the fact that because n_1 and n_2 are integers $(-1)^{2n_1n_2}=1$ and moreover that because J'_1, J'_2, and I

form a triangle that $(-1)^{2(J_1'+J_2'-I)}=1$.) Thus the matrix element we wish to evaluate (the right-hand side of equation (1.102) is just a phase factor times the matrix element given by equation (1.101) provided that in equation (1.101) one makes the replacements

$$j_1 \leftrightarrow j_2$$

$$J_1 \leftrightarrow J_2$$

$$\beta_1 \leftrightarrow \beta_2$$

$$n_1 \leftrightarrow n_2$$

$$J_1' \leftrightarrow J_2'$$

$$\beta_1' \leftrightarrow \beta_2'$$

Therefore, if one takes into account the symmetry properties of the $E_K(j_1j_2;j_3j_4)$ given by equation (1.46) one sees that

$$[(j_1)^{n_1-1}_{J_1'\beta_1'}\times(j_2)^{n_2+1}_{J_2'\beta_2'}]_{IM}|\,H\,|[(j_1)^{n_1}_{J_1\beta_1}\times(j_2)^{n_2}_{J_2\beta_2}]_{IM}\rangle=\langle V_3\rangle+\langle V_4\rangle \quad (1.103)$$

where

$$\begin{aligned}\langle V_3\rangle = {} & n_2\sqrt{\{(n_1(n_2+1)(2J_1+1)(2J_2+1)(2J_2'+1))/2\}}\\ & \times\sum_{J_3\beta_3K}(-1)^{j_2+J_2-J_2'-K+n_2}\sqrt{(2K+1)}\langle j_2^{n_2-1}J_3\beta_3,j_2|\}j_2^{n_2}J_2\beta_2\rangle\\ & \times\langle j_1^{n_1-1}J_1'\beta_1',j_1|\}j_1^{n_1}J_1\beta_1\rangle\langle j_2^{n_2-1}J_3\beta_3,j_2^2K|\}j_2^{n_2+1}J_2'\beta_2'\rangle\\ & \times W(J_2j_1IJ_1';J_2'J_1)W(J_3j_2J_2'j_1;J_2K)E_K(j_2j_2;j_1j_2) \end{aligned}\quad (1.104)$$

and

$$\begin{aligned}\langle V_4\rangle = {} & (-1)^{n_2+j_2+J_2-J_2'}(n_1-1)\sqrt{\{(n_1(n_2+1)(2J_1+1)(2J_1'+1)(2J_2'+1))/2\}}\\ & \times\sum_{J_4\beta_4K}\sqrt{(2K+1)}\langle j_2^{n_2}J_2\beta_2,j_2|\}j_2^{n_2+1}J_2'\beta_2'\rangle\\ & \times\langle j_1^{n_1-2}J_4\beta_4,j_1|\}j_1^{n_1-1}J_1'\beta_1'\rangle\langle j_1^{n_1-2}J_4\beta_4,j_1^2K|\}j_1^{n_1}J_1\beta_1\rangle\\ & \times W(j_2j_1J_1J_4;KJ_1')W(J_2j_2IJ_1';J_2'J_1)E_K(j_1j_2;j_1j_1). \end{aligned}\quad (1.105)$$

Finally to compute the matrix element for the case $n_1'=n_1+2$, $n_2'=n_2-2$ a double-parentage decomposition is made for the state $(j_2)^{n_2}_{J_2\beta_2}$.

Thus

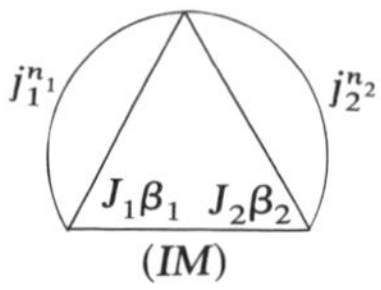

$$= \sum_{J_4\beta_4 K} (-1)^{J_4+K-J_2} \langle j_2^{n_2-2} J_4\beta_4, j_2^2 K | \} j_2^{n_2} J_2\beta_2 \rangle$$

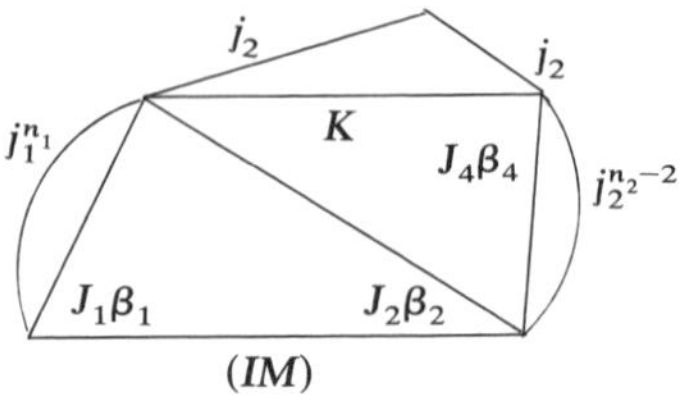

$$= \sum_{J_4\beta_4 KL} (-1)^{J_4+K-J_2} \langle j_2^{n_2-2} J_4\beta_4, j_2^2 K | \} j_2^{n_2} J_2\beta_2 \rangle$$

$$\times \sqrt{\{(2L+1)(2J_2+1)\}} W(J_1 K I J_4; L J_2)$$

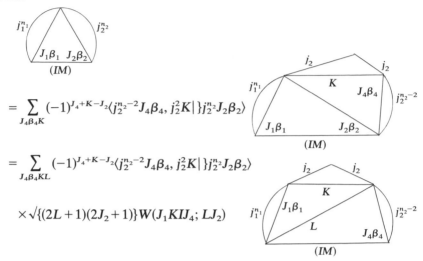

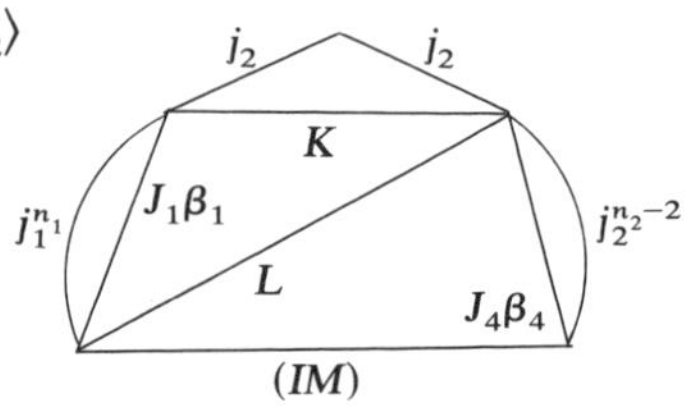

To have a non-vanishing overlap with the state $[(j_1)^{n_1+2}_{J_1'\beta_1'} \times (j_2)^{n_2-2}_{J_2'\beta_2'}]_{IM}$, V must destroy the two interacting particles in the state j_2 and recreate them in the state j_1. Therefore, if one makes a double-parentage decomposition analogous to that in equation (1.97), for the state $(j_1)^{n_1+2}_{J_1'\beta_1}$ one easily shows that

$$\begin{aligned}
&\langle [(j_1)^{n_1+2}_{J_1'\beta_1'} \times (j_2)^{n_2-2}_{J_2'\beta_2'}]_{IM} | \, H \, | [(j_1)^{n_1}_{J_1\beta_1} \times (j_2)^{n_2}_{J_2\beta_2}]_{IM} \rangle \\
&\quad = \tfrac{1}{2}\sqrt{\{n_2(n_2-1)(n_1+2)(n_1+1)(2J_1'+1)(2J_2+1)\}} \\
&\qquad \times \sum_K (-1)^{J_2'+K-J_2} W(J_1 K I J_2'; J_1' J_2) \\
&\qquad \times \langle j_2^{n_2-2} J_2'\beta_2', j_2^2 K | \} j_2^{n_2} J_2\beta_2 \rangle \langle j_1^{n_1} J_1\beta_1, j_1^2 K | \} j_1^{n_1+2} J_1'\beta_1' \rangle \\
&\qquad \times E_K(j_1 j_1; j_2 j_2). \qquad (1.106)
\end{aligned}$$

To have all the operators involving the orbit j_1 together one must bring $A^\dagger_{J_1}(j_1, n_1)$ through the operator that destroys a pair of nucleons in the state j_2. Thus Rule 2 gives a phase factor of $+1$. The counting factor, $\frac{1}{2}\sqrt{\{n_2(n_2-1)(n_1+2)(n_1+1)\}}$ of course arises from Rule 1.

In a similar manner

$$\begin{aligned}
&\langle [(j_1)^{n_1-2}_{J_1'\beta_1'} \times (j_2)^{n_2+2}_{J_2'\beta_2'}]_{IM} | \, H \, | [(j_1)^{n_1}_{J_1\beta_1} \times (j_2)^{n_2}_{J_2\beta_2}]_{IM} \rangle \\
&\quad = \tfrac{1}{2}\sqrt{\{n_1(n_1-1)(n_2+2)(n_2+1)(2J_1+1)(2J_2'+1)\}} \\
&\qquad \times \sum_K (-1)^{J_2+K-J_2'} W(J_1' K I J_2, J_1 J_2') \\
&\qquad \times \langle j_1^{n_1-2} J_1'\beta_1', j_1^2 K | \} j_1^{n_1} J_1\beta_1 \rangle \langle j_2^{n_2} J_2\beta_2, j_2^2 K | \} j_2^{n_2+2} J_2'\beta_2' \rangle \\
&\qquad \times E_K(j_2 j_2; j_1 j_1). \qquad (1.107)
\end{aligned}$$

4.4. Spectrum of $^{19}_{8}O_{11}$

To illustrate the use of the expressions for the Hamiltonian matrix elements just derived, we consider the calculation of the spectrum of $^{19}_{8}O_{11}$. For the low-lying states, the model space will be taken to be the same as that previously used for $^{18}_{8}O_{10}$; i.e. $^{16}_{8}O_8$ will be taken to be a doubly closed shell and the three valence neutrons will be confined to the $0d_{\frac{5}{2}}$ and $1s_{\frac{1}{2}}$ orbits.

By the methods discussed in sections 1.1–1.3 one can show that the allowable spin states and the configurations that give rise to them are

$$I=\tfrac{1}{2}:[(d_{\frac{5}{2}})^2_0\times s_{\frac{1}{2}}]_{\frac{1}{2}m}$$
$$I=\tfrac{3}{2}:(d_{\frac{5}{2}})^3_{\frac{3}{2}m};\;[(d_{\frac{5}{2}})^2_2\times s_{\frac{1}{2}}]_{\frac{3}{2}m}$$
$$I=\tfrac{5}{2}:(d_{\frac{5}{2}})^3_{\frac{5}{2}m};\;[(d_{\frac{5}{2}})^2_2\times s_{\frac{1}{2}}]_{\frac{5}{2}m};\;[d_{\frac{5}{2}}\times(s_{\frac{1}{2}})^2_0]_{\frac{5}{2}m}$$
$$I=\tfrac{7}{2}:[(d_{\frac{5}{2}})^2_4\times s_{\frac{1}{2}}]_{\frac{7}{2}m}$$
$$I=\tfrac{9}{2}:(d_{\frac{5}{2}})^3_{\frac{9}{2}m};\;[(d_{\frac{5}{2}})^2_4\times s_{\frac{1}{2}}]_{\frac{9}{2}m}.$$

Thus the energy matrices to be diagonalized are at most 3×3 and we shall now construct them. Expressions for the $I=\frac{1}{2}$, $\frac{3}{2}$ and $\frac{5}{2}$ matrices have been given by Pandya (1963).

$I=\frac{1}{2}$

Within this model space there is only one way to construct the state, and the energy is therefore given by equations (1.92) and (1.96) with $j_1=\frac{5}{2}$, $j_2=\frac{1}{2}$, $J_1=J_1'=0$, $J_2=J_2'=\frac{1}{2}$, $n_1=2$, and $n_2=1$. Since the $d_{\frac{5}{2}}$ orbit contains only two particles, the double-parentage coefficient in equation (1.92) has $J=0$ (because $n_1-2=0$ corresponds to the vacuum for nucleons with $j_1=\frac{5}{2}$) and $K=0$. Moreover, both single-nucleon-parentage coefficients in equation (1.96) are unity and have $J_3=\frac{5}{2}$ for the $d_{\frac{5}{2}}$ configuration and $J_4=0$ for the $s_{\frac{1}{2}}$ state (when one particle is 'split-off' from a one-particle state the parent is the vacuum state for that orbit and hence has angular momentum zero). Thus

$$\langle H\rangle_{I=\frac{1}{2}}=2\varepsilon_d+\varepsilon_s+E_0(\tfrac{5}{2}\tfrac{5}{2};\tfrac{5}{2}\tfrac{5}{2})+4\sum_{LK}(2L+1)(2K+1)$$
$$\times\{W(0\tfrac{1}{2}\tfrac{1}{2}0;L\tfrac{1}{2})\}^2\{W(\tfrac{5}{2}\tfrac{5}{2}L\tfrac{1}{2};0K)\}^2E_K(\tfrac{5}{2}\tfrac{1}{2};\tfrac{5}{2}\tfrac{1}{2}).$$

Since the Racah coefficient with one index equal to zero has the simple form given by equation (A4.14),

$$W(0\tfrac{1}{2}\tfrac{1}{2}0;L\tfrac{1}{2})=\tfrac{1}{2}\delta_{L\frac{1}{2}}$$

and

$$W(\tfrac{5}{2}\tfrac{5}{2}\tfrac{1}{2}\tfrac{1}{2};0K)=(-1)^{3-K}/\sqrt{12}$$

it follows that

$$\langle H\rangle_{I=\frac{1}{2}}=2\varepsilon_d+\varepsilon_s+E_0(\tfrac{5}{2}\tfrac{5}{2};\tfrac{5}{2}\tfrac{5}{2})+\tfrac{5}{6}E_2(\tfrac{5}{2}\tfrac{1}{2};\tfrac{5}{2}\tfrac{1}{2})+\tfrac{7}{6}E_3(\tfrac{5}{2}\tfrac{1}{2};\tfrac{5}{2}\tfrac{1}{2}).$$

As a check on the algebra one may take the limit that the range of the residual interaction tends to infinity and the strength tends to unity. In this case

$$\begin{aligned}E_K(j_1j_2;j_3j_4)&\rightarrow\langle\psi_{KM}(j_1j_2)\mid\psi_{KM}(j_3j_4)\rangle\\&=\delta_{j_1j_3}\delta_{j_2j_4}\end{aligned}$$

and the matrix element of $\sum_{i<j}V_{ij}$ then gives the number of pairs multiplied by the orthonormality integral for the states involved. For diagonal matrix elements the sum of the coefficients of E_K should be $n(n-1)/2=3$.

$I=\frac{3}{2}$

There are two ways to make this spin

$$\begin{aligned}\chi_{\frac{3}{2}m}(1)&=(d_{\frac{5}{2}})^3_{\frac{3}{2}m}\\\chi_{\frac{3}{2}m}(2)&=[(d_{\frac{5}{2}})^2_2\times s_{\frac{1}{2}}]_{\frac{3}{2}m}\end{aligned}$$

so one must diagonalize a 2×2 matrix. To calculate

$$\langle H_{11}\rangle_{I=\frac{3}{2}}=\langle\chi_{\frac{3}{2}m}(1)|\,H\,|\chi_{\frac{3}{2}m}(1)\rangle$$

we use equation (1.70) together with the one nucleon $j=\frac{5}{2}$ c.f.p's. given in Table A5.1. Thus

$$\langle H_{11}\rangle_{I=\frac{3}{2}}=3\varepsilon_d+\tfrac{15}{7}E_2(\tfrac{5}{2}\tfrac{5}{2};\tfrac{5}{2}\tfrac{5}{2})+\tfrac{6}{7}E_4(\tfrac{5}{2}\tfrac{5}{2};\tfrac{5}{2}\tfrac{5}{2}).$$

The other diagonal matrix element

$$\langle H_{22}\rangle_{I=\frac{3}{2}}=\langle\chi_{\frac{3}{2}m}(2)|\,H\,|\chi_{\frac{3}{2}m}(2)\rangle$$

is computed as was $\langle H\rangle_{I=\frac{1}{2}}$ except that in this case $J_1'=J_1=2$. Thus by use of equations (1.92) and (1.96)

$$\begin{aligned}\langle H_{22}\rangle_{I=\frac{3}{2}}&=2\varepsilon_d+\varepsilon_s+E_2(\tfrac{5}{2}\tfrac{5}{2};\tfrac{5}{2}\tfrac{5}{2})+20\sum_{LK}(2L+1)(2K+1)\\&\quad\times\{W(2\tfrac{1}{2}\tfrac{3}{2}0;L\tfrac{1}{2})\}^2\{W(\tfrac{5}{2}\tfrac{5}{2}L\tfrac{1}{2};2K)\}^2E_K(\tfrac{5}{2}\tfrac{1}{2};\tfrac{5}{2}\tfrac{1}{2})\\&=2\varepsilon_d+\varepsilon_s+E_2(\tfrac{5}{2}\tfrac{5}{2};\tfrac{5}{2}\tfrac{5}{2})+\tfrac{4}{3}E_2(\tfrac{5}{2}\tfrac{1}{2};\tfrac{5}{2}\tfrac{1}{2})+\tfrac{2}{3}E_3(\tfrac{5}{2}\tfrac{1}{2};\tfrac{5}{2}\tfrac{1}{2}).\end{aligned}$$

In obtaining the final numerical expression use has been made of the table of Racah coefficients (Rotenberg *et al.* 1959).

The off-diagonal matrix element

$$\langle H_{12}\rangle_{I=\frac{3}{2}}=\langle\chi_{\frac{3}{2}m}(1)|\,H\,|\chi_{\frac{3}{2}m}(2)\rangle$$

is computed using equation (1.101) with $n_1=2$, $n_2=1$, $J_1=2$, $J_2=\frac{1}{2}$, $J_1'=\frac{3}{2}$, and $J_2'=0$. Since the identical nucleon configuration $\frac{5}{2}^2$ can only have $I=0$, 2, and 4, and since $\frac{5}{2}$, $\frac{1}{2}$ can only couple to 2 and 3, the only allowable value for K is $K=2$. If we make use of the relationship between the one- and two-nucleon c.f.p's for the case $n=3$ (see equation (A5.15))

$$\langle j^1 j, j^2 J | \} j^3 I\alpha \rangle = (-1)^{J+j-I} \langle j^2 J, j | \} j^3 I\alpha \rangle$$

one finds that

$$\langle H_{12} \rangle_{I=\frac{3}{2}} = \frac{20\sqrt{105}}{7} W(2\tfrac{1}{2}\tfrac{3}{2}0;\tfrac{3}{2}\tfrac{1}{2}) W(\tfrac{5}{2}\tfrac{5}{2}\tfrac{3}{2}\tfrac{1}{2};22) E_2(\tfrac{5}{2}\tfrac{5}{2};\tfrac{5}{2}\tfrac{1}{2})$$

$$= \frac{2\sqrt{35}}{7} E_2(\tfrac{5}{2}\tfrac{5}{2};\tfrac{5}{2}\tfrac{1}{2}).$$

Since $\langle H_{12} \rangle_{I=\frac{3}{2}} = \langle H_{21} \rangle_{I=\frac{3}{2}}$, the energy matrix to be diagonalized is

$$\langle H \rangle_{I=\frac{3}{2}} = \begin{bmatrix} 3\varepsilon_d + \frac{15}{7} E_2(\frac{5}{2}\frac{5}{2};\frac{5}{2}\frac{5}{2}) + \frac{6}{7} E_4(\frac{5}{2}\frac{5}{2};\frac{5}{2}\frac{5}{2}) & \frac{2\sqrt{35}}{7} E_2(\frac{5}{2}\frac{5}{2};\frac{5}{2}\frac{1}{2}) \\ \frac{2\sqrt{35}}{7} E_2(\frac{5}{2}\frac{5}{2};\frac{5}{2}\frac{1}{2}) & 2\varepsilon_d + \varepsilon_s + E_2(\frac{5}{2}\frac{5}{2};\frac{5}{2}\frac{5}{2}) + \frac{4}{3} E_2(\frac{5}{2}\frac{1}{2};\frac{5}{2}\frac{1}{2}) + \frac{2}{3} E_3(\frac{5}{2}\frac{1}{2};\frac{5}{2}\frac{1}{2}) \end{bmatrix}.$$

For the diagonal matrix elements, the sum of the coefficients of E_K is 3. In addition, since $E_2(\frac{5}{2}\frac{5}{2};\frac{5}{2}\frac{1}{2}) \to 0$ when the range of V becomes infinite, the matrix satisfies the orthonormality conditions discussed for $I=\frac{1}{2}$.

$I=\frac{5}{2}$

The three ways of making a spin $\frac{5}{2}$ state are

$$\chi_{\frac{5}{2}m}(1) = (d_{\frac{5}{2}})^3_{\frac{5}{2}m}$$

$$\chi_{\frac{5}{2}m}(2) = [(d_{\frac{5}{2}})^2_2 \times s_{\frac{1}{2}}]_{\frac{5}{2}m}$$

$$\chi_{\frac{5}{2}m}(3) = [d_{\frac{5}{2}} \times (s_{\frac{1}{2}})^2_0]_{\frac{5}{2}m}$$

The calculation proceeds exactly as before and leads to the 3×3 energy matrix $\langle H\rangle_{I=\frac{5}{2}}$ the matrix elements of which are

$$\langle H_{11}\rangle_{I=\frac{5}{2}} = 3\varepsilon_d + \tfrac{2}{3}E_0(\tfrac{5}{2}\tfrac{5}{2};\tfrac{5}{2}\tfrac{5}{2}) + \tfrac{5}{6}E_2(\tfrac{5}{2}\tfrac{5}{2};\tfrac{5}{2}\tfrac{5}{2}) + \tfrac{3}{2}E_4(\tfrac{5}{2}\tfrac{5}{2};\tfrac{5}{2}\tfrac{5}{2})$$

$$\langle H_{22}\rangle_{I=\frac{5}{2}} = 2\varepsilon_d + \varepsilon_s + E_2(\tfrac{5}{2}\tfrac{5}{2};\tfrac{5}{2}\tfrac{5}{2}) + \tfrac{1}{2}E_2(\tfrac{5}{2}\tfrac{1}{2};\tfrac{5}{2}\tfrac{1}{2}) + \tfrac{3}{2}E_3(\tfrac{5}{2}\tfrac{1}{2};\tfrac{5}{2}\tfrac{1}{2})$$

$$\langle H_{33}\rangle_{I=\frac{5}{2}} = \varepsilon_d + 2\varepsilon_s + E_0(\tfrac{1}{2}\tfrac{1}{2};\tfrac{1}{2}\tfrac{1}{2}) + \tfrac{5}{6}E_2(\tfrac{5}{2}\tfrac{1}{2};\tfrac{5}{2}\tfrac{1}{2}) + \tfrac{7}{6}E_3(\tfrac{5}{2}\tfrac{1}{2};\tfrac{5}{2}\tfrac{1}{2})$$

$$\langle H_{12}\rangle_{I=\frac{5}{2}} = \langle H_{21}\rangle_{I=\frac{5}{2}} = \frac{\sqrt{5}}{2\sqrt{3}}\,E_2(\tfrac{5}{2}\tfrac{5}{2};\tfrac{5}{2}\tfrac{1}{2})$$

$$\langle H_{13}\rangle_{I=\frac{5}{2}} = \langle H_{31}\rangle_{I=\frac{5}{2}} = \sqrt{\tfrac{2}{3}}E_0(\tfrac{5}{2}\tfrac{5}{2};\tfrac{1}{2}\tfrac{1}{2})$$

$$\langle H_{23}\rangle_{I=\frac{5}{2}} = \langle H_{32}\rangle_{I=\frac{5}{2}} = -\sqrt{\tfrac{5}{6}}E_2(\tfrac{5}{2}\tfrac{5}{2};\tfrac{5}{2}\tfrac{1}{2}).$$

$I=\frac{7}{2}$

Since there is only one way to make the state the procedure used to calculate the energy is identical to that used for $\langle H\rangle_{I=\frac{1}{2}}$. By using equations (1.92) and (1.96) we find that

$$\langle H\rangle_{I=\frac{7}{2}} = 2\varepsilon_d + \varepsilon_s + E_4(\tfrac{5}{2}\tfrac{5}{2};\tfrac{5}{2}\tfrac{5}{2}) + \tfrac{5}{3}E_2(\tfrac{5}{2}\tfrac{1}{2};\tfrac{5}{2}\tfrac{1}{2}) + \tfrac{1}{3}E_3(\tfrac{5}{2}\tfrac{1}{2};\tfrac{5}{2}\tfrac{1}{2}).$$

$I=\frac{9}{2}$

There are two ways to make spin $\frac{9}{2}$:

$$\chi_{\frac{9}{2}m}(1) = (d_{\frac{5}{2}})^3_{\frac{9}{2}m}$$

$$\chi_{\frac{9}{2}m}(2) = [(d_{\frac{5}{2}})^2_4 \times s_{\frac{1}{2}}]_{\frac{9}{2}m}.$$

The 2×2 energy matrix that must be diagonalized to find the energies of the states is

$$\langle H\rangle_{I=\frac{9}{2}} = \begin{bmatrix} 3\varepsilon_d + \frac{9}{14}E_2(\frac{5}{2}\frac{5}{2};\frac{5}{2}\frac{5}{2}) + \frac{33}{14}E_4(\frac{5}{2}\frac{5}{2};\frac{5}{2}\frac{5}{2}) & \dfrac{\sqrt{3}}{2\sqrt{7}}E_2(\frac{5}{2}\frac{5}{2};\frac{5}{2}\frac{1}{2}) \\ \dfrac{\sqrt{3}}{2\sqrt{7}}E_2(\frac{5}{2}\frac{5}{2};\frac{5}{2}\frac{1}{2}) & 2\varepsilon_d + \varepsilon_s + E_4(\frac{5}{2}\frac{5}{2};\frac{5}{2}\frac{5}{2}) + \frac{1}{6}E_2(\frac{5}{2}\frac{1}{2};\frac{5}{2}\frac{1}{2}) + \frac{11}{6}E_3(\frac{5}{2}\frac{1}{2};\frac{5}{2}\frac{1}{2}) \end{bmatrix}$$

To obtain numerical values for the energies of these states, we take (as in section 2.1 of this Chapter) the residual nucleon–nucleon interaction to be the surface-delta potential whose matrix elements are given by equation (1.59). When V_0R_0, ε_d, and ε_s are taken to have the values previously used (1·057, −4·143, and −3·272 MeV, respectively) the resulting theoretical spectrum is shown in Fig. 1.10. The binding energy relative to ${}^{16}_{8}O_8$ is quite well reproduced by the calculation, but the predicted excitation energies compared with the experimental results of

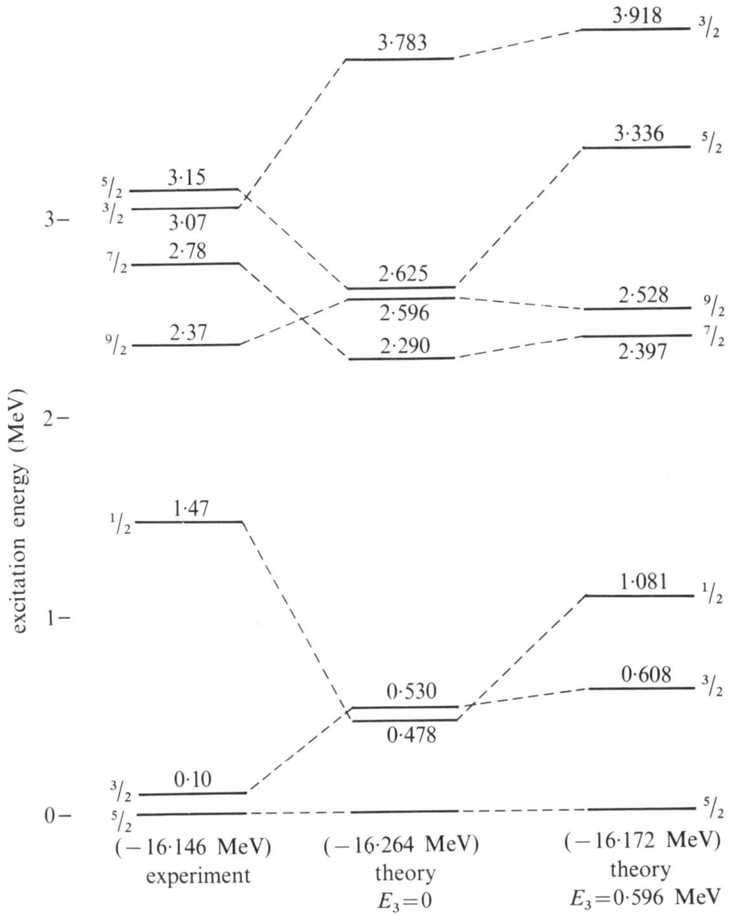

Fig. 1.10. Spectrum of $^{19}_{8}O_{11}$. The ground-state energy is given in MeV relative to the $^{16}_{8}O_8$ core. The excitation energies are taken from the work of Crozier *et al.* (1972). The theoretical predictions are based on the $(d_{\frac{5}{2}}, s_{\frac{1}{2}})$ model space with the single-particle energies of equations (1.9) and (1.10). In the centre column the results obtained with the surface-delta interaction ($V_0R_0 = 1{\cdot}057$ MeV) are shown. On the right the theoretical spectrum is shown when $E_3(\frac{5}{2}\frac{5}{2};\, \frac{5}{2}\frac{1}{2}) = 0{\cdot}596$ MeV and all other matrix elements are those of the surface-delta interaction.

Crozier *et al.* (1972) show an r.m.s. error of 0·61 MeV. This is only slightly worse than the 0·41 MeV r.m.s. error that arose when the same potential was used to calculate $^{18}_{8}O_{10}$, and is typical of the kind of agreement one generally obtains when a delta-function potential is used.

Some improvement in the calculated spectrum can be obtained if one uses a finite-range interaction. For example, the two neutrons making up the $I=3$ state $[d_{\frac{5}{2}} \times s_{\frac{1}{2}}]_{3M}$ can couple their orbital angular momenta to at most $L=2$ $(L=l_1+l_2)$. Thus to have $I=3$, they must be in the symmetric $S=1$ state. Since the total wave function must be antisymmetric, the spatial $L=2$ state must be antisymmetric to interchange of the two particles, and consequently in this I state the two nucleons can never be at the same point in space. For this reason the delta-function potential gives a vanishing value for $E_3(\frac{5}{2}\frac{1}{2};\frac{5}{2}\frac{1}{2})$. It turns out that the 3^+ state is the worst predicted in ${}^{18}_{8}O_{10}$. On the other hand, a finite-range potential can lead to a non-vanishing value for this matrix element and if we somewhat arbitrarily set

$$E_3(\tfrac{5}{2}\tfrac{1}{2};\tfrac{5}{2}\tfrac{1}{2}) = +0{\cdot}596 \text{ MeV}$$

(the value needed to obtain the correct excitation energy of the 3^+ state in ${}^{18}_{8}O_{10}$) the spectrum shown on the right-hand side of Fig. 1.10 is obtained. Clearly some improvement has been obtained in this way and the r.m.s. error in any calculated excitation energy is now 0·47 MeV. However, in general when a simple finite-range interaction is used in conjunction with harmonic-oscillator eigenfunctions, for the radial wave functions one does not predict excitation energies of multi-particle systems to much better than 300 keV. For a discussion of the 'best' finite-range potential to be used in shell-model calculations over a wide range of nuclei, see the work of Schiffer and True (1976).

5. Seniority

In this Section we define the seniority quantum number for the identical particle configuration j^n and discuss the experimental evidence that indicates this is a good quantum number for single-closed-shell nuclei. Some selection rules on nucleon transfer and gamma decay are given for seniority eigenfunctions and a simple expression is deduced for the energy of the lowest state of j^n. Finally the generalization of this quantum number to single-closed-shell nuclei in which the valence nucleons fill several single-particle orbitals simultaneously is discussed.

5.1. The seniority quantum number

If the effective interaction between the valence nucleons can be approximated by a delta-function potential

$$V(\underset{\sim}{r}_1-\underset{\sim}{r}_2) = -4\pi V_0 \delta(\underset{\sim}{r}_1-\underset{\sim}{r}_2)$$

it follows from equation (A2.29) that the interaction energy between two identical particles in the state j coupled to angular momentum $I=0$ is

$$E_0(jj;jj)=-(2j+1)\frac{V_0\bar{R}}{2}$$

where

$$\bar{R}=\int R_j^4 r^2\,\mathrm{d}r$$

and R_j is the radial wave function for a nucleon in the state j. For $I\neq 0$ one can use the special form of the Clebsch–Gordan coefficient when $(m_1m_2M)=(\frac{1}{2}-\frac{1}{2}0)$ (see equations (A1.27) and (A1.28)) together with the fact that

$$\frac{K!}{\{(K/2)!\}^2}=\frac{2^K(K-1)!!}{K!!}$$

to write

$$E_I(jj;jj)=-\frac{V_0\bar{R}}{4}\{1+(-1)^I\}\frac{(2j+1+I)!!(2j-I)!!}{(2j+I)!!(2j-1-I)!!}\left\{\frac{(I-1)!!}{I!!}\right\}^2.$$

(This equation can be used for all values of I provided one defines $(I-1)!!/I!!=1$ for $I=0$.) From the ratio

$$\frac{E_{I+2}(jj;jj)}{E_I(jj;jj)}=\left(\frac{I+1}{I+2}\right)^2\left\{1-\frac{(2I+3)}{(2j-I)(2j+2+I)}\right\}$$

it is clear that the interaction energy becomes less attractive as I increases.

In Fig. 1.11 we have plotted the ratio $E_I(jj;jj)/E_0(jj;jj)$ when $(j)_I^2=(0g_{\frac{9}{2}})_I^2$. Curves are given for three different potentials—the delta-function interaction, the Yukawa potential

$$V_Y=(e^{-\mu r})/\mu r$$

with $\mu=0{\cdot}7\ \mathrm{fm}^{-1}$ $\underline{r}=|\underline{r}_1-\underline{r}_2|$ and the Gaussian potential

$$V_G=e^{-\mu^2r^2}$$

with $\mu=0{\cdot}6\ \mathrm{fm}^{-1}$. Although E_I/E_0 falls off more rapidly for the delta function, the fall off still persists when a finite-range potential is used. Thus for a short-range residual two-body potential the $I=0$ state is predicted to lie considerably lower in energy than the $I=2, 4, \ldots$ states and this is precisely what is seen experimentally (see Figs. 1.4, 1.7, and 1.9).

When two nucleons couple to $I=0$ their spins are said to be paired.

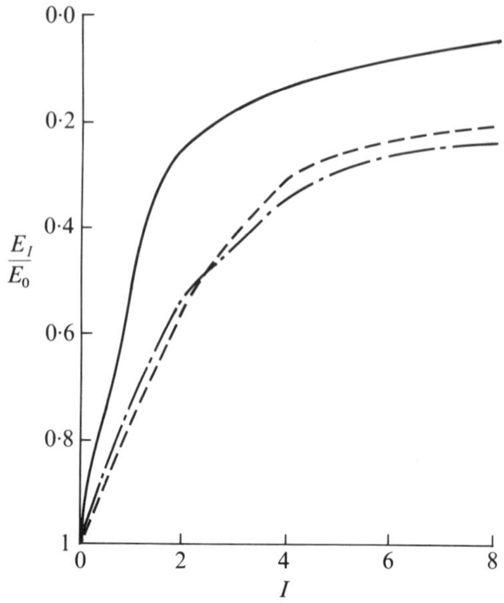

Fig. 1.11. Plot of $E_I(jj; jj)/E_0(jj; jj)$ as a function of I when j is the $0g_{\frac{9}{2}}$ orbit. The solid curve is for the delta-function potential, - - - - for the Gaussian potential $e^{-\mu^2r^2}$ with $\mu = 0{\cdot}6\ \text{fm}^{-1}$ and —-— for the Yukawa interaction $(e^{-\mu r})/\mu r$ with $\mu = 0{\cdot}7\ \text{fm}^{-1}$.

Furthermore, since

$$(jjmm' \mid 00) = (-1)^{j-m}(2j+1)^{-\frac{1}{2}}\delta_{m,-m'}$$

it follows that the normalized wave function describing this state is

$$\Psi_{00}(j^2) = \{2(2j+1)\}^{-\frac{1}{2}} \sum_m (-1)^{j-m} a^\dagger_{jm} a^\dagger_{j-m} |0\rangle$$
$$= \surd\{2/(2j+1)\} S_+(j) |0\rangle \qquad (1.108)$$

where

$$S_+(j) = \sum_{m>0} (-1)^{j-m} a^\dagger_{jm} a^\dagger_{j-m} \qquad (1.109)$$

is the zero-coupled pair-creation operator. Because $S_+(j)$ creates a state in which the z-components of angular momentum of the two particles are oppositely oriented, it is convenient to depict the situation in terms of an 'arrow' diagram

$$S_+(j) |0\rangle \Rightarrow \uparrow\!\downarrow. \qquad (1.110a)$$

On the other hand, the two-particle creation operator $A^{\dagger}_{IM}(jj)$ given in equation (1.42) with $I \neq 0$ does not pair spins and is pictured as

$$A^{\dagger}_{IM}(jj)\,|0\rangle \Rightarrow \uparrow\uparrow \qquad (1.110b)$$

and the excitation energy at which such states are seen experimentally is 1·5–3 MeV.

On the basis of energy considerations alone one would therefore expect the following hierarchy of states of the configuration j^n with n even:

(a) the lowest energy state would correspond to all nucleons having their spins paired and the eigenfunction would be

$$\{S_+(j)\}^{n/2}\,|0\rangle \Rightarrow \uparrow\downarrow\uparrow\downarrow\ldots\uparrow\downarrow \qquad (1.111a)$$

(b) in the energy region 1·5–3 MeV one would expect to see states described by wave functions having two unpaired particles

$$A^{\dagger}_{IM}(jj)\{S_+(j)\}^{(n-2)/2}\,|0\rangle \Rightarrow \uparrow\uparrow\uparrow\downarrow\ldots\uparrow\downarrow \qquad (1.111b)$$

(c) above 3 MeV, states with four unpaired nucleons should appear

$$B^{\dagger}_{IM}\{S_+(j)\}^{(n-4)/2}\,|0\rangle \Rightarrow \uparrow\uparrow\uparrow\uparrow\ldots\uparrow\downarrow \qquad (1.111c)$$

where $B^{\dagger}_{IM}$ stands for a four-particle creation operator in which no particles are paired, etc.

Clearly the number of unpaired particles is likely to be a useful labelling device for eigenfunctions of the configuration j^n. Thus we define the seniority v of a nuclear state as the number of unpaired nucleons in the eigenfunction describing the state (Racah 1943, 1949). From this definition it follows that equations (1.111a), (1.111b), and (1.111c) describe seniority-zero, -two, and -four states, respectively.

For an odd number of particles, the ground state of the system is expected to have one unpaired nucleon and hence be a seniority-one eigenfunction:

$$a^{\dagger}_{jm}\{S_+(j)\}^{(n-1)/2}\,|0\rangle \Rightarrow \uparrow\uparrow\downarrow\uparrow\downarrow\ldots\uparrow\downarrow. \qquad (1.112)$$

Since all the nucleons but one have their spins paired, it is clear that a seniority-one eigenfunction describes a state with $I = j$. The other possible states of the odd-n system would correspond to seniority 3, 5, . . . , n in which 3, 5, . . . , n particles are unpaired.

In section 5.3 we shall show how to construct seniority eigenfunctions explicitly. However, even without a detailed knowledge of their form one can deduce certain selection rules from the 'arrow' diagrams. For example, the general form of the single-particle operator given by equation

(1.51) corresponds to the destruction of a particle in the state (jm) followed by its recreation in the state (jm'). Thus the single-particle operator $Q_{\lambda\mu}$ operating on a seniority-v eigenfunction gives

$$
\begin{aligned}
Q_{\lambda\mu}\uparrow\uparrow\ldots\uparrow\uparrow\uparrow\downarrow\ldots\uparrow\downarrow \Rightarrow\ & \uparrow\uparrow\ldots\uparrow\uparrow\uparrow\downarrow\ldots\uparrow\downarrow\ (\Delta v=0)\\
& \uparrow\uparrow\ldots\uparrow\uparrow\uparrow\uparrow\ldots\uparrow\downarrow\ (\Delta v=+2)\\
& \uparrow\uparrow\ldots\uparrow\downarrow\uparrow\downarrow\ldots\uparrow\downarrow\ (\Delta v=-2).
\end{aligned}
$$

Therefore any operator which is the sum of single-particle operators has non-vanishing matrix elements only between states that differ in seniority by zero or ± 2 units. Consequently in the nucleus ${}^{52}_{24}\mathrm{Cr}_{28}$ the 2^+ state with seniority 4, which is predicted by use of equations (1.80) and (1.83) to lie at an excitation energy of

$$
\begin{aligned}
\Delta E &= \tfrac{1}{6}(E_2-E_0)+\tfrac{51}{22}(E_4-E_0)-\tfrac{65}{66}(E_6-E_0)\\
&= 3{\cdot}194\ \mathrm{MeV},
\end{aligned}
$$

should have no gamma decay to the ground state.

Other important selection rules that may be derived simply from the 'arrow' diagrams concern the possible final states that can be populated in direct-transfer reactions. We shall discuss these in more detail in Chapter 2 (section 5) and Chapter 4 (section 4). These reactions are considered to take place near the nuclear surface and the transferred nucleons are put into or taken from the target without disturbing the other nucleons in the target. Thus in single-nucleon direct-transfer experiments ((p, d), (d, p), etc.) the states populated in the final nucleus can only differ from the target ground state by one unit of seniority. Similarly, in two-nucleon transfer ((t, p), (p, t), etc.) the difference in seniority between the initial and final nuclear states is zero or ± 2 units. An example of the operation of these selection rules is provided by the ${}^{43}_{20}\mathrm{Ca}_{23}(d, p){}^{44}_{20}\mathrm{Ca}_{24}$ and ${}^{42}_{20}\mathrm{Ca}_{22}(t, p){}^{44}_{20}\mathrm{Ca}_{24}$ reactions. In the former the target ground state is $(\nu f_{\frac{7}{2}})^3_{I=\frac{7}{2}}$ with seniority 1. Thus in a single-nucleon direct-reaction process only seniority-zero and -two states in ${}^{44}_{20}\mathrm{Ca}_{24}$ can be populated. In the latter reaction the target has $(\nu f_{\frac{7}{2}})^2_{I=0}$ with $v=0$ and hence only seniority-zero and -two states in ${}^{44}_{20}\mathrm{Ca}_{24}$ can be reached. In ${}^{44}_{20}\mathrm{Ca}_{24}$ two low-lying 4^+ states have been observed (at 2·283 MeV and 3·044 MeV excitation energies), and these can be interpreted, in first approximation, as the seniority-two and four 4^+ states of the configuration $(\nu f_{\frac{7}{2}})^4$. Since the 3·044 MeV level is populated much more strongly than the 2·283 MeV state in both the (d, p) and (t, p) reactions (Bjerregaard *et al.* 1967), it is clear that the upper one of the two states is predominantly $v=2$. This is precisely what one would find if the proton–proton interaction energies of equation (1.83) were used to calculate the $(\nu f_{\frac{7}{2}})^4$ spectrum. Although these two-body matrix elements should be corrected for Coulomb effects before they are used in a neutron–neutron calculation this correction will not affect the predicted excitation energies very much. The reason for this is that the Coulomb interaction, because of its long range, modifies each of the two-body matrix elements by about the same amount. Since the addition of a constant to all two-body matrix elements does not change predicted excitation energies, it follows that one would expect the seniority-four state of $(\nu f_{\frac{7}{2}})^4_{I=4}$ to lie lower than the seniority-two level.

5.2. *How good is the seniority quantum number?*

As shown in Appendix A3 seniority is a good quantum number for the identical-particle configuration j^n if the residual interaction is a delta-function potential. Since the residual interaction is expected to have a range of the order of the π-meson Compton wavelength ($r = \hbar/m_\pi c \simeq 1{\cdot}4$ fm) and the nuclear radius is $R \simeq 1{\cdot}2 \times A^{\frac{1}{3}}$ fm it follows that $r/R \simeq 1{\cdot}2/A^{\frac{1}{3}}$. Thus for any but the lightest nuclei this ratio is considerably less than unity, and the delta-function approximation should be reasonable. Consequently one would expect to find experimentally that seniority is a good quantum number within the configuration j^n for single closed-shell nuclei.

To check this experimentally we examine the condition that the two-body matrix elements must satisfy for no seniority mixing to occur. As shown by French (1960), if

$$\sum_{K>0} \{(2K+1)W(jjjj;JK)+\tfrac{1}{2}\delta_{JK}\}E_K(jj;jj) = \frac{2}{(2j-1)(2j+1)}\sum_{K>0}(2K+1)E_K(jj;jj) \quad (1.113)$$

for all even $J>0$, seniority is a good quantum number within j^n. For $j=\frac{1}{2}$, $\frac{3}{2}$, and $\frac{5}{2}$ there is only one state of j^n for given I, so that no possibility exists for testing seniority conservation. The first place where mixing can occur is in the $f_{\frac{7}{2}}$ configuration with $n=4$. In this case there are two $I=2$ and two $I=4$ states. However, as can be shown by substituting numerical values for the Racah coefficients, equation (1.113) is automatically satisfied and so seniority is always conserved in the $f_{\frac{7}{2}}$ configuration for any two-body interaction (Racah and Talmi 1952, Schwartz and de Shalit 1954, and equation (A3.42)).

If we use the shorthand notation

$$E_K = E_K(\tfrac{9}{2}\tfrac{9}{2};\tfrac{9}{2}\tfrac{9}{2})$$

explicit calculation shows that for the $j=\frac{9}{2}$ orbit equation (1.113) can only be satisfied if

$$\Delta=\frac{1}{20\sqrt{429}}(65E_2-315E_4+403E_6-153E_8) \qquad (1.114)$$

vanishes. Since there are no nuclei where the $g_{\frac{9}{2}}$ level is isolated from all other single-particle orbits, one must make a careful analysis of the experimental data to see whether or not Δ (defined by equation (1.114) is indeed small. The nuclei that yield information about this are those in which $N=50$, $Z\geq 39$; i.e. ${}^{89}_{39}Y_{50}$, ${}^{90}_{40}Zr_{50}$, ${}^{91}_{41}Nb_{50}$, ${}^{92}_{42}Mo_{50}$, ${}^{93}_{43}Tc_{50}$, and ${}^{94}_{44}Ru_{50}$. Since the nucleus ${}^{89}_{39}Y_{50}$ shows a paucity of low-lying states (below 1·5 MeV excitation energy only the $\frac{1}{2}^-$ ground state and $\frac{9}{2}^+$ level at 915 keV are seen) $Z=38$ appears to be a very stable structure. Thus a suitable model for analyzing the data on these nuclei should be one in which ${}^{88}_{38}Sr_{50}$ is taken to be an inert core and the valence protons move in the $1p_{\frac{1}{2}}$ and $0g_{\frac{9}{2}}$ orbitals. Eleven parameters are needed to specify completely the shell-model Hamiltonian within this model space. They are two single-particle energies ε_p and ε_g and nine matrix elements $E_I(j_1j_2;j_3j_4)$ of the residual two-body interaction V. The reason there are so few E_I is because V is parity conserving so that the only E_I are

$$E_I(\tfrac{1}{2}\tfrac{1}{2};\tfrac{1}{2}\tfrac{1}{2}) \qquad I=0$$

$$E_I(\tfrac{9}{2}\tfrac{9}{2};\tfrac{9}{2}\tfrac{9}{2}) \qquad I=0,2,4,6,\text{ and }8$$

$$E_I(\tfrac{1}{2}\tfrac{1}{2};\tfrac{9}{2}\tfrac{9}{2}) \qquad I=0$$

$$E_I(\tfrac{9}{2}\tfrac{1}{2};\tfrac{9}{2}\tfrac{1}{2}) \qquad I=4\text{ and }5.$$

Gloeckner and Serduke (1974) have shown that the 45 energy levels with established spins and parities that are known for these nuclei can be reproduced with an r.m.s. error of only 52·2 keV if $\Delta\equiv 0$. If Δ is allowed to vary freely, a best fit to the data gives $\Delta=23$ keV and the r.m.s. error in each energy is 51·9 keV. Thus from spectra alone one would conclude that seniority is conserved in the $g_{\frac{9}{2}}$ configuration to a high degree.

If one requires that these model-space wave functions also reproduce the observed $E2$ transition rates Δ cannot be taken equal to zero (Gloeckner *et al.* 1972) but has the value $\Delta=-0{\cdot}162$ MeV. However, it is easy to show that this value of Δ induces only small seniority mixing. To see this we construct the energy matrix for the two $I=\frac{9}{2}$ states of the $(g_{\frac{9}{2}})^3$ configuration. By use of equation (1.70) and the $g_{\frac{9}{2}}$ c.f.p's given in Table

A5.4 we find

$$\langle H\rangle = \begin{bmatrix} 3\varepsilon_g + \frac{1}{20}\{16E_0+5E_2 + 9E_4+13E_6+17E_8\} & -\Delta \\ -\Delta & 3\varepsilon_g+\frac{1}{8580}\{845E_2+11025E_4 + 12493E_6+1377E_8\} \end{bmatrix}$$

where H_{11} and H_{22} are the diagonal seniority-one and -three matrix elements, respectively. The off-diagonal matrix element is given precisely by Δ in equation (1.114) and this is the reason for our choice of the overall multiplicative constant in this equation. Using E_I determined by Gloeckner *et al.* (1972)

$$\begin{aligned} E_0(\tfrac{9}{2}\tfrac{9}{2};\tfrac{9}{2}\tfrac{9}{2}) &= -1{\cdot}702\ \text{MeV} \\ E_2(\tfrac{9}{2}\tfrac{9}{2};\tfrac{9}{2}\tfrac{9}{2}) &= -0{\cdot}644\ \text{MeV} \\ E_4(\tfrac{9}{2}\tfrac{9}{2};\tfrac{9}{2}\tfrac{9}{2}) &= 0{\cdot}243\ \text{MeV} \\ E_6(\tfrac{9}{2}\tfrac{9}{2};\tfrac{9}{2}\tfrac{9}{2}) &= 0{\cdot}351\ \text{MeV} \\ E_8(\tfrac{9}{2}\tfrac{9}{2};\tfrac{9}{2}\tfrac{9}{2}) &= 0{\cdot}588\ \text{MeV} \end{aligned} \tag{1.115}$$

one finds that the eigenfunctions of $\langle H\rangle$ are

$$\Phi_{\frac{9}{2}}(1) = 0{\cdot}995(g_{\frac{9}{2}})^3_{I=\frac{9}{2},v=1} - 0{\cdot}104(g_{\frac{9}{2}})^3_{I=\frac{9}{2},v=3}$$

$$\Phi_{\frac{9}{2}}(2) = 0{\cdot}104(g_{\frac{9}{2}})^3_{I=\frac{9}{2},v=1} + 0{\cdot}995(g_{\frac{9}{2}})^3_{I=\frac{9}{2},v=3}$$

where $\Phi_{\frac{9}{2}}(1)$ describes the lower energy state.

Thus the residual nucleon–nucleon interaction induces only about a 1% seniority mixing in the three-particle case and a similarly small amount in states with $n>3$. Consequently within the $g_{\frac{9}{2}}$ configuration seniority is an excellent quantum number in agreement with what is expected if the residual interaction between identical particles has a short range.

Although there is always an energy gap of more than 1 MeV between the $v=0$ ($I=0$) and $v=2$ ($I\geq 2$) levels in the even-A nuclei (see, for example, Figs. 1.7 and 1.9 and the spectra shown by Gloeckner and Serduke 1974), such a large separation does not occur between $v=1$ and $v=3$ states in an odd-A nucleus. In Fig. 1.8 one sees that the seniority-three $I=\frac{5}{2}^-$ state of $(f_{\frac{7}{2}})^3$ lies only about 300 keV above the seniority-one ground state. The same is true for $(g_{\frac{9}{2}})^3$. If one uses Table A5.4 one easily shows that

$$E_{\frac{7}{2}}(n=3,\ v=3)-E_{\frac{9}{2}}(n=3,\ v=1)=\tfrac{1}{8580}\{11375(E_2-E_0)$$
$$-261(E_4-E_0)-5525(E_6-E_0)+1275(E_8-E_0)\}.$$

When the interaction energies of equation (1.115) are used, the $v=3$ $\frac{7}{2}^+$ state of this configuration is predicted to lie only 362 keV above the seniority-one $I=j=\frac{9}{2}^+$ level. When we discuss quasi-particle calculations in Chapter 6 it will be important to realize that in the odd-n nuclei with $n\geq 3$ the state $I=j-1$ always lies low in excitation energy.

5.3. Seniority eigenfunctions

To write down normalized seniority eigenfunctions it is convenient to make use of the fact that the zero-coupled pair-creation operator given in equation (1.109) and its Hermitian adjoint

$$\mathrm{S}_-(j)=\sum_{m>0}(-1)^{j-m}a_{j-m}a_{jm} \tag{1.116}$$

satisfy the commutation relationship

$$\begin{aligned}[\mathrm{S}_-(j),\mathrm{S}_+(j)]&=\mathrm{S}_-(j)\mathrm{S}_+(j)-\mathrm{S}_+(j)\mathrm{S}_-(j)\\ &=(2j+1)/2-\sum_m a^\dagger_{jm}a_{jm}\\ &=\{\Omega_j-\mathrm{N}_j\}\end{aligned} \tag{1.117}$$

where N_j, given by equation (1.41), is the number operator for the state j and

$$\Omega_j=(2j+1)/2 \tag{1.118}$$

is the number of pairs that the orbit j can accommodate. Although we shall not explicitly use the result here, it follows that if we define

$$\begin{aligned}\mathrm{S}_z(j)&=\tfrac{1}{2}\left\{\sum_m a^\dagger_{jm}a_{jm}-(2j+1)/2\right\}\\ &=\tfrac{1}{2}\{\mathrm{N}_j-\Omega_j\}\end{aligned}$$

the three operators $\mathrm{S}_+(j)$, $\mathrm{S}_-(j)$, and $\mathrm{S}_z(j)$ satisfy the same commutation relationship as the ordinary angular momentum operator $\underset{\sim}{\mathrm{J}}$,

$$\underset{\sim}{\mathrm{S}}(j)\times\underset{\sim}{\mathrm{S}}(j)=i\underset{\sim}{\mathrm{S}}(j). \tag{1.119}$$

For this reason $\underset{\sim}{\mathrm{S}}$ is often called the quasi-spin operator (Kerman 1961), and in Appendix 3 we exploit its properties to deduce simple expressions for the number dependence of nuclear matrix elements within the configuration j^n.

We now use equation (1.117) to find the normalization constant Λ_n for the n-particle seniority-zero eigenfunction depicted in equation (1.111a). If we write

$$\Psi_{00}(j^n) = \Lambda_n\{S_+(j)\}^{n/2}\,|0\rangle$$

it follows by use of equation (1.117) that

$$\begin{aligned}\langle\Psi_{00}(j^n)\mid\Psi_{00}(j^n)\rangle &= \Lambda_n^2\langle 0|\,\{S_-(j)\}^{n/2}\{S_+(j)\}^{n/2}\,|0\rangle\\ &= \Lambda_n^2\langle 0|\,\{S_-(j)\}^{(n-2)/2}[\Omega_j-(n-2)\\ &\qquad +S_+(j)S_-(j)]\{S_+(j)\}^{(n-2)/2}\,|0\rangle.\end{aligned}$$

Repeated application of equation (1.117) to the second term in this equation will finally give a term in which the pair-destruction operator $S_-(j)$ operates on the vacuum and hence vanishes. Thus

$$\begin{aligned}\langle\Psi_{00}(j^n)\mid\Psi_{00}(j^n)\rangle &= \Lambda_n^2[\{\Omega_j-(n-2)\}+\{\Omega_j-(n-4)\}+\ldots+\Omega_j]\\ &\qquad\times\langle 0|\,\{S_-(j)\}^{(n-2)/2}\{S_+(j)\}^{(n-2)/2}\,|0\rangle\\ &= \Lambda_n^2\left(\frac{n}{4}[2j+1-(n-2)]\right)\\ &\qquad\times\langle 0|\,\{S_-(j)\}^{(n-2)/2}\{S_+(j)\}^{(n-2)/2}\,|0\rangle.\end{aligned}$$

By continuing in this manner one finally obtains the expression

$$\begin{aligned}\langle\Psi_{00}(j^n)|\,\Psi_{00}(j^n)\rangle &= \Lambda_n^2\left(\frac{n}{4}\{(2j+1)-(n-2)\}\frac{n-2}{4}\{(2j+1)-(n-4)\}\right.\\ &\qquad\left.\times\ldots\tfrac{2}{4}\{2j+1\}\right)\langle 0\mid 0\rangle\\ &= \Lambda_n^2\left\{\frac{n!!(2j+1)!!}{2^n(2j+1-n)!!}\right\}.\end{aligned}$$

Thus the normalized seniority-zero wave function of the configuration j^n is

$$\Psi_{00}(j^n) = \left\{\frac{2^n(2j+1-n)!!}{n!!(2j+1)!!}\right\}^{\frac{1}{2}}\{S_+(j)\}^{n/2}\,|0\rangle. \qquad (1.120)$$

The normalization constant for the seniority-two eigenfunction of equation (1.111b) can be obtained in the same way or by using the following simple argument. The two unpaired nucleons in the seniority-two wave function occupy particular m substates, say m_1 and m_2. Since these particles are not to be paired, the substates $-m_1$ and $-m_2$ are blocked as far as putting nucleons into them. Thus the remaining $(n-2)/2$ zero-coupled pairs have only $(2j-3)$ substates at their disposal. Therefore to find the normalization constant in this case one need only make the replacement

$$(2j+1) \to (2j-3)$$
$$n \to n-2$$

in the normalization constant of equation (1.120). Consequently

$$\Psi_{IM,v=2}(j^n) = \left\{\frac{2^{n-2}(2j-1-n)!!}{(n-2)!!(2j-3)!!}\right\}^{\frac{1}{2}} A^{\dagger}_{IM}(jj)\{S_+(j)\}^{(n-2)/2}\,|0\rangle. \quad (1.121)$$

where $A^{\dagger}_{IM}(jj)$ is given by equation (1.42). Furthermore, since the two nucleons in $A^{\dagger}_{IM}(jj)$ must not be paired the only $v=2$ states are those with $I=2, 4, \ldots, (2j-1)$.

In general, if one wants to construct states with angular momentum I and $v>2$ one proceeds via the method outlined in section 1.2. First, all possible v-particle states

$$\phi_M(i) = a^{\dagger}_{jm_1} a^{\dagger}_{jm_2} \ldots a^{\dagger}_{jm_v}\,|0\rangle$$

with $M=I$ are written down. A linear combination of these $\phi_M(i)$ is taken

$$\Phi_{IM=I} = \sum_i \alpha_i \phi_{M=I}(i) \quad (1.122)$$

with coefficients α_i chosen so that

$$\mathrm{J}_+\Phi_{IM=I} = 0 \quad (1.123a)$$
$$\mathrm{S}_-(j)\Phi_{IM=I} = 0 \quad (1.123b)$$

and

$$\sum_i \alpha_i^2 = 1.$$

If α_i can be found so that equation (1.123a) is satisfied, $\Phi_{I,M=I}$ is guaranteed to have angular momentum I because of Theorem 1, section 1.1. Furthermore, if equation (1.123b) can be satisfied a v-particle state with no zero-coupled pairs has been constructed and hence by definition is a seniority-v eigenfunction. (In the same way that not all possible angular-momentum states can be realized in a given configuration (see section 1.2) it is clear that it will not always be possible to obtain a v-particle state with angular momentum I and seniority $v>2$.) However, if Φ_{IM} can be found that satisfies equations (1.123) the normalized n-particle seniority-v eigenfunction is

$$\Psi_{IM,v}(j^n) = \left\{\frac{2^{n-v}(2j+1-v-n)!!}{(n-v)!!(2j+1-2v)!!}\right\}^{\frac{1}{2}} \Phi_{IM}\{\mathrm{S}_+(j)\}^{(n-v)/2}\,|0\rangle \quad (1.124)$$

where the normalization constant is determined by the same arguments as used in deducing equation (1.121).

Equation (1.124) can be used to construct seniority eigenfunctions for the odd-n system. In particular the seniority-one eigenfunction has

$$\Phi_{IM} = a^{\dagger}_{jm}\,|0\rangle$$

so that

$$\Psi_{jm,v=1}(j^n) = \left\{\frac{2^{n-1}(2j-n)!!}{(n-1)!!(2j-1)!!}\right\}^{\frac{1}{2}} a^{\dagger}_{jm}\{S_+(j)\}^{(n-1)/2}\,|0\rangle. \quad (1.125)$$

5.4. *The Talmi binding-energy formula*

In Appendix 3, section 3.1 we deduce explicit expressions for the double-parentage coefficients associated with the seniority-zero wave function given in equation (1.120). Thus according to equation (A3.27) and (A3.28)

$$\langle j^{n-2}J=0v=0,\, j^2K=0|\}j^nI=0v=0\rangle = \left\{\frac{2j+3-n}{(n-1)(2j+1)}\right\}^{\frac{1}{2}} \quad (1.126a)$$

$$\langle j^{n-2}Jv=2,\, j^2K=J|\}j^nI=0v=0\rangle$$
$$= -\left\{\frac{2(n-2)(2I+1)}{(n-1)(2j+1)(2j-1)}\right\}^{\frac{1}{2}}\left\{\frac{1+(-1)^I}{2}\right\}. \quad (1.126b)$$

One may use these values in conjunction with equation (1.73) to write down the energy associated with the $(j)^n_{I=0v=0}$ configuration which will be the ground state of the even–even nucleus:

$$E_{I=0,v=0}(j^n) = \frac{n(2j+3-n)}{2(2j+1)}E_0 + \frac{n(n-2)}{(2j+1)(2j-1)}\sum_{K>0}(2K+1)E_K$$
$$= \frac{n(n-1)}{2}\left\{\frac{2(j+1)\bar{E}-E_0}{2j+1}\right\} + \frac{n}{2}\left\{\frac{2(j+1)(E_0-\bar{E})}{2j+1}\right\} \quad (1.127)$$

where

$$\bar{E} = \frac{\sum_{K>0}(2K+1)E_K}{\sum_{K>0}(2K+1)} = \frac{\sum_{K>0}(2K+1)E_K}{(j+1)(2j-1)}. \quad (1.128)$$

(In these expressions all energies E_K are understood to be $E_K(jj;jj)$.)

One is, of course, also interested in the energy of the seniority-one ground state of the odd-A nucleus and this may be directly calculated from equation (1.70) when one makes use of the fact that

$$\langle j^{n-1}J=jv=1,\, j|\}j^nI=0v=0\rangle = 1.$$

Thus

$$E_{I=j,v=1}(j^{n-1}) = \frac{n-2}{n} E_{I=0,v=0}(j^n)$$
$$= \frac{(n-1)(n-2)}{2}\left\{\frac{2(j+1)\bar{E} - E_0}{2j+1}\right\}$$
$$+ \frac{n-2}{2}\left\{\frac{2(j+1)(E_0 - \bar{E})}{2j+1}\right\}. \quad (1.129)$$

From equations (1.127) and (1.128) it is clear that the two-body interaction energies in the seniority-zero and -one states of j^n depend on only the two parameters $E_0(jj; jj)$, the energy when two nucleons couple to $J=0$, and $\bar{E}$, the average interaction in the $v=2$ states of j^2. This two-parameter dependence holds no matter how many J states can be realized within j^2. Thus relative to the closed shell in which the state j is empty the energy of the $I=v=0$ and $I=j$, $v=1$ states of the identical-particle configuration j^n is given by (Talmi 1962)

$$\mathrm{BE}(j^n) = n\varepsilon_j + \frac{n(n-1)}{2}\alpha + \left[\frac{n}{2}\right]\beta \quad (1.130)$$

where

$$\left[\frac{n}{2}\right] = \frac{n}{2} \quad \text{if } n \text{ is even}$$
$$= \frac{n-1}{2} \quad \text{if } n \text{ is odd} \quad (1.131)$$

ε_j is the single-particle energy relative to the closed shell and

$$\alpha = \frac{2(j+1)\bar{E} - E_0}{2j+1} \quad (1.132)$$

$$\beta = \frac{2(j+1)(E_0 - \bar{E})}{2j+1} \quad (1.133)$$

with $\bar{E}$ given by equation (1.128).

As an application of equation (1.130) we consider the energies of the calcium isotopes ${}^{41}_{20}Ca_{21}$ to ${}^{48}_{20}Ca_{28}$ relative to the closed-shell nucleus ${}^{40}_{20}Ca_{20}$. In Table 1.2 the theoretical results that emerge when one uses the best-fit parameters (see equation (1.82) for details of the procedure)

$$\varepsilon_{\frac{7}{2}} = -8{\cdot}42 \text{ MeV}$$
$$\alpha_\nu = +0{\cdot}23 \text{ MeV} \quad (1.134)$$
$$\beta_\nu = -3{\cdot}23 \text{ MeV}$$

TABLE 1.2

Comparison of the theoretical value of BE *from equation* (1.130) *to the values given by Wapstra and Gove* (1971)

Nucleus	BE Experiment	Theory	Nucleus	BE Experiment	Theory
$^{41}_{20}Ca_{21}$	−8·36	−8·42	$^{49}_{21}Sc_{28}$	−9·62	−9·69
$^{42}_{20}Ca_{22}$	−19·84	−19·84	$^{50}_{22}Ti_{28}$	−21·79	−21·75
$^{43}_{20}Ca_{23}$	−27·77	−27·80	$^{51}_{23}V_{28}$	−29·84	−29·86
$^{44}_{20}Ca_{24}$	−38·90	−38·76	$^{52}_{24}Cr_{28}$	−40·35	−40·34
$^{45}_{20}Ca_{25}$	−46·32	−46·26	$^{53}_{25}Mn_{28}$	−46·91	−46·87
$^{46}_{20}Ca_{26}$	−56·72	−56·76	$^{54}_{26}Fe_{28}$	−55·77	−55·77
$^{47}_{20}Ca_{27}$	−64·00	−63·80	$^{55}_{27}Co_{28}$	−60·82	−60·72
$^{48}_{20}Ca_{28}$	−73·94	−73·84	$^{56}_{28}Ni_{28}$	−68·01	−68·04

Note that BE is the negative of the binding energy of the nucleus relative to the closed shell. The theoretical values are based on the parameters of equation (1.134) for neutrons and on those of equation (1.136) for protons.

are compared with the experimental values (Wapstra and Gove 1971). The r.m.s. error in any one of the fitted energies is only 100 keV, so that clearly the three-parameter fit to these experimental data is excellent.

From the definitions of α and β in equations (1.132) and (1.133) it follows that

$$E_0 = E_0(\nu) = -3{\cdot}00\text{ MeV}$$
$$\bar{E} = \bar{E}(\nu) = -0{\cdot}13\text{ MeV}. \tag{1.135}$$

Thus the interaction energy in the $I = v = 0$ state is much greater than the average in the seniority-two states. The fact that α_ν is small is consistent with a short range for the residual two-body interaction. To see this we note that when the delta function is used it follows from equation (A2.29) that

$$E_J = \frac{-V_0\bar{R}(2j+1)^2}{4(2J+1)}[1+(-1)^J](jj\tfrac{1}{2}-\tfrac{1}{2}\mid J0)^2.$$

Thus

$$\begin{aligned}\sum_{J>0}(2J+1)E_J &= \frac{-V_0\bar{R}(2j+1)^2}{4}\left(\sum_{J>0}[1+(-1)^J](jj\tfrac{1}{2}-\tfrac{1}{2}\mid J0)^2\right)\\ &= \frac{-V_0\bar{R}(2j+1)^2}{4}\left(\sum_J (jj\tfrac{1}{2}-\tfrac{1}{2}\mid J0)\{(jj\tfrac{1}{2}-\tfrac{1}{2}\mid J0)\right.\\ &\quad\left. -(jj-\tfrac{1}{2}\tfrac{1}{2}\mid J0)\} - 2/(2j+1)\right)\\ &= \frac{-V_0\bar{R}(2j+1)(2j-1)}{4}.\end{aligned}$$

In writing the second line of this equation we have made use of the fact that $(jj\frac{1}{2}-\frac{1}{2}\mid J0)=(-1)^{J+1}(jj-\frac{1}{2}\frac{1}{2}\mid J0)$ and that the sum can be taken over all J provided the contribution from $J=0$ is subtracted. The final answer is a direct consequence of the summation relationship in equation (A1.24). Since

$$\sum_{J>0}(2J+1)=(j+1)(2j-1)$$

it follows that

$$\frac{2(j+1)\sum_{J>0}(2J+1)E_J}{\sum_{J>0}(2J+1)}=\frac{-V_0\bar{R}}{2}(2j+1)$$
$$\equiv E_0.$$

Thus α is zero for the delta-function potential, and the fact that empirically it is found to be small is consistent with a short range for the residual nucleon–nucleon interaction.

A similar analysis can be carried out for the proton $f_{\frac{7}{2}}$ nuclei ${}^{49}_{21}Sc_{28}$, ${}^{50}_{22}Ti_{28}$, ${}^{51}_{23}V_{28}$, ${}^{52}_{24}Cr_{28}$, ${}^{53}_{25}Mn_{28}$, ${}^{54}_{26}Fe_{28}$, ${}^{55}_{27}Co_{28}$, and ${}^{56}_{28}Ni_{28}$. The experimental values tabulated in Table 1.2 are the negative of the binding energies measured relative to the ${}^{48}_{20}Ca_{28}$ core (Wapstra and Gove 1971). The data is best fitted when

$$\varepsilon_{\frac{7}{2}}(\pi)=-9{\cdot}69\text{ MeV}$$
$$\alpha_\pi=+0{\cdot}79\text{ MeV}$$
$$\beta_\pi=-3{\cdot}16\text{ MeV} \tag{1.136}$$

and with these parameters the r.m.s. deviation in any energy is 50 keV.

As might be expected the values of the parameters needed in this fit are somewhat different to those for the valence neutron nuclei. The proton single-particle energy is with respect to the ${}^{48}_{20}Ca_{28}$ core whereas that for the neutron was with respect to the ${}^{40}_{20}Ca_{20}$. In addition, the residual Coulomb interaction between the $f_{\frac{7}{2}}$ protons will make $E_0(\pi)$ and $\bar{E}(\pi)$ less attractive than the values obtained for neutrons. From equations (1.136) it follows that

$$E_0(\pi)=-2{\cdot}37\text{ MeV}$$
$$\bar{E}(\pi)=+0{\cdot}44\text{ MeV}. \tag{1.137}$$

A comparison of these values with those obtained for neutrons (equations (1.135)) shows that

$$\Delta E_0=E_0(\pi)-E_0(\nu)=0{\cdot}63\text{ MeV} \tag{1.138a}$$
$$\Delta\bar{E}=\bar{E}(\pi)-\bar{E}(\nu)=0{\cdot}57\text{ MeV}. \tag{1.138b}$$

If the nuclear forces are assumed to be charge independent, one would conclude that the Coulomb repulsion in the $J=0$ state is about 630 keV. Approximately the same number is obtained for the average value of this interaction in the seniority-two levels and this is consistent with our contention in section 5.1 that even when the Coulomb repulsion is included the same excitation energies will be predicted for both the proton and neutron $f_{\frac{7}{2}}$ nuclei.

5.5. Many single-particle levels

When valence nucleons occupy several single-particle states the basis wave functions in terms of which the Hamiltonian is diagonalized can be written as (see section 1.3)

$$\Phi_{IM}=[[\ldots[[(j_1)^{n_1}_{J_1v_1}\times(j_2)^{n_2}_{J_2v_2}]_{J_{12}}\times(j_3)^{n_3}_{J_3v_3}]_{J_{23}} \ldots\times(j_{m-1})^{n_{m-1}}_{J_{m-1}v_{m-1}}]_{J_{m-2,m-1}}\times(j_m)^{n_m}_{J_mv_m}]_{IM}. \quad (1.139)$$

The seniority of this wave function is defined as the sum of the seniorities of its constituent parts

$$v=\sum_{i=1}^{m} v_i. \quad (1.140)$$

We shall now argue that only under very special conditions will the physically observed states have pure seniority given by equation (1.140).

Although the delta-function interaction has vanishing matrix elements between states with different seniority of the configuration j^n this does not happen when more than one shell is involved. For example, one may easily show that the off-diagonal matrix element between the seniority-one state $(f_{\frac{7}{2}})^3_{I=\frac{7}{2}M}$ and the seniority-three eigenfunction $[f_{\frac{7}{2}}\times(p_{\frac{3}{2}})^2_2]_{I=\frac{7}{2}M}$ does not vanish. The relevant formula for calculating this matrix element is given by equation (1.106). When this is combined with the fact that the double-parentage coefficients for the $n=3$ system are related to the one-particle c.f.p's (see equation (A5.15)) one may use Table A5.2 to show that

$$\langle(f_{\frac{7}{2}})^3_{I=\frac{7}{2}M}|V|[f_{\frac{7}{2}}\times(p_{\frac{3}{2}})^2_2]_{I=\frac{7}{2}M}\rangle=\frac{-\sqrt{5}}{2\sqrt{3}}E_2(\tfrac{7}{2}\tfrac{7}{2};\tfrac{3}{2}\tfrac{3}{2}).$$

When E_2 is calculated using the delta-function interaction in equation (A2.29) one finds that

$$E_2(\tfrac{7}{2}\tfrac{7}{2};\tfrac{3}{2}\tfrac{3}{2})=\frac{-2\sqrt{2}}{\sqrt{21}}V_0\bar{R}$$

where

$$\bar{R}=\int R^2_{\frac{7}{2}}R^2_{\frac{3}{2}}r^2\,\mathrm{d}r$$

and R_j is the radial wave function for the state j. Since E_2 does not vanish this matrix element will lead to seniority mixing.

In general, whether v (given by equation (1.140)) is a good quantum number or not will depend sensitively on two conditions:

(i) the energy separation between the basis states defined by equation (1.139), which are used to diagonalize the Hamiltonian
(ii) the magnitude of the off-diagonal matrix element between these states.

There are two cases where the seniority of the physical state is likely to be quite pure and these are the 0^+ ground state and the yrast 2^+ state of the even-A s.c.s. nuclei. This is because the only 0^+ levels that can mix into the seniority-zero ground state are those with at least seniority four; i.e. states with a broken pair in j_1 and a broken pair in j_2. For example, in the four-particle case the admixed state will have the form

$$\Phi_{00} = [(j_1)^2_{J_1} \times (j_2)^2_{J_1}]_{00}.$$

Since the energy to break a pair is $\sim 1{\cdot}5$ MeV (the excitation energy of the yrast 2^+ level of the configuration j_1^2) the energy separation of the $v=0$ and $v=4$ states is ~ 3 MeV. Consequently, unless one has a large off-diagonal matrix element between the states, the mixing will be small.

The same argument applies to the yrast 2^+ level. The energy separation between a state with one broken pair ($v=2$) and two broken pairs ($v=4$) is $\sim 1{\cdot}5$ MeV. Thus, although the seniority mixing is expected to be stronger in this case it should still be small.

On the other hand, from Figs. 1.4, 1.7, and 1.9 it is clear that the yrast levels with $I>2$ lie at an excitation energy of about twice that of the yrast 2^+ state. Thus, for example, the seniority-two 4^+ level will have an energy comparable with that of the state

$$\Phi_{4M} = [(j_1)^2_2 \times (j_2)^2_2]_{4M}$$

and consequently considerable seniority mixing is likely to occur.

In Table 1.3 we present the results that emerge when the theoretical wave functions of Cohen *et al.* (1967) for ${}^{62}_{28}Ni_{34}$ are decomposed into states of good seniority. The model used in these calculations was one in which ${}^{56}_{28}Ni_{28}$ was taken to be an inert core and the six valence particles were confined to the $1p_{\frac{3}{2}}$, $0f_{\frac{5}{2}}$, and $1p_{\frac{1}{2}}$ single-particle states. The residual two-body interaction between the valence nucleons was chosen so as best to reproduce all the known energy-level data of the nickel isotopes. From Table 1.3 it is clear that the ground state and yrast 2^+ level are fairly pure seniority-zero and -two states, respectively. Although the other states

TABLE 1.3

Seniority decomposition of the various states in $^{62}_{28}Ni_{34}$

State	Excitation energy (MeV) Theory	Experiment	Percentage seniority v: $v=0$	$v=2$	$v=4$	$v=6$
0_1^+	0	0	99·7		0·3	
0_2^+	2·01	2·05	87·3		12·7	
1_1^+	3·57			24·7	70·0	5·3
2_1^+	1·53	1·17		99·4	0·5	0·1
2_2^+	2·25	2·30		89·1	10.7	0·2
3_1^+	2·84			40·6	59·3	0·1
4_1^+	2·20	2·34		92·9	7·0	0·1
4_2^+	2·76			41·3	58·3	0·4

The notation I_k means the kth level of spin I. The theoretical excitation energies are those of Cohen *et al.* (1967) and the experimental energies are taken from the compilation of Verheul (1974).

exhibit seniority mixing, the wave functions of the first two states of each even I are mainly a combination of $v=0$, $v=2$, and $v=4$ basis states.

For the odd-A s.c.s. nuclei we have already pointed out that there is no appreciable gap between the seniority-1 and -3 states. Consequently one would expect much more seniority mixing in these nuclei than in the even-A case. In Table 1.4 we list the seniority decomposition of the

TABLE 1.4

Seniority decomposition of the various states in $^{61}_{28}Ni_{33}$

State	Excitation energy (MeV) Theory	Experiment	Percentage seniority v: $v=1$	$v=3$	$v=5$
$(\frac{1}{2})_1$	0·02	0·28	96·9	2·9	0·2
$(\frac{1}{2})_2$	1·02		24·1	74·5	1·4
$(\frac{3}{2})_1$	0	0	92·4	7·0	0·6
$(\frac{3}{2})_2$	1·03	0·66	31·2	65·7	3·1
$(\frac{5}{2})_1$	0·12	0·07	97·1	2·7	0·2
$(\frac{5}{2})_2$	0·93	0·91	24·3	71·3	4·4
$(\frac{7}{2})_1$	0·92	1·02		94·9	5·1
$(\frac{9}{2})_1$	1·00			99·3	0·7

The notation I_k means the kth level of spin I. The theoretical excitation energies are those of Cohen *et al.* (1967) and the experimental energies are taken from the compilation of Auble (1975).

states in ${}^{61}_{28}Ni_{33}$ (Cohen *et al.* 1967). As anticipated there is substantial seniority mixing. Since single-nucleon pick-up from an even-A s.c.s. nucleus with seniority-zero ground state can only populate the seniority-one part of the wave function of the odd-A nucleus (see section 5.1) this fragmentation of $v = 1$ strength into states other than the lowest for given I explains why several states with the same spin are populated in the ${}^{60}_{28}Ni_{32}(d, p){}^{61}_{28}Ni_{33}$ reaction (Cosman *et al.* 1967).

2

NEUTRON–PROTON PROBLEMS

In Chapter 1 we dealt with nuclei in which either the neutrons or the protons formed an inert closed shell (s.c.s. nuclei). We shall now discuss the various methods for calculating the spectra of nuclei in which both protons and neutrons are active.

1. The residual nucleon–nucleon interaction

To start with we shall review the properties that the effective interaction is likely to possess in these problems and introduce the concept of isospin, which often simplifies nuclear-structure calculations.

1.1. Charge symmetry and charge independence

Because of the Coulomb repulsion, the total potential acting between two protons is different from that between two neutrons. One might hope to find out about the specifically nuclear proton–proton interaction from low-energy scattering data. However, Sauer (1974) has shown that the separation of the scattering length into Coulomb and nuclear parts depends sensitively on the short-range nature of the nucleon–nucleon interaction. Since we know little about the behaviour of the nuclear interaction at short distances one cannot make the separation unambiguously. On the other hand, if one uses a 'reasonable' form for the short range behaviour the experimental evidence reviewed by Henley (1969) indicates that

$$\frac{V_{\mathrm{nn}}-V_{\mathrm{pp}}}{\frac{1}{2}(V_{\mathrm{nn}}+V_{\mathrm{pp}})}\,0\pm 0{\cdot}8\%$$

where V_{nn} and V_{pp} are the nuclear neutron-neutron and proton–proton forces, respectively. From an analysis of Coulomb energy differences Negele (1971) concluded that V_{nn} is slightly more attractive than V_{pp}. However, it is difficult to be very precise about the magnitude of the difference and as a consequence we shall assume $V_{\mathrm{nn}}=V_{\mathrm{pp}}$, that is, nuclear forces are charge symmetric.

As we have already discussed the Coulomb force has a long range and therefore its matrix elements should not vary much with the total angular

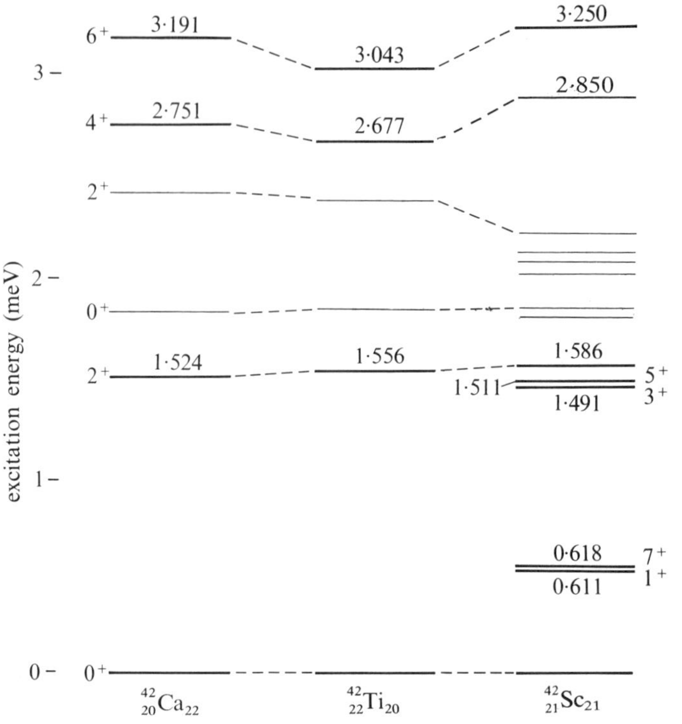

Fig. 2.1. Experimental spectrum of $^{42}_{20}Ca_{22}$ and $^{42}_{21}Sc_{21}$ taken from Hartmann *et al.* (1973) and that for $^{42}_{22}Ti_{20}$ obtained by Cox *et al.* (1973). The levels that can be attributed to the $(f_{\frac{7}{2}})^2$ configuration are shown with heavy lines and their observed excitation energies are given on the diagram. The assignment of 6^+ for the 3·043 MeV state in $^{42}_{22}Ti_{20}$ is tentative. The states that arise from configurations outside the $f_{\frac{7}{2}}$ model space are shown with lighter lines.

momentum. Consequently if $V_{nn} = V_{pp}$ one would expect that the observed excitation energies of nuclei with two valence protons should be the same as those with two valence neutrons. In Fig. 2.1 we show the experimental data for $^{42}_{20}Ca_{22}$ and $^{42}_{22}Ti_{20}$; the former is interpreted as two $f_{\frac{7}{2}}$ neutrons moving outside an inert $^{40}_{20}Ca_{20}$ core and the latter as two $f_{\frac{7}{2}}$ protons. The excitation energies are quite similar and in agreement with what one would expect for charge symmetric forces. However, the total interaction energy in the $I = 0$ ground states of these two nuclei should differ because of the Coulomb repulsion. From the binding-energy tables of Wapstra and Gove (1971) it follows that

$$E_0(\pi, \pi) = \mathrm{BE}(^{42}_{22}\mathrm{Ti}_{20}) - \mathrm{BE}(^{40}_{20}\mathrm{Ca}_{20}) - 2\{\mathrm{BE}(^{41}_{21}\mathrm{Sc}_{20}) - \mathrm{BE}(^{40}_{20}\mathrm{Ca}_{20})\}$$
$$= -2{\cdot}684\ \mathrm{MeV} \qquad (2.1)$$

whereas

$$E_0(\nu, \nu) = \mathrm{BE}(^{42}_{20}\mathrm{Ca}_{22}) - \mathrm{BE}(^{40}_{20}\mathrm{Ca}_{20}) - 2\{\mathrm{BE}(^{41}_{20}\mathrm{Ca}_{21}) - \mathrm{BE}(^{40}_{20}\mathrm{Ca}_{20})\}$$
$$= -3{\cdot}109\ \mathrm{MeV}. \tag{2.2}$$

Thus the Coulomb repulsion between two $f_{\frac{7}{2}}$ protons with $I=0$ is about 425 keV.

The Pauli exclusion principle requires that the wave function describing two protons or two neutrons is antisymmetric to interchange of the particles

$$\Psi_{IM} = \{2(1+\delta_{j_1j_2})\}^{-\frac{1}{2}} \sum_{m_1m_2} (j_1j_2m_1m_2 \mid IM)\{\phi_{j_1m_1}(1)\phi_{j_2m_2}(2) - \phi_{j_1m_1}(2)\phi_{j_2m_2}(1)\}. \tag{2.3}$$

Although this is not a requirement that the neutron–proton wave function must satisfy, it is certainly a possible state for the n–p system. The neutron–proton interaction (V_{np}) in this antisymmetric space-spin state was originally deduced from an analysis of the binding energies of $^2_1\mathrm{H}_1$, $^3_2\mathrm{He}_1$, and $^4_2\mathrm{He}_2$ (Feenberg and Knipp 1935), and from the experimental cross-section for the scattering of low-energy neutrons off hydrogen (Fermi and Amaldi 1936). If a reasonable form is used for the short range part of the nucleon–nucleon interaction one finds that V_{pp} is almost the same as V_{np}. Consequently Breit *et al.* (1936) proposed that nuclear forces be charge independent; in other words

$$V_{\mathrm{pp}} = V_{\mathrm{nn}} = V_{\mathrm{np}} \tag{2.4}$$

where V_{np} is the neutron–proton interaction in the antisymmetric space-spin state.

Over the past 40 years considerable effort has gone into checking the validity of this hypothesis, and it is now well established that the above equality is only approximately true. Henley (1969) has reviewed the data and finds that V_{np} is about 2% stronger than either V_{pp} or V_{nn}

$$\frac{V_{\mathrm{np}} - V_{\mathrm{nn}}}{\frac{1}{2}(V_{\mathrm{np}} + V_{\mathrm{nn}})} = 2{\cdot}13 \pm 0{\cdot}52\%.$$

The spectra and binding energies of complex nuclei also seem to bear this out (Altman and MacDonald 1962, Sherr and Talmi 1975). For example, by use of the Wapstra–Gove tables (1971) one finds that the interaction energy associated with a neutron and proton in the $f_{\frac{7}{2}}$ configuration when $I=0$ is

$$E_0(\pi, \nu) = \mathrm{BE}(^{42}_{21}\mathrm{Sc}_{21}) - \mathrm{BE}(^{40}_{20}\mathrm{Ca}_{20}) - \{\mathrm{BE}(^{41}_{21}\mathrm{Sc}_{20}) - \mathrm{BE}(^{40}_{20}\mathrm{Ca}_{20})\}$$
$$- \{\mathrm{BE}(^{41}_{20}\mathrm{Ca}_{21}) - \mathrm{BE}(^{40}_{20}\mathrm{Ca}_{20})\}$$
$$= -3{\cdot}174\ \mathrm{MeV}. \tag{2.5}$$

When this is compared with $E_0(\nu, \nu)$ in equation (2.2) one finds that

$$\frac{E_0(\pi, \nu) - E_0(\nu, \nu)}{|\frac{1}{2}\{E_0(\pi, \nu) + E_0(\nu, \nu)\}|} = \frac{-0{\cdot}065}{3{\cdot}142} = -0{\cdot}021, \tag{2.6}$$

i.e. the n–p interaction appears to be about 2% stronger than V_{nn}. One must be careful, however, in interpreting the above effect as being due entirely to a difference in V_{nn} and V_{np}, the free particle–particle interactions. The reason for this is that one always uses a truncated model space in making any shell-model calculation, and as we shall now show the potential needed is model-space dependent. To illustrate this we consider the small effect on the ${}^{42}_{20}\mathrm{Ca}_{22}$ and ${}^{42}_{21}\mathrm{Sc}_{21}$ binding energies brought about by extending the model space to include both the $0f_{\frac{7}{2}}$ and $1p_{\frac{3}{2}}$ orbitals. In this enlarged model space the energy of the ${}^{42}_{20}\mathrm{Ca}_{22}$ ground state is given by the lower eigenvalue of the 2×2 matrix

$$\langle H \rangle = \begin{pmatrix} E_0(\tfrac{7}{2}\tfrac{7}{2};\tfrac{7}{2}\tfrac{7}{2}) + 2\varepsilon_{\frac{7}{2}}(\nu) & E_0(\tfrac{7}{2}\tfrac{7}{2};\tfrac{3}{2}\tfrac{3}{2}) \\ E_0(\tfrac{7}{2}\tfrac{7}{2};\tfrac{3}{2}\tfrac{3}{2}) & E_0(\tfrac{3}{2}\tfrac{3}{2};\tfrac{3}{2}\tfrac{3}{2}) + 2\varepsilon_{\frac{3}{2}}(\nu) \end{pmatrix}$$

where

$$E_0(jj; j_1j_1) = \langle (j^2)_0 |\, V\, | (j_1^2)_0 \rangle$$

is the interaction energy in the $I = 0$ state and $\varepsilon_j(\nu)$ is the neutron single-particle energy in the state j. For ${}^{42}_{21}\mathrm{Sc}_{21}$ $2\varepsilon_j(\nu)$ must be replaced by $\varepsilon_j(\nu) + \varepsilon_j(\pi)$.

Values for the single-particle energies can be obtained from the spectrum of ${}^{41}_{20}\mathrm{Ca}_{21}$ and ${}^{41}_{21}\mathrm{Sc}_{21}$ (Endt and Van der Leun 1973)

$$\varepsilon_{\frac{3}{2}}(\nu) - \varepsilon_{\frac{7}{2}}(\nu) = 1{\cdot}943 \text{ MeV}$$

$$\varepsilon_{\frac{3}{2}}(\pi) - \varepsilon_{\frac{7}{2}}(\pi) = 1{\cdot}716 \text{ MeV}.$$

When these are combined with the charge-independent $f_{\frac{7}{2}}$–$p_{\frac{3}{2}}$ interaction energies found by Gloeckner *et al.* (1973),

$$E_0(\tfrac{7}{2}\tfrac{7}{2};\tfrac{7}{2}\tfrac{7}{2}) = -2{\cdot}551 \text{ MeV}$$

$$E_0(\tfrac{7}{2}\tfrac{7}{2};\tfrac{3}{2}\tfrac{3}{2}) = -1{\cdot}463 \text{ MeV}$$

$$E_0(\tfrac{3}{2}\tfrac{3}{2};\tfrac{3}{2}\tfrac{3}{2}) = -1{\cdot}568 \text{ MeV}$$

one obtains

$$E_0({}^{42}_{20}\mathrm{Ca}_{22}) - E_0({}^{40}_{20}\mathrm{Ca}_{20}) = 2\varepsilon_{\frac{7}{2}}(\nu) - 2{\cdot}957 \text{ MeV}$$

whereas

$$E_0({}^{42}_{21}\mathrm{Sc}_{21}) - E_0({}^{40}_{20}\mathrm{Ca}_{20}) = \varepsilon_{\frac{7}{2}}(\nu) + \varepsilon_{\frac{7}{2}}(\pi) - 2{\cdot}974 \text{ MeV}.$$

Thus 17 keV of the observed 65keV can be attributed to a difference in the neutron and proton $p_{\frac{3}{2}}$ single-particle excitation energies which is brought about primarily by the Coulomb interaction. Clearly other neglected configurations could modify the energy to be attributed to purely nuclear effects so it is therefore extremely difficult to obtain a quantitative estimate for the small difference between V_{np} and V_{nn} from the properties of complex nuclei. This result has already been pointed out by Blin–Stoyle and Nair (1963).

Because of the near equality of V_{nn} and V_{np} one would expect the excitation energies of the 2^+, 4^+, and 6^+ states in ${}^{42}_{21}Sc_{21}$ to be the same as those of the analogous states in ${}^{42}_{20}Ca_{22}$ and from Fig. 2.1 one sees that this is indeed true. Thus since shell-model calculations cannot be relied upon to predict energies to better than a few hundred kilovolts, it is clear that effects of the order of 50 keV brought about by the fact that $V_{nn} \neq V_{np}$ are unimportant. Consequently unless otherwise stated we shall take nuclear forces to be charge independent; i.e. we shall assume that they satisfy equation (2.4). (There may, of course, be cases where the assumption of charge independence imposes a selection rule on a process, and then it may be important to take into account this small difference.)

The neutron–proton system can also exist in the symmetric space-spin state

$$\Phi_{IM} = \{2(1+\delta_{j_1j_2})\}^{-\frac{1}{2}} \sum_{m_1m_2} (j_1j_2m_1m_2 \mid IM)\{\phi_{j_1m_1}(1)\phi_{j_2m_2}(2) + \phi_{j_1m_1}(2)\phi_{j_2m_2}(1)\}. \tag{2.7}$$

Because of the symmetry property given by equation (A1.18),

$$(j_1j_2m_1m_2 \mid IM) = (-1)^{j_1+j_2-I}(j_2j_1m_2m_1 \mid IM)$$

it follows that the only symmetric states that can be realized for the configuration $(j)^2$ have odd values of I $(I = 1, 3, 5, \ldots, (2j+1))$. These states have been identified in ${}^{42}_{21}Sc_{21}$ and are shown in Fig. 2.1. Clearly these states have no counterpart, either theoretically or experimentally, in ${}^{42}_{20}Ca_{22}$ and ${}^{42}_{22}Ti_{20}$. In the same way that the interaction energies in the even angular-momentum states are given by matrix elements of V_{np}, those in the odd I-states will be given by matrix elements of $\tilde{V}_{np}$ where $\tilde{V}_{np}$ is the neutron–proton interaction in symmetric space-spin states. Because the bound state of the deuteron has a symmetric space-spin wave function whereas the antisymmetric state is unbound and appears as a resonance in low-energy neutron scattering, $\tilde{V}_{np}$ is more attractive than V_{np}.

1.2. Isospin

For calculation it is often convenient to consider neutrons and protons as two different charge states of the same particle. Thus one introduces a fifth coordinate t_z in addition to the four space-spin variables. The new coordinate takes on only two values, namely $t_z = +\frac{1}{2}$ when we refer to a neutron and $t_z = -\frac{1}{2}$ for a proton (Heisenberg 1932, Cassen and Condon 1936). Consequently we write the wave function of a neutron in the state (jm) as

$$\begin{aligned}\phi_{jm}(\underline{r}, \underline{s}, t_z = +\tfrac{1}{2}) &= \phi_{jm}(\underline{r}, \underline{s})\begin{pmatrix}1\\0\end{pmatrix}\\ &= \phi_{jm}(\underline{r}, \underline{s})\zeta_{\frac{1}{2}}\end{aligned} \tag{2.8}$$

and of a proton

$$\begin{aligned}\phi_{jm}(\underline{r}, \underline{s}, t_z = -\tfrac{1}{2}) &= \phi_{jm}(\underline{r}, \underline{s})\begin{pmatrix}0\\1\end{pmatrix}\\ &= \phi_{jm}(\underline{r}, \underline{s})\zeta_{-\frac{1}{2}}.\end{aligned} \tag{2.9}$$

where $\underline{r}$ and $\underline{s}$ are the space and spin coordinates, respectively. The eigenfunctions ζ_μ of this fifth variable are formally identical to the ordinary spin eigenvectors χ_μ of equation (1.6) and as a consequence this coordinate is called the isospin of the particle. The isospin matrices can be introduced in analogy to normal spin, and using the same representation one writes

$$\begin{aligned}\tau_x &= \begin{pmatrix}0 & 1\\ 1 & 0\end{pmatrix}\\ \tau_y &= \begin{pmatrix}0 & -i\\ i & 0\end{pmatrix}\\ \tau_z &= \begin{pmatrix}1 & 0\\ 0 & -1\end{pmatrix}.\end{aligned} \tag{2.10}$$

It is to be understood that these operate only on the ζ_μ coordinate of the nucleon. The single-particle isospin operator is then defined as

$$\mathrm{t}_k = \tfrac{1}{2}\tau_k \tag{2.11}$$

and these have exactly the commutation properties of ordinary angular momentum (see equation (A1.1))

$$\begin{aligned}[\mathrm{t}_k, \mathrm{t}_m] &= \mathrm{t}_k\mathrm{t}_m - \mathrm{t}_m\mathrm{t}_k\\ &= i\mathrm{t}_n \quad (k, m, n, \text{cyclic}).\end{aligned} \tag{2.12}$$

Thus

$$\begin{aligned}\mathrm{t}^2\zeta_\mu &= t(t+1)\zeta_\mu\\ &= (\tfrac{1}{2})(\tfrac{3}{2})\zeta_\mu\end{aligned} \tag{2.13a}$$

and

$$\mathrm{t}_z\zeta_\mu = \mu\zeta_\mu. \tag{2.13b}$$

In other words, ζ_μ is an eigenfunction of the isospin operators t^2 and t_z with eigenvalue $\frac{1}{2}$ and μ ($\mu = \pm\frac{1}{2}$).

One may define the step-up and step-down operators for isospin

$$\mathrm{t}_+ = \tfrac{1}{2}(\tau_x + i\tau_y) \tag{2.14a}$$

$$\mathrm{t}_- = \tfrac{1}{2}(\tau_x - i\tau_y). \tag{2.14b}$$

The former has the properties

$$\begin{aligned} \mathrm{t}_+\zeta_{\frac{1}{2}} &= 0 \\ \mathrm{t}_+\zeta_{-\frac{1}{2}} &= \zeta_{\frac{1}{2}} \end{aligned} \tag{2.15a}$$

so that t_+ changes a proton to a neutron. For the latter operator

$$\begin{aligned} \mathrm{t}_-\zeta_{\frac{1}{2}} &= \zeta_{-\frac{1}{2}} \\ \mathrm{t}_-\zeta_{-\frac{1}{2}} &= 0 \end{aligned} \tag{2.15b}$$

and one sees that t_- changes a neutron to a proton. In addition the operators

$$\begin{aligned} \mathrm{t}_\nu &= (\tfrac{1}{2} + \mathrm{t}_z) \\ &= \tfrac{1}{2}(1 + \tau_z) \end{aligned} \tag{2.16a}$$

and

$$\begin{aligned} \mathrm{t}_\pi &= (\tfrac{1}{2} - \mathrm{t}_z) \\ &= \tfrac{1}{2}(1 - \tau_z) \end{aligned} \tag{2.16b}$$

are the neutron- and proton-projection operators, respectively,

$$\begin{aligned} \mathrm{t}_\nu\zeta_{\frac{1}{2}} &= \zeta_{\frac{1}{2}} \\ \mathrm{t}_\nu\zeta_{-\frac{1}{2}} &= 0 \\ \mathrm{t}_\pi\zeta_{\frac{1}{2}} &= 0 \\ \mathrm{t}_\pi\zeta_{-\frac{1}{2}} &= \zeta_{-\frac{1}{2}}. \end{aligned} \tag{2.16c}$$

To exploit fully the isospin concept we generalize the exclusion principle and require that any acceptable many-particle wave function must be antisymmetric with respect to the interchange of all coordinates (space, spin, and isospin) of any two particles. Clearly for the two-particle system this implies that the space-spin symmetric state Φ_{IM} of equation (2.7) can only have an antisymmetric isospin eigenfunction associated with it:

$$\Phi_{IM;00} = \Phi_{IM}\zeta_{00}(1, 2)$$

where

$$\zeta_{00}(1, 2) = 2^{-\frac{1}{2}}\{\zeta_{\frac{1}{2}}(1)\zeta_{-\frac{1}{2}}(2) - \zeta_{\frac{1}{2}}(2)\zeta_{-\frac{1}{2}}(1)\}. \tag{2.17}$$

On the other hand, the space-spin antisymmetric state (Ψ_{IM} of equation (2.3)) goes with a symmetric isospin eigenfunction

$$\Psi_{IM;1\mu} = \Psi_{IM}\zeta_{1\mu}(1,2)$$

where $\zeta_{1\mu}(1,2)$ is either

$$\zeta_{11}(1,2) = \zeta_{\frac{1}{2}}(1)\zeta_{\frac{1}{2}}(2) \tag{2.18a}$$

or

$$\zeta_{10}(1,2) = 2^{-\frac{1}{2}}\{\zeta_{\frac{1}{2}}(1)\zeta_{-\frac{1}{2}}(2) + \zeta_{\frac{1}{2}}(2)\zeta_{-\frac{1}{2}}(1) \tag{2.18b}$$

or

$$\zeta_{1-1}(1,2) = \zeta_{-\frac{1}{2}}(1)\zeta_{-\frac{1}{2}}(2) \tag{2.18c}$$

depending on whether the two particles are two neutrons, a neutron and a proton, or two protons.

The significance of the subscripts to $\zeta(1,2)$ and the additional pair of subscripts to Ψ and Φ is that they label the total isospin and the third component of isospin of the state. This follows because the total isospin operator

$$\mathrm{T}_{\mu} = \sum_{i} (\mathrm{t}_i)\mu \tag{2.19}$$

where (t_i) is the operator for particle number i, has the properties

$$\mathrm{T}_z\zeta_{1\mu}(1,2) = \mu\zeta_{1\mu}(1,2)$$
$$\mathrm{T}^2\zeta_{1\mu}(1,2) = \{(\mathrm{t}_1)^2 + (\mathrm{t}_2)^2 + 2\underline{\mathrm{t}}_1 \cdot \underline{\mathrm{t}}_2\}\zeta_{1\mu}(1,2)$$
$$= 2\zeta_{1\mu}(1,2).$$

Thus $\zeta_{1\mu}(1,2)$ is an eigenfunction of T^2 and T_z with eigenvalue $T=1$ and $T_z = \mu$. In a similar way

$$\mathrm{T}^2\zeta_{00}(1,2) = 0$$

$$\mathrm{T}_z\zeta_{00}(1,2) = 0$$

so that $\zeta_{00}(1,2)$ has $T = T_z = 0$.

If the residual nucleon–nucleon interaction is taken to be chare independent and the Coulomb potential and the neutron–proton mass difference are neglected, the Hamiltonian will commute with $\underline{\mathrm{T}}$:

$$[H, \underline{\mathrm{T}}] = 0$$

or

$$[H, \mathrm{T}^2] = 0. \tag{2.20}$$

Thus in this approximation nuclear states have a definite isospin. This means in the two-particle case that the $T=0$ and $T=1$ states just defined do not mix. In general, a system with N neutrons and Z protons will have

$$T_z = (N-Z)/2.$$

Since the total isospin cannot be less than T_z, the possible T values will be

$$T = (N-Z)/2, (N-Z+2)/2, \ldots, (N+Z)/2$$

and as long as equation (2.20) holds true these various isospin states will not mix. Because the nucleon isospin is $\frac{1}{2}$ a system with an odd number of particles will have half-integral isospin whereas one with an even number will have integral values. Except when $T_z = 0$, it is found experimentally that the ground state of the (N, Z) nucleus has $T = (N-Z)/2$. It should be emphasized that whether or not we use the isospin classification the same physical results emerge. This is obviously true in the two-particle case and can be shown to hold in many-particle configurations (see, for example, Bayman (1966) and section 2, Chapter 2).

Because the isospin quantum number provides a useful book-keeping device, it is convenient to introduce creation and destruction operators compatible with this concept. We therefore generalize the creation operator $a^\dagger_{jm}$ by introducing the additional subscripts $(\frac{1}{2}\mu)$ where μ can take on the values $\pm\frac{1}{2}$. Thus $a^\dagger_{jm;\frac{1}{2}\mu}$ creates a neutron in the state (jm) when $\mu = +\frac{1}{2}$ and a proton when $\mu = -\frac{1}{2}$. Since the entities created by these operators obey Fermi statistics

$$\begin{aligned}\{a_{jm;\frac{1}{2}\mu}, a_{j'm';\frac{1}{2}\mu'}\} &= a_{jm;\frac{1}{2}\mu}a_{j'm';\frac{1}{2}\mu'} + a_{j'm';\frac{1}{2}\mu'}a_{jm;\frac{1}{2}\mu} \\ &= 0 \\ \{a^\dagger_{jm;\frac{1}{2}\mu}, a^\dagger_{j'm'\frac{1}{2}\mu'}\} &= 0 \end{aligned} \tag{2.21}$$

$$\{a^\dagger_{jm;\frac{1}{2}\mu}, a_{j'm'\frac{1}{2}\mu'}\} = \delta_{jj'}\delta_{mm'}\delta_{\mu\mu'}. \tag{2.21}$$

The normalized two-particle creation operator with angular momentum IM and isospin TT_z may therefore be written as

$$\begin{aligned} A^\dagger_{IM;TT_z}(j_1j_2) = (1+\delta_{j_1j_2})^{-\frac{1}{2}} \sum_{m_1m_2} \sum_{\mu_1\mu_2} (j_1j_2m_1m_2 \mid IM)(\tfrac{1}{2}\tfrac{1}{2}\mu_1\mu_2 \mid TT_z) \\ \times a^\dagger_{j_1m_1;\frac{1}{2}\mu_1} a^\dagger_{j_2m_2;\frac{1}{2}\mu_2}. \end{aligned} \tag{2.22}$$

The destruction operator for a (j_3j_4) pair with $(IM; TT_z)$ is, of course, just the Hermitian adjoint of the creation operator. Thus

$$\begin{aligned} A_{IM;TT_z}(j_3j_4) = (1+\delta_{j_3j_4})^{-\frac{1}{2}} \sum_{m_3m_4} \sum_{\mu_3\mu_4} (j_3j_4m_3m_4 \mid IM)(\tfrac{1}{2}\tfrac{1}{2}\mu_3\mu_4 \mid TT_z) \\ \times a_{j_4m_4;\frac{1}{2}\mu_4} a_{j_3m_3;\frac{1}{2}\mu_3}. \end{aligned} \tag{2.23}$$

In terms of these operators the shell-model Hamiltonian takes the form

$$H=\sum_{jm\mu}\varepsilon_j a^{\dagger}_{jm;\frac{1}{2}\mu}a_{jm;\frac{1}{2}\mu}+\sum_{JMTT_z}\sum_{\{j_1j_2\}}\sum_{\{j_3j_4\}}E_{JT}(j_1j_2;j_3j_4)$$
$$\times A^{\dagger}_{JM;TT_z}(j_1j_2)A_{JM;TT_z}(j_3j_4) \tag{2.24}$$

where the interaction energy

$$E_{JT}(j_1j_2;j_3j_4)=\langle\Psi_{JM;TT_z}(j_1j_2)|\,V\,|\Psi_{JM;TT_z}(j_3j_4)\rangle \tag{2.25}$$

is the matrix element of the residual two-body potential evaluated using the normalized antisymmetric states created by $A^{\dagger}_{JM;TT_z}(j_1j_2)$ and destroyed by $A_{JM;TT_z}(j_3j_4)$. The notation $\{j_ij_k\}$ implies that in the summation $j_i\geq j_k$. This restriction, which is necessary in order to avoid overcounting, can be removed if the Hamiltonian is written in a form analogous to equation (1.47).

The isospin formalism provides a convenient way of expressing certain exchange operators commonly encountered in nuclear physics. Since

$$\mathrm{T}^2\zeta_{T\mu}(1,2)=T(T+1)\zeta_{T\mu}(1,2)$$
$$=\tfrac{1}{2}\{3+\underset{\sim}{\tau}_1\cdot\underset{\sim}{\tau}_2\}\zeta_{T\mu}(1,2)$$

where

$$\underset{\sim}{\tau}_1\cdot\underset{\sim}{\tau}_2=(\tau_1)_x(\tau_2)_x+(\tau_1)_y(\tau_2)_y+(\tau_1)_z(\tau_2)_z$$

one sees that

$$\underset{\sim}{\tau}_1\cdot\underset{\sim}{\tau}_2\zeta_{T\mu}(1,2)=\{2T(T+1)-3\}\zeta_{T\mu}(1,2). \tag{2.26}$$

Consequently the operator

$$P_{\mathrm{H}}=-\tfrac{1}{2}\{1+\underset{\sim}{\tau}_1\cdot\underset{\sim}{\tau}_2\} \tag{2.27}$$

has the property that it gives -1 when operating on a $T=1$ state and $+1$ when applied to a $T=0$ eigenfunction. Since the overall wave function must be antisymmetric in the combination of space-spin and isospin, P_{H} gives $+1$ when operating on a space-spin symmetric state and -1 when acting on a space-spin antisymmetric wave function. For this reason P_{H} is called the space-spin exchange operator or Heisenberg exchange operator (Heisenberg 1932).

The spin operators $\underset{\sim}{\sigma}$ satisfy the same type of relationship:

$$\underset{\sim}{\sigma}_1\cdot\underset{\sim}{\sigma}_2\chi_{SM}(1,2)=\{2S(S+1)-3\}\chi_{SM}(1,2) \tag{2.28}$$

where $\chi_{SM}(1,2)$ is the two-particle spin wave function

$$\chi_{SM}(1,2)=\sum_{\nu_1\nu_2}(\tfrac{1}{2}\tfrac{1}{2}\nu_1\nu_2\,|\,SM)\chi_{\nu_1}(1)\chi_{\nu_2}(2).$$

Thus the Bartlett operator (Bartlett 1936)

$$P_B = \tfrac{1}{2}\{1 + \underset{\sim}{\sigma}_1 \cdot \underset{\sim}{\sigma}_2\}, \tag{2.29}$$

which has the property that it gives +1 when operating on the symmetric $S = 1$ state and -1 when acting on the antisymmetric $S = 0$ eigenfunction, is called the spin-exchange operator.

Finally, because of the overall antisymmetry of wave functions, the operator

$$\begin{aligned} P_M &= P_H P_B \\ &= -\tfrac{1}{4}\{1 + \underset{\sim}{\tau}_1 \cdot \underset{\sim}{\tau}_2\}(1 + \underset{\sim}{\sigma}_1 \cdot \underset{\sim}{\sigma}_2) \end{aligned} \tag{2.30}$$

gives +1 or -1 when operating on a spatially symmetric or antisymmetric state. Consequently this operator is called the space exchange or Majorana operator (Majorana 1933).

If one neglects tensor forces and any velocity dependence of the residual two-body interaction, the most general potential that satisfies equation (2.20) can be written in terms of P_H, P_B, P_M, and the unit operator

$$V = V_W(\underset{\sim}{r}_1\underset{\sim}{r}_2) + V_H(\underset{\sim}{r}_1\underset{\sim}{r}_2)P_H + V_B(\underset{\sim}{r}_1\underset{\sim}{r}_2)P_B + V_M(\underset{\sim}{r}_1\underset{\sim}{r}_2)P_M \tag{2.31}$$

where the $V(\underset{\sim}{r}_1\underset{\sim}{r}_2)$ are arbitrary scalar functions of $\underset{\sim}{r}_1$ and $\underset{\sim}{r}_2$ which satisfy Galilean invariance. The interaction $V_W(\underset{\sim}{r}_1\underset{\sim}{r}_2)$, which has no exchange operator associated with it, is known as the Wigner interaction (Wigner 1933).

There is yet another form that is often used in writing the residual two-body interaction:

$$V = \sum_{ST} V_{ST}(\underset{\sim}{r}_1\underset{\sim}{r}_2)P_{ST} \tag{2.32}$$

where the P_{ST} are the spin–isospin projection operators

$$\begin{aligned} P_{11} &= \tfrac{1}{16}\{3 + \underset{\sim}{\sigma}_1 \cdot \underset{\sim}{\sigma}_2\}\{3 + \underset{\sim}{\tau}_1 \cdot \underset{\sim}{\tau}_2\} \\ P_{10} &= \tfrac{1}{16}\{3 + \underset{\sim}{\sigma}_1 \cdot \underset{\sim}{\sigma}_2\}\{1 - \underset{\sim}{\tau}_1 \cdot \underset{\sim}{\tau}_2\} \\ P_{01} &= \tfrac{1}{16}\{1 - \underset{\sim}{\sigma}_1 \cdot \underset{\sim}{\sigma}_2\}\{3 + \underset{\sim}{\tau}_1 \cdot \underset{\sim}{\tau}_2\} \\ P_{00} &= \tfrac{1}{16}\{1 - \underset{\sim}{\sigma}_1 \cdot \underset{\sim}{\sigma}_2\}\{1 - \underset{\sim}{\tau}_1 \cdot \underset{\sim}{\tau}_2\}. \end{aligned} \tag{2.33}$$

From equations (2.26) and (2.28) it follows that P_{11} gives +1 when operating on an $S = 1$, $T = 1$ state and vanished when applied to any other (ST) configuration. Thus it is the $S = 1\ T = 1$ projection operator. Similarly P_{10}, P_{01}, and P_{00} are the $S = 1\ T = 0$, the $S = 0\ T = 1$, and the $S = 0\ T = 0$ projection operators, respectively. It is straightforward to relate the V_{ST} to the Wigner, Heisenberg, Bartlett, and Majorana interactions of equation (2.31):

$$\begin{aligned} V_{11}(\underline{r}_1\underline{r}_2) &= V_W(\underline{r}_1\underline{r}_2) - V_H(\underline{r}_1\underline{r}_2) + V_B(\underline{r}_1\underline{r}_2) - V_M(\underline{r}_1\underline{r}_2) \\ V_{10}(\underline{r}_1\underline{r}_2) &= V_W(\underline{r}_1\underline{r}_2) + V_H(\underline{r}_1\underline{r}_2) + V_B(\underline{r}_1\underline{r}_2) + V_M(\underline{r}_1\underline{r}_2) \\ V_{01}(\underline{r}_1\underline{r}_2) &= V_W(\underline{r}_1\underline{r}_2) - V_H(\underline{r}_1\underline{r}_2) - V_B(\underline{r}_1\underline{r}_2) + V_M(\underline{r}_1\underline{r}_2) \\ V_{00}(\underline{r}_1\underline{r}_2) &= V_W(\underline{r}_1\underline{r}_2) + V_H(\underline{r}_1\underline{r}_2) - V_B(\underline{r}_1\underline{r}_2) - V_M(\underline{r}_1\underline{r}_2). \end{aligned} \tag{2.34}$$

The potential $V_{11}(\underline{r}_1\underline{r}_2)$ acts only on the $S=1$ (spin-triplet), $T=1$ (isospin-triplet) state. Thus when this interaction is operative the two nucleons are in an antisymmetric spatial state and consequently have only odd values for their relative angular momentum. For this reason V_{11} is usually referred to as the triplet–odd interaction. By similar reasoning V_{10} is the triplet–even, V_{01} the singlet-even, and V_{00} the singlet-odd interaction.

The interactions V_{nn}, V_{pp}, and V_{pn}, which are all taken to be equal when charge independence is assumed, can be written in terms of the spin–isospin projection operators as

$$\begin{aligned} V_{pn} &= V_{pp} = V_{nn} \\ &= \tfrac{1}{16}[V_{11}(\underline{r}_1\underline{r}_2)\{3+\underline{\sigma}_1 \cdot \underline{\sigma}_2\} + V_{01}(\underline{r}_1\underline{r}_2)\{1-\underline{\sigma}_1 \cdot \underline{\sigma}_2\}]\{3+\underline{\tau}_1 \cdot \underline{\tau}_2\} \end{aligned} \tag{2.35}$$

and $\tilde{V}_{pn}$, the interaction in the space-spin symmetric state, is

$$\tilde{V}_{pn} = \tfrac{1}{16}[V_{10}(\underline{r}_1\underline{r}_2)\{3+\underline{\sigma}_1 \cdot \underline{\sigma}_2\} + V_{00}(\underline{r}_1\underline{r}_2)\{1-\underline{\sigma}_1 \cdot \underline{\sigma}_2\}]\{1-\underline{\tau}_1 \cdot \underline{\tau}_2\}. \tag{2.36}$$

1.3. How good is the isospin quantum number?

It has been shown (Bethe and Bacher 1936, Swamy and Green 1958) that for $Z>10$ the total Coulomb energy of a nucleus with charge Z is given to good approximation by the expression

$$\begin{aligned} V_c(A, Z) &= \frac{3}{5}\frac{e^2Z^2}{R}\left(1-\frac{0{\cdot}7636}{Z^{\frac{2}{3}}}\right) \\ &= \frac{0{\cdot}864Z^2}{R}\left(1-\frac{0{\cdot}7636}{Z^{\frac{2}{3}}}\right) \end{aligned} \tag{2.37}$$

where R is the nuclear radius in fermis and $V_c(A, Z)$ is given in MeV. For intermediate and heavy nuclei this energy is large; for example, with $R=1{\cdot}25A^{\frac{1}{3}}$ one finds 334 MeV for ${}^{116}_{50}Sn_{66}$ and 753 MeV for ${}^{208}_{82}Pb_{126}$. Because this interaction is operative only between protons it violates charge independence and consequently can lead to mixing of isospin in nuclear states. Moreover, because of the magnitude of the energies involved, it was originally felt that isospin would provide a useful classification scheme for only very light nuclei. However, in the (p, n) reaction on several intermediate-weight nuclei, Anderson and Wong (1961) observed that on a target with $(Z-1)$ protons and $(N+1)$ neutrons $(N+Z=A)$ a sharp neutron group was seen with an energy loss given almost precisely by

$$\Delta V_c(A, Z) = V_c(A, Z) - V_c(A, Z-1) \tag{2.38}$$

where $V_c(A, Z)$ is calculated from equation (2.37) with $R = 1{\cdot}25A^{\frac{1}{3}}$. In other words, the final nuclear state populated in the reaction was identical to the target ground state except that one neutron had been changed to a proton. This state is called the isobaric analogue (I.A.S.) of the ground state of the nucleus ${}^{\;A}_{Z-1}\mathrm{X}_{N+1}$ in the nucleus ${}^{A}_{Z}\mathrm{Y}_{N}$. By definition its excitation energy in the nucleus ${}^{A}_{Z}\mathrm{Y}_{N}$ is

$$\Delta E(A, Z) = \mathrm{BE}({}^{\;A}_{Z-1}\mathrm{X}_{N+1}) - \mathrm{BE}({}^{A}_{Z}\mathrm{Y}_{N}) + \Delta V_c(A, Z) \tag{2.39}$$

where BE is the negative of the total binding energy tabulated by Wapstra and Gove (1971) and $\Delta V_c(A, Z)$ is given by equation (2.38). If ψ denotes the eigenfunction of the target ground state, the I.A.S. is described by the wave function

$$\phi = \mathrm{T}_-\psi / \langle\psi|\, \mathrm{T}_+\mathrm{T}_-\, |\psi\rangle \tag{2.40}$$

where

$$\mathrm{T}_- = \sum_i (\mathrm{t}_i)_-$$

is the operator that changes a neutron to a proton and the denominator merely ensures that the wave function is normalized. The fact that a sharp outgoing neutron group is observed implies that ϕ is quite accurately an eigenfunction of the total Hamiltonian. Clearly a sufficient condition that this is true is that isospin is a good quantum number. Because of these experiments people were led to re-examine the question of isospin purity of nuclear levels.

MacDonald (1954) showed that because of its long range, the major effect of the Coulomb interaction could be approximated by adding to the usual shell-model potential the Coulomb repulsion felt by a proton moving in the presence of a uniform charge distribution of radius R. This added potential for $r<R$ is given by

$$V_c(r) = \frac{(Z-1)e^2}{2R}\left(3 - \frac{r^2}{R^2}\right). \tag{2.41}$$

For $r>R$, $V_c(r)$ has the form $(Z-1)e^2/r$. However, for all but very loosely bound states the nuclear wave functions drop off sufficiently rapidly that equation (2.41) can be used to approximate $V_c(r)$ for all values of r. Thus if we assume the neutron shell-model Hamiltonian is

$$H_\nu = \frac{p^2}{2m} + \tfrac{1}{2}m\omega^2 r^2$$

the single-particle motion of the proton will be governed by

$$H_\pi = \frac{p^2}{2m} + \tfrac{1}{2}m\omega^2 r^2 + \frac{(Z-1)e^2}{2R}\left(3 - \frac{r^2}{R^2}\right).$$

The eigenfunction describing a neutron with total angular momentum (jm) and energy $E = (2n + l + \tfrac{3}{2})\hbar\omega$ is

$$\phi_{nljm} = R_{nl}(\alpha r) \sum_{k\mu} (l\tfrac{1}{2}k\mu \mid jm) Y_{lk}(\theta, \phi)\chi_\mu \tag{2.42}$$

where the extra indices (nl) have been attached to the single particle wave function, eq (1.4), and $R_{nl}(\alpha r)$ is the harmonic oscillator wave function discussed in detail in Chapter 4 and Appendix 6, with

$$\alpha^2 = \frac{m\omega}{\hbar}.$$

Apart from an additive constant, the only effect of the Coulomb potential is to replace α^2 by

$$\alpha_1^2 = \alpha^2\{1 - \delta\}^{\frac{1}{2}}$$

with

$$\delta = \frac{(Z-1)e^2 m}{R^3\hbar^2\alpha^4}. \tag{2.43}$$

Thus the proton wave function for the state (jm) is identical to the neutron eigenfunction except that α is replaced by α_1

$$\phi'_{nljm} = R_{nl}(\alpha_1 r) \sum_{k\mu} (l\tfrac{1}{2}k\mu \mid jm) Y_{lk}(\theta, \phi)\chi_\mu. \tag{2.44}$$

Consequently the main effect of the Coulomb interaction is to give a uniform dilation of the proton wave function relative to the neutron eigenfunction, and we shall now estimate the isospin mixing in nuclear states brought about by this.

To simplify estimating this effect we use perturbation theory to expand the proton wave functions in terms of the neutron eigenfunctions. Thus

$$\phi'_{nljm}(\alpha_1 r) = \phi_{nljm}(\alpha r) + \sum_{n'} \beta_{n';nl}\phi_{n'ljm}(\alpha r) \tag{2.45}$$

where

$$\beta_{n';nl} = \frac{\alpha^2\hbar^2\delta}{2m} \frac{\langle R_{n'l}(\alpha r)|\, \alpha^2 r^2\, |R_{nl}(\alpha r)\rangle}{(E_{n'l} - E_{nl})} \tag{2.46}$$

and because of the properties of the oscillator wave functions n' can be at most $n+1$. The many-particle wave function of the system is a Slater determinant (equation (1.18)) of neutron and proton eigenfunctions with $T_z = (N-Z)/2 = T_0$. We now expand this determinant and keep, at most, terms linear in δ. This expansion is illustrated in Fig. 2.2 for the case that

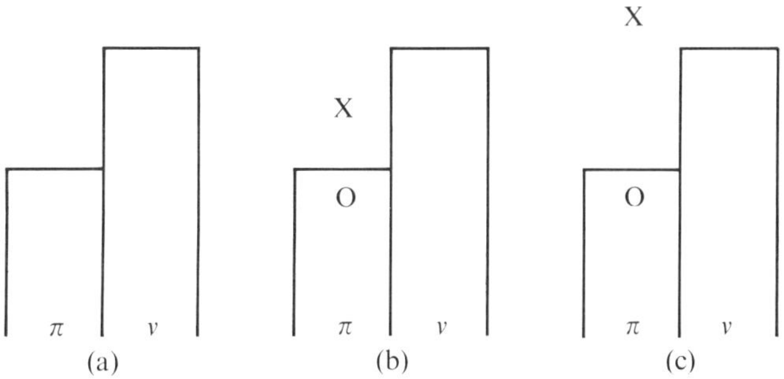

Fig. 2.2. The effect of Coulomb forces on a many-particle wave function that describes a nucleus with a neutron excess. In (a) the configuration is shown when the proton and neutron oscillator constants are the same. In (b) and (c) the effect on the proton configuration due to the Coulomb change in the oscillator constant is illustrated. X and 0 denote a proton particle-hole pair.

the nucleus has a neutron excess. The first term shown in Fig. 2.2a is the result of replacing α_1 by α. Since every occupied proton state is also occupied by a neutron

$$T_+\Phi_{T_z=T_0} = 0.$$

Thus from Theorem 1 of Chapter 1 it follows that this part of the wave function has $T = T_0$. The remaining terms are proportional to $\beta_{n';nl}$ and describe a situation in which the proton single-particle state $(n'ljm)$ is occupied but the state $(nljm)$ is not; in other words a particle-hole pair. The possible types of particle-hole pairs are illustrated in Figs. 2.2b and 2.2c. The contribution to the wave function illustrated in Fig. 2.2b corresponds to the excitation of a proton from the state $(nljm)$ to the state $(n+1ljm)$ with the neutron state $(n+1ljm)$ occupied. Consequently the T_+ operation applied to this part of the wave function gives zero, so that this component has precisely the same isospin $(T = T_0)$ as the unperturbed configuration.

On the other hand, excitation of a proton to a state not occupied by a neutron (as illustrated in Fig. 2.2c) can give rise to a state with $T > T_0$ because T_+ operating on the configuration does not vanish. However, as we shall now show, only a small part of the configuration illustrated in Fig. 2.2c corresponds to a state with $T = T_0 + 1$. To see this we note that the unperturbed nucleons in Fig. 2.2c have $T_z = (N+1-Z)/2 = T_0 + \frac{1}{2}$ and since T_+ operating on the unperturbed configuration vanishes, $T = T_z = T_0 + \frac{1}{2}$. The extracore proton can couple its isospin of $\frac{1}{2}$ with the core isospin to give T_0 or $T_0 + 1$. If we denote the isospin eigenfunction of the core by $\phi_{T_0+\frac{1}{2},T_0+\frac{1}{2}}$ then

$$\zeta_{\frac{1}{2}-\frac{1}{2}}\phi_{T_0+\frac{1}{2},T_0+\frac{1}{2}} = \sum_{T'} (\tfrac{1}{2}, T_0+\tfrac{1}{2}, -\tfrac{1}{2}, T_0+\tfrac{1}{2} \mid T'T_0)\Phi_{T'T_0}$$

$$= \frac{1}{\sqrt{(2T_0+2)}}\Phi_{T_0+1,T_0} - \left\{\frac{(2T_0+1)}{(2T_0+2)}\right\}^{\frac{1}{2}}\Phi_{T_0T_0} \tag{2.47}$$

where $\Phi_{T'T_0}$ is the isospin eigenfunction of the composite system and in writing the second line of this equation the value of the isospin Clebsch–Gordon coefficient has been taken from Table A1.1. Therefore, because of geometrical considerations the probability that the configuration shown in Fig. 2.2c has isospin (T_0+1) is only

$$\frac{1}{2T_0+2} = \frac{1}{N-Z+2}.$$

Thus there are two effects that cut down the isospin mixing. The first is the Pauli principle which forbids many of the protons (those of Figs. 2.2a and 2.2b) from contributing and the second is that even those protons that can contribute have their mixing cut down by the geometrical factor $1/(N-Z+2)$. From these considerations it follows, in first-order perturbation theory, that the probability γ^2 that a state with isospin T_0+1 will be mixed into the $T = T_0$ state is

$$\gamma^2 = \frac{1}{N-Z+2}\sum_{nlj}{}' N_{nlj}\beta^2_{n+1;nl}$$

$$= \frac{\delta^2(\hbar\omega)^2}{8(N-Z+2)}\sum_{nlj}{}' \frac{(n+1)(2l+2n+3)}{(E_{n+1l}-E_{nl})^2}N_{nlj} \tag{2.48}$$

where N_{nlj} is the number of protons in the orbit (nlj) and the prime on the summation means that only those (nlj) protons for which the $(n+1lj)$ neutron state is not occupied are to be considered. In writing the second line of this equation we have made use of the fact that with ocillator wave functions the radial matrix element has the form given by equation

(1.11a)

$$\langle R_{n+1l}(\alpha r)|\, \alpha^2 r^2\, |R_{nl}(\alpha r)\rangle = -\tfrac{1}{2}\sqrt{\{2(n+1)(2l+2n+3)\}}$$

and our convention for n, the radial quantum number, is that n is equal to zero the first time a given angular-momentum state occurs. These arguments for quenching of the isospin admixtures were originally given by Lane and Soper (1962), and they later showed that the result is close to being correct even when perturbation theory is not valid (Soper 1969).

If one sets $(E_{n+1l} - E_{nl}) = 2\hbar\omega$, equation (2.48) leads to the prediction that $\gamma^2 = 0{\cdot}022$ for ${}^{40}_{20}Ca_{20}$ and $0{\cdot}027$ for ${}^{208}_{82}Pb_{126}$, i.e. of the order of 3% isospin mixing even for very heavy nuclei (Soper 1969). There is, however, one additional effect that cuts the isospin mixing down even further (particularly for heavy nuclei), and this is due to the fact that the energy denominator in equation (2.48) is actually considerably larger than $2\hbar\omega$ (Bohr *et al.* 1967). To understand this we note that even when nuclear forces are taken to be charge independent there is a difference in the neutron–neutron and neutron–proton interaction because the latter pair can interact in both a $T=0$ and a $T=1$ state. Therefore the shell-model potential, which approximates the two-body interaction between the valence and core nucleons, should have an isospin-dependent part. Since the only scalar operator linear in $\underset{\sim}{t}$ and $\underset{\sim}{T}_c$ (the isospin of the extra-core nucleon and the core, respectively) is $\underset{\sim}{t} \cdot \underset{\sim}{T}_c$ one would expect this added potential (often referred to as the Lane potential) to have the form (Lane 1962)

$$V'(r) = \frac{V_a(r)}{A}\, \underset{\sim}{t} \cdot \underset{\sim}{T}_c. \tag{2.49}$$

The $1/A$ dependence follows from the fact that the nucleon–nucleon interaction is short range so two particles can only affect each other when they are close together. Thus this interaction should be inversely proportional to the nuclear volume, $(4\pi/3)r_0^3 A$.

A magnitude for $V_a(r)$ can be obtained by looking at nuclei with a single proton outside a core with isospin T_c. In this case the particle can be in a state with isospin $T = T_c + \frac{1}{2}$ or $T = T_c - \frac{1}{2}$. Since the usual shell-model potential has no isospin dependence the splitting between these two states is due entirely to the Lane potential. In the same way that equation (2.26) was derived we find

$$\langle \Phi_{TT_z}|\, V'(r)\, |\Phi_{TT_z}\rangle = \langle V'(r)\rangle_T$$
$$= \frac{\langle V_a(r)\rangle}{2A}\{T(T+1) - T_c(T_c+1) - \tfrac{3}{4}\}.$$

Thus

$$\langle V'(r)\rangle_{T=T_c+\frac{1}{2}} - \langle V'(r)\rangle_{T=T_c-\frac{1}{2}} = \frac{\langle V_a(r)\rangle}{A}(T_c + \tfrac{1}{2}). \tag{2.50}$$

The observed positions of the $p_{\frac{3}{2}}$ states in ${}^{49}_{21}Sc_{28}$ can now be used in conjunction with equation (2.50) to estimate $\langle V_a(r)\rangle$. In this case the core is ${}^{48}_{20}Ca_{28}$ with $T_c=4$ and the $p_{\frac{3}{2}}$ states with $T=\frac{7}{2}$ and $\frac{9}{2}$ have been identified by Bloom *et al.* (1973) to lie at 3·08 and 11·56 MeV excitation energy, respectively. Thus from these data one would conclude that $\langle V_a(r)\rangle = 92{\cdot}34$ MeV. To simplify the discussion we take $\langle V_a(r)\rangle$ to be independent of r and to have the value

$$\langle V_a(r)\rangle = 100\ \text{MeV}. \tag{2.51}$$

To find the effect the Lane term will have on the energy denominator in equation (2.48) we may use equation (2.50). Because the extra-core proton in the (N, Z) nucleus moves in the field generated by N neutrons and $(Z-1)$ protons it follows that $T_c=\{N-(Z-1)\}/2=T_0+\frac{1}{2}$ so that

$$E_{n+1l}-E_{nl}=2\hbar\omega+\frac{\langle V_a(r)\rangle}{A}(T_0+1).$$

Thus in perturbation theory the probability of mixing states with isospin (T_0+1) into the state T_0 is

$$\gamma^2=\frac{\delta^2}{32(N-Z+2)}\sum_{nlj}{}^{'}\frac{N_{nlj}(n+1)(2l+2n+3)}{\left\{1+\dfrac{\langle V_a(r)\rangle}{4\hbar\omega A}(N-Z+2)\right\}^2}. \tag{2.52}$$

In summary, there are three separate effects that cut down the isospin mixing naively expected in a state

(a) *Pauli quenching.* Because of the Pauli principle not all occupied proton states can contribute to the isospin mixing. In fact only those (nlj) for which the neutron state $(n+1\ lj)$ is unoccupied are important. For light nuclei this effect is small; e.g. in ${}^{40}_{20}Ca_{20}$ only excitation from the $0s$ to the $1s$ orbit is blocked; a 5% effect. However, for heavy nuclei this quenching is more important and in ${}^{208}_{82}Pb_{126}$ only 51%, $(\frac{42}{82})$, of the protons contribute.

(b) *Geometrical quenching.* The fact that a single proton outside a core with $T=T_z=T_0+\frac{1}{2}$ has only a small probability of being in a (T_0+1) state is the most important factor in suppressing isospin mixing. Although the factor $(N-Z+2)$ only cuts the mixing down by a factor of 2 in ${}^{40}_{20}Ca_{20}$ it provides a factor of 46 for ${}^{208}_{82}Pb_{126}$.

(c) *Analogue quenching.* This is the name generally applied to the increase in the single-particle energies brought about by the Lane potential. If one uses $\langle V_a(r)\rangle = 100$ MeV and $\hbar\omega = 41/A^{\frac{1}{3}}$ this leads to an increase of 10% in the energy denominator of equation (2.52) for ${}^{40}_{20}Ca_{20}$ and 80% for ${}^{208}_{82}Pb_{126}$.

When these three effects are taken into account the predicted isospin mixing (Soper 1969) is small for light nuclei ($\gamma^2 = 0{\cdot}11\%$ for ${}^{16}_{8}O_8$) but rises rapidly to a maximum of about 2% in ${}^{40}_{20}Ca_{20}$. Since stable nuclei beyond ${}^{40}_{20}Ca_{20}$ have a substantial neutron excess the geometrical quenching factor becomes important, and the isospin mixing actually decreases for heavy nuclei falling to about 1% in ${}^{208}_{82}Pb_{126}$. The same magnitude for this mixing is also predicted when sum-rule techniques are used (Lane and Mekjian 1973).

Thus on the basis of these simple considerations one would expect isospin to be a good quantum number even for the heaviest nuclei. One source of experimental information concerning the validity of this expectation comes from the data on $0^+ \rightarrow 0^+$ beta decays. According to the theory of beta decay (see, for example, Konopinski 1966, Wu and Moszkowski 1966) the operator governing allowed Fermi transitions is

$$F = \sum_i (\mathrm{t}_\pm) i = \mathrm{T}_\pm. \tag{2.53}$$

Since $\mathrm{T}_\pm$ cannot change the total isospin of a state, the matrix elements of this operator vanish identically if the two 0^+ states involved have different isospin. The data concerning these $\Delta T = 1$ decays have recently been collected by Raman *et al.* (1975), and in no case do they find the amount of isospin mixing needed to explain the transition rates exceeds a few tenths of a per cent.

A second way of obtaining information about isospin purity involves a study of the $1^-(T=0) \leftrightarrow 0^+(T=0)$ gamma decays. As we shall show in Chapter 5, these transitions are strictly forbidden in the long-wavelength limit if the states involved are rigorously $T = 0$. From the compilations of Endt and Van der Leun (1974, 1974a) one sees that these transitions do proceed more slowly than the isospin-allowed ones. By far the fastest of these is the one observed in ${}^{40}_{20}Ca_{20}$ between the 6·95 MeV 1^- $T=0$ state and the ground state, and even in this case the observed strength requires only about a 0·5% $T = 1$ admixture into the 1^- state (Gloeckner and Lawson 1975).

There are, of course, exceptional cases that occur when two levels of the same spin but different isospin are almost degenerate. In this case it is possible to have rather large mixing. The classic example of this is the 16·6 and 16·9 MeV 2^+ states in ${}^{8}_{4}Be_4$ which both appear to be 50–50 mixtures of $T = 0$ and $T = 1$ (Stephenson and Marion 1966). However, in general, one does not expect and indeed does not find more than a few tenths of a per cent admixture of $(T_0 + 1)$ into low-lying T_0 states.

1.4. What about seniority?

In Chapter 1, section 5, we defined the seniority quantum number for the identical particle system and showed that within the configuration j^n

seniority was conserved to a very high degree. The question of course arises 'is the same thing true for the neutron–proton system?' French (1960) has shown that the conditions that the two-body interaction energies must satisfy in order that seniority (more exactly symplectic symmetry) is a good quantum number is that

$$\sum_J (-1)^J(2J+1)W(jjjj;JJ_0)E_J(jj;jj)=C \tag{2.54}$$

where C is a constant independent of J_0 and the equation must hold for all even $J_0>0$. Since J_0 can take on the values $2, 4, \ldots, (2j-1)$ equation (2.54) imposes one constraint on E_J for $j=\frac{5}{2}$, *two conditions when* $j=\frac{7}{2}$ etc.

It is difficult to find values for E_J when $j=\frac{5}{2}$ since there is very substantial mixing of the $0d_{\frac{5}{2}}$, $1s_{\frac{1}{2}}$ and $0d_{\frac{3}{2}}$ orbits in the $(0d, 1s)$ shell. On the other hand, values of the $E_J(\frac{7}{2}\frac{7}{2};\frac{7}{2}\frac{7}{2})$ matrix elements may be taken directly from the spectrum of ${}^{42}_{21}Sc_{21}$ (Fig. 2.1). By use of the tables of Racah coefficients (Rotenberg *et al.* 1959) one can show that the conditions that the $f_{\frac{7}{2}}$ two-body matrix elements must satisfy so that symplectic symmetry is a good quantum number is that Δ and Δ' must both be identically equal to zero where

$$\Delta=(E_4-E_6)+\tfrac{1}{273}(143E_1-455E_3+242E_5+70E_7)$$

and

$$\Delta'=(E_2-E_6)+\tfrac{1}{1001}(858E_1-819E_3-1782E_5+1743E_7).$$

With the ${}^{42}_{21}Sc_{21}$ matrix elements one concludes that $\Delta=-1{\cdot}067$ MeV and $\Delta'=-3{\cdot}974$ MeV. The degree of seniority mixing depends not only on Δ and Δ' but also on the difference in the diagonal matrix elements of the Hamiltonian for the various seniority states. Thus to assess the degree to which seniority is mixed one must carry out an explicit calculation. The wave function for the lowest $I=\frac{7}{2}T=\frac{1}{2}$ state of the configuration $(f_{\frac{7}{2}})^3$ (the ground state of ${}^{43}_{21}Sc_{22}$) can be computed by use of the $f_{\frac{7}{2}}$ isospin coefficients of fractional parentage (given by Hubbard 1971 and Shlomo 1972) and the methods outlined in the next section. When this calculation is carried out one predicts that the ground-state wave function of ${}^{43}_{21}Sc_{22}$ is 83% seniority one and 17% seniority three. Thus energy considerations alone tell us that physically observed states do not have good symplectic symmetry for the neutron–proton system. It should be noted that this result is not at variance with the assumption of a short-range potential for the (n–p) interaction. Although equation (1.113) (the condition analogous to equation (2.54) for the (n–n) or (p–p) case) is automatically fulfilled with a delta-function potential, equation (2.54) is not satisfied when the matrix elements of a delta-function (n–p) interaction are used.

A second check on whether seniority is a good quantum number is provided by transfer reactions as discussed in Chapter 1, section 5.1. If the target is known to have seniority-zero, single-nucleon transfer should only populate the seniority-one level. $^{50}_{22}\text{Ti}_{28}$ provides such a target; the eight $f_{\frac{7}{2}}$ neutrons form a closed shell and in the ground state the two $f_{\frac{7}{2}}$ protons are paired to $I=0$. Thus if seniority is conserved, the direct reaction $^{50}_{22}\text{Ti}_{28}(d, t)^{49}_{22}\text{Ti}_{27}$ should populate mainly the $T=\frac{5}{2}$ $I=\frac{7}{2}^-$ state with only a small cross section to the $T=\frac{7}{2}$ $I=\frac{7}{2}^-$ state (see section (5.3) of this Chapter) which is expected to lie at about 9 MeV excitation energy. Experimentally (Yntema 1962, Lawson and Zeidman 1962) two strong peaks corresponding to $f_{\frac{7}{2}}$ pick-up are seen below 3 MeV and only about 50% of the strength goes to the ground state. Thus the direct reaction data also show that seniority mixing in the $f_{\frac{7}{2}}$ nuclei is quite strong so that the experimentally observed levels are not states with well-defined seniority.

2. Isospin and non-isospin methods of calculation

In this Section we give two examples of calculations carried out using isospin eigenfunctions and show that the results are identical to those obtained when the neutron–proton formalism is used. To carry out the isospin calculations we must first generalize the diagrammatic recoupling results given in Chapter 1, section 4.1.

2.1. Some diagrammatic rules for recoupling including isospin

When the isospin formalism is used nuclear wave functions are endowed with two new quantum numbers μ and T, the z component and total isospin, respectively. Thus if one wishes to couple $\Psi_{J_1M_1;T_1\mu_1}$ and $\Phi_{J_2M_2;T_2\mu_2}$ to total angular momentum $(J_3M_3; T_3\mu_3)$ one must do the vector addition in both ordinary and isospin space (see equation (2.22)). Thus the analogue of equation (1.84), which depicts the coupling of J only, is

$$[\Psi_{J_1T_1} \times \Phi_{J_2T_2}]_{J_3M_3;T_3\mu_3} = \overset{\displaystyle J_1T_1 \diagup \diagdown J_2T_2}{\underset{(J_3M_3;\,T_3\mu_3)}{\rule{3cm}{0.4pt}}} \qquad (2.55)$$

where the symbol $[\times]_{J_3M_3;T_3\mu_3}$ denotes coupling in both J and T space. Furthermore, because of the symmetry property of the Clebsch–Gordan coefficients given by equation (A1.18),

$$\overset{\displaystyle J_1T_1 \diagup \diagdown J_2T_2}{\underset{(J_3M_3;\,T_3\mu_3)}{\rule{3cm}{0.4pt}}} = (-1)^{J_1+J_2-J_3+T_1+T_2-T_3}\; \overset{\displaystyle J_2T_2 \diagup \diagdown J_1T_1}{\underset{(J_3M_3;\,T_3\mu_3)}{\rule{3cm}{0.4pt}}}. \qquad (2.56)$$

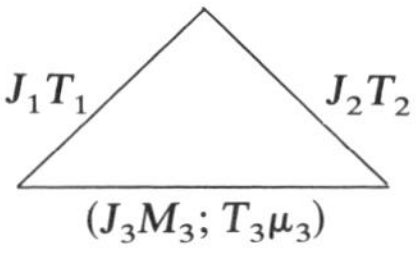

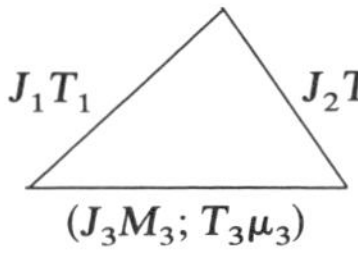

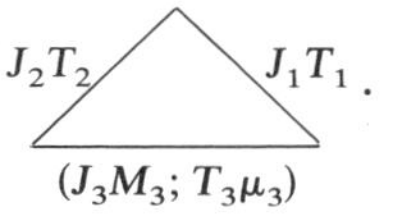

The above rules are readily extended to the coupling of three angular momenta:

$$[[\Psi_{J_1T_1}\times\Phi_{J_2T_2}]_{JT}\times\chi_{J_3T_3}]_{J_4M_4;T_4\mu_4}$$

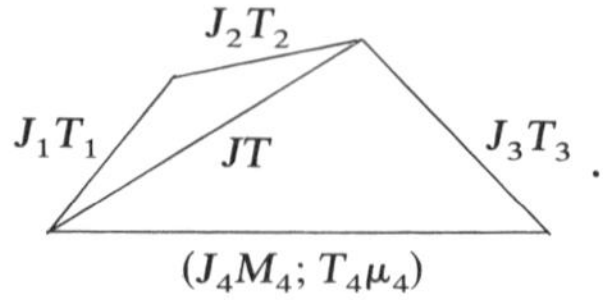

In analogy to equations (1.88) and (1.89) we therefore obtain the recoupling relationships

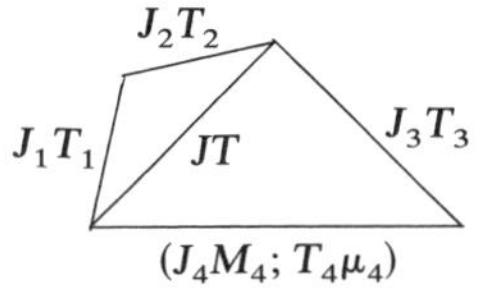

$$=\sum_{KT'}\sqrt{\{(2J+1)(2K+1)(2T+1)(2T'+1)\}}$$

$$\times W(J_1J_2J_4J_3;\, JK)W(T_1T_2T_4T_3;\, TT') \tag{2.57}$$

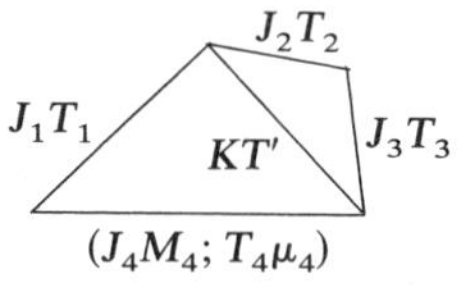

and

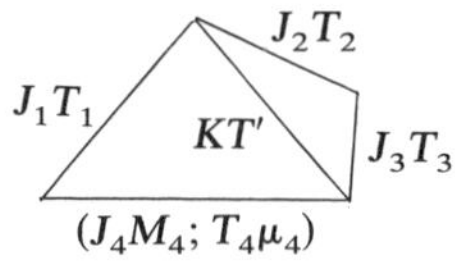

$$=\sum_{JT}\sqrt{\{(2J+1)(2K+1)(2T+1)(2T'+1)\}}$$

$$\times W(J_1J_2J_4J_3;\, JK)W(T_1T_2T_4T_3;\, TT') \tag{2.58}$$

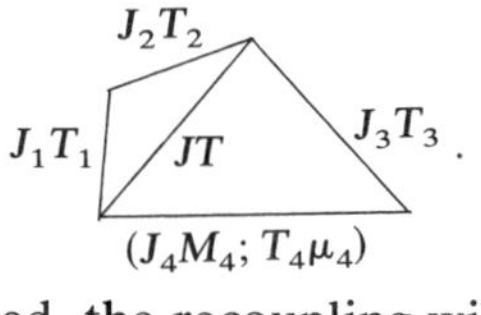

When four angular momenta are involved, the recoupling will lead to two $9j$ symbols, one for ordinary space and one for isospin:

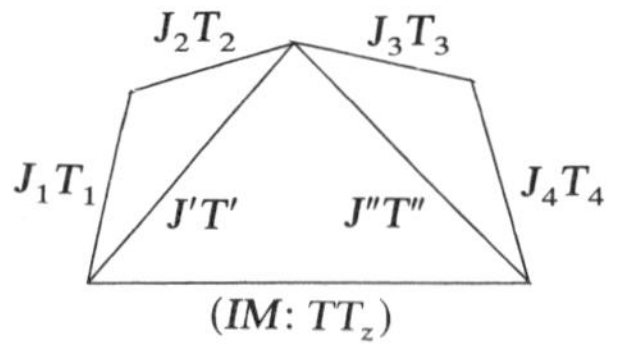

$$=\sum_{K'K''}\sum_{T_1'T_1''}\sqrt{\{(2J'+1)(2J''+1)(2K'+1)(2K''+1)\}}$$

$$\times\sqrt{\{(2T'+1)(2T''+1)(2T_1'+1)(2T_1''+1)\}}$$

$$\times\begin{Bmatrix} J_1 & J_2 & J' \\ J_3 & J_4 & J'' \\ K' & K'' & I \end{Bmatrix}\begin{Bmatrix} T_1 & T_2 & T' \\ T_3 & T_4 & T'' \\ T_1' & T_1'' & T \end{Bmatrix} \tag{2.59}$$

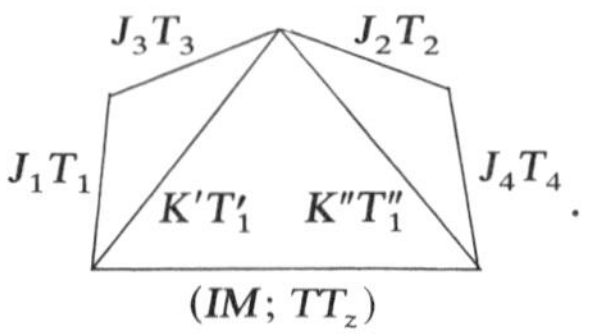

The isospin fractional-parentage coefficients can be defined (see equation (A5.29)) so that

$$(j^n)_{IM;TT_z\alpha} = \sum_{JT'\beta} \langle j^{n-1}JT'\beta, j\tfrac{1}{2}|\}j^nIT\alpha\rangle \times[\Phi_{JT'\beta}(1,\ldots,n-1)\times\phi_{j\frac{1}{2}}(n)]_{IM;TT_z}$$

where we have made use of the fact that the nucleon has $t=\frac{1}{2}$. In terms of diagrams this equation becomes

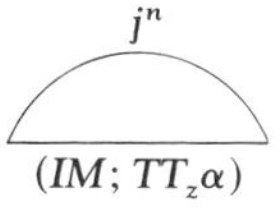

$$= \sum_{JT'\beta} \langle j^{n-1}JT'\beta, j\tfrac{1}{2}|\}j^nIT\alpha\rangle \qquad (2.60)$$

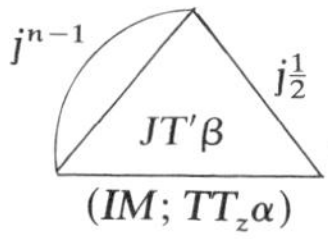

Tables of isospin c.f.p's have been given by Glaudemans *et al.* (1964) for the $j=\frac{1}{2}$ and $\frac{3}{2}$ shells, Towner and Hardy (1969a) for the $j=\frac{3}{2}$ and $\frac{5}{2}$ levels, Hubbard (1971) for $j=\frac{7}{2}$ $(n\le 5)$ and Shlomo (1972) for $j=\frac{1}{2}, \frac{3}{2}, \frac{5}{2}$, and $\frac{7}{2}$ ($n\le 4$ in this latter orbit).

The two-particle isospin c.f.p. is defined by equation (A5.32) and in terms of diagrams is

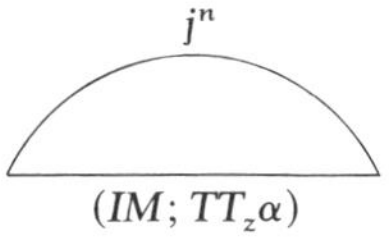

$$= \sum_{J_3T_3\beta_3}\sum_{KT_1} \langle j^{n-2}J_3T_3\beta_3, j^2KT_1|\}j^nIT\alpha\rangle \qquad (2.61)$$

$\times$

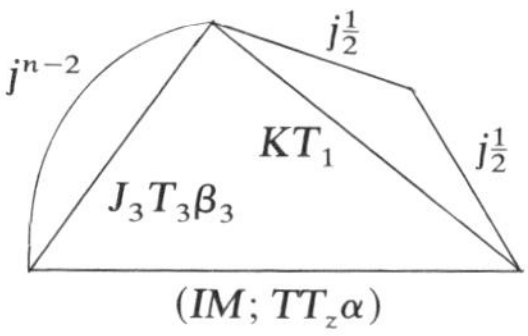

For the case $n=3$ the single- and double-parentage coefficients are related, as may be seen by comparing equations (2.60) and (2.61), by

$$\langle j^2JT_1, j\tfrac{1}{2}|\}j^3IT\alpha\rangle = (-1)^{J+j-I+T_1+\frac{1}{2}-T}\langle j^1 j\tfrac{1}{2}, j^2JT_1|\}j^3IT\alpha\rangle. \qquad (2.62)$$

Algebraic formulae for the two-particle isospin c.f.p's for seniority-zero and one states have been given by Towner and Hardy (1969). For higher-seniority states one can calculate them using equation (A5.39).

2.2 *Isospin calculation for* $^{35}_{17}Cl_{18}$

As an application of the isospin formalism we first consider the spectrum of $^{35}_{17}Cl_{18}$. To simplify the calculation we assume that $^{32}_{16}S_{16}$ forms an inert core and that the valence nucleons are confined to the $d_{\frac{3}{2}}$ orbit. Thus to calculate the energy levels of this nucleus one must know the neutron and proton $d_{\frac{3}{2}}$ single-particle energies and the matrix elements of the residual nucleon–nucleon interaction in the four possible $(d_{\frac{3}{2}})^2$ angular-momentum states $J=0$, 1, 2, and 3. Since we deal with a neutron and a proton all possible J states consistent with the triangle inequality $|j_1-j_2|\leq J\leq (j_1+j_2)$ are allowed. The even-J values have space-spin antisymmetric states and hence are associated with $T=1$ configurations, whereas the odd-J eigenfunctions are space-spin symmetric and hence have $T=0$.

The single-particle energies relative to the closed ${}^{32}_{16}S_{16}$ core can be found from the binding-energy tables of Wapstra and Gove (1971):

$$\begin{aligned}\varepsilon_\pi &= \mathrm{BE}({}^{33}_{17}\mathrm{Cl}_{16})-\mathrm{BE}({}^{32}_{16}\mathrm{S}_{16})\\ &=-2\cdot277\ \mathrm{MeV}\end{aligned} \tag{2.63a}$$

$$\begin{aligned}\varepsilon_\nu &= \mathrm{BE}({}^{33}_{16}\mathrm{S}_{17})-\mathrm{BE}({}^{32}_{16}\mathrm{S}_{16})\\ &=-8\cdot643\ \mathrm{MeV}.\end{aligned} \tag{2.63b}$$

As far as the two-body matrix elements are concerned, we take them from the spectrum of ${}^{34}_{17}Cl_{17}$. The ground state of this nucleus has $J=0$. Consequently the residual two-body interaction in the $J=0$, $T=1$ state of $(d_{\frac{3}{2}})^2$ is

$$\begin{aligned}E_{01} &= \mathrm{BE}({}^{34}_{17}\mathrm{Cl}_{17})-\mathrm{BE}({}^{32}_{16}\mathrm{S}_{16})-\varepsilon_\pi-\varepsilon_\nu\\ &=-2\cdot865\ \mathrm{MeV}\end{aligned} \tag{2.63c}$$

where E_{01} stands for $E_{01}(\frac{3}{2}\frac{3}{2};\frac{3}{2}\frac{3}{2})$ and the subscript 01 is the JT value of the state for which this is the interaction energy. There are several 2^+ states seen in ${}^{34}_{17}Cl_{17}$ (Endt and Van der Leun 1973). The lowest one with $T=1$ occurs at 2·158 MeV excitation energy and we assume this state corresponds to the configuration $(d_{\frac{3}{2}})^2_{J=2;T=1}$ so that

$$E_{21}=-2\cdot865+2\cdot158=-0.707\ \mathrm{MeV}. \tag{2.63d}$$

Only one low-lying 3^+ level is seen in ${}^{34}_{17}Cl_{17}$ and this lies at an excitation energy of 0·146 MeV. We assume the configuration of this state is $(d_{\frac{3}{2}})^2_{J=3;T=0}$ and as a consequence

$$E_{30}=-2\cdot719\ \mathrm{MeV}. \tag{2.63e}$$

Finally two $J=1^+$ states occur in this nucleus, one at 0·461 MeV and the second at 0·665 MeV. The latter shows up much more strongly with $l=2$ in the ${}^{33}_{16}S_{17}(t,d){}^{34}_{17}Cl_{17}$ reaction, and so we assume this state to be the $(d_{\frac{3}{2}})^2_{J=1;T=0}$ level (see section 5 of this chapter).

$$E_{10}=-2\cdot200\ \mathrm{MeV}.$$

The three-particle isospin fractional-parentage coefficients for $j=\frac{3}{2}$ have been given by Glaudemans *et al.* (1964) and are reproduced in

TABLE 2.1

One-particle isospin coefficients of fractional parentage for the configuration $(d_{\frac{3}{2}})^3$

	J_1T_1			
IT	0 1	2 1	1 0	3 0
$\frac{1}{2}\frac{1}{2}$	0	$\frac{1}{\sqrt{2}}$	$\frac{1}{\sqrt{2}}$	0
$\frac{3}{2}\frac{1}{2}$	$\sqrt{\frac{5}{12}}$	$\sqrt{\frac{1}{12}}$	$-\sqrt{\frac{3}{20}}$	$-\sqrt{\frac{7}{20}}$
$\frac{5}{2}\frac{1}{2}$	0	$\frac{1}{\sqrt{2}}$	$-\sqrt{\frac{7}{30}}$	$\sqrt{\frac{4}{15}}$
$\frac{7}{2}\frac{1}{2}$	0	$\frac{1}{\sqrt{2}}$	0	$-\frac{1}{\sqrt{2}}$
$\frac{3}{2}\frac{3}{2}$	$\frac{1}{\sqrt{6}}$	$-\sqrt{\frac{5}{6}}$	0	0

Table 2.1. A straightforward extension of equation (1.73) to the case that isospin is included together with the fact that the double-parentage coefficients are identical in magnitude to the one-particle c.f.p's for $n=3$ (see equation (2.62)) leads to the equation

$$\begin{bmatrix} \varepsilon_{\frac{3}{2}\frac{1}{2}} \\ \varepsilon_{\frac{1}{2}\frac{1}{2}} \\ \varepsilon_{\frac{5}{2}\frac{1}{2}} \\ \varepsilon_{\frac{7}{2}\frac{1}{2}} \\ \varepsilon_{\frac{3}{2}\frac{3}{2}} \end{bmatrix} = \varepsilon_\pi + 2\varepsilon_\nu + \begin{bmatrix} \frac{5}{4} & \frac{1}{4} & \frac{9}{20} & \frac{21}{20} \\ 0 & \frac{3}{2} & \frac{3}{2} & 0 \\ 0 & \frac{3}{2} & \frac{7}{10} & \frac{4}{5} \\ 0 & \frac{3}{2} & 0 & \frac{3}{2} \\ \frac{1}{2} & \frac{5}{2} & 0 & 0 \end{bmatrix} \begin{bmatrix} E_{01} \\ E_{21} \\ E_{10} \\ E_{30} \end{bmatrix} \tag{2.64}$$

where ε_{IT} is the total energy of the three-particle state with IT relative to the closed ${}^{32}_{16}S_{16}$ core and matrix multiplication is implied.

In Fig. 2.3 we compare the experimental results of Vignon *et al.* (1971) with the theoretical predictions that result when the energies of equations (2.63) are used in conjunction with equation (2.64). As can be seen, the $(d_{\frac{3}{2}})^3$ model predicts $T=\frac{1}{2}$, $I=\frac{5}{2}$ and $\frac{7}{2}$ states at 2·827 and 2·464 MeV excitation energy, respectively. Thus the observed $I=\frac{5}{2}$ state at 3·001 MeV and the $I=\frac{7}{2}$ level at 2·646 MeV undoubtedly have wave functions that are dominated by the $(d_{\frac{3}{2}})^3_{IT}$ configuration. On the other hand, the low-lying $T=\frac{1}{2}$, $I=\frac{1}{2}$ and $\frac{5}{2}$ states observed at 1·219 and 1·763 MeV, respectively, are not accounted for by the model and are undoubtedly due to excitation out of the assumed inert $(\pi d_{\frac{5}{2}})^6(\nu d_{\frac{5}{2}})^6(\pi s_{\frac{1}{2}})^2$ $(\nu s_{\frac{1}{2}})^2$ core. In addition the $d_{\frac{3}{2}}$ model predicts that the $I=\frac{3}{2}$, $T=\frac{3}{2}$ level (which is the isobaric analogue of the ${}^{35}_{16}S_{19}$ ground state) should occur at

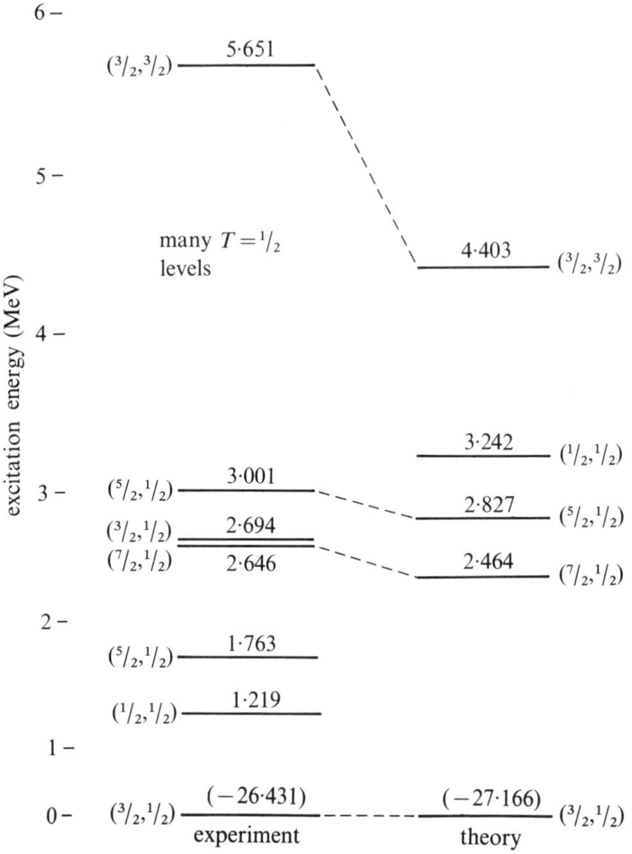

Fig. 2.3. Low-lying states of $^{35}_{17}Cl_{18}$. The experimental data are taken from Vignon *et al.* (1971). The theoretical results are for the $(d_{\frac{3}{2}})^3$ configuration with the energies given in equations (2.63).

4·403 MeV excitation energy. This is approximately 1.2 MeV lower than observed.

Thus from this simple calculation one learns that only two excited states (the first $I=\frac{7}{2}$, $T=\frac{1}{2}$ and second $I=\frac{5}{2}$, $T=\frac{1}{2}$ levels) can be adequately accounted for by the $(d_{\frac{3}{2}})^3$ model. If one is to obtain a fit to all the low-lying states a larger model space must be used. Vignon *et al.* (1971) have carried out a calculation in which the $s_{\frac{1}{2}}$ single-particle state is included, and when this is done one can explain the experimentally observed structure of $^{35}_{17}Cl_{18}$.

2.3. Neutron–proton calculation for $^{35}_{17}Cl_{18}$

To illustrate that the same results are obtained whether or not one uses the isospin formalism we now calculate, using the same model space as in the previous section, the spectrum of $^{35}_{17}\mathrm{Cl}_{18}$ by treating neutrons and protons as distinguishable particles.

Because of the exclusion principle, the two $d_{\frac{3}{2}}$ neutrons can only have $J_N=0$ or 2. These angular-momentum values can be coupled with the proton spin of $\frac{3}{2}$, and as a consequence the basis states used in the calculation are

$\psi_{IM}=$ 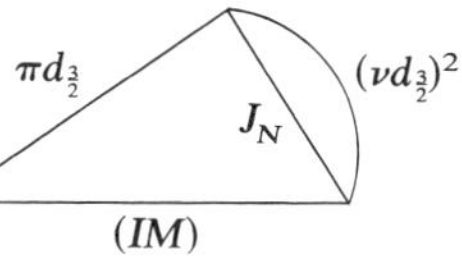

Since J_N takes on the values 0 and 2, there are two ways to obtain $I=\frac{3}{2}$ but only one way to arrive at each of the spins $I=\frac{1}{2}$, $\frac{5}{2}$, and $\frac{7}{2}$. The Hamiltonian matrix elements can be calculated by use of equation (1.96) with $n_1=1$, $n_2=2$, $j_1=j_2=J_1=J_1'=J_4=\frac{3}{2}$, and $J_3=0$. Since the Racah coefficient with one index equal to zero has the simple form given in equation (A4.14) and since the $2\to1$ and $1\to0$ one-nucleon c.f.p's are both unity, it follows that

$$\langle[\pi d_{\frac{3}{2}}\times(\nu d_{\frac{3}{2}})^2_{J_N'}]_{IM}|\,H\,|[\pi d_{\frac{3}{2}}\times(\nu d_{\frac{3}{2}})^2_{J_N}]_{IM}\rangle$$
$$=\{\varepsilon_\pi+2\varepsilon_\nu+E'_{J_N}\}\delta_{J_N'J_N}+2(-1)^{J_N'-J_N}\surd\{(2J_N+1)(2J_N'+1)\}$$
$$\times\sum_K(2K+1)W(\tfrac{3}{2}\tfrac{3}{2}I\tfrac{3}{2};\,KJ_N)W(\tfrac{3}{2}\tfrac{3}{2}I\tfrac{3}{2};\,KJ_N')E_K.$$

In writing this equation we have denoted the neutron–neutron interaction in the state J_N by E'_{J_N} and the proton–neutron matrix element by E_K.

By use of the tables of Racah coefficients (Rotenberg *et al.* 1959) one finds the Hamiltonian matrix for the $I=\frac{3}{2}$ state to be

$$\langle H\rangle_{I=\frac{3}{2}}=\begin{bmatrix}\begin{array}{l}\varepsilon_\pi+2\varepsilon_\nu+E_0'\\ \quad+\frac{1}{8}(E_0+3E_1+5E_2+7E_3)\end{array} & \frac{\surd5}{40}(5E_0+3E_1-15E_2+7E_3)\\[2ex] \frac{\surd5}{40}(5E_0+3E_1-15E_2+7E_3) & \begin{array}{l}\varepsilon_\pi+2\varepsilon_\nu+E_2'\\ \quad+\frac{1}{40}(25E_0+3E_1+45E_2+7E_3)\end{array}\end{bmatrix}$$

$$=\begin{bmatrix}\begin{array}{l}\varepsilon_\pi+2\varepsilon_\nu+(E_0'-E_0)\\ \quad+\frac{1}{8}(9E_0+3E_1+5E_2+7E_3)\end{array} & 0\\[2ex] 0 & \begin{array}{l}\varepsilon_\pi+2\varepsilon_\nu+(E_2'-E_2)\\ \quad+\frac{1}{8}(9E_0+3E_1+5E_2+7E_3)\end{array}\end{bmatrix}$$

$$+\frac{(5E_0+3E_1-15E_2+7E_3)}{10}\times\begin{bmatrix}0 & \frac{\surd5}{4}\\ \frac{\surd5}{4} & -1\end{bmatrix}. \qquad (2.65)$$

(Again we have the limiting checks on the algebra as discussed in Chapter 1, i.e. if the potential becomes long range all $E_K \to 1$ and the sum of the coefficients of E_K must be $n(n-1)/2 = 3$ for the diagonal matrix elements and zero for the off-diagonal ones.)

If nuclear forces are charge independent $E_0' = E_0$ and $E_2' = E_2$. When this is true the diagonal matrix in equation (2.65) becomes a multiple of the unit matrix and the eigenvectors of the $I = \frac{3}{2}$ states are determined by diagonalizing the second matrix in this equation. Since apart from a multiplication constant this matrix is independent of E_K, the eigenvectors will be independent of E_K. Thus

$$\Psi_{\frac{3}{2}M} = \sqrt{\tfrac{5}{6}}[\pi d_{\frac{3}{2}} \times (\nu d_{\frac{3}{2}})_0^2]_{\frac{3}{2}M} + \frac{1}{\sqrt{6}}[\pi d_{\frac{3}{2}} \times (\nu d_{\frac{3}{2}})_2^2]_{\frac{3}{2}M}$$

and

$$\Phi_{\frac{3}{2}M} = \frac{1}{\sqrt{6}}[\pi d_{\frac{3}{2}} \times (\nu d_{\frac{3}{2}})_0^2]_{\frac{3}{2}M} - \sqrt{\tfrac{5}{6}}[\pi d_{\frac{3}{2}} \times (\nu d_{\frac{3}{2}})_2^2]_{\frac{3}{2}M}$$

are the eigenvectors of this matrix. The former is associated with the eigenvalue

$$\varepsilon = \varepsilon_\pi + 2\varepsilon_\nu + \tfrac{5}{4}E_0 + \tfrac{9}{20}E_1 + \tfrac{1}{4}E_2 + \frac{21}{20}E_3$$

which is identical to $\varepsilon_{I=\frac{3}{2}, T=\frac{1}{2}}$ obtained from equation (2.64) provided one identifies the E_K of this Section with the E_{IT} of Equation (2.63). Thus $\Psi_{\frac{3}{2},M}$ corresponds (in the isospin language) to the $T = \frac{1}{2}$ eigenfunction. In a similar way the eigenvalue associated with $\Phi_{\frac{3}{2},M}$ is

$$\tilde{\varepsilon} = \varepsilon_\pi + 2\varepsilon_\nu + \tfrac{1}{2}E_0 + \tfrac{5}{2}E_2$$

and this is identical to $\varepsilon_{I=\frac{3}{2}, T=\frac{3}{2}}$ of equation (2.64). Consequently the state described by $\Phi_{\frac{3}{2},M}$ corresponds to the $T = \frac{3}{2}$ level.

The Hamiltonian matrices for the $I = \frac{1}{2}, \frac{5}{2}$, and $\frac{7}{2}$ states are all (1×1) and by explicitly substituting values for the Racah coefficients one can show that the expectation value of H is the same as given by equation (2.64). Thus the isospin formalism and the neutron–proton method give the same results.

The number of allowable angular-momentum states goes up extremely rapidly for increasing n and j, and it is for this reason that the existing tables of isospin c.f.p's do not go beyond $j = \frac{7}{2}$, $n = 5$. Consequently if one wants to do calculations in the $f_{\frac{7}{2}}$ shell for $n > 5$ one is forced either to compute the isospin c.f.p's or to proceed by the methods outlined in this section. This latter method has the disadvantage that one has to construct and diagonalize larger energy matrices since all isospin states for given n are calculated at once, whereas in the isospin formalism the matrix partitions into 'isospin blocks' if T is a good quantum number. However, the advantage of not having to calculate the isospin c.f.p's outweighs the disadvantage of constructing a larger matrix and consequently most calculations reported in the literature proceed along the lines outlined in this section (see, for example, McCullen *et al.* 1964).

2.4. ${}^{39}_{18}Ar_{21}$ *spectrum using the neutron–proton formalism*

As a second example of the calculation of nuclear spectra we consider the nucleus ${}^{39}_{18}Ar_{21}$. We shall first calculate the spectrum by treating neutrons and protons as distinguishable particles. To make the calculations as simple as possible we assume that ${}^{36}_{16}S_{20}$ forms an inert closed core and that the two valence protons and the valence neutrons are restricted to the $d_{\frac{3}{2}}$ and $f_{\frac{7}{2}}$ orbits, respectively. The single-particle energies associated with these states can be obtained from the binding-energy tables of Wapstra and Gove (1971):

$$\begin{aligned}\varepsilon_{d\pi} &= \mathrm{BE}({}^{37}_{17}\mathrm{Cl}_{20}) - \mathrm{BE}({}^{36}_{16}\mathrm{S}_{20}) \\ &= -8{\cdot}385\ \mathrm{MeV}\end{aligned} \tag{2.66a}$$

$$\begin{aligned}\varepsilon_{f\nu} &= \mathrm{BE}({}^{37}_{16}\mathrm{S}_{21}) - \mathrm{BE}({}^{36}_{16}\mathrm{S}_{20}) \\ &= -4{\cdot}313\ \mathrm{MeV}.\end{aligned} \tag{2.66b}$$

The $(d_{\frac{3}{2}})^2$ interaction energies in the $J=0$ and 2 states can be taken from the data on ${}^{38}_{18}Ar_{20}$. From the binding energy we find

$$\begin{aligned}E_0 &= E_0(\tfrac{3}{2}\tfrac{3}{2};\tfrac{3}{2}\tfrac{3}{2}) \\ &= \mathrm{BE}({}^{38}_{18}\mathrm{Ar}_{20}) - \mathrm{BE}({}^{36}_{16}\mathrm{S}_{20}) - 2\varepsilon_{d\pi} \\ &= -1{\cdot}857\ \mathrm{MeV}.\end{aligned}$$

(Part of the reason this value is different to that given in equation (2.63c) is that in ${}^{38}_{18}Ar_{20}$ we deal with the interaction of two protons which includes the Coulomb repulsion, whereas in ${}^{34}_{17}Cl_{17}$—the nucleus used to obtain the value given in equation (2.63c)—the interaction is between a neutron and a proton.) The $T=1$ $I=2^+$ state in ${}^{38}_{18}Ar_{20}$ lies at 2·168 MeV excitation energy (Endt and Van der Leun 1973). This value is almost identical to the energy of the 2^+ state in ${}^{34}_{17}Cl_{17}$ (2·158 MeV) and bears out the fact that Coulomb matrix elements are not very sensitive to the angular momentum of the state. These results lead to the $(d_{\frac{3}{2}})^2$ interaction energies in Table 2.2.

TABLE 2.2

Interaction energies in the various angular-momentum states of $(\pi d_{\frac{3}{2}})^2$ *and* $(\pi d_{\frac{3}{2}}, \nu f_{\frac{7}{2}})$

	Angular momentum					
Configuration	0	1	2	3	4	5
$(d_{\frac{3}{2}})^2$	−1·857		+0·311			
$(d_{\frac{3}{2}}, f_{\frac{7}{2}})$			−1·797	−1·042	−0·488	−1·126

Blanks indicate an angular-momentum state cannot occur. Interaction energies are in MeV.

Values for the $(d_{\frac{3}{2}}, f_{\frac{7}{2}})$ energies can be taken from ${}^{38}_{17}\mathrm{Cl}_{21}$. The ground-state spin is 2^- so that the residual two-body interaction in this state is

$$\tilde{E}_2 = E_2(\tfrac{3}{2}\tfrac{7}{2}; \tfrac{3}{2}\tfrac{7}{2})$$

$$\tilde{E}_2 = \mathrm{BE}({}^{38}_{17}\mathrm{Cl}_{21}) - \mathrm{BE}({}^{36}_{16}\mathrm{S}_{20}) - \varepsilon_{d\pi} - \varepsilon_{f\nu}$$

$$= -1{\cdot}797\ \mathrm{MeV}.$$

Excited states with spins 5^-, 3^-, and 4^- are seen at 0·671, 0·755, and 1·309 MeV, respectively (Endt and Van der Leun 1973), and these lead to the $(d_{\frac{3}{2}}, f_{\frac{7}{2}})$ entries given in Table 2.2.

Within the $(\pi d_{\frac{3}{2}} - \nu f_{\frac{7}{2}})$ model space the possible spin values are

$$I = \tfrac{3}{2}^-, \tfrac{5}{2}^-, (\tfrac{7}{2}^-)^2, \tfrac{9}{2}^-, \text{ and } \tfrac{11}{2}^-$$

where the superscript 2 indicates that $I = \frac{7}{2}^-$ can be realized in two ways: $[(\pi d_{\frac{3}{2}})^2_0 \times (\nu f_{\frac{7}{2}})]_{\frac{7}{2}M}$ and $[\pi d_{\frac{3}{2}})^2_2 \times (\nu f_{\frac{7}{2}})]_{\frac{7}{2}M}$. The Hamiltonian matrix can be calculated by use of equation (1.96) with $n_1 = 2$, $n_2 = 1$, $j_1 = J_3 = \frac{3}{2}$, $j_2 = J_2 = J_2' = \frac{7}{2}$, and $J_4 = 0$. Because of the special form of the Racah coefficient with one angular momentum equal to zero (equation (A4.14)) and because the $2 \to 1$ and $1 \to 0$ c.f.p's are both unity, one finds that

$$\langle H \rangle = \{2\varepsilon_{d\pi} + \varepsilon_{f\nu} + E_J\}\delta_{JJ'} + 2\sqrt{\{(2J+1)(2J'+1)\}}$$
$$\times \sum_K (2K+1) W(\tfrac{3}{2}\tfrac{3}{2}I\tfrac{7}{2}; JK) W(\tfrac{3}{2}\tfrac{3}{2}I\tfrac{7}{2}; J'K)\tilde{E}_K.$$

Explicit substitution of values for the Racah coefficients (Rotenberg *et al.* 1959) leads to

$$\langle H \rangle_{I=\frac{3}{2}} = 2\varepsilon_{d\pi} + \varepsilon_{f\nu} + E_2 + \tfrac{1}{2}\tilde{E}_2 + \tfrac{3}{2}\tilde{E}_3 \qquad (2.67\text{a})$$

$$\langle H \rangle_{I=\frac{5}{2}} = 2\varepsilon_{d\pi} + \varepsilon_{f\nu} + E_2 + \tfrac{1}{28}\{24\tilde{E}_2 + 7\tilde{E}_3 + 25\tilde{E}_4\} \qquad (2.67\text{b})$$

$$\langle H\rangle_{I=\frac{7}{2}}=\begin{bmatrix} 2\varepsilon_{d\pi}+\varepsilon_{f\nu}+E_0 & \frac{\sqrt{5}}{80\sqrt{21}}(75\tilde{E}_2 \\ +\frac{1}{16}(5\tilde{E}_2+7\tilde{E}_3+9\tilde{E}_4+11\tilde{E}_5) & -35\tilde{E}_3-117\tilde{E}_4+77\tilde{E}_5) \\ \frac{\sqrt{5}}{80\sqrt{21}}(75\tilde{E}_2 & 2\varepsilon_{d\pi}+\varepsilon_{f\nu}+E_2 \\ & +\frac{1}{1680}(1125\tilde{E}_2+175\tilde{E}_3 \\ -35\tilde{E}_3-117\tilde{E}_4+77\tilde{E}_5) & +1521\tilde{E}_4+539\tilde{E}_5) \end{bmatrix} \tag{2.67c}$$

$$\langle H\rangle_{I=\frac{9}{2}}=2\varepsilon_{d\pi}+\varepsilon_{f\nu}+E_2+\tfrac{1}{60}(55\tilde{E}_3+9\tilde{E}_4+56\tilde{E}_5) \tag{2.67d}$$

$$\langle H\rangle_{I=\frac{11}{2}}=2\varepsilon_{d\pi}+\varepsilon_{f\nu}+E_2+\tfrac{1}{10}(7\tilde{E}_4+13\tilde{E}_5). \tag{2.67e}$$

Again one has a minimal check on the arithmetic in that the sum of the coefficients of E_J and $\tilde{E}_K$ must total $n(n-1)/2=3$ for diagonal matrix elements and zero for off-diagonal ones.

The theoretical spectrum that results when the single-particle energies of equations (2.66) and the interaction energies of Table 2.2 are used is shown in Fig. 2.4. With the exception of the 1·267 MeV $\frac{3}{2}^-$ and

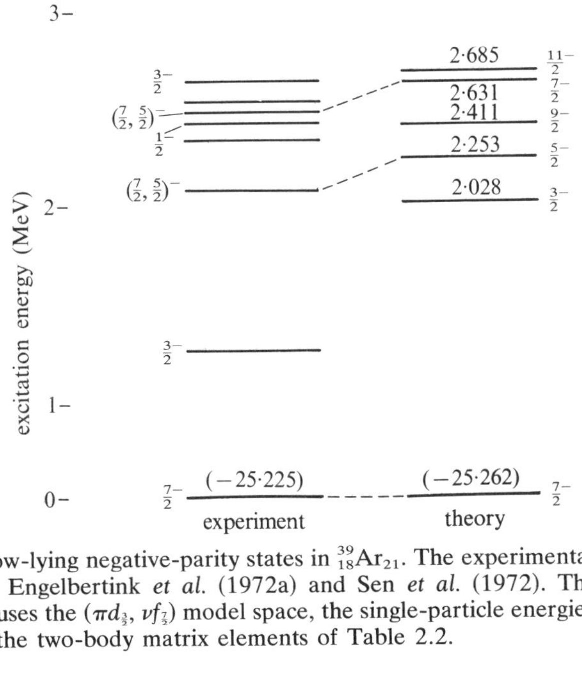

Fig. 2.4. Low-lying negative-parity states in $^{39}_{18}Ar_{21}$. The experimental spectrum is taken from Engelbertink *et al.* (1972a) and Sen *et al.* (1972). The theoretical calculation uses the $(\pi d_{\frac{3}{2}}, \nu f_{\frac{7}{2}})$ model space, the single-particle energies of equation (2.66) and the two-body matrix elements of Table 2.2.

2·423 MeV $\frac{1}{2}^-$ states, this model is able to account for all the levels below 2·7 MeV excitation energy that have so far been identified as having negative parity. The 1·267 MeV state shows up strongly with $l=1$ in the ${}^{38}_{18}\mathrm{Ar}_{20}(d,p){}^{39}_{18}\mathrm{Ar}_{21}$ reaction (Sen *et al.* 1972) and as we shall show in section 5 of this chapter this implies that the wave function for the state has a substantial $1p_{\frac{3}{2}}$ component. When the model space is extended to include the $1p_{\frac{3}{2}}$ orbital (Gloeckner *et al.* 1973) this state can be accounted for and moreover the $[(\pi d_{\frac{3}{2}})^2 \times \nu f_{\frac{7}{2}}]_{\frac{3}{2}}$ level (which is predicted by this calculation to lie at 2·028 MeV) is pushed up closer to the observed position of the second $\frac{3}{2}^-$ level seen at 2·63 MeV.

Thus by use of this simple model one is able to account for many of the observed properties of ${}^{39}_{18}\mathrm{Ar}_{21}$ including its binding energy relative to ${}^{36}_{16}\mathrm{S}_{20}$. There are a number of low-lying positive-parity states seen in this nucleus, and these correspond to excitation of a $d_{\frac{3}{2}}$ nucleon to the $f_{\frac{7}{2}}$ or $p_{\frac{3}{2}}$ orbits (see the discussion concerning core-excited states in ${}^{18}_{8}\mathrm{O}_{10}$ in Chapter 1, section 2.2). We shall discuss the positive-parity states observed at the end of the *ds* shell in some detail in section 4.2 of this Chapter.

2.5. ${}^{39}_{18}\mathrm{Ar}_{21}$ using isospin

An alternative way of calculating the properties ${}^{39}_{18}\mathrm{Ar}_{21}$ would be to consider ${}^{32}_{16}\mathrm{S}_{16}$ to be a closed shell and to use the eigenfunctions

$$[(d_{\frac{3}{2}})^6_{J'T'} \times (f_{\frac{7}{2}})_{\frac{7}{2}\frac{1}{2}}]_{IM;TT_z} = \qquad (2.68)$$

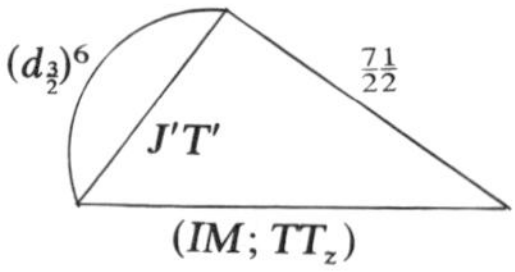

to diagonalize the Hamiltonian. The six $d_{\frac{3}{2}}$ nucleons have $T'_z=1$ and since the neutron $d_{\frac{3}{2}}$ orbit is full a proton cannot be changed into a neutron (i.e. $T_+\phi=0$) so that the configuration has $T'=1$. The only possible J' values are $J'=0$ and 2 consequently, as before,

$$I=\tfrac{3}{2}^-, \tfrac{5}{2}^-, (\tfrac{7}{2}^-)^2, \tfrac{9}{2}^-, \text{ and } \tfrac{11}{2}^-$$

with total isospin $T=T_z=\frac{3}{2}$.

To calculate the Hamiltonian matrix when the basis states of equation (2.68) are used, one has to compute the energy of $(d_{\frac{3}{2}})^6_{J'T'}$ and the interaction between the $d_{\frac{3}{2}}$ and $f_{\frac{7}{2}}$ orbits. The energy of the $(j^n)_{IT}$ configuration is given by

$$\langle (j^n)_{IT\alpha}|\,H\,|(j^n)_{IT\alpha}\rangle = n_\pi\tilde{\varepsilon}_\pi + n_\nu\tilde{\varepsilon}_\nu + \frac{n(n-1)}{2}\sum_{JT'\beta K\tau}\langle j^{n-2}JT'\beta, j^2K\tau|\}j^nIT\alpha\rangle^2 E_{K\tau} \qquad (2.69)$$

where n_π and n_ν are the numbers of protons and neutrons in the orbit $(n_\pi + n_\nu = n)$, $\tilde{\varepsilon}_n$ and $\tilde{\varepsilon}_\pi$ are the neutron and proton single-particle energies, and $E_{K\tau} = E_{K\tau}(jj; jj)$ are the diagonal matrix elements of the residual interaction V in the state $(j^2)_{K\tau}$. If the one-nucleon c.f.p's are known for j^n and j^{n-1}, the double-parentage coefficients can be calculated from equation (A5.39). Alternatively one can use the bootstrap procedure of equation (1.70) and make use of the fact that once the energies

$$E_{JT\alpha\beta}(j^{n-1}) = \langle (j^{n-1})_{JT\alpha} | \, V \, | (j^{n-1})_{JT\beta} \rangle$$

are known the interaction in the n-particle states can be calculated by use of the equation

$$E_{IT\alpha\beta}(j^n) = \frac{n}{n-2} \sum_{J\tau\alpha'\beta'} \langle j^{n-1} J\tau\alpha', j\tfrac{1}{2} | \} j^n IT\alpha \rangle \times \langle j^{n-1} J\tau\beta', j\tfrac{1}{2} | \} j^n IT\beta \rangle E_{J\tau\alpha'\beta'}(j^{n-1}). \qquad (2.70)$$

For the $d_{\frac{3}{2}}$ configuration the one-nucleon isospin c.f.p's have been given by Glaudemans *et al.* (1964) and these can be used with either of the methods just described to yield

$$\begin{aligned} E_{J1}(d_{\frac{3}{2}}^6) &= \langle (d_{\frac{3}{2}})^6_{JT=1} | \, H \, | (d_{\frac{3}{2}})^6_{JT=1} \rangle \\ &= E_{J1} + 2\{\tilde{\varepsilon}_{d\pi} + \tfrac{1}{4}(E_{01} + 3E_{10} + 5E_{21} + 7E_{30})\} \\ &\quad + (4\tilde{\varepsilon}_{d\nu} + E_{01} + 5E_{21}) \end{aligned} \qquad (2.71)$$

where $\tilde{\varepsilon}_{d\pi}$ and $\tilde{\varepsilon}_{d\nu}$ are the single-particle energies relative to the ${}^{32}_{16}S_{16}$ core and

$$E_{KT} = E_{KT}(\tfrac{3}{2}\tfrac{3}{2}; \tfrac{3}{2}\tfrac{3}{2}).$$

Our reason for grouping the terms as we have will become clear later.

To calculate the $d_{\frac{3}{2}} - f_{\frac{7}{2}}$ interaction a one-nucleon fractional-parentage decomposition of the $(d_{\frac{3}{2}})^6_{JT}$ configuration must be carried out and then the $d_{\frac{3}{2}}$ and $f_{\frac{7}{2}}$ nucleons have to be coupled to a resultant spin $K\tau$. By use of equations (2.58) and (2.60) one finds

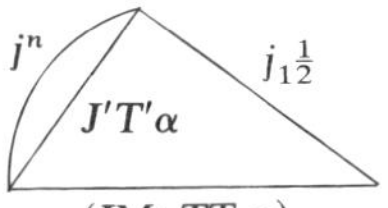

$$= \sum_{J_3T_3\beta K\tau} \sqrt{\{(2J'+1)(2K+1)(2T'+1)(2\tau+I)\}} \times \langle j^{n-1} J_3 T_3 \beta, j\tfrac{1}{2} | \} j^n J'T'\alpha \rangle \times W(J_3 j I j_1; J'K) W(T_3 \tfrac{1}{2} T \tfrac{1}{2}; T'\tau)$$

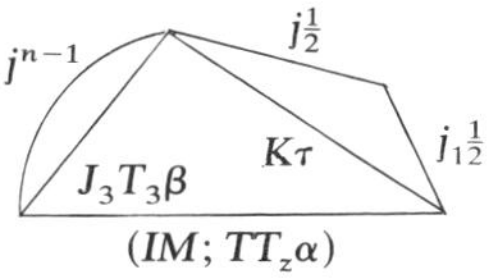

.

If V_{df} denotes the $d_{\frac{3}{2}}f_{\frac{7}{2}}$ interaction one easily shows that

$$E_{df}(JJ') = \langle[(j^n)_{J'T'\alpha'} \times (j_1\tfrac{1}{2})]_{IM;TT_z\alpha'}|\, V_{df}\, |[(j^n)_{J''T''\alpha''} \times (j_1\tfrac{1}{2})]_{IM;TT_z\alpha''}\rangle$$
$$= n\sqrt{\{(2J'+1)(2J''+1)(2T'+1)(2T''+1)\}} \sum_{J_3T_3\beta K\tau} (2K+1)(2\tau+1)$$
$$\times \langle j^{n-1}J_3T_3\beta, j\tfrac{1}{2}|\}j^nJ''T''\alpha''\rangle \langle j^{n-1}J_3T_3\beta, j\tfrac{1}{2}|\}j^nJ'T'\alpha'\rangle$$
$$\times W(J_3jIj_1;\, J'K)W(J_3jIj_1;\, J''K)W(T_3\tfrac{1}{2}T\tfrac{1}{2};\, T'\tau)W(T_3\tfrac{1}{2}T\tfrac{1}{2};\, T''\tau)$$
$$\times \tilde{E}_{K\tau}(jj_1;\, jj_1). \qquad (2.72)$$

The one nucleon c.f.p's for the configuration $(d_{\frac{3}{2}})^6$ are given in Table 2.3. When these are used in conjunction with the values for the Racah coefficients one can calculate the matrix elements of V_{df}. The Hamiltonian matrix $\langle H_I\rangle_{JJ'}$ has the form

$$\langle H_I\rangle_{JJ'} = \{E_{J1}(d^6_{\frac{3}{2}}) + \tilde{\varepsilon}_{f\nu}\}\delta_{JJ'} + E_{df}(JJ')$$

where $E_{J1}(d^6_{\frac{3}{2}})$ is given by equation (2.71), $E_{df}(JJ')$ by equation (2.72), and $\tilde{\varepsilon}_{f\nu}$ is the neutron $f_{\frac{7}{2}}$ single-particle energy relative to the ${}^{32}_{16}S_{16}$ core.

TABLE 2.3

One-nucleon coefficients of fractional-parentage $\langle j^5J'T', j\frac{1}{2}|\}j^6JT\rangle$ for the $(d_{\frac{3}{2}})^6_{JT}$ configuration

	$J'T'$				
JT	$\frac{3}{2}\ \frac{1}{2}$	$\frac{1}{2}\ \frac{1}{2}$	$\frac{5}{2}\ \frac{1}{2}$	$\frac{7}{2}\ \frac{1}{2}$	$\frac{3}{2}\ \frac{3}{2}$
0 1	$\frac{\sqrt{5}}{3}$				$\frac{2}{3}$
2 1	$-\sqrt{\frac{1}{45}}$	$-\sqrt{\frac{1}{15}}$	$-\sqrt{\frac{1}{5}}$	$\sqrt{\frac{4}{15}}$	$\frac{2}{3}$

If one sets

$$\tilde{E}_K = \tfrac{1}{2}(\tilde{E}_{K0}(\tfrac{3}{2}\tfrac{7}{2};\, \tfrac{3}{2}\tfrac{7}{2}) + \tilde{E}_{K1}(\tfrac{3}{2}\tfrac{7}{2};\, \tfrac{3}{2}\tfrac{7}{2})) \qquad (2.73)$$

$$\varepsilon_{f\nu} = \tilde{\varepsilon}_{f\nu} + \tfrac{1}{8}\sum_K (2K+1)\tilde{E}_{K1}(\tfrac{3}{2}\tfrac{7}{2};\, \tfrac{3}{2}\tfrac{7}{2}) \qquad (2.74)$$

$$\varepsilon_{d\pi} = \tilde{\varepsilon}_{d\pi} + \tfrac{1}{4}\sum_{K\tau} (2K+1)E_{K\tau}(\tfrac{3}{2}\tfrac{3}{2};\, \tfrac{3}{2}\tfrac{3}{2}) \qquad (2.75)$$

one finds that $\langle H_I\rangle_{JJ'}$ calculated in this way is identical to the result given in equation (2.67), except that each diagonal matrix element has the added term

$$\varepsilon_0 = 4\tilde{\varepsilon}_{d\nu} + \sum_K (2K+1)E_{K1}(\tfrac{3}{2}\tfrac{3}{2};\, \tfrac{3}{2}\tfrac{3}{2}). \qquad (2.76)$$

One may easily show that these replacements are exactly what one would expect. To start with, in the neutron–proton formulation of section 2.4, the two-particle wave functions used to calculate V_{df} were

$$
\begin{aligned}
\psi_{KM}(jj_1) &= \sum_{mm_1} (jj_1mm_1 \mid KM) a^\dagger_{jm;\frac12-\frac12} a^\dagger_{j_1m_1;\frac12\frac12} |0\rangle \\
&= \tfrac12 \sum_{mm_1} (jj_1mm_1 \mid KM)(\{a^\dagger_{jm;\frac12\frac12} a^\dagger_{j_1m_1;\frac12-\frac12} \\
&\quad + a^\dagger_{jm;\frac12-\frac12} a^\dagger_{j_1m_1;\frac12\frac12}\} \\
&\quad - \{a^\dagger_{jm;\frac12\frac12} a^\dagger_{j_1m_1;\frac12-\frac12} - a^\dagger_{jm;\frac12-\frac12} a^\dagger_{j_1m_1;\frac12\frac12}\}) |0\rangle \\
&= 2^{-\frac12}\{A^\dagger_{KM;T=1,T_z=0}(jj_1) - A^\dagger_{KM;T=0,T_z=0}(jj_1)\} |0\rangle
\end{aligned} \tag{2.77}
$$

where the two-particle creation operators $A^\dagger_{KM;TT_z}$ are given by equation (2.22). Since

$$\tilde{E}_K(\tfrac32\tfrac72;\tfrac32\tfrac72) = \langle \psi_{KM}(\tfrac32\tfrac72)| V |\psi_{KM}(\tfrac32\tfrac72)\rangle$$

equation (2.73) follows provided V commutes with T^2.

As far as the $f_{\frac72}$ single-particle energies are concerned, $\varepsilon_{f\nu}$ in section 2.4 was measured relative to the ${}^{36}_{16}S_{20}$ core whereas $\tilde{\varepsilon}_{f\nu}$ is with respect to ${}^{32}_{16}S_{16}$. Since ${}^{36}_{16}S_{20}$ has four additional $d_{\frac32}$ neutrons it follows that $\varepsilon_{f\nu}$ should be identical to $\tilde{\varepsilon}_{f\nu}$ plus the added attraction between the $f_{\frac72}$ neutron and the four $d_{\frac32}$ neutrons. Because a closed shell has angular momentum zero and since the one-nucleon c.f.p. $\langle j^{2j} j, j|\}j^{2j+1}0\rangle = 1$ one sees that

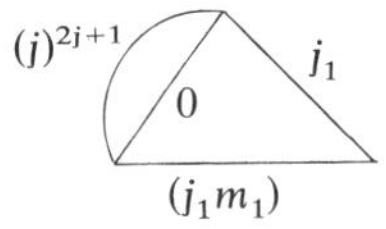

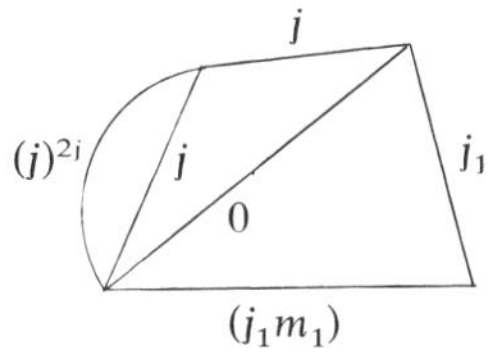

$$\sum_K (-1)^{j+j_1-K}\left\{\frac{(2K+1)}{(2j+1)(2j_1+1)}\right\}^{\frac12}$$

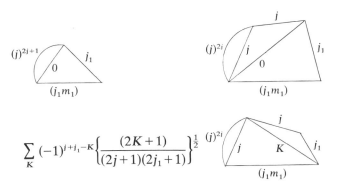

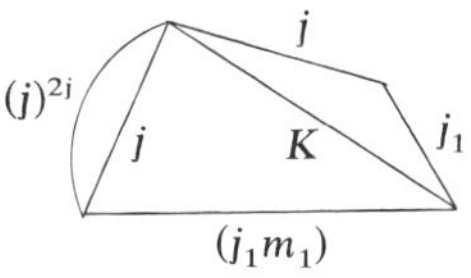

Since there are $(2j+1)$ core particles the energy that should be added to take into account the interaction between j_1 and the core is

$$\langle[(j)_0^{2j+1} \times j_1]_{j_1m_1}| V_{jj_1} |[(j)_0^{2j+1} \times j_1]_{j_1m_1}\rangle = \frac{1}{2j_1+1} \sum_K (2K+1) E_K(jj_1; jj_1). \tag{2.78}$$

In this example both the $f_{\frac{7}{2}}$ and $d_{\frac{3}{2}}$ particles are neutrons so the added energy should only depend on the $T=1$ matrix elements. For the extra-core $d_{\frac{3}{2}}$ proton the same argument applies except that in this case the particles can interact in both $T=0$ and $T=1$ states. Thus the single-particle energy replacements of equations (2.74) and (2.75) are precisely what one would expect.

Finally in section 2.4 we calculated energies relative to ${}^{36}_{16}S_{20}$ whereas in this section the energies are relative to ${}^{32}_{16}S_{16}$. Thus one would expect every computed energy in this section to differ from the analogous quantity of section 2.4 by the energy of the configuration $(\nu d_{\frac{3}{2}})^4_0$. By use of the two-particle c.f.p's for the seniority-zero closed shell given in equation (A5.53) one finds

$$\langle (j)_0^{2j+1} | H | (j)_0^{2j+1} \rangle = (2j+1)\varepsilon_j + \sum_K (2K+1) E_{K1}(jj;jj) \qquad (2.79)$$

and this is precisely the energy ε_0 of equation (2.76) when j is set equal to $\frac{3}{2}$.

Thus the two methods of calculation give identical results when one interprets the physics correctly. Of course, for the negative parity states in ${}^{39}_{18}Ar_{21}$ the method outlined in section 2.4 is easier because the $d_{\frac{3}{2}}$ neutrons form a closed shell. However, there are situations where the methods given in this section would be more useful. For example, if one wanted to calculate the positive parity states in this nucleus one would have to deal with a nucleon excited out of the $d_{\frac{3}{2}}$ level to the $f_{\frac{7}{2}}$ orbit; in other words the configuration of interest would be $(d_{\frac{3}{2}})^5(f_{\frac{7}{2}})^2$. Since either a neutron or a proton can be excited, the calculation is most easily done by classifying the $d_{\frac{3}{2}}$ states by $(d_{\frac{3}{2}})^5_{JT}$ and the $f_{\frac{7}{2}}$ nucleons by $(f_{\frac{7}{2}})^2_{J'T'}$.

3. Weak-coupling model

Since nuclear forces have a short range, one would expect that the interaction between nucleons in the same shell would be stronger than that between particles in different shells. To see that this expectation is borne out experimentally we compare the interaction between two $(0d, 1s)$ shell nucleons with that between a $(0d, 1s)$ and a $0p$ shell particle. In section 2.2, Chapter 1 we estimated this latter interaction (see equation (1.64)) and found that to fit the observed excitation energy of the $\frac{1}{2}^-$ level in ${}^{19}_{9}F_{10}$ the average interaction between a $p_{\frac{1}{2}}$ hole and an (s, d) particle was

$$E_{p-ds} = 0{\cdot}21 \text{ MeV}.$$

The interaction of two particles in the $(0d, 1s)$ shell can be estimated from the binding energy of ${}^{18}_{8}O_{10}$. If one takes out the effect of the single-particle energies one finds

$$E_{ds} = -3{\cdot}90 \text{ MeV}.$$

Of course, this is the strongest $T=1$ matrix element, but even if one considers the 3·55 MeV 4^+ level in ${}^{18}_{8}O_{10}$ one finds $E_{ds} = -0{\cdot}35$ MeV which is almost a factor of 2 greater than E_{p-ds}.

Because of this, a convenient way to approach problems in which one group of nucleons in a given shell interacts with a second group in a different shell is first to diagonalize the Hamiltonian of each group separately and then to use the resulting eigenfunctions as basis states to solve the entire problem. That is

$$H = H_1 + H_2 + H_{12} \tag{2.80}$$

would be diagonalized using the basis functions

$$\Psi_{IM} = [\phi_{J_1\alpha} \times \psi_{J_2\beta}]_{IM} \tag{2.81}$$

where

$$H_1\phi_{J_1M_1\alpha} = E_{J_1\alpha}\phi_{J_1M_1\alpha}$$

$$H_2\psi_{J_2M_2\beta} = E_{J_2\beta}\psi_{J_2M_2\beta}$$

and H_{12} is the interaction between the two shells. In order to diagonalize H one uses the basis states of equation (2.81) to calculate matrix elements of H_{12}, and in principle this may be done by the methods given in Chapter 1, section 4 or those of the preceding section of this chapter.

However, the number of basis states that can be generated by use of all $\phi_{J_1M_1\alpha}$ and $\psi_{J_2M_2\beta}$ can become extremely large as soon as one gets away from any but the simplest nuclei. For example, if one wishes to calculate the negative-parity states in ${}^{19}_{9}F_{10}$ that arise when one nucleon is excited out of the $0p$ shell up to the $(0d_{\frac{5}{2}}, 1s_{\frac{1}{2}}, 0d_{\frac{3}{2}})$ shell, a complete calculation would involve coupling either the $p_{\frac{1}{2}}$ or $p_{\frac{3}{2}}$ hole to all the allowable states of $(0d, 1s)^4$. The states of the $(sd)^n$ configuration have been enumerated by Harvey and Sebe (1968), and by use of the simple counting procedure given in Chapter 1, section 1.3, one concludes that for $I \leq \frac{9}{2}$ the number of $T=\frac{1}{2}$ states is

$$I = (\tfrac{1}{2}^-)^{329}, (\tfrac{3}{2}^-)^{565}, (\tfrac{5}{2}^-)^{651}, (\tfrac{7}{2}^-)^{596}, (\tfrac{9}{2}^-)^{451}$$

where the superscript indicates the dimensionality of the Hamiltonian matrix that must be constructed and diagonalized! Clearly an approximation procedure is called for and the logical one is to restrict the basis states by considering only those $\phi_{J_1M_1\alpha}$ and $\psi_{J_2M_2\beta}$ that correspond to the lowest energy states of the first and second groups of particles. How many states one must consider depends on the number of states with given IM one wishes to describe. In this section we consider results that can be obtained when a severe truncation of the basis is assumed.

3.1. *Centre-of-gravity theorem*

The simplest situation corresponds to a single nucleon moving outside a core whose low-lying states are described by the configuration $(j^n)_{J\beta}$ with energies $E_{J\beta}$. In this case the eigenfunctions of the composite system are

$\psi_{IM}(j_1, J\beta) =$ 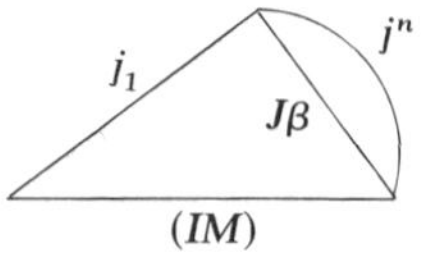

and the diagonal matrix elements of the Hamiltonian associated with these $\Psi_{IM}(j_1, J\beta)$ can be calculated by use of equation (1.96)

$$\begin{aligned} E_I(j_1, J\beta) &= \langle \Psi_{IM}(j_1, J\beta)|\, H\, |\Psi_{IM}(j_1, J\beta)\rangle \\ &= \varepsilon_{j_1} + E_{J\beta} + n(2J+1) \sum_{J_3\beta_3 K} (2K+1)\langle j^{n-1}J_3\beta_3, j|\}j^nJ\beta\rangle^2 \\ &\quad \times \{W(j_1 jIJ_3; KJ)\}^2 E_K(j_1 j; j_1 j) \end{aligned}$$

where $E_{J\beta}$ is the energy of the configuration $(j)^n_{J\beta}$ including the single-particle energy. If every pair of states $\Psi_{IM}(j_1, J\beta)$ and $\Psi_{IM}(j_1', J'\beta')$ has the property that

$$|E_I(j_1, J\beta) - E_I(j_1', J'\beta')| \gg |\langle \Psi_{IM}(j_1, J\beta)|\, H\, |\Psi_{IM}(j_1', J'\beta')\rangle| \tag{2.82}$$

then, to a good approximation, $E_I(j_1, J\beta)$ will be an eigenvalue of H. This relationship is, however, only likely to hold if n is even (the states of j^n for n odd lie close in energy; see, for example, Fig. 1.8) and if the states j_1 that the extra-core nucleon can move in are well separated in energy.

If use is made of the fact that

$$\sum_I (2I+1) = (2j_1+1)(2J+1)$$

together with the summation relationship in equation (A4.16) for the Racah coefficients and the normalization condition for the fractional-parentage coefficients in equation (A5.4) one finds that

$$\begin{aligned} \bar{E}(j_1, J\beta) &= \frac{\sum_I (2I+1) E_I(j_1, J\beta)}{\sum_I (2I+1)} \\ &= \varepsilon_{j_1} + E_{J\beta} + \frac{n}{(2j+1)(2j_1+1)} \sum_K (2K+1) E_K(j_1 j; j_1 j) \\ &\quad \times \sum_{J_3\beta_3} \langle j^{n-1}J_3\beta_3, j|\}j^nJ\beta\rangle^2 \\ &= \varepsilon_{j_1} + E_{J\beta} + \frac{n}{(2j+1)(2j_1+1)} \sum_K (2K+1) E_K(j_1 j; j_1 j) \end{aligned} \tag{2.83}$$

where $\bar{E}(j_1, J\beta)$ is called the centre-of-gravity of the states $E_I(j_1, J\beta)$. Since the sum over K in this equation is independent of $(J\beta)$ it follows that

$$\bar{E}(j_1, J'\beta') - \bar{E}(j_1, J\beta) = E_{J'\beta'} - E_{J\beta}; \tag{2.84}$$

in other words the difference in energy between the centre-of-gravity of the levels described by $[j_1 \times (j)^n_{J'\beta'}]_{IM}$ and $[j_1 \times (j)^n_{J\beta}]_{IM}$ is identical to the energy difference between the two n particle states to which j_1 is coupled (Lawson and Uretsky 1957, De Shalit 1961).

Although we have explicitly derived this theorem for the case of a single nucleon in the state j_1, the derivation goes through for $(j_1)^{n'}_{J_1\beta_1}$ and in this case gives

$$\bar{E}(J_1'\beta_1', J'\beta') - \bar{E}(J_1\beta_1, J\beta) = (E_{J_1'\beta_1'} - E_{J_1\beta_1}) + (E_{J'\beta'} - E_{J\beta}). \tag{2.85}$$

For $n' = 2j_1$ there is only one state of the configuration $(j_1)^{n'}$. In consequence $(J_1'\beta_1') = (J_1\beta_1)$ and this result becomes identical to equation (2.84). However, for $n' \neq 1$ or $2j_1$ the result has little practical importance. For n' odd the states of $(j_1)^{n'}_{J_1\beta_1}$ are usually not sufficiently separated in energy to make equation (2.82) valid. Moreover, for n and n' both even there is little difference in the energy of the state $[(j_1)^{n'}_{J_1} \times (j)^n_J]_{IM}$ and the state $[(j_1)^{n'}_J \times (j)^n_{J_1}]_{IM}$ when J and J_1 correspond to low-lying states.

There are not many situations where the core states $J\beta$ and $J'\beta'$ of equation (2.84) can actually be considered to arise from the configuration j^n. In the more complicated case where the n particles are distributed over several single-particle states, one can generalize the fractional-parentage decomposition

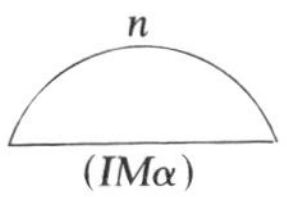

$$\sum_{J_3\beta_3 j'} \langle (n-1)J_3\beta_3, j' | \} nI\alpha \rangle$$

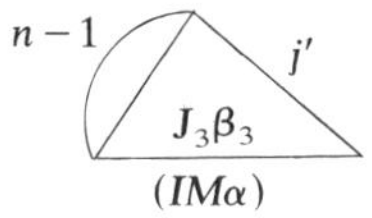

and the generalized-parentage coefficients satisfy the orthonormality condition

$$\sum_{J_3\beta_3 j'} \langle (n-1)J_3\beta_3, j' | \} nI\alpha \rangle \langle (n-1)J_3\beta_3, j' | \} nI\alpha' \rangle = \delta_{\alpha\alpha'}. \tag{2.86}$$

In this case equation (2.83) becomes

$$\bar{E}(j_1, J\beta) = \varepsilon_{j_1} + E_{J\beta} + \frac{n}{(2j_1+1)} \sum_K (2K+1) \times \sum_{J_3\beta_3 j} \frac{\langle (n-1)J_3\beta_3, j | \} nJ\beta \rangle^2}{(2j+1)} E_K(j_1 j; j_1 j) \tag{2.87}$$

and because of the sum over j, the last term in the equation does not cancel out in the energy difference $\bar{E}(j_1, J'\beta') - \bar{E}(j_1, J\beta)$. However, if this term does not vary appreciably with $J\beta$, the centre-of-gravity relationship in equation (2.84) should be approximately correct. In the next two sections we present experimental data that indicate that this is often true.

3.2. Negative-parity states in $^{19}_{9}F_{10}$

As discussed in section 2.2, Chapter 1, the $\frac{1}{2}^-$ state at 110 keV excitation energy in $^{19}_{9}F_{10}$ can be interpreted as a $p_{\frac{1}{2}}$- proton coupled to the $^{20}_{10}Ne_{10}$ ground state (see Fig. 1.6). Since this is the case, there should also be negative-parity doublets corresponding to the $p_{\frac{1}{2}}$-particle coupling to various excited states of the $^{20}_{10}Ne_{10}$ core. In the extreme weak-coupling

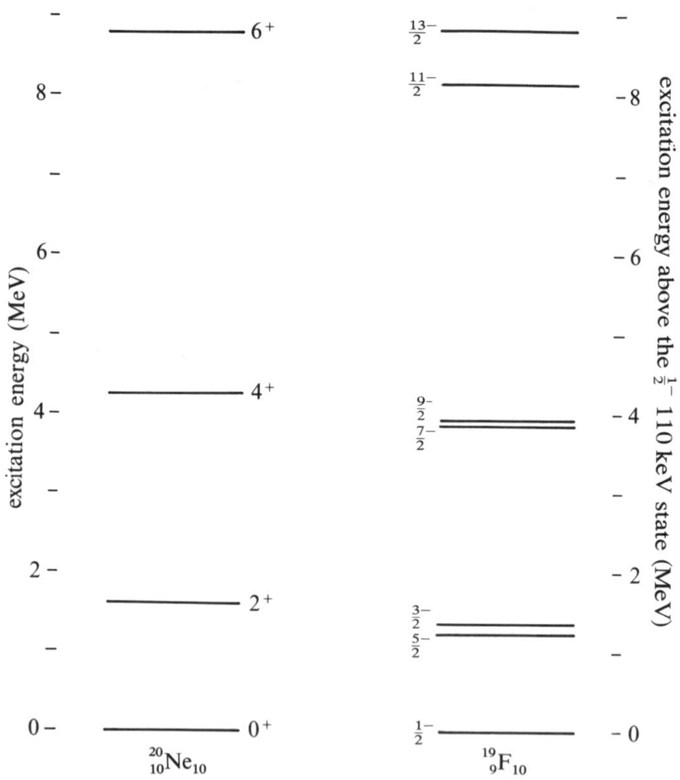

Fig. 2.5. The ground-state rotational band in $^{20}_{10}Ne_{10}$ and the yrast negative-parity states in $^{19}_{9}F_{10}$. For illustrative purposes the spectra have been plotted so that the 110 keV $\frac{1}{2}^-$ state in $^{19}_{9}F_{10}$ and the 0^+ ground state of $^{20}_{10}Ne_{10}$ have the same energy. The data have been taken from the compilation of Ajzenberg-Selove (1972).

limit the wave functions of these states would be

$$\Psi_{IM} = [\pi p_{\frac{1}{2}} \times \phi_J({}^{20}_{10}\text{Ne}_{10})]_{I=J\pm\frac{1}{2},M}. \tag{2.88}$$

In Fig. 2.5 we show the 0^+, 2^+, 4^+, and 6^+ members of the ground-state rotational band of ${}^{20}_{10}\text{Ne}_{10}$ and on the same figure the yrast negative-parity states in ${}^{19}_{9}\text{F}_{10}$ (Ajzenberg-Selove 1972). Clearly there are negative-parity doublets of the appropriate spins with excitation energies quite near to those of the 2^+, 4^+, and 6^+ states (Zamick 1965, Arima *et al.* 1967).

In Table 2.4 we compare the excitation energies of the centre-of-gravity of the various multiplets with the theoretical prediction based on

TABLE 2.4

The yrast negative-parity states in ${}^{19}_{9}\text{F}_{10}$

State in ${}^{20}_{10}\text{Ne}_{10}$		Observed doublet		Centre-of-gravity excitation energy (MeV)	
Spin	Excitation energy, E_J (MeV)	Spin	Excitation energy (MeV)	Observed	Calculated
2^+	1·63	$\frac{5}{2}^-$ $\frac{3}{2}^-$	1·35 1·46	1·39	1·74
4^+	4·25	$\frac{7}{2}^-$ $\frac{9}{2}^-$	4·00 4·04	4·02	4·36
6^+	8·79	$\frac{11}{2}^-$ $\frac{13}{2}^-$	8·25 8·91	8·61	8·90

Since the $\frac{1}{2}^-$ state lies at 110 keV the predicted value of the excitation energy of the centre of gravity of a doublet is $\bar{E}(p_{\frac{1}{2}}, J) = (E_J + 0{\cdot}110)$ MeV.

equation (2.84). The fact that the observed levels do not satisfy precisely the centre-of-gravity rule is, of course, to be expected since

(a) the yrast positive-parity levels are unlikely to have the pure configuration $(d_{\frac{5}{2}})^4_J$ so that the last term in equation (2.87) is not independent of J and hence the centre-of-gravity rule can only be expected to be approximate.
(b) a drastic truncation of the model space has been made. For example, we have assumed that the yrast $\frac{5}{2}^-$ can be described by the one-component wave function of equation (2.88) with $J = 2^+$ whereas the complete wave function would have 651 components!

Of course, one can improve the agreement between theory and experiment by extending the model space, and this has been done by Benson and Flowers (1969). However, the qualitative features and, more important, a simple understanding of the structure of the yrast negative-parity levels can be obtained from this simple model.

3.3. Inelastic scattering; the positive-parity septuplet in $^{209}_{83}\mathrm{Bi}_{126}$

The probability that inelastic scattering ((*p*, *p*′), (*d*, *d*′) etc.) will populate a final state with spin λ in an even–even nucleus depends on the nuclear matrix element

$$\langle \psi_{J_c=\lambda M'} |\, \mathrm{G}_{\lambda\mu} \,| \psi_{J_c=0M} \rangle$$

where $\mathrm{G}_{\lambda\mu}$ is a tensor operator of rank λ which is a function of the nuclear coordinates. The scattering cross-section will be proportional to the square of this matrix element summed over the final spin states and averaged over initial spin states. From the definition of the reduced matrix element given in equation (A2.7), it follows that

$$\frac{\mathrm{d}\sigma}{\mathrm{d}\Omega}(0 \rightarrow \lambda) \sim \sum_{MM'\mu} |\langle \psi_{J_c=\lambda M'} |\, \mathrm{G}_{\lambda\mu} \,| \psi_{J_c=0M} \rangle|^2$$

$$= (2\lambda+1)\, |\langle \psi_{J_c=\lambda} \|\, \mathrm{G}_\lambda \,\| \psi_{J_c=0} \rangle|^2. \qquad (2.89)$$

In the neighbouring odd-even nucleus there may be low-lying states corresponding to the coupling of a single particle in the orbit j to the various core states J_c. In the extreme weak-coupling limit the wave functions of these levels will be

$\Psi_{IM} =$ 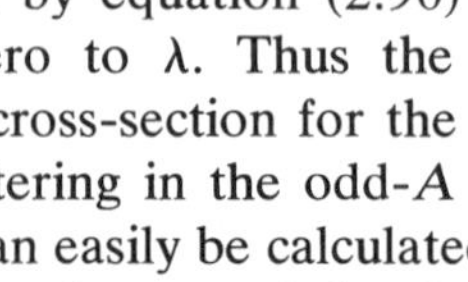 (2.90)

If we assume that the ground state of the odd–even nucleus has $J_c = 0$, the excitation of states described by equation (2.90) with $J_c = \lambda$ will again involve changing J_c from zero to λ. Thus the same reduced matrix element that comes into the cross-section for the even–even system will also govern the inelastic scattering in the odd-*A* nucleus. The angular-momentum factors involved can easily be calculated diagrammatically. To carry out this calculation we make use of the definition of the reduced matrix element in terms of diagrams given in equation (1.54). Since the operator $\mathrm{G}_{\lambda\mu}$ that induces the inelastic scattering operates on the 'core

coordinates' only

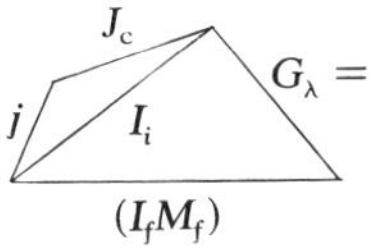

$$G_\lambda = \sum_{J_c'} \sqrt{\{(2I_i+1)(2J_c'+1)\}}\, W(jJ_cI_f\lambda\,;\,I_iJ_c')$$

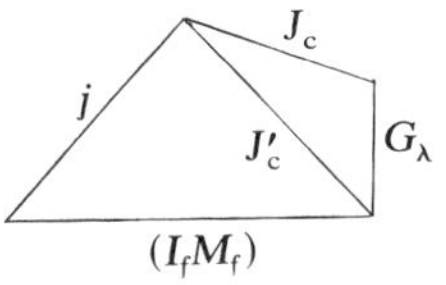

$$= \sum_{J_c'} \sqrt{\{(2I_i+1)(2J_c'+1)\}}\, W(jJ_cI_f\lambda\,;\,I_iJ_c')\langle\psi_{J_c'}\|G_\lambda\|\,\psi_{J_c}\rangle$$

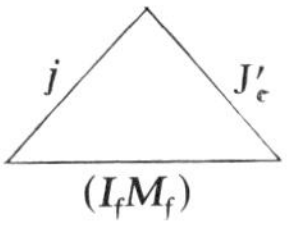

$$\quad . \qquad (2.91)$$

Since we are interested in calculating the transition probability for going from the initial state in which $J_c = 0$ in equation (2.90) to one in which $J_c = \lambda$, it follows that the reduced matrix element for the transition is precisely the coefficient of

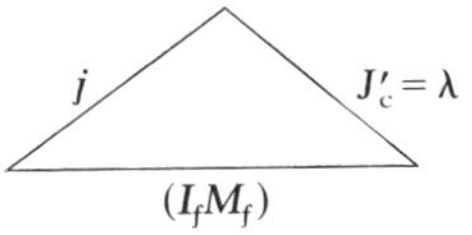

in equation (2.91). Thus

$$\begin{aligned}
\langle[\phi_j \times \psi_{J_c=\lambda}]_{I_fM_f}|\, \mathrm{G}_{\lambda\mu}\, |[\phi_j \times \psi_{J_c=0}]_{I_iM_i}\rangle
&= [\sqrt{\{2I_i+1)(2\lambda+1)\}}\, W(j0I_f\lambda\,;\,I_i\lambda)\langle\psi_{J_c'=\lambda}\|\,\mathrm{G}_\lambda\,\|\psi_{J_c=0}\rangle] \\
&\quad \times (I_i\lambda M_i\mu \mid I_fM_f) \\
&= \langle\psi_{J_c'=\lambda}\|\,\mathrm{G}_\lambda\,\|\psi_{J_c=0}\rangle\delta_{jI_i}(I_i\lambda M_i\mu \mid I_fM_f).
\end{aligned}$$

To calculate the cross-section one must sum over final spin states and average over initial spin states. By use of the sum rule for the Clebsch-Gordan coefficients (equation (A1.23)) one finds

$$\frac{\mathrm{d}\sigma}{\mathrm{d}\Omega}\{j \to (j\lambda)I_f\} \sim \frac{(2I_f+1)}{(2j+1)}\,|\langle\psi_{J_c'=\lambda}\|\,\mathrm{G}_\lambda\,\|\psi_{J_c=0}\rangle|^2.$$

In the limit that the states of the multiplet are degenerate and have excitation energy equal to that in the even–even nucleus the phase-space factors that come into the cross-section for the even-A and odd-A nuclei will be the same so that

$$\frac{\mathrm{d}\sigma}{\mathrm{d}\Omega}\{j \to (j\lambda)I_f\} = \left\{\frac{(2I_f+1)}{(2j+1)(2\lambda+1)}\right\}\frac{\mathrm{d}\sigma}{\mathrm{d}\Omega}(0 \to \lambda). \qquad (2.92)$$

Even when the levels are not degenerate this relationship should remain true to good approximation, provided the energy of the incident particle is much greater than the energy splitting of the levels from their unperturbed position. Since

$$\sum_{I_f=|\lambda-j|}^{\lambda+j} (2I_f+1)=(2j+1)(2\lambda+1)$$

it follows that the sum of the cross-sections to the multiplet in the odd-A nucleus is equal to the cross-section in the neighbouring even–even system. Furthermore, since the scattering to the state I is proportional to $(2I_f+1)$, a measurement of the cross-section allows a determination of the final-state spin.

This result has been used to interpret data in several regions of the periodic table. Probably the most spectacular example of its use has been for nuclei one particle removed from the doubly magic $Z=82$, $N=126$ core. The first excited state of ${}^{208}_{82}\text{Pb}_{126}$ is a 3^- level at 2.615 MeV. Since the ground state of ${}^{209}_{83}\text{Bi}_{126}$ is an $h_{\frac{9}{2}}$ level, one would expect a septuplet of positive-parity levels to occur at approximately 2·615 MeV. The spins of these states would range from $\frac{3}{2}^+$ to $\frac{15}{2}^+$. In Table 2.5 we tabulate the

TABLE 2.5

Excitation energies and assumed spins of the $[h_{\frac{9}{2}}\times 3^-]_{I_f}$ *septuplet in* ${}^{209}_{83}\text{Bi}_{126}$

Excitation energy (MeV)	2·489	2·560	2·580	2·597	2·615	2·739
Assumed spin	$\frac{3}{2}^+$	$\frac{9}{2}^+$	$\frac{7}{2}^+$	$\frac{11}{2}^+$ and $\frac{13}{2}^+$	$\frac{5}{2}^+$	$\frac{15}{2}^+$
$R_{\text{th}}(I_f)$	5·7	14·3	11·4	17·1+20·0	8·6	22·9
$R_{\text{exp}}(I_f)$	4·1	15·1	12·7	37·7	8·7	21·7

$R_{\text{th}}(I_f)$ is the theoretical value of the relative cross-section to the various states given by equation (2.93a). This quantity should be compared with the experimental value $R_{\text{exp}}(I_f)$ given by equation (2.93b).

results of Hafele and Woods (1966) who measured the (p, p') cross-sections to these levels using 21 MeV protons. In this table the quantity

$$R_{\text{th}}(I_f)=\left\{\frac{(2I_f+1)}{(2j+1)(2\lambda+1)}\right\}\times 100=\left\{\frac{(2I_f+1)}{70}\right\}\times 100 \qquad (2.93\text{a})$$

and

$$R_{\text{exp}}(I_f)=\left\{\frac{(\mathrm{d}\sigma/\mathrm{d}\Omega)(\frac{9}{2}^-\to I_f)}{\sum_I (\mathrm{d}\sigma/\mathrm{d}\Omega)(\frac{9}{2}^-\to I)}\right\}\times 100. \qquad (2.93\text{b})$$

Since an anomalously large value was obtained for the inelastic cross-section to the 2·597 MeV state, Hafele and Woods assumed the state was a doublet with spins of $\frac{11}{2}^+$ and $\frac{13}{2}^+$. Subsequent work by Hertel *et al.* (1969) in which the multiplet was Coulomb excited by ^{16}O ions and the gamma decay of the states was observed, confirmed this level was a doublet split by approximately 2 keV with the $\frac{13}{2}^+$ level lying above the $\frac{11}{2}^+$ state.

The centre of gravity of the $^{209}_{83}Bi_{126}$ septuplet is 2·618 MeV. This is in excellent agreement with the energy (2·615 MeV) of the 3^- state in $^{208}_{82}Pb_{126}$. Other properties of these states have been measured, and Hamamoto (1969, 1974) has found that the weak-coupling model gives an excellent description of the data.

3.4. *More detailed calculations for* $^{63}_{29}Cu_{34}$

So far we have presented results based on the extreme weak-coupling model in which the wave function describing a nuclear state is assumed to have only one component. We shall now calculate the properties of $^{63}_{29}Cu_{34}$ using a model space that includes several proton single-particle states and several neutron core states. We take $Z = 28$ to be a closed shell and restrict the one valence proton to the $1p_{\frac{3}{2}}$, $0f_{\frac{5}{2}}$ or $1p_{\frac{1}{2}}$ orbit. In addition, it is assumed that the low-lying states of $^{62}_{28}Ni_{34}$ are due to neutron excitation alone, and that it is to these states that the valence proton couples. Thus in equations (2.80) and (2.81) H_1, $E_{J_1\alpha}$, and $\phi_{J_1M_1\alpha}$ describe the single-particle proton states ϕ_{jm} with energies ε_j, and H_2, $E_{J_2\beta}$, and $\psi_{J_2M_2\beta}$ characterize the neutron core states.

The interaction H_{12} between the proton and neutron core is written as

$$H_{12} = \sum_{n\lambda} \sqrt{(2\lambda+1)}\, V_\lambda(r_p, r_n)[\underset{\sim}{G}_\lambda(p) \times \underset{\sim}{G}_\lambda(n)]_{00}$$

$$= \sum_{n\lambda} V_\lambda(r_p, r_n) \underset{\sim}{G}_\lambda(p) \,.\, \underset{\sim}{G}_\lambda(n)$$

where p stands for the space-spin coordinate of the proton and n for any one of the core neutrons. $\underset{\sim}{G}_\lambda(p)(\underset{\sim}{G}_\lambda(n))$ is a tensor operator of rank λ that acts on the proton (neutron) coordinates, $V_\lambda(r_p, r_n)$ is the strength of the interaction in the multipole mode λ and the dot product $\underset{\sim}{G}_\lambda(p) \,.\, \underset{\sim}{G}_\lambda(n)$ is defined as

$$\underset{\sim}{G}_\lambda(p) \,.\, \underset{\sim}{G}_\lambda(n) = (-1)^\lambda \sqrt{(2\lambda+1)}[\underset{\sim}{G}_\lambda(p) \times \underset{\sim}{G}_\lambda(n)]_{00}.$$

In order to make a minimum number of assumptions about the structure of the core states we consider only the $\lambda = 0$, 1, and 2 components of H_{12} and parameterize them as follows: The $\lambda = 0$ part (known as the monopole part of the force) has diagonal matrix elements only. If the matrix elements of $\underset{\sim}{G}_0(n)$ are approximately the same for all the core states (this would be rigorously true if the states arose from the configuration (j^n)), the effect of the monopole interaction could be lumped into the energies ε_j. We make this assumption and pass on to the $\lambda=1$ multipole known as the dipole part of the force. From a macroscopic point of view, the characteristic vector associated with the core is its angular momentum $\underline{\mathrm{J}}_\mathrm{c}$. Therefore for the dipole–dipole potential we take the form $\underline{\mathrm{j}}\,.\,\underline{\mathrm{J}}_\mathrm{c}$, where $\underline{\mathrm{j}}$ is the angular-momentum operator for the particle. In a similar way the $\lambda=2$ tensor associated the core (which is best described macroscopically) is the quadrupole operator. Thus for the interaction Hamiltonian we take the form first used by Thankappan and True (1965)

$$\begin{aligned} H_{12} &= -\zeta \underline{\mathrm{j}}\,.\,\underline{\mathrm{J}}_\mathrm{c} - \eta \underline{\mathrm{Q}}_\mathrm{p}\,.\,\underline{\mathrm{Q}}_\mathrm{c} \\ &= -\zeta \underline{\mathrm{j}}\,.\,\underline{\mathrm{J}}_\mathrm{c} - \eta\sqrt{5}[\mathrm{Q}_\mathrm{p}\times\mathrm{Q}_\mathrm{c}]_{00} \end{aligned} \tag{2.94}$$

where ζ and η are numbers to be determined, $(\mathrm{Q}_\mathrm{c})_\mu = \sum_n r_n^2 Y_{2\mu}(\theta_n, \phi_n)$ with $Y_{2\mu}$ the spherical harmonic given by equation (1.5) and a similar definition holds for $(\mathrm{Q}_\mathrm{p})_\mu$.

Matrix elements of the Hamiltonian can be calculated simply by graphical methods. Thus by use of equation (1.90) we find

$\underline{\mathrm{Q}}_\mathrm{p}\,.\,\underline{\mathrm{Q}}_\mathrm{c}$ 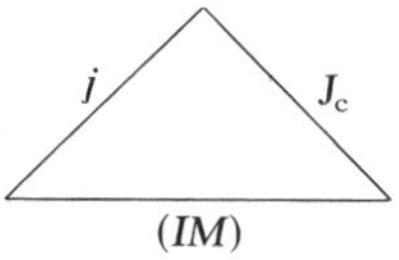

$=\sqrt{5}$ 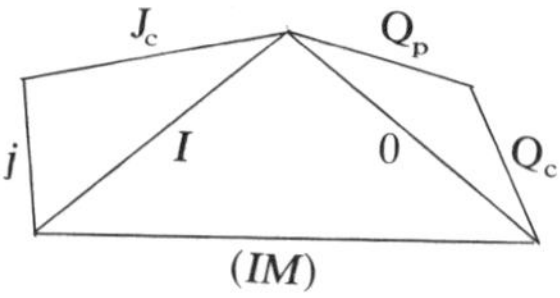

$$=\sqrt{\{5(2I+1)\}}\sum_{j'J_\mathrm{c}'}\sqrt{\{(2j'+1)(2J_\mathrm{c}'+1)\}}\begin{Bmatrix} j & J_\mathrm{c} & I \\ 2 & 2 & 0 \\ j' & J_\mathrm{c}' & I \end{Bmatrix}$$

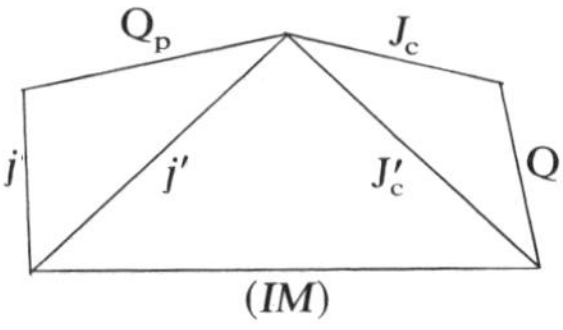

$$=\sum_{j'J_\mathrm{c}'}(-1)^{j'+J_\mathrm{c}-I}\sqrt{\{(2j'+1)(2J_\mathrm{c}'+1)\}}\,W(jJ_\mathrm{c}j'J_\mathrm{c}';\,I2)$$

$$\times\langle j'\|\,\mathrm{Q}_\mathrm{p}\,\|j\rangle\,\langle J_\mathrm{c}'\|\,\mathrm{Q}_\mathrm{c}\,\|J_\mathrm{c}\rangle$$ 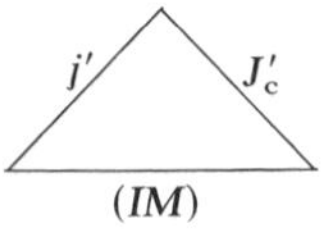

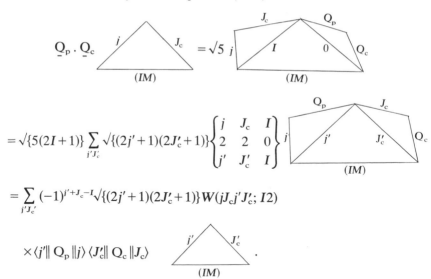

.

In writing the last line of this equation we have made use of the diagrammatic definition of the reduced matrix element given in equation (1.54), and the simple form the $9j$ coefficient takes when one argument is zero given by equation (A4.33). Since the total angular-momentum operator is $\underline{\mathrm{I}}=\underline{\mathrm{j}}+\underline{\mathrm{J}}_\mathrm{c}$, the $\underline{\mathrm{j}}\,.\,\underline{\mathrm{J}}_\mathrm{c}$ term can be evaluated by the method used in deducing equation (2.26). Thus

$$\langle[\phi_{j'}\times\psi_{J_c'}]_{IM}|\,H\,\|[\phi_j\times\psi_{J_c}]_{IM}\rangle$$
$$=\left(\varepsilon_j+E_{J_c}-\frac{\zeta}{2}\{I(I+1)-j(j+1)-J_c(J_c+1)\}\right)\delta_{j'j}\delta_{J_c'J_c}$$
$$-(-1)^{j'+J_c-I}\eta\sqrt{\{(2j'+1)(2J_c'+1)\}}W(jJ_cj'J_c';I2)$$
$$\times\langle j'\|\,Q_p\,\|j\rangle\langle J_c'\|\,Q_c\,\|J_c\rangle. \tag{2.95}$$

De Jager and Boeker (1973) have used this model to analyze all the known properties of the levels below 1·6 MeV excitation energy in $^{63}_{29}Cu_{34}$. They assume only two neutron core states need be considered—the 0^+ ground state and 2^+ state at 1·173 MeV (Fanger *et al.* 1970) of $^{62}_{28}Ni_{34}$. The matrix element of the quadrupole moment operator for the proton was calculated by use of equation (A2.23) and the radial matrix element was evaluated using harmonic-oscillator wave functions with $\hbar\omega=41/A^{\frac{1}{3}}$ (see equation (1.11a))

$$\int R_{0f}r^2R_{0f}r^2\,dr=\int R_{1p}r^2R_{1p}r^2\,dr=\tfrac{9}{2}\hbar/m\omega$$

$$\int R_{0f}r^2R_{1p}r^2\,dr=-\sqrt{14}\hbar/m\omega.$$

If one does not consider the ground-state binding energy, then only five parameters remain to be determined, namely $(\varepsilon_{\frac{5}{2}}-\varepsilon_{\frac{3}{2}})$, $(\varepsilon_{\frac{1}{2}}-\varepsilon_{\frac{3}{2}})$, ζ, $\eta\langle\psi_{J_c=2}\|\,Q_c\,\|\psi_{J_c=0}\rangle$, and $\eta\langle\psi_{J_c=2}\|\,Q_c\,\|\psi_{J_c=2}\rangle$. These five parameters were varied so as to reproduce not only the observed excitation energies but also the spectroscopic factors for one-nucleon transfer, static moments and gamma-ray transition probabilities. A best fit to all the data is obtained when

$$\varepsilon_{\frac{5}{2}}-\varepsilon_{\frac{3}{2}}=1{\cdot}40\text{ MeV}$$
$$\varepsilon_{\frac{1}{2}}-\varepsilon_{\frac{3}{2}}=1{\cdot}01\text{ MeV}$$
$$\zeta=0{\cdot}13\text{ MeV} \tag{2.96}$$
$$\sqrt{5}\eta\langle\psi_{J_c=2}\|\,Q_c\,\|\psi_{J_c=0}\rangle=0{\cdot}33\text{ MeV fm}^{-2}$$
$$\sqrt{5}\eta\langle\psi_{J_c=2}\|\,Q_c\,\|\psi_{J_c=2}\rangle=0{\cdot}37\text{ MeV fm}^{-2}.$$

The spectrum that one calculates using these values is shown in Fig. 2.6 and reproduces the experimentally known states (Markham and Fulbright 1973) extremely well. However, other calculated nuclear properties are only in fair agreement with experiment. Thus if one wishes to reproduce all the theoretical properties of the levels below 1·6 MeV one must

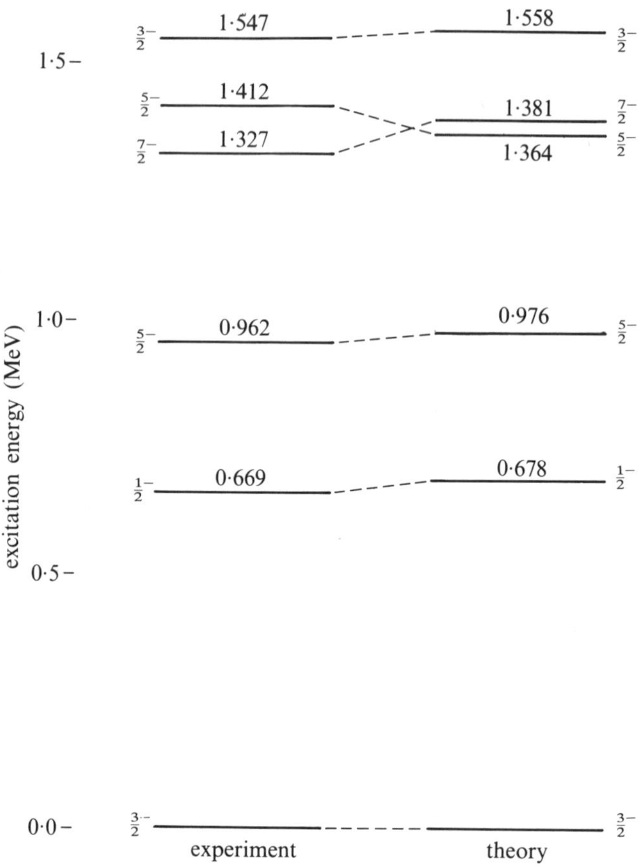

Fig. 2.6. Low-lying states of $^{63}_{29}Cu_{34}$. The experimental results are those of Markham and Fulbright (1973). The theoretical spectrum is based on the Hamiltonian whose matrix elements are given by equation (2.95) with the parameters of equation (2.96).

include more basis states in the calculation. On the assumption that $Z = 28$ is inert, the extension would involve using additional core states. however, the next states in $^{62}_{28}Ni_{34}$ are an almost degenerate 0^+, 2^+, and 4^+ triplet at energies of 2·047, 2·293, and 2·336 MeV, respectively (Fanger *et al.* 1970). Consequently on the basis of energies there is little to choose between these states and all three should be considered. However, if this is done one then has all the quadrupole properties of the core states at one's disposal (instead of the two parameters proportional to η in equation (2.96) there will be 10), and with so many parameters a meaningful fit to the data cannot be made.

Of course, the empirically determined core quadrupole properties are related to measurable quantities. As we shall show in Chapter 5 the off-diagonal matrix elements of Q_c are proportional to the observed $B(E2)$ values and the diagonal matrix elements to the quadrupole moment $Q(J_c)$:

$$Q(J_c)=e\left(\frac{16\pi}{5}\right)^{\frac{1}{2}}\langle\psi_{J_cM=J_c}|\sum_n r_n^2 Y_{20}(\theta_n,\phi_n)\,|\psi_{J_cM=J_c}\rangle$$
$$=e\left(\frac{16\pi}{5}\right)^{\frac{1}{2}}(J_c2J_c0\mid J_cJ_c)\langle\psi_{J_c}\|\,Q_c\,\|\psi_{J_c}\rangle$$

$$B(E2;J_c\rightarrow J_c')=\frac{(2J_c'+1)}{(2J_c+1)}\,e^2\,|\langle\psi_{J_c'}\|\,Q_c\,\|\psi_{J_c}\rangle|^2.$$

Thus, if all the quadrupole properties of the first five states in $^{62}_{28}Ni_{34}$ were known, only the parameter η and the signs of the off-diagonal matrix elements would be at our disposal. In this case a fit to the data including all states could be attempted. However, at the moment this is not so, and if one wishes to extend the model one must reduce the number of parameters by making some assumption about the structure of the neutron core states (for example, one might use the microscopic structure of the Ni states determined theoretically by Cohen *et al.* 1967).

3.5. Failure of weak coupling through violation of the Pauli principle

There are situations where blind application of the weak-coupling model can lead to results in gross disagreement with experiment. This occurs when the Pauli principle is violated. As an example, consider the negative-parity states in the nucleus $^{91}_{41}Nb_{50}$. As a model for such states one might consider the coupling of a $p_{\frac{1}{2}}$ particle to the various states of the $^{90}_{40}Zr_{50}$ core. Thus the lowest $\frac{1}{2}^-$ state would arise from a $p_{\frac{1}{2}}$ proton attached to the $J_c=0$ ground state of $^{90}_{40}Zr_{50}$; states with spins of $\frac{3}{2}^-$ and $\frac{5}{2}^-$ would occur when the $p_{\frac{1}{2}}$ particle is coupled to the 2·18 MeV 2^+ state etc. States of these spins and parities are observed in the spectrum of $^{91}_{41}Nb_{50}$. The $\frac{1}{2}^-$ is seen at an excitation energy of 0·1045 MeV and the $\frac{5}{2}^-$ and $\frac{3}{2}^-$ are seen at 1·187 and 1·313 MeV, respectively (Verheul and Ewbank 1972). According to equation (2.84) the centre of gravity of this doublet should be 2·18 MeV above the 0·1045 MeV level. Experimentally it is at 1·237 MeV in gross disagreement with theory.

The origin of this discrepancy can easily be understood. The $^{90}_{40}Zr_{50}(d,{}^3He)^{89}_{39}Y_{50}$ pick-up data (Yntema 1964) imply that the ground-state wave function of $^{90}_{40}Zr_{50}$ is

$$\psi_{0^+}=0{\cdot}8(p_{\frac{1}{2}})^2_0\pm0{\cdot}6(g_{\frac{9}{2}})^2_0.$$

Therefore by attempting to explain the lowest $\frac{1}{2}^-$ state in ${}^{91}_{41}\mathrm{Nb}_{50}$ as the coupling of a $p_{\frac{1}{2}}$ proton to this 0^+ level we violate the Pauli principle badly by trying to force three protons into the $p_{\frac{1}{2}}$ orbit. On the other hand, Gloeckner *et al.* (1972) have found that the $(p_{\frac{1}{2}}, g_{\frac{9}{2}})$ model explains the data on the $N=50, Z\geq 39$ nuclei quite well. On the basis of this model, negative-parity states in ${}^{91}_{41}\mathrm{Nb}_{50}$ can only arise when one $p_{\frac{1}{2}}$ proton is coupled to a pair of $g_{\frac{9}{2}}$ particles. Furthermore, since the only states of $g_{\frac{9}{2}}^2$ are $J_c=0$, 2, 4, 6, and 8, it follows that there is only one way to make a negative-parity state of given spin. Thus, according to equation (2.84) the energy difference between the centre of gravity of the $\frac{3}{2}^-$, $\frac{5}{2}^-$ doublet and the $\frac{1}{2}^-$ state should be equal to the difference in the interaction energy when two $g_{\frac{9}{2}}$ protons couple to spin 2 and zero. These latter energies are given in equation (1.115) and yield 1·058 MeV for the splitting in good agreement with the experimental result 1·237−0·1045 = 1·1325 MeV.

4. Weak coupling including isospin

In sections 2.3 and 2.5 of this chapter we showed how the isospin formalism is used to calculate nuclear spectra. In this section we consider the energy associated with the state described by the wave function

$$\Psi_{IM;TT_z} = \qquad (2.97)$$

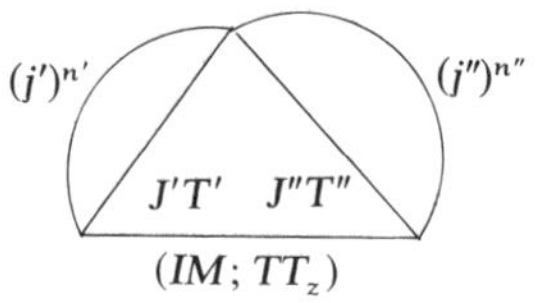

in the special case that either

(a) J' and/or $J''=0$
(b) one is interested in the centre of gravity of the states whose wave functions are given by equation (2.97).

(For convenience we have suppressed the additional quantum numbers α' and α'' needed to give a complete specification of the states.)

The main problem in calculating H is the evaluation of the interaction between j' and j''. However, since this interaction is a scalar in spin and isospin space it takes on a very simple form when J' or $J''=0$. In this case the only scalar quantities that one can make that depend simultaneously on the two configurations is $\mathrm{N}_{j'}\mathrm{N}_{j''}$ and $\underline{\mathrm{T}}'\cdot\underline{\mathrm{T}}''$, where $\mathrm{N}_{j'}$ is the number operator for the state j' (see equation (1.41)) and $\underline{\mathrm{T}}'$ and $\underline{\mathrm{T}}''$ are the isospin operators associated with $(j')^{n'}$ and $(j'')^{n''}$, respectively. The total isospin operator of the composite system is

$$\underset{\sim}{T} = \underset{\sim}{T}' + \underset{\sim}{T}''$$

so that the energy $\bar{\varepsilon}$ associated with this special case is

$$\begin{aligned}\bar{\varepsilon} &= E_{J'T'}\{(j')^{n'}\} + E_{J''T''}\{(j'')^{n''}\} \\ &\quad + \langle \Psi_{IM;TT_z} |\, \gamma \mathrm{N}_{j'}\mathrm{N}_{j''} + \beta \underset{\sim}{T}' \,.\, \underset{\sim}{T}'' \,| \Psi_{IM;TT_z} \rangle \\ &= E_{J'T'}\{(j')^{n'}\} + E_{J''T''}\{(j'')^{n''}\} + \gamma n' n'' \\ &\quad + \frac{\beta}{2}(T(T+1) - T'(T'+1) - T''(T''+1)) \end{aligned} \tag{2.98}$$

where γ and β are constants (Bansal and French 1964, Zamick 1965).

The same argument goes through if we consider the centre of gravity of the states described by equation (2.97). In this case we are interested in

$$\sum_I (2I+1)\langle \Psi_{IM;TT_z}|\, H \,|\Psi_{IM;TT_z}\rangle = \sum_{IM} \langle \Psi_{IM;TT_z}|\, H \,|\Psi_{IM;TT_z}\rangle.$$

Summing over I and M is essentially the same as averaging over all orientation in space so that again only $\mathrm{N}_{j'}\mathrm{N}_{j''}$ and $\underset{\sim}{T}' \,.\, \underset{\sim}{T}''$ remain to characterize the interaction energy.

In section 4.1 we give a formal proof of equation (2.98) and find the relationship between γ and β and the two-body interaction energies (see equations (2.106)). In section 4.2 we use the result to study the $d_{\frac{3}{2}}$ -hole states near $A = 40$.

4.1. *The Bansal–French formula*

In order to calculate the j', j'' interaction energy it is convenient to write $V_{j'j''}$ of equation (2.24) in the form

$$\begin{aligned} V_{j'j''} &= \sum_{JT} \surd\{(2J+1)(2T+1)\} E_{JT}(j'j''; j'j'') \\ &\quad \times [A^\dagger_{JT}(j'j'') \times \tilde{A}_{JT}(j'j'')]_{00;00} \end{aligned}$$

where $A^\dagger_{JM;TT_z}(j'j'')$ is given by equation (2.22),

$$\begin{aligned} \tilde{A}_{J-M;T-T_z}(j'j'') &= (-1)^{J-M+T-T_z}\{A^\dagger_{JM;TT_z}(j'j'')\}^\dagger \\ &= -[\tilde{a}_{j';\frac{1}{2}} \times \tilde{a}_{j'';\frac{1}{2}}]_{J-M;T-T_z} \end{aligned} \tag{2.99}$$

and

$$\tilde{a}_{j-m;\frac{1}{2}-\mu} = (-1)^{j-m+\frac{1}{2}-\mu} a_{jm;\frac{1}{2}\mu}. \tag{2.100}$$

In appendix 2, section 1, we showed that $\tilde{A}_{J-M}$ is a tensor operator of rank J in space-spin. By the same arguments $\tilde{a}_{j-m;\frac{1}{2}-\mu}$ and $\tilde{A}_{J-M;T-Tz}$ can be shown to be tensors of rank $(j, \frac{1}{2})$ and (J, T) in space-spin-isospin. Since

$a^{\dagger}_{j''m'';\frac{1}{2}\mu''}$ and $a^{\dagger}_{j'm';\frac{1}{2}\mu'}$ anticommute (equation (2.21)) it follows that

$$V_{j'j''} = -\sum_{JT} \sqrt{\{(2J+1)(2T+1)\}} E_{JT}(j'j''; j'j'')$$ 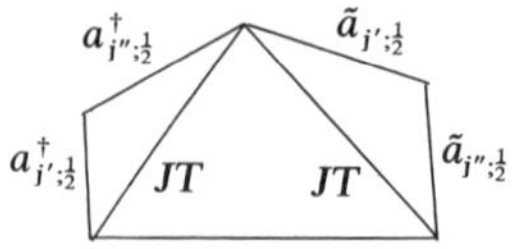

$$= \sum_{JTK\tau} (2K+1)(2\tau+1)(2J+1)(2T+1)\sqrt{\{(2J+1)(2T+1)\}} E_{JT}(j'j''; j'j'')$$

$$\times \begin{Bmatrix} j' & j'' & J \\ j' & j'' & J \\ K & K & 0 \end{Bmatrix} \begin{Bmatrix} \frac{1}{2} & \frac{1}{2} & T \\ \frac{1}{2} & \frac{1}{2} & T \\ \tau & \tau & 0 \end{Bmatrix}$$ 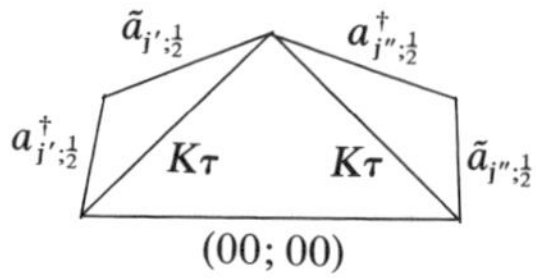

$$= \sum_{K\tau} \sqrt{\{(2K+1)(2\tau+1)\}} \tilde{E}_{K\tau}(j'j''; j'j'')$$ 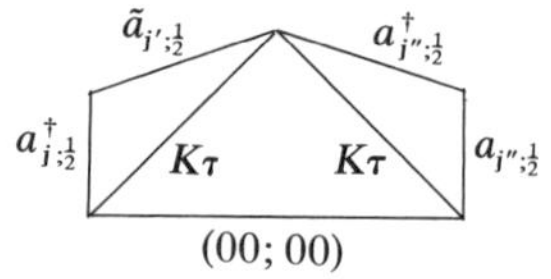

$$= \sum_{K\tau} \sqrt{\{(2K+1)(2\tau+1)\}} \tilde{E}_{K\tau}(j'j''; j'j'')$$
$$\times [[a^{\dagger}_{j';\frac{1}{2}} \times \tilde{a}_{j';\frac{1}{2}}]_{K\tau} \times [a^{\dagger}_{j'';\frac{1}{2}} \times \tilde{a}_{j'';\frac{1}{2}}]_{K\tau}]_{00;00} \tag{2.101a}$$

where

$$\tilde{E}_{K_\tau}(j'j''; j'j'') = \sum_{JT} (-1)^{j'-j''-J-K-T-\tau}(2J+1)(2T+1)W(j'j''j'j''; JK)$$
$$\times W(\tfrac{1}{2}\tfrac{1}{2}\tfrac{1}{2}\tfrac{1}{2}; T\tau)E_{JT}(j'j''; j'j''). \tag{2.101b}$$

In making the recoupling involved in this equation we have made use of the diagrammatic definition of the $9j$ coefficient (see equation (2.59) and Appendix 4, section 2 together with the fact that this coefficient has a particularly simple form when one argument is zero (see equation A4.33)).

Finally, in order to calculate the effect of $V_{j'j''}$ it is necessary to define the reduced matrix element when isospin is included and furthermore to know the value of the reduced matrix element of $[a^{\dagger}_{j;\frac{1}{2}} \times \tilde{a}_{j;\frac{1}{2}}]_{Kk;\tau\mu}$ when $K = k = 0$. According to equation (A2.6) the ordinary reduced matrix element is simply the matrix element of a tensor operator divided by its M dependence. Consequently when isospin is included one must also factor out the T_z dependence. Since the quantities of interest are also tensors in isospin space it follows that the T_z dependence will involve a

Clebsch-Gordan coefficient. Thus

$$\langle \Psi_{I'M';T'T_z'} | S_{Kk;\tau\mu} | \Psi_{IM;TT_z} \rangle = \langle \Psi_{I';T'} ||| S_{K;\tau} ||| \Psi_{I;T} \rangle \times (IKMk \mid I'M')(T\tau T_z\mu \mid T'T_z') \qquad (2.102a)$$

or

$$\sum_{MkT_z\mu} (IKMk \mid I'M')(T\tau T_z\mu \mid T'T_z') S_{Kk;\tau\mu} \Psi_{IM;TT_z} = \langle \Psi_{I';T'} ||| S_{K;\tau} ||| \Psi_{I;T} \rangle \Psi_{I'M';T'T_z'} \qquad (2.102b)$$

where $S_{Kk;\tau\mu}$ is any tensor of rank K in ordinary space and rank τ in isospin space and $\langle ||| \;\; ||| \rangle$ is the reduced matrix element.

The operator $[a^\dagger_{j;\frac{1}{2}} \times \tilde{a}_{j;\frac{1}{2}}]_{Kk;\tau\mu}$ is a tensor operator of rank $(K\tau)$ and when $K = k = 0$ it takes the simple form

$$[a^\dagger_{j;\frac{1}{2}} \times \tilde{a}_{j;\frac{1}{2}}]_{00;\tau\mu} = (2j+1)^{-\frac{1}{2}} \sum_{m\nu\nu'} (-1)^{j-m} (\tfrac{1}{2}\tfrac{1}{2}\nu\nu' \mid \tau\mu) a^\dagger_{jm;\frac{1}{2}\nu} \tilde{a}_{j-m;\frac{1}{2}\nu'}$$
$$= (2j+1)^{-\frac{1}{2}} \sum_{m\nu\nu'} (-1)^{\frac{1}{2}+\nu'} (\tfrac{1}{2}\tfrac{1}{2}\nu\nu' \mid \tau\mu) a^\dagger_{jm;\frac{1}{2}\nu} a_{jm;\frac{1}{2}-\nu'} \qquad (2.103)$$

where use has been made of equation (2.100) and the form of the Clebsch–Gordan coefficient when two angular momenta couple to zero, equation (A1.21). When $\tau = 0$ this equation becomes

$$[a^\dagger_{j;\frac{1}{2}} \times \tilde{a}_{j;\frac{1}{2}}]_{00;00} = \{2(2j+1)\}^{-\frac{1}{2}} \sum_{jm} (a^\dagger_{jm;\frac{1}{2}\frac{1}{2}} a_{jm;\frac{1}{2}\frac{1}{2}} + a^\dagger_{jm;\frac{1}{2}-\frac{1}{2}} a_{jm;\frac{1}{2}-\frac{1}{2}}).$$

Thus this operator is proportional to the sum of the number operator for neutrons plus that for protons (see equation (1.41)) and as such gives the total number of particles in the orbit j. Consequently

$$\langle \Psi_{JM;TT_z}(j^n) | [a^\dagger_{j;\frac{1}{2}} \times \tilde{a}_{j;\frac{1}{2}}]_{00;00} | \Psi_{JM;TT_z}(j^n) \rangle = n/\sqrt{\{2(2j+1)\}}$$
$$= \langle \Psi_{J;T}(j^n) ||| [a^\dagger_{j;\frac{1}{2}} \times \tilde{a}_{j;\frac{1}{2}}]_{0;0} ||| \Psi_{J;T}(j^n) \rangle (J0M0 \mid JM) \times (T0T_z0 \mid TT_z).$$

Since the Clebsch–Gordan coefficients in this equation are both unity it follows that

$$\langle \Psi_{J;T}(j^n) ||| [a^\dagger_{j;\frac{1}{2}} \times \tilde{a}_{j;\frac{1}{2}}]_{0;0} ||| \Psi_{J;T}(j^n) \rangle = n/\sqrt{\{2(2j+1)\}}. \qquad (2.104)$$

When $\tau = 1$ and $\mu = 0$ use of Table A1.1 allows us to write

$$[a^\dagger_{j;\frac{1}{2}} \times \tilde{a}_{j;\frac{1}{2}}]_{00;10} = \{2(2j+1)\}^{-\frac{1}{2}} \sum_{jm} (a^\dagger_{jm;\frac{1}{2}\frac{1}{2}} a_{jm;\frac{1}{2}\frac{1}{2}} - a^\dagger_{jm;\frac{1}{2}-\frac{1}{2}} a_{jm;\frac{1}{2}-\frac{1}{2}}).$$

In other words this operator is proportional to the number of neutrons minus the number of protons; i.e. to the T_z of the state. Therefore

$$\begin{aligned}\langle \Psi_{JM;TT_z}(j^n)|\,[a^\dagger_{j;\frac12}\times \tilde a_{j;\frac12}]_{00;10}\,|\Psi_{JM;TT_z}(j^n)\rangle \\ &= \left\{\frac{2}{2j+1}\right\}^{\frac12} T_z \\ &= \langle \Psi_{J;T}(j^n)|||\,[a^\dagger_{j;\frac12}\times \tilde a_{j;\frac12}]_{0;1}\,|||\Psi_{J;T}(j^n)\rangle (J0M0\,|\,JM) \\ &\quad \times (T1T_z0\,|\,TT_z).\end{aligned}$$

The space-spin Clebsch–Gordan coefficient in this equation is unity whereas the isospin one, according to Table A1.2, has the value $T_z/\sqrt{\{T(T+1)\}}$. Thus

$$\langle \Psi_{J;T}(j^n)|||\,[a^\dagger_{j;\frac12}\times \tilde a_{j;\frac12}]_{0;1}\,|||\Psi_{J;T}(j^n)\rangle = \left\{\frac{2T(T+1)}{2j+1}\right\}^{\frac12}. \tag{2.105}$$

With these results established we now turn to the evaluation of the effect of $V_{j'j''}$ on $\Psi_{IM;TT_z}$ of equation (2.97). When $V_{j'j'}$ is written in the form given by equations (2.101) one sees that

$$V_{j'j''}\Psi_{IM;TT_z} = \sum_{K\tau} \sqrt{\{(2K+1)(2\tau+1)\}}\tilde E_{K\tau}(j'j'';j'j'')$$

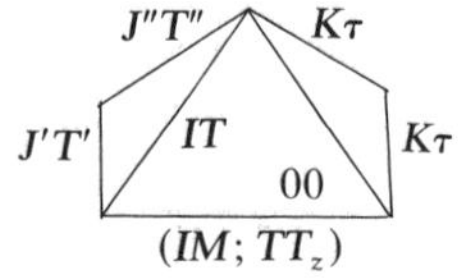

$$= \sum_{J_3'T_3'J_3''T_3''K\tau} \sqrt{\{(2I+1)(2K+1)(2J_3'+1)(2J_3''+1)(2T+1)}$$

$$\times (2\tau+1)(2T_3'+1)(2T_3''+1)\}$$

$$\times \tilde E_{K\tau}(j'j'';j'j'')\begin{Bmatrix} J' & J'' & I \\ K & K & 0 \\ J_3' & J_3'' & I\end{Bmatrix}\begin{Bmatrix} T' & T'' & T \\ \tau & \tau & 0 \\ T_3' & T_3'' & T\end{Bmatrix}$$

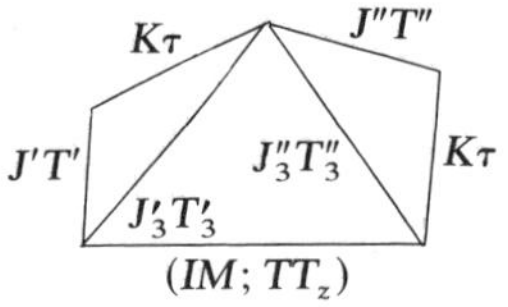

where it is to be understood that in this diagram

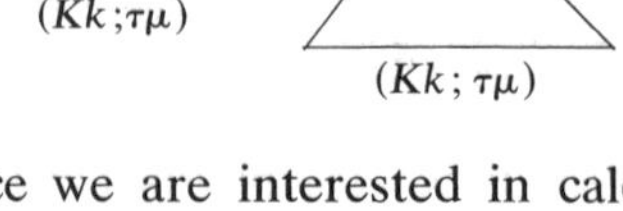

with $j = j'$ or j''. Since we are interested in calculating diagonal matrix elements of $V_{j'j''}$, $(J_3'T_3') = (J'T')$ and $(J_3''T_3'') = (J''T'')$. Use of this together with the definition of the reduced matrix element given in equation (2.102b) and the simple form of the $9j$ coefficient, equation (A4.33), leads to

$$
\begin{aligned}
&\langle[(j')^{n'}_{J'T'}\times(j'')^{n''}_{J''T''}]_{IM;TT_z}|\,H\,|[(j')^{n'}_{J'T'}\times(j'')^{n''}_{J''T''}]_{IM;TT_z}\rangle\\
&=E_{J'T'}\{(j')^{n'}\}+E_{J''T''}\{(j'')^{n''}\}+\sqrt{\{(2J'+1)(2J''+1)}\\
&\quad\times(2T'+1)(2T''+1)\}\sum_{K\tau}W(J'IKJ'';J''J')\\
&\quad\times W(T'T\tau T'';T''T')\langle(j')^{n'}_{J'T'}|||[a^{\dagger}_{j';\frac{1}{2}}\times\tilde{a}_{j';\frac{1}{2}}]_{K;\tau}|||(j')^{n'}_{J'T'}\rangle\\
&\quad\times\langle(j'')^{n''}_{J''T''}|||[a^{\dagger}_{j'';\frac{1}{2}}\times\tilde{a}_{j'';\frac{1}{2}}]_{K;\tau}|||(j'')^{n''}_{J''T''}\rangle\tilde{E}_{K\tau}(j'j'';j'j'').
\end{aligned}
$$

We are particularly interested in this energy when either J' and/or $J''=0$ or when the centre of gravity of the levels is to be calculated. We define $\bar{\varepsilon}$ to be the energy in these cases. When J' and/or $J''=0$ one sees from equations (A4.13) and (A4.14) that

$$\sqrt{\{(2J'+1)(2J''+1)\}}W(J'IKJ'';J''J')=\delta_{K0}.$$

Alternatively when the centre of gravity of the levels is the quantity of interest one obtains by use of equations (A4.13), (A4.14), and (A4.15)

$$
\begin{aligned}
&\sqrt{\{(2J'+1)(2J''+1)\}}\sum_{I}(2I+1)W(J'IKJ'';J''J')\\
&\quad=(-1)^K(2J'+1)(2J''+1)\sum_{I}(2I+1)W(J'J'J''J'';KI)W(J'J'J''J'';0I)\\
&\quad=(-1)^K(2J'+1)(2J''+1)\delta_{K0}\\
&\quad=(-1)^K\sum_{I}(2I+1)\delta_{K0}.
\end{aligned}
$$

Thus

$$
\begin{aligned}
\bar{\varepsilon}&=E_{J'T'}\{(j')^{n'}\}+E_{J''T''}\{(j'')^{n''}\}\\
&\quad+\sqrt{\{(2T'+1)(2T''+1)\}}\sum_{\tau}W(T'T\tau T'';T''T')\\
&\quad\times\tilde{E}_{0\tau}(j'j'';j'j'')\langle(j')^{n'}_{J'T'}|||[a^{\dagger}_{j';\frac{1}{2}}\times\tilde{a}_{j';\frac{1}{2}}]_{0;\tau}|||(j')^{n'}_{J'T'}\rangle\\
&\quad\times\langle(j'')^{n''}_{J''T''}|||[a^{\dagger}_{j'';\frac{1}{2}}\times\tilde{a}_{j'';\frac{1}{2}}]_{0;\tau}|||(j'')^{n''}_{J''T''}\rangle.
\end{aligned}
$$

This expression can be simplified if one uses the form of the reduced matrix elements given by equations (2.104) and (2.105) together with the value of the Racah coefficient when $\tau=0$ (equation (A4.14) or when $\tau=1$ given in Table A4.2:

$$
\begin{aligned}
\bar{\varepsilon}&=E_{J'T'}\{(j')^{n'}\}+E_{J''T''}\{(j'')^{n''}\}+\left[\frac{\tilde{E}_{00}(j'j'';j'j'')}{2\sqrt{\{(2j'+1)(2j''+1)\}}}\right]n'n''\\
&\quad-\left[\frac{\tilde{E}_{01}(j'j'';j'j'')}{\sqrt{\{(2j'+1)(2j''+1)\}}}\right]\{T(T+1)-T'(T'+1)-T''(T''+1)\}\\
&=E_{J'T'}\{(j')^{n'}\}+E_{J''T''}\{(j'')^{n''}\}+\gamma n'n''\\
&\quad+\frac{\beta}{2}\{T(T+1)-T'(T'+1)-T''(T''+1)\}
\end{aligned}
$$

which is identical to the result given in equation (2.98) arrived at earlier on the basis of physical arguments. From the definition of $\tilde{E}_{0\tau}(j'j'';j'j'')$ given in equation (2.101b) one can obtain expressions for γ and β in terms of the two-body interaction energies

$$\gamma=\frac{\tilde{E}_{00}(j'j'';j'j'')}{2\sqrt{\{(2j'+1)(2j''+1)\}}}$$

$$=\frac{1}{4(2j'+1)(2j''+1)}\sum_{JT}(2J+1)(2T+1)E_{JT}(j'j'';j'j'') \qquad (2.106a)$$

and

$$\beta=\frac{-2\tilde{E}_{01}(j'j'';j'j'')}{\sqrt{\{(2j'+1)(2j''+1)\}}}$$

$$=\frac{1}{(2j'+1)(2j''+1)}\sum_{J}(2J+1)\{E_{J1}(j'j'';j'j'')-E_{J0}(j'j'';j'j'')\}. \qquad (2.106b)$$

In deriving the expression for $\langle V_{j'j''}\rangle$ we did not take into account the Coulomb interaction which acts only between protons. If n'_π and n''_π denote the number of protons in j' and j'', respectively, and ε_c is the average Coulomb interaction energy, it follows that

$$\tilde{\varepsilon}=E_{J'T'}\{(j')^{n'}\}+E_{J''T''}\{(j'')^{n''}\}+\gamma n'n''+n'_\pi n''_\pi\varepsilon_c$$

$$+\frac{\beta}{2}\{T(T+1)-T'(T'+1)-T''(T''+1)\} \qquad (2.107)$$

where $\tilde{\varepsilon}$ is the energy of the configuration $[(j')^{n'}_{J'T'}\times(j'')^{n''}_{J''T''}]_{IM;TT_z}$ when either J' and/or $J''=0$ or the energy of the centre-of-gravity of the levels is the quantity of interest (Bansal and French 1964). By analogy with equation (2.106a) one sees that

$$\varepsilon_c=\frac{1}{(2j'+1)(2j''+1)}\sum_{J}(2J+1)\langle(j'j'')_J|\frac{e^2}{r}|(j'j'')_J\rangle \qquad (2.108)$$

where the matrix element is the expectation value of the Coulomb potential in the two-particle state $(j'j'')_J$.

4.2. $d_{\frac{3}{2}}$ hole states near $A=40$

The first excited state in the odd-A Sc isotopes ${}^{43}_{21}Sc_{22}$, ${}^{45}_{21}Sc_{24}$, and ${}^{47}_{21}Sc_{26}$ is a $\frac{3}{2}^+$ level that occurs at 150, 12, and 765 keV, respectively (Endt and Van der Leun 1973). Since these states are strongly excited with $l=2$ by the $(d,{}^3He)$ reaction on the titanium isotopes thay have been interpreted (Yntema and Satchler 1964) as the $d_{\frac{3}{2}}$ hole states; i.e. a $d_{\frac{3}{2}}$ hole coupled to the $J''=0$ ground state of the even-A titanium core. One may use equation (2.107) together with the experimental excitation energies of these levels to predict the position of the $(d_{\frac{3}{2}})^{-2}$ states in the neighbouring calcium nuclei. As an example we treat in detail the $({}^{47}_{21}\mathrm{Sc}_{26}, {}^{46}_{20}\mathrm{Ca}_{26})$ pair.

The configurations of the ground state and the $\frac{3}{2}^{+}$ hole state in ${}^{47}_{21}\mathrm{Sc}_{26}$ are illustrated in Fig. 2.7a. The energy of the seven $f_{\frac{7}{2}}$ nucleons relative to the ${}^{40}_{20}\mathrm{Ca}_{20}$ core can be obtained from the binding-energy tables of Wapstra and Gove (1971). If we denote this energy by E_{gs}, then

$$\begin{aligned} E_{\mathrm{gs}} &= \mathrm{BE}({}^{47}_{21}\mathrm{Sc}_{26}) - \mathrm{BE}({}^{40}_{20}\mathrm{Ca}_{20}) \\ &= -65{\cdot}202\ \mathrm{MeV}. \end{aligned}$$

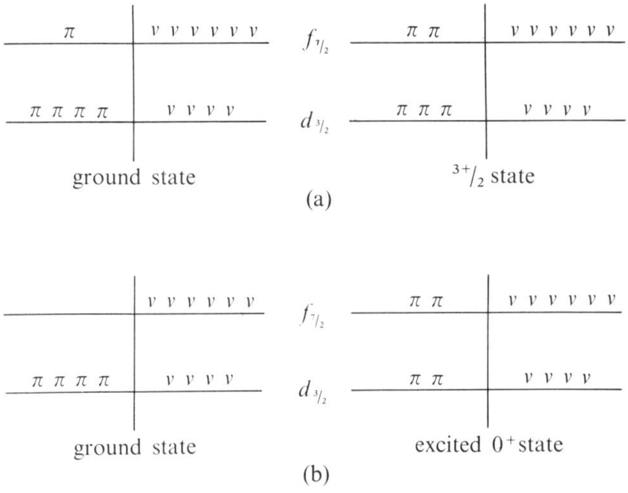

Fig. 2.7. In (a) the configurations of the ground state and $\frac{3}{2}^{+}$ hole state in ${}^{47}_{21}\mathrm{Sc}_{26}$ are shown. In (b) the ground state and the structure of the excited 0^{+} state of ${}^{46}_{20}\mathrm{Ca}_{26}$ are illustrated. In these diagrams π denotes a proton and ν a neutron.

Since the hole state is strongly seen in the ${}^{48}_{22}\mathrm{Ti}_{26}(d, {}^{3}\mathrm{He}){}^{47}_{21}\mathrm{Sc}_{26}$ reaction its wave function is assumed to have the form

$$\Psi_{IM;TT_z} = [(d_{\frac{3}{2}})^{-1}_{\frac{3}{2};\frac{1}{2}} \times (f_{\frac{7}{2}})^{8}_{0;2}]_{I=\frac{3}{2}M;T=\frac{5}{2}T_z=\frac{5}{2}}. \tag{2.109}$$

Thus the configuration is similar to that of equation (2.97) except that one deals with a $d_{\frac{3}{2}}$ hole instead of a $d_{\frac{3}{2}}$ particle. However, as we shall show in Chapter 3, one can equally well formulate shell-model problems in terms of holes or particles so that equation (2.107) can be used provided one interprets γ, β, and ε_{c} as arising from the particle-hole interaction. Since $n'_{\pi} = 1$ and $n''_{\pi} = 2$ it follows that $E_{\frac{3}{2}}$, the energy of the $\frac{3}{2}^{+}$ state relative to the ${}^{40}_{20}\mathrm{Ca}_{20}$ core, is

$$
\begin{aligned}
E_{\frac{3}{2}} &= E_{\frac{3}{2};\frac{1}{2}}\{(d_{\frac{3}{2}})^{-1}\} + E_{02}\{(f_{\frac{7}{2}})^{8}\} + 8\gamma + 2\varepsilon_{c} + \beta \\
&= \{\mathrm{BE}({}^{39}_{19}\mathrm{K}_{20}) - \mathrm{BE}({}^{40}_{20}\mathrm{Ca}_{20})\} + \{\mathrm{BE}({}^{48}_{22}\mathrm{Ti}_{26}) - \mathrm{BE}({}^{40}_{20}\mathrm{Ca}_{20})\} \\
&\quad + 8\gamma + 2\varepsilon_{c} + \beta \\
&= (-68{\cdot}317 + 8\gamma + 2\varepsilon_{c} + \beta)\ \mathrm{MeV}.
\end{aligned}
$$

Since the $\frac{3}{2}^{+}$ state is known to lie 765 keV above the ground state, it follows that

$$8\gamma + 2\varepsilon_{c} + \beta = 3{\cdot}880\ \mathrm{MeV}. \tag{2.110}$$

By the same arguments one may obtain an expression for the energy of the excited 0^{+} state

$$\Phi_{IM;TT_z} = [(d_{\frac{3}{2}})^{-2}_{0;1} \times (f_{\frac{7}{2}})^{8}_{0;2}]_{I=0M;T=3T_z=3}$$

which occurs in ${}^{46}_{20}\mathrm{Ca}_{26}$ and is illustrated in Fig. 2.7b. The energy of this state relative to the ${}^{46}_{20}\mathrm{Ca}_{26}$ ground state is

$$
\begin{aligned}
\Delta E &= \mathrm{BE}({}^{38}_{18}\mathrm{Ar}_{20}) + \mathrm{BE}({}^{48}_{22}\mathrm{Ti}_{26}) - \mathrm{BE}({}^{40}_{20}\mathrm{Ca}_{20}) - \mathrm{BE}({}^{46}_{20}\mathrm{Ca}_{26}) \\
&\quad + 16\gamma + 4\varepsilon_{c} + 2\beta \\
&= \{-5{\cdot}217 + 2(8\gamma + 2\varepsilon_{c} + \beta)\}\ \mathrm{MeV}.
\end{aligned}
$$

If one uses the value given by equation (2.110) for the $(d_{\frac{3}{2}})^{-1} - f_{\frac{7}{2}}$ interaction energies one predicts that the eight-particle two-hole $I = 0$, $T = 3$ state illustrated in Fig. 2.7b should lie at an excitation energy of 2·543 MeV in ${}^{46}_{20}\mathrm{Ca}_{26}$. This is in good agreement with the excitation energy of the first excited 0^{+} state observed in ${}^{46}_{20}\mathrm{Ca}_{26}$ at 2·424 MeV (Endt and Van der Leun 1973).

One can make a complete analysis of the many $d_{\frac{3}{2}}$ hole states observed in this mass region and this has been done by Sherr, Kouzes, and del Vecchio (1974). There are some cases where the analysis differs slightly from that carried out in the preceding paragraph and this occurs when proton excitation alone does not produce a state with good isospin. To illustrate how one proceeds in this case we consider the excitation energy of the $I = \frac{3}{2}^{+}$, $T = \frac{5}{2}$ level in ${}^{39}_{18}\mathrm{Ar}_{21}$. This state, which is the isobaric analogue of the ${}^{39}_{17}\mathrm{Cl}_{22}$ ground state, has the structure shown in Fig. 2.8. The factors $\sqrt{\frac{2}{5}}$ and $\sqrt{\frac{3}{5}}$ are the isospin Clebsch–Gordan coefficients $(\frac{3}{2}1\frac{3}{2}0 \mid \frac{5}{2}\frac{3}{2})$ and $(\frac{3}{2}1\frac{1}{2}1 \mid \frac{5}{2}\frac{3}{2})$ which ensure $T = \frac{5}{2}$. From this diagrammatic representation of the state it follows that its energy relative to the ${}^{40}_{20}\mathrm{Ca}_{20}$ core is

$$
\begin{aligned}
E_{\frac{3}{2}} &= \tfrac{2}{5}(\mathrm{BE}({}^{42}_{21}\mathrm{Sc}_{21}) + \mathrm{BE}({}^{37}_{17}\mathrm{Cl}_{20}) - 2\mathrm{BE}({}^{40}_{20}\mathrm{Ca}_{20}) + 3\varepsilon_{c}) \\
&\quad + \tfrac{3}{5}(\mathrm{BE}({}^{42}_{20}\mathrm{Ca}_{22}) + \mathrm{BE}({}^{37}_{18}\mathrm{Ar}^{*}_{19}) - 2\mathrm{BE}({}^{40}_{20}\mathrm{Ca}_{20})) + 6\gamma + \tfrac{3}{2}\beta
\end{aligned}
$$

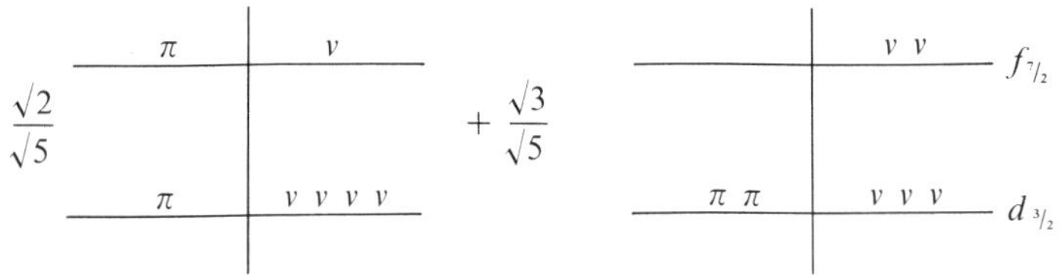

Fig. 2.8. Structure of the $I=\frac{3}{2}^{+}$ $T=\frac{5}{2}$ state in ${}^{39}_{18}\mathrm{Ar}_{21}$. This state is the isobaric analogue in the nucleus ${}^{39}_{18}\mathrm{Ar}_{21}$ of the ground state of ${}^{39}_{17}\mathrm{Cl}_{22}$. The coefficients multiplying the two components are the isospin Clebsch–Gordan coefficients $(\frac{3}{2}1\frac{3}{2}0\,|\,\frac{5}{2}\frac{3}{2})$ and $(\frac{3}{2}1\frac{1}{2}1\,|\,\frac{5}{2}\frac{3}{2})$ as determined from Table (A1.2).

where $\mathrm{BE}({}^{37}_{18}\mathrm{Ar}^{*}_{19})$ is the negative of the binding energy of the $I=\frac{3}{2}^{+}$ $T=\frac{3}{2}$ state in ${}^{37}_{18}\mathrm{Ar}_{19}$ which is known to lie at 4·989 MeV excitation energy. Thus the excitation energy of the isobaric analogue state in ${}^{39}_{18}\mathrm{Ar}_{21}$ is

$$\delta E=\tfrac{2}{5}(\mathrm{BE}({}^{42}_{21}\mathrm{Sc}_{21})+\mathrm{BE}({}^{37}_{17}\mathrm{Cl}_{20}))+\tfrac{3}{5}(\mathrm{BE}({}^{42}_{20}\mathrm{Ca}_{22})+\mathrm{BE}({}^{37}_{18}\mathrm{Ar}^{*}_{19}))$$
$$-\mathrm{BE}({}^{40}_{20}\mathrm{Ca}_{20})-\mathrm{BE}({}^{39}_{18}\mathrm{Ar}_{21})+6\gamma+\tfrac{3}{2}\beta+\tfrac{6}{5}\varepsilon_{c}.$$

In their analysis of this region, Sherr *et al.* (1974) found that

$$\gamma=0{\cdot}25\ \mathrm{MeV}$$
$$\beta=2{\cdot}74\ \mathrm{MeV} \tag{2.111}$$
$$\varepsilon_{c}=-0{\cdot}29\ \mathrm{MeV}.$$

When these values are used the excitation energy in ${}^{39}_{18}\mathrm{Ar}_{21}$ of the isobaric analogue of the ${}^{39}_{17}\mathrm{Cl}_{22}$ ground state is predicted to be $\delta E=9{\cdot}100$ MeV, and this is in almost exact agreement with the observed position of the state, 9·09 MeV (Endt and Van der Leun 1973).

According to equations (2.106) the values of γ and β should be related to the observed particle-hole interaction energies. If one interprets the lowest $2^{-}-5^{-}$ $T=0$ and $T=1$ states seen in ${}^{40}_{20}\mathrm{Ca}_{20}$ as belonging to the $[(d_{\frac{3}{2}})^{-1}_{\frac{3}{2};\frac{1}{2}}\times(f_{\frac{7}{2}})^{1}_{\frac{7}{2};\frac{1}{2}}]_{JM;TT_z}$ configuration one can calculate the expected value of these quantities. In Table 2.6 we list the excitation energies of these states (Endt and Van der Leun 1973). Since β involves only the difference in the $T=0$ and $T=1$ energies it can be calculated directly and one finds

$$\beta=3{\cdot}219\ \mathrm{MeV}.$$

That this value is somewhat larger than the result found by Sherr *et al.* (1974) given in equation (2.111) can be traced to the fact that the lowest 2^{-}–5^{-} levels in ${}^{40}_{20}\mathrm{Ca}_{20}$, particularly those with $T=0$, are not pure $[d_{\frac{3}{2}})^{-1}_{\frac{3}{2};\frac{1}{2}}\times(f_{\frac{7}{2}})^{1}_{\frac{7}{2};\frac{1}{2}}]_{JM;TT_z}$ states. This follows from the experimental observation that more than one $T=0$ state of each of these spins is populated with $l=3$ in the $^{39}_{19}K_{20}(^3He, d)^{40}_{20}Ca_{20}$ reaction (see section 5.3 of this chapter for a detailed discussion of this point). Thus the 'pure' $[(d_{\frac{3}{2}})^{-1}_{\frac{3}{2};\frac{1}{2}} \times (f_{\frac{7}{2}})^{1}_{\frac{7}{2};\frac{1}{2}}]_{JM;T=T_z=0}$ states lie at somewhat higher excitation energies than those given in Table 2.6 and consequently use of only the lowest states will give an overestimate of β.

TABLE 2.6

Observed excitation energies in MeV of the lowest 2^-–5^- states with $T=0$ and $T=1$ in $^{40}_{20}Ca_{20}$

T \ J	2^-	3^-	4^-	5^-
0	6·025	3·737	5·614	4·492
1	8·474	7·696	7·659	8·551

Since γ in equation (2.106a) depends on both the $T=0$ and $T=1$ matrix elements additively one has to look at the structure of the states in more detail to estimate its value. In Fig. 2.9 we illustrate these states

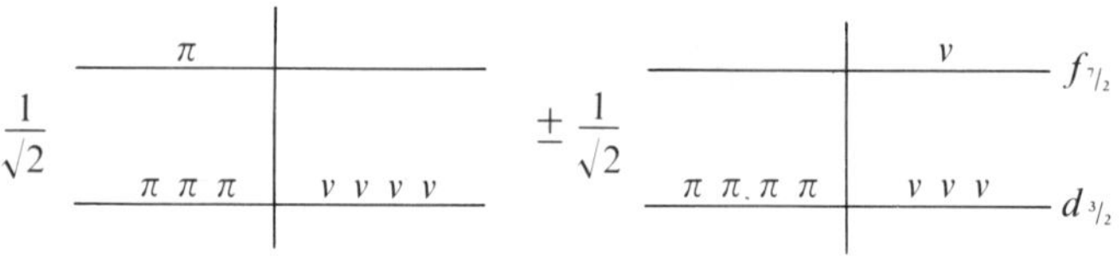

Fig. 2.9. The $d_{\frac{3}{2}}$ hole states in $^{40}_{20}Ca_{20}$. When both coefficients are $+1/\sqrt{2}$ the state has $T=0$ whereas when the phase of the two components in opposite the state has $T=1$.

schematically, and from this diagram it is clear that the excitation energy $\bar{E}_{JT}$ of $[(d_{\frac{3}{2}})^{-1}_{\frac{3}{2};\frac{1}{2}} \times (f_{\frac{7}{2}})^{1}_{\frac{7}{2};\frac{1}{2}}]_{JM;TT_z}$ relative to the $^{40}_{20}Ca_{20}$ ground state is

$$\bar{E}_{JT} = \tfrac{1}{2}\{\varepsilon^{-1}_{d\pi} + \varepsilon_{f\pi} + (\varepsilon_c)_J\} + \tfrac{1}{2}(\varepsilon^{-1}_{d\nu} + \varepsilon_{f\nu}) + E_{JT}$$

where $\varepsilon^{-1}_{d\pi}$ and $\varepsilon^{-1}_{d\nu}$ are the $d_{\frac{3}{2}}$ hole state energies for the proton and neutron, respectively, $\varepsilon_{f\pi}$ and $\varepsilon_{f\nu}$ are the corresponding $f_{\frac{7}{2}}$ single-particle energies, $(\varepsilon_c)_J$ is the Coulomb particle-hole interaction when the particle and hole couple to spin J, and E_{JT} is the particle-hole nuclear interaction. One may use the Wapstra–Gove (1971) binding-energy tables to deduce the single-particle and single-hole energies

$$\varepsilon_{d\pi}^{-1}+\varepsilon_{f\pi}=\mathrm{BE}({}^{39}_{19}\mathrm{K}_{20})+\mathrm{BE}({}^{41}_{21}\mathrm{Sc}_{20})-2\mathrm{BE}({}^{40}_{20}\mathrm{Ca}_{20})$$
$$=7{\cdot}245\ \mathrm{MeV}$$
$$\varepsilon_{d\nu}^{-1}+\varepsilon_{f\nu}=\mathrm{BE}({}^{39}_{20}\mathrm{Ca}_{19})+\mathrm{BE}({}^{41}_{20}\mathrm{Ca}_{21})-2\mathrm{BE}({}^{40}_{20}\mathrm{Ca}_{20})$$
$$=7{\cdot}275\ \mathrm{MeV}$$

so that

$$E_{JT}+\tfrac{1}{2}(\varepsilon_c)_J=(\bar{E}_{JT}-7{\cdot}260)\ \mathrm{MeV}.$$

From equations (2.106a) and (2.108) it follows that only the value of $(\gamma+\frac{1}{2}\varepsilon_c)$ can be deduced from the position of the particle-hole states in ${}^{40}_{20}\mathrm{Ca}_{20}$. By using the values given in Table 2.6 for $\bar{E}_{JT}$ one finds

$$\gamma+\tfrac{1}{2}\varepsilon_c=0{\cdot}036\ \mathrm{MeV}$$

compared with the empirical value 0·105 MeV given by equation (2.111). Since the pure $T=0$ $[(d_{\frac{3}{2}})^{-1}_{\frac{3}{2};\frac{1}{2}}\times(f_{\frac{7}{2}})^{1}_{\frac{7}{2};\frac{1}{2}}]_{JM;T=T_z=0}$ states lie at higher excitation energy than the values given in Table 2.6 it is clear that use of these energies will lead to too small an estimated value for $(\gamma+\frac{1}{2}\varepsilon_c)$.

The analysis carried out by Sherr *et al.* (1974) explained quite nicely the excitation energies of states with stretched isospin (i.e. $T=T_d+T_f$). However, it led to the unexpected result that the energies of states with $T<T_d+T_f$ could not be understood on the basis of the $[(d_{\frac{3}{2}})^{-n'}_{J';T_d}\times(f_{\frac{7}{2}})^{n''}_{J''T_f}]_{IM;TT_z}$ model. As an example of this, consider the energy of the $I=0$ $T=1$ state in ${}^{40}_{19}\mathrm{K}_{21}$ arising from the configuration

$$\Psi_{IM;TT_z}=[(d_{\frac{3}{2}})^{-2}_{0;1}\times(f_{\frac{7}{2}})^{2}_{0;1}]_{00;11}.$$

From the pictorial representation of this state given in Fig. 2.10 one easily sees that its energy relative to the ${}^{40}_{20}\mathrm{Ca}_{20}$ core is

$$E_0=\tfrac{1}{2}(BE({}^{38}_{18}Ar_{20})+\mathrm{BE}({}^{42}_{21}\mathrm{Sc}_{21})+\mathrm{BE}({}^{38}_{19}\mathrm{K}^{*}_{19})+\mathrm{BE}({}^{42}_{20}\mathrm{Ca}_{22}))$$
$$-2\mathrm{BE}({}^{40}_{20}\mathrm{Ca}_{20})+4\gamma-\beta+\varepsilon_c$$

where $\mathrm{BE}({}^{38}_{19}\mathrm{K}^{*}_{19})$ is the negative of the total binding energy of the $J=0$ $T=1$ state in ${}^{38}_{19}\mathrm{K}_{19}$ which lies at 0·131 MeV excitation energy (Endt and Van der Leun 1973). Thus the excitation energy ΔE_0 of this state relative to the ${}^{40}_{19}\mathrm{K}_{21}$ ground state

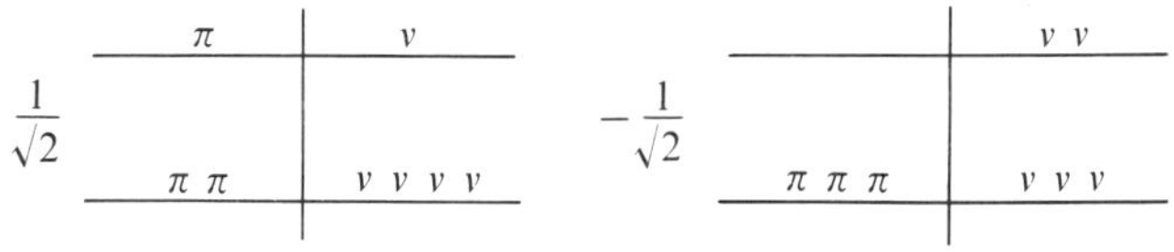

Fig. 2.10. The $I=0^{+}$, $T=1$ two-particle two-hole state in ${}^{40}_{19}\mathrm{K}_{21}$.

is

$$\begin{aligned}\Delta E_0 &= \tfrac{1}{2}(\mathrm{BE}({}^{38}_{18}\mathrm{Ar}_{20})+\mathrm{BE}({}^{42}_{21}\mathrm{Sc}_{21})+\mathrm{BE}({}^{38}_{19}\mathrm{K}^{*}_{19})+\mathrm{BE}({}^{42}_{20}\mathrm{Ca}_{22}))\\ &\quad -\mathrm{BE}({}^{40}_{20}\mathrm{Ca}_{20})-\mathrm{BE}({}^{40}_{19}\mathrm{K}_{21})+4\gamma-\beta+\varepsilon_{\mathrm{c}}\\ &= 1{\cdot}368+4\gamma-\beta+\varepsilon_{\mathrm{c}}\\ &= -0{\cdot}662\ \mathrm{MeV}.\end{aligned}$$

Thus when the values of γ, β, and ε_c given by equation (2.111) are used, a negative excitation energy results; *in other words this $I=0$ $T=1$ state is predicted to lie below the known ground state of the nucleus.* The fact that too low an excitation energy is predicted when one deals with the non-stretched isospin configuration does not seem to have a simple explanation.

From this result, however, it is clear that one must test as many consequences as possible of a configuration assignment before one believes it to be true. Another example where many nuclear properties can be explained by a wrong configuration assignment is the oxygen isotopes. If one considers only the excitation and binding energies of ${}^{17}_{8}\mathrm{O}_9$, ${}^{18}_{8}\mathrm{O}_{10}$, ${}^{19}_{8}\mathrm{O}_{11}$, and ${}^{20}_{8}\mathrm{O}_{12}$ one can explain all the low-lying states on the assumption that ${}^{16}_{8}\mathrm{O}_8$ is an inert core and that the valence neutrons are restricted to the $0d_{\frac{5}{2}}$ and $1s_{\frac{1}{2}}$ single-particle states (Cohen *et al.* 1964). However, once other properties such as gamma decay and nucleon-transfer experiments are considered one can detect the inadequacy of the model and show the need for the collective intruder state discussed in Chapter 1, section 2.2 (Brown 1964, Engeland 1965). Thus it is important to check quantities other than the energies of states and in particular one should compare the predicted properties of the non-yrast levels with experiment since it often happens that these are a sensitive indicator of an inadequate model space.

5. Single-nucleon transfer

A powerful tool for learning about the structure of nuclear wave functions is provided by the single-nucleon-transfer process. If the process takes place near the nuclear surface a particle may be either picked up from or transferred to the target nucleus without appreciably disturbing the other nucleons in the nucleus and such a process is called a direct reaction. The types of processes we shall consider are illustrated in Fig. 2.11 and are embodied in the equations

$$\mathrm{A}+\mathrm{a}\rightleftarrows\mathrm{B}+\mathrm{b}$$

$$\mathrm{a}=\mathrm{b}+\mathrm{x}$$

$$\mathrm{B}=\mathrm{A}+\mathrm{x}$$

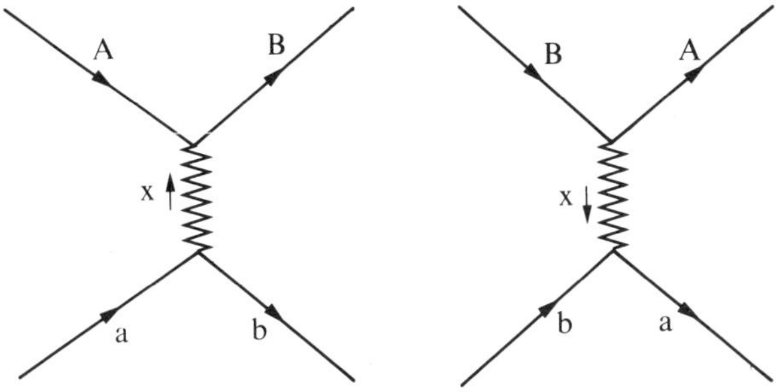

Fig. 2.11. The stripping process ($A+a \rightarrow B+b$) and pick-up process ($B+b \rightarrow A+a$). In the first case the nucleon x is transferred from the incident projectile a to the target A and in the second the nucleon x is picked up by the incident projectile b.

where x is the transferred particle and is either a neutron or proton. In the case that x is transferred from the projectile to the target the reaction is called stripping (i.e. (d, p), (^{3}He, d), (t, d) etc.) and when x is transferred to the projectile from the target we deal with pick-up (i.e. (p, d), (d, ^{3}He), (d, t) etc.).

In this transfer process conservation of linear momentum implies that when the target is at rest

$$\underline{p}_i = \underline{p}_t + \underline{p}_o$$

where $\underline{p}_i$, $\underline{p}_t$, and $\underline{p}_o$ are the momenta of the incident, transferred, and outgoing particles, respectively. Thus if θ is the scattering angle shown in Fig. 2.12

$$\begin{aligned} p_t^2 &= p_i^2 + p_o^2 - 2p_i p_o \cos\theta \\ &= 2p_i^2(1-\cos\theta)(1-\delta/p_i) + \delta^2 \end{aligned}$$

where we have set

$$p_o = p_i - \delta.$$

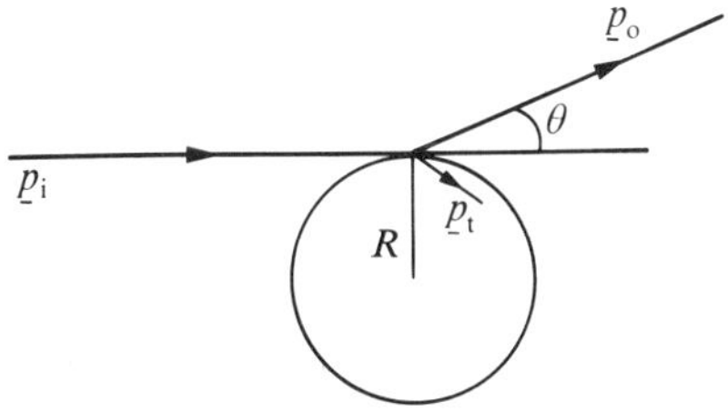

Fig. 2.12. Linear momentum conservation for a grazing collision.

Furthermore, if the process takes place at the nuclear radius R, $(\underline{\mathrm{R}}\times \mathrm{p_t})$ will be the transferred orbital angular momentum l. Thus for a grazing collision

$$p_t \simeq \hbar\sqrt{\{l(l+1)\}}/R$$

and

$$\cos\theta \simeq 1-\left\{\frac{\hbar^2 l(l+1)-\delta^2 R^2}{2p_i^2 R^2(1-\delta/p_i)}\right\}.$$

Provided the momentum mismatch δ is small and that the assumption that the reaction takes place entirely at the nuclear surface is reasonable, the transfer reaction will be a sensitive l-meter (Butler 1951). For example, when 12 MeV deuterons are incident on ^{40}Ca this equation implies, when δ is neglected, that for $l=0$ transfer $\theta=0°$, for $l=1$ $\theta\simeq 18°$, for $l=2$ $\theta\simeq 32°$ etc. Thus if one observes a strong peak in the forward direction $l=0$ transfer is involved. On the other hand, if the forward scattering vanishes but peaks near 18°, $l=1$ transfer is taking place etc.

This type of process is generally analyzed using the distorted wave Born approximation (DWBA) and is discussed in detail by Satchler (1966), Austern (1970), and Hodgson (1971). Optical model wave functions are used to describe the relative motion of the (A, a) and (B, b) systems and the interaction causing the transfer of x is assumed to be sufficiently weak that it can be treated in perturbation theory. In addition to the scattering mechanism there are two other quantities on which the cross-section depends. The first is the probability that the system a will be found as b+x. For reactions initiated by light projectiles ($A\leq 4$) this factor has been computed explicitly and incorporated in the DWBA programmes. The second factor contains the nuclear-structure information, namely the probability that the nuclear system B will be found as A+x. The experimental value for this quantity is extracted from the data by dividing the experimental differential cross-section at forward-scattering angles by $(\mathrm{d}\sigma/\mathrm{d}\Omega)$ computed using one of the standard DWBA codes. This gives a quantity $G_{jl}(\mathrm{A},\mathrm{B})$ known as the strength of the transition,

$$G_{jl}(\mathrm{A},\mathrm{B})=\frac{(\mathrm{d}\sigma/\mathrm{d}\Omega)_{\mathrm{expt}}}{(\mathrm{d}\sigma/\mathrm{d}\Omega)}. \tag{2.112}$$

This strength is related to the shell-model probability by the relationship

$$G_{jl}(\mathrm{A},\mathrm{B})=\frac{1}{(2I_i+1)}\sum_{M_A M_B m}|\langle\Psi_{I_B M_B;T_B T_z'}|\,a^{\dagger}_{jm;\frac{1}{2}\mu}\,|\Psi_{I_A M_A;T_A T_z}\rangle|^2 \tag{2.113}$$

where $\Psi_{I_AM_A;T_AT_z'}$ and $\Psi_{I_BM_B;T_BT_z}$ are the many-particle wave functions describing the nuclear states involved in the transition, $a^{\dagger}_{jm;\frac{1}{2}\mu}$ creates a nucleon in the state $(jlm;\frac{1}{2}\mu)$ (equation (2.21)), and as usual the l quantum number has been suppressed.

Clearly $G_{jl}(\mathrm{A, B})$ should be a constant independent of angle, and this would be true of the experimental value if the theoretical DWBA cross-section reproduced experiment throughout the entire forward-angle scattering range. In practice this is not the case, and to minimize the error in $G_{jl}(\mathrm{A, B})$ the ratio in equation (2.112) is taken at the angle where the experimental cross-section has its maximum. For strong transitions the value of $G_{jl}(\mathrm{A, B})$ extracted in this way can usually be trusted to within about 20%. However, for weak transitions $G_{jl}(\mathrm{A, B})$ can be in error by about a factor of 2.

Generally the strength itself is not quoted, but instead the spectroscopic factor $\mathscr{S}_{jl}(\mathrm{A, B})$ or $C^2\mathscr{S}_{jl}(\mathrm{A, B})$ is the quantity one finds in the literature. This quantity is related to $G_{jl}(\mathrm{A, B})$ by the equation

$$G_{jl}(\mathrm{A, B}) = \frac{2I_\mathrm{B}+1}{2I_\mathrm{i}+1}(T_\mathrm{A}\tfrac{1}{2}T_z\mu \mid T_\mathrm{B}T_z')^2\mathscr{S}_{jl}(\mathrm{A, B}) \tag{2.114}$$

and $\mathscr{S}_{jl}(\mathrm{A, B})$ is the nuclear probability

$$\mathscr{S}_{jl}(\mathrm{A, B}) = |\langle\Psi_{I_\mathrm{B};T_\mathrm{B}}|||\, a^{\dagger}_{j;\frac{1}{2}}\, |||\Psi_{I_\mathrm{A};T_\mathrm{A}}\rangle|^2. \tag{2.115}$$

In equation (2.115) the triple-barred matrix element is the matrix element reduced in both ordinary and isospin space given in equation (2.102a). Because of the M sums in the definition of $G_{jl}(\mathrm{A, B})$, the Clebsch–Gordan coefficients involving I_A, I_B, and j sum to $(2I_\mathrm{B}+1)$ (see equation (A1.23)). However, the isospin Clebsch–Gordan remains and often it is the product

$$C^2\mathscr{S}_{jl}(\mathrm{A, B}) = (T_\mathrm{A}\tfrac{1}{2}T_z\mu \mid T_\mathrm{B}T_z')^2\mathscr{S}_{jl}(\mathrm{A, B}) \tag{2.116}$$

that is quoted. *Note that* $\mathscr{S}_{jl}(\mathrm{A, B})$ *is defined as the matrix element of the creation operator* $a^{\dagger}_{jm;\frac{1}{2}\mu}$ *independently of whether the reaction is stripping or pick-up.*

If the nuclear states I_A and I_B involved in the transfer process have the same parity, then according to equation (1.40) only even l values can be transferred. On the other hand, if their parities are different only odd values of l are involved. Because the cross-section decreases rapidly with increasing l $(\sigma_{l+2}/\sigma_l \simeq \frac{1}{10})$ the transfer process is not only a sensitive l meter but also a sensitive indicator of low l contaminants in a nuclear state (Bethe and Butler 1952). For example, the low-lying states in the calcium isotopes are generally assumed to be describable in terms of the

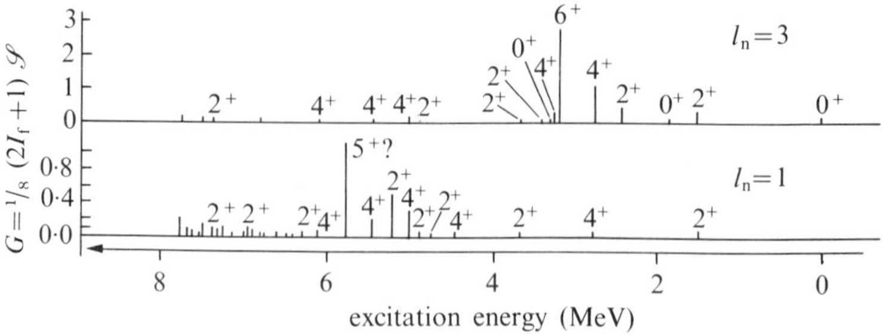

Fig. 2.13. Strength distribution in the ${}^{41}_{20}Ca_{21}(d, p){}^{42}_{20}Ca_{22}$ reaction. l_n is the orbital angular-momentum transfer and the experimental data are taken from Hansen *et al.* (1975).

$(\nu f_{\frac{7}{2}})^n$ configuration and as such only $l=3$ transfer should exist. In Fig. 2.13 we show the experimental data of Hansen *et al.* (1975) for the ${}^{41}_{20}Ca_{21}(d, p){}^{42}_{20}Ca_{22}$ reaction populating states below 5 MeV excitation energy. If the pure $f_{\frac{7}{2}}$ model were valid for describing the yrast 0^+, 2^+, 4^+, and 6^+ states there should be no $l=1$ transfer to the 2^+ (1·524 MeV) and 4^+ (2·752 MeV) states. Clearly non-vanishing $l=1$ transfer is seen and Hansen *et al.* report

$$\mathscr{S}_{jl=3}(2^+)=0{\cdot}62$$

$$\mathscr{S}_{jl=1}(2^+)=0{\cdot}05$$

$$\mathscr{S}_{jl=3}(4^+)=1{\cdot}06$$

$$\mathscr{S}_{jl=1}(4^+)=0{\cdot}04.$$

Thus despite the small values of $\mathscr{S}_{jl=1}$, the $l=1$ contaminant in these states can be detected.

We shall now examine what these data tell us about the structure of the yrast states of ${}^{42}_{20}Ca_{22}$. To do this we assume that the ground state of ${}^{41}_{20}Ca_{21}$ is a single $f_{\frac{7}{2}}$ neutron outside a closed shell. If we write the ${}^{42}_{20}Ca_{22}$ wave functions as

$$\Psi_{IM;11}=\alpha_I(f^2_{\frac{7}{2}})_{IM;11}+\beta_I(f_{\frac{7}{2}}p_{\frac{3}{2}})_{IM;11}+\dots$$

it follows that

$$\mathscr{S}_{\frac{7}{2}l=3}(I^+)=\alpha_I^2\langle(f^2_{\frac{7}{2}})_{I;1}|||\,a^\dagger_{\frac{7}{2};\frac{1}{2}}\,|||(f_{\frac{7}{2}})_{\frac{7}{2};\frac{1}{2}}\rangle^2 \tag{2.117a}$$

$$\mathscr{S}_{\frac{3}{2}l=1}(I^+)=\beta_I^2\langle(f_{\frac{7}{2}}p_{\frac{3}{2}})_{I;1}|||\,a^\dagger_{\frac{3}{2};\frac{1}{2}}\,|||(f_{\frac{7}{2}})_{\frac{7}{2};\frac{1}{2}}\rangle^2. \tag{2.117b}$$

The reduced matrix element in equation (2.117a) is just $\sqrt{2}$ times the two-particle to one-particle c.f.p. given by equation (A5.64), and since this c.f.p. is unity

$$\mathscr{S}_{\frac{7}{2}l=3}(I^+)=2\alpha_I^2. \tag{2.118a}$$

Since

$$\begin{aligned}&\langle(f_{\frac{7}{2}}p_{\frac{3}{2}})_{IM;11}|\, a^{\dagger}_{\frac{3}{2}m';\frac{1}{2}\frac{1}{2}}\,|(f_{\frac{7}{2}})_{\frac{7}{2}m;\frac{1}{2}\frac{1}{2}}\rangle\\&-(\tfrac{7}{2}\tfrac{3}{2}mm'\mid IM)(\tfrac{1}{2}\tfrac{1}{2}\tfrac{1}{2}\tfrac{1}{2}\mid 11)\langle(f_{\frac{7}{2}}p_{\frac{3}{2}})_{I;1}|||\, a^{\dagger}_{\frac{3}{2};\frac{1}{2}}\,|||(f_{\frac{7}{2}})_{\frac{7}{2};\frac{1}{2}}\rangle\\&=\sum_{m_1m_2}(\tfrac{7}{2}\tfrac{3}{2}m_1m_2\mid IM)(\tfrac{1}{2}\tfrac{1}{2}\tfrac{1}{2}\tfrac{1}{2}\mid 11)\\&\quad\times\langle 0|\, a_{\frac{3}{2}m_2;\frac{1}{2}\frac{1}{2}}a_{\frac{7}{2}m_1;\frac{1}{2}\frac{1}{2}}a^{\dagger}_{\frac{3}{2}m';\frac{1}{2}\frac{1}{2}}a^{\dagger}_{\frac{7}{2}m;\frac{1}{2}\frac{1}{2}}\,|0\rangle\\&=-(\tfrac{7}{2}\tfrac{3}{2}mm'\mid IM)(\tfrac{1}{2}\tfrac{1}{2}\tfrac{1}{2}\tfrac{1}{2}\mid 11)\end{aligned}$$

it follows that the reduced matrix element is -1 so that

$$\mathcal{S}_{\frac{3}{2}l=1}(I^{+})=\beta_I^2. \tag{2.118b}$$

Therefore to fit the observed values of the spectroscopic factors

$$\Psi_{2M;11}=0\cdot 56(f_{\frac{7}{2}}^2)_{2M;11}\pm 0\cdot 22(f_{\frac{7}{2}}p_{\frac{3}{2}})_{2M;11}+\dots$$

$$\Psi_{4M;11}=0\cdot 73(f_{\frac{7}{2}}^2)_{4M;11}\pm 0\cdot 20(f_{\frac{7}{2}}p_{\frac{3}{2}})_{4M;11}+\dots.$$

Consequently the yrast 2^+ state in ${}^{42}_{20}\mathrm{Ca}_{22}$ appears to be only about 30% $(f_{\frac{7}{2}}^2)$ and the yrast 4^+ about 50%. Although the $p_{\frac{3}{2}}$ admixture in these states is only about 5%, this contaminant can be observed because single-nucleon transfer favours low-l-values.

A further interesting aspect of single-nucleon transfer is its ability to detect whether a level is empty or is full. For example, the simplest model of ${}^{40}_{20}\mathrm{Ca}_{20}$ assumes the ground state is a doubly closed shell with no nucleons in the $0f_{\frac{7}{2}}$ orbit. If this is true $l=3$ pick-up on ${}^{40}_{20}\mathrm{Ca}_{20}$ should vanish. Thus since a $0^+\to\frac{7}{2}^-$ transition can only proceed via the transfer of $l=3$ and $j=\frac{7}{2}$, the ${}^{40}_{20}\mathrm{Ca}_{20}(p,d){}^{39}_{20}\,\mathrm{Ca}_{19}$ strength to the 2·79 MeV $\frac{7}{2}^-$ state should vanish. Since this is a weak transition it is not surprising that different experiments give different results; the (p, d) work of Kozub (1968) gives

$$\mathcal{S}_{j=\frac{7}{2}l=3}(\tfrac{7}{2}^-)=0\cdot 58, \tag{2.119a}$$

whereas the more recent work of Martin *et al.* (1972) yields

$$\mathcal{S}_{j=\frac{7}{2}l=3}(\tfrac{7}{2}^-)=0.21. \tag{2.119b}$$

However, both experiments do show $l=3$ pick-up and consequently there must be some $f_{\frac{7}{2}}$ nucleons present in the ${}^{40}_{20}\mathrm{Ca}_{20}$ ground state.

A similar situation exists for adding an $l=2$ nucleon to ${}^{40}_{20}\mathrm{Ca}_{20}$; if the $0d_{\frac{3}{2}}$ and $0d_{\frac{5}{2}}$ orbits are full then $l=2$ transfer can only populate the $1d$ orbit which should lie at 10–20 MeV excitation energy. However, the 2·01 MeV $0d_{\frac{3}{2}}$ hole state in ${}^{41}_{20}\mathrm{Ca}_{21}$ is populated in the ${}^{40}_{20}\mathrm{Ca}_{20}(d,p){}^{40}_{20}\mathrm{Ca}_{21}$ reaction (Seth *et al.* 1973) with

$$\mathcal{S}_{j=\frac{3}{2}l=2}(\tfrac{3}{2}^+)=0.05. \tag{2.120}$$

Thus it is clear that ${}^{40}_{20}Ca_{20}$ does not form a closed shell, and as we shall see in section 5.2 the observed (d, p) and (p, d) results indicate about a 20% admixture of the configuration $[(d_{\frac{3}{2}}^{-2})_{JT} \times (f_{\frac{7}{2}}^{2})_{JT}]_{00;00}$ in the ground state. This same situation is seen in other nuclei. Consequently the geometrical relationships (relationships involving only Racah and fractional-parentage coefficients) that would exist between the spectra of neighbouring nuclei if the configurations were pure seem to persist to a high degree of accuracy even when the configurations involved have a 20% contaminant. In actual fact the situation may be much worse. In a study of the hypothetical pseudonium nuclei it was shown in special cases that the geometrical relationships hold even when the configurations are only 20% pure (Cohen *et al.* 1966, Lawson and Soper 1966, Soper 1970). Thus the entities we have been calling 'single particles' are really quite complicated. One might describe them as 'clothed nucleons' or 'quasi-particles' that behave under many conditions *as if* they were a particle with angular momentum j. Through detailed studies of one- and two-nucleon transfer processes, static nuclear moments, gamma decays, beta decays, electron scattering, etc. one can investigate situations where this *as if* description breaks down and thus one can probe the 'clothing' or structure of these entities.

5.1. *Spectroscopic factors for the configuration j^n*

If the states A and B involved in the transfer process belong to the configuration j^n it is clear from the definition of $\mathcal{S}_{jl}(\mathrm{A}, \mathrm{B})$ given in equation (2.115) that

$$\mathcal{S}_{jl}\{(j^n)_{IT\alpha} \rightleftarrows (j^{n+1})_{I'T'\alpha'}\} = (n+1)\, |\langle j^n IT\alpha, j\tfrac{1}{2}|\}j^{n+1}I'T'\alpha'\rangle|^2 \tag{2.121}$$

where $\langle j^n IT\alpha, j\frac{1}{2}|\}j^{n+1}I'T'\alpha'\rangle$ is the isospin coefficient of fractional parentage defined by equation (A5.31). Tabulations of these coefficients exist for $j \leq \frac{5}{2}$ and for $j = \frac{7}{2}$, $n \leq 5$ (Glaudemans *et al.* 1964, Towner and Hardy 1969a, Hubbard 1971, Shlomo 1972), and analytic expressions for the seniority-zero and -one c.f.p.s can be found in Grayson and Nordheim (1956) and deShalit and Talmi (1963).

For the identical nucleon problem the isospin Clebsch–Gordan coefficient is unity. Furthermore only $T = (T' - \frac{1}{2})$ parents exist for the state T', and the isospin c.f.p. is equal to the identical nucleon c.f.p. defined by equation (A5.6). Thus in this case

$$\mathcal{S}_{jl}\{j^n)_{I\alpha} \rightleftarrows (j^{n+1})_{I'\alpha'}\} = (n+1)\, |\langle j^n I\alpha, j|\}j^{n+1}I'\alpha'\rangle|^2. \tag{2.122}$$

The quasi-spin formalism discussed in Appendix 3 can be used to write down the explicit n dependence of these coefficients, and provided the

TABLE 2.7

Spectroscopic factors for one-nucleon transfer within the identical nucleon configuration j^n

n even	n odd
Stripping	
$\mathscr{S}_{jl}=(n+1)\langle j^n 0, j\| \} j^{n+1} I\rangle^2$	$\mathscr{S}_{jl}=(n+1)\langle j^n j, j\| \} j^{n+1} I\rangle^2$
$=\left(\dfrac{2j+1-n}{2j+1}\right)\delta_{Ij}\delta_{v1}$	$=(n+1)$ if $I=0$ $v=0$
	$=\left(\dfrac{2j-n}{2j-1}\right)(1+(-1)^I)$ if $I\neq 0, v=2$
Pick-up	
$\mathscr{S}_{jl}=n\langle j^{n-1} I, j\| \} j^n 0\rangle^2$	$\mathscr{S}_{jl}=n\langle j^{n-1} I, j\| \} j^n j\rangle^2$
$=n\delta_{Ij}\delta_{v1}$	$=\dfrac{2j+2-n}{2j+1}$ if $I=0$ $v=0$
	$=\left(\dfrac{(n-1)(2I+1)}{(2j+1)(2j-1)}\right)(1+(-1)^I)$ if $I\neq 0, v=2$

n stands for the numbers of nucleons with angular momentum j in the target ground state. For n even the target has spin zero and v (the seniority quantum number discussed in Chapter 1, section 5) equal to zero. For n odd the target spin is j and $v=1$.

target is in the $I=0$, $v=0$ or $I=j$, $v=1$ state the simple formulae given in Table 2.7 are obtained for the spectroscopic factors (see equations (A5.54)–(A5.56) for the relevant c.f.p's).

One can use these relationships to check the consistency of the $(\pi f_{\frac{7}{2}})^n$ assignment which so successfully explained the spectrum of ${}^{49}_{21}Sc_{28}$, ${}^{50}_{22}Ti_{28}$, ${}^{51}_{23}V_{28}$ and ${}^{52}_{24}Cr_{28}$ (see Chapter 1, section 3.2). Armstrong and Blair (1965) studied the ${}^{51}_{23}V_{28}({}^3He, d){}^{52}_{24}Cr_{28}$ reaction, and according to Table 2.7 the ratio of the spectroscopic factors for populating the ground state to the $I=2^+, v=2$ state at 1·434 MeV should be

$$\frac{\mathscr{S}_{\frac{7}{2}l=3}(0^+)}{\mathscr{S}_{\frac{7}{2}l=3}(2^+)}=\frac{(n+1)(2j-1)}{2(2j-n)}$$
$$=3.$$

Experimentally this ratio is 3·7. Thus if one assumes that the ground states of the ${}^{51}_{23}V_{28}$ target and ${}^{52}_{24}Cr_{28}$ daughter are $(\pi f_{\frac{7}{2}})^n$ with $n=3$ and 4, respectively, the yrast 2^+ state in ${}^{52}_{24}Cr_{28}$ is only 81% $(\pi f_{\frac{7}{2}})^4$. Again this implies a 20% 'other-configuration' impurity despite the rather good overall fit to the energy-level data given by the pure $f_{\frac{7}{2}}$ model.

As we have already discussed in Chapter 1, section 5.1, single-nucleon transfer can only change the seniority quantum number by ±1 unit. Thus as direct calculation shows $\mathcal{S}_{jl}$ vanishes for $\Delta v > 1$. As we saw in Chapter 1, section 3.2 there are two 4^+ levels observed in ${}^{52}_{24}\mathrm{Cr}_{28}$, and according to our $f_{\frac{7}{2}}$ fit to the energy-level data the lower one at 2·370 MeV has pure $v = 4$. Thus we predict there should be no transition to the lower 4^+ state. Armstrong and Blair have measured the spectroscopic factors for these transitions and find that

$$\mathcal{S}_{j=\frac{7}{2}l=3}(4^+, 2{\cdot}370\ \mathrm{MeV}) = 0{\cdot}51$$

$$\mathcal{S}_{j=\frac{7}{2}l=3}(4^+, 2{\cdot}766\ \mathrm{MeV}) = 0{\cdot}81.$$

The sum of the spectroscopic factors is 1·32 in good agreement with the $(f^4_{\frac{7}{2}})$ prediction of 1·33. Thus the observed states are consistent with pure $(f^4_{\frac{7}{2}})$ but with mixed seniority.

$$\begin{aligned}\psi_{4M}(2{\cdot}370\ \mathrm{MeV}) &= 0{\cdot}62(\pi f_{\frac{7}{2}})^4_{4Mv=2} \pm 0{\cdot}78(\pi f_{\frac{7}{2}})^4_{4Mv=4}\\ \psi_{4M}(2{\cdot}766\ \mathrm{MeV}) &= 0{\cdot}78(\pi f_{\frac{7}{2}})^4_{4Mv=2} \mp 0{\cdot}62(\pi f_{\frac{7}{2}})^4_{4Mv=4}.\end{aligned} \quad (2.123)$$

(Gamma decay involving these levels leads to almost the same conclusion regarding the mixing; 0·62 is replaced by 0·56 and 0·78 by 0·83.) In Appendix 3 and in Chapter 3, section 2.3, we show that within the identical nucleon configuration j^n, seniority must be a good quantum number if the residual interaction is a two-body potential. Consequently for mixing to occur other configurations (such as $[f^3_{\frac{7}{2}} \times p_{\frac{3}{2}}]_{4M}$) must be involved. However, we also show in Chapter 3, section 2.3, that because of their near degeneracy the two $(\pi f_{\frac{7}{2}})^4$ levels can be thoroughly mixed without introducing large components of the other configuration. Therefore the fact that the sum of the spectroscopic factors is 1·32 is in agreement with an 'almost pure' $f_{\frac{7}{2}}$ description of the levels.

Stripping and pick-up reactions can often be used to determine where a given pure configuration state would lie if it had not had any interaction with other model space states. Let us suppose we are interested in the unperturbed energy E' of the n particle state

$$\psi_{IM;TT_z} = \Lambda[\phi_{J;T'} \times a^\dagger_{j;\frac{1}{2}}]_{IM;TT_z}$$

where Λ is a normalization constant and $\phi_{JM';T'T_z'}$ describes the ground state of the neighboring $(n-1)$ particle nucleus. For example, if $\phi_{JM';T'T_z'}$ corresponds to a closed-shell wave function, $\Lambda = 1$ and E' is the energy of the single-particle state (jm). Alternatively $\phi_{JM';T'T_z'}$ might represent the seniority-zero ground state of the configuration $(j)^{n-1}$, in which case E' would be the energy of the seniority-1 state of the configuration $(j)^n$ and Λ would be obtained from equations (1.120) and (1.125).

If one denotes the wave function of the kth observed state with angular momentum $(IM;\, TT_z)$ by $\Psi_{IM;TT_z}(k)$, then

$$\Psi_{IM;TT_z}(k) = \alpha_{1k}\psi_{IM;TT_z} + \sum_{i=2}^{a} a_{ik}\Phi_{IM;TT_z}(i)$$

where $\psi_{IM;TT_z}$ and $\Phi_{IM;TT_z}(i)$ are the normalized basis states used to diagonalize H. Provided $\langle\Phi_{I;T}|||\, a^{\dagger}_{j;\frac{1}{2}}\, |||\phi_{J;T'}\rangle = 0$ the spectroscopic factor for single-nucleon transfer to any one of the observed states will be

$$\mathscr{S}_{jl}(k) = \alpha^2_{1k}\langle\psi_{I;T}|||\, a^{\dagger}_{j;\frac{1}{2}}\, |||\phi_{J;T'}\rangle^2.$$

The state ψ_{IM,TT_z} can of course be expressed in terms of the $\Psi_{IM;TT_z}(k)$, and since the α_{ik} form an orthogonal matrix

$$\psi_{IM;TT_z} = \sum_k \alpha_{1k}\Psi_{IM;TT_z}(k).$$

Thus

$$\begin{aligned} E' &= \Big\langle \sum_k \alpha_{1k}\Psi_{IM;TT_z}(k) \Big| H \Big| \sum_{k'} \alpha_{1k'}\Psi_{IM;TT_z}(k')\Big\rangle \\ &= \sum_k \alpha^2_{1k} E_k \end{aligned}$$

where the E_k are the observed energies of the various states $\Psi_{IM;TT_z}(k)$, and the cross-terms $\langle\Psi_{IM;TT_z}(k)|\, H\, |\Psi_{IM;TT_z}(k')\rangle$ vanish because $\Psi_{IM;TT_z}(k)$ is an eigenfunction of H. Consequently

$$\begin{aligned} E' &= \frac{\sum_k \mathscr{S}_{jl}(k)E_k}{\langle\psi_{I;T}|||\, a^{\dagger}_{j;\frac{1}{2}}\, |||\phi_{J;T'}\rangle^2} \\ &= \frac{\sum_k \mathscr{S}_{jl}(k)E_k}{\sum_k \mathscr{S}_{jl}(k)} \end{aligned} \tag{2.124}$$

where the last equality follows from the fact that $\sum_k \alpha^2_{1k} = 1$.

One may use equation (2.124) in several different ways. For example, if one wishes to know the energy of the single-particle state relative to a closed shell (or at least what one assumes to be a closed shell) the (d, p) reaction to states with the spin of interest will give the information. Although the l value of the transferred particle can be simply obtained for strong transitions from the angle at which the cross-section peaks, the j value is more difficult to determine. One way of doing this is to use a polarized deuteron beam and observe the asymmetry of the outgoing protons. Kocher and Haeberli (1972) have studied the ${}^{40}_{20}Ca_{20}(d, p){}^{41}_{20}Ca_{21}$ reaction using polarized deuterons and find $l=1$ transfer to eight states:

$$\mathcal{S}_{j=\frac{3}{2}l=1}(1{\cdot}95\ \text{MeV})=0{\cdot}70$$

$$\mathcal{S}_{j=\frac{3}{2}l=1}(2{\cdot}47\ \text{MeV})=0{\cdot}25$$

$$\mathcal{S}_{j=\frac{3}{2}l=1}(4.62\ \text{MeV})=0{\cdot}04$$

$$\mathcal{S}_{j=\frac{3}{2}l=1}(5{\cdot}49\ \text{MeV})=0{\cdot}03$$

and

$$\mathcal{S}_{j=\frac{1}{2}l=1}(3{\cdot}62\ \text{MeV})=0{\cdot}11$$

$$\mathcal{S}_{j=\frac{1}{2}l=1}(3{\cdot}95\ \text{MeV})=0{\cdot}67$$

$$\mathcal{S}_{j=\frac{1}{2}l=1}(4{\cdot}77\ \text{MeV})=0{\cdot}19$$

$$\mathcal{S}_{j=\frac{1}{2}l=1}(5{\cdot}46\ \text{MeV})=0{\cdot}03.$$

Thus the energies of the single-particle $p_{\frac{3}{2}}$ and $p_{\frac{1}{2}}$ states relative to the ${}^{41}_{20}Ca_{21}$ ground state are

$$E'_{p_{\frac{3}{2}}}=2{\cdot}29\ \text{MeV}$$

$$E'_{p_{\frac{1}{2}}}=4{\cdot}11\ \text{MeV}.$$

A second place where such results are useful is in unravelling the residual two-body interaction. Obviously when we take say the $I=0$ interaction from an analysis of the binding energies, it is an effective interaction which includes the influence of many other configurations. One can remove the effects of at least the nearest configurations by using equation (2.124). In the ${}^{41}_{20}Ca_{21}(d, p){}^{42}_{20}Ca_{22}$ work of Hansen *et al.* (1975) three 0^+ states are seen below 4 MeV excitation energy with

$$\mathcal{S}_{j=\frac{7}{2}l=3}(0^+,\ \text{ground state})=1{\cdot}0$$

$$\mathcal{S}_{j=\frac{7}{2}l=3}(0^+,\ 1{\cdot}836\ \text{MeV})=0{\cdot}23$$

$$\mathcal{S}_{j=\frac{7}{2}l=3}(0^+,\ 3{\cdot}295\ \text{MeV})=0{\cdot}08.$$

If one assumes that ${}^{41}_{20}Ca_{21}$ is a single $f_{\frac{7}{2}}$ neutron outside a closed core and that the observed $\mathcal{S}_{jl}$ exhaust all the $f_{\frac{7}{2}}$ strength to the $I=0^+$ levels, one can use equation (2.124) to deduce that the pure $(f^2_{\frac{7}{2}})$ $I=0$ state lies $0{\cdot}524$ MeV above the ground state. Thus instead of the value given by equation (2.2) a more realistic value for the two-body interaction between the $f_{\frac{7}{2}}$ neutrons themselves is $E_0=-3{\cdot}109+0{\cdot}524=-2{\cdot}585$ MeV.

Of course, the difficulty with this method is the fact that very weak states at high excitation energy play an important role, and it is precisely these values of $\mathcal{S}_{jl}$ that are the most uncertain.

5.2. Spectroscopic factors when more than one level is involved

In general, the valence nucleons will occupy several different single-particle orbits and in this section we outline the method of calculating $\mathscr{S}_{jl}$ in this case. Let us suppose that the nuclear states involved in the transition are

$$\Psi_{IM;TT_z} = \sum_{J_1T_1J_2T_2} \alpha_{IT}(J_1T_1; J_2T_2)[(j_1)^{n_1}_{J_1;T_1} \times (j_2)^{n_2}_{J_2;T_2}]_{IM;TT_z}$$

$$= \sum_{J_1T_1J_2T_2} \alpha_{IT}(J_1T_1; J_2T_2 \qquad (2.125a)$$

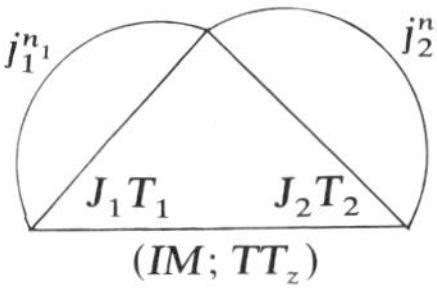

and

$$\Psi_{I'M';T'T_z'} = \sum_{J_1'T_1'J_2'T_2'} \beta_{I'T'}(J_1'T_1'; J_2'T_2') \qquad (2.125b)$$

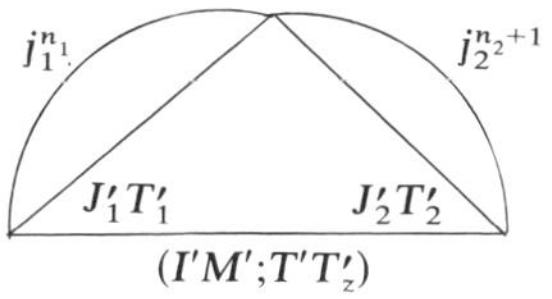

where the expansion coefficients α_{IT} and $\beta_{I'T'}$ are determined by diagonalizing the shell-model Hamiltonian. From the structure of the wave functions it is clear that we are interested in the transfer of a (j_2l_2) nucleon so that we must calculate the effect of $a^{\dagger}_{j_2m;\frac{1}{2}\mu}$ operating on $\Psi_{IM;TT_z}$. On a typical term in equation (2.125a) this operation gives

$$a^{\dagger}_{j_2m;\frac{1}{2}\mu}[(j_1)^{n_1}_{J_1T_1} \times (j_2)^{n_2}_{J_2T_2}]_{IM;TTz}$$

$$= \sum_{I''M''T''T_z''} (Ij_2Mm \mid I''M'')(T\tfrac{1}{2}T_z\mu \mid T''T_z'')$$

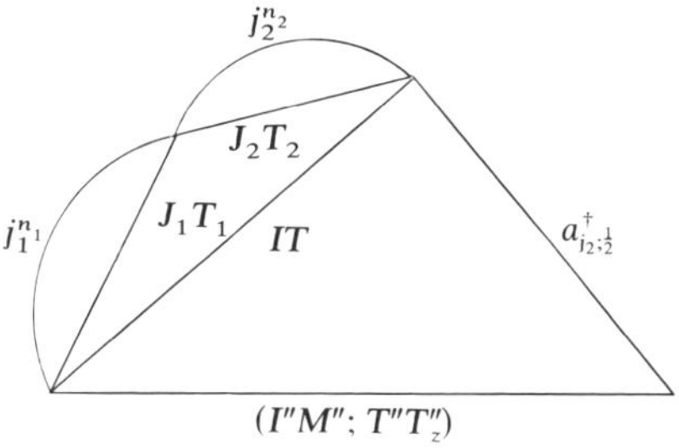

$$= \sqrt{\{(2I+1)(2T+1)\}} \sum_{J_2''T_2''I''M''T''T_z''} \sqrt{\{(2J_2''+1)(2T_2''+1)\}}$$

$$\times W(J_1J_2I''j_2; IJ_2'')W(T_1T_2T''\tfrac{1}{2}; TT_2'')$$
$$\times (Ij_2Mm \mid I''M'')(T\tfrac{1}{2}T_z\mu \mid T''T_z'')$$

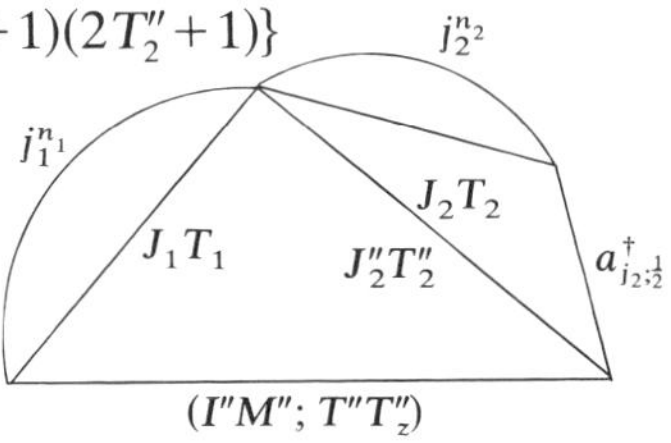

$$= \sqrt{\{(2I+1)(2T+1)\}} \sum_{J_2''T_2''I''M''T''T_z''} \sqrt{\{(2J_2''+1)(2T_2''+1)\}} W(J_1J_2I''j_2; IJ_2'')$$

$$\times W(T_1T_2T''\tfrac{1}{2}; TT_2'') < (j_2)^{n_2+1}_{J_2'';T_2''} \| a^{\dagger}_{j_2;\frac{1}{2}} \| (j_2)^{n_2}_{J_2;T_2} >$$
$$\times (Ij_2Mm \mid I''M'')(T\tfrac{1}{2}T_z\mu \mid T''T_z'')$$

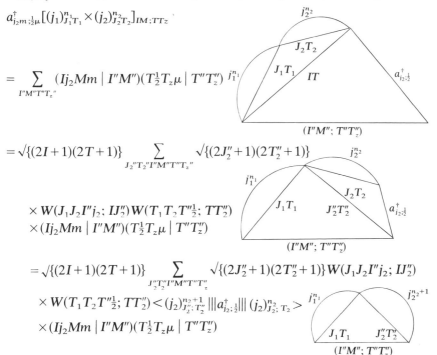

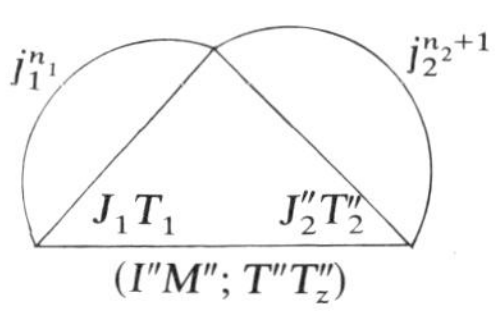

The final-state wave function given in equation (2.125b) picks out the term in which $(I''M''T''T''_z)=(I'M'T'T'_z)$, and from the definition of the reduced matrix element, equation (2.102b), it follows that the coefficient of $(Ij_2Mm \mid I'M')(T\frac{1}{2}T_z\mu \mid T'T'_z)$ is precisely the spin–isospin reduced matrix element. Moreover, since the reduced matrix element of $a^{\dagger}_{j_2;\frac{1}{2}}$ is $\sqrt{(n_2+1)}$ times the isospin c.f.p. given in equation (A5.31) the spectroscopic factor for one-nucleon transfer connecting the states of equations (2.125) is

$$\mathscr{S}_{jl}(I'T' \rightleftarrows IT)=(n_2+1)(2I+1)(2T+1)\Bigg|\sum_{J_1T_1J_2T_2J_2'T_2'} \alpha_{IT}(J_1T_1;J_2T_2)$$
$$\times \beta_{I'T'}(J_1T_1;J_2'T_2')\sqrt{\{(2J_2'+1)(2T_2'+1)\}}\langle j_2^{n_2}J_2T_2,j_2\tfrac{1}{2}|\}j_2^{n_2+1}J_2'T_2'\rangle$$
$$\times W(J_1J_2I'j_2;IJ_2')W(T_1T_2T'\tfrac{1}{2};TT_2')\Bigg|^2. \qquad (2.126)$$

From the work of Seth *et al.* (1973) we know that in the ${}^{40}_{20}\mathrm{Ca}_{20}(d,p){}^{41}_{20}\mathrm{Ca}_{21}$ reaction the 2·01 MeV $d_{\frac{3}{2}}$ hole state is populated with the spectroscopic factor given by equation (2.120). Thus the ${}^{40}_{20}\mathrm{Ca}_{20}$ ground state must at least contain a two-particle two-hole component and we take it to have the form

$$\Psi_{00;00}=\alpha_0(d_{\frac{3}{2}})^8_{00;00}+\sum_{J_1T_1}\alpha_{00}(J_1T_1;J_1T_1)[(f_{\frac{7}{2}})^2_{J_1;T_1}\times(d_{\frac{3}{2}})^6_{J_1;T_1}]_{00;00} \qquad (2.127)$$

Furthermore, if we assume the 2·01 MeV $\frac{3}{2}^+$ state has the structure

$$\Psi_{\frac{3}{2}m;\frac{1}{2}\frac{1}{2}}=\sum_{J_1'T_1'}\beta_{\frac{3}{2}\frac{1}{2}}(J_1'T_1';\tfrac{3}{2}\tfrac{1}{2})[(f_{\frac{7}{2}})^2_{J_1';T_1'}\times(d_{\frac{3}{2}})^7_{\frac{3}{2};\frac{1}{2}}]_{\frac{3}{2}m;\frac{1}{2}\frac{1}{2}} \qquad (2.128)$$

comparison of these wave functions with equations (2.125) implies that to calculate the spectroscopic factor one must set $I=T=0$, $I'=J_2'=j_2=\frac{3}{2}$, $T'=T_2'=\frac{1}{2}$, $n_2=6$, and $J_1=J_2$ $T_1=T_2$ in equation (2.126). When the special form of the Racah coefficient with one index equal to zero is used (equation A(4.14)) one finds

$$\mathscr{S}_{j=\frac{3}{2}l=2}({}^{40}_{20}\mathrm{Ca}_{20}\rightarrow{}^{41}_{20}\mathrm{Ca}_{21})$$
$$=7\Bigg|\sum_{J_1T_1}(-1)^{J_1+T_1}\frac{\langle(\frac{3}{2})^6J_1T_1,\frac{3}{2}\frac{1}{2}|\}(\frac{3}{2})^7\frac{3}{2}\frac{1}{2}\rangle}{\sqrt{\{(2J_1+1)(2T_1+1)\}}}\alpha_{00}(J_1T_1;J_1T_1)$$
$$\times\beta_{\frac{3}{2}\frac{1}{2}}(J_1T_1;\tfrac{3}{2}\tfrac{1}{2})\Bigg|^2.$$

According to equation (A5.67)

$$\langle j^{4j}JT,j\tfrac{1}{2}|\}j^{4j+1}j\tfrac{1}{2}\rangle=-\left\{\frac{(2J+1)(2T+1)}{4(4j+1)(2j+1)}\right\}^{\frac{1}{2}}\{1-(-1)^{J+T}\}$$

and consequently when $j=\frac{3}{2}$

$$\mathcal{S}_{j=\frac{3}{2}l=2}({}^{40}_{20}\mathrm{Ca}_{20}\to{}^{41}_{20}\mathrm{Ca}_{21})=\tfrac{1}{16}\left|\sum_{J_1T_1}\{1-(-1)^{J_1+T_1}\}\alpha_{00}(J_1T_1;J_1T_1)\times\beta_{\frac{3}{2}\frac{1}{2}}(J_1T_1;\tfrac{3}{2}\tfrac{1}{2})\right|^2.$$

If we assume the extreme weak-coupling model for the $\frac{3}{2}^+$ hole state, i.e.

$$\beta_{\frac{3}{2}\frac{1}{2}}(J_1T_1;\tfrac{3}{2}\tfrac{1}{2})=\delta_{J_10}\,\delta_{T_11} \tag{2.129}$$

the stripping data of Seth *et al.* (1973) given in equation (2.120) implies that

$$\alpha^2_{00}(00;00)=0.20; \tag{2.130}$$

in other words there is a 20% probability that the ground state of ${}^{40}_{20}\mathrm{Ca}_{20}$ has the two-particle two-hole component $[(f^2_{\frac{7}{2}})_{0;0}\times(d^6_{\frac{3}{2}})_{0;0}]_{00;00}$.

Since $f_{\frac{7}{2}}$ pick-up cannot occur unless there are some $f_{\frac{7}{2}}$ nucleons to be picked up, the ${}^{40}_{20}\mathrm{Ca}_{20}(p,d){}^{39}_{20}\mathrm{Ca}_{19}$ strength to the 2.79 MeV $\frac{7}{2}^-$ state also gives information about the two-particle two-hole components in the ${}^{40}_{20}\mathrm{Ca}_{20}$ ground state. The simplest wave function for the $\frac{7}{2}^-$ state corresponds to one nucleon excited from the $d_{\frac{3}{2}}$ orbit to the $f_{\frac{7}{2}}$ shell so that

$$\Psi_{\frac{7}{2}m;\frac{1}{2}\frac{1}{2}}=\sum_{J_2'T_2'}\beta_{\frac{7}{2}\frac{1}{2}}(\tfrac{7}{2}\tfrac{1}{2};J_2'T_2')[(f_{\frac{7}{2}})_{\frac{7}{2};\frac{1}{2}}\times(d^6_{\frac{3}{2}})_{J_2';T_2'}]_{\frac{7}{2}m;\frac{1}{2}\frac{1}{2}}. \tag{2.131}$$

When this ansatz is combined with the wave function of the ${}^{40}_{20}\mathrm{Ca}_{20}$ ground state given by equation (2.127) one finds from equation (2.126) that

$$\mathcal{S}_{j=\frac{7}{2}l=3}({}^{40}_{20}\mathrm{Ca}_{20}\to{}^{39}_{20}\mathrm{Ca}_{19})=2\left|\sum_{J_1T_1}(-1)^{J_1+T_1}\alpha_{00}(J_1T_1;J_1T_1)\beta_{\frac{7}{2}\frac{1}{2}}(\tfrac{7}{2}\tfrac{1}{2};J_1T_1)\right|^2$$

where use has been made of the fact that the two-particle to one-particle c.f.p. is unity.

The (p,d) work of Kozub (1968) gives 0·58 for this spectroscopic factor (equation (2.119a)), so that if we assume the extreme weak-coupling model for the $\frac{7}{2}^-$ state, i.e.

$$\beta_{\frac{7}{2}\frac{1}{2}}(\tfrac{7}{2}\tfrac{1}{2};J_1T_1)=\delta_{J_10}\,\delta_{T_11}$$

one finds

$$\alpha^2_{00}(00;00)=0.29. \tag{2.132a}$$

On the other hand, the results of Martin *et al.* (1972) give $\mathcal{S}_{jl}=0{\cdot}21$ (equation (2.119b)) which would imply that

$$\alpha^2_{00}(00;00)=0.105. \tag{2.132b}$$

Thus within a factor of 2 the (p, d) estimate of $\alpha_{00}(00; 00)$ is consistent with the (d, p) value given by equation (2.130), and the implication is that there is about a 20% probability of the two-particle two-hole configuration $[(f^{2}_{\frac{7}{2}})_{0;0}\times d^{6}_{\frac{3}{2}})_{0;0}]_{00;00}$ in the $^{40}_{20}\mathrm{Ca}_{20}$ ground state.

5.3. *Sum rules*

The strength function $G_{jl}(\mathrm{A}, \mathrm{B})$ of equation (2.113) obeys certain sum rules which allow us to extract useful nuclear-structure information (French and Macfarlane 1961). We first consider a stripping reaction in which a nucleon is added to the initial state $(I_\mathrm{A}M_\mathrm{A}T_\mathrm{A}T_z)=(I_\mathrm{i}M_\mathrm{i}T_\mathrm{i}T_z)$. If we sum over all final states $(I_\mathrm{B}M_\mathrm{B}T_\mathrm{B}T'_z)=(I_\mathrm{f}M_\mathrm{f}T_\mathrm{f}T'_z)$ and make use of the fact that the final states form a complete set

$$\sum_{I_\mathrm{f}M_\mathrm{f}T_\mathrm{f}T_z'} |\Psi_{I_\mathrm{f}M_\mathrm{f};T_\mathrm{f}T_{z'}}\rangle\langle\Psi_{I_\mathrm{f}M_\mathrm{f};T_\mathrm{f}T_{z'}}|=1$$

one sees that

$$\sum_{\{\mathrm{f}\}} G_{jl}(I_\mathrm{i}T_\mathrm{i}T_z\,;\,I_\mathrm{f}T_\mathrm{f}T'_z)=\frac{1}{2I_\mathrm{i}+1}\sum_{mM_\mathrm{i}}\langle\Psi_{I_\mathrm{i}M_\mathrm{i};T_\mathrm{i}T_z}|\,a_{jm;\frac{1}{2}\mu}a^{\dagger}_{jm;\frac{1}{2}\mu}\,|\Psi_{I_\mathrm{i}M_\mathrm{i};T_\mathrm{i}T_z}\rangle$$

$$=\frac{1}{2I_\mathrm{i}+1}\sum_{mM_\mathrm{i}}\langle\Psi_{I_\mathrm{i}M_\mathrm{i};T_\mathrm{i}T_z}|\,1-a^{\dagger}_{jm;\frac{1}{2}\mu}a_{jm;\frac{1}{2}\mu}\,|\Psi_{I_\mathrm{i}M_\mathrm{i};T_\mathrm{i}T_z}\rangle$$

where use has been made of the anticommutation relationship given in equation (2.21) in writing the last line of this equation. The operator $\sum_m a^{\dagger}_{jm;\frac{1}{2}\mu}a_{jm;\frac{1}{2}\mu}$ is just the number operator for nucleons of type μ (neutrons when $\mu=+\frac{1}{2}$, protons when $\mu=-\frac{1}{2}$) given in equation (1.41). Since $(2j+1)$ is the total number of neutrons or protons an orbit can accommodate, it follows that $\sum_m(1-a^{\dagger}_{jm;\frac{1}{2}\mu}a_{jm;\frac{1}{2}\mu})=(2j+1)-\sum_m a_{jm;\frac{1}{2}\mu}a^{\dagger}_{jm;\frac{1}{2}\mu}$ is precisely the number of holes in the orbit j. Thus for stripping

$$\sum_{\{\mathrm{f}\}} G_{jl}(I_\mathrm{i}T_\mathrm{i}T_z\,;\,I_\mathrm{f}T_\mathrm{f}T'_z)=\bar{N}_j \text{ when a neutron is transferred}$$

$$=\bar{Z}_j \text{ when a proton is transferred} \qquad (2.133)$$

where $\bar{N}_j$ and $\bar{Z}_j$ are the number of neutron and proton holes, respectively, in the orbit j.

Since the strength is related to the spectroscopic factor, this relationship implies that $\mathcal{S}_{jl}$ satisfies certain sum rules. From equations (2.113) and (2.115) one sees that

$$\sum_{\{f\}} G_{jl}(I_iT_iT_z\,;\,I_fT_fT_z') = \frac{1}{(2I_i+1)} \sum_{mM_iM_fI_fT_fT_z'} (I_ijM_im \mid I_fM_f)^2$$
$$\times (T_i\tfrac{1}{2}T_z\mu \mid T_fT_z')^2 \mathscr{S}_{jl}(I_iT_i;\,I_fT_f)$$
$$= \sum_{I_fT_fT_z'} \left(\frac{2I_f+1}{2I_i+1}\right)(T_i\tfrac{1}{2}T_z\mu \mid T_fT_z')^2 \mathscr{S}_{jl}(I_iT_i;\,I_fT_f) \tag{2.134}$$

where use has been made of equation (A1.23) to carry out the M-sums. Comparison of equations (2.133) and (2.134) leads to the stripping sum rules

$$\bar{N}_j = \sum_{I_fT_f} \left(\frac{2I_f+1}{2I_i+1}\right)(T_i\tfrac{1}{2}T_z\tfrac{1}{2} \mid T_fT_z+\tfrac{1}{2})^2 \mathscr{S}_{jl}(I_iT_i;\,I_fT_f)$$
$$\bar{Z}_j = \sum_{I_fT_f} \left(\frac{2I_f+1}{2I_i+1}\right)(T_i\tfrac{1}{2}T_z-\tfrac{1}{2} \mid T_fT_z-\tfrac{1}{2})^2 \mathscr{S}_{jl}(I_iT_i;I_fT_f).$$

By substituting the explicit values for the isospin Clebsch–Gordan coefficients given in Table A1.2, one may rewrite these equations for a target with a neutron excess and $T_z = T_i$ as

$$\bar{N}_j = \sum_{I_f} \left(\frac{2I_f+1}{2I_i+1}\right)\mathscr{S}_{jl}(I_iT_i;\,I_fT_i+\tfrac{1}{2}) \tag{2.135a}$$

$$\bar{Z}_j = \sum_{I_f} \left(\frac{2I_f+1}{2I_i+1}\right)\left\{\left(\frac{2T_i}{2T_i+1}\right)\mathscr{S}_{jl}(I_iT_i;\,I_fT_i-\tfrac{1}{2}) + \left(\frac{1}{2T_i+1}\right)\mathscr{S}_{jl}(I_iT_i;\,I_fT_i+\tfrac{1}{2})\right\}.$$
2.135b)

Thus stripping (in which a neutron is added to an already neutron-rich target nucleus) can only lead to states with isospin half a unit greater than that of the target. On the other hand, when a proton is added to the target, states with isospin half a unit greater and half a unit less can be populated. From equations (2.135) it follows that in a proton-stripping reaction *the sum of all the transition strengths to the T–upper states is*

$$\sum_{I_f} G_{jl}(I_iT_iT_z\,;\,I_fT_f = T_i+\tfrac{1}{2}T_z' = T_i-\tfrac{1}{2}) = \frac{\bar{N}_j}{2T_i+1} \tag{2.136}$$

Since the strength and the spectroscopic factor are always defined as the matrix element of the creation operator, the roles of the A and B quantum numbers in equation (2.113) are interchanged when pick-up is to be described. Thus for pick-up $(I_A M_A T_A T_z) = (I_f M_f T_f T_z')$ and $(I_B M_B T_B T_z') = (I_i M_i T_i T_z)$. In the same way that equations (2.135) were deduced one can show that the pick-up sum rules are

$$N_j = \sum_{I_f T_f} (T_f \tfrac{1}{2} T_z - \tfrac{1}{2} \tfrac{1}{2} \mid T_i T_z)^2 \mathcal{S}_{jl}(I_i T_i;\, I_f T_f)$$

$$Z_j = \sum_{I_f T_f} (T_f \tfrac{1}{2} T_z + \tfrac{1}{2} - \tfrac{1}{2} \mid T_i T_z)^2 \mathcal{S}_{jl}(I_i T_i;\, I_f T_f)$$

where N_j and Z_j are the number of neutrons and protons in the level j. If one puts in the explicit values for the isospin Clebsch–Gordan coefficients one finds for $T_z = T_i$ that

$$N_j = \sum_{I_f} \left\{ \mathcal{S}_{jl}(I_i T_i;\, I_f T_i - \tfrac{1}{2}) + \left(\frac{1}{2T_i + 2}\right) \mathcal{S}_{jl}(I_i T_i;\, I_f T_i + \tfrac{1}{2}) \right\} \quad (2.137a)$$

$$Z_j = \left(\frac{2T_i + 1}{2T_i + 2}\right) \sum_{I_f} \mathcal{S}_{jl}(I_i T_i;\, I_f T_i + \tfrac{1}{2}). \quad (2.137b)$$

Thus proton pick-up on an already neutron-rich target leads to a nucleus which has an even larger neutron excess; in other words only states with $T_f = T_i + \frac{1}{2}$ can be populated. However, neutron pick-up can populate states with $T_f = T_i \pm \frac{1}{2}$. From equations (2.137) it follows that in a neutron pick-up reaction *the sum of all the transition strengths to the T-upper states is*

$$\sum_{I_f} G_{jl}(I_i T_i T_z;\, I_f T_f = T_i + \tfrac{1}{2} T_z' = T_i - \tfrac{1}{2}) = \frac{Z_j}{2T_i + 1} \quad (2.138)$$

Since the sum rules tell us the number of neutron and proton particles or holes in a shell, it is clear that stripping and pick-up experiments can be used to detect the filling of the various single-particle orbits throughout the periodic table. This has been extensively carried out by Cohen (1968), and we shall discuss this aspect of single-nucleon transfer further in Chapter 6, section 2.3.

In section 4.2 of this chapter we discussed the $d_{\frac{3}{2}}$-hole states in ${}^{40}_{20}Ca_{20}$ i.e. the states $[(d_{\frac{3}{2}})^{-1}_{\frac{3}{2};\frac{1}{2}} \times (f_{\frac{7}{2}})^{1}_{\frac{7}{2};\frac{1}{2}}]_{IM;T0}$. These states should be strongly populated by $l = 2$ pick-up in the ${}^{41}_{20}Ca_{21}(d, t){}^{40}_{20}Ca_{20}$ reaction. We now examine what the sum rules tell us about this reaction and compare the predicted results with experiment. If we assume that the ground state of ${}^{41}_{20}Ca_{21}$ has a full $d_{\frac{3}{2}}$ shell then $Z_{j=\frac{3}{2}} = 4$ and equation (2.138) implies that the sum of the strengths to the T upper states $(T = 1)$ should be

$$\sum_{I_f} G_{j=\frac{3}{2} l=2}(\tfrac{7}{2}\tfrac{1}{2}\tfrac{1}{2};\, I_f T_f = 1 T_z' = 0) = \frac{4}{2T_i + 1}$$
$$= 2.$$

Furthermore, since the total strength for neutron pick-up must be equal to $N_{j=\frac{3}{2}}=4$ (equation (2.137a)) it follows that the total strength to the T-lower states is also 2. Since the initial and final states involved in the transition have the structure

$$[(f_{\frac{7}{2}})_{\frac{7}{2};\frac{1}{2}} \times (d_{\frac{3}{2}})^{8}_{0;0}]_{\frac{7}{2}m;\frac{1}{2}\frac{1}{2}}$$

and

$$[(f_{\frac{7}{2}})_{\frac{7}{2};\frac{1}{2}} \times (d_{\frac{3}{2}})^{7}_{\frac{3}{2};\frac{1}{2}}]_{IM;T0}$$

it follows from equations (2.125) that the appropriate values of the angular momenta to be used for calculating $\mathscr{S}_{jl}$, from equation (2.126) are $J_1=\frac{7}{2}$, $T_1=\frac{1}{2}$, $j_2=J_2=\frac{3}{2}$, $T_2=\frac{1}{2}$, $n_2=7$, $J_2'=T_2'=0$, $I'=\frac{7}{2}$, and $T'=\frac{1}{2}$. The full shell c.f.p. is unity so that

$$\mathscr{S}_{j=\frac{3}{2}l=2}(\tfrac{7}{2}\tfrac{1}{2}\rightarrow IT)=\frac{(2I+1)(2T+1)}{16}.$$

In the pick-up reaction $I_{\mathrm{B}}=I_{\mathrm{i}}$ so that the strength for pick-up given by equation (2.114) is

$$G_{j=\frac{3}{2}l=2}(\tfrac{7}{2}\tfrac{1}{2}\tfrac{1}{2};\, IT0)=(T\tfrac{1}{2}0\tfrac{1}{2}\mid\tfrac{1}{2}\tfrac{1}{2})^2\mathscr{S}_{j=\frac{3}{2}l=2}(\tfrac{7}{2}\tfrac{1}{2}\rightarrow IT)$$

and consequently when the isospin Clebsch–Gordan coefficients are taken into account

$$G_{j=\frac{3}{2}l=2}(\tfrac{7}{2}\tfrac{1}{2}\tfrac{1}{2};\, IT0)=\frac{(2I+1)}{16} \tag{2.139}$$

(Since I can be 2, 3, 4, and 5 the strength to each T group given by equation (2.139) is 2 and is in agreement with the sum rule results.)

Betts *et al.* (1975) have studied this reaction and their $l=2$ strengths together with the observed excitation energies are given in Table 2.8. One should remember that even with careful experiments and data analysis the experimental value of $C^2\mathscr{S}$ is uncertain to about 20%. Consequently, except for the $T=0$ 3^- transition, the individual strengths are in good agreement with theory. As we have already mentioned, observation at forward angles does not differentiate between $j=l-\frac{1}{2}$ and $j=l+\frac{1}{2}$ pick-up unless polarization measurements are made. Thus it is possible that the $l=2$ strength to the 7·11 MeV 3^- level may contain a substantial $d_{\frac{5}{2}}$ component, and in fact if one assumes all this strength corresponds to $j=\frac{5}{2}$ pick-up then $\sum(C^2\mathscr{S})=0{\cdot}49$ for the other two 3^- states in good agreement with the theoretical expectation.

TABLE 2.8

Strengths of the $l = 2$ transitions observed in the ${}^{41}_{20}Ca_{21}(d, t){}^{40}_{20}Ca_{20}$ reaction

Isospin T	Angular momentum I	Excitation energy (MeV)	$(C^2\mathscr{S})_{expt}$	$\sum(C^2\mathscr{S})_{expt}$	$(C^2\mathscr{S})_{theory}$
0	2^-	6·03	0·10		
		6·75	0·18	0·28	0·31
	3^-	3·74	0·22		
		6·58	0·27		
		7·11	0·23	0·72	0·44
	4^-	5·61	0·50	0·50	0·56
	5^-	4·49	0·58	0·58	0·69
1	2^-	8·42	0·24	0·24	0·31
	3^-	7·69	unresolved	1·2	0·44
	4^-	7·66			0·56
	5^-	8·55	0·68	0·68	0·69

The experimental results are those of Betts *et al.* (1975). The theoretical values of $C^2\mathscr{S}$ are given by equation (2.139).

Since $l = 2$ pick-up is seen to two 2^- and two 3^- states (we shall now assume that the $l = 2$ pick-up to the 7·11 MeV level is indeed $d_{\frac{5}{2}}$), it follows from equation (2.124) that the 'true' positions of the $[(d_{\frac{3}{2}})^{-1}_{\frac{3}{2};\frac{1}{2}} \times (f_{\frac{7}{2}})^{1}_{\frac{7}{2};\frac{1}{2}}]_{IM;00}$ states with $I = 2$ and 3 are

$$E_2 = 6{\cdot}49 \text{ MeV}$$

$$E_3 = 5{\cdot}30 \text{ MeV}.$$

If these values are used instead of the 2^- and 3^- energies given in Table 2.6 one finds that β, the interaction parameter to be used in the $d_{\frac{3}{2}} - f_{\frac{7}{2}}$ particle-hole calculation (see equations (2.106b) and (2.107)), is 2·80 MeV. This is in close agreement with the value given by Sherr *et al.* (1974) which is given in equation (2.111).

From equations (2.136) and (2.138) it is clear that the proton-stripping strength and neutron pick-up strength to the T upper states decreases rapidly with increasing isospin of the target. This result is, of course, also true for the individual T upper states. For example, the ${}^{50+N}_{50}Sn_N(d, {}^3He){}^{49+N}_{49}In_N$ reaction populates many states in the indium nucleus with isospin $T = T_i + \frac{1}{2}$, where $T_i = (N - 50)/2$ is the isospin of the tin ground state. The strength to a particular final state in the indium nucleus will be, according to equation (2.114)

$$G_{jl}(I_f T_f = T_i + \tfrac{1}{2} T'_z = T_i + \tfrac{1}{2};\ I_i T_i T_z = T_i)$$
$$= (T_i + \tfrac{1}{2} \tfrac{1}{2}\ T_i + \tfrac{1}{2}\ -\tfrac{1}{2} \mid T_i T_i)^2 \mathscr{S}_{jl}(I_f T_f;\ I_i T_i)$$
$$= \frac{(2T_i+1)}{(2T_i+2)} \mathscr{S}_{jl}(I_f T_f;\ I_i T_i)$$

where we have made use of the fact that in the pick-up reaction B stands for the quantum numbers of the initial state and A for those of the final nucleus. If we now consider the (d, t) reaction on the same target there will be strength to states with $T = T_i - \frac{1}{2}$ and also to states with $T = T_i + \frac{1}{2}$. In particular, the states with $T = T_i + \frac{1}{2}$ observed in the $^{49+N}_{50}\mathrm{Sn}_{N-1}$ nucleus will be the isobaric analogues of the $^{49+N}_{49}\mathrm{In}_N$ states; i.e. as discussed in section 1.3 of this chapter, the T upper states in tin have exactly the same structure as the indium states except that a proton has been changed to a neutron. The strength for populating a particular isobaric analogue state by the (d, t) reaction is

$$G_{jl}(I_f T_f = T_i + \tfrac{1}{2} T'_z = T_i - \tfrac{1}{2};\ I_i T_i T_z = T_i)$$
$$= (T_i + \tfrac{1}{2} \tfrac{1}{2}\ T_i - \tfrac{1}{2} \tfrac{1}{2} \mid T_i T_i)^2 \mathscr{S}_{jl}(I_f T_f;\ I_i T_i)$$
$$= \frac{1}{(2T_i+2)} \mathscr{S}_{jl}(I_f T_f;\ I_i T_i).$$

Since the spectroscopic factor itself has no T_z dependence (this is all contained in the isospin Clebsch–Gordan coefficient) it follows that if we consider the ratio of the $(d, {}^3\mathrm{He})$ strength to a particular final state compared with the (d, t) strength to its isobaric analogue

$$\frac{G_{jl}(I_f T_f = T_i + \tfrac{1}{2} T'_z = T_i + \tfrac{1}{2};\ I_i T_i T_z = T_i)}{G_{jl}(I_f T_f = T_i + \tfrac{1}{2} T'_z = T_i - \tfrac{1}{2};\ I_i T_i T_z = T_i)} = (2T_i + 1). \qquad (2.140)$$

This rule is dramatically illustrated by the pick-up experiments on the tin isotopes. If one compares the spectroscopic factor for the $(d, {}^3\mathrm{He})$ obtained by Weiffenbach and Tickle (1971) with the (d, t) spectroscopic factors for the isobaric analogue states determined by Sekiguchi *et al.* (1977) one obtains the following results:

$$\frac{G_{j=\frac{9}{2} l=4}(I_f = \frac{9}{2} T_f = \frac{23}{2} T'_z = \frac{23}{2};\ I_i = 0 T_i = 11 T_z = 11)}{G_{j=\frac{9}{2} l=4}(I_f = \frac{9}{2} T_f = \frac{23}{2} T'_z = \frac{21}{2};\ I_i = 0 T_i = 11 T_z = 11)} = 26{\cdot}7$$

$$\frac{G_{j=\frac{3}{2} l=1}(I_f = \frac{3}{2} T_f = \frac{23}{2} T'_z = \frac{23}{2};\ I_i = 0 T_i = 11 T_z = 11)}{G_{j=\frac{3}{2} l=1}(I_f = \frac{3}{2} T_f = \frac{23}{2} T'_z = \frac{21}{2};\ I_i = 0 T_i = 11 T_z = 11)} = 22{\cdot}0$$

$$\frac{G_{j=\frac{1}{2} l=1}(I_f = \frac{1}{2} T_f = \frac{23}{2} T'_z = \frac{23}{2};\ I_i = 0 T_i = 11 T_z = 11)}{G_{j=\frac{1}{2} l=1}(I_f = \frac{1}{2} T_f = \frac{23}{2} T'_z = \frac{21}{2};\ I_i = 0 T_i = 11 T_z = 11)} = 24.4$$

where the levels involved in the (d, ^{3}He) reaction are the yrast $\frac{9}{2}^+$, $\frac{3}{2}^-$, and $\frac{1}{2}^-$ states in $^{121}_{49}\mathrm{In}_{72}$. Considering that the strengths to the analogue states (which lie at about 16 MeV excitation energy in $^{121}_{50}\mathrm{Sn}_{71}$) are small, the agreement between the theoretical expectation ($2T_i+1=23$) and experiment is excellent and confirms that these highly excited states are indeed isobaric analogues of the low-lying indium states.

3

PARTICLES AND HOLES

One often encounters situations in which a nuclear shell is nearly full, and then it is convenient to deal with holes instead of particles. In this Chapter we shall discuss how these calculations are made and illustrate, with several examples, how one can check the purity of an assumed configuration assignment by comparing the properties of nuclei that appear to belong to the configuration j^n and j^{-n}.

1. Particle–particle and particle–hole spectra

In this section we prove the Pandya particle–hole theorem which relates the spectrum of a nucleus with one nucleon in each of the single-particle levels j' and j to the spectrum of the nucleus with one particle in j' and one hole in j. This relationship will then be used to correlate the spectra observed in various nuclei.

1.1. The Pandya transformation

The fractional-parentage coefficients for the configuration $(j)_j^{-1}=(j)_j^{2j}$ are deduced in Appendix 5 and according to equation (A5.52) have the values

$$\langle j^{-2}Jv, j|\}j^{-1}j\rangle = \langle j^{2j-1}Jv, j|\}j^{2j}j\rangle$$
$$= (-1)^{v/2}\left\{\frac{2J+1}{4j(2j+1)}\right\}^{\frac{1}{2}}(1+(-1)^J) \qquad (3.1)$$

where v is the seniority of the two-hole wavefunction ($v=0$ when $J=0$, $v=2$ when $J\neq 0$). With the aid of these coefficients one may easily calculate the energy $\varepsilon_I(j'j^{-1}; j'j^{-1})$ associated with the state

$\Phi_{IM}(j', j^{-1}) =$ 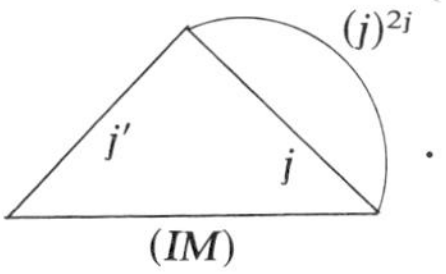

$\cdot \qquad (3.2)$

If one denotes the interaction between nucleons in the orbit j' and those in j by $V_{j'j}$ it follows that

$$\varepsilon_I(j'j^{-1}; j'j^{-1}) = \varepsilon_{j'} + 2j\varepsilon_j + E_j + \langle\Phi_{IM}(j', j^{-1})|\, V_{j'j}\, |\Phi_{IM}(j', j^{-1})\rangle \quad (3.3)$$

where E_j is the residual two-body interaction in the state $(j)_j^{2j}$. Since the second term in equation (2.79) gives the interaction energy for the configuration $(j)_0^{2j+1}$, it follows from equation (1.70) that

$$E_j = \left(\frac{2j-1}{2j+1}\right) \sum_K (2K+1) E_{K1}(jj;jj). \tag{3.4}$$

To calculate the matrix elements of $V_{j'j}$ one must get j' and j into the same triangle. Thus

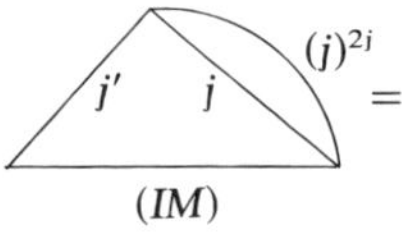

$$= \sum_{Jv} \langle j^{2j-1} Jv, j | \} j^{2j} j \rangle$$

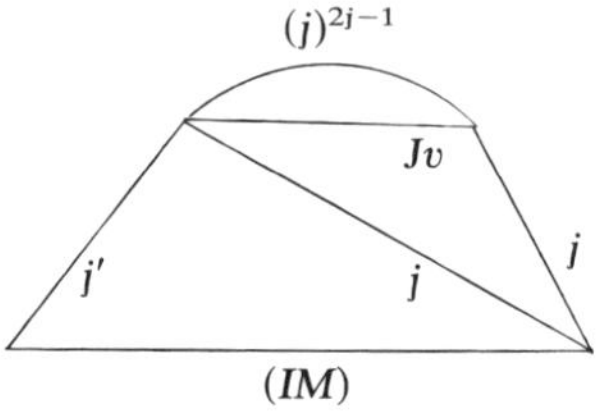

$$= \sum_{Jv} (-1)^{v/2} \left\{ \frac{2J+1}{4j(2j+1)} \right\}^{\frac{1}{2}} \{1+(-1)^J\}$$

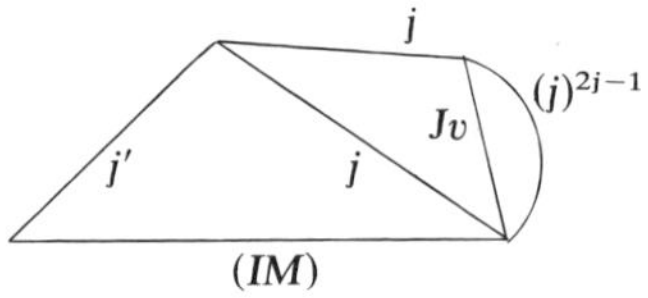

$$= \sum_{JKv} (-1)^{v/2} \left\{ \frac{(2J+1)(2K+1)}{4j} \right\}^{\frac{1}{2}} \{1+(-1)^J\} \times W(j'jIJ;Kj)$$

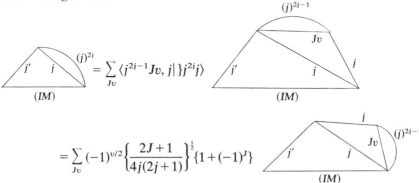

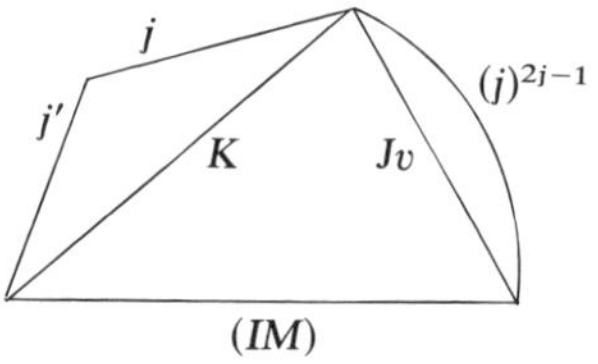

(3.5)

The effect of $V_{j'j}$ operating on the wave function in equation (3.5) is merely to multiply each term in the K sum by $E_K(j'j;j'j)$, the interaction energy when j' and j couple to spin K. Since there are $2j$ possible particles that can interact with the nucleon in j', it follows that

$$\langle \Phi_{IM}(j', j^{-1}) | V_{j'j} | \Phi_{IM}(j', j^{-1}) \rangle = \sum_{JK} (2J+1)(2K+1)\{1+(-1)^J\}$$
$$\times \{W(j'jIJ;Kj)\}^2 E_K(j'j;j'j). \tag{3.6}$$

The sum over J in this equation can be carried out by using equations

(A4.12), (A4.16), and (A4.17) so that one may finally write

$$\varepsilon_I(j'j^{-1};j'j^{-1})=\left\{2j\varepsilon_j+\left(\frac{2j-1}{2j+1}\right)\sum_K(2K+1)E_{K1}(jj;jj)\right\}$$
$$+\left\{\varepsilon_{j'}+\frac{1}{2j'+1}\sum_K(2K+1)E_K(j'j;j'j)\right\}$$
$$-\sum_K(2K+1)W(jj'j'j;IK)E_K(j'j;j'j). \tag{3.7}$$

From the foregoing discussion it is clear that the first group of terms in this equation is the energy of the configuration $(j)_j^{2j}$. Furthermore, if $\varepsilon_{j'}$ is the single-particle energy when there are no particles in the shell j, then as shown in equation (2.78)

$$\bar{\varepsilon}_{j'}=\varepsilon_{j'}+\frac{1}{(2j'+1)}\sum_K(2K+1)E_K(j'j;j'j) \tag{3.8}$$

gives the energy of the single-particle state j' when the shell j is full. The last term

$$E_I(j'j^{-1};j'j^{-1})=-\sum_K(2K+1)W(jj'j'j;IK)E_K(j'j;j'j) \tag{3.9}$$

represents the additional energy due to the fact there is a hole in the shell j (i.e. represents the particle–hole interaction energy). If one is interested in calculating excitation energies alone, then only equation (3.9) needs to be evaluated; the additional terms in equation (3.7) are common to each state and hence cancel. Equation (3.9) is often referred to as the Pandya particle–hole transformation (Pandya 1956).

1.2. The $^{38}_{17}Cl_{21}$—$^{40}_{19}K_{21}$ spectra

As a first application of these ideas consider the relationship between the spectra of $^{38}_{17}Cl_{21}$ and $^{40}_{19}K_{21}$. The first of these is described by the particle–particle configuration $(\pi d_{\frac{3}{2}}, \nu f_{\frac{7}{2}})$ whereas the second is the hole–particle configuration $(\pi d_{\frac{3}{2}}^{-1}, \nu f_{\frac{7}{2}})$. Substituting the values for the Racah coefficients into equation (3.9) we obtain equations for the hole-particle interaction energies in terms of the particle–particle matrix elements. In matrix notation

$$\begin{bmatrix}E_2(\pi d_{\frac{3}{2}}^{-1},\nu f_{\frac{7}{2}})\\E_3(\pi d_{\frac{3}{2}}^{-1},\nu f_{\frac{7}{2}})\\E_4(\pi d_{\frac{3}{2}}^{-1},\nu f_{\frac{7}{2}})\\E_5(\pi d_{\frac{3}{2}}^{-1},\nu f_{\frac{7}{2}})\end{bmatrix}=-\begin{bmatrix}-\frac{1}{56}&\frac{1}{8}&-\frac{27}{56}&\frac{11}{8}\\\frac{5}{56}&-\frac{11}{24}&\frac{51}{56}&\frac{11}{24}\\-\frac{15}{56}&\frac{17}{24}&\frac{131}{280}&\frac{11}{120}\\\frac{5}{8}&\frac{7}{24}&\frac{3}{40}&\frac{1}{120}\end{bmatrix}\begin{bmatrix}E_2(\pi d_{\frac{3}{2}},\nu f_{\frac{7}{2}})\\E_3(\pi d_{\frac{3}{2}},\nu f_{\frac{7}{2}})\\E_4(\pi d_{\frac{3}{2}},\nu f_{\frac{7}{2}})\\E_5(\pi d_{\frac{3}{2}},\nu f_{\frac{7}{2}})\end{bmatrix}.$$

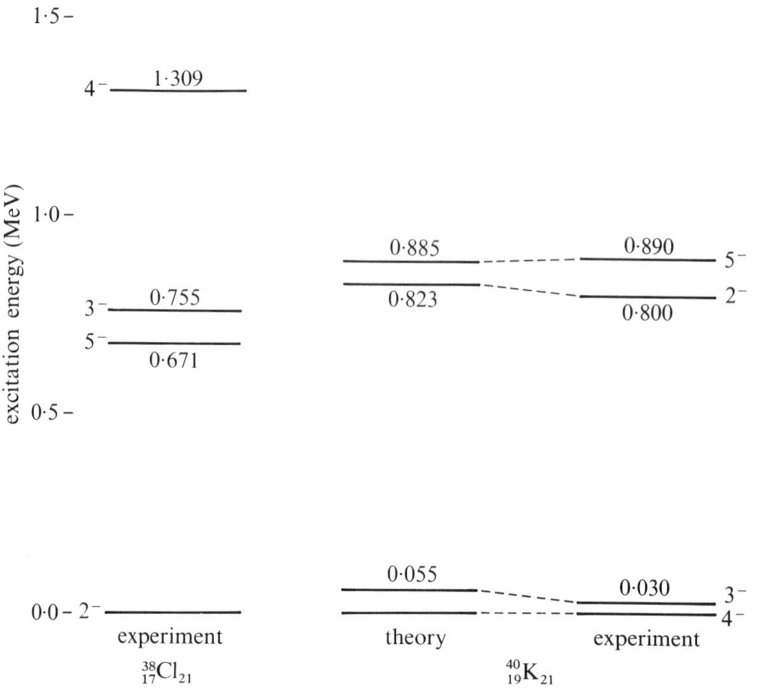

Fig. 3.1. The experimental spectra of ${}^{38}_{17}\mathrm{Cl}_{21}$ and ${}^{40}_{19}\mathrm{K}_{21}$ together with the theoretical predictions for ${}^{40}_{19}\mathrm{K}_{21}$. The theoretical results are derived by use of the particle-hole relationship (equation (3.9)) under the assumption that the yrast levels in ${}^{38}_{17}\mathrm{Cl}_{21}$ arise from $[\pi d_{\frac{3}{2}} \times \nu f_{\frac{7}{2}}]_{IM}$ and those in ${}^{40}_{19}\mathrm{K}_{21}$ from $[(\pi d_{\frac{3}{2}})^{-1} \times \nu f_{\frac{7}{2}}]_{IM}$.

When the ${}^{38}_{17}\mathrm{Cl}_{21}$ energies $E_I(\pi d_{\frac{3}{2}}, \nu f_{\frac{7}{2}})$ are used to predict the spectrum of ${}^{40}_{19}\mathrm{K}_{21}$ the results shown in Fig. 3.1 are in excellent agreement with experiment; the r.m.s. error in any of the excitation energies is 20 keV. Thus energy considerations alone imply that the assumed configuration assignment seems to be quite close to reality (Goldstein and Talmi 1956, Pandya 1956).

A numerical check on the evaluation of the Racah coefficients in equation (3.9) arises from the following considerations. The total number of interacting $(j'j)$ pairs is $2j$. Since the interaction with the closed shell (the second term in equation (3.8)) involves $(2j+1)$ pairs it follows that the sum of the coefficients of E_K in equation (3.9) must be -1.

1.3. *The ${}^{42}_{21}\mathrm{Sc}_{21}$—${}^{48}_{21}\mathrm{Sc}_{27}$ spectra*

Although the neutrons and protons are in the same shell, one may also use equation (3.9) to relate the spectrum of ${}^{42}_{21}\mathrm{Sc}_{21}$ to that in ${}^{48}_{21}\mathrm{Sc}_{27}$

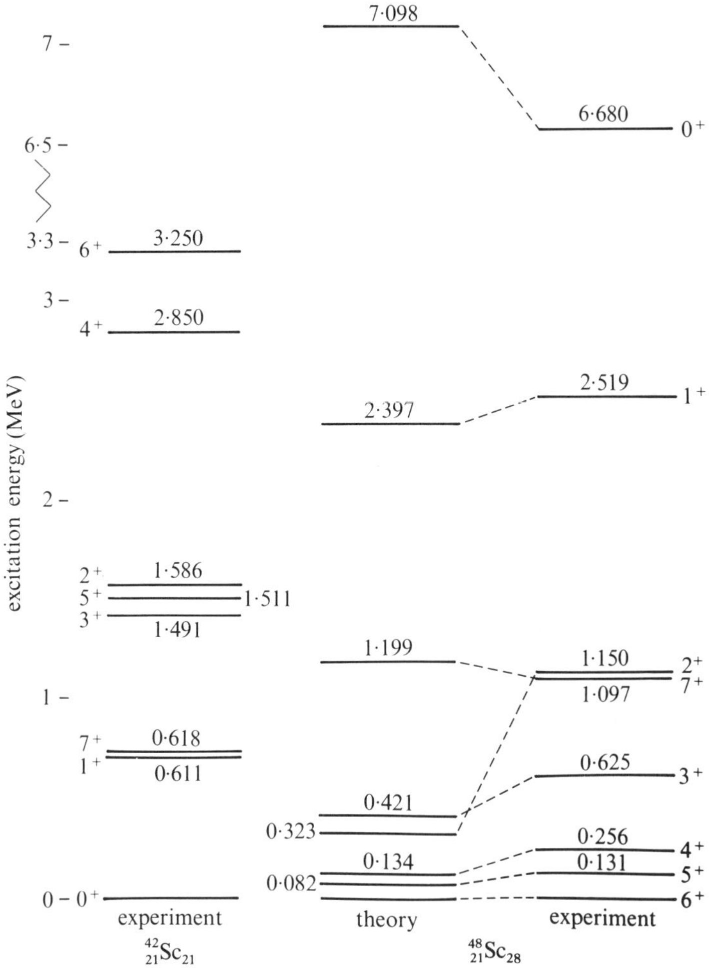

Fig. 3.2. The experimental spectra of $^{42}_{21}\mathrm{Sc}_{21}$ and $^{48}_{21}\mathrm{Sc}_{27}$ together with the theoretical predictions for $^{48}_{21}\mathrm{Sc}_{27}$. The theoretical results are derived by use of the particle-hole relationship (equation (3.9)) under the assumption that the yrast levels in $^{42}_{21}\mathrm{Sc}_{21}$ arise from $[\pi f_{\frac{7}{2}} \times \nu f_{\frac{7}{2}}]_{IM}$ and those in $^{48}_{21}\mathrm{Sc}_{27}$ from $[(\pi f_{\frac{7}{2}}) \times (\nu f_{\frac{7}{2}})^{-1}]_{IM}$.

(Schwartz and Watson 1969). This follows, of course, from the fact that in carrying out shell-model calculations it is not necessary to use the isospin formalism. If we denote by E_I the energy of the state $[\pi f_{\frac{7}{2}} \times \nu f_{\frac{7}{2}}]_{IM}$ and by E'_I the energy of the configuration $[\pi f_{\frac{7}{2}} \times (\nu f_{\frac{7}{2}})^{-1}]_{IM}$, equation (3.9) takes the numerical form

$$\begin{bmatrix} E'_0 \\ E'_1 \\ E'_2 \\ E'_3 \\ E'_4 \\ E'_5 \\ E'_6 \\ E'_7 \end{bmatrix} = - \begin{bmatrix} -\frac{1}{8} & \frac{3}{8} & -\frac{5}{8} & \frac{7}{8} & -\frac{9}{8} & \frac{11}{8} & -\frac{13}{8} & \frac{15}{8} \\ \frac{1}{8} & -\frac{59}{168} & \frac{85}{168} & -\frac{13}{24} & \frac{23}{56} & -\frac{11}{168} & -\frac{13}{24} & \frac{35}{24} \\ -\frac{1}{8} & \frac{17}{56} & -\frac{7}{24} & \frac{1}{24} & \frac{3}{8} & -\frac{121}{168} & \frac{13}{24} & \frac{7}{8} \\ \frac{1}{8} & -\frac{13}{56} & \frac{5}{168} & \frac{31}{88} & -\frac{303}{616} & -\frac{1}{56} & \frac{221}{264} & \frac{35}{88} \\ -\frac{1}{8} & \frac{23}{168} & \frac{5}{24} & -\frac{101}{264} & -\frac{1}{8} & \frac{103}{168} & \frac{13}{24} & \frac{35}{264} \\ \frac{1}{8} & -\frac{1}{56} & -\frac{55}{168} & -\frac{1}{88} & \frac{309}{616} & \frac{363}{728} & \frac{53}{264} & \frac{35}{1144} \\ -\frac{1}{8} & -\frac{1}{8} & \frac{5}{24} & \frac{119}{264} & \frac{3}{8} & \frac{53}{312} & \frac{1}{24} & \frac{5}{1144} \\ \frac{1}{8} & \frac{7}{24} & \frac{7}{24} & \frac{49}{264} & \frac{7}{88} & \frac{7}{312} & \frac{1}{264} & \frac{1}{3432} \end{bmatrix} \begin{bmatrix} E_0 \\ E_1 \\ E_2 \\ E_3 \\ E_4 \\ E_5 \\ E_6 \\ E_7 \end{bmatrix}$$

When the ${}^{42}_{21}\mathrm{Sc}_{21}$ matrix elements of Fig. 2.1 are used (with E_0 arbitrarily set equal to zero) the predicted spectrum of ${}^{48}_{21}\mathrm{Sc}_{27}$ is as shown in Fig. 3.2. The agreement with experiment for all but the 2^+ state is satisfactory; the r.m.s. error in the excitation energy of any level (except the 2^+) is 182 keV. On the other hand, the 2^+ state is off by 816 keV. One can question whether this concentration of error in the 2^+ state implies that only one of the levels in ${}^{42}_{21}\mathrm{Sc}_{21}$ is badly predicted by the ${}^{48}_{21}\mathrm{Sc}_{27}$ data. The inverse of equation (3.9) can be obtained by using equation (A4.15) and yields

$$E_K(j'j;\, j'j) = -\sum_I (2I+1)\, W(jj'j'j;\, IK) E_I(j'j^{-1};\, j'j^{-1}). \qquad (3.10)$$

If one substitutes the experimental ${}^{48}_{21}\mathrm{Sc}_{27}$ matrix elements in this equation one obtains the results listed in Table 3.1, and it is seen that the error is

TABLE 3.1

Comparison of experimental two-body matrix elements in ${}^{42}_{21}\mathrm{Sc}_{21}$ *with those predicted from the* ${}^{48}_{21}\mathrm{Sc}_{27}$ *data.*

	0	1	2	3	4	5	6	7
Experiment	0	0·611	1·586	1·491	2·85	1·511	3·25	0·618
Theory	0	−0·003	1·305	1·074	2·183	1·253	2·392	−0·158

All energies are in MeV. The interaction energy in the $I=0$ state has been arbitrarily normalized to zero. A negative value for the $I=1$ and 7 states means they are predicted to lie below the $I=0$ level.

distributed over all levels. Thus either ${}^{42}_{21}\mathrm{Sc}_{21}$ or ${}^{48}_{21}\mathrm{Sc}_{27}$ is not particularly well described by the pure $f_{\frac{7}{2}}$ configuration. A careful analysis of the data suggests that ${}^{42}_{21}\mathrm{Sc}_{21}$ has the less pure configuration (Moinester *et al.* 1969).

1.4. *The* ${}^{92}_{41}Nb_{51}$—${}^{96}_{41}Nb_{55}$ *spectra*

In deriving equation (3.9) there was no need for j' to be a single-particle state. The same derivation goes through if j' describes a group of particles coupled to spin j' provided $E_K(j'j; j'j)$ stands for the interaction energy in the state K between the single particle j and the group with spin j'. The reason this result is not generally of much practical use is the following: In the n-particle configuration $(j')^n_{JM}$ (n odd) the lowest energy state will have $J = j'$. However, the state with $J = j' - 1$ is usually quite close in energy. For example, in ${}^{51}_{23}V_{28}$ the $\frac{5}{2}^-$ state lies only 320 keV above the $\frac{7}{2}^-$

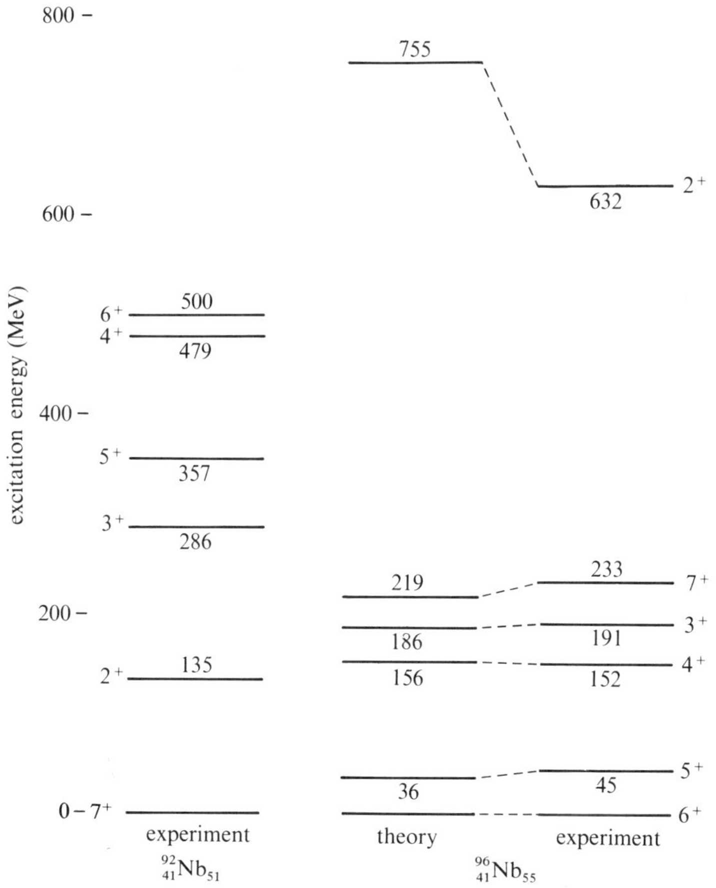

Fig. 3.3. The experimental spectra of ${}^{92}_{41}Nb_{51}$ and ${}^{96}_{41}Nb_{55}$ together with the theoretical predictions for ${}^{96}_{41}Nb_{55}$. The theoretical results are derived by use of the particle-hole relationship (equation (3.9)) under the assumption that the yrast levels in ${}^{92}_{41}Nb_{51}$ arise from the configuration $[(\pi J = \frac{9}{2}^+) \times (\nu d_{\frac{5}{2}})]_{IM}$ and those in ${}^{96}_{41}Nb_{55}$ from $[(\pi J = \frac{9}{2}^+) \times (\nu d_{\frac{5}{2}})^{-1}]_{IM}$.

ground state (see Fig. 1.8). Thus the energy of the configuration $[(j')^n_{j'}\times j]_{IM}$ cannot be taken directly from the observed spectrum because for most values of I there will be at least two ways of making the physically observed state: $[(j')^n_{j'}\times j]_{IM}$ and $[(j')^n_{j'-1}\times j]_{IM}$.

There is, however, one region where the generalization of the particle-hole theorem may be applicable. The properties of nuclei with $N=50$ and $Z>38$ are quite well described by the assumption that $N=50$, $Z=38$ forms a closed shell and the valence protons occupy the $1p_{\frac{1}{2}}$ and $0g_{\frac{9}{2}}$ orbits. If one uses the two-body matrix elements of Gloeckner *et al.* (1972), the ground state of ${}^{91}_{41}\mathrm{Nb}_{50}$ is described by the wave function

$$\psi_{\frac{9}{2}m}=0{\cdot}831(\pi p_{\frac{1}{2}})^2_0(\pi g_{\frac{9}{2}})-0{\cdot}556(\pi g_{\frac{9}{2}})^3_{\frac{9}{2}v=1}$$
$$-0{\cdot}034(\pi g_{\frac{9}{2}})^3_{\frac{9}{2}v=3}. \tag{3.11}$$

Because of the strong mixing between the $p_{\frac{1}{2}}$ and $g_{\frac{9}{2}}$ orbitals the ground $\frac{9}{2}^+$ state is considerably pushed down in energy and no other positive-parity states are seen in the nucleus below 1·5 MeV. Because of this, it is logical to interpret the low-lying 2^+–7^+ multiplet in ${}^{92}_{41}\mathrm{Nb}_{51}$ as due to the coupling of a $d_{\frac{5}{2}}$ neutron to this $\frac{9}{2}^+$ state. If the $N=55$ system can be considered to be a $d_{\frac{5}{2}}$ neutron hole and if the addition of four extra neutrons does not appreciably change the structure of the $\frac{9}{2}^+$ state given by equation (3.11), the Pandya transformation can be used to predict the spectrum of ${}^{96}_{41}\mathrm{Nb}_{55}$ by use of the observed levels in ${}^{92}_{41}\mathrm{Nb}_{51}$. In Fig. 3.3 we compare the predicted and observed spectrum of ${}^{96}_{41}\mathrm{Nb}_{55}$ (Comfort *et al.* 1970). Since the r.m.s. error in any predicted excitation energy is only 56 keV it would appear, on the basis of excitation energies alone, that ${}^{92}_{41}\mathrm{Nb}_{51}$ and ${}^{96}_{41}\mathrm{Nb}_{55}$ are particle-hole conjugates.

2. The particle–hole conjugation operator

Since it is cumbersome to construct the many-hole wave functions explicitly, it is convenient to define the particle–hole conjugation operator and to derive theorems concerning nuclei using only the properties of this operator (Bell 1959). If Γ is to be the particle–hole conjugation operator it must have the property that

$$\Gamma\psi_{IM}(n\text{-particles})=\psi_{IM}(n\text{-holes}). \tag{3.12}$$

Since Γ is to transform one complete orthonormal set of states (the states of j^n) into another complete orthonormal set (the states of j^{-n}), it follows that Γ is unitary. Thus

$$\Gamma^\dagger=\Gamma^{-1}$$

and

$$\Gamma^\dagger\Gamma=\Gamma\Gamma^\dagger=1. \tag{3.13}$$

From its defining equation (equation (3.12)) $\Gamma|0\rangle$ must be the state $(j)_0^{2j+1}$ whose wave function is given by equation (1.120). Consequently

$$\Gamma|0\rangle=\frac{2^{\Omega}}{(2j+1)!!}\{S_+\}^{\Omega}|0\rangle$$

$$=\frac{1}{\Omega!}\{S_+\}^{\Omega}|0\rangle \tag{3.14}$$

where

$$\Omega=(2j+1)/2 \tag{3.15}$$

is the pair degeneracy of the state j,

$$S_+=\sum_{m>0}(-1)^{j-m}a_m^\dagger a_{-m}^\dagger \tag{3.16a}$$

and for convenience the index j on the creation operator has been suppressed. In addition, $\Gamma a_m^\dagger|0\rangle$ must be the seniority-1 state $(j)_j^{2j}$ given by equation (1.125):

$$\Gamma a_m^\dagger|0\rangle=\frac{2^{\Omega-1}}{(2j-1)!!}a_m^\dagger\{S_+\}^{\Omega-1}|0\rangle$$

$$=\frac{1}{(\Omega-1)!}a_m^\dagger\{S_+\}^{\Omega-1}|0\rangle.$$

From the same arguments used in determining the normalization constant for the seniority-zero state (equation (1.120) one may show that for $p\le q$ and $q\le\Omega$

$$\{S_-\}^p\{S_+\}^q|0\rangle=\frac{q!(\Omega-q+p)!}{(q-p)!(\Omega-q)!}\{S_+\}^{q-p}|0\rangle \tag{3.17}$$

where

$$S_-=\sum_{m>0}(-1)^{j-m}a_{-m}a_m. \tag{3.16b}$$

Thus $\{S_+\}^{\Omega-1}|0\rangle$ can be written as

$$\{S_+\}^{\Omega-1}|0\rangle=\frac{1}{\Omega}S_-\{S_+\}^{\Omega}|0\rangle$$

so that

$$\Gamma a_m^\dagger|0\rangle=\frac{1}{\Omega!}a_m^\dagger S_-\{S_+\}^{\Omega}|0\rangle$$

$$=a_m^\dagger S_-\Gamma|0\rangle$$

where use has been made of equation (3.14) in writing the second line of this equation. Finally, from the anticommutation relationship, equation

(1.31), one sees that

$$a_m^\dagger S_- - S_- a_m^\dagger = (-1)^{j+m} a_{-m} \\ = \tilde{a}_m \tag{3.18}$$

where $\tilde{a}_m$ is the modified Hermitian adjoint operator given by equation (A2.9) and is an irreducible tensor operator of rank j. Since $\Gamma\,|0\rangle$ denotes the full shell it follows that $a_m^\dagger \Gamma\,|0\rangle \equiv 0$. When this result is combined with equation (3.13) one sees that

$$\Gamma a_m^\dagger\,|0\rangle = \Gamma a_m^\dagger \Gamma^\dagger \Gamma\,|0\rangle \\ = \tilde{a}_m \Gamma\,|0\rangle.$$

In other words, the particle–hole conjugation operator has the property that

$$\Gamma a_m^\dagger \Gamma^\dagger = (-1)^{j+m} a_{-m} \\ = \tilde{a}_m \tag{3.19a}$$

$$\Gamma a_m \Gamma^\dagger = (-1)^{j+m} a_{-m}^\dagger \tag{3.19b}$$

and

$$\Gamma^\dagger a_m^\dagger \Gamma = (-1)^{j-m} a_{-m} \\ = -\tilde{a}_m \tag{3.20a}$$

$$\Gamma^\dagger a_m \Gamma = (-1)^{j-m} a_{-m}^\dagger \tag{3.20b}$$

where equations (3.20) are obtained from equations (3.19) through multiplication by $\Gamma^\dagger$ from the left and Γ from the right.

By making use of the properties that the particle–hole conjugation operator must satisfy (equations (3.12), (3.19), and (3.20)) one can deduce many useful theorems in nuclear spectroscopy. However, before doing this we shall construct explicitly the operator Γ. The reader who is uninterested in this discussion can move immediately to section 2.2.

2.1. Explicit form of the particle–hole conjugation operator

Although one need only know the properties of Γ given by equations (3.12), (3.19), and (3.20), it is instructive to obtain the explicit form of the operator (Müller-Arnke 1973). Since the operator must be unitary it is convenient to try an exponential form

$$\Gamma = e^{i\alpha \bar{S}}. \tag{3.21}$$

This operator is obviously unitary if $\bar{S} = \bar{S}^\dagger$ and α is real. The effect of Γ on an arbitrary operator $Q_{\lambda\mu}$ can be calculated by making a Taylor's series expansion of $\Gamma Q_{\lambda\mu} \Gamma^\dagger$ in powers of α. If we set $x = i\alpha$ and define

$$f(x) = e^{x\bar{S}} Q_{\lambda\mu}\, e^{-x\bar{S}}$$

it follows that

$$\Gamma Q_{\lambda\mu}\Gamma^{\dagger}=\sum_{n=0}^{\infty}\frac{(\mathrm{i}\alpha)^n}{n!}\left(\frac{\partial^n f}{\partial x^n}\right)_{x=0}$$

From the definition of $f(x)$ one sees that

$$\frac{\partial f}{\partial x}=[\bar{\mathrm{S}}, f(x)]$$

$$\frac{\partial^2 f}{\partial x^2}=[\bar{\mathrm{S}},[\bar{\mathrm{S}}, f(x)]]$$

and in general the nth derivative of $f(x)$ gives the n-fold commutator of $\bar{\mathrm{S}}$ and $f(x)$. Since $f(0)=Q_{\lambda\mu}$ it follows that

$$\Gamma Q_{\lambda\mu}\Gamma^{\dagger}=Q_{\lambda\mu}+\mathrm{i}\alpha[\bar{\mathrm{S}}, Q_{\lambda\mu}]+\frac{(\mathrm{i}\alpha)^2}{2!}[\bar{\mathrm{S}},[\bar{\mathrm{S}}, Q_{\lambda\mu}]]+\ldots \tag{3.22}$$

To make a guess about the form of $\bar{\mathrm{S}}$ we note that it is possible to extract the number dependence of matrix elements involving the configuration j^n by using the quasi-spin formalism (see Appendix 3). The generators of rotations in quasi-spin are the operators S_+ and S_- (equations (3.16)) together with S_z where

$$\begin{aligned}\mathrm{S}_z&=\tfrac{1}{2}\Big\{\sum_m a_m^{\dagger}a_m-\Omega\Big\}\\&=\tfrac{1}{2}\{\mathrm{N}_j-\Omega\}\end{aligned}$$

and N_j is the number operator for the state j given by equation (1.41). A rotation through an angle π around the y axis in this space changes S_z to $-\mathrm{S}_z$; i.e. it changes an n-particle system to one with n holes. Thus one would expect that Γ would be proportional to $\bar{\Gamma}$ where

$$\begin{aligned}\bar{\Gamma}&=\exp(\mathrm{i}\pi\mathrm{S}_y)\\&=\exp\left(\frac{\pi}{2}\{\mathrm{S}_+-\mathrm{S}_-\}\right).\end{aligned}$$

To demonstrate that this form for $\bar{\Gamma}$ satisfies equations (3.19) and (3.20) we use the commutation relationship for $a_{jm}^{\dagger}$ and a_{jm} given by equation (1.31) to show that

$$[\mathrm{S}_y, a_m]=\frac{\mathrm{i}}{2}(-1)^{j-m}a_{-m}^{\dagger} \tag{3.23a}$$

$$[\mathrm{S}_y, a_m^{\dagger}]=\frac{\mathrm{i}}{2}(-1)^{j-m}a_{-m} \tag{3.23b}$$

and that

$$[S_y, [S_y, a_m]] = (\tfrac{1}{2})^2 a_m \tag{3.24a}$$

$$[S_y, [S_y, a_m^\dagger]] = (\tfrac{1}{2})^2 a_m^\dagger. \tag{3.24b}$$

Thus if either $a_{jm}^\dagger$ or a_{jm} is substituted for $Q_{\lambda\mu}$ in equation (3.22) it follows that the terms with an even number of commutators will just give a constant times $Q_{\lambda\mu}$ and those with an odd number of commutators will give the modified adjoint operator $\tilde{Q}_{\lambda\mu}$. Explicitly

$$\begin{aligned}\bar{\Gamma} a_m \bar{\Gamma}^\dagger &= a_m\left\{1 - \left(\frac{\pi}{2}\right)^2 \Big/ 2! + \left(\frac{\pi}{2}\right)^4 \Big/ 4! + \ldots\right\} \\ &\quad - (-1)^{j-m} a_{-m}^\dagger\left\{\left(\frac{\pi}{2}\right) - \left(\frac{\pi}{2}\right)^3 \Big/ 3! + \left(\frac{\pi}{2}\right)^5 \Big/ 5! + \ldots\right\} \\ &= a_m \cos\frac{\pi}{2} + (-1)^{j+m} a_{-m}^\dagger \sin\frac{\pi}{2} \\ &= (-1)^{j+m} a_{-m}^\dagger \end{aligned} \tag{3.25}$$

which shows that equation (3.19b) is satisfied. In the same way one can show that the other transformation properties given in equations (3.19a) and (3.20) are reproduced by this choice of $\bar{\Gamma}$.

Finally, one must deduce the effect of $\bar{\Gamma}$ on the vacuum state $|0\rangle$ to see if equation (3.14) is satisfied. The appropriate technique for doing this is one used by Macfarlane (1966) in a different context. Because of the anticommutation properties of $a_m^\dagger$ and a_m given by equations (1.28), (1.30), and (1.31), the operators

$$S_+(m) = (-1)^{j-m} a_m^\dagger a_{-m}^\dagger \tag{3.26a}$$

and

$$S_-(m) = (-1)^{j-m} a_{-m} a_m \tag{3.26b}$$

satisfy the commutation relationship

$$[S_+(m), S_-(m')] = 0 \quad \text{if} \quad m \neq m'.$$

Since

$$S_+ = \sum_{m>0} S_+(m) \tag{3.27}$$

$\bar{\Gamma}$ can be written as

$$\begin{aligned}\bar{\Gamma} &= \prod_{m>0} \bar{\Gamma}_m \\ &= \prod_{m>0} \exp\left[\frac{\pi}{2}\{S_+(m) - S_-(m)\}\right]. \end{aligned} \tag{3.28}$$

In addition, since these operators create fermions,

$$\{S_+(m)\}^q\,|0\rangle = 0 \quad \text{if} \quad q > 1 \tag{3.29}$$

$$S_-(m)S_+(m)\,|0\rangle = |0\rangle \tag{3.30}$$

and because $|0\rangle$ is the vacuum for particles in the state j

$$S_-(m)\,|0\rangle = 0. \tag{3.31}$$

Because of equations (3.29)–(3.31)

$$\begin{aligned}
&\{S_+(m) - S_-(m)\}^n\,|0\rangle \\
&\quad = \{S_+(m) - S_-(m)\}^{n-2}\{S_+(m)^2 - S_+(m)S_-(m) - S_-(m)S_+(m) + S_-(m)^2\}\,|0\rangle \\
&\quad = -\{(S_+(m) - S_-(m)\}^{n-2}\,|0\rangle.
\end{aligned}$$

By repeatedly taking off a pair of $S_\pm(m)$ operators one sees that

$$\{S_+(m) - S_-(m)\}^n\,|0\rangle = (-1)^{n/2}\,|0\rangle \quad \text{for } n \text{ even} \tag{3.32a}$$

and

$$\{S_+(m) - S_-(m)\}^n\,|0\rangle = (-1)^{(n-1)/2}S_+(m)\,|0\rangle \quad \text{for } n \text{ odd.} \tag{3.32b}$$

Therefore when one expands the exponential in equation (3.28) one finds

$$\begin{aligned}
\bar{\Gamma}_m\,|0\rangle &= \sum_{\substack{n=0\\(n\text{ even})}}^{\infty}\left((-1)^{n/2}\frac{1}{n!}\left\{\frac{\pi}{2}\right\}^n + S_+(m)\sum_{\substack{n=1\\(n\text{ odd})}}^{\infty}(-1)^{(n-1)/2}\frac{1}{n!}\left\{\frac{\pi}{2}\right\}^n\right)|0\rangle \\
&= \left(\cos\frac{\pi}{2} + S_+(m)\sin\frac{\pi}{2}\right)|0\rangle \\
&= S_+(m)\,|0\rangle.
\end{aligned}$$

Consequently

$$\begin{aligned}
\bar{\Gamma}\,|0\rangle &= \prod_{m>0}\bar{\Gamma}_m\,|0\rangle \\
&= \prod_{m>0} S_+(m)\,|0\rangle.
\end{aligned} \tag{3.33}$$

To make the final comparison with equation (3.14) we note that because of equation (3.29) each value of m can occur only once in the product $\{S_+\}^\Omega$. Moreover, since the first m value can be chosen from any of the Ω S_+ factors, the second from any of the remaining $(\Omega - 1)$ S_+ factors etc.

$$\begin{aligned}
\{S_+\}^\Omega\,|0\rangle &= \left\{\sum_{m>0} S_+(m)\right\}^\Omega |0\rangle \\
&= \Omega!\prod_{m>0} S_+(m)\,|0\rangle.
\end{aligned} \tag{3.34}$$

Therefore equation (3.33) can be written as

$$\bar{\Gamma}\,|0\rangle = \frac{1}{\Omega!}\{S_+\}^\Omega\,|0\rangle,$$

i.e. $\bar{\Gamma}$ operating on the vacuum produces exactly the full-shell wave function of equation (3.14). Consequently Γ, the particle–hole conjugation operator, is identically equal to $\bar{\Gamma}$ so that

$$\Gamma = \exp\{i\pi S_y\} \\ = \exp\left\{\frac{\pi}{2}(S_+ - S_-)\right\}. \tag{3.35}$$

There is one case where caution must be exercised and this is when the shell is half filled. There would be no trouble if Γ operating on the wave function gave the eigenfunction back again. However, as we shall now show there is a seniority-dependent phase factor that arises from the Γ operation. To see this consider the n particle seniority-v wave function (see Chapter 1, section 5.3)

$$\Psi_{IMv}(j^n) = \Lambda\phi_{IM}(j^v)\{S_+\}^q\,|0\rangle \tag{3.36}$$

with Λ, the normalization constant, given by equation (1.124) and

$$n = 2q + v. \tag{3.37}$$

According to equation (1·123b) for this to be a seniority v state

$$S_-\phi_{IM}(j^v)\,|0\rangle = 0.$$

$\phi_{IM}(j^v)$ will be a sum of terms each having v creation operators. If

$$\phi_M(k)\,|0\rangle = a^\dagger_{m_1}a^\dagger_{m_2}\ldots a^\dagger_{m_v}\,|0\rangle \qquad \left(M = \sum_i m_i\right) \tag{3.38}$$

denotes these possible occupation number states, $\phi_{IM}(j^v)$ can be written in terms of them as

$$\phi_{IM}(j^v) = \sum_k \alpha_{Ik}\phi_M(k) \tag{3.39}$$

where the coefficients α_{IK} must be chosen so that $J_+\phi_{IM=I}(j^v) = 0$.

In order to find the effect of Γ on equation (3.36) there are several relationships we need to know. From the anticommutation rule of equation (1.31) one sees that

$$[(-1)^{j+m}a_{-m}, \{S_+\}^p] = (p)a^\dagger_m\{S_+\}^{p-1} \tag{3.40}$$

and from equation (3.19a) it follows that

$$\Gamma(S_+)^p\Gamma^\dagger=(-1)^p(S_-)^p. \tag{3.41}$$

Thus

$$\begin{aligned}\Gamma\Psi_{IMv}(j^n)&=\Lambda\Gamma\phi_{IM}(j^v)\Gamma^\dagger\Gamma\{S_+\}^q\Gamma^\dagger\Gamma\,|0\rangle\\&=\frac{(-1)^q}{\Omega!}\Lambda\Gamma\phi_{IM}(j^v)\Gamma^\dagger\{S_-\}^q\{S_+\}^\Omega\,|0\rangle\\&=(-1)^q\frac{q!}{(\Omega-q)!}\Lambda\Gamma\phi_{IM}(j^v)\Gamma^\dagger\{S_-\}^{\Omega-q}\,|0\rangle\end{aligned} \tag{3.42}$$

where use has been made of equations (3.14) and (3.41) in writing the second line of this equation and of equation (3.17) to write the final expression. By using equation (3.19a) one sees that

$$\Gamma\phi_{IM}(j^v)\Gamma^\dagger=\sum_k\alpha_{Ik}\tilde{\phi}_M(k)$$

where

$$\tilde{\phi}_M(k)=(-1)^{j+m_1+j+m_2+\ldots+j+m_v}a_{-m_1}a_{-m_2}\ldots a_{-m_v}.$$

Finally, since $a_m\,|0\rangle=0$ one may use equation (3.40) to write

$$\begin{aligned}\Gamma\phi_M(k)\Gamma^\dagger\{S_+\}^{\Omega-q}\,|0\rangle&=\tilde{\phi}_M(k)\{S_+\}^{\Omega-q}\,|0\rangle\\&=(\Omega-q)(-1)^{j+m_1\ldots+j+m_{v-1}}a_{-m_1}\ldots a_{-m_{v-1}}\\&\quad\times a^\dagger_{m_v}\{S_+\}^{\Omega-q-1}\,|0\rangle\\&=(\Omega-q)(\Omega-q-1)\ldots(\Omega-q-v+1)\\&\quad\times a^\dagger_{m_1}\ldots a^\dagger_{m_v}\{S_+\}^{\Omega-q-v}\,|0\rangle\\&=\frac{(\Omega-q)!}{(\Omega-q-v)!}\phi_M(k)\{S_+\}^{\Omega-q-v}\,|0\rangle.\end{aligned} \tag{3.43}$$

When this expression is inserted into equation (3.42) one finds that

$$\Gamma\Psi_{IMv}(j^n)=(-1)^q\frac{q!}{(\Omega-q-v)!}\Lambda\phi_{IM}(j^v)\{S_+\}^{\Omega-q-v}\,|0\rangle.$$

By comparison with equation (1.124) one sees that $(q!/(\Omega-q-v)!)\Lambda$ is precisely the numerical constant necessary to make this the normalized seniority-v $(2j+1-n)$-particle wavefunction of equation (1.124). Consequently when use is made of equation (3.37) one sees that

$$\Gamma\Psi_{IMv}(j^n)=(-1)^{\tilde{v}}\Psi_{IMv}(j^{2j+1-n}) \tag{3.44}$$

where

$$\tilde{v}=(n-v)/2 \tag{3.45}$$

This result, which was originally arrived at by Racah (1943), shows that if one insists on the phase convention of equation (1.124) the $(2j+1-n)$-particle wavefunction must be obtained from the n-particle one by the operation

$$\Psi_{IM\upsilon}(j^{2j+1-n})=(-1)^{\tilde{\upsilon}}\Gamma\Psi_{IM\upsilon}(j^n) \tag{3.44a}$$

Since the overall sign of the wavefunction is unimportant the phase factor $(-1)^{\tilde{\upsilon}}$ can usually be neglected. However, there is one place where it must be taken into account and this is when we deal with the half filled shell. In this case the effect of Γ operating on $\Psi_{IM\upsilon}(j^{\Omega})$ gives the same wavefunction back again but multiplied by the phase factor $(-1)^{(\Omega-\upsilon)/2}$. Consequently in order to be consistent in this case one *must* use $(-1)^{\tilde{\upsilon}}\Gamma\Psi_{IM\upsilon}(j^{\Omega})$ and not $\Gamma\Psi_{IM\upsilon}(j^{\Omega})$ alone.

2.2. *Hole–hole interaction energies*

In the preceding Section we have considered the particle–hole conjugation operator for a single j shell so that Γ should really have had a subscript j attached to it. When several single-particle levels are involved the particle–hole conjugation operator for the system becomes

$$\Gamma=\prod_j \Gamma_j \tag{3.46}$$

With this in mind, we now wish to find the relationship between the interaction energies in the particle state

$$\Psi_{IM}\{(j)^n_J(j')^{n'}_{J'}\}=[(j)^n_J\times(j')^{n'}_{J'}]_{IM}$$

and the hole state

$$\Phi_{IM}\{(j)^{-n}_J(j')^{-n'}_{J'}\}=[(j)^{-n}_J\times(j')^{-n'}_{J'}]_{IM}.$$

Since Γ satisfies equation (3.12)

$$\Phi_{IM}\{(j)^{-n}_J(j')^{-n'}_{J'}\}=\Gamma_j\Gamma_{j'}\Psi_{IM}\{(j)^n_J(j')^{n'}_{J'}\} \tag{3.47}$$

so that

$$\langle\Phi_{IM}\{(j)^{-n}_J(j')^{-n'}_{J'}\}|\,V\,|\Phi_{IM}\{(j)^{-n}_J(j')^{-n'}_{J'}\}\rangle$$
$$=\langle\Psi_{IM}\{(j)^n_J(j')^{n'}_{J'}\}|\,\Gamma^\dagger_{j'}\Gamma^\dagger_j V\Gamma_j\Gamma_{j'}\,|\Psi_{IM}\{(j)^n_J(j')^{n'}_{J'}\}\rangle.$$

Thus one may calculate the matrix element of V in the hole-state configuration $\Phi_{IM}\{(j)^{-n}_J(j')^{-n'}_{J'}\}$ by use of the particle wave function $\Psi_{IM}\{(j)^n_J(j')^{n'}_{J'}\}$ once one knows the form of

$$V'=\Gamma^\dagger_{j'}\Gamma^\dagger_j V\Gamma_j\Gamma_{j'}. \tag{3.48}$$

Since

$$V = \sum_{KMjj'm_i} \frac{E_K(jj'; jj')}{(1+\delta_{jj'})} (jj'm_1m_2 \mid KM)(jj'm_3m_4 \mid KM) \times a^\dagger_{jm_1} a^\dagger_{j'm_2} a_{j'm_4} a_{jm_3}$$

it follows from equations (3.20) that

$$V' = \sum_{KMjj'm_i} \frac{E_K(jj'; jj')}{(1+\delta_{jj'})} (jj'm_1m_2 \mid KM)(jj'm_3m_4 \mid KM) \times a_{j-m_1} a_{j'-m_2} a^\dagger_{j'-m_4} a^\dagger_{j-m_3}$$

where we have made use of the fact that $m_1+m_2=m_3+m_4=M$ so that

$$(-1)^{2j+2j'-2M}=1.$$

Because of the anticommutation properties of a_{jm} in equations (1.28), (1.30), and (1.31), V' can be written as

$$\begin{aligned} V' = \sum_{KMjj'm_i} & \frac{E_K(jj'; jj')}{(1+\delta_{jj'})} (jj'm_1m_2 \mid KM)(jj'm_3m_4 \mid KM) \\ & \times (a^\dagger_{j-m_3} a^\dagger_{j'-m_4} a_{j'-m_2} a_{j-m_1} + \delta_{m_1m_3}\delta_{m_2m_4} - \delta_{jj'}\delta_{m_1m_4}\delta_{m_2m_3} \\ & - \delta_{m_2m_4} a^\dagger_{j-m_3} a_{j-m_1} - \delta_{m_1m_3} a^\dagger_{j'-m_4} a_{j'-m_2} \\ & + \delta_{jj'}\delta_{m_1m_4} a^\dagger_{j-m_3} a_{j-m_2} + \delta_{jj'}\delta_{m_2m_3} a^\dagger_{j-m_4} a_{j-m_1}). \end{aligned}$$

By using the symmetry property in equation (A1.19) of the Clebsch–Gordan coefficients together with the fact that m_i and M are dummy indices, one sees that the first term in V' is identical to V itself. Furthermore, the other terms can all be simplified by use of equations (A1.18)–(A1.24) so that

$$\begin{aligned} V' &= \Gamma^\dagger_{j'}\Gamma^\dagger_j V \Gamma_j \Gamma_{j'} \\ &= V + \sum_{Kjj'} (2K+1) E_K(jj'; jj') \left(1 - \frac{\mathrm{N}_j}{2j+1} - \frac{\mathrm{N}_{j'}}{2j'+1}\right) \end{aligned} \tag{3.49}$$

where

$$\mathrm{N}_j = \sum_m a^\dagger_{jm} a_{jm}$$

is the number operator for the shell j.

Thus apart from a constant I-dependent term the matrix element $\langle [(j)^{-n}_J \times (j')^{-n'}_{J'}]_{IM} | \, V \, | [(j)^{-n}_J \times (j')^{-n'}_{J'}]_{IM} \rangle$ is identical to

$$\langle [(j)^n_J \times (j')^{n'}_{J'}]_{IM} | \, V \, | [(j)^n_J \times (j')^{n'}_{J'}]_{IM} \rangle$$

so that corresponding particle–particle and hole–hole excitation energies should be the same.

As a first example of this theorem we compare the spectrum of $(j)^n$ with that of $(j)^{-n}$ when j is the proton $0f_{\frac{7}{2}}$ orbit and $n=2$. The nuclei

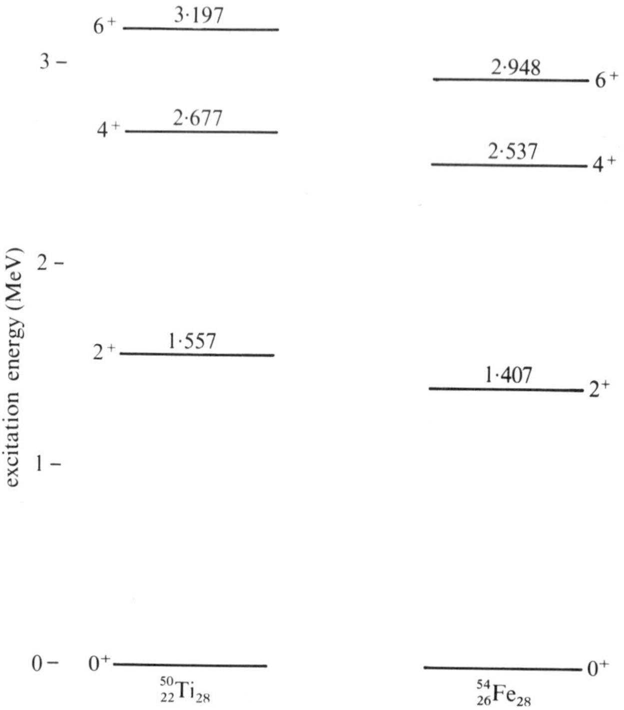

Fig. 3.4. The experimental excitation energies of the yrast levels in $^{50}_{22}Ti_{28}$ and $^{54}_{26}Fe_{28}$. If the states are $(\pi f_{\frac{7}{2}})^2$ and $(\pi f_{\frac{7}{2}})^{-2}$ their excitation energies should be the same.

involved in this case are $^{50}_{22}Ti_{28}$ and $^{54}_{26}Fe_{28}$ and their yrast 0^+, 2^+, 4^+, and 6^+ states are compared in Fig. 3.4. It is clear that the spectra are quite similar so that as far as energies are concerned these two nuclei are well described by the proton configurations $(\pi f_{\frac{7}{2}})^2$ and $(\pi f_{\frac{7}{2}})^{-2}$.

Another place where equation (3.49) can be used to determine the validity of a configuration assignment is in the nuclei $^{38}_{17}Cl_{21}$ and $^{46}_{19}K_{27}$. We have already seen (section 1.2 of this Chapter) that the spectra of $^{40}_{19}K_{21}$ and $^{38}_{17}Cl_{21}$ are consistent with a $[\pi d_{\frac{3}{2}} \times \nu f_{\frac{7}{2}}]_{IM}$ assignment for the first four levels in $^{38}_{17}Cl_{21}$. One might hope that the yrast 2^-, 3^-, 4^-, and 5^- states in $^{46}_{19}K_{27}$ could be described as $[(\pi d_{\frac{3}{2}})^{-1} \times (\nu f_{\frac{7}{2}})^{-1}]_{IM}$. Although some of the spins are not definitely known, the excitation energies in $^{46}_{19}K_{27}$ are well established (Dupont *et al.* 1973, Daehnick and Sherr 1973) and are compared in Fig. 3.5 with $^{38}_{17}Cl_{21}$. Clearly the energies are quite different so that with no more work one may unambiguously conclude that $^{46}_{19}K_{27}$ is not the hole–hole analogue of $^{38}_{17}Cl_{21}$; i.e. the yrast states of $^{46}_{19}K_{27}$ cannot be described by the configuration $[(\pi d_{\frac{3}{2}})^{-1} \times (\nu f_{\frac{7}{2}})^{-1}]_{IM}$.

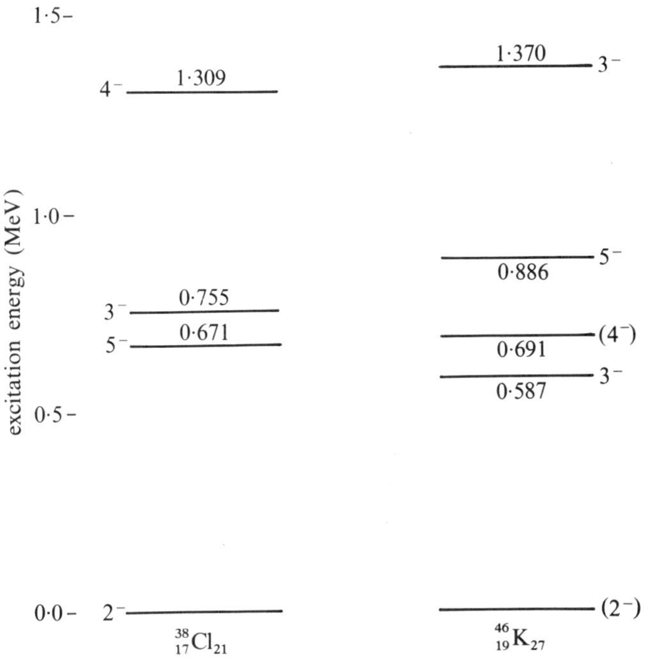

Fig. 3.5. The experimental states below 1·5 MeV excitation energy in ${}^{38}_{17}\mathrm{Cl}_{21}$ and ${}^{46}_{19}\mathrm{K}_{27}$. If the nuclei are particle–particle hole–hole conjugates their spectra should be the same. Uncertain spins are bracketed.

Finally, as was shown in the preceding Section, one must be careful when calculating matrix elements involving the half-filled shell. According to equation (3.44) and the discussion following it, the results obtained when one calculates with the wave function $\Psi_{IMv}(j^{\Omega})$ must be identical to those obtained by use of $(-1)^{\tilde{v}}\Gamma_j\Psi_{IMv}(j^{\Omega})$ where $\tilde{v}$ is given by equation (3.45). Thus

$$\langle\Psi_{IMv'}(j^{\Omega})|\, V\,|\Psi_{IMv}(j^{\Omega})\rangle = (-1)^{\tilde{v}+\tilde{v}'}\langle\Psi_{IMv'}(j^{\Omega})|\,\Gamma_j^{\dagger}V\Gamma_j\,|\Psi_{IMv}(j^{\Omega})\rangle.$$

Since

$$N_j\Psi_{IMv}(j^{\Omega}) = \Omega\Psi_{IMv}(j^{\Omega})$$

it follows that only V in the transformed potential of equation (3.49) can contribute to the matrix element so that

$$\langle\Psi_{IMv'}(j^{\Omega})|\, V\,|\Psi_{IMv}(j^{\Omega})\rangle = (-1)^{\tilde{v}+\tilde{v}'}\langle\Psi_{IMv'}(j^{\Omega})|\, V\,|\Psi_{IMv}(j^{\Omega})\rangle. \tag{3.50}$$

Because $\tilde{v}$ and $\tilde{v}'$ are both integers it follows that

$$(-1)^{\tilde{v}+\tilde{v}'}=(-1)^{2\tilde{v}}(-1)^{(v-v')/2}=(-1)^{(v-v')/2}.$$

Consequently for the half-full shell only states with $(v-v')=4, 8, \ldots$ can be mixed. This immediately explains why there is no seniority mixing in the identical nucleon $f_{\frac{7}{2}}$ shell; in this case both the two $I=2$ and two $I=4$ states have $v-v'=2$ so that no two-body force can mix seniority in the pure $(\frac{7}{2})^4$ configuration.

2.3. *Transformation of single-particle operators*

It is often important to know the matrix element of an operator $Q_{\lambda\mu}$ which is a sum of single-particle operators $Q_{\lambda\mu}(i)$:

$$Q_{\lambda\mu}=\sum_i Q_{\lambda\mu}(i).$$

For example, the quadrupole moment of a nucleus is the sum of the quadrupole moments of its constituent nucleons or, when a nucleus undergoes gamma decay, any one of the nucleons in the nucleus can emit the observed photon. Since

$$\langle\phi_{j'm'}(i)|\, Q_{\lambda\mu}(i)\, |\phi_{jm}(i)\rangle=\langle\phi_{j'}(i)\|\, Q_{\lambda}(i)\, \|\phi_j(i)\rangle(j\lambda m\mu \mid j'm')$$

it follows that $Q_{\lambda\mu}$, written in terms of annihilation and creation operators, has the form given by equation (1.51)

$$Q_{\lambda\mu}=\sum_{jmj'm'} \langle\phi_{j'}\|\, Q_{\lambda}\, \|\phi_j\rangle(j\lambda m\mu \mid j'm')a^{\dagger}_{j'm'}a_{jm} \tag{3.51}$$

where $\langle\phi_{j'}\|\, Q_{\lambda}\, \|\phi_j\rangle$ is the single-particle reduced matrix element.

In this section we investigate the relationship between the matrix element $\langle[(j)^{n+1}_{J_1}\times(j')^{m-1}_{J_1'}]_{I'M'}|\, Q_{\lambda\mu}\, \|[(j)^{n}_{J}\times(j')^{m}_{J'}]_{IM}\rangle$ and

$$\langle[(j)^{-n-1}_{J_1}\times(j')^{-m+1}_{J_1'}]_{I'M'}|\, Q_{\lambda\mu}\, \|[(j)^{-n}_{J}\times(j')^{-m}_{J'}]_{IM}\rangle.$$

Because of equation (3.12) it is clear we can use the particle wave function to determine the matrix elements of $Q_{\lambda\mu}$ in the hole–state configurations provided we make the appropriate transformation of the operator. That is

$$\langle[(j)^{-n-1}_{J_1}\times(j')^{-m+1}_{J_1'}]_{I'M'}|\, Q_{\lambda\mu}\, \|[(j)^{-n}_{J}\times(j')^{-m}_{J'}]_{IM}\rangle$$
$$=\langle[(j)^{n+1}_{J_1}\times(j')^{m-1}_{J_1'}]_{I'M'}|\, Q'_{\lambda\mu}\, \|[(j)^{n}_{J}\times(j')^{m}_{J'}]_{IM}\rangle \tag{3.52}$$

provided

$$Q'_{\lambda\mu}=\Gamma^{\dagger}_{j}\Gamma^{\dagger}_{j'}Q_{\lambda\mu}\Gamma_{j'}\Gamma_{j}. \tag{3.53}$$

In order to determine how $Q_{\lambda\mu}$ transforms we use equations (3.20). Thus

$$Q'_{\lambda\mu}=\sum_{jmj'm'}\langle\phi_{j'}\|\,Q_\lambda\,\|\phi_j\rangle(j\lambda m\mu\mid j'm')(-1)^{j+j'-m-m'}a_{j'-m'}a^{\dagger}_{j-m}.$$

$$=\sum_{jmj'm'}(-1)^{j+j'+\lambda}\left(\frac{2j'+1}{2j+1}\right)^{\frac{1}{2}}\langle\phi_{j'}\|\,Q_\lambda\,\|\phi_j\rangle(j'\lambda m'\mu\mid jm)$$

$$\times\{a^{\dagger}_{jm}a_{j'm'}-\delta_{jj'}\delta_{mm'}\}$$

$$=-\sum_{jmj'm'}\langle\phi_j\|\,\tilde{Q}_\lambda\,\|\phi_{j'}\rangle(j'\lambda m'\mu\mid jm)\{a^{\dagger}_{jm}a_{j'm'}-\delta_{jj'}\delta_{mm'}\}$$

where in writing the second line use has been made of equations (A1.18) and (A1.20), the fact that the m's are dummy indices and the anticommutation relationship, equation (1.31). In the final expression equation (A2.10), which relates the reduced matrix element of an operator and its Hermitian adjoint, has been used.

The single-particle operators $Q_{\lambda\mu}$ of interest have the form of the spherical harmonics given in equation (A1.8) or of the spin operators in (A2.15); in all cases they have the property that when λ is an integer

$$\tilde{Q}_{\lambda\mu}=(-1)^{\lambda}Q_{\lambda\mu}.$$

Furthermore, by using equations (A1.20) and (A1.23) one can show that

$$\sum_{mm'}(j\lambda m'\mu\mid jm)\delta_{jj'}\delta_{mm'}=\delta_{jj'}\sum_{mm'}(j\lambda m'\mu\mid jm)(j0m'0\mid jm)=(2j+1)\delta_{\lambda 0}\delta_{jj'}.$$

Therefore

$$Q'_{\lambda\mu}=(-1)^{1+\lambda}Q_{\lambda\mu}+\sum_{jj'}(2j+1)\delta_{\lambda 0}\delta_{jj'}\langle\phi_{j'}\|\,Q_\lambda\,\|\phi_j\rangle. \tag{3.54}$$

Thus *except when* $\lambda=0$ *the most that can happen to a matrix element when one transforms from particles to holes is that it changes sign.*

It is often of interest to compare the magnetic dipole moment ($\lambda=1$) and the electric quadrupole moment ($\lambda=2$) associated with the configurations j^n and j^{-n}. From equation (3.54) it follows that

(a) the magnetic moment of the state $(j)^n_I$ is identical to that of $(j)^{-n}_I$;
(b) the quadrupole moment of the state $(j)^n_I$ is equal in magnitude but opposite in sign to that of the state $(j)^{-n}_I$.

One may use these rules to test the assumption that the $\frac{3}{2}^+$ ground states of ${}^{37}_{17}Cl_{20}$ and ${}^{39}_{19}K_{20}$ are particle–hole conjugates. Since ${}^{39}_{19}K_{20}$ has a single hole in the $(0d, 1s)$ shell its ground state is expected to be $(\pi d_{\frac{3}{2}})^{-1}$. The simplest model for ${}^{37}_{17}Cl_{20}$ is a single $d_{\frac{3}{2}}$ proton outside a ${}^{36}_{16}S_{20}$ core. If this is the case the magnetic dipole moments of the two nuclei should be the same whereas their quadrupole moments should be equal but of opposite sign. Experimentally (Endt and van der Leun 1973) for ${}^{37}_{17}Cl_{20}$ the magnetic moment μ and the quadrupole moment Q are

$$\begin{aligned}\mu &= 0{\cdot}68411(e\hbar/2mc)\\ &= 0{\cdot}68411\,\mu_N\\ Q &= -0{\cdot}06213e\times 10^{-24}\ \text{cm}^2\\ &= -0{\cdot}06213e\ \text{barns},\end{aligned}\tag{3.55}$$

where μ_N is the nuclear magneton and e is the proton charge. For ${}^{39}_{19}K_{20}$

$$\begin{aligned}\mu &= 0{\cdot}39142\mu_N\\ Q &= +0{\cdot}049e\ \text{barns}.\end{aligned}\tag{3.56}$$

Although the sign expectations are borne out the magnitudes, in each case, are somewhat different. Thus if ${}^{39}_{19}K_{20}$ indeed looks like $(\pi d_{\frac{3}{2}})^{-1}$ it follows that ${}^{37}_{17}Cl_{20}$ cannot be considered to have the configuration $(\pi d_{\frac{3}{2}})$. In Chapter 5 we shall show that the observed properties of ${}^{39}_{19}K_{20}$ are closer to those expected for a $\pi d_{\frac{3}{2}}$ nucleon.

As a second application of equation (3.54) we consider the calculation of gamma-ray transition rates. As we shall show in Chapter 5, $E2$ transitions between nuclear states depend on the value of $B(E2;I_i\to I_f)$ where

$$B(E2;I\to I')=\frac{2I'+1}{2I+1}\,|\langle\Psi_{I'}\|\,Q_2\,\|\Psi_I\rangle|^2$$

$$Q_{2\nu}=\sum_i e_i r_i^2 Y_{2\nu}(\theta_i\phi_i)$$

and e_i is the electric charge on the ith particle. Thus $E2$ transition strengths between states of the configuration j^n should be the same as those between corresponding states of j^{-n}. One may use this fact to test further the validity of the $f_{\frac{7}{2}}$ configuration assignment for the nuclei ${}^{50}_{22}Ti_{28}$ and ${}^{54}_{26}Fe_{28}$ discussed in the preceding section. The experimental data (Bizetti 1971) concerning the levels that appear to belong to $(\pi f_{\frac{7}{2}})^{\pm 2}$ (see Fig. 3.4) are as follows: for ${}^{50}_{22}Ti_{28}$

$$\begin{aligned}B(E2;2^+\to 0^+)&=77\pm 15e^2\ \text{fm}^4\\ B(E2;6^+\to 4^+)&=33.6\pm 1{\cdot}1e^2\ \text{fm}^4.\end{aligned}\tag{3.57}$$

whereas for ${}^{54}_{26}Fe_{28}$

$$\begin{aligned}B(E2;2^+\to 0^+)&=120\pm 10e^2\ \text{fm}^4\\ B(E2;6^+\to 4^+)&=39{\cdot}9\pm 0{\cdot}8e^2\ \text{fm}^4.\end{aligned}\tag{3.58}$$

Thus although on the basis of energies alone these states act *as if* they belong to the configuration $(\pi f_{\frac{7}{2}})^{\pm 2}$, once transition rates are studied it is clear that at least the 2^+ and/or 0^+ levels in the two nuclei have quite different structure. *This result once more points out the necessity of studying experimental data other than energies if one is to learn about configuration purity.*

As pointed out in the previous section one must be careful when one calculates nuclear properties of the half-filled shell. In the same way that equation (3.50) was deduced, one may show that for $\lambda \neq 0$

$$\langle \Psi_{I'M'v'}(j^{\Omega})| \, Q_{\lambda\mu} \, |\Psi_{IMv}(j^{\Omega})\rangle = (-1)^{1+\lambda+\bar{v}+\bar{v}'} \langle \Psi_{I'M'v'}(j^{\Omega})| \, Q_{\lambda\mu} \, |\Psi_{IMv}(j^{\Omega})\rangle$$
$$= (-1)^{\{1+\lambda+(v-v')/2\}} \langle \Psi_{I'M'v'}(j^{\Omega})| \, Q_{\lambda\mu} \, |\Psi_{IMv}(j^{\Omega})\rangle. \quad (3.59)$$

Since a single-particle operator can only change the seniority of a state by zero or ±2 units (see Chapter 1, section 5.1) this leads to the centre-of-the-shell selection rules:

(i) All static multipole moments with λ even vanish for the half-filled shell; i.e. the quadrupole moment, the 2^4-pole moment, etc. are identically zero.
(ii) Magnetic multipole transitions (λ odd) can only take place between states of the same seniority. (Actually this selection rule is always true within j^n. See Appendix 3 for the proof.)
(iii) Electric multipole transitions (λ even) can only take place between states differing in seniority by two units.

The nucleus ${}^{52}_{24}Cr_{28}$ provides a case where rule (iii) should be applicable if the low-lying states are indeed $(\pi f_{\frac{7}{2}})^4$. The first 6^+ level in this nucleus is observed at 3·1146 MeV and according to the calculations in Chapter 1, section 3.2 is described as the seniority-2 state of $(\pi f_{\frac{7}{2}})^4$. There are two 4^+ states observed and according to our calculations the upper one at 2·7688 MeV is pure $v = 2$ and the lower at 2·3703 MeV is $v = 4$. From equation (3.50) it follows that no residual two-body interaction can mix these two states as long as the configuration is pure. Thus according to equation (3.59) there should be no $E2$ gamma decay from the 6^+ state to the upper 4^+ level. Experimentally (Freedman *et al.* 1966) it is found that

$$\frac{B(E2; 6^+ v = 2(3{\cdot}1146 \text{ MeV}) \rightarrow 4^+ v = 4(2{\cdot}3703 \text{ MeV}))}{B(E2; 6^+ v = 2(3{\cdot}1146 \text{ MeV}) \rightarrow 4^+ v = 2(2{\cdot}7688 \text{ MeV}))} = 2{\cdot}2.$$

Assuming that the 6^+ state is seniority-2 this would imply that

$$\Psi_{I=4,M}(2{\cdot}7688 \text{ MeV}) = 0{\cdot}83(\pi f_{\frac{7}{2}})^4_{I=4,v=2} \pm 0{\cdot}56(\pi f_{\frac{7}{2}})^4_{I=4,v=4}$$

and

$$\Psi_{I=4,M}(2{\cdot}3703 \text{ MeV}) = 0{\cdot}56(\pi f_{\frac{7}{2}})^4_{I=4,v=2} \mp 0{\cdot}83(\pi f_{\frac{7}{2}})^4_{I=4,v=4}$$

if the experimental data are to be explained using only the $(\pi f_{\frac{7}{2}})^4$ configuration.

This result shows that the states have mixed seniority, and because of equation (3.50) this implies that configurations other than $(\pi f_{\frac{7}{2}})^4$ must be included to mediate the mixing. However, one must be careful in drawing conclusions about the amount of 'other configuration' mixing needed to explain the data. As we shall now show, because the two $(f_{\frac{7}{2}})^4$ $I=4$ levels are predicted to lie close in energy, only a small amount of mixing is needed to explain experiment.

To see this take the limiting case that the two $(f_{\frac{7}{2}})^4$ $I=4$ states are degenerate and assume their mixing is brought about by the action of a third 4^+ level that lies at an energy ΔE above the $(f_{\frac{7}{2}})^4_{I=4}$ states. If ε_2 and ε_4 are the off-diagonal matrix elements of the seniority-2 and -4 states with this third one, the Hamiltonian matrix describing the situation in this extended model space is

$$\langle H\rangle = \begin{pmatrix} E_0 & 0 & \varepsilon_2 \\ 0 & E_0 & \varepsilon_4 \\ \varepsilon_2 & \varepsilon_4 & E_0+\Delta E \end{pmatrix}.$$

The eigenvalues of this matrix are

$$E=E_0 \quad \text{and} \quad E_0+\frac{[\Delta E \pm \sqrt{\{(\Delta E)^2+4(\varepsilon_2^2+\varepsilon_4^2)\}}]}{2}$$

and independently of the value of ΔE the eigenfunction corresponding to the unperturbed eigenvalue E_0 is

$$\psi(E_0)=\frac{\varepsilon_4}{\sqrt{(\varepsilon_2^2+\varepsilon_4^2)}}(\pi f_{\frac{7}{2}})^4_{I=4,v=2}-\frac{\varepsilon_2}{\sqrt{(\varepsilon_2^2+\varepsilon_4^2)}}(\pi f_{\frac{7}{2}})^4_{I=4,v=4}.$$

Thus even a minute off-diagonal matix element can give a large mixing of the originally degenerate states without introducing any admixture of states out of the shell.

In a similar way the second eigenfunction can become thoroughly mixed with only a small 'other configuration contaminant' provided $\varepsilon_i/\Delta E \ll 1$. In this case the second eigenvalue of the matrix is

$$E_2 \simeq E_0-\frac{(\varepsilon_2^2+\varepsilon_4^2)}{\Delta E}=E_0-\Delta$$

and

$$\psi(E_2)=\frac{1}{\sqrt{\{1+(\varepsilon_4/\varepsilon_2)^2+(\Delta/\varepsilon_2)^2\}}}\left((\pi f_{\frac{7}{2}})^4_{I=4,v=2}+\frac{\varepsilon_4}{\varepsilon_2}(\pi f_{\frac{7}{2}})^4_{I=4,v=4}-\frac{\Delta}{\varepsilon_2}\phi_{I=4}\right)$$

where $\phi_{I=4}$ is the wave function of the other configuration. Thus the almost complete seniority mixing of the two lowest 4^+ levels in ${}^{52}_{24}Cr_{28}$ cannot be used to infer a gross breakdown of the $(f_{\frac{7}{2}})^4$ configuration assignment because of the near degeneracy of the two levels.

2.4. General particle–hole theorem

In section 1 we discussed the particle–hole theorem for the particular case that only diagonal matrix elements are to be calculated. It often happens, of course, that one has particles in several orbits interacting with holes in several other orbits and then the theorem must be generalized. for example, as we have already mentioned, the properties of nuclei with $N=50$, $Z>38$ (${}^{89}_{39}T_{50}$, ${}^{90}_{40}Zr_{50}$, ${}^{91}_{41}Nb_{50}$, ${}^{92}_{42}Mo_{50}$, ${}^{93}_{43}Tc_{50}$, and ${}^{94}_{44}Ru_{50}$) can be readily interpreted using a model in which $N=50$, $Z=38$ forms an inert core and the extra-core protons occupy the $0g_{\frac{9}{2}}$ and $1p_{\frac{1}{2}}$ single-particle levels (for a detailed discussion see Gloeckner and Serduke 1974). If one wants to discuss the isotopes of these nuclei with $N=49$ a convenient model is one in which the protons (treated as particles) interact with a neutron hole in either the $0g_{\frac{9}{2}}$ or $1p_{\frac{1}{2}}$ orbit (Serduke *et al.* 1976). In this calculation the proton single-particle energies are obtained from the binding energy of the $p_{\frac{1}{2}}$ ground state and the energy of the excited $g_{\frac{9}{2}}$ state of ${}^{89}_{39}Y_{50}$ (Johns *et al.* 1970) relative to the ${}^{88}_{38}Sr_{50}$ closed shell (see Chapter 1, section 2 where this type of analysis was carried out for the oxygen isotopes). As far as the neutron hole energies are concerned they are obtained from the states of ${}^{87}_{38}Sr_{49}$. Since the ground state of ${}^{87}_{38}Sr_{49}$ is $\frac{9}{2}^+$ and the excited $\frac{1}{2}^-$ level lies at 388 keV (Verheul 1971) it follows from the tables of Wapstra and Gove (1971) that

$$\begin{aligned}\varepsilon_{\frac{9}{2}}^{-1} &= \mathrm{BE}({}^{87}_{38}\mathrm{Sr}_{49}) - \mathrm{BE}({}^{88}_{38}\mathrm{Sr}_{50}) \\ &= 11{\cdot}113\ \mathrm{MeV}\end{aligned}$$

and

$$\begin{aligned}\varepsilon_{\frac{1}{2}}^{-1} &= 11{\cdot}113 + 0{\cdot}388 \\ &= 11{\cdot}501\ \mathrm{MeV}.\end{aligned}$$

(Note that the single-hole energies are positive and in part this is due to the $(-1)^{1+\lambda}$ factor that occurs in the transformation of a single-particle operator given in equation (3.54); $\lambda=0$ since energy is a scalar.) The particle–hole interaction energies can be determined to give a best fit to the data in much the same way as discussed in Chapter 1, section 3. If one wishes to express these particle–hole energies in terms of particle–particle matrix elements one must know the generalization of equation (3.9) and we shall now investigate this.

In the general case we are interested in evaluating matrix elements involving the hole–state configurations

$$\begin{aligned}\tilde{\phi} &= (j_1)^{n_1}(j_2)^{-n_2}(j_3)^{n_3}(j_4)^{-n_4}\\ \tilde{\phi}' &= (j_1)^{n_1+1}(j_2)^{-(n_2-1)}(j_3)^{n_3-1}(j_4)^{-(n_4+1)}\end{aligned} \tag{3.60}$$

in terms of particle eigenfunctions

$$\begin{aligned}\phi &= (j_1)^{n_1}(j_2)^{n_2}(j_3)^{n_3}(j_4)^{n_4}\\ \phi' &= (j_1)^{n_1+1}(j_2)^{n_2-1}(j_3)^{n_3-1}(j_4)^{n_4+1}.\end{aligned} \tag{3.61}$$

Since $\Gamma_{j_2}\Gamma_{j_4}\phi = \tilde{\phi}$ and $\Gamma_{j_2}\Gamma_{j_4}\phi' = \tilde{\phi}'$

$$\langle\tilde{\phi}'|\, V\,|\tilde{\phi}\rangle = \langle\phi'|\,\bar{V}\,|\phi\rangle \tag{3.62}$$

where

$$\bar{V} = \Gamma^{\dagger}_{j_4}\Gamma^{\dagger}_{j_2}V\Gamma_{j_2}\Gamma_{j_4} \tag{3.63}$$

i.e. matrix elements involving the hole configurations should be calculated using the transformed potential $\bar{V}$.

In order to calculate $\bar{V}$ in equation (3.63) it is convenient to express V in terms of diagrams. From the definition of the modified Hermitian adjoint operator given in equation (A2.9) it follows that

$$\begin{aligned}\tilde{A}_{J-M}(j_3j_4) &= (-1)^{J-M}A_{JM}(j_3j_4)\\ &= \sum_{m_3m_4}(j_3j_4m_3m_4\,|\,J-M)\tilde{a}_{j_4m_4}\tilde{a}_{j_3m_3}\\ &= (-1)^{j_3+j_4-J}[\tilde{a}_{j_4}\times\tilde{a}_{j_3}]_{J-M}.\end{aligned}$$

Furthermore, since we are interested in the case that we have particles in the orbits j_1 and j_3 and holes in the orbits j_2 and j_4, it follows that not all the quantum numbers associated with j_1 will be the same as those associated with j_2 and a similar statement holds for j_3 and j_4. Thus the factors $(1+\delta_{j_1j_2})^{-\frac{1}{2}}$ in equation (1.42) and $(1+\delta_{j_3j_4})^{-\frac{1}{2}}$ in equation (1.43) are both unity. Consequently

$$\begin{aligned}V &= \sum_{JMj_i}(-1)^{J-M}E_J(j_1j_2;\, j_3j_4)A^{\dagger}_{JM}(j_1j_2)\tilde{A}_{J-M}(j_3j_4)\\ &= \sum_{Jj_i}(-1)^{j_3+j_4-J}\sqrt{(2J+1)}E_J(j_1j_2;\, j_3j_4)[[a^{\dagger}_{j_1}\times a^{\dagger}_{j_2}]_J\times[\tilde{a}_{j_4}\times\tilde{a}_{j_3}]_J]_{00}\\ &= \sum_{Jj_i}(-1)^{j_3+j_4-J}\sqrt{(2J+1)}E_J(j_1j_2;\, j_3j_4)\end{aligned}$$

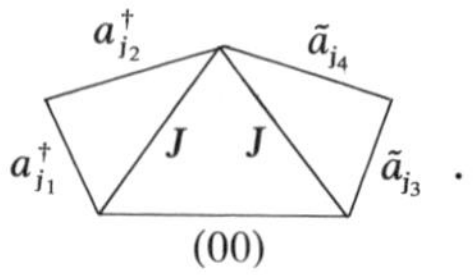

. (3.64)

Therefore when equations (3.20) are used one sees that

$$\bar{V} = -\sum_{Jj_i} (-1)^{j_3+j_4-J}\sqrt{(2J+1)}E_J(j_1j_2;\, j_3j_4)$$

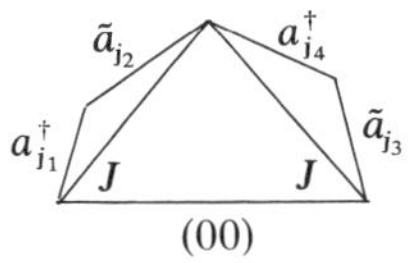

To write $\bar{V}$ in a form in which the two creation operators stand together the recoupling described by equation (1.90) must be carried out. Because $a_j^\dagger$ and a_j satisfy the anticommutation relationship of equation (1.31) one has a sign change together with an extra term that arises when $j_2 = j_4$, $m_2 = m_4$. Thus

$$\bar{V} = V' + V'' \tag{3.65}$$

where

$$V' = \sum_{JKj_i} (-1)^{j_3+j_4-J}(2J+1)(2K+1)\sqrt{(2J+1)}E_J(j_1j_2;\, j_3j_4) \times \begin{Bmatrix} j_1 & j_2 & J \\ j_4 & j_3 & J \\ K & K & 0 \end{Bmatrix}$$

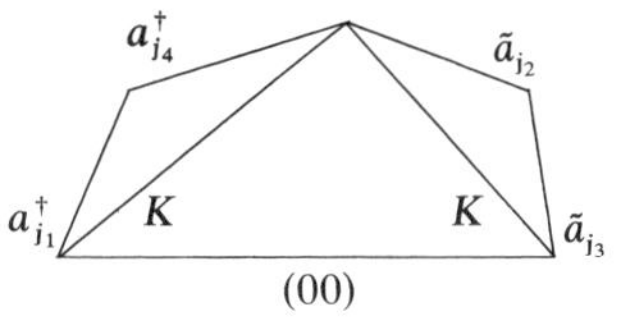

and

$$V'' = \sum_{JMj_im_i} \delta_{j_2j_4}\delta_{m_2m_4}E_J(j_1j_2;\, j_3j_4) \times (j_1j_2m_1m_2 \mid JM)(j_3j_4m_3m_4 \mid JM)a^\dagger_{j_1m_1}a_{j_3m_3}.$$

By using equations (A1.20) and (A1.23) one easily shows that

$$V'' = \sum_{j_i} \delta_{j_2j_4}\delta_{j_1j_3}\left(\sum_J \frac{(2J+1)}{(2j_1+1)} E_J(j_1j_2;\, j_3j_4)\right)\mathrm{N}_{j_1}. \tag{3.66}$$

In addition the simple form that the $9j$ coefficient takes when one argument is zero (equation (A4.33) means that V' becomes

$$\begin{aligned} V' &= \sum_{Kj_i} (-1)^{j_2+j_3-K}\sqrt{(2K+1)}F_K(j_1j_4;\, j_3j_2) \\ &= \sum_{KMj_i} F_K(j_1j_4;\, j_3j_2)A^\dagger_{KM}(j_1j_4)A_{KM}(j_3j_2) \end{aligned} \tag{3.67}$$

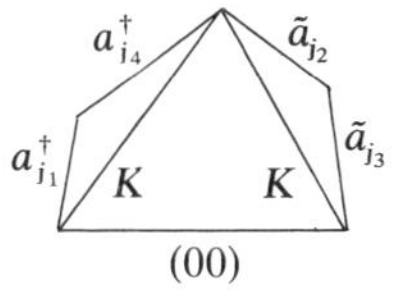

where

$$\begin{aligned} F_K(j_1j_4;\, j_3j_2) &= E_K(j_1j_4^{-1};\, j_3j_2^{-1}) \\ &= -(-1)^{j_1+j_2+j_3+j_4}\sum_J (2J+1)W(j_1j_2j_4j_3;\, JK)E_J(j_1j_2;\, j_3j_4). \end{aligned} \tag{3.68}$$

Thus the potential $\bar{V}$ to be used in calculating the particle–hole interaction is given by equations (3.65)–(3.68) and F_K in equation (3.68) is the general expression for the particle–hole interaction. For the special case $j_1=j_3=j'$, $j_2=j_4=j$, equation (3.68) becomes identical to equation (3.9).

When we deal with off-diagonal matrix elements what appear at first glance to be some surprising relationships are obtained. For example, in the $N=49$ isotones discussed at the beginning of this section the off-diagonal matrix element

$$\begin{aligned}F_K(\tfrac{1}{2}\tfrac{1}{2};\tfrac{9}{2}\tfrac{9}{2}) &= \langle[(\pi p_{\frac{1}{2}})\times(\nu p_{\frac{1}{2}})^{-1}]_{KM}|\, V\,|[(\pi g_{\frac{9}{2}})\times(\nu g_{\frac{9}{2}})^{-1}]_{KM}\rangle\\ &= -\sum_J (2J+1)W(\tfrac{1}{2}\tfrac{9}{2}\tfrac{1}{2}\tfrac{9}{2};JK)\\ &\qquad \langle[(\pi p_{\frac{1}{2}})\times(\nu g_{\frac{9}{2}})]_{JM}|\, V\,|[(\pi g_{\frac{9}{2}})\times(\nu p_{\frac{1}{2}})]_{JM}\rangle \end{aligned} \quad (3.69)$$

describing the interaction between a particle and a hole (both with negative parity) and a particle and a hole (both with positive parity) involves the actual two-body interaction evaluated in negative-parity states. The reason for this is, of course, clear when one remembers that the shorthand notation used in equation (3.69) stands for

$$\begin{aligned}&\langle[(\pi p_{\frac{1}{2}})\times(\nu p_{\frac{1}{2}})^{-1}]_{KM}|\, V\,|[(\pi g_{\frac{9}{2}})\times(\nu g_{\frac{9}{2}})^{-1}]_{KM}\rangle\\ &\qquad = \langle[(\pi p_{\frac{1}{2}})\times(\nu p_{\frac{1}{2}})]_{KM}(\nu g_{\frac{9}{2}})^{10}_{0}|\, V\,|[(\pi g_{\frac{9}{2}})\times(\nu g_{\frac{9}{2}})^{9}_{\frac{9}{2}}]_{KM}(\nu p_{\frac{1}{2}})^{2}_{0}\rangle.\end{aligned}$$

Thus the only way a non-vanishing value can arise for this matrix element is if the annihilation operators in V destroy a $g_{\frac{9}{2}}$ proton and a $p_{\frac{1}{2}}$ neutron and the $a^{\dagger}_{jm}$ in V create a $p_{\frac{1}{2}}$ proton and a $g_{\frac{9}{2}}$ neutron.

On the left-hand side of the particle–particle matrix element in equation (3.69) the proton is in a $p_{\frac{1}{2}}$ state and the neutron in $g_{\frac{9}{2}}$ whereas on the right-hand side the situation is reversed. According to equation (2.22)

$$[(\pi p_{\frac{1}{2}})\times(\nu g_{\frac{9}{2}})]_{JM} = 2^{-\frac{1}{2}}\{\Psi_{JM;T=1,T_z=0}(p_{\frac{1}{2}}, g_{\frac{9}{2}}) - \Psi_{JM;T=0,T_z=0}(p_{\frac{1}{2}}, g_{\frac{9}{2}})\}$$

and

$$\begin{aligned}&[(\pi g_{\frac{9}{2}})\times(\nu p_{\frac{1}{2}})]_{JM}\\ &\qquad = -(-1)^{\frac{1}{2}+\frac{9}{2}-J}[2^{-\frac{1}{2}}\{\Psi_{JM;T=1,T_z=0}(p_{\frac{1}{2}}, g_{\frac{9}{2}}) + \Psi_{JM;T=0,T_z=0}(p_{\frac{1}{2}}, g_{\frac{9}{2}})\}]\end{aligned}$$

Thus if V is an isospin-conserving interaction it follows that

$$\begin{aligned}&\langle[(\pi p_{\frac{1}{2}})\times(\nu p_{\frac{1}{2}})^{-1}]_{KM}|\, V\,|[(\pi g_{\frac{9}{2}})\times(\nu g_{\frac{9}{2}})^{-1}]_{KM}\rangle\\ &\qquad = -\tfrac{1}{2}\sum_J (-1)^J(2J+1)W(\tfrac{1}{2}\tfrac{9}{2}\tfrac{1}{2}\tfrac{9}{2};JK)\{E_{JT=1}(\tfrac{1}{2}\tfrac{9}{2};\tfrac{1}{2}\tfrac{9}{2}) - E_{JT=0}(\tfrac{1}{2}\tfrac{9}{2};\tfrac{1}{2}\tfrac{9}{2})\}\end{aligned}$$

where

$$E_{JT}(\tfrac{1}{2}\tfrac{9}{2};\tfrac{1}{2}\tfrac{9}{2}) = \langle\Psi_{JM;TT_z}(p_{\frac{1}{2}}, g_{\frac{9}{2}})|\, V\,|\Psi_{JM;TT_z}(p_{\frac{1}{2}}, g_{\frac{9}{2}})\rangle.$$

3. Particle–hole transformation including isospin

Although it is not necessary to introduce isospin it is often useful to do so if for no other reason than to reduce the dimensionality of the energy matrix that must be diagonalized. In this section we consider the particle–hole transformation when isospin is included.

By analogy with the non-isospin case given in equation (3.12) we look for a unitary transformation Γ_t that has the property

$$\Gamma_t \Psi_{IM;TT_z}(n\text{-particles}) = \Psi_{IM;TT_z}(n\text{-holes}). \tag{3.70}$$

Consequently $\Gamma_t\,|0\rangle$ is the full shell with $(2j+1)$ neutrons and $(2j+1)$ protons defined by

$$\Gamma_t\,|0\rangle = a^\dagger_{jj;\frac{1}{2}\frac{1}{2}} \ldots a^\dagger_{j-j;\frac{1}{2}\frac{1}{2}} a^\dagger_{jj;\frac{1}{2}-\frac{1}{2}} \ldots a^\dagger_{j-j;\frac{1}{2}-\frac{1}{2}}\,|0\rangle.$$

If we define the one-hole state $\psi_{jj;\frac{1}{2}\frac{1}{2}}$ as

$$\psi_{jj;\frac{1}{2}\frac{1}{2}} = a^\dagger_{jj;\frac{1}{2}\frac{1}{2}} \ldots a^\dagger_{j-j;\frac{1}{2}\frac{1}{2}} a^\dagger_{jj;\frac{1}{2}-\frac{1}{2}} \ldots a^\dagger_{j-j+1;\frac{1}{2}-\frac{1}{2}}\,|0\rangle$$

then by application of the T_- operator one sees that

$$\psi_{jj;\frac{1}{2}-\frac{1}{2}} = -a^\dagger_{jj;\frac{1}{2}\frac{1}{2}} \ldots a^\dagger_{j-j+1;\frac{1}{2}\frac{1}{2}} a^\dagger_{jj;\frac{1}{2}-\frac{1}{2}} \ldots a^\dagger_{j-j;\frac{1}{2}-\frac{1}{2}}\,|0\rangle.$$

Consequently if $\Gamma_t a^\dagger_{jm;\frac{1}{2}\mu}\,|0\rangle$ is to represent the one hole state with $(jm;\frac{1}{2}\mu)$ it follows that

$$\begin{aligned}\Gamma_t a^\dagger_{jm;\frac{1}{2}\mu}\,|0\rangle &= \Gamma_t a^\dagger_{jm;\frac{1}{2}\mu}\Gamma_t^\dagger\Gamma_t\,|0\rangle \\ &= -\tilde{a}_{jm;\frac{1}{2}\mu}\Gamma_t\,|0\rangle\end{aligned}$$

where $a^\dagger_{jm;\frac{1}{2}\mu}$ satisfy the anticommutation relationship of equation (2.21) and

$$\tilde{a}_{jm;\frac{1}{2}\mu} = (-1)^{j+m+\frac{1}{2}+\mu} a_{j-m;\frac{1}{2}-\mu} \tag{3.71}$$

is the modified adjoint operator which is a tensor of rank (jm) in ordinary space and $(\frac{1}{2}\mu)$ in isospin. Thus Γ_t must have properties similar to those of equations (3.19) and (3.20):

$$\begin{aligned}\Gamma_t a^\dagger_{jm;\frac{1}{2}\mu}\Gamma_t^\dagger &= -(-1)^{j+m+\frac{1}{2}+\mu} a_{j-m;\frac{1}{2}-\mu} \\ &= -\tilde{a}_{jm;\frac{1}{2}\mu}\end{aligned} \tag{3.72a}$$

$$\Gamma_t a_{jm;\frac{1}{2}\mu}\Gamma_t^\dagger = -(-1)^{j+m+\frac{1}{2}+\mu} a^\dagger_{j-m;\frac{1}{2}-\mu} \tag{3.72b}$$

and

$$\begin{aligned}\Gamma_t^\dagger a^\dagger_{jm;\frac{1}{2}\mu}\Gamma_t &= -(-1)^{j-m+\frac{1}{2}-\mu} a_{j-m;\frac{1}{2}-\mu} \\ &= -\tilde{a}_{jm;\frac{1}{2}\mu}\end{aligned} \tag{3.73a}$$

$$\Gamma_t^\dagger a_{jm;\frac{1}{2}\mu}\Gamma_t = -(-1)^{j-m+\frac{1}{2}-\mu} a^\dagger_{j-m;\frac{1}{2}-\mu} \tag{3.73b}$$

To calculate the particle–hole interaction one must, as was done in section 2.4, evaluate the transformed potential

$$\bar{V} = (\Gamma_t^\dagger)_{j_4}(\Gamma_t^\dagger)_{j_2} V (\Gamma_t)_{j_2}(\Gamma_t)_{j_4}$$

where

$$V = \sum_{JMTT_z j_i} E_{JT}(j_1 j_2; j_3 j_4) A^\dagger_{JM;TT_z}(j_1 j_2) A_{JM;TT_z}(j_3 j_4)$$

$$= \sum_{JTj_i} (-1)^{j_3+j_4-J+1-T} \surd\{(2J+1)(2T+1)\} E_{JT}(j_1 j_2; j_3 j_4)$$

$\times$ 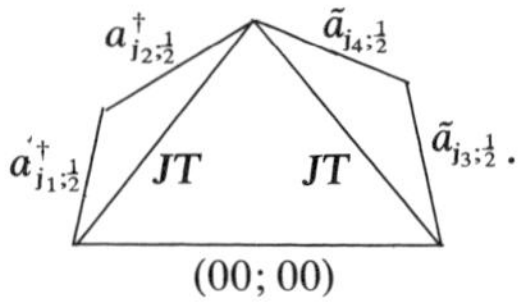

.

In V, $A^\dagger_{JM;TT_z}$ and $A_{JM;TT_z}$ are defined by equations (2.22) and (2.23) and $E_{JT}(j_1 j_2; j_3 j_4)$ is the interaction in the spin state J with isospin T. Finally, since the particle–hole transformation will not involve a level which simultaneously has a particle and a hole, $j_1 \neq j_2$ and $j_3 \neq j_4$ so that the factor $(1+\delta_{j_1 j_2})^{-\frac{1}{2}}$ can be neglected. By exactly the same methods as employed in the last section one can use equations (3.73) to show that

$$\bar{V} = V_t' + V_t'' \tag{3.74}$$

where

$$V_t'' = \sum_{j_i} \delta_{j_1 j_3} \delta_{j_2 j_4} \sum_{JT} \frac{(2J+1)(2T+1)}{2(2j_1+1)} E_{JT}(j_1 j_2; j_3 j_4)$$
$$\times \sum_{m\mu} a^\dagger_{j_1 m;\frac{1}{2}\mu} a_{j_1 m;\frac{1}{2}\mu} \tag{3.75}$$

and in this expression $\sum_{m\mu} a^\dagger_{j_1 m;\frac{1}{2}\mu} a_{j_1 m;\frac{1}{2}\mu}$ gives the total number of nucleons (neutrons plus protons) in the level j_1.

The two-body part of equation (3.74), V_t', has the form

$$V_t' = \sum_{K\tau j_i} (-1)^{j_2+j_3-K+1-\tau} \surd\{(2K+1)(2\tau+1)\} F_{K\tau}(j_1 j_4; j_3 j_2)$$

$\times$ 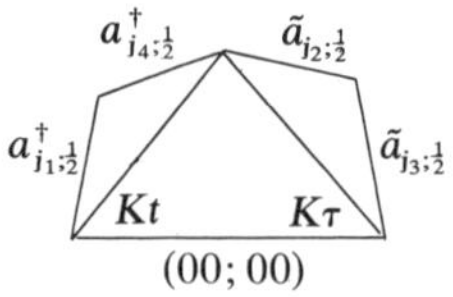

$$= \sum_{KM\tau\tau_z j_i} F_{K\tau}(j_1 j_4; j_3 j_2) A^\dagger_{KM;\tau\tau_z}(j_1 j_4) A_{KM;\tau\tau_z}(j_3 j_2) \tag{3.76}$$

with

$$F_{K\tau}(j_1j_4;j_3j_2)=E_{K\tau}(j_1j_4^{-1};j_3j_2^{-1})$$
$$=-(-1)^{j_1+j_2+j_3+j_4}\sum_{JT}(2J+1)(2T+1)W(j_1j_2j_4j_3;JK)$$
$$\times W(\tfrac{1}{2}\tfrac{1}{2}\tfrac{1}{2}\tfrac{1}{2};T\tau)E_{JT}(j_1j_2;j_3j_4). \tag{3.77}$$

Consequently in the isospin formalism, the generalized particle–hole theorem is given by equation (3.77). Furthermore, if one inserts the values of the isospin Racah coefficients from Table (A4.2) and equation (A4.14) one finds that

$$E_{K0}(j_1j_4^{-1};j_3j_2^{-1})=\tfrac{1}{2}(-1)^{j_1+j_2+j_3+j_4}\sum_J(2J+1)W(j_1j_2j_4j_3;JK)$$
$$\times\{E_{J0}(j_1j_2;j_3j_4)-3E_{J1}(j_1j_2;j_3j_4)\} \tag{3.78}$$
$$E_{K1}(j_1j_4^{-1};j_3j_2^{-1})=-\tfrac{1}{2}(-1)^{j_1+j_2+j_3+j_4}\sum_J(2J+1)W(j_1j_2j_4j_3;JK)$$
$$\times\{E_{J0}(j_1j_2;j_3j_4)+E_{J1}(j_1j_2;j_3j_4)\}. \tag{3.79}$$

3.1. The yrast negative-parity states in $^{40}_{20}Ca_{20}$

As an application of the particle–hole transformation including isospin we consider the yrast 2^-–5^- states in $^{40}_{20}Ca_{20}$. In the first approximation these states may be considered to arise from the excitation of a $d_{\frac{3}{2}}$ particle to the $f_{\frac{7}{2}}$ state. Relative to the $^{40}_{20}Ca_{20}$ ground state, the energy it takes to produce such an excitation is

$$\varepsilon_\pi=\varepsilon_\pi(d_{\frac{3}{2}}^{-1})+\varepsilon_\pi(f_{\frac{7}{2}})$$
$$=\{BE(^{39}_{19}K_{20})-BE(^{40}_{20}Ca_{20})\}+\{BE(^{41}_{21}Sc_{20})-BE(^{40}_{20}Ca_{20})\}$$
$$=7\cdot244\text{ MeV}$$

for a proton and

$$\varepsilon_\nu=\varepsilon_\nu(d_{\frac{3}{2}}^{-1})+\varepsilon_\nu(f_{\frac{7}{2}})$$
$$=\{BE(^{39}_{20}Ca_{19})-BE(^{40}_{20}Ca_{20})\}+\{BE(^{41}_{20}Ca_{21})-BE(^{40}_{20}Ca_{20})\}$$
$$=7\cdot269\text{ MeV}.$$

for a neutron. Since both the neutron and the proton particle–hole excitation energies are almost identical we shall take the value 7·25 MeV for both, and consequently in the absence of any particle–hole interaction the $T=0$ and $T=1$ 2^-–5^- states would be degenerate and lie at this excitation energy.

To simplify the calculation we shall assume that the residual two-body interaction is the delta-function potential whose matrix elements are given by equation (A2.29). By using equations (A4.19), (A4.20), and

(A4.22) it follows that

$$-\sum_J (2J+1)W(j_1j_2j_4j_3;JK)E_{JT}(j_1j_2;j_3j_4)$$
$$=\frac{V_T\bar{R}\sqrt{\{(2j_1+1)(2j_2+1)(2j_3+1)(2j_4+1)\}}}{2(2K+1)\sqrt{\{(1+\delta_{j_1j_2})(1+\delta_{j_3j_4})\}}}\left\{\frac{1+(-1)^{l_1+l_2+l_3+l_4}}{2}\right\}$$
$$\times((-1)^{l_2+l_4}(j_1j_4\tfrac{1}{2}\tfrac{1}{2}\mid K1)(j_3j_2\tfrac{1}{2}\tfrac{1}{2}\mid K1)$$
$$+(-1)^{j_2-j_4}\{1+(-1)^T+(-1)^{K+T+l_2+l_3}\}(j_1j_4\tfrac{1}{2}-\tfrac{1}{2}\mid K0)(j_3j_2\tfrac{1}{2}-\tfrac{1}{2}\mid K0)) \quad (3.80)$$

where V_T is the strength of the interaction in the isospin state T and

$$\bar{R}=\int R_{j_1}R_{j_2}R_{j_3}R_{j_4}r^2\,\mathrm{d}r$$

with R_j the radial wave function for the state j.

Equation (3.80) may now be used in conjunction with equations (3.78) and (3.79) to calculate the $(d_{\frac{3}{2}}, f_{\frac{7}{2}})$ particle–hole interaction energies. One can show that the interaction strengths

$$V_0\bar{R}=0{\cdot}90\ \text{MeV}$$
$$V_1\bar{R}=0{\cdot}54\ \text{MeV} \quad (3.81)$$

gives a good fit to the ${}^{38}_{17}\text{Cl}_{21}$ spectrum and we assume this same strength for ${}^{40}_{20}\text{Ca}_{20}$. In Fig. 3.6 the calculated $(d_{\frac{3}{2}}^{-1}, f_{\frac{7}{2}})$ spectrum is compared with experiment (Endt and Van der Leun 1973). The $T=1$ states are quite well reproduced by this potential; however, the experimental $T=0$ levels all occur at lower excitation energies.

Zamick (1973) has shown that if all possible one-particle excitations from the *ds* shell to the *fp* shell are considered the $T=1$ states are hardly affected whereas the $T=0$, $I=3^-$, and $I=5^-$ levels are substantially depressed as is required to fit experiment. For example, in the full-shell-model space the 3^- state can be realized in nine different ways—$(f_{\frac{7}{2}}, d_{\frac{3}{2}}^{-1})$, $(f_{\frac{7}{2}}, s_{\frac{1}{2}}^{-1})$, $(f_{\frac{7}{2}}, d_{\frac{5}{2}}^{-1})$, $(p_{\frac{3}{2}}, d_{\frac{3}{2}}^{-1})$, $(p_{\frac{3}{2}}, d_{\frac{5}{2}}^{-1})$, $(p_{\frac{1}{2}}, d_{\frac{5}{2}}^{-1})$, $(f_{\frac{5}{2}}, d_{\frac{3}{2}}^{-1})$, $(f_{\frac{5}{2}}, s_{\frac{1}{2}}^{-1})$, $(f_{\frac{5}{2}}, d_{\frac{5}{2}}^{-1})$—and consequently a full one-particle, one-hole calculation requires the construction and diagonalization of a 9×9 matrix (18×18 if one works in a non-isospin formalism). If one takes $\bar{R}$ equal to a constant (i.e. the surface-delta form of the interaction) and uses the strengths given by equation (3.81) the spectrum labelled 'full shell' in Fig. 3.6 emerges when the single-particle $p_{\frac{3}{2}}$, $p_{\frac{1}{2}}$, and $f_{\frac{5}{2}}$ states are taken to lie 2·0 MeV, 4·25 MeV, and 6·0 MeV above the $f_{\frac{7}{2}}$ level and the $s_{\frac{1}{2}}^{-1}$ and $d_{\frac{5}{2}}^{-1}$ states are taken 2·5 MeV and 5·0 MeV above the $d_{\frac{3}{2}}$ hole state. Although some improvement is apparent, with the exception of the 3^- level, all $T=0$

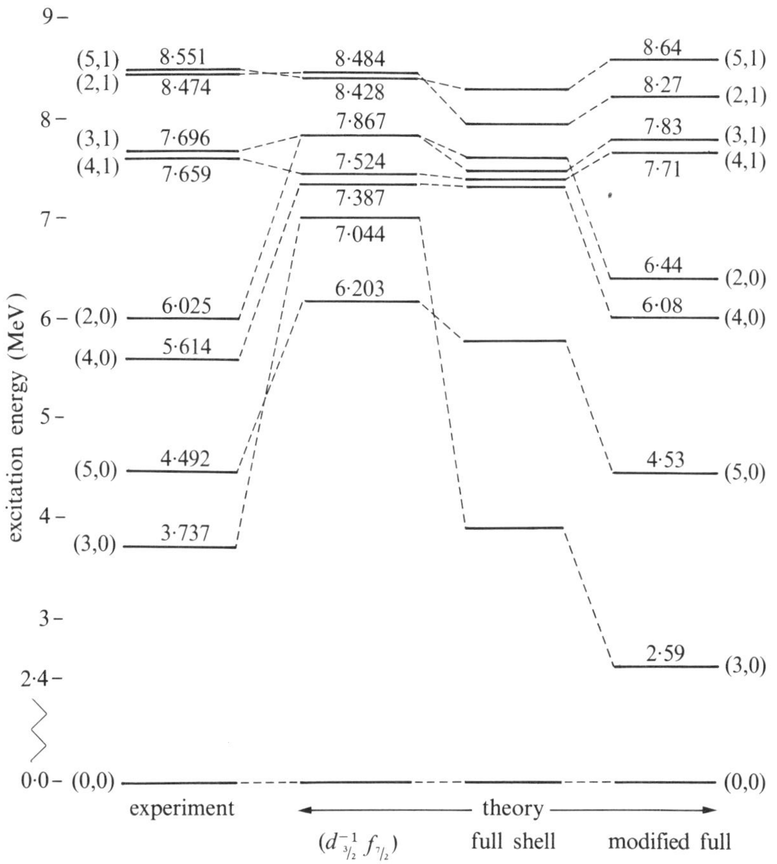

Fig. 3.6. Comparison of theory and experiment for the yrast 2^-–5^- states in ${}^{40}_{20}Ca_{20}$. In the absence of any residual particle–hole interaction all theoretical states would be degenerate and lie at 7·25 MeV excitation energy. The values in braces are the spin and isospin. See text for more details.

states still lie at too high an excitation energy and all $T=1$ states are still too low.

Glaudemans *et al.* (1967) noted that a substantial improvement in the splitting of the $T=0$ and $T=1$ levels could be obtained by using the modified surface-delta interaction

$$V=\sum_T\{-4\pi V_T\delta(\Omega_{12})+B_T\} \tag{3.82}$$

where B_T is a constant that depends only on the isospin of the two interacting particles. Since the added terms have no space-spin dependence they have only diagonal matrix elements. According to equations

(3.78) and (3.79) one must therefore add to each $T=0$ particle-hole interaction an amount

$$\Delta E_{K0}(j_1 j_4^{-1};\, j_3 j_2^{-1}) = \tfrac{1}{2}(B_0 - 3B_1)\delta_{j_1 j_3}\delta_{j_2 j_4}$$

and to each $T=1$ an energy

$$\Delta E_{K1}(j_1 j_4^{-1};\, j_3 j_2^{-1}) = -\tfrac{1}{2}(B_0 + B_1)\delta_{j_1 j_3}\delta_{j_2 j_4}.$$

In deriving these values for ΔE_{KT} we have made use of equations (A4.13), (A4.14), and (A4.17) to show that

$$\begin{aligned}\sum_J (2J+1)W(j_1 j_2 j_2 j_1;\, JK) &= \sqrt{\{(2j_1+1)(2j_2+1)\}}\sum_J (-1)^{j_1+j_2-J}(2J+1)W(j_1 j_2 j_2 j_1;\, JK)W(j_1 j_2 j_1 j_2;\, J0)\\ &= 1.\end{aligned}$$

The results that one obtains when $B_0 = -1{\cdot}05$ MeV and $B_1 = 0{\cdot}505$ MeV are shown in the last column of Fig. 3.6, and it is apparent that theory and experiment are now in quite reasonable agreement.

4

HARMONIC OSCILLATOR WAVE-FUNCTIONS

With the exception of the delta-function interaction we have always taken the matrix elements of the residual two-body force from experimental data. This procedure is not always possible and there may also be instances when we wish to evaluate these matrix elements from some assumed finite-range interaction. In general this can be done by making a Slater decomposition (Slater 1929) of the interaction potential. For simplicity we outline the method of treatment when the potential is spin independent. We assume that V is Galilean invariant and expand $V = V(|\underline{r}_1 - \underline{r}_2|)$ in terms of the cosine of the angle between $\underline{r}_1$ and $\underline{r}_2$, that is in terms of Legendre polynomials. Thus

$$\begin{aligned} V(|\underline{r}_1-\underline{r}_2|) &= \sum_l F_l(r_1, r_2) P_l(\cos\theta_{12}) \\ &= 4\pi \sum_{lm} \frac{(-1)^m F_l(r_1, r_2)}{(2l+1)} Y_{l-m}(\theta_1, \phi_1) Y_{lm}(\theta_2, \phi_2) \\ &= 4\pi \sum_l \frac{(-1)^l F_l(r_1, r_2)}{\sqrt{(2l+1)}} [Y_l(\theta_1, \phi_1) \times Y_l(\theta_2, \phi_2)]_{00} \end{aligned} \tag{4.1}$$

and in writing the last line of this equation we have made use of equation (A1.21) to write $(-1)^m = (-1)^l \sqrt{(2l+1)}(llm-m \mid 00)$. Since the P_l are orthogonal $F_l(r_1, r_2)$ is given by

$$F_l(r_1, r_2) = \frac{2l+1}{2} \int_0^\pi P_l(\cos\theta_{12}) V(|\underline{r}_1 - \underline{r}_2|) \sin\theta_{12}\, d\theta_{12}. \tag{4.2}$$

For example, the Gaussian potential

$$V = \bar{V}_0 e^{-\beta^2(\underline{r}_1-\underline{r}_2)^2} = \bar{V}_0 e^{-\beta^2(r_1{}^2+r_2{}^2)} e^{2\beta^2(\underline{r}_1\cdot\underline{r}_2)} \tag{4.3}$$

has

$$F_l(r_1, r_2) = \bar{V}_0(2l+1) i^l e^{-\beta^2(r_1{}^2+r_2{}^2)} j_l(-2i\beta^2 r_1 r_2) \tag{4.4}$$

where j_l is the spherical Bessel function (Morse and Feshbach 1953) and $\bar{V}_0$ is a constant. Note that the quadrupole–quadrupole interaction used in equation (2.94) is just the Gaussian F_2 taken in the limit that β is small.

Once the Slater decomposition has been carried out, the matrix elements of the potential can be calculated by the same method as was used to deduce equation (2.95). Thus

$$\begin{aligned} E_I(j_c j_d\,;\, j_a j_b) &= \langle \Phi_{IM}(j_c j_d)|\, V\, |\Phi_{IM}(j_a j_b)\rangle \\ &= \sqrt{\{(2j_a+1)(2j_b+1)\}} \sum_l (-1)^l \left\{\frac{1+(-1)^{l_a+l_c+l}}{2}\right\} \\ &\quad \times \left\{\frac{1+(-1)^{l_b+l_d+l}}{2}\right\} (j_a l \tfrac{1}{2} 0 \mid j_c \tfrac{1}{2})(j_b l \tfrac{1}{2} 0 \mid j_d \tfrac{1}{2}) \\ &\quad \times W(j_c l I j_b\,;\, j_a j_d) \\ &\quad \times \iint r_1^2\, \mathrm{d}r_1 r_2^2\, \mathrm{d}r_2 R^*_{n_c l_c}(r_1) R^*_{n_d l_d}(r_2) F_l(r_1, r_2) R_{n_a l_a}(r_1) R_{n_b l_b}(r_2) \end{aligned} \tag{4.5}$$

where $R_{nl}(r)$ is the radial wave function for the state (njl).

Equation (4.5) gives the value of the matrix element when

$$\Phi_{IM}(j_1 j_2) = [\phi_{j_1 l_1}(1) \times \phi_{j_2 l_2}(2)]_{IM} \tag{4.6}$$

where the ϕ_j are the space-spin angular-momentum eigenfunctions given by equation (1.4). This wave function is antisymmetric only when $(j_1 l_1) = (j_2 l_2)$ and I is even. Consequently, when this condition is not fulfilled one must explicitly antisymmetrize the wave functions used to calculate the values of $E_{IT=1}(j_c j_d\,;\, j_a j_b)$ to be used when the potential is expressed in terms of annihilation and creation operators and one must symmetrize the Φ_{IM} used to calculate $E_{IT=0}(j_c j_d\,;\, j_a j_b)$.

When the $R_{nl}(r)$ are harmonic-oscillator functions one need not make the Slater decomposition because a simpler procedure exists. We shall now discuss how calculations are carried out in this case and derive explicit expressions for the matrix elements of the various commonly used forms of the residual two-body interaction.

1. Energy matrix elements calculated with oscillator functions

The Hamiltonian describing the independent motion of two particles in a harmonic-oscillator potential can be separated in (r_1, r_2) space and also in the relative and centre-of-mass coordinate system of the two particles:

$$\begin{aligned} H &= \frac{p_1^2}{2m} + \tfrac{1}{2} m\omega^2 r_1^2 + \frac{p_2^2}{2m} + \tfrac{1}{2} m\omega^2 r_2^2 \\ &= \frac{p^2}{2m} + \tfrac{1}{2} m\omega^2 r^2 + \frac{P^2}{2m} + \tfrac{1}{2} m\omega^2 R^2 \end{aligned} \tag{4.7}$$

where the latter coordinates and their canonically conjugate momenta are

$$\begin{aligned} \underline{r} &= \frac{1}{\sqrt{2}}(\underline{r}_1 - \underline{r}_2) \\ \underline{R} &= \frac{1}{\sqrt{2}}(\underline{r}_1 + \underline{r}_2) \\ \underline{p} &= \frac{1}{\sqrt{2}}(\underline{p}_1 - \underline{p}_2) \\ \underline{P} &= \frac{1}{\sqrt{2}}(\underline{p}_1 + \underline{p}_2). \end{aligned} \tag{4.8}$$

Talmi (1952) showed that the oscillator is the only potential that can be separated in both these coordinate systems and pointed out that because the residual nucleon–nucleon interaction is a function of the relative coordinate $\underline{r}$ an enormous simplification results in the calculation of the energy matrix elements. Because of this separability we may write the spatial part of the two-particle wave function as

$$\begin{aligned} \psi_{\lambda\mu}\{n_1l_1(1), n_2l_2(2)\} &= [\phi_{n_1l_1}(\underline{r}_1)\times\phi_{n_2l_2}(\underline{r}_2)]_{\lambda\mu} \\ &= \sum_{nlNL} M_\lambda(nlNL; n_1l_1n_2l_2)[\phi_{nl}(\underline{r})\times\phi_{NL}(\underline{R})]_{\lambda\mu} \end{aligned} \tag{4.9}$$

where the notation $n_1l_1(1)$, $n_2l_2(2)$ means particle number 1 is in the state n_1l_1 and particle number 2 is in the n_2l_2 orbit. The $\phi_{n'l'}(\underline{r}')$ describe states with energy $(2n'+l'+\frac{3}{2})\hbar\omega$ and have the form

$$\phi_{n'l'}(\underline{r}') = R_{n'l'}(\alpha r')Y_{l'm'}(\theta', \phi') \tag{4.10}$$

with

$$\begin{aligned} R_{nl}(\alpha r) = {} & \left[\frac{2^{l-n+2}(2l+2n+1)!!\alpha^{2l+3}}{\sqrt{\pi}n![(2l+1)!!]^2}\right]^{\frac{1}{2}}\left(\exp -\frac{\alpha^2}{2}r^2\right)r^l \\ & \times \sum_{k=0}^{n} \frac{(-1)^k 2^k n!(2l+1)!!(\alpha^2r^2)^k}{k!(n-k)!(2l+2k+1)!!} \end{aligned} \tag{4.11a}$$

and

$$\alpha^2 = m\omega/\hbar. \tag{4.11b}$$

Since both sides of equation (4.9) must describe a state with the same energy it follows that

$$2n_1 + l_1 + 2n_2 + l_2 = 2n + l + 2N + L. \tag{4.12}$$

Thus the $(nlNL)$ sum in equation (4.9) is finite.

Moshinsky (1959) deduced the form of the matrix elements of the central, tensor, and spin-orbit interactions in terms of the coefficients

$M_\lambda(nlNL; n_1l_1n_2l_2)$, and for this reason they are often called Moshinsky coefficients. In addition he gave an explicit expression for them when $n_1 = n_2 = 0$ (essentially the same as given in equation (A6.16)) and deduced a recursion relationship that could be used to find the coefficients for other values of n_1 and n_2. Since then many different formulae and procedures have been proposed for evaluating M_λ. For n_1 and/or $n_2 \neq 0$ the formula given by Bakri (1967) seems to be the one best suited for numerical evaluation. Even so, this formula (which is given by equation (A6.19)) contains a sum over five independent dummy variables. In view of this complexity it is fortunate that a tabulation of these coefficients exists (Brody and Moshinsky 1960).

Because of the strong one-body spin-orbit force in nuclei, which leads to the required shell structure, we have always dealt with j–j coupled wave functions. To exploit the simplicity afforded by the oscillator functions it is necessary to change from this coupling scheme to the L–S scheme. This is easily done in terms of diagrams. From the definition of the $9j$ coefficient given in equations (1.90) or (A4.8) it follows that

$$[\phi_{j_1l_1} \times \phi_{j_2l_2}]_{IM} =$$

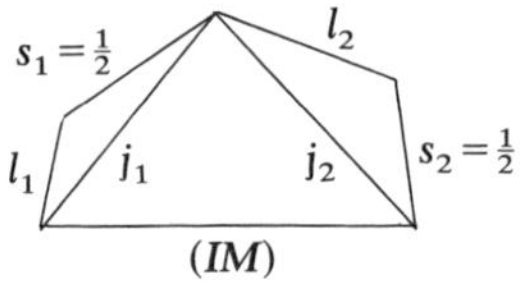

$$= \sum_{\lambda S} \sqrt{\{(2j_1+1)(2j_2+1)(2\lambda+1)(2S+1)\}} \begin{Bmatrix} l_1 & \frac{1}{2} & j_1 \\ l_2 & \frac{1}{2} & j_2 \\ \lambda & S & I \end{Bmatrix}$$

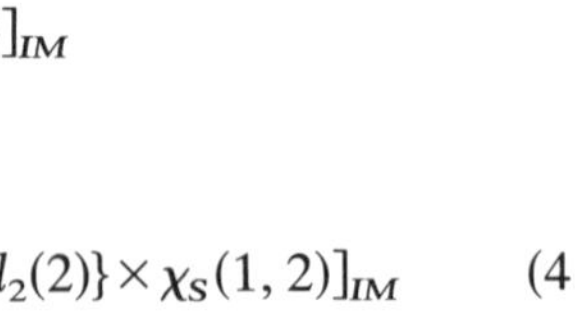

Thus

$$\Phi_{IM}\{j_1n_1l_1(1), j_2n_2l_2(2)\} = [\phi_{j_1l_1}(1) \times \phi_{j_2l_2}(2)]_{IM}$$
$$= \sum_{\lambda S} \gamma^{(I)}_{\lambda S}(j_1l_1; j_2l_2)$$
$$\times [\psi_\lambda\{n_1l_1(1), n_2l_2(2)\} \times \chi_S(1, 2)]_{IM} \tag{4.13}$$

where $\chi_{SM'}(1, 2)$ is the two-particle spin wave function

$$\chi_{SM'}(1, 2) = \sum_{\nu\nu'} (\tfrac{1}{2}\tfrac{1}{2}\nu\nu' \mid SM')\chi_\nu(1)\chi_{\nu'}(2) \tag{4.14}$$

and $\psi_{\lambda\mu}$ is the two-particle wave function given by (4.9). The coefficient $\gamma_{\lambda S}^{(I)}$ is the j–j to L–S transformation coefficient

$$\gamma_{\lambda S}^{(I)}(j_1 l_1; j_2 l_2) = \sqrt{\{(2j_1+1)(2j_2+1)(2S+1)(2\lambda+1)\}} \begin{Bmatrix} l_1 & \frac{1}{2} & j_1 \\ l_2 & \frac{1}{2} & j_2 \\ \lambda & S & I \end{Bmatrix} \quad (4.15)$$

and when squared it gives the probability that the state $[\psi_\lambda \times \chi_S]_{IM}$ will be realized in the original j–j coupled eigenfunction. Since S can take on the values 0 and 1, it follows that $\lambda = I,\ I \pm 1$.

The Brody–Moshinsky tabulation of $M_\lambda(nlNL; n_1 l_1 n_2 l_2)$ and the formulae given in Appendix 6 are for the wave function in equation (4.13). However, for $(j_1 n_1 l_1) \neq (j_2 n_2 l_2)$ this eigenvector is neither symmetric nor antisymmetric to interchange of the particles. On the other hand, the $T=1$ interaction energies must be calculated using space-spin antisymmetric states and those for $T=0$ by use of space-spin symmetric functions. Thus instead of equation (4.13) the appropriate wave function to use is

$$\begin{aligned}\tilde{\Phi}_{IM}(j_1 n_1 l_1, j_2 n_2 l_2; T) = {} & \frac{1}{\sqrt{\{2(1+\delta_{j_1 j_2}\delta_{l_1 l_2}\delta_{n_1 n_2})\}}} \\ & \times \sum_{\lambda S} \gamma_{\lambda S}^{(I)}(j_1 l_1; j_2 l_2) \\ & \times ([\psi_\lambda\{n_1 l_1(1), n_2 l_2(2)\} \times \chi_S(1,2)]_{IM} \\ & + (-1)^T [\psi_\lambda\{n_1 l_1(2), n_2 l_2(1)\} \times \chi_S(2,1)]_{IM}).\end{aligned} \quad (4.16)$$

As a consequence of the defining equation (4.14)

$$\chi_{SM'}(2,1) = (-1)^{1+S}\chi_{SM'}(1,2)$$

and from equation (4.9) it follows that

$$\psi_{\lambda\mu}\{n_1 l_1(2), n_2 l_2(1)\} = (-1)^{l_1+l_2-\lambda}\psi_{\lambda\mu}\{n_2 l_2(1), n_1 l_1(2)\}.$$

Furthermore, from equation (A6.22)

$$M_\lambda(nlNL; n_1 l_1 n_2 l_2) = (-1)^{L+\lambda} M_\lambda(nlNL; n_2 l_2 n_1 l_1).$$

If we combine these three results together with the fact that

$$(-1)^{l_1+l_2-L} = (-1)^l,$$

which follows from equation (4.12), we see that $\tilde{\Phi}_{IM}$ may be written as

$$\tilde{\Phi}_{IM}(j_1n_1l_1, j_2n_2l_2; T) = \frac{1}{\sqrt{\{2(1+\delta_{j_1j_2}\delta_{l_1l_2}\delta_{n_1n_2})\}}} \times \sum_{\lambda S}\sum_{nlNL} \gamma^{(I)}_{\lambda S}(j_1l_1; j_2l_2)\{1-(-1)^{S+T+l}\} \times M_\lambda(nlNL; n_1l_1n_2l_2) \times [[\phi_{nl}(\underset{\sim}{r})\times\phi_{NL}(\underset{\sim}{R})]_\lambda \times \chi_S(1,2)]_{IM}. \tag{4.17}$$

The term involving the phase factor shows that even l wave functions, which are symmetric under the interchange of the two particles, must go with $S=1$, $T=0$ and $S=0$, $T=1$, whereas antisymmetric spatial states (l odd) are associated with $S=1$, $T=1$ and $S=0$, $T=0$. Thus if we merely replace $\gamma^{(I)}_{\lambda S}$ by

$$\tilde{\gamma}^{(I)}_{\lambda S}(j_1l_1; j_2l_2) = \frac{\{1-(-1)^{S+T+l}\}}{\sqrt{\{2(1+\delta_{j_1j_2}\delta_{l_1l_2}\delta_{n_1n_2})\}}}\gamma^{(I)}_{\lambda S}(j_1l_1; j_2l_2) \tag{4.18}$$

we may calculate energy matrix elements corresponding to physical states using equation (4.13). We shall now deduce expressions for the matrix elements of commonly encountered potentials.

1.1. Central spin-dependent interaction

The spin operator for the two-nucleon system is

$$\underset{\sim}{S} = \tfrac{1}{2}(\underset{\sim}{\sigma}_1 + \underset{\sim}{\sigma}_2). \tag{4.19}$$

By squaring this equation it follows that

$$\langle \chi_{SM}|\, \underset{\sim}{\sigma}_1 \cdot \underset{\sim}{\sigma}_2\, |\chi_{SM}\rangle = 2S(S+1)-3.$$

Thus the spin singlet projection operator

$$P_0 = \tfrac{1}{4}(1-\underset{\sim}{\sigma}_1\cdot\underset{\sim}{\sigma}_2) \tag{4.20a}$$

has the property that it gives 1 and zero when operating on the singlet and triplet spin states, respectively.

$$P_1 = \tfrac{1}{4}(3+\underset{\sim}{\sigma}_1\cdot\underset{\sim}{\sigma}_2) \tag{4.20b}$$

has just the opposite effect; it gives zero and 1 when operating on the singlet and triplet spin states, respectively.

Thus a general potential which is the sum of a spin-independent (Wigner) force and spin-dependent part can be written as

$$V = \sum_S P_S V_S(\beta\,|\underset{\sim}{r}_1 - \underset{\sim}{r}_2|) \tag{4.21}$$

where V_0 and V_1 are the singlet and triplet potentials which are functions of $|\underline{r}_1-\underline{r}_2|$, have strength V_S, and range parameter β (for a Gaussian potential the form would be given by equation (4.3)).

Matrix elements of this potential between basis states given by equation (4.17) have the form

$$\langle\tilde{\Phi}_{IM}(j_1n_1l_1, j_2n_2l_2; T)|\, V\, |\tilde{\Phi}_{IM}(j_3n_3l_3, j_4n_4l_4; T)\rangle$$

$$= \sum_{n'l'N'L'}\ \sum_{nlNL}\ \sum_{\lambda S}\ \sum_{\lambda' S'}\ \sum_{\mu\mu'}\ \sum_{M_SM_S'}\ \sum_{S''} \tilde{\gamma}^{(I)}_{\lambda' S'}(j_1l_1; j_2l_2)\tilde{\gamma}^{(I)}_{\lambda S}(j_3l_3; j_4l_4)$$

$$\times(\lambda S\mu M_S \mid IM)(\lambda' S'\mu' M_S' \mid IM)\langle\chi_{S'M_S'} \mid P_{S''}\, |\chi_{SM_S}\rangle$$

$$\times M_{\lambda'}(n'l'N'L'; n_1l_1n_2l_2)M_\lambda(nlNL; n_3l_3n_4l_4)$$

$$\times\langle[\phi_{n'l'}(\underline{r})\times\phi_{N'L'}(\underline{R})]_{\lambda'\mu'}|\, V_{S''}\, |[\phi_{nl}(\underline{r})\times\phi_{NL}(\underline{R})]_{\lambda\mu}\rangle.$$

Since the V_S are scalars independent of the centre-of-mass coordinate, it follows that $(lNL)=(l'N'L')$, $\mu=\mu'$, and $\lambda=\lambda'$. Also the P_S are scalars so that $M_S=M_S'$ and $S=S'$. The sums over the magnetic quantum numbers can be carried out and give unity. Thus

$$\langle\tilde{\Phi}_{IM}(j_1n_1l_1, j_2n_2l_2; T)|\, V\, |\tilde{\Phi}_{IM}(j_3n_3l_3, j_4n_4l_4; T)\rangle$$

$$= \sum_{\lambda S}\ \sum_{n'nlNL} \tilde{\gamma}^{(I)}_{\lambda S}(j_1l_1; j_2l_2)\tilde{\gamma}^{(I)}_{\lambda S}(j_3l_3; j_4l_4)$$

$$\times M_\lambda(n'lNL; n_1l_1n_2l_2)M_\lambda(nlNL; n_3l_3n_4l_4)$$

$$\times\langle R_{n'l}(x)|\, V_S(\kappa x)\, |R_{nl}(x)\rangle \tag{4.22}$$

where

$$x=\alpha r$$

$$\kappa=\sqrt{2}\beta/\alpha. \tag{4.23}$$

The reason the factor κ has a $\sqrt{2}$ is because of our definition of the relative and centre-of-mass coordinate for the oscillator transformation given in equation (4.8). This leads to symmetry in the wave functions but the potential is written conventionally as a function of $\underline{r}_1-\underline{r}_2=\sqrt{2}\underline{r}$.

1.2. *Tensor force*

To express the two-body tensor force

$$V_t=\frac{1}{r^2}V_t(\beta\,|\underline{r}_1-\underline{r}_2|)\{3(\underline{\sigma}_1\cdot\underline{r})(\underline{\sigma}_2\cdot\underline{r})-(\underline{\sigma}_1\cdot\underline{\sigma}_2)r^2\} \tag{4.24}$$

in a convenient form for evaluation we make use of the fact that according to equations (A2.12) and (A2.15)

$$\underset{\sim}{\sigma} \cdot \underset{\sim}{r} = \left(\frac{\sigma_x + i\sigma_y}{\sqrt{2}}\right)\left(\frac{x - iy}{\sqrt{2}}\right) + \left(\frac{\sigma_x - i\sigma_y}{\sqrt{2}}\right)\left(\frac{x + iy}{\sqrt{2}}\right) + \sigma_z z$$

$$= \left(\frac{4\pi}{3}\right)^{\frac{1}{2}} r \sum_{\mu} (-1)^{\mu} \sigma_{\mu} Y_{1-\mu}(\theta, \phi). \tag{4.25}$$

From this result it follows that

$$3(\underset{\sim}{\sigma}_1 \cdot \underset{\sim}{r})(\underset{\sim}{\sigma}_2 \cdot \underset{\sim}{r}) = 4\pi r^2 \sum_{\mu\mu'} (-1)^{\mu+\mu'} \sigma_{1\mu}\sigma_{2\mu'} Y_{1-\mu} Y_{1-\mu'}$$

$$= 3\sqrt{(4\pi)} r^2 \sum_{\mu\mu' LM} (-1)^{\mu+\mu'}$$

$$\times \frac{(1100 \mid L0)(11 - \mu - \mu' \mid L - M)}{\sqrt{(2L+1)}} \sigma_{1\mu}\sigma_{2\mu'} Y_{L-M}$$

where the two spherical harmonics have been combined by use of equation (A2.11). If we define the spin operator

$$X_{SM_S} = \sum_{\nu\nu'} (11\nu\nu' \mid SM_S) \sigma_{1\nu}\sigma_{2\nu'} \tag{4.26}$$

it follows that

$$3(\underset{\sim}{\sigma}_1 \cdot \underset{\sim}{r})(\underset{\sim}{\sigma}_2 \cdot \underset{\sim}{r}) = 3\sqrt{(4\pi)} r^2 \sum_{SM_S LM} (-1)^M \frac{(1100 \mid L0)}{\sqrt{(2L+1)}} X_{SM_S} Y_{L-M}$$

$$\times \sum_{\mu\mu'} (11\mu\mu' \mid SM_S)(11 - \mu - \mu' \mid L - M)$$

$$= 3\sqrt{(4\pi)} r^2 \sum_{LM} (-1)^{L-M} \frac{(1100 \mid L0)}{\sqrt{(2L+1)}} X_{LM} Y_{L-M} \tag{4.27}$$

where use has been made of equations (A1.19) and (A1.23) in writing the final form of this result. When $L = 0$ straightforward substitution yields

$$3(\underset{\sim}{\sigma}_1 \cdot \underset{\sim}{r})(\underset{\sim}{\sigma}_2 \cdot \underset{\sim}{r})_{L=0} - (\underset{\sim}{\sigma}_1 \cdot \underset{\sim}{\sigma}_2) r^2 = 0.$$

Furthermore because the parity Clebsch–Gordan coefficient $(1100 \mid L0)$ vanishes when $L = 1$ (see Table A1.2), the only allowable L value is $L = 2$. Since $(1100 \mid 20) = \sqrt{\frac{2}{3}}$ it follows that

$$V_t = \sqrt{(24\pi)}\, V_t(\beta\, |\underset{\sim}{r}_1 - \underset{\sim}{r}_2|)[X_2 \times Y_2]_{00} \tag{4.28}$$

where $V_t(\beta\, |\underset{\sim}{r}_1 - \underset{\sim}{r}_2|)$ is a scalar.

In deducing the 'three-to-two' sum rule for Racah coefficients we evaluated the matrix element of an operator with precisely the form of V_t

(see equation (A4.6)). Consequently once the reduced matrix element of X_{2M} is known the problem is solved. From equation (4.26) it follows that

$$\begin{aligned}X_{20} &= \sum_{\nu} (11\nu-\nu \mid 20)\sigma_{1\nu}\sigma_{2-\nu} \\ &= \frac{1}{\sqrt{6}}\{2\sigma_{1z}\sigma_{2z} - \sigma_{1x}\sigma_{2x} - \sigma_{1y}\sigma_{2y}\} \\ &= \sqrt{\tfrac{2}{3}}(3S_z^2 - \underset{\sim}{S}^2).\end{aligned} \tag{4.29}$$

Since neither S_z nor $\underset{\sim}{S}^2$ can change the spin

$$\begin{aligned}\langle \chi_{S'M_S} | X_{20} | \chi_{SM_S} \rangle &= \langle \chi_{S'} \| X_2 \| \chi_S \rangle (S2M_S0 \mid S'M_S) \\ &= \sqrt{\tfrac{2}{3}}\{3M_S^2 - S(S+1)\}\delta_{SS'}\delta_{S1}\end{aligned}$$

so that

$$\langle \chi_{S'} \| X_2 \| \chi_S \rangle = \frac{2\sqrt{5}}{\sqrt{3}}\,\delta_{SS'}\delta_{S1}. \tag{4.30}$$

Inserting this value for the reduced matrix element together with the reduced matrix element of the spherical harmonic Y_2, given by equation (A2.11), into equation (A4.6) we arrive at the result

$$\begin{aligned}&\langle \tilde{\Phi}_{IM}(j_1n_1l_1, j_2n_2l_2; T) | \, V_t \, | \tilde{\Phi}_{IM}(j_3n_3l_3, j_4n_4l_4; T) \rangle \\ &= 2\sqrt{30} \sum_{nln'l'} \sum_{NL} \sum_{\lambda\lambda'} (-1)^{l-L+1-I} \sqrt{\{(2l+1)(2\lambda+1)(2\lambda'+1)\}} \\ &\quad \times \tilde{\gamma}^{(I)}_{\lambda'1}(j_1l_1; j_2l_2)\tilde{\gamma}^{(I)}_{\lambda 1}(j_3l_3; j_4l_4)(l200 \mid l'0) \\ &\quad \times M_{\lambda'}(n'l'NL; n_1l_1n_2l_2)M_{\lambda}(nlNL; n_3l_3n_4l_4)W(\lambda\lambda'11; 2I) \\ &\quad \times W(ll'\lambda\lambda'; 2L)\langle R_{n'l'}(x) | \, V_t(\kappa x) \, | R_{nl}(x) \rangle\end{aligned} \tag{4.31}$$

where x and κ are given by equation (4.23).

1.3. *Two-body spin-orbit potential*

Equation (A4.6) can also be used to write down the matrix elements of the spin-orbit potential:

$$\begin{aligned}V_{so} &= V_s(\beta \, |\underset{\sim}{r}_1 - \underset{\sim}{r}_2|)\underset{\sim}{S} \cdot \underset{\sim}{\ell} \\ &= -\sqrt{3}\,V_s(\beta \, |\underset{\sim}{r}_1 - \underset{\sim}{r}_2|)[S \times \ell]_{00}\end{aligned} \tag{4.32}$$

where $\underset{\sim}{S}$ is the spin operator of equation (4.19) and $\underset{\sim}{\ell}$ is the relative angular momentum operator

$$\underset{\sim}{\ell} = \underset{\sim}{r} \times \underset{\sim}{p} \tag{4.33}$$

where × stands for the cross product encountered in ordinary vector analysis. Since $\underset{\sim}{\ell}$ is a tensor operator of rank 1

$$\langle Y_{l'm}|\,\ell_z\,|Y_{lm}\rangle = \langle Y_{l'}\|\,\ell\,\|Y_l\rangle(l1m0\mid l'm) = m\delta_{ll'}$$

it follows from Table A1.2 that

$$\langle Y_{l'}\|\,\ell\,\|Y_l\rangle = \delta_{ll'}\sqrt{\{l(l+1)\}} \tag{4.34}$$

and a similar expression holds for the matrix element of $\underset{\sim}{S}$.

By use of these relationships in conjunction with equation (A4.6) one finds that

$$\begin{aligned}&\langle\tilde{\Phi}_{IM}(j_1n_1l_1, j_2n_2l_2; T)|\,V_{so}\,|\tilde{\Phi}_{IM}(j_3n_3l_3, j_4n_4l_4; T)\rangle\\ &= \sum_{nn'lNL}\sum_{\lambda\lambda'}(-1)^{l-L-I}\sqrt{\{6l(l+1)(2l+1)(2\lambda+1)(2\lambda'+1)\}}\\ &\quad\times\tilde{\gamma}^{(I)}_{\lambda'1}(j_1l_1; j_2l_2)\tilde{\gamma}^{(I)}_{\lambda 1}(j_3l_3; j_4l_4)M_{\lambda}(n'lNL; n_1l_1n_2l_2)\\ &\quad\times M_{\lambda}(nlNL; n_3l_3n_4l_4)W(ll\lambda\lambda'; 1L)W(\lambda\lambda'11; 1I)\\ &\quad\times\langle R_{n'l}(x)|\,V_s(\kappa x)\,|R_{nl}(x)\rangle.\end{aligned} \tag{4.35}$$

2. The $0p$ Shell

The properties of nuclei with N and Z between 2 and 8 have been investigated theoretically by many authors on the assumption that the valence nucleons are filling the $0p_{\frac{3}{2}}$ and $0p_{\frac{1}{2}}$ levels outside an inert $(0s_{\frac{1}{2}})^4$ core. Cohen and Kurath (1965) critically surveyed the existing data and made a least-squares fit of the matrix elements of the residual two-body force to selected states that appeared to arise from the $0p$ configuration. Using the coupling rules of Chapter 1, section 1.1, and Chapter 2, sections 1.1 and 1.2 it follows that the possible two-particle states $(j_1j_2)_I$ that can arise in this shell are

$T=1$

$(p_{\frac{3}{2}})^2_0;\ (p_{\frac{1}{2}})^2_0;\ (p_{\frac{3}{2}}p_{\frac{1}{2}})_1;\ (p_{\frac{3}{2}})^2_2;\ (p_{\frac{3}{2}}p_{\frac{1}{2}})_2$

$T=0$

$(p_{\frac{3}{2}})^2_1;\ (p_{\frac{3}{2}}p_{\frac{1}{2}})_1;\ (p_{\frac{1}{2}})^2_1;\ (p_{\frac{3}{2}}p_{\frac{1}{2}})_2;\ (p_{\frac{3}{2}})^2_3.$

Thus the $T=0$ interaction energies are characterized by eight parameters; namely five diagonal and three off-diagonal $I=1$ matrix elements. Similarly the $T=1$ potential can be characterized by seven parameters. In this section we shall treat the Cohen–Kurath matrix elements as if they were experimental data and examine how well they can be fitted by a general potential whose matrix elements are calculated using harmonic-oscillator functions.

Since the Cohen–Kurath matrix elements are given in j–j coupling we must first calculate the j–j to L–S transformation coefficients given by equation (4.15). Values for the $9j$ symbols may be calculated directly from equation (A4.32) or taken from tables (Kennedy *et al.* 1954, Smith and Stephenson 1957). The resulting values of $\gamma_{\lambda S}^{(I)}$ are given in Table 4.1.

TABLE 4.1

j–j to L–S transformation coefficients for two p-shell nucleons calculated by using equation (4.15)

Configuration	Angular momentum I	Spin singlet $\gamma_{I,0}^{(I)}$	Spin triplet $\gamma_{I-1,1}^{(I)}$	$\gamma_{I,1}^{(I)}$	$\gamma_{I+1,1}^{(I)}$
$(p_{\frac{1}{2}})^2$	0	$1/\sqrt{3}$	0	0	$\sqrt{(\frac{2}{3})}$
$(p_{\frac{3}{2}})^2$	0	$\sqrt{(\frac{2}{3})}$	0	0	$-1/\sqrt{3}$
$(p_{\frac{1}{2}})^2$	1	$\sqrt{2}/3$	$-\frac{1}{3\sqrt{3}}$	0	$\frac{2\sqrt{5}}{3\sqrt{3}}$
$(p_{\frac{3}{2}}p_{\frac{1}{2}})$	1	$\frac{1}{3}$	$-\frac{2\sqrt{2}}{3\sqrt{3}}$	$1/\sqrt{2}$	$\frac{-\sqrt{5}}{3\sqrt{6}}$
$(p_{\frac{3}{2}})^2$	1	$\sqrt{5}/3$	$\frac{\sqrt{10}}{3\sqrt{3}}$	0	$-\frac{\sqrt{2}}{3\sqrt{3}}$
$(p_{\frac{3}{2}}p_{\frac{1}{2}})$	2	$1/\sqrt{3}$	$-1/\sqrt{6}$	$1/\sqrt{2}$	0
$(p_{\frac{3}{2}})^2$	2	$1/\sqrt{3}$	$\sqrt{(\frac{2}{3})}$	0	0
$(p_{\frac{3}{2}})^2$	3	0	1	0	0

The $\tilde{\gamma}_{\lambda S}^{(I)}$ that appear in the expressions for the energy matrix elements (equations (4.22), (4.31), and (4.35)) are related to these coefficients by equation (4.18). As a check on the numerics, note that $\sum_{\lambda S} \gamma_{\lambda S}^{(I)}(i)\gamma_{\lambda S}^{(I)}(k) = \delta_{ik}$.

The relative centre-of-mass transformation coefficients M_λ can be looked up directly in the Brody–Moshinsky tables (1960) or can be calculated from equation (A6.16). When this is done one finds

$$\begin{aligned}[\phi_{01}(\underline{r}_1)\times\phi_{01}(\underline{r}_2)]_{2\mu} &= M_2(0002;0101)[\phi_{00}(\underline{r})\times\phi_{02}(\underline{R})]_{2\mu} \\ &\quad + M_2(0200;0101)[\phi_{02}(\underline{r})\times\phi_{00}(\underline{R})]_{2\mu} \\ &= 2^{-\frac{1}{2}}\{[\phi_{00}(\underline{r})\times\phi_{02}(\underline{R})]_{2\mu} \\ &\quad -[\phi_{02}(\underline{r})\times\phi_{00}(\underline{R})]_{2\mu}\} \end{aligned} \tag{4.36a}$$

$$\begin{aligned}[\phi_{01}(\underline{r}_1)\times\phi_{01}(\underline{r}_2)]_{1\mu} &= M_1(0101;0101)[\phi_{01}(\underline{r})\times\phi_{01}(\underline{R})]_{1\mu} \\ &= [\phi_{01}(\underline{r})\times\phi_{01}(\underline{R})]_{1\mu}\end{aligned} \tag{4.36b}$$

$$[\phi_{01}(\underline{r}_1)\times\phi_{01}(\underline{r}_2)]_{00}=M_0(0010;0101)[\phi_{00}(\underline{r})\times\phi_{10}(\underline{R})]_{00}$$

$$+M_0(1000;0101)[\phi_{10}(\underline{r})\times\phi_{00}(\underline{R})]_{00}$$

$$=2^{-\frac{1}{2}}\{[\phi_{00}(\underline{r})\times\phi_{10}(\underline{R})]_{00}$$

$$-[\phi_{10}(\underline{r})\times\phi_{00}(\underline{R})]_{00}\} \tag{4.36c}$$

To simplify writing the expressions for the matrix elements it is convenient to introduce the notation

$$I^{(iT)}_{n'l';nl}=\int R_{n'l'}(x)V_{iT}(\kappa x)R_{nl}(x)x^2\,\mathrm{d}x \tag{4.37}$$

where x and κ are defined by equation (4.23) and i can take on the values 0, 1, t, and s corresponding to the spin singlet and triplet potentials of equation (4.21), the tensor potential of equation (4.24), and the spin-orbit potential of equation (4.32), respectively. T is the isospin index which is zero or 1 and for $(n'l')=(nl)$ we abbreviate

$$I^{(iT)}_{nl;nl}=I^{(iT)}_{nl}. \tag{4.37a}$$

In terms of the $I^{(iT)}_{n'l';nl}$ expressions for the interaction energies can be simply written down by use of equations (4.22), (4.31), and (4.35).

2.1. *The $T=1$ interaction in the $0p$ shell*

The central and two-body spin-orbit interactions, whose matrix elements are given by equations (4.22) and (4.35), can have non-vanishing values when n and n', the radial quantum numbers of the relative motion, are different. In addition, the tensor interaction can lead to matrix elements off-diagonal in both n and l. However, in all cases the (NL) quantum number characterizing the centre-of-mass motion of the two particles must be the same on both sides of the matrix element. When this is combined with the form of the $0p$ wave functions in equations (4.36) it follows that in all cases $n=n'$ and $l=l'$.

In terms of the radial integrals of equation (4.37a) the $T=1$ interaction energies

$$E_{I1}(j_1j_2;j_3j_4)=\langle\tilde{\Phi}_{IM}(j_1n_1l_1,j_2n_2l_2;1)|\,V\,|\tilde{\Phi}_{IM}(j_3n_3l_3,j_4n_4l_4;1)\rangle$$

are given by

$$
\begin{bmatrix}
E_{01}(\frac{1}{2}\frac{1}{2};\frac{1}{2}\frac{1}{2}) \\
E_{01}(\frac{3}{2}\frac{3}{2};\frac{3}{2}\frac{3}{2}) \\
E_{01}(\frac{3}{2}\frac{3}{2};\frac{1}{2}\frac{1}{2}) \\
E_{11}(\frac{3}{2}\frac{1}{2};\frac{3}{2}\frac{1}{2}) \\
E_{21}(\frac{3}{2}\frac{3}{2};\frac{3}{2}\frac{3}{2}) \\
E_{21}(\frac{3}{2}\frac{1}{2};\frac{3}{2}\frac{1}{2}) \\
E_{21}(\frac{3}{2}\frac{3}{2};\frac{3}{2}\frac{1}{2})
\end{bmatrix}
=
\begin{bmatrix}
\frac{1}{6} & 0 & \frac{2}{3} & -\frac{2}{3} & \frac{4}{3} \\
\frac{1}{3} & 0 & \frac{1}{3} & -\frac{1}{3} & \frac{2}{3} \\
\frac{\sqrt{2}}{6} & 0 & -\frac{\sqrt{2}}{3} & \frac{\sqrt{2}}{3} & -\frac{2\sqrt{2}}{3} \\
0 & 0 & 1 & -\frac{1}{2} & -1 \\
0 & \frac{1}{6} & \frac{2}{3} & \frac{1}{3} & \frac{2}{15} \\
0 & \frac{1}{3} & \frac{1}{3} & \frac{1}{6} & \frac{1}{15} \\
0 & \frac{\sqrt{2}}{6} & -\frac{\sqrt{2}}{3} & -\frac{\sqrt{2}}{6} & -\frac{\sqrt{2}}{15}
\end{bmatrix}
\begin{bmatrix}
I_{00}^{(01)}+I_{10}^{(01)} \\
I_{00}^{(01)}+I_{02}^{(01)} \\
I_{01}^{(11)} \\
I_{01}^{(s1)} \\
I_{01}^{(t1)}
\end{bmatrix}
\tag{4.38}
$$

where matrix multiplication is implied.

As a check on the numerics of the calculation of the central force matrix elements we know that when $V_{S=0,T=1}$ and $V_{S=1,T=1}$ of equation (4.21) both equal unity the I_{nl} become unity and we are left with the orthonormality integrals. Therefore the sum of the coefficients of $I_{nl}^{(01)}$ and $I_{nl}^{(11)}$ must be unity or zero depending on whether we deal with a diagonal or non-diagonal matrix element.

In Table 4.2 we present the results of a least-squares fit (similar to that carried out in Chapter 1, section 3.2) of the $I_{nl}^{(iT)}$ to the Cohen-Kurath matrix elements. The results given in column 3 are the best fit possible when only two parameters are used. These two parameters, $(I_{00}^{(01)}+I_{10}^{(01)})$ and $(I_{00}^{(01)}+I_{02}^{(01)})$, are those that characterize the $S=0$ $T=1$ interaction. Columns 4, 5, and 6 give the best fit when one more, two more, and finally three more parameters are added. To see whether the addition of one more parameter is meaningful we have included in each column a quantity $\mathscr{F}$ which we call the 'figure of merit':

$$
\mathscr{F}=\left[\left\{\sum_{i=1}^{\bar{N}}(E_i^{(\text{expt})}-E_i^{(\text{theory})})^2\right\}\Big/(\bar{N}-\bar{n})\right]^{\frac{1}{2}}. \tag{4.39}
$$

where $E_i^{(\text{expt})}$ is the Cohen–Kurath value for the energy and $E_i^{(\text{theory})}$ is the value of the matrix element computed using the values of the $I_{nl}^{(i1)}$ given in that column, $\bar{n}$ is the number of parameters in the fit and $\bar{N}$ is the total number of data points. For the addition of one more parameter to be meaningful $\mathscr{F}$ should decrease by roughly a factor of $\sqrt{2}$. Thus from a comparison of columns 5 and 6 one would conclude that there is no evidence for a two-body spin-orbit part of the effective $T=1$ interaction. It should be noted that with the addition of each new parameter the $I_{nl}^{(iT)}$ change slightly but their qualitative features remain the same; the central

TABLE 4.2

Comparison of the Cohen–Kurath $T=1$ matrix elements with those calculated from a potential using oscillator wave functions

Matrix Element	Experiment Cohen–Kurath	Value of matrix element (MeV): Number of parameters $\bar{n}$ 2	3	4	5
$E_{01}(\frac{1}{2}\frac{1}{2};\frac{1}{2}\frac{1}{2})$	−0·26	−1·931	−0·944	−0·148	−0·057
$E_{01}(\frac{3}{2}\frac{3}{2};\frac{3}{2}\frac{3}{2})$	−3·19	−3·861	−3·614	−3·415	−3·393
$E_{01}(\frac{3}{2}\frac{3}{2};\frac{1}{2}\frac{1}{2})$	−4·86	−2·730	−3·776	−4·621	−4·717
$E_{11}(\frac{3}{2}\frac{1}{2};\frac{3}{2}\frac{1}{2})$	0·92	0	1·726	0·750	0·920
$E_{21}(\frac{3}{2}\frac{3}{2};\frac{3}{2}\frac{3}{2})$	−0·17	−0·687	0·300	0·284	0·057
$E_{21}(\frac{3}{2}\frac{1}{2};\frac{3}{2}\frac{1}{2})$	−0·96	−1·373	−1·126	−1·130	−1·187
$E_{21}(\frac{3}{2}\frac{3}{2};\frac{3}{2}\frac{1}{2})$	−1·92	−0·971	−2·017	−2·000	−1·760
$I_{00}^{(01)}+I_{10}^{(01)}$		−11·583	−12·569	−13·365	−13·456
$I_{00}^{(01)}+I_{02}^{(01)}$		−4·119	−5·105	−5·089	−4·862
$I_{01}^{(11)}$			1·726	1·540	1·394
$I_{01}^{(s1)}$					−0·468
$I_{01}^{(t1)}$				0·790	0·708
Figure of merit $\mathcal{F}$		1·412	0·826	0·361	0·340

The various columns give the best-fit matrix elements as a function of the number of parameters describing the potential. The figure of merit $\mathcal{F}$ is defined by equation (4.39).

$S=0$ interaction is strongly attractive while the central $S=1$ and tensor interactions are repulsive.

Let us now see what these results imply about the form of the interaction in the $S=0$ state, i.e. the potential $V_{01}(\beta\,|\underline{r}_1-\underline{r}_2|)$ of equation (4.21). To carry out this investigation it is convenient to express $I_{nl}^{(iT)}$ in terms of 'moments' of the potential, usually called Talmi integrals. As long as the residual two-body interaction is Galilean invariant and conserves parity, the only $I_{n'l';nl}^{(iT)}$ that can arise are those involving states in which l and l' have the same parity. From the form of the oscillator wave functions it therefore follows that

$$I_{n'l';nl}^{(iT)}=\sum_q B(n'l'nl;q)I_q^{(iT)}$$

where $B(n'l'nl;q)$ are numerical constants and $I_q^{(iT)}$ is the Talmi integral:

$$I_q^{(iT)}=\frac{2^{q+2}}{\sqrt{\pi}(2q+1)!!}\int_0^\infty x^{2q+2}e^{-x^2}V_{iT}(\kappa x)\,\mathrm{d}x. \tag{4.40}$$

The advantage of introducing these 'moments' is that if $V_{iT}(\kappa x)$ has the same sign for all values of x, all $I_q^{(iT)}$ are attractive if V_{iT} is attractive and repulsive if V_{iT} is repulsive. From the form of the oscillator wave functions in equations (4.11) it follows that

$$W = \frac{I_{00}^{(01)} + I_{10}^{(01)}}{I_{00}^{(01)} + I_{02}^{(01)}} = \frac{\frac{5}{2}I_0^{(01)} - 3I_1^{(01)} + \frac{5}{2}I_2^{(01)}}{I_0^{(01)} + I_2^{(01)}} \tag{4.41}$$

or

$$(W - \tfrac{5}{2})I_0^{(01)} + 3I_1^{(01)} + (W - \tfrac{5}{2})I_2^{(01)} = 0. \tag{4.42}$$

Thus if W is found empirically to be $\geq 2{\cdot}5$ it follows that the spin-singlet part of the residual two-body force cannot be taken to be purely attractive. Since $\mathscr{F}$ decreases by more than a factor of 2 in going from $\bar{n} = 3$ to $\bar{n} = 4$ it would appear that the effective $T = 1$ interaction has a tensor component and in this case $W = 2{\cdot}626$. Thus the $S = 0$ $T = 1$ interaction must be a sum of a repulsive and an attractive interaction with different ranges.

$I_{00}^{(01)}$ and $I_{10}^{(01)}$ are integrals of the potential when the two interacting particles are in a relative s-state ($l = 0$). Since in a relative s-state two particles may sit on top of each other, it is clear that these integrals depend sensitively on the short-range nature of the potential. On the other hand, $I_{02}^{(01)}$ is the integral of $V_{01}(\kappa x)$ when the two particles are in a relative d-state ($l = 2$). Consequently $I_{02}^{(01)}$ is more sensitive to the long-range part of the force. Thus if we take $V_{01}(\kappa x)$ to be the sum of a short-range attractive and long-range repulsive potential we will be able to force W to have the empirically observed value. To demonstrate this explicitly, we consider the Gaussian potential

$$V_{01}(\beta\,|\underline{r}_1 - \underline{r}_2|) = -V_{01}\mathrm{e}^{-\beta_{01}^2(\underline{r}_1-\underline{r}_2)^2} + \tilde{V}_{01}\mathrm{e}^{-\tilde{\beta}_{01}^2(\underline{r}_1-\underline{r}_2)^2}.$$

In this case the Talmi integrals become

$$I_q^{(01)} = \frac{-V_{01}}{\left(1 + \dfrac{2\beta_{01}^2}{\alpha^2}\right)^{q+\frac{3}{2}}} + \frac{\tilde{V}_{01}}{\left(1 + \dfrac{2\tilde{\beta}_{01}^2}{\alpha^2}\right)^{q+\frac{3}{2}}}. \tag{4.44}$$

It is clear that the data do not allow us to determine all of the parameters of equation (4.43). We therefore take

$$\frac{\beta_{01}}{\alpha} = 1{\cdot}4 \tag{4.45a}$$

the value chosen by Kurath (1956) in his early studies of the p-shell nuclei. If we arbitrarily choose

$$\frac{\tilde{\beta}_{01}}{\alpha} = 1 \tag{4.45b}$$

the potential strengths of equation (4.43) needed to fit the $T=1$ $S=0$ integrals when $\bar{n}=4$ are

$$V_{01} = 174{\cdot}00 \text{ MeV}$$
$$\tilde{V}_{01} = 53{\cdot}84 \text{ MeV}. \tag{4.46}$$

When the two nucleons couple to $S=1$ the interaction is mediated by $V_{11}(\beta_1\,|\underline{r}_1-\underline{r}_2|)$ and $V_t(\beta_t\,|\underline{r}_1-\underline{r}_2|)$. Since two $0p$-shell nucleons can only be in a state with relative orbital angular momentum equal to unity (equation (4.36b)), it follows that a single integral of each of these potentials is all that is determined by the data. Thus if we assume the Gaussian shape

$$V_{i1}(\beta_i\,|\underline{r}_1-\underline{r}_2|) = V_{i1}\mathrm{e}^{-\beta_i^2(\underline{r}_1-\underline{r}_2)^2} \tag{4.47}$$

only the product

$$I_{01}^{(i1)} = \frac{V_{i1}}{\left(1+\dfrac{2\beta_i^2}{\alpha^2}\right)^{\frac{5}{2}}}. \tag{4.48}$$

of V_{i1} and β_i is determined. Consequently all one can say is that if $V_{i1}(\beta_i\,|\underline{r}_1-\underline{r}_2|)$ is a single term (and not a sum as with the $S=0$ interaction) V_{11} and V_{t1} are positive and hence correspond to a repulsion.

One may ask how the effective interaction for the $0p$ shell compares with the free nucleon-nucleon potential (Hamada and Johnston 1962, Reid 1968). The most important part as far as the shell model is concerned is the central $S=0$ interaction and this must be attractive in relative s-states. The Hamada–Johnston potential, which is fitted to the free nucleon–nucleon scattering data and the deuteron binding energy, is the sum of an infinite repulsive core with radius 0·485 fm plus an attractive potential outside the hard core. Because the shell model uses uncorrelated wave functions one must, of course, be careful in treating the repulsive core. If one uses the reaction-matrix formalism to take care of this, Kuo and Brown (1966) have shown that the net effect of this interaction is to give attraction in the low even l states. However, in order to get sufficient attraction, particularly in the $I=0$ ground state of even–even nuclei, Kuo and Brown found that one must allow a rather large amount of core excitation and also excitation of the valence nucleons to states outside the model space. (For more details about these calculations see the review articles by Barrett and Kirson 1973, Kuo 1974, Ellis and Osnes 1977). Thus the free nucleon–nucleon interaction has the same effect as our empirical one. However, the free and effective potentials are bound to be different for the following reasons.

(a) In the shell model uncorrelated wave functions are used so that only a delta-function hard core can be tolerated.
(b) The effective interaction deduced here is to be used in a model space that consists of the $0p$ shell alone. The free nucleon–nucleon interaction, because of its infinite repulsive core, has very high momentum components in its Fourier transform, and as a consequence many oscillator shells must be included when it is to be used.

The effective $0p$ shell tensor and spin-orbit potential has the same sign as the Hamada–Johnston interaction whereas the central $S=1$ part has the opposite sign to that given by the free nucleon–nucleon data. As we have seen, these parts of V are less important in determining the interaction matrix elements, and their values are undoubtedly very sensitive to the model space used in the shell-model calculation.

Finally one can compare our potential with the Schiffer–True interaction (1976) which was chosen to give a best fit to all the shell-model data between $A=12$ and $A=210$. The most important part of the potential (the $S=0$ central interaction) has the same structure as they found, namely an attractive interaction followed by a longer-range repulsion. However, the $S=1$ part (which as we have seen leads to 'fine tuning') differs from that of Schiffer and True.

2.2. The $T=0$ interaction in the $0p$ shell

Since the $T=0$ tensor interaction acts in spatially symmetric states it follows from equation (4.31) and the form of the $0p$-shell wave functions in equations (4.36) that the off-diagonal matrix element

$$I^{(t,0)}_{10;02} = \int R_{10}(\alpha r) V_t(\beta_t\, |\underline{r}_1 - \underline{r}_2|) R_{02}(\alpha r) r^2 \, \mathrm{d}r$$

$$= \int R_{10}(x) V_t(\kappa x) R_{02}(x) x^2 \, \mathrm{d}x$$

can give a non-vanishing contribution to the $T=0$ energies. In terms of the integrals of the $T=0$ potential these interaction energies are given by

$$\begin{bmatrix} E_{30}(\frac{3}{2}\frac{3}{2};\frac{3}{2}\frac{3}{2}) \\ E_{20}(\frac{3}{2}\frac{1}{2};\frac{3}{2}\frac{1}{2}) \\ E_{10}(\frac{1}{2}\frac{1}{2};\frac{1}{2}\frac{1}{2}) \\ E_{10}(\frac{3}{2}\frac{1}{2};\frac{3}{2}\frac{1}{2}) \\ E_{10}(\frac{3}{2}\frac{3}{2};\frac{3}{2}\frac{3}{2}) \\ E_{10}(\frac{3}{2}\frac{1}{2};\frac{1}{2}\frac{1}{2}) \\ E_{10}(\frac{3}{2}\frac{3}{2};\frac{1}{2}\frac{1}{2}) \\ E_{10}(\frac{3}{2}\frac{3}{2};\frac{3}{2}\frac{1}{2}) \end{bmatrix} = \begin{bmatrix} \frac{1}{2} & 0 & 0 & 1 & -\frac{2}{7} & 0 \\ \frac{1}{2} & 0 & 0 & -\frac{1}{2} & 1 & 0 \\ \frac{10}{27} & \frac{1}{54} & \frac{2}{9} & -\frac{10}{9} & -\frac{20}{27} & -\frac{4\sqrt{10}}{27} \\ \frac{5}{54} & \frac{8}{27} & \frac{2}{9} & -\frac{5}{18} & -\frac{5}{27} & \frac{8\sqrt{10}}{27} \\ \frac{1}{27} & \frac{5}{27} & \frac{5}{9} & -\frac{1}{9} & -\frac{2}{27} & -\frac{4\sqrt{10}}{27} \\ -\frac{5}{27} & \frac{2}{27} & \frac{2}{9} & \frac{5}{9} & \frac{10}{27} & \frac{-7\sqrt{10}}{27} \\ -\frac{\sqrt{10}}{27} & -\frac{\sqrt{10}}{54} & \frac{\sqrt{10}}{9} & \frac{\sqrt{10}}{9} & \frac{2\sqrt{10}}{27} & \frac{22}{27} \\ \frac{\sqrt{10}}{54} & \frac{-2\sqrt{10}}{27} & \frac{\sqrt{10}}{9} & \frac{-\sqrt{10}}{18} & \frac{-\sqrt{10}}{27} & \frac{-2}{27} \end{bmatrix} \begin{bmatrix} I_{00}^{(10)}+I_{02}^{(10)} \\ I_{00}^{(10)}+I_{10}^{(10)} \\ I_{01}^{(00)} \\ I_{02}^{(s0)} \\ I_{02}^{(t0)} \\ I_{10;02}^{(t0)} \end{bmatrix} \tag{4.49}$$

where matrix multiplication is implied. As before, a partial check on the numerics is provided by the fact that the sum of the coefficients of the $I_{nl}^{(00)}$ and $I_{nl}^{(10)}$ must be unity for diagonal matrix elements and zero for off-diagonal elements.

All attempts to fit the Cohen–Kurath 'data' (shown in column 2 of Table 4.3), lead to the conclusion that the effective tensor interaction is extremely long range, and in fact a fit with a better figure of merit is obtained when the tensor force is assumed to have infinite range. If we assume a Gaussian for the spatial form of the tensor potential, the integrals characterizing this interaction are

$$I_{02}^{(t0)} = \frac{V_{t0}}{\left(1+\dfrac{2\beta_{t0}^2}{\alpha^2}\right)^{\frac{7}{2}}}.$$

and

$$I_{10;02}^{(t0)} = \frac{3V_{t0}}{\sqrt{10}\left(1+\dfrac{2\beta_{t0}^2}{\alpha^2}\right)^{\frac{7}{2}}}\left\{\left(1+\frac{2\beta_{t0}^2}{\alpha^2}\right)-\frac{5}{3}\right\}. \tag{4.50}$$

TABLE 4.3

Comparison of the Cohen–Kurath $T=0$ matrix elements with those calculated from a potential using harmonic oscillator wave functions

		Value of matrix element (MeV)			
Matrix element	Experiment Cohen–Kurath	Number of parameters $\bar{n}$ 2	3	4	5
$E_{30}(\frac{3}{2}\frac{3}{2};\frac{3}{2}\frac{3}{2})$	−7·23	−5·583	−5·899	−7·177	−7·178
$E_{20}(\frac{3}{2}\frac{1}{2};\frac{3}{2}\frac{1}{2})$	−4·00	−5·583	−4·491	−4·133	−4·129
$E_{10}(\frac{1}{2}\frac{1}{2};\frac{1}{2}\frac{1}{2})$	−4·15	−4·449	−4·927	−4·042	−4·056
$E_{10}(\frac{3}{2}\frac{1}{2};\frac{3}{2}\frac{1}{2})$	−6·22	−6·047	−6·727	−6·639	−6·646
$E_{10}(\frac{3}{2}\frac{3}{2};\frac{3}{2}\frac{3}{2})$	−3·58	−3·547	−3·196	−3·122	−3·171
$E_{10}(\frac{3}{2}\frac{1}{2};\frac{1}{2}\frac{1}{2})$	1·69	0·814	1·832	1·438	1·413
$E_{10}(\frac{3}{2}\frac{3}{2};\frac{1}{2}\frac{1}{2})$	1·56	2·299	1·958	1·647	1·605
$E_{10}(\frac{3}{2}\frac{3}{2};\frac{3}{2}\frac{1}{2})$	3·55	3·309	3·096	3·289	3·243
$I_{00}^{(10)}+I_{10}^{(10)}$		−16·920	−16·339	−16·564	−16·518
$I_{00}^{(10)}+I_{02}^{(10)}$		−11·166	−11·172	−11·662	−11·656
$I_{01}^{(00)}$					−0·105
$I_{02}^{(s0)}$				−1·004	−1·009
V_{t0}			1·095	1·196	1·195
Figure of merit $\mathscr{F}$		1·058	0·825	0·373	0·428

See Table 4.2 for further details.

In the limit the tensor force has infinite range $(\beta_{t0}\rightarrow 0)I_{02}^{(t0)}$ and $I_{10;02}^{(t0)}$ have opposite signs. In fitting the strength of the $T=0$ tensor interaction we shall take $\beta_{t0}=0$ so that $I_{02}^{(t0)}=V_{t0}$ and $I_{10;02}^{(t0)}=-\sqrt{\frac{2}{5}}\,V_{t0}$. With this assumption the problem is similar to the $T=1$ case in that there are five integrals which characterize the interaction.

In Table 4.3 we list the best two-, three-, four-, and five-parameter fits to the data. From an inspection of this table it is apparent that there is no evidence for an effective central $S=0$ $T=0$ interaction in the $0p$ shell. The best fit is a four-parameter potential which is the sum of an $S=1$ central, infinite range tensor, and spin-orbit interaction.

The central-force matrix elements $(I_{00}^{(10)}+I_{10}^{(10)})$ and $(I_{00}^{(10)}+I_{02}^{(10)})$ are remarkably constant as the number of parameters is increased. Furthermore since W in equation (4.41) (defined now in terms of the $I_{nl}^{(10)}$) is 1.42 for the best-fit four-parameter interaction, these matrix elements can be fitted by a single Gaussian interaction

$$V_{10}(\beta_{10}\,|\underline{r}_1-\underline{r}_2|)=-V_{10}\mathrm{e}^{-\beta_{10}^2(\underline{r}_1-\underline{r}_2)^2} \qquad (4.51)$$

A fit to W determines the potential range

$$\frac{\beta_{10}}{\alpha}=0\cdot823$$

and the value of $(I_{00}^{(10)}+I_{10}^{(10)})$ fixes V_{10} to be

$$V_{10}=35\cdot7 \text{ MeV}.$$

Thus the central $S=1$ $T=0$ interaction is attractive with a range slightly greater than the $0p$ shell nuclear size (if $\hbar\omega$ is taken to be $41/(16)^{\frac{1}{3}}=16\cdot27$ MeV, $1/\alpha=1\cdot596$ fm and $1/\beta_{10}=1\cdot939$ fm).

The fitted matrix element of the spin-orbit interaction requires that it be attractive and about twice as strong as the analogous $T=1$ force. Because there is only one matrix element of this potential within our model space only $I_{02}^{(s0)}$ is determined, and even when a potential form is assumed one cannot differentiate between strength and range.

Finally the tensor force is repulsive and in the foregoing calculations was taken to have infinite range. If a finite range is taken for this interaction the best-fit parameters are

$$\begin{aligned} I_{00}^{(10)}+I_{10}^{(10)}&=-16\cdot612 \text{ MeV}\\ I_{00}^{(10)}+I_{02}^{(10)}&=-11\cdot655 \text{ MeV}\\ I_{02}^{(s0)}&=-1\cdot027 \text{ MeV}\\ I_{02}^{(t0)}&=1\cdot314 \text{ MeV}\\ I_{10;02}^{(t0)}&=-0\cdot622 \text{ MeV}\end{aligned} \tag{4.52}$$

with a figure of merit of $0\cdot400$. Since $\mathscr{F}$ does not decrease with the addition of the range parameter, the difference between infinite and finite range is not significant. However, if one is troubled by the use of an infinite-range force these results can be used to see 'how long the infinite range is'. By using equations (4.50) one may deduce the value of β_{t0} necessary to satisfy equations (4.52). On the assumption that the force has a Gaussian character one finds

$$\frac{\beta_{t0}}{\alpha}=0\cdot290$$

i.e. almost three times that of the central interaction.

The most important part of the $T=0$ interaction is the central $S=1$ potential, $V_{10}(\beta_{10}\,|\underline{r}_1-\underline{r}_2|)$ given by equation (4.21), and for the $0p$ shell a two-parameter (V_{10} and β_{10}) attractive potential is required. This component of the Hamada-Johnston potential has an infinite repulsive core of radius $c=0\cdot485$ fm followed by an attractive part for $|\underline{r}_1-\underline{r}_2|>c$. When treated by the reaction matrix formalism, its net effect is to give attraction in the low l states (Kuo and Brown 1966). However, the central $T=0$ $S=1$ part of the Hamada–Johnston potential itself does not bind the deuteron, and it is only when the strong attractive tensor force is included that the required binding is obtained. Thus in the reaction matrix calculation the attractive tensor force gives an appreciable contribution to the shell-model matrix elements. In our calculation the tensor force is not so important and moreover has the opposite sign. Finally, the effective $0p$ shell spin-orbit interaction is attractive whereas the free nucleon-nucleon one is repulsive. Overall the free nucleon-nucleon interaction gives the effects required by the shell model, but once again the detailed form of the effective $0p$ shell interaction is quite different than the free nucleon-nucleon potential.

Our effective interaction agrees with the results of Schiffer and True (1976) in its most important part; namely the central $S=1$ interaction is attractive in low l states (Schiffer and True do find some improvement when this potential is taken to be the sum of an attractive plus longer range repulsive interaction). However, except for the sign of the spin-orbit interaction, the less important parts of the potential (for $T=0$ this means the tensor, and $S=0$ central interaction) differ in sign from the best-fit Schiffer–True interaction.

3. Spurious centre-of-mass motion

In any nuclear physics calculation one would like ideally to determine the eigenvalues and eigenvectors of the translationally invariant A-particle Hamiltonian

$$H=\sum_{i=1}^{A}\frac{p_i^2}{2m}+\sum_{i<j}V(|\underline{r}_i-\underline{r}_j|)-\frac{\tilde{P}^2}{2Am} \tag{4.53}$$

where

$$\tilde{\underline{P}}=\sum_{i=1}^{A}\underline{p}_i \tag{4.54}$$

is the centre-of-mass momentum of the nucleus with A nucleons. (That H in equation (4.53) is translationally invariant follows from the fact that

$$\sum_i p_i^2=\frac{1}{A}\left\{\tilde{P}^2+\sum_{i<j}(\underline{p}_i-\underline{p}_j)^2\right\} \tag{4.55}$$

so that H is a function of $|\underline{p}_i-\underline{p}_j|$ and $|\underline{r}_i-\underline{r}_j|$.) In practice, however, one is forced to approximate the Hamiltonian and in the shell model one assumes that the eigenvalues and eigenvectors of

$$H=\sum_{i=1}^{A}\left\{\frac{p_i^2}{2m}+V(r_i)+\underline{\sigma}_i\cdot\underline{\ell}_i f(r_i)\right\}+\sum_{i<j}V(|\underline{r}_i-\underline{r}_j|) \tag{4.56}$$

describe nuclear properties despite the fact the Hamiltonian is not translationally invariant. This, of course, immediately raises a question about the reliability of any results deduced on the basis of such a model. We shall now examine to what extent one would expect this model to correspond to observation.

We first show that if H is diagonalized using harmonic oscillator wave functions with $f(r_i)\equiv 0$, the antisymmetric shell-model wave functions can be factored into a function of the relative coordinates and the centre-of-mass coordinate (Elliott and Skyrme 1955). From equation (4.55) and its analogue for $\sum_i r_i^2$ one sees that

$$\begin{aligned}H_0&=\sum_i\left(\frac{p_i^2}{2m}+\tfrac{1}{2}m\omega^2r_i^2\right)\\&=\frac{1}{2mA}\tilde{P}^2+\tfrac{1}{2}mA\omega^2\tilde{R}^2+\frac{1}{A}\sum_{i<j}\left\{\frac{(\underline{p}_i-\underline{p}_j)^2}{2m}+\tfrac{1}{2}m\omega^2(\underline{r}_i-\underline{r}_j)^2\right\}\end{aligned} \tag{4.57}$$

where

$$\underline{\tilde{R}}=\frac{1}{A}\sum_{i=1}^{A}\underline{r}_i \tag{4.58}$$

is the centre-of-mass coordinate of the nucleus. Since H_0 is separable in relative and centre-of-mass coordinates it follows that if the many-particle shell-model wavefunction Ψ is a product of oscillator functions it can be written as $\Psi=\phi(\underline{\tilde{R}})\Pi_{i<j}\phi(|\underline{r}_i-\underline{r}_j|)$.

From equation (4.57) it follows that the wave function describing the centre-of-mass motion is a harmonic-oscillator function, and we shall now show that provided the A nucleons fill the lowest possible oscillator orbitals consistent with the exclusion principle, the centre-of-mass wave function is the $0s$ eigenfunction.

To demonstrate this we note that the single-particle oscillator eigenfunctions ϕ_{nl} satisfy the wave equation

$$H\phi_{nl}=\left(\frac{p^2}{2m}+\tfrac{1}{2}m\omega^2r^2\right)\phi_{nl}=(2n+l+\tfrac{3}{2})\hbar\omega\phi_{nl}.$$

Furthermore the operator

$$b_\mu^\dagger=\frac{1}{\sqrt{(2m\hbar\omega)}}(-ip_\mu+m\omega r_\mu) \tag{4.59}$$

with r_μ and p_μ the μth spherical components of the vector operators $\underline{r}$ and $\underline{p}$ respectively has the property that $b^\dagger_\mu \phi_{nl}$ is an eigenfunction of H with energy $(2n+l+\frac{5}{2})\hbar\omega$. This follows from the fact that

$$Hb^\dagger_\mu \phi_{nl} = [H, b^\dagger_\mu]\phi_{nl} + (2n+l+\tfrac{3}{2})b^\dagger_\mu \phi_{nl}$$

where $[H, b^\dagger_\mu]$, the commutator of H and $b^\dagger_\mu$, is

$$\begin{aligned}[H, b^\dagger_\mu] &= \frac{1}{\sqrt{(2m\hbar\omega)}}\left\{-\frac{i}{2}m\omega^2[r^2_\mu, p_\mu] + \frac{\omega}{2}[p^2_\mu, r_\mu]\right\}\\ &= \hbar\omega b^\dagger_\mu.\end{aligned}$$

Thus when $b^\dagger_\mu$ is applied to the oscillator state ϕ_{nl} it produces an oscillator eigenfunction with one additional unit of energy; i.e. $b^\dagger_\mu$ is the 'step-up' operator for the harmonic oscillator. In a similar way the operator

$$\begin{aligned} b_\mu &= (b^\dagger_\mu)^\dagger \\ &= \frac{1}{\sqrt{(2m\hbar\omega)}}(ip^\dagger_\mu + m\omega r^\dagger_\mu) \\ &= \frac{(-1)^\mu}{\sqrt{(2m\hbar\omega)}}(ip_{-\mu} + m\omega r_{-\mu}) \end{aligned} \qquad (4.60)$$

is the 'step-down' operator because when it is applied to the state ϕ_{nl} it produces an oscillator eigenfunction with one unit less energy.

By analogy with these single-particle operators one can define the 'step-up' and 'step-down' operators for the centre-of-mass motion by summing $b^\dagger_\mu$ and b_μ over all nucleons in the nucleus (Baranger and Lee 1961). Thus

$$B^\dagger_\mu = \frac{1}{\sqrt{A}}\sum_k b^\dagger_\mu(k) = \frac{1}{\sqrt{(2mA\hbar\omega)}}\sum_{k=1}^{A}\{-i(p_k)_\mu + m\omega(r_k)_\mu\} \qquad (4.61a)$$

$$B_\mu = \frac{1}{\sqrt{A}}\sum_k b_\mu(k) = \frac{(-1)^\mu}{\sqrt{(2mA\hbar\omega)}}\sum_{k=1}^{A}\{i(p_k)_{-\mu} + m\omega(r_k)_{-\mu}\}. \qquad (4.61b)$$

B_μ applied to a shell-model wave function Ψ (an antisymmetric product of single-particle oscillator eigenfunctions in which the A nucleons occupy the lowest possible orbits consistent with the exclusion principle) gives zero because each term in B_μ destroys a particle in its original state and recreates it in a state that is already occupied (see Fig. 4.1a). Therefore if we assume that oscillator wave functions are a good approximation to the nuclear problem it follows that states Ψ which do not possess any core excitation will always have centre-of-mass motion characteristic of an oscillator in its $0s$ ground state. If these states are used to diagonalize the physical Hamiltonian–i.e. the effects of $f(r_i)$ and

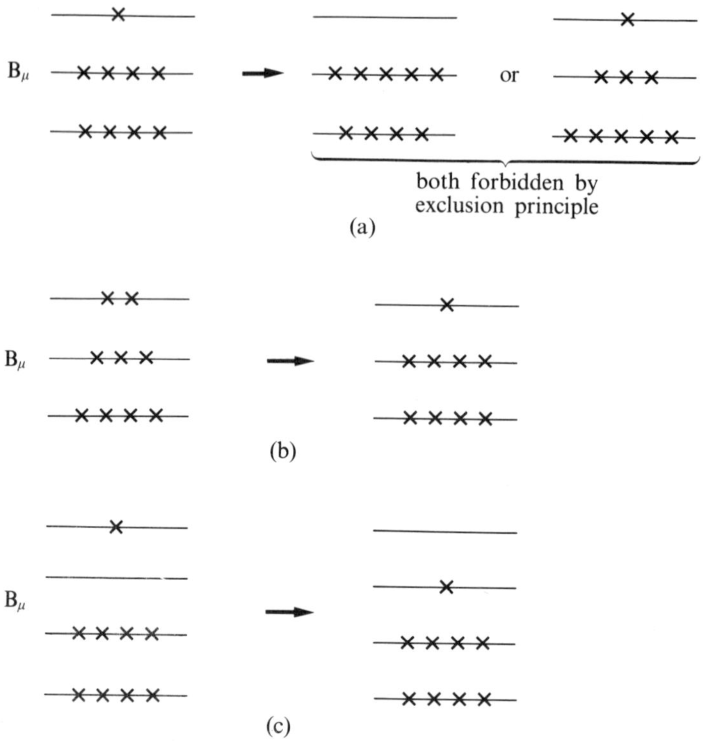

Fig. 4.1. The effect of the centre-of-mass step-down operator B_μ (equation (4.61b)) on several different configurations. For simplicity it has been assumed that an orbit can accommodate only four nucleons.

$V(|\underline{r}_i - \underline{r}_j|)$ are treated by perturbation theory—it follows that the low-lying excited nuclear states (of normal parity) will correspond to excitation of only internal degrees of freedom of the particles and the centre-of-mass always remains in its ground state. Since excited nuclear states *must* correspond to excitation of internal degrees of freedom and not changes in centre-of-mass motion the excitation energies computed by use of these functions are not likely to be grossly in error. Furthermore, since theory and experiment are in such good agreement it would appear that little error *is* introduced by the fact that the single-particle eigenfunctions are not really oscillator functions.

On the other hand, once excitation out of the core (Fig. 4.1b) or excitation of the valence nucleons from the lowest oscillator orbit to higher ones (Fig. 4.1c) is allowed, the centre-of-mass is no longer always in a 0s oscillator state. (Note that an oscillator orbit in this sense is defined as all the levels with the same single-particle energy.) On physical grounds one would expect that wave functions corresponding to centre-of-mass excitation should not be included in the diagonalization of the Hamiltonian, and in section 3.1 we shall outline a procedure for getting rid of these spurious states.

Before doing this, however, it is useful to have some idea about the number of spurious states (states corresponding to centre-of-mass excitation) that can occur in a given shell-model calculation. To determine this we note that $B_\mu^\dagger$ (the 'step-up' operator for centre-of-mass motion) carries one unit of angular momentum, and because it is a symmetric sum over all nucleons in the nucleus it carries isospin zero. Thus $B_\mu^\dagger$ operating on a non-spurious state with angular momentum J and isospin T will produce a spurious state with the same T and $I = J, J \pm 1$.

The set of states

$$\Phi_{IM\alpha;TT_z} = \sum_{M'\mu} (J'1M'\mu \mid IM) B_\mu^\dagger \Psi_{J'M'\gamma';TT_z} \tag{4.62}$$

obtained by allowing the 'step-up' operator to act on the non-spurious wave functions $\Psi_{J'M'\gamma';TT_z}$ are orthonormal and correspond to the possible states with one unit of centre-of-mass excitation energy. To see that these states are orthonormal we observe that

$$p_{\pm 1} = \mp \frac{1}{\sqrt{2}} (p_x \pm i p_y)$$

$$p_0 = p_z$$

together with the analogous expressions for r_μ implies that the commutator $[B_{\mu'}, B_\mu^\dagger]$ has the value

$$\begin{aligned}
[B_{\mu'}, B_\mu^\dagger] &= \frac{(-1)^{\mu'}}{2mA\hbar\omega} \left[\sum_{k'} \{ i(p_{k'})_{-\mu'} + m\omega (r_{k'})_{-\mu'} \}, \sum_k \{ -i(p_k)_\mu + m\omega (r_k)_\mu \} \right] \\
&= \frac{(-1)^{\mu'}}{2A\hbar} \sum_k \{ i[(p_k)_{-\mu'}, (r_k)_\mu] - i[(r_k)_{-\mu'}, (p_k)_\mu] \} \\
&= \delta_{\mu\mu'}.
\end{aligned} \tag{4.63}$$

Consequently

$$\begin{aligned}
\langle \Phi_{IM\alpha';TT_z} \mid \Phi_{IM\alpha;TT_z} \rangle &= \sum_{M'M''} \sum_{\mu\mu'} (J'1M'\mu \mid IM)(J''1M''\mu' \mid IM) \\
&\quad \times \langle \Psi_{J''M''\gamma'';TT_z} | [B_{\mu'}, B_\mu^\dagger] + B_\mu^\dagger B_{\mu'} | \Psi_{J'M'\gamma';TT_z} \rangle \\
&= \delta_{J'J''} \delta_{\gamma'\gamma''}
\end{aligned}$$

where use has been made of the summation property of the Clebsch–Gordan coefficients, equation (A1.23), together with the fact that $B_\mu \Psi_{JM\gamma;TT_z}=0$. Thus the states given by equation (4.62) are orthonormal and we may now use this result to calculate the number of $1\hbar\omega$ spurious states in a given model space.

First, there are certain types of core excitation that never give rise to spurious states. This happens because $B_\mu^\dagger$ and B_μ can only change j by at most one unit. Thus we have the general theorem that *there is no spurious component in the states that arise when the highest j of the lower oscillator shell is full and only the highest j of the excited shell has particles in it.* Thus all the core-excited states discussed in Chapter 2, section 4.2, which were attributed to excitation of nucleons from the $0d_{\frac{3}{2}}$ to the $0f_{\frac{7}{2}}$ level, had no spurious component. Moreover, the $(0g_{\frac{9}{2}}, 1p_{\frac{1}{2}})$ model used to interpret the data near $A=90$ has no spurious states (see the discussion in Chapter 1, section 5.2 and Chapter 3, section 2.4).

From equation (4.62) it follows that doubly closed-shell nuclei have only one spurious state with energy $\frac{5}{2}\hbar\omega$ and that is a state with spin parity $I=1^-$ and isospin equal to that of the core. Thus the ${}^{40}_{20}Ca_{20}$ 2^-–5^- states discussed in Chapter 3, section 3.1, also had no spurious components. On the other hand, non-closed-shell nuclei have many $\frac{5}{2}\hbar\omega$ spurious states with various spins and isospins. For example, the $A=14$ triad (${}^{14}_{6}C_8$, ${}^{14}_{7}N_7$, and ${}^{14}_{8}O_6$) has the following non-spurious states corresponding to two holes in the $0p$ shell:

$(0s_{\frac{1}{2}})^4(0p_{\frac{3}{2}})^8(0p_{\frac{1}{2}})^2$ has $J=0$, $T=1$ and $J=1$, $T=0$
$(0s_{\frac{1}{2}})^4(0p_{\frac{3}{2}})^7(0p_{\frac{1}{2}})^3$ has $J=1$ and 2 with $T=0$ and 1
$(0s_{\frac{1}{2}})^4(0p_{\frac{3}{2}})^6(0p_{\frac{1}{2}})^4$ has $J=0$ and 2, $T=1$ and $J=1$ and 3, $T=0$.

$B_\mu^\dagger$ operating on any one of these will produce a spurious state and all these states are orthogonal. Thus the spurious states with energy $\frac{5}{2}\hbar\omega$ are

$$T=0\ I=(0^-)^3,(1^-)^4,(2^-)^5,(3^-)^2,4^-$$
$$T=1\ I=0^-,(1^-)^5,(2^-)^3,(3^-)^2$$

where the superscript indicates the number of orthogonal states of that spin. All of these must be eliminated when the negative-parity states of these nuclei are investigated (Halbert and French 1957, Unna and Talmi 1958).

3.1. Elimination of spurious centre-of-mass states

One way to eliminate the states corresponding to centre-of-mass excitation is first to construct them all and then choose as basis states for diagonalizing the Hamiltonian only those wave functions that are orthogonal to the spurious ones. For $A=15$ and 17 Giraud (1965) has tabulated all the states corresponding to $1\hbar\omega$ centre-of-mass excitation energy, and for $A=16$ he has constructed all $I=0^+$ states with $2\hbar\omega$ centre-of-mass excitation. However, it is clear that in the general case this is a prohibitive procedure. Fortunately there is an alternative and simpler procedure and that is to diagonalize the modified operator

$$H' = H_{\text{SM}} + \beta\left\{\frac{\tilde{P}^2}{2Am} + \tfrac{1}{2}mA\omega^2\tilde{R}^2 - \tfrac{3}{2}\hbar\omega\right\} \tag{4.64}$$

where H_{SM} is the usual shell-model Hamiltonian. The added term gives zero when operating on the physical states. For spurious states a multiple of $\beta\hbar\omega$ is added to the energy, and if β is taken to be sufficiently large ($\sim 10^5$) the spurious states will lie at such a high excitation energy they will not be mixed into any physical state (Palumbo and Prosperi 1968, Gloeckner and Lawson 1974).

We shall now outline the procedure for handling the extra term in equation (4.64), and then in section 3.2 we shall show how this may be used to eliminate the spurious state in ${}^4_2\text{He}_2$. In making any shell-model calculation involving core excitation there will generally be some core orbits that will never be excited. For example, excitation out of the $(0s_{\frac{1}{2}})^4$ orbit in ${}^{208}_{82}\text{Pb}_{126}$ would not contribute until we were interested in spurious centre-of-mass states with energy $\frac{13}{2}\hbar\omega$. Let us denote the number of nucleons occupying these 'passive' orbits by N_c and the single-particle angular momenta associated with them by j_c. To examine the effect of the added centre-of-mass energy, we rewrite it explicitly separating out the role of the 'passive' orbits:

$$\begin{aligned} H_\beta &= \frac{\beta}{A}\left(\frac{\tilde{P}^2}{2m} + \tfrac{1}{2}mA^2\omega^2\tilde{R}^2\right) \\ &= \frac{\beta N_c}{A}\left(\frac{\tilde{P}_c^2}{2mN_c} + \tfrac{1}{2}mN_c\omega^2\tilde{R}_c^2\right) \\ &\quad + \frac{\beta}{A}\left\{\sum_{i<j=N_c+1}^{A}\left(\frac{\underline{p}_i\cdot\underline{p}_j}{m} + m\omega^2\underline{r}_i\cdot\underline{r}_j\right)\right. \\ &\quad + \sum_{i=N_c+1}^{A}\left(\frac{p_i^2}{2m} + \tfrac{1}{2}m\omega^2 r_i^2\right) \\ &\quad \left. + \sum_{i=1}^{N_c}\sum_{j=N_c+1}^{A}\left(\frac{\underline{p}_i\cdot\underline{p}_j}{m} + m\omega^2\underline{r}_i\cdot\underline{r}_j\right)\right\}. \end{aligned} \tag{4.65}$$

In the first term in this decomposition $\tilde{\underline{R}}_c$ and $\tilde{\underline{P}}_c$ are the centre-of-mass coordinate and momentum of the unperturbed core particles. Because these nucleons are in their lowest possible orbits it follows that the energy associated with this term is $\frac{3}{2}\beta\hbar\omega(N_c/A)$. The second term represents an added two-body interaction $V_\beta(i, k)$ between the valence nucleons:

$$V_\beta(i, k) = \frac{\beta}{A}\left(\frac{\underline{p}_i \cdot \underline{p}_k}{m} + m\omega^2 \underline{r}_i \cdot \underline{r}_k\right). \tag{4.66a}$$

To calculate matrix elements of this term it is convenient to express $V_\beta(i, k)$ in terms of the relative and centre-of-mass coordinates of the two interacting particles. From equations (4.8) it follows that

$$V_\beta(i, k) = \frac{\beta}{A}\left\{\left(\frac{P^2}{2m} + \tfrac{1}{2}m\omega^2 R^2\right) - \left(\frac{p^2}{2m} + \tfrac{1}{2}m\omega^2 r^2\right)\right\}. \tag{4.66b}$$

Thus in the same way that we evaluated the central-force matrix elements with equation (4.22), it follows that

$$\begin{aligned} E^{(\beta)}_{IT}(j_1j_2; j_3j_4) &= \langle\tilde{\Phi}_{IM}(j_1n_1l_1, j_2n_2l_2; T)|V_\beta|\tilde{\Phi}_{IM}(j_3n_3l_3, j_4n_4l_4; T)\rangle \\ &= \frac{\beta}{A}\hbar\omega \sum_{S\lambda}\sum_{nlNL} \tilde{\gamma}^{(I)}_{\lambda S}(j_1l_1; j_2l_2) \\ &\quad \times \tilde{\gamma}^{(I)}_{\lambda S}(j_3l_3; j_4l_4) \\ &\quad \times \{(2N+L)-(2n+l)\} M_\lambda(nlNL; n_1l_1n_2l_2) \\ &\quad \times M_\lambda(nlNL; n_3l_3n_4l_4). \end{aligned} \tag{4.67}$$

The third term in equation (4.65) is the single-particle oscillator energy of the valence nucleons and the last gives the change in this single-particle energy due to interaction with the N_c core nucleons. This core-particle interaction is exactly the quantity that was evaluated when we discussed the particle-hole transformation in Chapter 3, section 3. When this transformation was carried out, the term V_t'' in equation (3.75) corresponded to the interaction of the nucleons in the orbit j_1 with the closed shell $j_2 = j_c$. Thus in terms of nucleon creation and annihilation operators the last term in equation (4.65) can be written as

$$\frac{\beta}{A}\left(\sum_{i=1}^{N_c}\sum_{k=N_c+1}^{A}\frac{\underline{p}_i \cdot \underline{p}_k}{m} + m\omega^2 \underline{r}_i \cdot \underline{r}_k\right)$$

$$= \sum_{j_cJTj}\frac{(2J+1)(2T+1)E^{(\beta)}_{JT}(jj_c; jj_c)}{2(2j+1)}\sum_{m\mu} a^\dagger_{jm;\frac{1}{2}\mu}a_{jm;\frac{1}{2}\mu}$$

where $E^{(\beta)}_{JT}(jj_c; jj_c)$ is given by equation (4.67) and $\sum_{m\mu} a^\dagger_{jm;\frac{1}{2}\mu}a_{jm;\frac{1}{2}\mu}$ gives the number of nucleons in the orbit j.

Consequently the procedure to be followed in evaluating the matrix elements of the modified shell-model Hamiltonian H' of equation (4.64) is as follows:

(i) Having decided on a model space, carry out the usual shell-model calculation.

(ii) Subtract from each diagonal matrix element the energy

$$\Delta E = \tfrac{3}{2}\beta\hbar\omega\left(1-\frac{N_c}{A}\right). \tag{4.68}$$

(iii) Replace each single-particle energy ε_j by

$$\varepsilon_j = \tilde{\varepsilon}_j + (2n+l+\tfrac{3}{2})\frac{\beta}{A}\hbar\omega + \frac{1}{2(2j+1)}\sum_{j_c IT}(2I+1)(2T+1)E^{(\beta)}_{IT}(jj_c;\, jj_c) \tag{4.69}$$

where $\tilde{\varepsilon}_j$ is the empirically determined single-particle energy, $(2n+l+\tfrac{3}{2})\hbar\omega$ is the oscillator energy for a nucleon in the state (nlj) and the last term represents the interaction with the unperturbed core.

(iv) Whenever $V(|\underline{r}_i-\underline{r}_j|)$ occurs it must be replaced by $V(|\underline{r}_i-\underline{r}_j|)+V_\beta$ where the matrix elements of V_β are given by equation (4.67).

Once these modifications have been made and the limit that β becomes large is taken, the spurious states will be eliminated from the calculation without ever having to construct them explicitly. This is true no matter how many units of $\hbar\omega$ spurious centre-of-mass excitation one wants to eliminate.

3.2. *Negative-parity states in* $^4_2\mathrm{He}_2$

To illustrate the procedure for getting rid of spurious centre-of-mass excitation and to show the troubles one encounters if it is not eliminated, we consider the negative-parity states in $^4_2\mathrm{He}_2$. In the 20–30 MeV excitation-energy region seven negative-parity states are observed in this nucleus (Fiarman and Meyerhof 1973), namely three $T=0$ levels with spins 0^-, 1^-, and 2^- and four $T=1$ states with angular momenta 0^-, 1^-, 1^-, and 2^-. If one considers a model in which a single nucleon is excited from the $0s_{\frac{1}{2}}$ shell to the $0p_{\frac{3}{2}}$ and $0p_{\frac{1}{2}}$ orbitals the allowable states are $I=1^-$ and 2^- with $T=0$ and 1 from the configuration $[p_{\frac{3}{2}}\times(s_{\frac{1}{2}})^3]_{IM;TT_z}$ and $I=0^-$ and 1^- with $T=0$ and 1 from $[p_{\frac{1}{2}}\times(s_{\frac{1}{2}})^3]_{IM;TT_z}$. The energies of the observed states are close to where one would expect the $(0p, 0s^{-1})$ states to lie. In the same way that the particle-hole excitation energy was estimated in section 3.1 of Chapter 3, one finds by using the Wapstra-Gove tables (1971) that

$$\begin{aligned}\varepsilon(0p)+\varepsilon(0s^{-1}) &= \{\mathrm{BE}(^5_2\mathrm{He}_3)-\mathrm{BE}(^4_2\mathrm{He}_2)\}+\{\mathrm{BE}(^3_2\mathrm{He}_1)-\mathrm{BE}(^4_2\mathrm{He}_2)\}\\ &= 21{\cdot}466\ \mathrm{MeV}.\end{aligned}$$

Furthermore, as we showed at the beginning of this section, a closed-shell nucleus with $T=0$ will have one $1\hbar\omega$ spurious state which is a 1^- $T=0$ level.

Thus the simple $(0p, 0s^{-1})$ model accounts for both the number of states observed and their approximate excitation energies (de Shalit and Walecka 1966, Barrett 1967). Since we deal with a particle-hole excitation, the formalism developed in Chapter 3, section 3 can be used to set up the Hamiltonian matrix in this model space. From equations (3.75) and (3.79) it follows that

$$\langle[0p_j\times(0s_{\frac{1}{2}})^3]_{IM;\tau 0}|\,H\,|[0p_{j'}\times(0s_{\frac{1}{2}})^3]_{IM;\tau 0}\rangle$$
$$=\left\{3\varepsilon_{s_{\frac{1}{2}}}+E_{\frac{1}{2}\frac{1}{2}}(s^3_{\frac{1}{2}})+\varepsilon_{p_j}+\sum_{JT}\frac{(2J+1)(2T+1)}{2(2j+1)}E_{JT}(j\tfrac{1}{2};j\tfrac{1}{2})\right\}\delta_{jj'}$$
$$-\tfrac{1}{2}(-1)^{j+j'+\tau}\sum_{J}(2J+1)W(j\tfrac{1}{2}\tfrac{1}{2}j';JI)\Big(E_{J0}(j\tfrac{1}{2};j'\tfrac{1}{2})$$
$$+(\delta_{\tau 1}-3\delta_{\tau 0})E_{J1}(j\tfrac{1}{2};j'\tfrac{1}{2})\Big) \qquad (4.70)$$

where $\varepsilon_{s_{\frac{1}{2}}}$ and ε_{p_j} are the $0s_{\frac{1}{2}}$ and $0p_j$ single-particle energies and $E_{\frac{1}{2}\frac{1}{2}}(s^3_{\frac{1}{2}})$ denotes the mutual interaction between the three $s_{\frac{1}{2}}$ nucleons. The remaining two terms have the following origin. The first is precisely the energy V''_t in equation (3.74) that arises from the particle-hole transform of the potential V_t. This term, given by equation (3.75), represents the interaction energy of the $0p_j$ nucleon with the full $0s_{\frac{1}{2}}$ shell. The last term in equation (4.70) is just the particle-hole interaction given by equations (3.78) and (3.79), and as discussed in Chapter 3 takes care of the fact that the $0p_j$ nucleon does not have a full shell with which to interact.

To calculate the two-body matrix elements of V by the methods given in section 1 one must know the j–j to L–S transformation coefficients and these are given in Table 4.4. In addition, the relative centre-of-mass transformation coefficients for the two-particle wave functions must be known. Because of the simple form of the $(0p, 0s)$ function

$$R_{01}(\alpha r_1)R_{00}(\alpha r_2)\sim r_1\exp\{-\alpha^2(r_1^2+r_2^2)/2\}$$
$$\sim\frac{r+R}{\sqrt{2}}\exp\{-\alpha^2(r^2+R^2)/2\}$$

they can be written down without calculation

$$[\phi_{01}(\underline{r}_1)\times\phi_{00}(\underline{r}_2)]_{1\mu}=\frac{1}{\sqrt{2}}[\phi_{01}(\underline{r})\times\phi_{00}(\underline{R})]_{1\mu}$$
$$+\frac{1}{\sqrt{2}}[\phi_{00}(\underline{r})\times\phi_{01}(\underline{R})]_{1\mu}. \qquad (4.71)$$

TABLE 4.4
j–j to L–S transformation coefficients, computed by use of eq (4.15) for the configurations $[p_j \times s_{\frac{1}{2}}]_{IM}$.

Configuration	I	$\gamma_{10}^{(I)}$	$\gamma_{11}^{(I)}$
$[p_{\frac{1}{2}} \times s_{\frac{1}{2}}]_{IM}$	0	0	1
	1	$-1/\sqrt{3}$	$\sqrt{\frac{2}{3}}$
$[p_{\frac{3}{2}} \times s_{\frac{1}{2}}]_{IM}$	1	$\sqrt{\frac{2}{3}}$	$1/\sqrt{3}$
	2	0	1

To simplify the analysis we assume a central interaction. As a consequence the matrix elements of V are given by equation (4.22). In addition, from equation (4.71) it is clear that

$$2N+L=1 \quad \text{implies that} \quad 2n+l=0$$

and

$$2N+L=0 \quad \text{implies that} \quad 2n+l=1.$$

When this is used in conjunction with equation (4.67) one finds that

$$\begin{aligned}
&\langle [0p_j \times 0s_{\frac{1}{2}}]_{IM;TT_z}| \, V+V_\beta \, \| [0p_{j'} \times 0s_{\frac{1}{2}}]_{IM;TT_z}\rangle \\
&= \tfrac{1}{4} \sum_{lS} \gamma_{1S}^{(I)}(0p_j, 0s_{\frac{1}{2}}) \gamma_{1S}^{(I)}(0p_{j'}, 0s_{\frac{1}{2}}) \{1-(-1)^{S+T+l}\}^2 \\
&\quad \times \{\langle \phi_{0l} | \, V \, | \phi_{0l}\rangle + (-1)^l (\beta \hbar\omega/4)\} \\
&= \tfrac{1}{2} \sum_{lS} \gamma_{1S}^{(I)}(0p_j, 0s_{\frac{1}{2}}) \gamma_{1S}^{(I)}(0p_{j'}, 0s_{\frac{1}{2}}) \{1-(-1)^{S+T+l}\} \\
&\quad \times \{I_l^{(ST)} + (-1)^l (\beta \hbar\omega/4)\}
\end{aligned} \tag{4.72}$$

where $I_l^{(ST)}$ is the Talmi integral in equation (4.40).

These matrix elements can be used in conjunction with equation (4.70) to write down the matrix of the modified Hamiltonian

$$H = H_{SM} + V_\beta$$

for the various $(I\tau)$ states. Thus

$$\langle H\rangle_{I=0,\tau}=3\varepsilon_{s_{\frac{1}{2}}}+E_{\frac{1}{2}\frac{1}{2}}(s_{\frac{1}{2}}^{3})+\varepsilon_{p_{\frac{3}{2}}}+a_{\tau}-\frac{\beta\hbar\omega}{4} \quad (4.73a)$$

$$\langle H\rangle_{I=2,\tau}=3\varepsilon_{s_{\frac{1}{2}}}+E_{\frac{1}{2}\frac{1}{2}}(s_{\frac{1}{2}}^{3})+\varepsilon_{p_{\frac{3}{2}}}+a_{\tau}-\frac{\beta\hbar\omega}{4} \quad (4.73b)$$

$$\langle H\rangle_{I=1,\tau}=\left\{3\varepsilon_{s_{\frac{1}{2}}}+E_{\frac{1}{2}\frac{1}{2}}(s_{\frac{1}{2}}^{3})+\varepsilon_{p_{\frac{1}{2}}}+b_{\tau}-\frac{\beta\hbar\omega}{4}\right\}$$

$$\times\begin{pmatrix}1 & 0\\ 0 & 1\end{pmatrix}+\begin{pmatrix}0 & -\sqrt{2}c_{\tau}\\ -\sqrt{2}c_{\tau} & c_{\tau}+\Delta\end{pmatrix} \quad (4.73c)$$

where

$$a_0=\tfrac{1}{2}\{2I_0^{(10)}+I_1^{(00)}+3I_1^{(11)}\} \quad (4.74a)$$

$$b_0=\tfrac{1}{6}\{7I_0^{(10)}+2I_1^{(00)}+6I_1^{(11)}+3I_0^{(01)}+2\beta\hbar\omega\} \quad (4.74b)$$

$$c_0=\tfrac{1}{6}\{I_0^{(10)}-I_1^{(00)}-3I_1^{(11)}+3I_0^{(01)}+2\beta\hbar\omega\} \quad (4.74c)$$

$$a_1=\tfrac{1}{2}\{I_0^{(10)}+I_0^{(01)}+4I_1^{(11)}\} \quad (4.74d)$$

$$b_1=\tfrac{1}{6}\{2I_0^{(10)}+I_1^{(00)}+11I_1^{(11)}+4I_0^{(01)}\} \quad (4.74e)$$

$$c_1=\tfrac{1}{6}\{-I_0^{(10)}+I_1^{(00)}-I_1^{(11)}+I_0^{(01)}\} \quad (4.74f)$$

$$\Delta=\varepsilon_{p_{\frac{3}{2}}}-\varepsilon_{p_{\frac{1}{2}}}. \quad (4.74g)$$

These expressions give the conventional shell-model results plus the effect of the V_β term in equation (4.66b). We must also take into account the additional diagonal energies given by equations (4.68) and (4.69). Since $N_c=0$ these relationships imply that we must subtract from each diagonal matrix element $\frac{3}{2}\beta\hbar\omega$ and modify the single-particle energies by adding $(2n+l+\frac{3}{2})\beta\hbar\omega/4$ to each of them. Since we have three $0s_{\frac{1}{2}}$ and one $0p$ nucleons, the total energy $\Delta\varepsilon$ that must be added to each diagonal matrix element is

$$\Delta\varepsilon=-\tfrac{3}{2}\beta\hbar\omega+\tfrac{14}{8}\beta\hbar\omega$$

$$=\frac{\beta\hbar\omega}{4}.$$

When $\Delta\varepsilon$ is included in $\langle H\rangle$ the $\beta\hbar\omega$ terms cancel out in all places except in b_0 and c_0 which characterize the 1^- $\tau=0$ states. We now turn to a closer inspection of these states.

Because the first matrix in equation (4.73c) is the unit matrix, the eigenvalues and eigenfunctions of the $\tau=0$ 1^- states are determined by diagonalizing the second term. The eigenvalues of this matrix are

$$\lambda_{\pm}=\frac{c_0}{2}[(1+\Delta/c_0)\pm\sqrt{\{(9+2\Delta/c_0+(\Delta/c_0)^2)\}}].$$

In the limit that $\beta\hbar\omega$ becomes large $c_0 \gg \Delta$ and if we neglect terms of the order of Δ/c_0 we find

$$\lambda_+=2c_0+\tfrac{2}{3}\Delta$$
$$\lambda_-=-c_0+\tfrac{1}{3}\Delta.$$

When these are added to the first term in equation (4.73c) the $I=1^-$ $\tau=0$ energies are obtained

$$E'_{10}=3\varepsilon_{s_{\frac{1}{2}}}+E_{\frac{1}{2}\frac{1}{2}}(s_{\frac{1}{2}}^3)+\tfrac{2}{3}\varepsilon_{p_{\frac{3}{2}}}+\tfrac{1}{3}\varepsilon_{p_{\frac{1}{2}}}+\tfrac{3}{2}\{I_0^{(10)}+I_0^{(01)}\}+\beta\hbar\omega \tag{4.75a}$$
$$E_{10}=3\varepsilon_{s_{\frac{1}{2}}}+E_{\frac{1}{2}\frac{1}{2}}(s_{\frac{1}{2}}^3)+\tfrac{1}{3}\varepsilon_{p_{\frac{3}{2}}}+\tfrac{2}{3}\varepsilon_{p_{\frac{1}{2}}}+a_0 \tag{4.75b}$$

with the corresponding eigenvectors

$$\psi'_{1M}=\frac{1}{\sqrt{3}}[p_{\frac{1}{2}}\times s_{\frac{1}{2}}^{-1}]_{1M}-\sqrt{\tfrac{2}{3}}[p_{\frac{3}{2}}\times s_{\frac{1}{2}}^{-1}]_{1M} \tag{4.76a}$$
$$\psi_{1M}=\sqrt{\tfrac{2}{3}}[p_{\frac{1}{2}}\times s_{\frac{1}{2}}^{-1}]_{1M}+\frac{1}{\sqrt{3}}[p_{\frac{3}{2}}\times s_{\frac{1}{2}}^{-1}]_{1M}. \tag{4.76b}$$

Clearly ψ'_M is the spurious state as it contains the entire centre-of-mass excitation energy; ψ_{1M} has no spurious part and the energy associated with this state is given by equation (4.75b).

To obtain a numerical estimate for the splitting of the negative-parity states we use one of the interactions suggested by de Shalit and Walecka in their original analysis of this system:

$$V=\left(V_0\frac{e^{-\gamma|\underline{r}_1-\underline{r}_2|}}{\gamma|\underline{r}_1-\underline{r}_2|}P_0+V_1\frac{e^{-\gamma'|\underline{r}_1-\underline{r}_2|}}{\gamma'|\underline{r}_1-\underline{r}_2|}P_1\right)\left(\frac{1+P_M}{2}\right) \tag{4.77a}$$

with

$$V_0=-46{\cdot}9\text{ MeV}$$
$$V_1=-52{\cdot}1\text{ MeV} \tag{4.77b}$$
$$\gamma=0{\cdot}855\text{ fm}^{-1}$$
$$\gamma'=0{\cdot}726\text{ fm}^{-1}. \tag{4.77c}$$

P_0 and P_1 are the spin singlet and triplet projection operators of equation (4.20) and P_M is the Majorana space-exchange operator of equation (2.30):

$$P_M=-\tfrac{1}{4}(1+\underline{\tau}_1\cdot\underline{\tau}_2)(1+\underline{\sigma}_1\cdot\sigma_2).$$

Since

$$\underset{\sim}{\sigma}_1 \cdot \underset{\sim}{\sigma}_2 \chi_{SM} = \{2S(S+1)-3\}\chi_{SM}$$
$$\underset{\sim}{\tau}_1 \cdot \underset{\sim}{\tau}_2 \zeta_{T\nu} = \{2T(T+1)-3\}\zeta_{T\nu}$$

it follows that

$$\tfrac{1}{2}(1+P_{\mathrm{M}})\chi_{SM}\zeta_{T\nu} = \tfrac{1}{2}[1-\{1-T(T+1)\}\{1-S(S+1)\}]\chi_{SM}\zeta_{T\nu}.$$

Therefore $\frac{1}{2}(1+P_{\mathrm{M}})$ gives zero when operating on $T=0$, $S=0$ or $T=1$, $S=1$ states and unity when $T=0$, $S=1$ or $T=1$, $S=0$ are involved. Since the matrix elements of the potential in equation (4.72) are proportional to $\{1-(-1)^{S+T+l}\}$ one sees that this interaction has non-vanishing values only in even l states. An interaction possessing this property is referred to as a Serber interaction. Thus the only non-vanishing Talmi integral is the one with $l=0$:

$$I_0^{(ST)} = V_{\mathrm{T}}\left[\frac{2}{\kappa\sqrt{\pi}} - \mathrm{e}^{(\kappa^2/4)}\left\{1-\frac{2}{\sqrt{\pi}}\int_0^{\kappa/2} \mathrm{e}^{-x^2}\,\mathrm{d}x\right\}\right]$$

where κ is given by equation (4.23).

An analysis of the electron-scattering data on ${}^4_2\mathrm{He}_2$ yields the value

$$\hbar\omega = 21{\cdot}8\ \mathrm{MeV}$$

or

$$\alpha = 0{\cdot}725\ \mathrm{fm}^{-1}.$$

Thus

$$I_0^{(01)} = -9{\cdot}3\ \mathrm{MeV}$$
$$I_0^{(10)} = -14{\cdot}3\ \mathrm{MeV}.$$

When these values are combined with the spin-orbit splitting taken by de Shalit and Walecka

$$\varepsilon_{\mathrm{p}_{\frac{1}{2}}} - \varepsilon_{\mathrm{p}_{\frac{3}{2}}} = 3{\cdot}2\ \mathrm{MeV},$$

one can use equations (4.73), (4.74), and (4.75b) to calculate the relative spacings of the negative-parity states. Their excitation energy above the ${}^4_2\mathrm{He}_2$ ground state depends on

$$\bar{E} = \varepsilon_{\mathrm{p}_{\frac{3}{2}}} - \varepsilon_{\mathrm{s}_{\frac{1}{2}}} + E_{\frac{1}{2}\frac{1}{2}}(s^3_{\frac{1}{2}}) - E_{00}(s^4_{\frac{1}{2}})$$

where $E_{00}(s^4_{\frac{1}{2}})$ is the interaction energy of the four $s_{\frac{1}{2}}$ nucleons in the α particle ground state. We arbitrarily choose $\bar{E}$ so that the computed position of the 2^- $T=1$ state agrees with experiment. When this is done the spectrum shown in Fig. 4.2 is obtained.

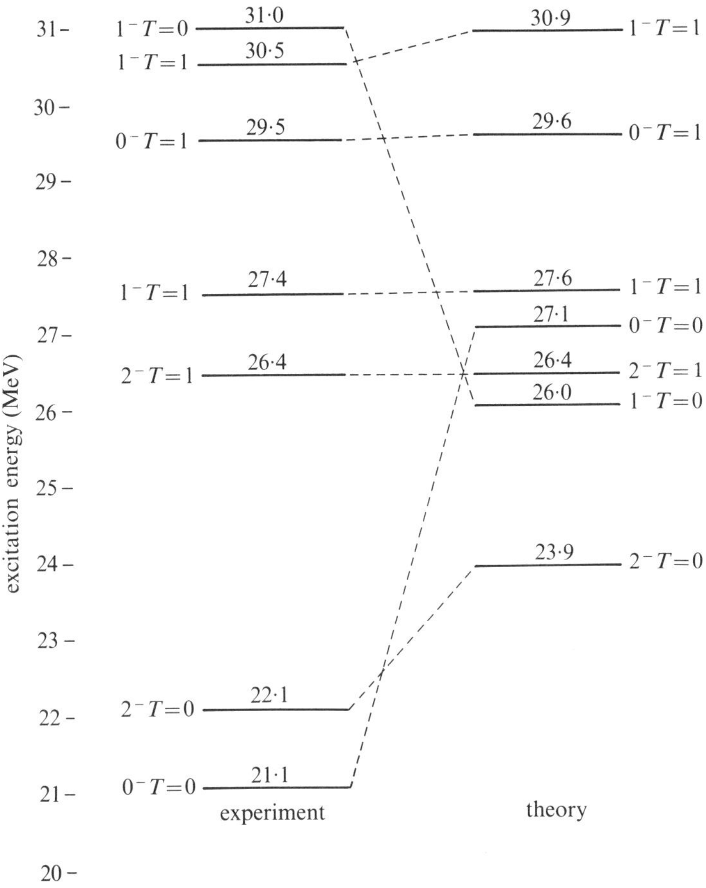

Fig. 4.2. Negative-parity states in ${}^{4}_{2}He_2$. The experimental data are taken from the compilation of Fiarman and Meyerhof (1973). The theoretical results are based on the $0s_{\frac{1}{2}}^{-1}-0p$ shell-model space and the Serber interaction given in equation (4.77).

Clearly the simple Serber interaction of equation (4.77) does quite well in fitting the $T=1$ levels. On the other hand, if one is to explain the relative position of the $T=0$ states a more sophisticated potential is needed (Barrett 1967). However, our main interest in this section was not so much a detailed fit to the data but instead to show how the spurious centre-of-mass contaminant can be removed. To demonstrate the importance of removing the spurious basis state we consider what happens if $\beta\hbar\omega=0$. When this is done the lower $T=0$ 1^- eigenvalue lies 22·53 MeV *below* the $T=1$ 2^- state. In other words, with this choice of parameters the lower 1^- $T=0$ state would be predicted to lie at an excitation energy of only 3·87 MeV! Thus it is imperative that the spurious states be removed in shell-model calculations. This has also been found in ${}^{16}_{8}O_8$ (Elliott and Flowers 1957) and even in the heavy nucleus ${}^{208}_{82}Pb_{126}$ (True *et al.* 1971). In Chapter 5, section 8 we discuss why the spurious 1^- state is always strongly depressed.

The value of β to be used in any numerical shell-model calculation depends on how much spurious contaminant one is willing to allow in the wave function. In perturbation theory the probability of a spurious-state admixture would be

$$\eta = \left|\frac{\langle\psi_{sp}| H |\psi_{ph}\rangle}{\beta\hbar\omega}\right|^2$$

where $\langle\psi_{sp}| H |\psi_{ph}\rangle$ is the off-diagonal matrix element connecting the physical state ψ_{ph} with the spurious state ψ_{sp}. If the matrix elements of V are computed by use of harmonic-oscillator wave functions from a potential that is a function of $(\underline{r}_1-\underline{r}_2)$, $\langle\psi_{sp}| V(\underline{r}_1-\underline{r}_2) |\psi_{ph}\rangle=0$ since the physical state always has the centre-of-mass in the $0s$ state. On the other hand, if the matrix elements of V are taken to have empirical values that best fit the data, there is, of course, no guarantee that they will be consistent with oscillator matrix elements of $V(\underline{r}_1-\underline{r}_2)$. When the one-body spin-orbit interaction of the shell model is included, the matrix element automatically has a non-vanishing value. For example, if we use the wave functions of equation (4.76)

$$\langle\psi'_{1M}| \sum_{i=1}^{A} \left\{\frac{p_i^2}{2m}+\tfrac{1}{2}m\omega^2 r_i^2+f(r_i)\underline{\sigma}_i\cdot\underline{\ell}_i\right\} |\psi_{1M}\rangle=\frac{\sqrt{2}}{3}\{\varepsilon_{p_{\frac{1}{2}}}-\varepsilon_{p_{\frac{3}{2}}}\}.$$

Since the order of magnitude of this off-diagonal matrix element is 1 MeV, one would expect the spurious-state amplitude in the wave function to be of the order of $1/\beta\hbar\omega$. One often wishes to calculate $E1$ and $M2$ transition rates with these wave functions and as we shall show in Chapter, 5, sections 4 and 5, these transitions are often hindered. Thus to obtain a reliable estimate for them one should take $\beta \gtrsim 10^5$.

3.3. *Amount of spurious state in a given wave function*

In many published shell-model calculations the spurious states have not been eliminated. One is therefore faced with the question as to whether the properties predicted with a given wave function can be trusted (i.e. is the state mainly non-spurious). Although the complete set of $1\hbar\omega$ spurious states may be tedious to find, it is often simple to calculate the amount of spuriosity in a wave function without explicitly constructing the spurious ones. In order to do this we must know the single-particle matrix elements of the 'step-up' operator $b_\mu^\dagger$ of equation (4.59). Since $p_\mu = -i\hbar\nabla_\mu$, $b_\mu^\dagger$ can be written in the form

$$b_\mu^\dagger = \frac{1}{\sqrt{\{2m\hbar\omega\}}}\{-\hbar\nabla_\mu + m\omega r_\mu\}.$$

Consequently by use of the methods of Appendix 2, sections 3 and 4 together with equation (A2.23) it follows that

$$\langle\phi_{j'l'm'}|\, b_\mu^\dagger\, |\phi_{jlm}\rangle$$
$$= (-1)^{j+j'}\left\{\frac{(2j+1)}{2(2j'+1)}\right\}^{\frac{1}{2}}(j1\tfrac{1}{2}0\mid j'\tfrac{1}{2})(j1m\mu\mid j'm')\left\{\frac{1-(-1)^{l+l'}}{2}\right\}$$
$$\times\int_0^\infty R_{n'l'}(x)\left\{-\frac{\mathrm{d}}{\mathrm{d}x}-\frac{2+l(l+1)-l'(l'+1)}{2x}+x\right\}R_{nl}(x)x^2\,\mathrm{d}x \tag{4.78}$$

where $x = \alpha r$, $R_{nl}(x)$ are the harmonic-oscillator functions and ϕ_{jlm} is the single-particle wave function of equation (1.4) with the orbital angular-momentum quantum number explicitly displayed. From the form of the oscillator functions given in equation (4.11a) one can show that

$$\left\{-\frac{\mathrm{d}}{\mathrm{d}x}-\frac{2+l(l+1)-l'(l'+1)}{2x}+x\right\}R_{nl}(x)$$
$$= \sqrt{\{2(2n+2l+3)\}}R_{n,l+1}(x)\delta_{l',l+1}$$
$$-2\sqrt{(n+1)}R_{n+1,l-1}(x)\delta_{l',l-1}.$$

Thus, as shown by Gartenhaus and Schwartz (1957),

$$\langle\phi_{j'l'm'}|\, b_\mu^\dagger\, |\phi_{jlm}\rangle = \langle n'l'j'\|\, b^\dagger\, \|nlj\rangle(j1m\mu\mid j'm')$$
$$= (-1)^{j+j'}\left(\frac{2j+1}{2j'+1}\right)^{\frac{1}{2}}(j1\tfrac{1}{2}0\mid j'\tfrac{1}{2})$$
$$\times(j1m\mu\mid j'm')\left\{\frac{1-(-1)^{l+l'}}{2}\right\}\{\sqrt{(2n+2l+3)}\delta_{n',n}\delta_{l',l+1}$$
$$-\sqrt{\{2(n+1)\}}\delta_{n',n+1}\delta_{l',l-1}\}. \tag{4.79}$$

As we shall see, in order to calculate the amount of spurious state it will be necessary to evaluate matrix elements of $B_\mu^\dagger$ (equation (4.61a)) between many-particle shell-model states. In order to keep the counting and phase factors straight (see Chapter 1, section 4.2) it is convenient to write $B_\mu^\dagger$ in terms of creation and destruction operators so that

$$B_\mu^\dagger = \frac{1}{\sqrt{A}}\sum_{nljm\nu}\sum_{n'l'j'm'}\langle n'l'j'\|\, b^\dagger\, \|nlj\rangle(j1m\mu\mid j'm')a^\dagger_{j'm';\frac{1}{2}\nu}a_{jm;\frac{1}{2}\nu} \tag{4.80}$$

where the reduced matrix element $\langle n'l'j'\| b^\dagger \|nlj\rangle$ is given by equation (4.79) and the $a^\dagger_{jm;\frac{1}{2}\nu}$ are the neutron and proton creation operators of equation (2.21) with the (nl) quantum number suppressed.

With these results at our disposal we now turn to calculating the amount of spurious state encountered in some simple wave functions. McGrory and Wildenthal (1973) studied the properties of the $A = 18$, 19 and 20 nuclei based on the assumption that ${}^{12}_{6}C_6$ is the $(0s_{\frac{1}{2}})^4(0p_{\frac{3}{2}})^8$ closed shell and that the valence nucleons occupy the $(0p_{\frac{1}{2}}, 0d_{\frac{5}{2}}, 1s_{\frac{1}{2}})$ levels. In this calculation the spurious states were not removed and

$$\psi_{\frac{1}{2}} = (1s_{\frac{1}{2}})(0p_{\frac{3}{2}})^8(0s_{\frac{1}{2}})^4$$

$$\psi_{\frac{5}{2}} = (0d_{\frac{5}{2}})(0p_{\frac{3}{2}})^8(0s_{\frac{1}{2}})^4$$

were taken to be single-particle states. It is, of course, important to know how much spurious state there is in each of these functions. The complete set of normalized spurious states can be constructed by allowing the centre-of-mass step-up operator, $B^\dagger_\mu$ of equation (4.61a), to operate on each of the non-spurious $A = 13$ states (see equation (4.62)). Although there are many of these non-spurious states in the 13-particle system, the only one which can be connected with $\psi_{\frac{1}{2}}$ and $\psi_{\frac{5}{2}}$ by the $B^\dagger_\mu$ operator is

$$(0p_{\frac{1}{2}})(0p_{\frac{3}{2}})^8(0s_{\frac{1}{2}})^4.$$

Because $B^\dagger_\mu$ can only change j by 1 it follows that $\psi_{\frac{5}{2}}$ is entirely non-spurious. (This is just a special case of the maximum j theorem given in section 3.) On the other hand, the probability that the $\frac{1}{2}^+$ state is spurious is

$$w' = |\langle\psi_{\frac{1}{2}M}| \sum_{\mu M_0} (\tfrac{1}{2}1M_0\mu \mid \tfrac{1}{2}M)B^\dagger_\mu\, (0p_{\frac{1}{2}})_{M_0}(0p_{\frac{3}{2}})^8(0s_{\frac{1}{2}})^4\rangle|^2$$

$$= \left|\frac{1}{\sqrt{13}} \sum_{\mu M_0} (\tfrac{1}{2}1M_0\mu \mid \tfrac{1}{2}M)\langle(1s_{\frac{1}{2}})_M|\, b^\dagger_\mu\, |(0p_{\frac{1}{2}})_{M_0}\rangle\right|^2.$$

When the value of the single-particle matrix element given by equation (4.79) is used, the M sum involving the Clebsch–Gordan coefficients gives unity and

$$w' = \tfrac{2}{13}(\tfrac{1}{2}1\tfrac{1}{2}0 \mid \tfrac{1}{2}\tfrac{1}{2})^2$$

$$= \tfrac{2}{39}.$$

Thus the $\frac{1}{2}^+$ state is 5·13% spurious (Elliott and Skyrme 1955).

One may also calculate the percentage spurious state in a wave function with excitation energy $2\hbar\omega$ above the ground state. In this case centre-of-mass excitation can arise in two ways. First, there are states in which the centre-of-mass has energy $\frac{5}{2}\hbar\omega$. These may be constructed by using equation (4.62) provided we replace $\Psi_{J'M'\gamma';TT_z}$ by $\tilde{\Psi}_{J'M'\gamma';TT_z}$ where the $\tilde{\Psi}$ are all the non-spurious states with energy $1\hbar\omega$ above the ground state. Secondly, there are states in which the centre-of-mass has energy $\frac{7}{2}\hbar\omega$. These may be found explicitly by use of the operator that produces two quanta of centre-of-mass excitation

$$D^{\dagger}_{KM}=\frac{1}{\sqrt{2}}\sum(11\mu\mu'\,|\,KM)B^{\dagger}_{\mu}B^{\dagger}_{\mu'}. \tag{4.81}$$

Because $(11\mu\mu'\,|\,KM)=(-1)^{K}(11\mu'\mu\,|\,KM)$ and since μ and μ' are dummy indices it follows that $K=0$ and 2 only. Thus in analogy with equation (4.62) the $2\hbar\omega$ spurious states are

$$\Phi''_{IM\alpha;TT_z}=\sum_{M'M''}(J'KM'M''\,|\,IM)D^{\dagger}_{KM''}\Psi_{J'M'\gamma';TT_z} \tag{4.82}$$

where the $\Psi_{J'M'\gamma';TT_z}$ are all states of the A-particle system with the nucleons in their lowest possible oscillator orbits and α stands for the quantum numbers $(J'\gamma'K)$. By use of equation (4.63) it follows that the $2\hbar\omega$ spurious states generated by this procedure are orthonormal; i.e.

$$\langle\Phi''_{IM\alpha';TT_z}\,|\,\Phi''_{IM\alpha;TT_z}\rangle=\delta_{J'J''}\delta_{\gamma'\gamma''}\delta_{KK'}.$$

Because of this, the probability w that a state $\Xi_{IM;TT_z}$, with energy $2\hbar\omega$ above the ground state, is spurious is

$$w=\sum_{J'\gamma'}{}'\,|\langle\Xi_{IM;TT_z}\,|\sum_{\mu M'}(J'1M'\mu\,|\,IM)B^{\dagger}_{\mu}\tilde{\Psi}_{J'M'\gamma';TT_z}\rangle|^{2}$$

$$+\sum_{J'\gamma'K}|\langle\Xi_{IM;TT_z}\,|\sum_{M'M''}(J'KM'M''\,|\,IM)D^{\dagger}_{KM''}\Psi_{J'M'\gamma';TT_z}\rangle|^{2} \tag{4.83}$$

where $\sum'$ means that in the sum only those states $\tilde{\Psi}$ with energy $1\hbar\omega$ above the ground state which are entirely non-spurious are to be considered.

As it stands this relationship is awkward because it implies we must construct explicitly the $1\hbar\omega$ states orthogonal to the spurious ones. One can circumvent this by a recoupling in the second term. Since $B^{\dagger}_{\mu}$ is a

tensor of rank-one it follows that

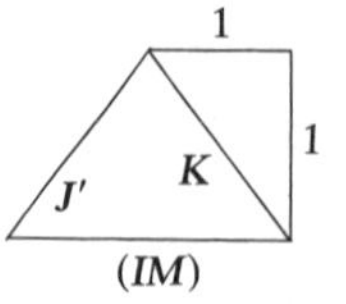

$$=\sum_{J_1}\sqrt{\{(2K+1)(2J_1+1)\}}\,W(J'1I1;J_1K)$$

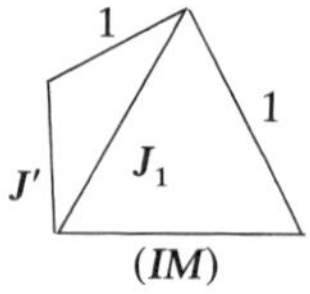

Thus

$$\sum_{M'M''}(J'KM'M''\mid IM)D^{\dagger}_{KM''}\Psi_{J'M'\gamma';TT_z}$$
$$=\frac{1}{\sqrt{2}}\sum_{J_1M_1\mu_1}\sqrt{\{(2J_1+1)(2K+1)\}}\,W(J'1I1;J_1K)(J_11M_1\mu_1\mid IM)$$
$$\times B^{\dagger}_{\mu_1}\Phi_{J_1M_1\gamma_1;TT_z}$$

where the $\Phi_{J_1M_1\gamma_1;TT_z}$ are given by equation (4.62). Because the second term in equation (4.83) is summed over all $(J'\gamma')$ these states are all the $1\hbar\omega$ spurious states; in other words they are precisely the states left out of $\sum'$. Since the square of this expression is summed over K, the second term in equation (4.83) is proportional to

$$\sqrt{\{(2J_1+1)(2J_1'+1)\}}\sum_{K}(2K+1)W(J'1I1;J_1K)W(J'1I1;J_1'K)=\delta_{J_1J_1}$$

where use has been made of equation (A4.15) to carry out the K sum. Thus

$$\sum_{J'\gamma'K}|\langle\Xi_{IM;TT_z}|\sum_{M'M''}(J'KM'M''\mid IM)D^{\dagger}_{KM''}\Psi_{J'M'\gamma';TT_z}\rangle|^2$$
$$=\tfrac{1}{2}\sum_{J_1\gamma_1}{}''|\langle\Xi_{IM;TT_z}|\sum_{\mu M_1}(J_11M_1\mu_1|\,IM)B^{\dagger}_{\mu}\Phi_{J_1M_1\gamma_1;TT_z}\rangle|^2$$

where $\sum''$ is the sum over all the spurious $1\hbar\omega$ states. Consequently $\sum'+\sum''$ is the sum over all $1\hbar\omega$ states and w in equation (4.83) may be written as

$$w=w_1-w_2$$
$$=\sum_{J_1\gamma_1}\left|\langle\Xi_{IM;TT_z}|\sum_{\mu M_1}(J_11M_1\mu\mid IM)B^{\dagger}_{\mu}\bar{\Psi}_{J_1M_1\gamma_1;TT_z}\rangle\right|^2$$
$$-\sum_{J'\gamma'K}\left|\langle\Xi_{IM;TT_z}|\sum_{M'M''}(J'KM'M''\mid IM)D^{\dagger}_{KM''}\Psi_{J'M'\gamma';TT_z}\rangle\right|^2 \tag{4.84}$$

where $\bar{\Psi}_{J_1M_1\gamma_1;TT_z}$ are *all* the states with $1\hbar\omega$ excitation energy and $\Psi_{J'M'\gamma';TT_z}$ are *all* the states with the particles in their lowest possible orbits.

To illustrate the use of this expression we calculate the percentage spurious state in the $A=18$, $T=1$ wave function $(0f_{\frac{7}{2}})^2_{J=0}(0p_{\frac{1}{2}})^4(0p_{\frac{3}{2}})^8(0s_{\frac{1}{2}})^4$. The only state which comes into the first terms in equation (4.84) is $(0f_{\frac{7}{2}}0d_{\frac{5}{2}})_{J'=1}(0p_{\frac{1}{2}})^4(0p_{\frac{3}{2}})^8(0s_{\frac{1}{2}})^4$. The effect of $B^{\dagger}_{\mu}$ on this must be to change the $d_{\frac{5}{2}}$ to $f_{\frac{7}{2}}$. Thus

$$\sum_{M\mu}(11M\mu\,|\,00)B^{\dagger}_{\mu}\bar{\Psi}_{1M;11}(0f_{\frac{7}{2}}, 0d_{\frac{5}{2}}) =$$

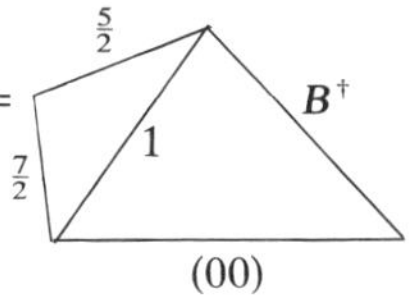

$$= \sqrt{\{(3)(8)\}}W(\tfrac{7}{2}\tfrac{5}{2}01;1\tfrac{7}{2})$$

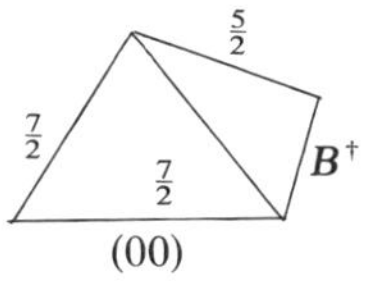

$$= \frac{1}{\sqrt{18}}\langle 0f_{\frac{7}{2}}\|\, b^{\dagger}\,\|0d_{\frac{5}{2}}\rangle$$

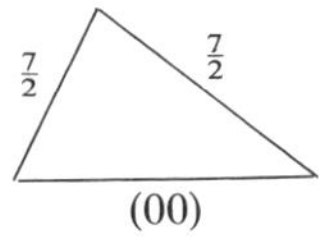

(4.85)

where use has been made of equation (A4.14) to evaluate the Racah coefficient. To make the $(0f_{\frac{7}{2}})^2$ wave function look like the right-hand side of equation (4.85) one must make a fractional-parentage decomposition. Although $\langle(f_{\frac{7}{2}}, f_{\frac{7}{2}}|\}(f_{\frac{7}{2}})^2I=0\rangle = 1$ the purpose of including this is to remind us that a counting factor ($\sqrt{2}$) must be included in the matrix element (Rule 1 in section 4.2 of Chapter 1). Thus by use of equation (4.79)

$$w_1 = \tfrac{7}{12}(\tfrac{5}{2}1\tfrac{1}{2}0\,|\,\tfrac{7}{2}\tfrac{1}{2})^2$$

$$= \tfrac{1}{3}.$$

To evaluate w_2 one need only consider the states $(0d_{\frac{5}{2}})^2_{J'}(0p_{\frac{1}{2}})^4(0p_{\frac{3}{2}})^8(0s_{\frac{1}{2}})^4$. Since $D^{\dagger}_K$ carries isospin zero and since $(0f_{\frac{7}{2}})^2_{I=0}$ must have $T=1$, it follows that $(0d_{\frac{5}{2}})^2$ is also $T=1$, so that $J'=0$ and 2 only. Thus according to equations (1.86) and (1.90)

$$\sum_{M'M''} (J'KM'M'' \mid 00) D^{\dagger}_{KM''} \Psi_{J'M';11}(0d^{2}_{\frac{5}{2}}) =$$

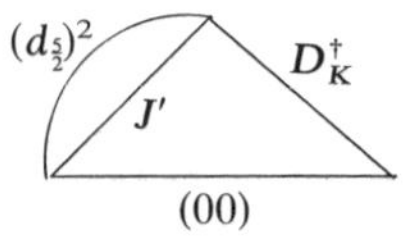

$$= \frac{1}{\sqrt{2}} \langle d_{\frac{5}{2}}, d_{\frac{5}{2}} | \} d^{2}_{\frac{5}{2}} J' \rangle \rangle$$

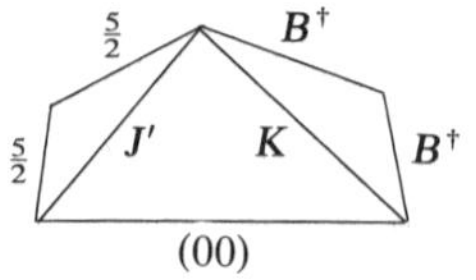

$$= 8 \left\{ \frac{(2J'+1)(2K+1)}{2} \right\}^{\frac{1}{2}} \langle d_{\frac{5}{2}}, d_{\frac{5}{2}} | \} d^{2}_{\frac{5}{2}} J' \rangle \begin{Bmatrix} \frac{5}{2} & \frac{5}{2} & J' \\ 1 & 1 & K \\ \frac{7}{2} & \frac{7}{2} & 0 \end{Bmatrix}.$$

$\times$ 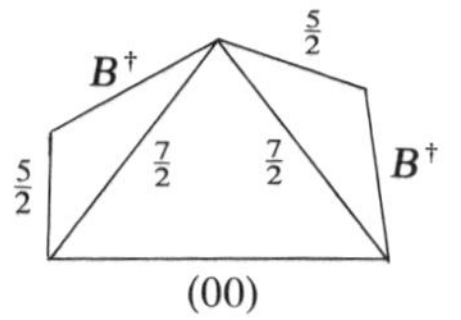

$$= \frac{(-1)^{J'}}{9} \delta_{J'K} \sqrt{(2J'+1)} \langle d_{\frac{5}{2}}, d_{\frac{5}{2}} | \} d^{2}_{\frac{5}{2}} J' \rangle$$
$$\times W(\tfrac{5}{2}\tfrac{5}{2}11; J'\tfrac{7}{2}) \langle 0f_{\frac{7}{2}} \| b^{\dagger} \| 0d_{\frac{5}{2}} \rangle^{2}$$

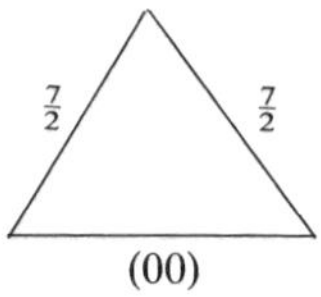

where use has been made of the simple form of the $9j$ coefficient, equation (A4.33). Again the fractional-parentage coefficient is trivially unity but reminds us of the $\sqrt{2}$ that must be included in the matrix element. Similarly a fractional-parentage decomposition must be made of the $(0f_{\frac{7}{2}})^2$ wave function and this again leads to the counting factor $\sqrt{2}$. Thus

$$w_2 = \tfrac{4}{81} \langle 0f_{\frac{7}{2}} \| b^{\dagger} \| 0d_{\frac{5}{2}} \rangle^{4} \sum_{J'=0,2} (2J'+1)\big(W(\tfrac{5}{2}\tfrac{5}{2}11; J'\tfrac{7}{2})\big)^{2}$$
$$= \tfrac{11}{378}.$$

Therefore

$$w = w_1 - w_2$$
$$= \tfrac{115}{378}$$

so that, as shown by Baranger and Lee (1961), the state $(0f_{\frac{7}{2}})^{2}_{I=0}(0p_{\frac{1}{2}})^{4}(0p_{\frac{3}{2}})^{8}(0s_{\frac{1}{2}})^{4}$ is 30·4% spurious.

4. Two-nucleon transfer

In this Section we discuss some of the spectroscopic information that can be obtained from the two-nucleon transfer process

$$A + a \rightleftarrows B + b$$

where

$$a = b + n_1 + n_2$$
$$B = A + n_1 + n_2 \tag{4.86}$$

and in particular to examine the role of the relative centre-of-mass transformation in describing this process. Examples would be the (t, p), $(^3\text{He}, p)$ and (α, d) reactions where the projectile to the left in the bracket is designated in equation (4.86) by a and that on the right by b.

Although some calculations (Ascuitto and Glendenning 1970) indicate that two-step processes may be important in analyzing the two-nucleon transfer data, we shall assume throughout this section that it is a one-step direct-reaction process that can be treated by the distorted wave Born approximation. We shall first summarize the DWBA treatment of the two-nucleon transfer process (for additional details the reader may consult Satchler 1964, Glendenning 1965, Towner and Hardy 1969; Broglia *et al.* (1973)) and then use the results to see what can be learned about nuclear structure from this reaction. According to DWBA the transition amplitude for the process is

$$T_{\text{DWBA}} = \langle \Psi^{(-)}_{\text{Bb}} | \, V \, | \Psi^{(+)}_{\text{Aa}} \rangle \tag{4.87}$$

where V is the interaction causing the transition. $\Psi^{(+)}_{\text{Aa}}$ is the solution of the Schrödinger equation

$$(H_{\text{a}} + H_{\text{A}} + U_{\text{Aa}})\Psi^{(+)}_{\text{Aa}} = E\Psi^{(+)}_{\text{Aa}}$$

with E the total energy of the system (A+a), H_{a} and H_{A} the Hamiltonians of the isolated systems a and A, and U_{Aa} the optical potential describing elastic scattering in the (A+a) channel. The superscript (+) on Ψ indicates that the wave function satisfies outgoing spherical wave boundary conditions. Similarly $\Psi^{(-)}_{\text{Bb}}$ is the DWBA solution to the problem in the (B+b) channel and satisfies the Schrödinger equation

$$(H_{\text{b}} + H_{\text{B}} + U_{\text{Bb}})\Psi^{(-)}_{\text{Bb}} = \text{E}\Psi^{(-)}_{\text{Bb}}$$

where U_{Bb} is the elastic optical potential for (B+b) and the superscript (−) implies an incoming spherical wave boundary condition.

The distorted wave solution $\Psi^{(+)}_{\text{Aa}}$ may be written in terms of the coordinates $\underset{\sim}{R}_{\text{Aa}}$, the internal coordinates $\underset{\sim}{\zeta}_{\text{a}}$ of the cluster a and $\underset{\sim}{\zeta}_{\text{A}}$ of A.

Thus

$$\Psi_{\mathrm{Aa}}^{(+)} = \psi_{\mathrm{Aa}}(\underset{\sim}{\zeta}_{\mathrm{A}}, \underset{\sim}{\zeta}_{\mathrm{a}})\phi^{(+)}(\underset{\sim}{k}_{\mathrm{a}}, \underset{\sim}{R}_{\mathrm{Aa}})$$
$$= \psi_{\mathrm{Aa}}(\underset{\sim}{\zeta}_{\mathrm{A}}, \underset{\sim}{\zeta}_{\mathrm{a}})\phi^{(+)}\left(\underset{\sim}{k}_{\mathrm{a}}, \frac{1}{\sqrt{2}}\underset{\sim}{R} + \frac{a-2}{a}\underset{\sim}{\rho}\right) \tag{4.88}$$

where $\phi^{(+)}(\underset{\sim}{k}_{\mathrm{a}}, \underset{\sim}{R}_{\mathrm{Aa}})$ satisfies the optical potential wave equation

$$\left(-\nabla_{\mathrm{Aa}}^2 + \frac{2\mu_{\mathrm{a}}}{\hbar^2} U_{\mathrm{Aa}}\right)\phi^{(+)}(\underset{\sim}{k}_{\mathrm{a}}, \underset{\sim}{R}_{\mathrm{Aa}}) = k_{\mathrm{a}}^2\phi^{(+)}(\underset{\sim}{k}_{\mathrm{a}}, \underset{\sim}{R}_{\mathrm{Aa}})$$

with $k_{\mathrm{a}}^2 = 2\mu_{\mathrm{a}}E/\hbar^2$ and μ_{a} is the reduced mass $M_{\mathrm{A}}M_{\mathrm{a}}/(M_{\mathrm{A}}+M_{\mathrm{a}})$ in this channel. In writing the second line of this equation use has been made of the coordinate system shown in Fig. 4.3. The coordinate vector $\underset{\sim}{R}_{\mathrm{Aa}}$ connecting the centre of the target A with the incident projectile a has been set equal to

$$\underset{\sim}{R}_{\mathrm{Aa}} = \frac{1}{\sqrt{2}}\underset{\sim}{R} + \frac{a-2}{a}\underset{\sim}{\rho}$$

where $\underset{\sim}{R}/\sqrt{2}$ is the distance between the centre of mass of A and the transferred pair $(\mathrm{n}_1, \mathrm{n}_2)$ and $\underset{\sim}{\rho}$ is the radius vector between the centres of

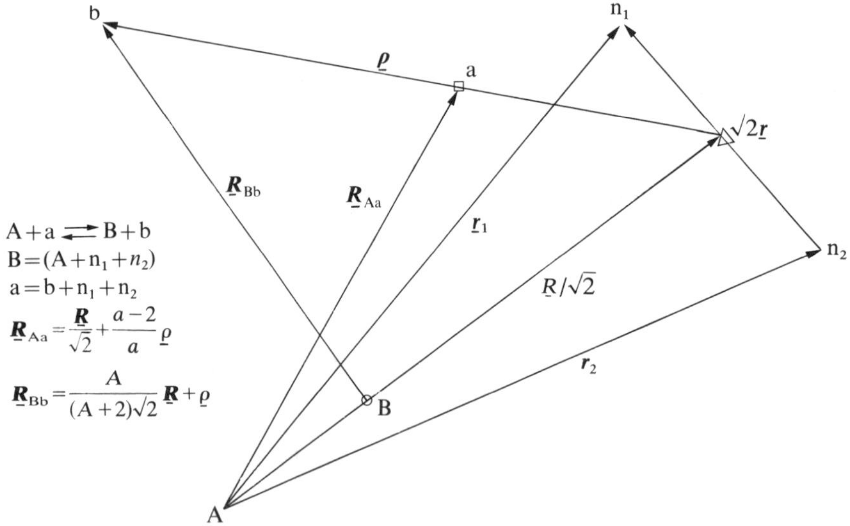

Fig. 4.3. Coordinate system used in the two-nucleon transfer process. ∘ is the centre-of-mass of $(\mathrm{A}+\mathrm{n}_1+\mathrm{n}_2)$ (the position of B), □ is the centre-of-mass of $(\mathrm{b}+\mathrm{n}_1+\mathrm{n}_2)$ (the position of a), and Δ is the centre-of-mass of the transferred pair $(\mathrm{n}_1+\mathrm{n}_2)$. The $\sqrt{2}$ are needed because of our definition of the relative and centre-of-mass coordinates given in equation (4.8). The DWBA wave function for the (A, a) channel is a function of $\underset{\sim}{R}_{\mathrm{Aa}}$ and for (B, b) is a function of $\underset{\sim}{R}_{\mathrm{Bb}}$.

mass of (n_1, n_2) and the outgoing cluster b. In this equation a stands for the number of nucleons in the cluster a. We have implicitly assumed that the optical model potential does not possess a one-body spin-orbit potential because the spin quantum numbers have been omitted from $\phi^{(+)}$ (and $\phi^{(-)}$ defined below). The treatment when this potential is included is discussed by Satchler (1964).

A similar equation can be written for $\Psi^{(-)}_{\mathrm{Bb}}$

$$\Psi^{(-)}_{\mathrm{Bb}} = \psi_{\mathrm{Bb}}(\underline{\zeta}_{\mathrm{B}}, \underline{\zeta}_{\mathrm{b}})\phi^{(-)}(\underline{k}_{\mathrm{b}}, \underline{R}_{\mathrm{Bb}})$$
$$= \psi_{\mathrm{Bb}}(\underline{\zeta}_{\mathrm{B}}, \underline{\zeta}_{\mathrm{b}})\phi^{(-)}\left(\underline{k}_{\mathrm{b}}, \frac{A}{(A+2)\sqrt{2}}\,\underline{R} + \underline{\rho}\right) \tag{4.89}$$

where the coordinates $\underline{R}$, $\underline{\rho}$ and $\underline{R}_{\mathrm{Bb}}$ are shown in Fig. 4.3,

$$\underline{R}_{\mathrm{Bb}} = \frac{A}{(A+2)\sqrt{2}}\,\underline{R} + \underline{\rho}$$

and A stands for the number of nucleons in the nucleus A.

Therefore T_{DWBA} can be written as

$$T_{\mathrm{DWBA}} = \mathcal{J}\iint \mathrm{d}\underline{R}\ \mathrm{d}\underline{\rho}\,\phi^{*(-)}\left(\underline{k}_{\mathrm{b}}, \frac{A}{(A+2)\sqrt{2}}\,\underline{R} + \underline{\rho}\right)$$
$$\times\langle\psi_{\mathrm{Bb}}(\underline{\zeta}_{\mathrm{B}}, \underline{\zeta}_{\mathrm{b}})|\ V\ |\psi_{\mathrm{Aa}}(\underline{\zeta}_{\mathrm{A}}, \underline{\zeta}_{\mathrm{a}})\rangle\phi^{(+)}\left(\underline{k}_{\mathrm{a}}, \frac{1}{\sqrt{2}}\,\underline{R} + \frac{a-2}{a}\,\underline{\rho}\right) \tag{4.90}$$

where $\mathcal{J}$ is the Jacobian of the transformation connecting the coordinate system $\underline{r}_1, \underline{r}_2, \ldots, \underline{r}_{\mathrm{A}}, \underline{r}_{\mathrm{A+1}}, \ldots, \underline{r}_{\mathrm{A+a}}$ with the one used to carry out the evaluation of the integral.

The potential V of equation (4.90) should be a sum over all possible pairs of particles. However, it is assumed that the introduction of the optical model potentials U_{Aa} and U_{Bb} takes into account the majority of the interaction and the only remaining part of V which causes the transition from (A, a) to (B, b) is

$$V = \sum_{i=1}^{b} (V_{\mathrm{in}_1} + V_{\mathrm{in}_2}) \tag{4.91}$$

i.e. V is the sum of the interactions between each nucleon in the cluster b with n_1 and with n_2.

In a direct reaction process the final state of the nucleus B must look like the target A in its initial state J_{A} with two added particles coupled to angular momentum J. Thus a double-parentage decomposition of the state $J_B T_B$ must be carried out:

$$\Psi_{J_BM_B;T_BT_{zB}} = \sum_{J_A'T_A'J'T'j_1j_2} \langle (n-2)J_A'T_A', (j_1j_2)J'T'| \} nJ_BT_B\rangle \times [\Psi_{J_A';T_A'} \times \Phi_{J';T'}]_{J_BM_B;T_BT_{zB}}$$

$$= \sum_{J_A'T_A'J'T'j_1j_2} \left\{\frac{2}{n(n-1)}\right\}^{\frac{1}{2}} \langle \Psi_{J_B;T_B} ||| A^{\dagger}_{J';T'}(j_1j_2) ||| \Psi_{J_A;T_A}\rangle \times [\Psi_{J_A';T_A'} \times \Phi_{J';T'}]_{J_BM_B;T_BT_{zB}}$$

$$= \sum_{J_A'T_A'J'T'j_1j_2} \left\{\frac{2}{n(n-1)}\right\}^{\frac{1}{2}} \{\mathcal{S}_{J'T'}(j_1j_2)\}^{\frac{1}{2}} . \times [\Psi_{J_A';T_A'} \times \Phi_{J';T'}]_{J_BM_B;T_BT_{zB}}$$

where

$$\mathcal{S}_{JT}(j_1j_2) = |\langle \Psi_{J_B;T_B} ||| A^{\dagger}_{J;T}(j_1j_2) ||| \Psi_{J_A;T_A}\rangle|^2 \tag{4.92}$$

and $A^{\dagger}_{JM;TT_z}(j_1j_2)$ is given by equation (2.22).

In writing this relationship we have generalized the notion of the double-parentage coefficient (see Appendix 5, section 2) to the case that the Ψ are antisymmetric wave functions with particles occupying several single-particle states $j_1, j_2, \ldots$. According to Rule 1 of Chapter 1, section 4.2, whenever a double-parentage coefficient arises the matrix element must be multiplied by $\sqrt{\{n(n-1)/2\}}$. Consequently once one integrates over the coordinates of A in equation (4.90) and takes this counting factor into account one finds that

$$T_{\text{DWBA}} = \mathcal{J} \sum_{JT} (J_A J M_A M \mid J_B M_B)(T_A T T_{zA} T_z \mid T_B T_{zB}) \times \sum_{j_1j_2} \{\mathcal{S}_{JT}(j_1j_2)\}^{\frac{1}{2}} \iint \mathrm{d}\underset{\sim}{R}\, \mathrm{d}\rho\, \phi^{*(-)}\left(\underset{\sim}{k}_b, \frac{A}{(A+2)\sqrt{2}}\underset{\sim}{R} + \underset{\sim}{\rho}\right) \times \langle \psi_b(\underset{\sim}{\zeta}_b)\Phi_{JM;TT_z}(\underset{\sim}{r}_1, \underset{\sim}{r}_2)|\, V\, |\psi_a(\underset{\sim}{\zeta}_a)\rangle \times \phi^{(+)}\left(\underset{\sim}{k}_a, \frac{1}{\sqrt{2}}\underset{\sim}{R} + \frac{a-2}{a}\underset{\sim}{\rho}\right)$$

$$= \sum_{j_1j_2} T_{\text{DWBA}}(j_1j_2) \tag{4.93}$$

where the notation $\langle|\, V\, |\rangle$ implies an integration over all coordinates and spin variables except $\underset{\sim}{R}$ and $\underset{\sim}{\rho}$.

For one-particle transfer the spectroscopic factor can be extracted from the data by dividing the peak experimental cross-section by the DWBA result. In two-particle reactions the cross-section, which is proportional to $|T_{\text{DWBA}}|^2$, is a coherent sum over the contributions from many (j_1, j_2) orbits. Consequently only in the special case that a single (j_1, j_2) state contributes is it possible to extract $\mathcal{S}$. On the other hand, one may check the coherence properties (phases) of shell-model wave functions by multiplying each $T_{\text{DWBA}}(j_1 j_2)$ by the predicted $\{\mathcal{S}_{JT}(j_1 j_2)\}^{\frac{1}{2}}$ and computing the cross-section using the coherent sum of the contributions.

Most microscopic codes which describe the (t, p), $(^3\text{He}, p)$ and (α, d) reactions have built into them the assumption that any pair of particles in the light nuclei is in a relative s-state. Thus the spatial wave function of the two nucleons is symmetric and since their space-spin-isospin eigenfunction must be antisymmetric it follows that $S+T=1$ for the transferred pair. For the three- and four-particle systems a convenient radial wave function for ψ_a is

$$\varphi_a \propto \exp\left(-\eta^2 \sum_{i<j} r_{ij}^2\right).$$

In terms of the coordinates $\underline{\rho}$ and $\underline{r}$ of Fig. 4.3 the exponential can be written for the triton as

$$\begin{aligned}\sum_{i<j} r_{ij}^2 &= (\sqrt{2}\underline{r})^2 + (\underline{\rho} + \underline{r}/\sqrt{2})^2 + (\underline{\rho} - \underline{r}/\sqrt{2})^2 \\ &= 3r^2 + 2\rho^2\end{aligned}$$

and for the α particle as

$$\sum_{i<j} r_{ij}^2 = 4(\zeta^2 + r^2 + \rho^2)$$

where the coordinate $\sqrt{2}\zeta$ is the distance between the two particles that make up b (see Fig. 4.4). Thus the normalized radial wave function describing the projectile a is

$$\begin{aligned}\varphi_a &= R_{00}(\sqrt{6}\eta r) R_{00}(2\eta\rho) && \text{for } ^3\text{He or } t \\ &= R_{00}(\sqrt{8}\eta r) R_{00}(\sqrt{8}\eta\rho) R_{00}(\sqrt{8}\eta\zeta) && \text{for } ^4\text{He}\end{aligned} \tag{4.94}$$

where $R_{00}(\gamma r)$ is the harmonic-oscillator wave function of equation (4.11a) with $n=l=0$ and α replaced by γ. The parameter η in these wave functions can be related to the r.m.s. radius of the three- and

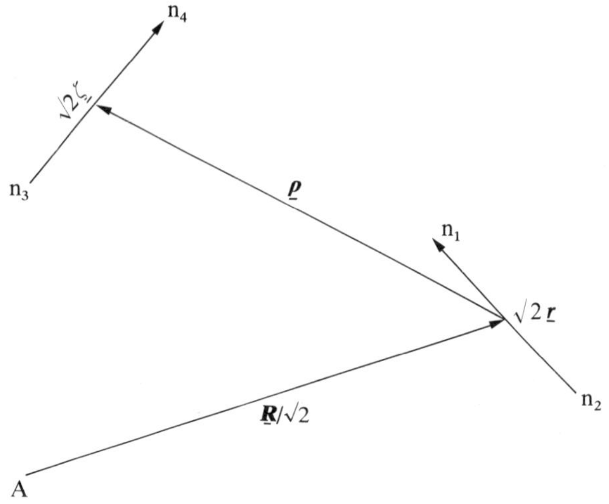

Fig. 4.4. Coordinate system used for the α-particle wave function in equation (4.94).

four-particle systems

$$\langle r^2\rangle = \frac{1}{6\eta^2} \quad \text{for } {}^3\text{He or t}$$

$$\langle r^2\rangle = \frac{9}{64\eta^2} \quad \text{for } {}^4\text{He}.$$

To fit the experimental r.m.s. radii requires $\eta = 0{\cdot}242\ \text{fm}^{-1}$, $0{\cdot}206\ \text{fm}^{-1}$, and $0{\cdot}233\ \text{fm}^{-1}$ for the triton, ^{3}He, and α particle, respectively (Glendenning 1965).

Since φ_a can be expressed as a function of $\underline{r}$ (Fig. 4.4), it is clear that a relative centre-of-mass decomposition of the transferred pair is called for, and this is where the Moshinsky coefficient comes into the two-particle transfer process. Thus

$$\begin{aligned}\mathcal{M} &= \langle \psi_b(\underline{\zeta}_b)\Phi_{JM;TT_z}(\underline{r}_1, \underline{r}_2)|\, V\, |\psi_a(\underline{\zeta}_a)\rangle \\ &= \sum_{S\lambda}\sum_{nlNL}\sum_{M_SM'} \tilde{\gamma}^{(J)}_{\lambda S}(j_1l_1; j_2l_2) \\ &\quad \times M_\lambda(nlNL; n_1l_1n_2l_2)(\lambda SM'M_S \mid JM) \\ &\quad \times \langle \varphi_b(\zeta)\chi_{S_bM_S';T_bT_{zb}}(\text{b})\chi_{SM_S;TT_z}(n_1n_2) \\ &\quad \times [\phi_{nl}(\underline{r})\times\phi_{NL}(\underline{R})]_{\lambda M'}|\, V|\, \varphi_a\chi_{S_aM_S'';T_aT_{za}}(\text{a})\rangle \qquad (4.95)\end{aligned}$$

where $\varphi_b(\zeta)=1$ if b is either a proton or a neutron and is the spatial deuteron wave function if we deal with the (α, d) process, φ_a is given by equation (4.94) and the integration is over all coordinates except $\underset{\sim}{R}$ and $\underset{\sim}{\rho}$. The functions $\chi_{S_aM_S;T_aT_{za}}(a)$ and $\chi_{S_bM_S;T_bT_{zb}}(b)$ are the spin-isospin wave functions of a and b, respectively, and $\chi_{SM_S:TT_z}(n_1n_2)$ is the analogous quantity for the transferred pair. $\bar{\gamma}^{(J)}_{\lambda S}(j_1l_1; j_2l_2)$, the modified j–j to L–S transformation coefficient, is given by equations (4.15) and (4.18). For the special case that $(j_1l_1)=(j_2l_2)$ the first and second rows of the $9j$ coefficient are the same. From the symmetry property of the $9j$ coefficients in equation (A4.29) it follows that $\mathcal{M}$ of equation (4.95) vanishes unless $\{2(l+\frac{1}{2}+j)+S+\lambda+j\}$ is even. Thus we have the selection rule that *the state* $(j^2)_J$ *can only be populated in the two-particle transfer reaction if* $S+\lambda+J$ is even.

Most DWBA analyses of the two-nucleon transfer data assume a spin-independent potential for V and in fact assume that

$$V=V(\underset{\sim}{\rho}) \tag{4.96}$$

i.e. V is a function only of the vector that connects the centres of mass of b and (n_1, n_2). *This assumption together with the assumption that each pair in a or b is in a relative s state leads to the result that only $l=0$ can contribute to the transfer process.* Furthermore, because of equation (4.12), it follows that

$$(-1)^{l_1+l_2}=(-1)^L.$$

Thus *only even (odd) values of L contribute to the cross-section if the parity of the final nuclear state is the same as (opposite to) the parity of the ground state of the target.*

To simplify the spin-isospin part of the calculation we make a double-parentage decomposition of a. For either the three- or four-nucleon system of spin $\frac{1}{2}$ particles it is simple to show that

$$\langle(\tfrac{1}{2})^{n-2}ST, (\tfrac{1}{2})^2S'T'|\}(\tfrac{1}{2})^nS_aT_a\rangle=\left\{\frac{2}{n(n-1)}\right\}^{\frac{1}{2}}\langle\chi_{S_a;T_a}|||A^{\dagger}_{S';T'}|||\chi_{S;T}\rangle$$

$$=\frac{1-(-1)^{S'+T'}}{2\sqrt{2}}.$$

Furthermore, according to Rule 1 of Chapter 1, section 4.2, whenever a double-parentage coefficient arises in a calculation the matrix element must be multiplied by the counting factor $\sqrt{\{n(n-1)/2\}}$ where $n=a$, the number of nucleons in the projectile a. Thus the "integration" over the spin variables in equation (4.95) gives

$$\mathcal{M}=\left\{\frac{a(a-1)}{2}\right\}^{\frac{1}{2}}\sum_{S\lambda}\sum_{nN}\sum_{M_SM'}\frac{\{1-(-1)^{S+T}\}}{2\sqrt{2}}\tilde{\gamma}^{(J)}_{\lambda S}(j_1l_1;j_2l_2)$$
$$\times M_\lambda(n0N\lambda;n_1l_1n_2l_2)(\lambda SM'M_S\mid JM)(S_bSM'_SM_S\mid S_aM''_S)$$
$$\times(T_bTT_{zb}T_z\mid T_aT_{za})\Omega_b\Omega_nV(\rho)$$
$$\times R_{00}(\sqrt{\{4(a-2)\}}\eta\rho)R_{N\lambda}(\alpha R)Y_{\lambda M'}(\theta,\Phi) \qquad (4.97)$$

where

$$\Omega_b=1 \qquad \text{for } a=3$$
$$=\int\zeta^2\,\mathrm{d}\zeta\varphi_b(\zeta)R_{00}(\sqrt{8}\eta\zeta) \quad \text{for } a=4. \qquad (4.98)$$

The remaining overlap integral

$$\Omega_n=\int r^2\,\mathrm{d}rR_{n0}(\alpha r)R_{00}\{\sqrt{(2a)}\eta r\}$$

can be evaluated when the explicit form of the oscillator wave functions in equation (4.11) is used

$$\Omega_n=\frac{\sqrt{(2n+1)!}}{2^nn!}(xy)^{\frac{3}{2}}(1-x)^n \qquad n=0,1,2,\ldots \qquad (4.99a)$$

where

$$x=\frac{2\alpha^2}{2a\eta^2+\alpha^2} \qquad y=\sqrt{(2a)}(\eta/\alpha). \qquad (4.99b)$$

The two-particle transfer cross-section is proportional to the square of T_{DWBA} summed over final-spin states and averaged over initial-spin states. Inserting equation (4.97) into equation (4.93) it follows that when the optical potential has no spin orbit part

$$Q_{\text{DWBA}}=\frac{1}{(2J_i+1)(2S_i+1)}\sum_{M_AM_BM}\sum_{M'_SM''_S}|T_{\text{DWBA}}|^2$$
$$=\frac{(2J_B+1)(2S_a+1)}{(2J_i+1)(2S_i+1)}\mathcal{J}^2\Omega_b^2\sum_{JST\lambda M'}(T_ATT_{zA}T_z\mid T_BT_{zB})^2b^2_{ST}$$
$$\times\left|\sum_{j_1j_2nN}\{\tilde{\mathcal{G}}_{JT}(j_1j_2\lambda SnN)\}^{\frac{1}{2}}\Omega_nB_{\lambda M'}(N)\right|^2. \qquad (4.100)$$

The terms in this equation have the following meanings:

(i) In the statistical factor multiplying the cross-section J_i is the angular momentum of the initial nuclear state, and S_i is the spin of the incident projectile. Thus

$$\frac{(2J_{\mathrm{B}}+1)(2S_{\mathrm{a}}+1)}{(2J_{\mathrm{i}}+1)(2S_{\mathrm{i}}+1)}=\frac{(2J_{\mathrm{B}}+1)}{(2J_{\mathrm{A}}+1)}\quad \begin{array}{l}\text{for the process}\\ \mathrm{A+a\rightarrow B+b}\end{array}$$
$$=\frac{(2S_{\mathrm{a}}+1)}{(2S_{\mathrm{b}}+1)}\quad \begin{array}{l}\text{for the process}\\ \mathrm{B+b\rightarrow A+a}\end{array}. \tag{4.101}$$

(ii) b_{ST} arises from the spin and isospin summations of the light projectiles a and b

$$b_{ST}^2=\left\{\frac{a(a-1)}{2}\right\}\frac{[1-(-1)^{S+T}]}{4(2S+1)}\quad (T_{\mathrm{b}}TT_{z\mathrm{b}}T_z \mid T_{\mathrm{a}}T_{z\mathrm{a}})^2$$
$$=\delta_{S0}\delta_{T1}\qquad \text{for } (t, p) \text{ or } (^3\mathrm{He}, n)$$
$$=\tfrac{1}{2}(\delta_{S0}\delta_{T1}+\delta_{S1}\delta_{T0})\qquad \text{for } (t, n) \text{ or } (^3\mathrm{He}, p)$$
$$=\delta_{S1}\delta_{T0}\qquad \text{for } (\alpha, d). \tag{4.102}$$

(iii) As we have already discussed, Ω_n is the overlap integral involving the relative coordinate of the transferred pair and its value is given by equations (4.99). This quantity is generally small for $n\neq 0$ because x in equation (4.99b) is close to unity. For example with $\hbar\omega=41/A^{\frac{1}{3}}$ one finds for the triton that

$$x=\frac{2}{1+0{\cdot}351A^{\frac{1}{3}}}=1{\cdot}061\quad \text{for}\quad A=16$$
$$=0{\cdot}909\quad \text{for}\quad A=40$$
$$=0{\cdot}649\quad \text{for}\quad A=208.$$

Thus for light nuclei $|1-x|\lesssim\frac{1}{10}$ and consequently $\Omega_{n+1}/\Omega_n\simeq\frac{1}{10}$. Therefore, provided $M_\lambda(n0N\lambda\,;n_1l_1n_2l_2)$ does not increase markedly with n, only $n=0$ need be considered. Glendenning (1968) has published tables of $\tilde{\gamma}^{(\lambda)}_{\lambda 0}(j_1l_1;\,j_2l_2)M_\lambda(n0N\lambda\,;n_1l_1n_2l_2)\Omega_n$ for various values of the oscillator constant. For light nuclei typical values are quoted in Table 4.5, and it is apparent that only $n=0$ need be considered for light systems. Thus the sum on n and N in equation (4.100) becomes a single term with

$$n=0$$
$$N=\tfrac{1}{2}(2n_1+l_1+2n_2+l_2-\lambda) \tag{4.103}$$

(iv) The modified spectroscopic factor $\tilde{\mathscr{S}}_{JT}$ is defined as

$$\{\tilde{\mathscr{S}}_{JT}(j_1j_2\lambda SnN)\}^{\frac{1}{2}}=\{\mathscr{S}_{JT}(j_1j_2)\}^{\frac{1}{2}}\tilde{\gamma}^{(J)}_{\lambda S}(j_1l_1;\,j_2l_2)$$
$$\times M_\lambda(n0N\lambda\,;n_1l_1n_2l_2). \tag{4.104}$$

TABLE 4.5

Values of $\tilde{\gamma}_{00}^{(0)}(j_1l_1; j_2l_2)M_0(n0N0; n_1l_1n_2l_2)\Omega_n$ *as a function of n for various nuclei*

			n			
Configuration	A	α^2	0	1	2	3
$(0p_{\frac{3}{2}})_0^2$	10	0·464	0·5690	0·0964		
$(0d_{\frac{5}{2}})_0^2$	16	0·397	0·3154	0·0429	0·0016	
$(0f_{\frac{7}{2}})_0^2$	40	0·292	0·1680	−0·0499	0·0051	−0·0002

α^2 has been taken equal to $1/A^{\frac{1}{3}}$ and the quoted numbers are for the (t, p) reaction.

The expression for $\tilde{\mathscr{S}}_{JT}$ simplifies when the transferred pair has $T=1$. Because only $l=0$ can contribute when the projectile wave function has the form given by equation (A4.94) and $V=V(\rho)$, it follows from equation (4.18) that $S=0$ for the transferred pair. Furthermore, according to equation (A4.33) the j–j to L–S transformation coefficient takes on a simple form when $S=0$. Thus

$$\{\tilde{\mathscr{S}}_{J=\lambda T=1}(j_1j_2\lambda 0nN)\}^{\frac{1}{2}} = (-1)^{l_2+j_1-\frac{1}{2}-\lambda}\left(\frac{1}{1+\delta_{j_1j_2}\delta_{l_1l_2}\delta_{n_1n_2}}\right)^{\frac{1}{2}}$$
$$\times\{\mathscr{S}_{JT=1}(j_1j_2)\}^{\frac{1}{2}}\times\sqrt{\{(2j_1+1)(2j_2+1)\}}$$
$$\times W(j_1l_1j_2l_2;\tfrac{1}{2}\lambda)M_\lambda(n0N\lambda; n_1l_1n_2l_2).$$

Furthermore, according to equation (4.103) only $n=0$ is important and when $n=l=0$ a simple form can be deduced for the Moshinsky coefficient (see Appendix 6, equation (A6.20))

$$M_\lambda(00N\lambda; n_1l_1n_2l_2) = (-1)^{n_1+n_2-N}(l_1l_200\,|\,\lambda 0)$$
$$\times\left[\frac{(2l_1+1)(2l_2+1)(2N)!!(2N+2\lambda+1)!!}{2^{2N+\lambda}(2\lambda+1)(2n_1)!!(2n_2)!!(2n_1+2l_1+1)!!(2n_2+2l_2+1)!!}\right]^{\frac{1}{2}}.$$

Thus $(\tilde{\mathscr{S}})^{\frac{1}{2}}$ is proportional to $(l_1l_200\,|\,\lambda 0)W(j_1l_1j_2l_2;\frac{1}{2}\lambda)$ and according to equation (A4.23) this combination can be written as a single Clebsch-Gordan coefficient

$$\frac{1}{\sqrt{(2j_2+1)}}(j_1\lambda\tfrac{1}{2}0\,|\,j_2\tfrac{1}{2}) = (-1)^{\frac{1}{2}+\lambda-j_2}\left\{\frac{(2l_1+1)(2l_2+1)}{2\lambda+1}\right\}^{\frac{1}{2}}(l_1l_200\,|\,\lambda 0)$$
$$\times W(j_1l_1j_2l_2;\tfrac{1}{2}\lambda).$$

When these results are combined it follows that

$$\{\tilde{\mathscr{S}}_{J=\lambda T=1}(j_1j_2\lambda 00N)\}^{\frac{1}{2}} = (-1)^{j_1-j_2+l_2+n_1+n_2-N}$$
$$\times\left(\frac{1}{1+\delta_{j_1j_2}\delta_{l_1l_2}\delta_{n_1n_2}}\right)^{\frac{1}{2}}$$
$$\times\{\mathscr{S}_{JT}(j_1j_2)\}^{\frac{1}{2}}\sqrt{(2j_1+1)}(j_1\lambda\tfrac{1}{2}0\mid j_2\tfrac{1}{2})$$
$$\times\left\{\frac{(2N)!!(2N+2\lambda+1)!!}{2^{2N+\lambda}(2n_1)!!(2n_2)!!(2n_1+2l_1+1)!!(2n_2+2l_2+1)!!}\right\}^{\frac{1}{2}}. \quad (4.105)$$

(v) The final term in the cross-section $B_{\lambda M'}(N)$ is a six-dimensional integral which must generally be evaluated numerically:

$$B_{\lambda M'}(N) = \frac{1}{\sqrt{(2\lambda+1)}}\iint \mathrm{d}\underline{R}\;\mathrm{d}\underline{\rho}\,\phi^{*(-)}\Big(\underline{k}_{\mathrm{b}}, \frac{A}{(A+2)\sqrt{2}}\,\underline{R}+\underline{\rho}\Big)V(\rho)$$
$$\times R_{00}(2\sqrt{\{a-2\}}\eta\rho)R_{N\lambda}(\alpha R)\,Y_{\lambda M'}(\theta,\Phi)\phi^{(+)}\Big(\underline{k}_{\mathrm{a}}, \frac{1}{\sqrt{2}}\,\underline{R}+\frac{a-2}{a}\,\underline{\rho}\Big) \quad (4.106)$$

where θ and Φ are the angular coordinates associated with $\underline{R}$. As in the case of single-nucleon transfer, it is important to have the correct exponential decay for $R_{N\lambda}(\alpha R)$.

At present there are two prescriptions that are commonly used.

(*a*) *The oscillator prescription* (*Glendenning* 1965). This is essentially the method we have used throughout our derivation. The wave function of the transferred pair is written with the aid of the Moshinsky coefficients in terms of relative and centre-of-mass coordinates; i.e. $M_\lambda(n0N\lambda; n_1l_1n_2l_2)$ is given by equation (A6.19). To take into account the correct exponential decay, $R_{N\lambda}(\alpha R)$ is replaced by a Hankel or Coulomb function beyond the nuclear surface. The wave number K in the exponential of the Hankel function (e^{-KR}) is given by

$$K^2 = 2\tilde{\mu}\tilde{E}/\hbar^2$$

where

$$\tilde{E} = |\mathrm{BE}(\mathrm{A}, J_{\mathrm{A}}) - \mathrm{BE}(\mathrm{B}, J_{\mathrm{B}})| \quad (4.107)$$

is the separation energy of the pair calculated by use of the total binding energies of the nuclei A and B in their states J_{A} and J_{B} and $\tilde{\mu}$ is the reduced mass of the pair.

(*b*) *The Bayman–Kallio prescription* (*Bayman and Kallio* 1967). In this case the factor $M_\lambda(n0N\lambda; n_1l_1n_2l_2)\Omega_n R_{N\lambda}(\alpha R)$ is generated from the wave function of two particles (j_1l_1) and (j_2l_2) moving in a spherical Woods-Saxon potential (equation (1.7)) with spin-orbit coupling. The well depth is adjusted so that each particle has half the energy $\tilde{E}$ of

TABLE 4.6

Relative values of the modified spectroscopic factor $\tilde{\mathscr{S}}$ for $T=1$ transfer when $J=\lambda=0$ and 2

	$\lambda=0$				$\lambda=2$		
Configuration	$\tilde{E}=11{\cdot}525$ Bayman–Kallio	$\tilde{E}=15{\cdot}525$ Bayman–Kallio	Oscillator	Configuration	$\tilde{E}=11{\cdot}525$ Bayman–Kallio	$\tilde{E}=15{\cdot}525$ Bayman–Kallio	Oscillator
$(1s_{\frac{1}{2}})^2$	1·000	1·000	1·000	$(0d_{\frac{5}{2}})^2$	0·272	0·268	0·229
$(0d_{\frac{3}{2}})^2$	0·317	0·302	0·320	$(0d_{\frac{5}{2}}0d_{\frac{3}{2}})$	0·111	0·107	0·114
$(0d_{\frac{5}{2}})^2$	0·700	0·671	0·480	$(0d_{\frac{5}{2}}1s_{\frac{1}{2}})$	1·000	1·000	1·000
				$(0d_{\frac{3}{2}})^2$	0·108	0·105	0·133
$(0f_{\frac{7}{2}})^2$	0·237	0·236	0·163	$(0d_{\frac{3}{2}}1s_{\frac{1}{2}})$	0·547	0·541	0·667
$(0f_{\frac{5}{2}})^2$	0·121	0·048	0·122				
$(1p_{\frac{3}{2}})^2$	1·000	1·000	1·000	$(0f_{\frac{7}{2}})^2$	0·152	0·157	0·097
$(1p_{\frac{1}{2}})^2$	0·373	0·385	0·500	$(0f_{\frac{7}{2}}0f_{\frac{5}{2}})$	0·019	0·017	0·023
				$(0f_{\frac{7}{2}}1p_{\frac{3}{2}})$	0.979	1·000	0·735
$(0g_{\frac{9}{2}})^2$	0·089	0·076	0·081	$(0f_{\frac{5}{2}})^2$	0·041	0·062	0·070
$(0g_{\frac{7}{2}})^2$	0·049	0·042	0·064	$(0f_{\frac{5}{2}}1p_{\frac{3}{2}})$	0·092	0·091	0·122
$(1d_{\frac{5}{2}})^2$	1·000	0·999	0·980	$(0f_{\frac{5}{2}}1p_{\frac{1}{2}})$	0·299	0·339	0·429
$(1d_{\frac{3}{2}})^2$	0·687	0·650	0·653	$(1p_{\frac{3}{2}})^2$	0·571	0·571	0·500
$(2s_{\frac{1}{2}})^2$	0·885	1·000	1·000	$(1p_{\frac{3}{2}}1p_{\frac{1}{2}})$	1·000	0·962	1·000

The values when the oscillator prescription is used are given by equation (4.105) with $\mathscr{S}_{JT}(j_1 j_2)=1$. For each oscillator shell and each value of λ the maximum value of $\tilde{\mathscr{S}}$ was arbitrarily normalized to unity. The Bayman–Kallio result was obtained by calculating the DWBA cross-section using the code TWOPAR. The optical model potential of Schlegel *et al.* (1973) was used and the calculation was carried out for two values of $\tilde{E}$ of (4.107).

equation (4.107). The two independent particle wave functions are then coupled to J and T. Because the spatial wave functions of a and b are assumed to be relative s-states, the integrals involved in projecting out the $l=0$ part can be carried out numerically without too much trouble.

In Table 4.6 we compare the values of $\tilde{\mathcal{S}}_{JT=1}$ obtained in these two different ways. In the Bayman-Kallio calculation the shell-model Woods-Saxon potential has a one-body spin-orbit part, and consequently the $j=l+\frac{1}{2}$ and $j=l-\frac{1}{2}$ states have different radial wave functions. Thus the ratio of $(0f_{\frac{7}{2}})^2$ to $(0f_{\frac{5}{2}})^2$ transfer differs somewhat from the oscillator prescription. The Bayman–Kallio results are given for two different separation energies $\tilde{E}$ in equation (4.107), and in both cases the results are qualitatively the same as the oscillator prediction.

In addition it is clear from this table that *two-nucleon transfer should be a sensitive indicator of low angular momentum admixtures into states that are predominantly made up of particles in a large j single-particle state.*

4.1. Wave functions for the 0^+ *states in* ${}^{18}_{8}\mathrm{O}_{10}$

In section 2.1 of Chapter 1 we discussed the structure of the states of ${}^{18}_{8}\mathrm{O}_{10}$ using the $(d_{\frac{5}{2}}, s_{\frac{1}{2}})$ model space, and in section 2.2 we showed that one should expect a low-lying core excited 0^+ state. Thus if we wish to describe the structure of the 0^+ states in this nucleus observed below 5·5 MeV excitation energy, we must at least take into account the three basis states

$$\begin{aligned}\phi_1 &= (d_{\frac{5}{2}})^2_0\\ \phi_2 &= (s_{\frac{1}{2}})^2_0 \qquad\qquad (4.108)\\ \phi_3 &= \psi_0\end{aligned}$$

where ψ_0 is the four-particle two-hole state shown in Fig. 1.5. Fortune and Headley (1974) have used the spectroscopic factors obtained from the ${}^{17}_{8}\mathrm{O}_9(d,p){}^{18}_{8}\mathrm{O}_{10}$ and ${}^{18}_{8}\mathrm{O}_{10}(d,t){}^{17}_{8}\mathrm{O}_9$ reactions together with the observed (t,p) cross-sections to determine the structure of the ground state and the 3·63 and 5·33 MeV excited 0^+ states in ${}^{18}_{8}\mathrm{O}_{10}$. To do this they assume

(a) that in its ground state ${}^{16}_{8}\mathrm{O}_8$ is a doubly closed-shell nucleus,
(b) that the $\frac{5}{2}^+$ ground state and the 871 keV $\frac{1}{2}^+$ in ${}^{17}_{8}\mathrm{O}_9$ are the single-particle $0d_{\frac{5}{2}}$ and $1s_{\frac{1}{2}}$ states, respectively,
(c) that the three 0^+ states can be described by the wave functions

$$\Psi(0_i^+) = \sum_k a_{ik}\phi_k \qquad (4.109)$$

where the ϕ_k are given by equation (4.108) and the coefficients a_{ik}

are real and form an orthogonal matrix satisfying the conditions

$$\sum_k a_{ik}a_{i'k} = \delta_{i'i}.$$

We first show that on the basis of these assumptions the one-nucleon transfer data determines the ground-state wave function but allows two possibilities for the structure of the 3·63 and 5·33 MeV states. We shall then show that the (t, p) data can be used to resolve this ambiguity.

According to equation (2.115) the spectroscopic factor $\mathcal{S}_{jl}$ for populating the ith 0^+ state via the ${}^{17}_{8}O_9(d, p){}^{18}_{8}O_{10}$ reaction is

$$\mathcal{S}_{j=\frac{5}{2},l=2}(0_i^+) = 2a_{i1}^2 \tag{4.110}$$

where a_{i1} is the coefficient of the $(d_{\frac{5}{2}})_0^2$ configuration in the state $\Psi(0_i^+)$. Furthermore, the spectroscopic factor for populating the $\frac{5}{2}^+$ and $\frac{1}{2}^+$ states in ${}^{17}_{8}O_9$ by the (d, t) pickup reaction is

$$\begin{aligned} \mathcal{S}_{j=\frac{5}{2},l=2}(\tfrac{5}{2}^+) &= 2a_{11}^2 \\ \mathcal{S}_{j=\frac{1}{2},l=0}(\tfrac{1}{2}^+) &= 2a_{12}^2 \end{aligned} \tag{4.111}$$

where a_{11} and a_{12} are the coefficients of $(d_{\frac{5}{2}})_0^2$ and $(s_{\frac{1}{2}})_0^2$ respectively in the ground-state wave function of ${}^{18}_{8}O_{10}$. To minimize the errors inherent in extracting spectroscopic factors, Fortune and Headley have fitted ratios of these quantities. From the (d, p) work of Wiza *et al.* (1966) they conclude that

$$\frac{\mathcal{S}_{j=\frac{5}{2},l=2}(0_2^+)}{\mathcal{S}_{j=\frac{5}{2},l=2}(0_1^+)} = \left(\frac{a_{21}}{a_{11}}\right)^2 = 0{\cdot}192 \tag{4.112a}$$

$$\frac{\mathcal{S}_{j=\frac{5}{2},l=2}(0_3^+)}{\mathcal{S}_{j=\frac{5}{2},l=2}(0_1^+)} = \left(\frac{a_{31}}{a_{11}}\right)^2 = 0{\cdot}066 \tag{4.112b}$$

and from the (d, t) study of Armstrong and Quisenberry (1961) that

$$\frac{\mathcal{S}_{j=\frac{1}{2},l=0}(\frac{1}{2}^+)}{\mathcal{S}_{j=\frac{5}{2},l=2}(\frac{5}{2}^+)} = \left(\frac{a_{12}}{a_{11}}\right)^2 = 0{\cdot}17. \tag{4.112c}$$

Since the expansion coefficients form an orthogonal matrix not only $\sum_k a_{ik}^2 = 1$ but also $\sum_i a_{i1}^2 = 1$. Thus the data given by equations (4.112a) and (4.112b) are sufficient to find the magnitudes of a_{11}, a_{21}, and a_{31}:

$$\begin{aligned} a_{11} &= 0{\cdot}892 \\ a_{21} &= \pm 0{\cdot}391 \\ a_{31} &= \pm 0{\cdot}229 \end{aligned} \tag{4.113}$$

where we have arbitrarily taken a_{11} to be positive. From the (d, t) data in equation (4.112c) one may deduce the magnitude of a_{12}. Since the phase of this coefficient relative to that of a_{11} should be the same as determined by a conventional calculation, it follows from equation (1.61) that if a_{11} is positive then so is a_{12}. Thus

$$a_{12} = 0{\cdot}368.$$

Finally, since $\sum_k a_{1k}^2 = 1$,

$$a_{13} = \pm 0{\cdot}265.$$

So far we have said nothing about the explicit form of ψ_0 expect that it is a four-particle two-hole state. We now assume that the phase associated with ψ_0 is such as to make a_{13} positive. Therefore the (d, p) and (d, t) data lead to the conclusion that the ground-state wave function of ${}^{18}_{8}O_{10}$ is

$$\Psi(0_1^+) = 0{\cdot}892(d_{\frac{5}{2}})_0^2 + 0{\cdot}368(s_{\frac{1}{2}})_0^2 + 0{\cdot}265\psi_0. \qquad (4.114)$$

In other words there is a 7 per cent probability of finding the four-particle two-hole state in the ground state of ${}^{18}_{8}O_{10}$ and a 79·5 per cent and 13·5 per cent probability of finding the $(d_{\frac{5}{2}})_0^2$ and $(s_{\frac{1}{2}})_0^2$ configurations respectively.

Since $\Psi(0_2^+)$ and $\Psi(0_3^+)$ must be orthogonal to $\Psi(0_1^+)$ it is convenient to express these two wave functions in terms of basis states that are mutually orthogonal and also orthogonal to equation (4.114). Clearly

$$\Phi_1 = 0{\cdot}381(d_{\frac{5}{2}})_0^2 - 0{\cdot}925(s_{\frac{1}{2}})_0^2$$

and

$$\Phi_2 = 0{\cdot}245(d_{\frac{5}{2}})_0^2 + 0{\cdot}101(s_{\frac{1}{2}})_0^2 - 0{\cdot}964\psi_0 \qquad (4.115)$$

are orthonormal and orthogonal to $\Psi(0_1^+)$ and if we take

$$\begin{aligned} \Psi(0_2^+) &= \Phi_1 \cos\alpha + \Phi_2 \sin\alpha \\ \Psi(0_3^+) &= -\Phi_1 \sin\alpha + \Phi_2 \cos\alpha \end{aligned} \qquad (4.116)$$

the three states $\Psi(0_1^+)$, $\Psi(0_2^+)$, and $\Psi(0_3^+)$ are orthonormal provided α is real. We now determine the possible values of α that are consistent with the (d, p) data. From the definitions of Φ_1 and Φ_2 one sees that equations (4.113) are satisfied if

$$0{\cdot}381 \cos\alpha + 0{\cdot}245 \sin\alpha = \pm 0{\cdot}391$$

$$-0{\cdot}381 \sin\alpha + 0{\cdot}245 \cos\alpha = \pm 0{\cdot}229.$$

When a_{21} and a_{31} are both taken to be positive, $\cos\alpha = 0{\cdot}999$ and

$\sin\alpha = 0{\cdot}042$ so that

$$\begin{aligned}\Psi(0_2^+) &= 0{\cdot}391(d_{\frac{5}{2}})_0^2 - 0{\cdot}920(s_{\frac{1}{2}})_0^2 - 0{\cdot}039\psi_0\\ \Psi(0_3^+) &= 0{\cdot}229(d_{\frac{5}{2}})_0^2 + 0{\cdot}138(s_{\frac{1}{2}})_0^2 - 0{\cdot}964\psi_0.\end{aligned} \tag{4.117}$$

Aside from a trivial overall phase factor the same solution is found when both a_{21} and a_{31} are taken to be negative. On the other hand, if a_{21} is positive and a_{31} is negative a second solution is found, namely $\cos\alpha = 0{\cdot}453$ and $\sin\alpha = 0{\cdot}892$ which leads to

$$\begin{aligned}\Psi(0_2^+) &= 0{\cdot}391(d_{\frac{5}{2}})_0^2 - 0{\cdot}329(s_{\frac{1}{2}})_0^2 - 0{\cdot}860\psi_0\\ \Psi(0_3^+) &= -0{\cdot}229(d_{\frac{5}{2}})_0^2 + 0{\cdot}870(s_{\frac{1}{2}})_0^2 - 0{\cdot}437\psi_0.\end{aligned} \tag{4.118}$$

If the signs of a_{21} and a_{31} are reversed the same solution is obtained.

Thus the (d, p) and (d, t) data lead to a unique result for the ground-state wave function, but do not distinguish between equations (4.117) and (4.118) for the 3·63 (0_2^+) and 5·33 (0_3^+) states. However, the predicted ${}^{16}_{8}\mathrm{O}_8(t, p){}^{18}_{8}\mathrm{O}_{10}$ cross-sections differ markedly in the two cases. If one neglects the difference in $\tilde{E}$ in equation (4.107) for the various states in ${}^{18}_{8}\mathrm{O}_{10}$ $B_{\lambda M'}(N)$ cancels out of equation (4.100) when we take ratios of cross-sections. Furthermore, if we assume ${}^{16}_{8}\mathrm{O}_8$ to be an inert core the only parts of the ${}^{18}_{8}\mathrm{O}_{10}$ wave functions involved in the transfer process are the $(d_{\frac{5}{2}})_0^2$ and $(s_{\frac{1}{2}})_0^2$ components. According to equation (4.103) $N=2$ so that it follows from equation (4.105) that

$$\{\tilde{\mathcal{G}}_{J=0,T=1}(\tfrac{5}{2}\tfrac{5}{2}0002)\}^{\frac{1}{2}} = 1/\sqrt{10}$$

and

$$\{\tilde{\mathcal{G}}_{J=0,T=1}(\tfrac{1}{2}\tfrac{1}{2}0002)\}^{\frac{1}{2}} = \sqrt{\tfrac{5}{24}}.$$

Consequently if $(\mathrm{d}\sigma/\mathrm{d}\Omega)_{0_i^+}$ is the differential cross-section to the state 0_i^+ one predicts that

$$\frac{\left(\dfrac{\mathrm{d}\sigma}{\mathrm{d}\Omega}\right)_{0_i^+}}{\left(\dfrac{\mathrm{d}\sigma}{\mathrm{d}\Omega}\right)_{0_1^+}} = \left|\frac{\dfrac{1}{\sqrt{10}}\,a_{i1} + \sqrt{\tfrac{5}{24}}\,a_{i2}}{\dfrac{1}{\sqrt{10}}\,a_{11} + \sqrt{\tfrac{5}{24}}\,a_{12}}\right|^2.$$

When the wave functions of equation (4.117) are used in conjunction with equation (4.114) one finds that

$$\frac{\left(\dfrac{\mathrm{d}\sigma}{\mathrm{d}\Omega}\right)_{0_2^+}}{\left(\dfrac{\mathrm{d}\sigma}{\mathrm{d}\Omega}\right)_{0_1^+}} = 0{\cdot}43 \quad\text{and}\quad \frac{\left(\dfrac{\mathrm{d}\sigma}{\mathrm{d}\Omega}\right)_{0_3^+}}{\left(\dfrac{\mathrm{d}\sigma}{\mathrm{d}\Omega}\right)_{0_1^+}} = 9{\cdot}1\times10^{-2}.$$

Alternatively equation (4.118) yields

$$\frac{\left(\frac{d\sigma}{d\Omega}\right)_{0_2^+}}{\left(\frac{d\sigma}{d\Omega}\right)_{0_1^+}} = 3\cdot5\times10^{-3} \quad \text{and} \quad \frac{\left(\frac{d\sigma}{d\Omega}\right)_{0_3^+}}{\left(\frac{d\sigma}{d\Omega}\right)_{0_1^+}} = 0\cdot52.$$

The (t, p) data of Middleton and Pullen (1964) give

$$\frac{\left(\frac{d\sigma}{d\Omega}\right)_{0_2^+}}{\left(\frac{d\sigma}{d\Omega}\right)_{0_1^+}} = 3\cdot2\times10^{-2} \quad \text{and} \quad \frac{\left(\frac{d\sigma}{d\Omega}\right)_{0_3^+}}{\left(\frac{d\sigma}{d\Omega}\right)_{0_1^+}} = 0\cdot43$$

for the ratios of these cross-sections at their peak experimental values. In other words, experiment gives a small cross-section for populating the 0_2^+ state, and this is only consistent with the theoretical predictions given by equations (4.118).

In their analysis Fortune and Headley use the Bayman–Kallio prescription to calculate the (t, p) cross-sections. When the wave functions of equations (4.118) are used they find

$$\frac{\left(\frac{d\sigma}{d\Omega}\right)_{0_2^+}}{\left(\frac{d\sigma}{d\Omega}\right)_{0_1^+}} = 3\cdot0\times10^{-2} \quad \text{and} \quad \frac{\left(\frac{d\sigma}{d\Omega}\right)_{0_3^+}}{\left(\frac{d\sigma}{d\Omega}\right)_{0_1^+}} = 0\cdot62$$

in satisfactory agreement with experiment. It is, of course, not surprising that the ratio involving the transition to 0_2^+ changes by an order of magnitude since in the oscillator calculation the $(s_{\frac{1}{2}})_0^2$ and $(d_{\frac{5}{2}})_0^2$ contributions cancel nearly exactly.

Thus the (t, p) data can be combined with the (d, p) and (d, t) results to pin down the structure of the three lowest 0^+ states in ${}^{18}_{8}O_{10}$, and all the data are well reproduced if one assumes the wave functions of these states are given by equations (4.114) and (4.118). Therefore the second excited 0^+ state is mainly core excited ($\sim$74 per cent). On the other hand, if one tries to fit only excitation energies one can carry out a calculation similar to that described for the $f_{\frac{7}{2}}$ nuclei in section 3, Chapter 1 and one finds that all the observed states below 4 MeV excitation energy in ${}^{18}_{8}O_{10}$, ${}^{19}_{8}O_{11}$, and ${}^{20}_{8}O_{12}$ can be reproduced with an r.m.s. error of about 100 keV by a model which implies that the 3·63 MeV state is mainly $(s_{\frac{1}{2}})_0^2$ (in other words, it has a wave function similar to that given by equation (4.117) (Cohen *et al.* 1964). Thus one sees once more that it is important to study properties other than energies if one is to make meaningful configuration assignments.

4.2. (t, p) reaction on the calcium isotopes

According to equation (4.104) when $S=0$ the only allowable value of λ is $\lambda = J$, where J is the angular momentum of the transferred pair. Thus the (t, p) reaction on a spin-zero target is mediated by a single λ value, namely the angular momentum of the final state. Consequently if one neglects Q-value effects and changes in the optical model potential, $B_{\lambda M'}(N)$ will be the same when we consider the $0 \to J$ transition in neighbouring nuclei filling the same major oscillator shell. Consequently cross-section ratios should be determined entirely by $\tilde{\mathscr{G}}_{JT=1}$ in equation (4.104).

An example of this is provided by the calcium nuclei. We shall assume that the low-lying states of these nuclei can be described as $f_{\frac{7}{2}}$ neutrons outside an inert ${}^{40}_{20}\mathrm{Ca}_{20}$ core and compare the theoretical predictions with experiment. When ratios of cross-sections are taken, the only part of $\tilde{\mathscr{G}}_{JT}$ that does not cancel out is $\mathscr{G}_{JT=1}(jj)$ given by equation (4.92). For the identical particle configuration, $\mathscr{G}_{JT=1}(jj)$ is

$$\mathscr{G}_{JT=1}(jj) = \frac{n(n-1)}{2} \langle j^{n-2}00, j^2J|\}j^nJv\rangle^2.$$

The double-parentage coefficients are given in Appendix 5 in equations (A5.57) and (A5.59) so that

$$\frac{\sigma(J, n)}{\sigma(J, 2)} = \frac{n(2j+3-n)}{2(2j+1)} \delta_{J,0}\delta_{v,0} \tag{4.119a}$$

$$= \frac{(2j+3-n)(2j+1-n)}{(2j+1)(2j-1)} \delta_{v,2} \tag{4.119b}$$

where $\sigma(J, n)$ is the cross-section for populating the state $(j^n)_J$.

In Table 4.7 we compare the results of Bjerregaard *et al.* (1967) with theory, and it is clear that the observed trend can be understood on the basis of an $f_{\frac{7}{2}}$ configuration assignment. In addition, the $f_{\frac{7}{2}}$ model predicts that the lower 4^+ state in ${}^{44}_{20}\mathrm{Ca}_{24}$ (observed at 2·283 MeV) should have seniority-4 and the upper one (seen at 3·044 MeV) should be $v=2$. *Since the direct two-particle transfer process can change v by at most two units,* only the upper 4^+ state should be populated in the (t, p) reaction. Experimentally

$$\frac{\sigma(4^+, 2{\cdot}283\ \mathrm{MeV})}{\sigma(4^+, 3{\cdot}044\ \mathrm{MeV})} = 0{\cdot}33$$

TABLE 4.7

Ratios of (t, p) cross-sections for the calcium nuclei

		$\sigma(0, n)/\sigma(0, 2)$		$\sigma(2, n)/\sigma(2, 2)$	
Reaction	n	Experiment	Theory	Experiment	Theory
${}^{40}_{20}Ca_{20}(t, p){}^{42}_{20}Ca_{22}$	2	1·00	1·00	1·00	1·00
${}^{42}_{20}Ca_{22}(t, p){}^{44}_{20}Ca_{24}$	4	1·72	1·50	0·50	0·50
${}^{44}_{20}Ca_{24}(t, p){}^{46}_{20}Ca_{26}$	6	1·44	1·50	0·27	0·17
${}^{46}_{20}Ca_{26}(t, p){}^{48}_{20}Ca_{28}$	8	1·06	1·00		

$\sigma(J, n)$ stands for the cross-section to the state J in the reaction ${}^{38+n}_{20}Ca_{18+n}(t, p){}^{40+n}_{20}Ca_{20+n}$. The experimental results are those of Bjerregaard *et al.* (1967) and the theoretical predictions are given by equation (4.119) with $j=\frac{7}{2}$.

so that the observed states do not obey the $f_{\frac{7}{2}}$ prediction. Since no two-body interaction can mix seniority in the $f_{\frac{7}{2}}$- shell (see Appendix 3, section 4.2) this implies admixtures of other configurations. However, as shown in Chapter 3, section 2.3, the admixture of other states need not be very large to produce the observed result because of the near degeneracy of the two $(f_{\frac{7}{2}})^4$ 4^+ states.

A further consequence of a pure $f_{\frac{7}{2}}$ assignment for the nuclei ${}^{40}_{20}Ca_{20}$—${}^{48}_{20}Ca_{28}$ is that ${}^{50}_{20}Ca_{30}$ should differ from ${}^{48}_{20}Ca_{28}$ by the addition of a $(1p_{\frac{3}{2}})^2_{J=0}$ pair. According to Table 4.6 $\tilde{\mathscr{G}}_{JT=1}$ for adding a $(1p_{\frac{3}{2}})^2_0$ pair is a factor of between 4 and 6 larger than the analogous quantity for adding a $(0f_{\frac{7}{2}})^2_0$ pair. Thus one would predict for the ground-state transitions

$$\frac{\sigma({}^{48}_{20}Ca_{28}(t, p){}^{50}_{20}Ca_{30})}{\sigma({}^{40}_{20}Ca_{20}(t, p){}^{42}_{20}Ca_{22})} \simeq 4-6.$$

Experimentally Bjerregaard *et al.* (1967) find this ratio to be 2·7. Thus once more one can qualitatively understand the observed results on a pure configuration picture.

Bayman and Hintz (1968) have examined the possibility of obtaining a more quantitative understanding of both the (t, p) and single-nucleon transfer data to the calcium ground states. They analyze the data using a full DWBA calculation taking into account changes in $\tilde{E}$ (equation (4.107)) in going from nucleus to nucleus and also changes in the optical-model parameters. When this is done they find that a satisfactory fit to the data requires about a 20 per cent admixture of configurations other than $f_{\frac{7}{2}}$. Thus on the basis of binding energies alone one would have concluded that the $f_{\frac{7}{2}}$ configuration must be fairly pure since the calcium binding energies could be fitted with an r.m.s. error of only 100 keV with the $(f_{\frac{7}{2}})^n$ model (see Table 1.2). However, once other nuclear properties are analyzed one gains important information about the shortcomings of the $f_{\frac{7}{2}}$ model.

4.3. *Comparison of the* (p, t) *and* $(p, {}^3\mathrm{He})$ *reactions*

The (p, t) process involves the transfer of a $T=1$ pair with $S=0$. Furthermore, for this reaction T_B of equation (4.100) is the initial isospin state and $T_\mathrm{A}=T_\mathrm{f}$. Thus for processes in which $T_\mathrm{i}=T_\mathrm{f}$ the isospin factors that come into this two-nucleon transfer process are

$$b_{ST}^2(T_\mathrm{A}TT_{z\mathrm{A}}T_z \mid T_\mathrm{B}T_{z\mathrm{B}})^2 = \delta_{S,0}(T_\mathrm{i}1T_\mathrm{i}-11 \mid T_\mathrm{i}T_\mathrm{i})^2$$
$$= \delta_{S,0}(T_\mathrm{i}+1)^{-1}.$$

On the other hand, for the $(p, {}^3\mathrm{He})$ reaction on the same target nucleus

$$b_{ST}^2(T_\mathrm{A}TT_{z\mathrm{A}}T_z \mid T_\mathrm{B}T_{z\mathrm{B}})^2 = \tfrac{1}{2}\{\delta_{S,0}(T_\mathrm{i}1T_\mathrm{i}0 \mid T_\mathrm{i}T_\mathrm{i})^2$$
$$+\delta_{S,1}(T_\mathrm{i}0T_\mathrm{i}0 \mid T_\mathrm{i}T_\mathrm{i})^2\}$$
$$= \frac{T_\mathrm{i}}{2(T_\mathrm{i}+1)}\delta_{S,0}+\tfrac{1}{2}\delta_{S,1}$$

where the value of b_{ST}^2 has been taken from equation (4.102).

According to equation (4.100) the contributions from $S=0$ and $S=1$ should be added incoherently so that

$$\left\{\frac{Q_{\mathrm{DWBA}}(p,t)}{Q_{\mathrm{DWBA}}(p,{}^3\mathrm{He})}\right\}_{\Delta T=0} \leq \frac{2}{T_\mathrm{i}}.$$

When this result is combined with the kinematic factors that come into the cross-section one finds

$$\left\{\frac{\dfrac{\mathrm{d}\sigma}{\mathrm{d}\Omega}(p,t)}{\dfrac{\mathrm{d}\sigma}{\mathrm{d}\Omega}(p,{}^3\mathrm{He})}\right\}_{\Delta T=0} \leq \frac{k_\mathrm{t}}{k_{{}^3\mathrm{He}}}\frac{2}{T_\mathrm{i}} \tag{4.120}$$

where k_t and $k_{{}^3\mathrm{He}}$ are the wave numbers of the outgoing triton and ${}^3\mathrm{He}$ respectively, and the equality in the equation holds for processes in which only $S=0$ contributes to the $(p, {}^3\mathrm{He})$ reaction.

As pointed out by Hardy *et al.* (1969), equation (4.120) can be used to test whether a given transition involves only the transfer of a pair in the same single-particle state. If the target has spin 0^+, then as we have already discussed only even values of λ can contribute to the transfer process if positive-parity states are populated. Consequently if the reaction involves the transfer of a $(j^2)_{JT=1}$ pair, only $S=0$ can contribute to the $(p, {}^3\mathrm{He})$ process ($\gamma_{\lambda S}^{(J)}(jl; jl)$ given by equation (4.15) vanishes when

TABLE 4.8

Ratio of the (p, t) *to* $(p, {}^3\mathrm{He})$ *cross-sections for populating analogue states*

	Final states			Cross-section ratio	
Target	Nucleus	Excitation Energy (MeV)	(J, T)	Experiment	Theory
${}^{30}_{14}\mathrm{Si}_{16}$	${}^{28}_{14}\mathrm{Si}_{14}$ ${}^{28}_{13}\mathrm{Al}_{15}$	10·70 1·35	0,1	$1{\cdot}85 \pm 0{\cdot}20$	1·84
${}^{38}_{18}\mathrm{Ar}_{20}$	${}^{36}_{18}\mathrm{Ar}_{18}$ ${}^{36}_{17}\mathrm{Cl}_{19}$	9·70 3·12	0,1	$2{\cdot}20 \pm 0{\cdot}70$	1·80
${}^{34}_{16}\mathrm{S}_{18}$	${}^{32}_{16}\mathrm{S}_{16}$ ${}^{32}_{15}\mathrm{P}_{17}$	7·005 0·078	2,1	$1{\cdot}20 \pm 0{\cdot}30$	1·82
${}^{38}_{18}\mathrm{Ar}_{20}$	${}^{36}_{18}\mathrm{Ar}_{18}$ ${}^{36}_{17}\mathrm{Cl}_{19}$	8·55 1·949	2,1	$1{\cdot}45 \pm 0{\cdot}20$	1·81
${}^{22}_{10}\mathrm{Ne}_{12}$	${}^{20}_{10}\mathrm{Ne}_{10}$ ${}^{20}_{9}\mathrm{F}_{11}$	10·275 g.s.	2,1	$2{\cdot}00 \pm 0{\cdot}20$	1·88
${}^{38}_{18}\mathrm{Ar}_{20}$	${}^{36}_{18}\mathrm{Ar}_{18}$ ${}^{36}_{17}\mathrm{Cl}_{19}$	6·612 g.s.	2,1	$1{\cdot}90 \pm 0{\cdot}20$	1·81

The theoretical ratio is for $(j^2)_{JT=1}$ transfer and the data are taken from Hardy *et al.* (1969). g.s. stands for ground state.

$\lambda = J$ and $S = 1$). Thus for a given target if we compare the (p, t) and $(p, {}^3\mathrm{He})$ cross-sections for populating isobaric analogue states in the final nuclei, one can learn whether or not the process is dominated by $(j^2)_{JT=1}$ transfer.

The data concerning this have been collected by Hardy *et al.* (1969) and are presented in Table 4.8. In the last column the predicted value for $(j^2)_{JT=1}$ transfer is given based on the assumption that only $S = 0$ transfer takes place (i.e. the equality in equation (4.120) holds true). When the final state has spin 0^+ the transferred pair must have the same single-particle angular momentum. Consequently no $S = 1$ transfer is possible, and the equality must be true if the process is a direct-reaction one. On the other hand, when $J = 2$, $(j_1 j_2)_{JT=1}$ pairs may be transferred and this may proceed with $S = 1$ when $j_1 \neq j_2$. Thus if the experimental ratio is smaller than the prediction given in the last column of Table 4.8 one has clear evidence that $(j_1 j_2)_{JT=1}$ transfer is taking place. From the experimental results one sees that the yrast 2^+ $T = 1$ state in ${}^{32}_{15}\mathrm{P}_{17}$ and the 1·949 MeV 2^+ $T = 1$ state in ${}^{36}_{17}\mathrm{Cl}_{19}$ have a substantial component $(j_1 j_2)_{J=2}$ in their wave functions. On the other hand, the last two cases quoted in the table exhibit cross-sections that look *as if* only $(j^2)_{J=2}$ transfer is involved.

5

ELECTROMAGNETIC PROPERTIES

In this Chapter we shall consider static and dynamic electromagnetic properties of nuclei and discuss how they may be used to check configuration assignments. In particular a detailed discussion will be given concerning magnetic dipole and electric quadrupole moments and the selection rules for $E1$, $M1$, $E2$, and $M2$ gamma decay. It will also be shown how core excitation can modify the form of the free-particle operators and lead to the concept of an effective operator.

1. Single-particle estimate for gamma decay

In the presence of an electromagnetic field the Hamiltonian describing the motion of a shell-model particle of charge e and magnetic moment $\mu(e\hbar/2mc)$ takes the form

$$H=\frac{1}{2m}\left(\underset{\sim}{p}-\frac{e}{c}\underset{\sim}{A}\right)^2-\mu\frac{e\hbar}{2mc}\underset{\sim}{\sigma}\,.\,(\underset{\sim}{\nabla}\times\underset{\sim}{A})+V(r)+f(r)\underset{\sim}{\sigma}\,.\left\{\underset{\sim}{r}\times\left(\underset{\sim}{p}-\frac{e}{c}\underset{\sim}{A}\right)\right\} \tag{5.1}$$

where $\underset{\sim}{A}=\underset{\sim}{A}(\underset{\sim}{r})$ is the divergence-free vector potential describing the electromagnetic field at the position of the particle and $\times$ stands for the cross product encountered in ordinary vector analysis. Since the electric and magnetic field strengths are related to $\underset{\sim}{A}$ by

$$\underset{\sim}{\mathscr{E}}=-\frac{1}{c}\frac{\partial\underset{\sim}{A}}{\partial t}$$

$$\underset{\sim}{\mathscr{H}}=\underset{\sim}{\nabla}\times\underset{\sim}{A}$$

it follows that the second term, $\underset{\sim}{\sigma}\,.\,\underset{\sim}{\mathscr{H}}$, is just the interaction of a particle with dipole moment $\mu(e\hbar/2mc)$ with the magnetic field. The last term, $-(e/c)f(r)\underset{\sim}{\sigma}\,.\,(\underset{\sim}{r}\times\underset{\sim}{A})$, is the additional electromagnetic interaction which arises because the shell-model potential must have a strong one-body spin-orbit part $f(r)\underset{\sim}{\sigma}\,.\,(\underset{\sim}{r}\times\underset{\sim}{p})$ to reproduce the observed shell structure and level sequence.

In general the vector potential $\underset{\sim}{A}(\underset{\sim}{r})$ is expanded in terms of the complete set of functions

$$u_{LM} = j_L(kr)Y_{LM}(\theta, \phi)$$

where $j_L(kr)$ is the spherical Bessel function (Morse and Feshbach 1953) and $\hbar k$ is the momentum of the emitted photon. The energy E of a photon with momentum $\hbar k$ is $k = E/(\hbar c)$. Consequently for a large nucleus (say ^{208}Pb)

$$kr \cong \frac{E}{\hbar c}(1{\cdot}2)(208)^{\frac{1}{3}} \times 10^{-13} = 0{\cdot}036E$$

when E is measured in MeV. Thus for normal gamma-ray transitions with energy of the order of a few MeV, $kr \ll 1$ and hence we may take the long wavelength limit and replace u_{LM} by

$$u_{LM} \cong \frac{(kr)^L}{(2L+1)!!} Y_{LM}(\theta, \phi). \tag{5.2}$$

Throughout the discussion this approximation will be made.

In this chapter we shall deal exclusively with nuclear decays involving the emission of a single gamma quantum. Consequently only terms linear in $\underset{\sim}{A}$ will be retained. In this approximation the total interaction Hamiltonian governing the emission of radiation is given by

$$H_{\text{int}} = -\sum_k \left[\frac{e_k}{mc} \underset{\sim}{p}_k \cdot \underset{\sim}{A}(\underset{\sim}{r}_k) + \mu_k \mu_N \underset{\sim}{\sigma}_k \cdot \{\underset{\sim}{\nabla}_k \times \underset{\sim}{A}(\underset{\sim}{r}_k)\} + \frac{e_k}{c} f(r_k) \underset{\sim}{\sigma}_k \cdot \{\underset{\sim}{r}_k \times \underset{\sim}{A}(\underset{\sim}{r}_k)\}\right]. \tag{5.3}$$

In this equation the sum is over all nucleons in the nucleus. $\underset{\sim}{p}_k$, $\underset{\sim}{r}_k$, and $\underset{\sim}{\sigma}_k$ stand for the momentum, position coordinate, and Pauli spin operator of the kth particle. The charge e_k is the electric charge e if k is a proton and is zero if k is a neutron. μ_k is the magnetic moment of the nucleon ($\mu_k = 2{\cdot}79$ $(-1{\cdot}91)$ if k is a proton (neutron)) and

$$\mu_N = \frac{e\hbar}{2mc} \tag{5.4}$$

is the nuclear magneton.

The problem of dividing this operator into parts corresponding to the emission of electric and magnetic multipole radiation is tedious but straightforward and is discussed in many places (Heitler 1936, DeBenedetti 1964, Moszkowski 1966, Rose and Brink 1967, Eisenberg and Greiner 1970). The electric and magnetic parts of the field (so named because they correspond closely to the fields of a vibrating electric and magnetic multipole) are given respectively by

$$A^{(\mathscr{E})}_{LM} = \frac{1}{k} \underset{\sim}{\nabla} \times \underset{\sim}{\ell} u_{LM}$$

$$A^{(\mathscr{M})}_{LM} = i \underset{\sim}{\ell} u_{LM}$$

where $\underset{\sim}{\ell}$ is the orbital angular-momentum operator

$$\underset{\sim}{\ell} = -i \underset{\sim}{r} \times \underset{\sim}{\nabla}.$$

In the long wavelength limit given by equation (5.2) the operator governing the emission of electric multipole radiation of order L with z component M from the kth particle is

$$(T^{(\mathscr{E})}_k)_{LM} = er^L Y_{LM}(\theta, \phi) + i\mu_k\mu_N \frac{E}{\hbar c}\left(\frac{1}{L+1}\right)(\underset{\sim}{\sigma} \times r) . \{\underset{\sim}{\nabla} r^L Y_{LM}(\theta, \phi)\}.$$

The second term in this expression can be simplified if one notes that

$$(\underset{\sim}{\sigma} \times \underset{\sim}{r})_\nu = -i\left(\frac{8\pi}{3}\right)^{\frac{1}{2}} r[\sigma \times Y_1]_{1\nu}.$$

When this is combined with the fact that $\underset{\sim}{a} . \underset{\sim}{b} = -\sqrt{3}[a \times b]_{00}$, it follows that

$$(\underset{\sim}{\sigma} \times \underset{\sim}{r}) . (\underset{\sim}{\nabla} r^L Y_{LM}(\theta, \phi)) = i\sqrt{(8\pi)} r$$

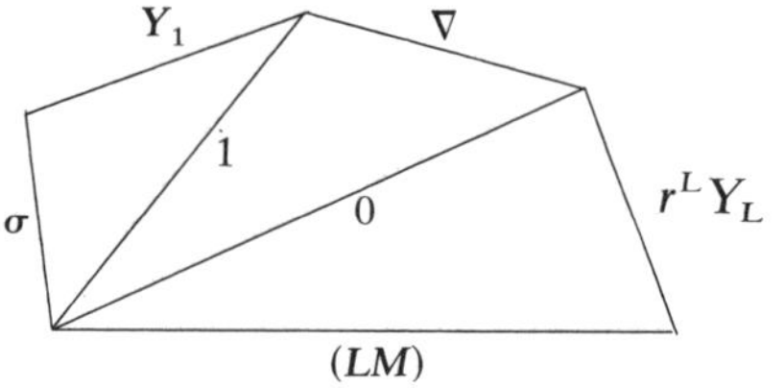

$$= i\sqrt{(8\pi)} r \sum_{L'} \sqrt{(2L'+1)}\, W(11LL; 0L') \langle Y_{L'} \| \nabla \| r^L Y_L \rangle$$

$$\times$$

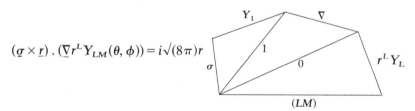

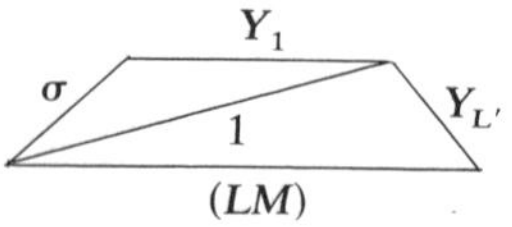

The value of the reduced matrix element in this equation can be obtained from equation (A2.14)

$$\langle Y_{L'} \| \nabla \| r^L Y_L \rangle = -(2L+1) r^{L-1} \left(\frac{L}{2L-1}\right)^{\frac{1}{2}} \delta_{L', L-1}$$

Thus

$$(\underset{\sim}{\sigma} \times \underset{\sim}{r}) \cdot (\nabla r^L Y_{LM}(\theta, \phi)) = -i\left\{\frac{8\pi L(2L+1)}{3}\right\}^{\frac{1}{2}} r^L$$

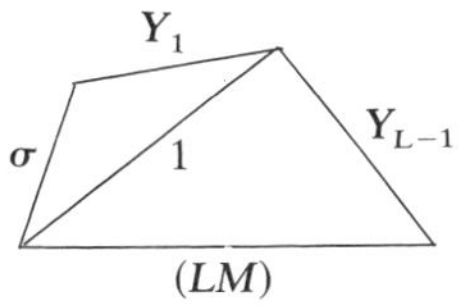

$$= -i\sqrt{\{8\pi L(2L+1)\}} r^L \sum_{L''} (-1)^{1+L''-L} \sqrt{(2L''+1)}\, W(11LL-1; 1L'')$$

$$\times \langle Y_{L''} \| Y_1 \| Y_{L-1} \rangle$$

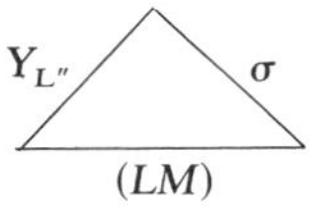

$$= i\sqrt{\{L(L+1)\}} \sum_{L''} \delta_{L'',L}$$

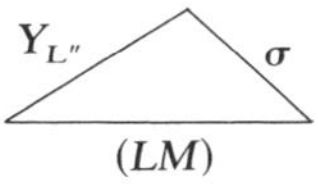

where use has been made of equation (A2.11) to evaluate the reduced matrix element and Table A4.2 to find the Racah coefficient. Consequently

$$(T_k^{(\mathscr{E})})_{LM} = er^L Y_{LM}(\theta, \phi) - \mu_k \mu_N \frac{E}{\hbar c} r^L \left(\frac{L}{L+1}\right)^{\frac{1}{2}} [Y_L(\theta, \phi) \times \sigma]_{LM} \tag{5.5}$$

where E is the gamma-ray energy which is equal to the difference in energy between the initial and final nulcear states involved in the transition.

The fact that there is no term proportional to $f(r)$ in the electric multipole operator is a consequence of gauge invariance. This result, which was originally demonstrated by Sachs and Austern (1951), is rigorously true in the long wavelength limit.

From equation (5.5) it follows that the emission of electric multipole radiation of order L can only take place between two nuclear states that satisfy the condition

$$|I_i - I_f| = \Delta I \leq L; \qquad \Delta(\text{parity}) = (-1)^L \quad \text{for } EL \text{ radiation.} \tag{5.6}$$

Furthermore, since the total operator governing the transition is a sum of the single-particle operators, this selection rule holds true not only for the total angular momentum and parity but also for the individual nucleons involved in the transition. Thus, for example, the transition

$$(\pi p_{\frac{1}{2}}, \pi g_{\frac{9}{2}})_{5-} \xrightarrow{E1} (\pi g_{\frac{9}{2}})^2_{4+}$$

would be allowed on the basis of overall angular-momentum conservation, but cannot occur because the $E1$ operator cannot change $p_{\frac{1}{2}}$ to $g_{\frac{9}{2}}$.

As we shall see later this type of selection rule allows us to tell whether a gross error has been made in a configuration assignment.

The second term in equation (5.5) is considerably smaller than the first. The ratio of the contribution from the second to the first is approximately

$$\mathcal{R} = \frac{\mu_k E}{2mc^2} \cong 1{\cdot}5 \times 10^{-3},$$

where the numerical value is given for a 1 MeV photon emitted by a proton. Consequently this term is usually dropped from further consideration.

The operator for the kth particle, which is responsible for the emission of magnetic multipole radiation, can be written in the form

$$(T_k^{(\mathcal{M})})_{LM} = \left\{\frac{e_k \hbar}{(L+1)mc}\right\} \underset{\sim}{\ell} \,.\, (\underset{\sim}{\nabla} r^L Y_{LM}) + \mu_k \mu_N \underset{\sim}{\sigma} \,.\, (\underset{\sim}{\nabla} r^L Y_{LM})$$

$$- \frac{e_k}{\hbar c}\left\{\frac{f(r)}{L+1}\right\} \underset{\sim}{\sigma} \,.\, \{\underset{\sim}{r} \times (\underset{\sim}{r} \times \underset{\sim}{\nabla} r^L Y_{LM})\}$$

and by the same arugments used in deducing equation (5.5) one may show that this can be written as

$$\begin{aligned}(T_k^{(\mathcal{M})})_{LM} &= \sqrt{\{L(2L+1)\}}\, r^{L-1} \\ &\quad \times \left\{\frac{1}{L+1}\left(\frac{e_k \hbar}{mc}\right)[Y_{L-1} \times \ell]_{LM} + \mu_k \mu_N [Y_{L-1} \times \sigma]_{LM}\right\} \\ &\quad + \frac{e_k}{\hbar c} f(r) \left\{\frac{L}{(L+1)(2L+1)}\right\}^{\frac{1}{2}} r^{L+1} (\sqrt{L}[Y_{L+1} \times \sigma]_{LM} \\ &\quad + \sqrt{(L+1)}[Y_{L-1} \times \sigma]_{LM}). \end{aligned} \quad (5.7)$$

From an inspection of this equation it is apparent that the selection rule governing the emission of magnetic multipole radiation of order L is

$$|I_i - I_f| = \Delta I \leq L; \qquad \Delta(\text{parity}) = (-1)^{L-1} \quad \text{for } ML \text{ radiation.} \quad (5.8)$$

There is no result like the theorem of Sachs and Austern (1951) that can be applied to magnetic radiation, and hence for protons there may be a contribution to the interaction proportional to the spin–orbit potential (Jensen and Mayer 1952). Qualitatively the ratio of this term to the $\mu_k = \mu_p$ term is

$$\mathcal{R}' \cong \left(\frac{2}{2L+1}\right) \frac{m}{\hbar^2} \frac{R^2 f(R)}{\mu_p}$$

where R is the nuclear radius. To estimate $\mathcal{R}'$ assume for simplicity that the central nuclear potential is an harmonic oscillator well and that the spin–orbit potential has the Thomas form (Thomas 1926, Inglis 1936); i.e.

$$f(r)\underset{\sim}{\sigma}\,.\,\underset{\sim}{\ell}=\lambda\frac{1}{r}\frac{\partial V}{\partial r}\underset{\sim}{\sigma}\,.\,\underset{\sim}{\ell}$$

with λ a constant and V the oscillator potential $V=\frac{1}{2}m\omega^2r^2$. Since

$$\begin{aligned}\underset{\sim}{\sigma}\,.\,\underset{\sim}{\ell}\phi_{jlm} &= l\phi_{jlm} &&\text{if } j=l+\tfrac{1}{2}\\ &= -(l+1)\phi_{jlm} &&\text{if } j=l-\tfrac{1}{2}\end{aligned}$$

it follows that the Thomas term leads to a splitting between spin–orbit partners of

$$\Delta E=\lambda m\omega^2(2l+1).$$

For ${}^{41}_{20}\mathrm{Ca}_{21}$, $\Delta E\simeq 6$ MeV for the $f_{\frac{7}{2}}$–$f_{\frac{5}{2}}$ splitting (Endt and Van der Leun 1973). Because $\omega\sim 1/A^{\frac{1}{3}}$ it follows that

$$f(r)=\lambda m\omega^2\simeq-\frac{6}{7}\left(\frac{41}{A}\right)^{\frac{2}{3}}=-10{\cdot}2/A^{\frac{2}{3}}\,\text{MeV}.$$

Since $R^2\cong(1{\cdot}2)^2A^{\frac{2}{3}}\times10^{-26}\,\text{cm}^2$, $\mathcal{R}'$ is roughly independent of A and has the value

$$\mathcal{R}'\cong\frac{0{\cdot}254}{(2L+1)}.$$

This correction, which is small and decreases rapidly with L, is generally omitted in calculations. In addition, as we shall see both meson exchange and core polarization effects lead to a term of similar form (see section 2.1). Consequently it is difficult to isolate the contribution of this term alone and therefore when we discuss magnetic multipole properties we shall use the conventional operator in which $f(r)=0$. That is

$$(T_k^{(\mathcal{M})})_{LM}=\sqrt{\{L(2L+1)\}}r^{L-1}\left\{\frac{1}{L+1}\left(\frac{e_k\hbar}{mc}\right)[Y_{L-1}\times\ell]_{LM}+\mu_k\mu_N[Y_{L-1}\times\sigma]_{LM}\right\}. \tag{5.9}$$

The gamma-ray transition probability is proportional to the square of the matrix element of $T_{LM}^{(\mathscr{E})}$ or $T_{LM}^{(\mathcal{M})}$ summed over final and averaged over initial magnetic substates. Since this transition probability is dependent on the energy of the emitted gamma ray it is convenient to define the reduced transition probability

$$\begin{aligned}B(\xi L;I_\mathrm{i}\to I_\mathrm{f})&=\frac{1}{(2I_\mathrm{i}+1)}\sum_{M_\mathrm{i}M_\mathrm{f}M}|\langle\Psi_{I_\mathrm{f}M_\mathrm{f}}|\,T_{LM}^{(\xi)}\,|\Psi_{I_\mathrm{i}M_\mathrm{i}}\rangle|^2\\ &=\left(\frac{2I_\mathrm{f}+1}{2I_\mathrm{i}+1}\right)|\langle\Psi_{I_\mathrm{f}}\|\,T_L^{(\xi)}\,\|\Psi_{I_\mathrm{i}}\rangle|^2\end{aligned} \tag{5.10}$$

with $T^{(\xi)}_{LM}$ given by equations (5.5) and (5.9) for electric and magnetic radiation, respectively. The reduced transition probability satisfies the relationship

$$(2I_1+1)B(\xi L; I_1 \to I_2) = (2I_2+1)B(\xi L; I_2 \to I_1). \tag{5.11}$$

The transition probability/unit time $\omega(\xi)$ for electromagnetic decay of type ξ is proportional to the reduced transition probability. The factors involved in the proportionality depend on the density of photon states and the normalization of the vector potential. When these factors are included one finds

$$w(\xi) = \frac{8\pi(L+1)}{L\{(2L+1)!!\}^2} \frac{1}{\hbar} \left(\frac{E}{\hbar c}\right)^{2L+1} B(\xi L; I_i \to I_f) \tag{5.12}$$

where E, the energy of the emitted photon, is the difference in energy between the initial and final nuclear states

$$E = E_i - E_f.$$

The mean lifetime of the state is then given by

$$\tau_m(\xi) = 1/w(\xi). \tag{5.13}$$

In Table 5.1 we tabulate the numerical values that relate $B(\xi L; I_i \to I_f)$ to the transition probability/unit time.

TABLE 5.1

The relationship between w, the transition probability/unit time, and $B(\xi L)$

Radiation multipolarity L	$w/B(\xi L; I_i \to I_f)$ Electric	Magnetic
1	$1{\cdot}59 \times 10^{15}\, E^3$	$1{\cdot}76 \times 10^{13}\, E^3$
2	$1{\cdot}23 \times 10^{9}\, E^5$	$1{\cdot}35 \times 10^{7}\, E^5$
3	$5{\cdot}72 \times 10^{2}\, E^7$	$6{\cdot}31 \times 10^{0}\, E^7$
4	$1{\cdot}70 \times 10^{-4}\, E^9$	$1{\cdot}88 \times 10^{-6}\, E^9$
5	$3{\cdot}47 \times 10^{-11}\, E^{11}$	$3{\cdot}82 \times 10^{-13}\, E^{11}$
6	$5{\cdot}12 \times 10^{-18}\, E^{13}$	$5{\cdot}65 \times 10^{-20}\, E^{13}$

In these formulae the gamma-ray energy is measured in MeV. For electric multipole transitions $B(EL)$ is to be taken in units of $e^2(\text{fermis})^{2L}$. For magnetic transitions the $B(ML)$ are in units of $(\text{nuclear magnetons})^2 \times (\text{fermis})^{2L-2} = \mu_N^2 \times (\text{fermis})^{2L-2}$. When these units are used $\tau_m = 1/w$ is calculated in seconds.

Before proceeding to a detailed discussion of the various multipole decays it is convenient to obtain the single-particle estimate for the transition rate. To do this we assume a single particle outside an inert core and calculate the transition probability/unit time for emission of radiation of multipolarity L. If we neglect the second term in equation (5.5) we obtain from equation (A2.23)

$$B_{\rm sp}(EL;j_{\rm i}\rightarrow j_{\rm f})=e^2\left(\frac{2L+1}{4\pi}\right)(j_{\rm i}L\tfrac{1}{2}0\mid j_{\rm f}\tfrac{1}{2})^2\left|\int R_{j_{\rm f}}(r)R_{j_{\rm i}}(r)r^{L+2}\,{\rm d}r\right|^2$$

$$=e^2\left(\frac{2L+1}{4\pi}\right)(j_{\rm i}L\tfrac{1}{2}0\mid j_{\rm f}\tfrac{1}{2})^2\left(\frac{3}{L+3}\right)^2R^{2L} \qquad (5.14)$$

for the emission of electric radiation. In writing the second line of this equation we have assumed that the radial wave functions of both the initial and final states are constant, $(3/R^3)^{\frac{1}{2}}$, out to the nuclear radius R and then drop discontinuously to zero. When this result is inserted into equations (5.12) and (5.13) the single-particle lifetime estimate is obtained

$$\frac{1}{\tau_{\rm sp}(EL)}=w_{\rm sp}(EL)=\frac{2(L+1)}{L\{(2L+1)!!\}^2}\left(\frac{3}{L+3}\right)^2\left(\frac{e^2}{\hbar c}\right)c\left(\frac{E}{\hbar c}\right)^{2L+1}$$

$$\times R^{2L}\mathcal{S}_L(j_{\rm i}\rightarrow j_{\rm f}) \qquad (5.15)$$

where $\mathcal{S}_L(j_{\rm i}\rightarrow j_{\rm f})$, the statistical factor, is defined by

$$\mathcal{S}_L(j_{\rm i}\rightarrow j_{\rm f})=(2L+1)(j_{\rm i}L\tfrac{1}{2}0\mid j_{\rm f}\tfrac{1}{2})^2. \qquad (5.16)$$

If one sets $\mathcal{S}_L(j_{\rm i}\rightarrow j_{\rm f})=1$ the Weisskopf single-particle estimate is obtained (Weisskopf 1951). On the other hand, if one uses the value of $\mathcal{S}_L(j_{\rm i}\rightarrow j_{\rm f})$ given by equation (5.16) one obtains the single-particle Moszkowski estimate (Moszkowski 1953). In Table 5.2 we list the values of $w_{\rm sp}(EL)$ on the assumption that the nuclear radius is $1{\cdot}2A^{\frac{1}{3}}$ fm. From this table it is clear that the transition probability decreases rapidly with increasing multipolarity and consequently the gamma-ray lifetime becomes longer with increasing multipolarity.

To evaluate the single-particle matrix element for magnetic multipole radiation we make use of the completeness relationship for the Clebsch–Gordan coefficients (equation (A1.24)) to write

$$T^{(\mathcal{M})}_{LM}\phi_{j_{\rm i}m_{\rm i}}=\sum_{j''m''}(j_{\rm i}Lm_{\rm i}M\mid j''m'')\sum_{m_{\rm i}'M'}(j_{\rm i}Lm_{\rm i}'M'\mid j''m'')T^{(\mathcal{M})}_{LM'}\phi_{j_{\rm i}m_{\rm i}'}$$

TABLE 5.2

Numerical values for the single-particle transition probability/unit time given by equations (5.15) *and* (5.20)

Radiation multipole L	$w_{sp}/\mathcal{S}_L(j_i \to j_f)$ Electric	Magnetic Weisskopf	Magnetic Moszkowski
1	$1{\cdot}0\times10^{14}\,A^{\frac{2}{3}}E^3$	$3{\cdot}1\times10^{13}\,E^3$	$2{\cdot}9\times10^{13}\,E^3$
2	$7{\cdot}3\times10^{7}\,A^{\frac{4}{3}}E^5$	$2{\cdot}2\times10^{7}\,A^{\frac{2}{3}}E^5$	$8{\cdot}4\times10^{7}\,A^{\frac{2}{3}}E^5$
3	$3{\cdot}4\times10^{1}\,A^{2}E^7$	$1{\cdot}0\times10^{1}\,A^{\frac{4}{3}}E^7$	$8{\cdot}7\times10^{1}\,A^{\frac{4}{3}}E^7$
4	$1{\cdot}1\times10^{-5}\,A^{\frac{8}{3}}E^9$	$3{\cdot}3\times10^{-6}\,A^{2}E^9$	$4{\cdot}8\times10^{-5}\,A^{2}E^9$
5	$2{\cdot}4\times10^{-12}\,A^{\frac{10}{3}}E^{11}$	$7{\cdot}4\times10^{-13}\,A^{\frac{8}{3}}E^{11}$	$1{\cdot}7\times10^{-11}\,A^{\frac{8}{3}}E^{11}$
6	$4{\cdot}0\times10^{-19}\,A^{4}E^{13}$	$1{\cdot}2\times10^{-19}\,A^{\frac{10}{3}}E^{13}$	$4{\cdot}0\times10^{-18}\,A^{\frac{10}{3}}E^{13}$

If E is the energy of the emitted gamma ray in MeV and A is the number of nucleons in the nucleus, $1/w_{sp}$ gives the mean lifetime for the decay in seconds. $\mathcal{S}_L(j_i \to j_f)$ is equal to one for the Weisskopf estimate and is given by equation (5.16) for the Moszkowski estimate. The nuclear radius $R = 1{\cdot}2A^{\frac{1}{3}}$ fm.

so that in terms of diagrams this becomes

$$T^{(\mathcal{M})}_{LM}\phi_{j_i m_i} = \sum_{j''m''} (j_i L m_i M \mid j''m'')$$ 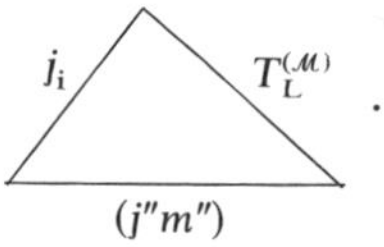

(5.17)

When this is combined with the fact that $[Y_{L-1}\times \ell]_{LM}$ may be written as

$$[Y_{L-1}\times \ell]_{LM} = [Y_{L-1}\times \mathrm{j}]_{LM} - \tfrac{1}{2}[Y_{L-1}\times \sigma]_{LM}$$

one sees that the reduced matrix element of the $[Y_{L-1}\times\sigma]_{LM}$ part of the operator can be written down at once by using equation (A2.24). On the other hand, to evaluate the $[Y_{L-1}\times \mathrm{j}]_{LM}$ part requires the use of equation (5.17). Thus

$$[Y_{L-1}\times \mathrm{j}]_{LM}\phi_{j_i m_i} = \sum_{j''m''} (j_i L m_i M \mid j''m'')$$ 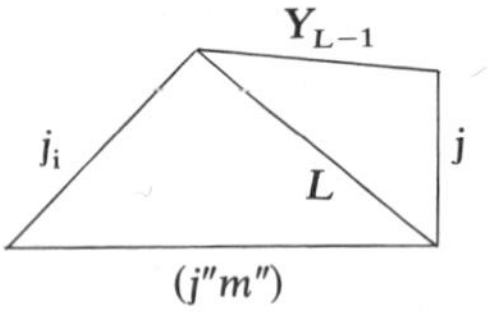

$$= \sum_{j''m''} \sqrt{\{j_i(j_i+1)(2j_i+1)(2L+1)\}}\,W(j_i 1 j'' L-1;\, j_i L)(j_i L m_i M \mid j''m'')$$ 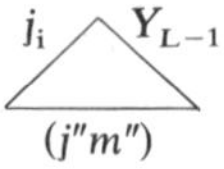

$$= \sum_{j''m''} (-1)^{l_i+l''+j_i-j''}(2j_i+1)\left\{\frac{(2L-1)(2L+1)j_i(j_i+1)}{4\pi(2j''+1)}\right\}^{\frac{1}{2}}(j_i L-1\tfrac{1}{2}0 \mid j''\tfrac{1}{2})$$

$$\times\, W(j_i 1 j'' L-1;\, j_i L)(j_i L m_i M \mid j''m'')$$ 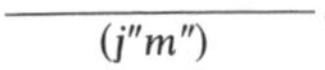

.

Since we are interested in $\langle\phi_{j_f m_f}|[Y_{L-1}\times j]_{LM}|\phi_{j_i m_i}\rangle$ it follows that the reduced matrix element is just the coefficient of $(j_i L m_i M \mid j'' m'')$ with $j''=j_f$, $m''=m_f$. Thus

$$\langle\phi_{j_f}\|\,T_L^{(\mathcal{M})}\,\|\phi_{j_i}\rangle=(-1)^{l_i+l_f+j_i-j_f}\left\{\frac{L(2L+1)(2j_i+1)}{4\pi(2j_f+1)}\right\}^{\frac{1}{2}}$$

$$\times\int R_{j_f}(r)r^{L-1}R_{j_i}(r)r^2\,\mathrm{d}r\left[\frac{1}{L+1}\left(\frac{e_k\hbar}{mc}\right)\right.$$

$$\times\sqrt{\{j_i(j_i+1)(2j_i+1)(2L-1)(2L+1)\}}$$

$$\times(j_i L-1\tfrac{1}{2}0\mid j_f\tfrac{1}{2})W(j_i 1 j_f L-1;\,j_i L)+\sqrt{L}\left\{\mu_k\mu_N-\frac{e_k\hbar}{2mc(L+1)}\right\}$$

$$\left.\times(j_i L\tfrac{1}{2}0\mid j_f\tfrac{1}{2})\left\{1+\frac{(-1)^{\frac{1}{2}+l_i-j_i}\eta_L(j_i,j_f)}{L}\right\}\right] \tag{5.18}$$

where $\eta_L(j_i, j_f)$ is

$$\eta_L(j_i,j_f)=\tfrac{1}{2}\{(2j_i+1)+(-1)^{j_i+j_f-L}(2j_f+1)\}.$$

Equation (5.18) is the general form for the single-particle matrix element. For the case

$$|j_i-j_f|=L;\qquad |l_i-l_f|=L-1,$$

the matrix element takes a particularly simple form and it is for this condition that the single-particle estimate is evaluated. In this case the first term in equation (5.18) does not contribute. Moreover, for these conditions to be satisfied, the only transitions that can take place are those for which

$$j_i=l_i+\tfrac{1}{2}\rightarrow j_f=l_f-\tfrac{1}{2}$$

or

$$j_i=l_i-\tfrac{1}{2}\rightarrow j_f=l_f+\tfrac{1}{2}.$$

In both cases

$$(-1)^{\frac{1}{2}+l_i-j_i}\eta_L(j_i,j_f)=\mathrm{L}.$$

Consequently if one assumes a constant radial wave function, $(3/R^3)^{\frac{1}{2}}$, out of the nuclear surface, one obtains for the proton single particle $B(ML)$

$$B_{\mathrm{sp}}(ML;\,j_i\rightarrow j_f)=\frac{1}{2j_i+1}\sum_{m_i m_f M}|\langle\phi_{j_f m_f}|\,T_{LM}^{(\mathcal{M})}\,|\phi_{j_i m_i}\rangle|^2$$

$$=(\mu_N^2/\pi)\left(\mu_p L-\frac{L}{L+1}\right)^2\left(\frac{3}{L+2}\right)^2 R^{2L-2}\mathcal{S}_L(j_i\rightarrow j_f) \tag{5.19}$$

where $\mathcal{S}_L(j_i \rightarrow j_f)$ is given by equation (5.16). By combining this result with equations (5.12) and (5.13) one obtains the Moszkowski estimate for the single-particle magnetic multipole lifetime

$$\frac{1}{\tau_{sp}(ML)} = w_{sp}(ML) = \frac{2(L+1)}{L\{(2L+1)!!\}^2}\left(\frac{e^2}{\hbar c}\right)c\left(\frac{E}{\hbar c}\right)^{2L+1}$$

$$\times\left(\frac{\hbar}{mc}\right)^2\left(\frac{3}{L+2}\right)^2\left(\mu_p L - \frac{L}{L+1}\right)^2 R^{2L-2}\mathcal{S}_L(j_i \rightarrow j_f). \quad (5.20)$$

If one replaces $\mathcal{S}_L(j_i \rightarrow j_f)$ by unity and further replaces

$$\left\{\frac{1}{L+2}\left(\mu_p L - \frac{L}{L+1}\right)\right\}^2 \quad \text{by} \quad \frac{10}{(L+3)^2}$$

one obtains the Weisskopf single-particles lifetime estimate.

In Table 5.2 we have tabulated the values of $w_{sp}(ML)$ for both the Weisskopf and Moszkowski estimates. Again with increasing multipolarity, the transition probability/unit time rapidly decreases in value.

One can use these single-particle estimates, together with the fact that the transition-rate operator is a sum of single-particle operators, to determine whether a particular configuration assignment is grossly in error. For example, in section 2.2 of Chapter 2 we calculated the spectrum of ${}^{35}_{17}Cl_{18}$ that would arise from the $(d_{\frac{3}{2}})^3$ configuration. States with $I = \frac{5}{2}^+$, $T = \frac{1}{2}$ and $I = \frac{7}{2}^+$, $T = \frac{1}{2}$ were predicted at excitation energies of 2·823 and 2·508 MeV, respectively, and levels with these spins have been observed in this nucleus at 3·001 and 2·646 MeV. A further test of the configuration assignment is provided by the gamma decay of the 3·163 MeV $\frac{7}{2}^-$ state to these levels. Since this latter state is seen strongly with $l = 3$ in the ${}^{34}_{16}S_{18}({}^3He, d){}^{35}_{17}Cl_{18}$ reaction, it follows that the gamma transition should be inhibited since the $E1$ operator cannot change an $f_{\frac{7}{2}}$ nucleon to a $d_{\frac{3}{2}}$ orbit. The experimental $E1$ lifetimes for these transitions are (Endt and Van der Leun 1974)

$$\tau_{expt}(E1; \tfrac{7}{2}^- \rightarrow \tfrac{5}{2}^+) = 2940 \times 10^{-12}\ \text{s},$$
$$\tau_{expt}(E1; \tfrac{7}{2}^- \rightarrow \tfrac{7}{2}^+) = 625 \times 10^{-12}\ \text{s}.$$

By using Table 5.1 it follows that the Weisskopf estimate for these lifetimes is

$$\tau_W(E1; \tfrac{7}{2}^- \rightarrow \tfrac{5}{2}^+) = 0{\cdot}22 \times 10^{-12}\ \text{s},$$
$$\tau_W(E1; \tfrac{7}{2}^- \rightarrow \tfrac{7}{2}^+) = 6{\cdot}8 \times 10^{-15}\ \text{s}.$$

Thus the transitions are inhibited by $1{\cdot}3\times10^4$ and $9{\cdot}2\times10^4$, respectively. Consequently the $E1$ gamma decay is consistent with the $(d_{\frac{3}{2}})^3$ description of the positive-parity states.

Except for the statistical factor, the Weisskopf and Moszkowski estimates are the same for electric radiation. On the other hand, for magnetic transitions the Moszkowski estimate for the transition probability is larger and consequently the single-particle lifetime based on this estimate is shorter than the Weisskopf limit. Thus when quoting a single-particle estimate it is important to state whether the Weisskopf or Moszkowski estimate is being given and to give the value of the nuclear radius R used in making the estimate. A nomogram to find both electric and magnetic lifetimes based on the Weisskopf estimate has been given by Wilkinson (1960).

In addition, there are two other frequently encountered ways of quoting lifetimes. The first of these is in terms of the radiative width Γ_γ of a state. If Γ_γ is given in electron volts, the mean lifetime of the state in seconds is

$$\tau_m(s) = 6{\cdot}58\times10^{-16}/\Gamma_\gamma(\text{eV}). \tag{5.21}$$

In addition, the half-life is often quoted and this is related to the mean lifetime by

$$\tau_{\frac{1}{2}} = 0{\cdot}693\tau_m. \tag{5.22}$$

2. Effective operators

In shell-model calculations one always deals with a truncated model space, and hence to make up for this fact one must introduce effective operators which take care of the mixing of configurations not explicitly included in the calculations. It is, of course, hoped that the model space used is a good approximation and that the neglected configuration mixing can be treated in perturbation theory. In this section we discuss the large effects weak-configuration mixing can have on calculated nuclear properties—a result first recognized by Blin Stoyle and Arima and Horie (Blin Stoyle 1953, Blin Stoyle and Perks 1954, Arima and Horie 1954, Noya *et al.* 1958). We also show the conditions under which these perturbations can give rise to state independent effective operators; i.e. operators that describe the properties of several adjoining nuclei and several transitions in a given nucleus.

2.1. Unlike-particle mixing

To start we consider the case that the properties of the low-lying states can be attributed to the motion of only one type of nucleon (say the neutrons), the other type forming an inert $J_c = 0$ core. We want to know what change occurs in the neutron single-particle operators when proton core excitation is considered. We assume the probability of proton excitation is small and consequently look for effects that are linear in the admixed configurations. If we are interested in calculating the matrix elements of $T^{(\xi)}_{\lambda\mu}$ between an initial state $\Phi_{J_c=0}\psi_{I_iM_i}$, where $\Phi_{J_c=0}$ is the proton core state and $\psi_{I_iM_i}$ is the neutron eigenfunction, and a final state $\Phi_{J_c=0}\psi_{I_fM_f}$, the perturbations that can contribute linearly in the mixing amplitudes are

$$\Psi_{I_iM_i} = \Phi_0\psi_{I_iM_i} + \sum_q \alpha_{I_iq}[\Phi_{\lambda q} \times \psi_{I_f}]_{I_iM_i} \tag{5.23a}$$

$$\Psi_{I_fM_f} = \Phi_0\psi_{I_fM_f} + \sum_q \alpha_{I_fq}[\Phi_{\lambda q} \times \psi_{I_i}]_{I_fM_f} \tag{5.23b}$$

where q denotes the various possible core states with angular momentum $J_c = \lambda$.

Since the operators of interest have integral values of λ and transform like the spherical harmonics, equation (A2.10) takes the form

$$\sqrt{(2J'+1)}(-1)^{J'-J}\langle\Psi_{J'}\| T^{(\xi)}_{\lambda} \|\Psi_J\rangle = \sqrt{(2J+1)}\langle\Psi_J\| T^{(\xi)}_{\lambda} \|\Psi_{J'}\rangle. \tag{5.24}$$

This result, together with some straightforward Racah algebra, leads to the expression for the reduced matrix element including terms linear in α_{Iq}

$$\langle\Psi_{I_f}\| T^{(\xi)}_{\lambda} \|\Psi_{I_i}\rangle = \langle\psi_{I_f}\| \sum_n (T^{(\xi)}_n)_\lambda \|\psi_{I_i}\rangle + \sum_q \left\{\left(\frac{2I_i+1}{2I_f+1}\right)^{\frac{1}{2}}\alpha_{I_iq} + (-1)^{I_f-I_i}\alpha_{I_fq}\right\}$$
$$\times(-1)^\lambda\langle\Phi_{\lambda q}\| \sum_n (T^{(\xi)}_n)_\lambda \|\Phi_0\rangle. \tag{5.25}$$

First assume the residual two-body force that induces the mixing is spin independent. In that case a Slater decomposition of the potential can be made (see equation (4.1)). Since we consider unlike-particle mixing, antisymmetrization is not important and consequently only the $l = \lambda$ multipole of the decomposition contributes to α_{Iq}. Thus

$$\langle[\Phi_\lambda \times \psi_{I'}]_{IM}|\, V\, |\Phi_0\psi_{IM}\rangle$$
$$= 4\pi(-1)^\lambda \sum_{kn} \frac{\langle|F_\lambda(r_k, r_n)|\rangle}{2\lambda+1} \sqrt{\left(\frac{2I'+1}{2I+1}\right)}\langle\Phi_{\lambda q}\| Y_\lambda(\theta_k, \phi_k) \|\Phi_0\rangle$$
$$\times\langle\psi_{I'}\| Y_\lambda(\theta_n, \phi_n) \|\psi_I\rangle \tag{5.26}$$

where $\langle|F_\lambda(r_k, r_n)|\rangle$ stands for $F_\lambda(r_k, r_n)$ integrated over the radial coordinates of the kth core and nth valence particle. If $E_\lambda(q)$ denotes the excitation energy of the core state, this matrix element divided by $[E_I - \{E_{I'} + E_\lambda(q)\}]$ is α_{Iq}. Consequently

$$\langle\Psi_{I_f}\|\, T_\lambda^{(\xi)}\,\|\Psi_{I_i}\rangle = \langle\psi_{I_f}\|\sum_n \{(T_n^{(\xi)})_\lambda + C_\lambda(r_n)Y_\lambda(\theta_n, \phi_n)\}\,\|\psi_{I_i}\rangle \tag{5.27}$$

where

$$C_\lambda(r_n) = \frac{4\pi}{2\lambda+1}\sum_q \left\{\frac{1}{E_{I_i} - E_{I_f} - E_\lambda(q)} + \frac{1}{E_{I_f} - E_{I_i} - E_\lambda(q)}\right\}$$
$$\times \langle\Phi_{\lambda q}\|\sum_k F_\lambda(r_k, r_n)Y_\lambda(\theta_k, \phi_k)\,\|\,\Phi_0\rangle\,\langle\Phi_{\lambda q}\|\sum_{k'}(T_{k'}^{(\xi)})_\lambda\,\|\,\Phi_0\rangle \tag{5.28}$$

and by including $F_\lambda(r_k, r_n)$ inside the reduced matrix element we imply an integration over r_k as well as the angular coordinates of k.

Thus in perturbation theory the operator $(T_n^{(\xi)})_{\lambda\mu}$ must be replaced by the effective operator

$$(T_n^{(\xi)})_{\lambda\mu} \to (T_n^{(\xi)})_{\lambda\mu} + C_\lambda(r_n)Y_{\lambda\mu}(\theta_n, \phi_n). \tag{5.29}$$

Furthermore, in the case that $|E_\lambda(q)| \gg |E_{I_i} - E_{I_f}|$ for all I_i and I_f of interest the sum of the two energy denominators may be approximated by $\{-2/E_\lambda(q)\}$ and consequently the effective operator given by equation (5.29) would become state independent.

From this it is clear that, for example, a neutron which has no electric charge will be endowed with an effective charge proportional to $C_\lambda(r_n)$ due to its polarizing effect on the proton core.

We may also have core polarization induced by a spin-dependent potential. In this case the added operator in equation (5.29) would also have a spin-dependent part

$$(T_n^{(\xi)})_{\lambda\mu} \to (T_n^{(\xi)})_{\lambda\mu} + C_\lambda(r_n)Y_{\lambda\mu}(\theta_n, \phi_n) + \sum_l C_{\lambda l}(r_n)[\sigma \times Y_l(\theta_n, \phi_n)]_{\lambda\mu} \tag{5.30}$$

where $C_{\lambda l}(r_n)$ is given by the multipole decomposition of the more complicated spin-dependent potential. The values of l that can enter are, of course, governed by parity considerations. Thus a term similar to the $f(r)$ contribution in equation (5.7) arises from core polarization.

2.2. Like-particle mixing

We again deal with the case that the properties of the states under consideration are determined by only one type of nucleon. However, we now want to consider core polarization involving nucleons of the same type as the valence particles. Thus we must antisymmetrize explicitly our core-excited and valence-nucleon wave functions.

There are two distinct cases to be studied. The first is the one in which the core nucleons are excited to orbits different from those occupied by the valence particles. In this case one may proceed in the same way as before and deduce equation (5.25). However, instead of calculating the admixture coefficients by using equation (5.26) we evaluate them using properly antisymmetrized functions. To do this, denote the valence orbits by j, j' etc. and the core and core-excited orbits by j_c, j_c' etc. The part of V which comes into the evaluation of α_I involves the destruction of a valence nucleon and a core nucleon and then their recreation. That is

$$V=-\sum_{Kjj'}\sum_{j_cj_c'}\sqrt{(2K+1)}E_K(j_c'j';j_cj)[[a_{j_c'}^\dagger\times a_{j'}^\dagger]_K\times[\tilde{a}_{j_c}\times\tilde{a}_j]_K]_{00}$$

$$=\sum_{Jjj'j_cj_c'}\sqrt{(2J+1)}\tilde{E}_J(j_c'j';j_cj)[[a_{j_c'}^\dagger\times\tilde{a}_{j_c}]_J\times[a_{j'}^\dagger\times\tilde{a}_j]_J]_{00} \tag{5.31}$$

where

$$\tilde{E}_J(j_c'j';j_cj)=\sum_K(-1)^{j+j_c'-K-J}(2K+1)E_K(j_c'j';j_cj)W(j'jj_c'j_c;JK) \tag{5.32}$$

and $E_K(j_c'j';j_cj)$ is the $T=1$ interaction in the angular momentum state K. This expression for V was deduced in exactly the same way that we arrived at equation (2.101) except that in this case there is no isospin recoupling.

Once the interaction has been written in this form it is simple to evaluate the matrix element that governs the mixing

$$\langle[\Phi_{\lambda q}\times\psi_{I'}]_{IM}|\,V\,|\Phi_0\psi_{IM}\rangle=\sqrt{\left(\frac{2I'+1}{2I+1}\right)}\sum_{jj'j_cj_c'}\tilde{E}_\lambda(j_c'j';j_cj)$$

$$\times\langle\Phi_{\lambda q}\|\,[a_{j_c'}^\dagger\times\tilde{a}_{j_c}]_\lambda\,\|\Phi_0\rangle\langle\psi_{I'}\|\,[a_{j'}^\dagger\times\tilde{a}_j]_\lambda\,\|\psi_I\rangle. \tag{5.33}$$

The operator $\sum_n(T_n^{(\xi)})_{\lambda\mu}$ may also be expressed in terms of annihilation and creation operators (see equation (1.53))

$$\sum_n(T_n^{(\xi)})_{\lambda\mu}=\sum_{jj'mm'}\langle\phi_{j'}\|\,T_\lambda^{(\xi)}\,\|\phi_j\rangle(j\lambda m\mu\mid j'm')a_{j'm'}^\dagger a_{jm}$$

$$=\sum_{jj'}\sqrt{\left(\frac{2j'+1}{2\lambda+1}\right)}\langle\phi_{j'}\|\,T_\lambda^{(\xi)}\,\|\phi_j\rangle[a_{j'}^\dagger\times\tilde{a}_j]_{\lambda\mu}. \tag{5.34}$$

When equations (5.33) and (5.34) are substituted into equation (5.25) and use is made of equation (5.24) one finds

$$\langle\Psi_{I_f}\|\,T_\lambda^{(\xi)}\,\|\Psi_{I_i}\rangle=\sum_{jj'}\langle\psi_{I_f}\|\,[a_{j'}^\dagger\times\tilde{a}_j]_\lambda\,\|\psi_{I_i}\rangle\Bigg[\sqrt{\left(\frac{2j'+1}{2\lambda+1}\right)}$$

$$\times\langle\phi_{j'}\|\,T_\lambda^{(\xi)}\,\|\phi_j\rangle+\sum_{j_cj_c'}D_\lambda(j_c',j_c)$$

$$\times\left\{\frac{\tilde{E}_\lambda(j_c'j';j_cj)}{E_{I_i}-E_{I_f}-E_\lambda(q)}+(-1)^{j-j'}\frac{\tilde{E}_\lambda(j_c'j;j_cj')}{E_{I_f}-E_{I_i}-E_\lambda(q)}\right\}\Bigg] \tag{5.35}$$

where

$$D_\lambda(j_c', j_c) = (-1)^{j_c'-j_c} \frac{\sqrt{(2j_c'+1)}}{2\lambda+1} \langle \phi_{j_c'} \| T_\lambda^{(\xi)} \| \phi_{j_c} \rangle$$

$$\times \sum_q \langle \Phi_0 \| [a_{j_c}^\dagger \times \tilde{a}_{j_c}]_\lambda \| \Phi_{\lambda q} \rangle \langle \Phi_{\lambda q} \| [a_{j_c'}^\dagger \times \tilde{a}_{j_c}]_\lambda \| \Phi_0 \rangle. \qquad (5.36)$$

If $|E_\lambda(q)| \gg |E_{I_i} - E_{I_f}|$ the added term is independent of the initial and final state angular momenta and the effect of the perturbation is to once more give rise to a state-independent effective operator.

One could, of course, have gone through exactly these same arguments for the unlike-particle mixing discussed in section 2.1. The only difference would be that the $E_K(j_c'j'; j_cj)$ which enter into the definition of $\tilde{E}_J$ in equation (5.32) would be a half of the sum of the $T=0$ and $T=1$ interactions (see equation (2.77)).

The final case, excitation of core particles to the valence orbits, will not lead to a state-independent effective operator. For example, in addition to the first term of equation (5.35) there will be a contribution of the form

$$\langle \Psi_{I_f} \| T_\lambda^{(\xi)} \| \Psi_{I_i} \rangle_{\text{add}} = \alpha_{I_f;I'J_c'} \sqrt{\left(\frac{2j_c+1}{2\lambda+1}\right)} \langle \phi_{j_c} \| T_\lambda^{(\xi)} \| \phi_j \rangle$$

$$\times \langle \Phi_0(m) \psi_{I_f}(n) \| [a_{j_c}^\dagger \times \tilde{a}_j]_\lambda \| [\Phi_{J_c'}(m-1) \times \psi_{I'}(n+1)]_{I_i} \rangle$$

where the notation indicates that initially there were m core particles and n nucleons in the valence orbits, and in the admixed states these numbers become $(m-1)$ and $(n+1)$. This matrix element depends on all the quantum numbers involved including I_f and I_i. Moreover, there is a different n dependence of this matrix element than that arising from the unperturbed states. This follows because according to the definition of the one-nucleon c.f.p. in equation (A5.6), $\langle \psi_{I_f}(n) \| \tilde{a}_j \| \psi_{I'}(n+1) \rangle$ is proportional to $\sqrt{(n+1)} \langle j^n I_f j | \} j^{n+1} I' \rangle$. Thus although one can get a linear contribution from this type of mixing it will not lead to a state-independent effective operator.

3. Matrix elements and phases

Before beginning a detailed discussion of electromagnetic properties and selection rules we shall derive some simple formulae for the general evaluation of matrix elements of one-particle operators and discuss the importance of consistently choosing the sign of off-diagonal energy matrix elements.

The effect of $\sum_n (T_n^{(\xi)})_{\lambda\mu}$ on the state $[\Phi_{J_1}\times\chi_{J_2}]_{I_iM_i}$ may be easily computed diagramatically by use of equation (5.17). Thus

$$\sum_n (T_n^{(\xi)})_{\lambda\mu}[\Phi_{J_1}\times\chi_{J_2}]_{I_iM_i}=\sum_{IM}(I_i\lambda M_i\mu\mid IM)$$ 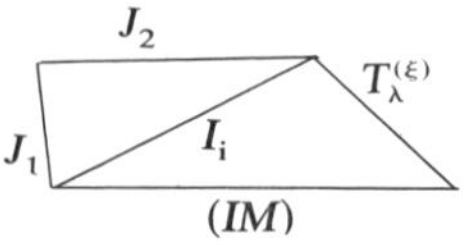

$$=\sum_{IM}(I_i\lambda M_i\mu\mid IM)\surd(2I_i+1)\Big(\sum_{J_2'}\surd(2J_2'+1)\langle J_2'\|\,T_\lambda^{(\xi)}\,\|J_2\rangle$$

$$\times W(J_1J_2I\lambda;\,I_iJ_2')$$ 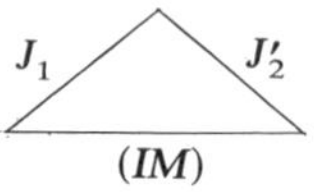

$$+\sum_{J_1'}(-1)^{J_1'-J_1+I_i-I}\surd(2J_1'+1)\langle J_1'\|\,T_\lambda^{(\xi)}\,\|J_1\rangle$$

$$\times W(J_2J_1I\lambda;\,I_iJ_1')$$ 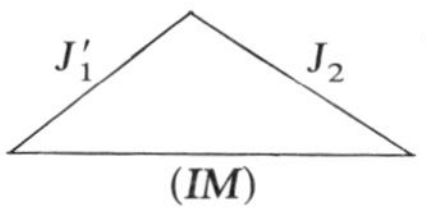 $\Big)$

(5.37)

where $\langle J'\|\,T_\lambda^{(\xi)}\,\|J\rangle$ is the many-particle reduced matrix element, the calculation of which we shall now discuss.

To be specific let us consider the state Φ_{JM} to arise from $(j^n)_{JM}$. The reduced matrix element may be calculated by making a fractional-parentage decomposition. Thus

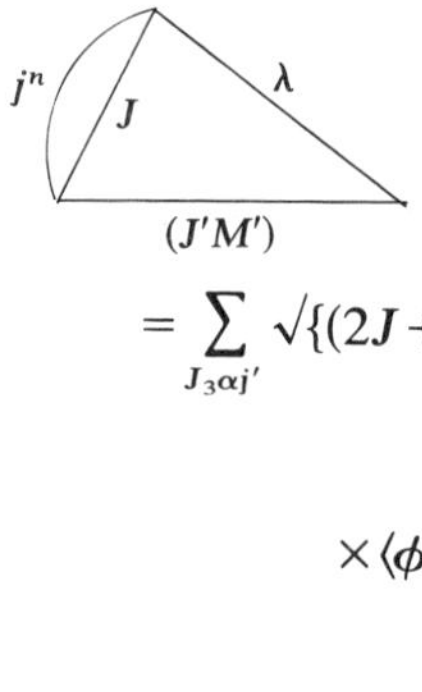

$$=\sum_{J_3\alpha}\langle j^{n-1}J_3\alpha,j\,|\,\}j^nJ\rangle$$ 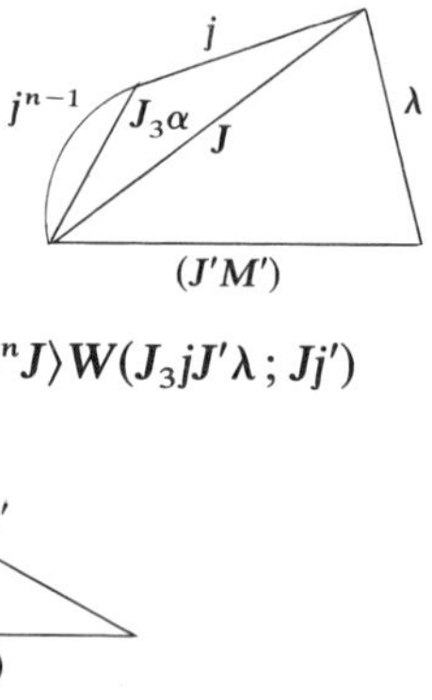

$$=\sum_{J_3\alpha j'}\surd\{(2J+1)(2j'+1)\}\langle j^{n-1}J_3\alpha,j|\,\}j^nJ\rangle W(J_3jJ'\lambda;\,Jj')$$

$$\times\langle\phi_{j'}\|\,T_\lambda^{(\xi)}\,\|\phi_j\rangle$$ 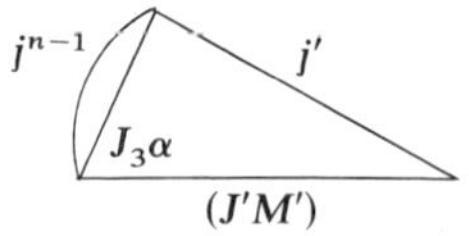 (5.38)

where α stands for any other quantum numbers necessary to specify completely the $(n-1)$ particle state (J_3M_3).

For $j'=j$ the many-particle reduced matrix element is

$$\langle (j)_{J'}^n\|\sum_i (T_i^{(\xi)})_\lambda\,\|(j)_J^n\rangle=n\sum_{J_3\alpha}\surd\{(2J+1)(2j+1)\}\langle j^{n-1}J_3\alpha,j|\,\}j^nJ\rangle$$
$$\times\langle j^{n-1}J_3\alpha,j|\,\}j^nJ'\rangle\,\langle\phi_j\|\,T_\lambda^{(\xi)}\,\|\phi_j\rangle W(J_3jJ'\lambda;\,Jj).\quad(5.39)$$

In this equation the factor n merely reflects the fact that any one of the n particles that make up the state j^n can emit the gamma ray (see Rule 1 in section 4.2, Chapter 1).

The only other state to which $(j^n)_{JM}$ can be connected, if $T^{(\xi)}_{\lambda\mu}$ is a sum of single-particle operators, is $[(j^{n-1})_{\bar{J}} \times j_1]_{J'M'}$. In this case Rule 1 in section 4.2, Chapter 1 tells us we must multiply the matrix element by $(-1)^{1-n}\sqrt{n}$. In addition, Rule 2 tells us an additional phase factor may arise. To calculate this factor forget for the moment about any angular-momentum coupling and consider

$$\begin{aligned} ME &= \langle (j^{n-1})_{\bar{J}} j_1 |\, a^{\dagger}_{j_1} a_j \,| (j^n)_J \rangle \\ &= \langle 0 |\, a_{j_1} (j^{n-1})_{\bar{J}} a^{\dagger}_{j_1} a_j (j^n)_J \,| 0 \rangle \\ &= (-1)^{n-1} \langle 0 |\, a_{j_1} a^{\dagger}_{j_1} \,| 0 \rangle \langle 0 |\, (j^{n-1})_{\bar{J}} a_j (j^n)_J \,| 0 \rangle . \end{aligned}$$

That is, to bring each of the operators a_j which go to make up the configuration $(j^{n-1})_J$ through a_{j_1} gives a factor (-1) or in total $(-1)^{n-1}$. Thus

$$\begin{aligned} \langle [(j^{n-1})_{\bar{J}} \times j_1]_{J'} \| \, T^{(\xi)}_{\lambda} \, \| (j^n)_J \rangle = \sqrt{\{n(2J+1)(2j_1+1)\}} \langle \phi_{j_1} \| \, T^{(\xi)}_{\lambda} \, \| \phi_j \rangle \\ \times \langle j^{n-1} \bar{J} \alpha, j | \} j^n J \rangle W(\bar{J} j J' \lambda; J j_1). \end{aligned} \tag{5.40a}$$

In the same way

$$\begin{aligned} \langle (j^n)_{J'} \| \, T^{(\xi)}_{\lambda} \, \| [(j^{n-1})_{\bar{J}} \times j_1]_J \rangle = \sqrt{\{n(2J+1)(2j+1)\}} \langle \phi_j \| \, T^{(\xi)}_{\lambda} \, \| \phi_{j_1} \rangle \\ \times W(\bar{J} j_1 J' \lambda; J j) \langle j^{n-1} \bar{J} \alpha, j | \} j^n J' \rangle . \end{aligned} \tag{5.40b}$$

3.1. Gamma transition between two-particle states; application to $^{18}_{8}O_{10}$

As an application of these ideas consider a gamma transition between an initial state

$$\psi_{I_i M_i} = \alpha (j^2)_{I_i M_i} + \beta [j \times j_1]_{I_i M_i} + \gamma (j_1^2)_{I_i M_i}$$

and a final state

$$\psi_{I_f M_f} = \tilde{\alpha} (j^2)_{I_f M_f} + \tilde{\beta} [j \times j_1]_{I_f M_f} + \tilde{\gamma} (j_1^2)_{I_f M_f} .$$

The two-particle-to-one-particle fractional-parentage coefficients are given by equation (A5.49) and have the values

$$\langle j^1 j, j | \} j^2 I \rangle = \begin{cases} 1 & \text{if } I \text{ is even} \\ 0 & \text{if } I \text{ is odd.} \end{cases}$$

Furthermore, one must remember that the notation $[j \times j_1]_{IM}$ stands for

$$[j \times j_1]_{IM} = \sum_{m m_1} (j j_1 m m_1 \mid IM) a^{\dagger}_{jm} a^{\dagger}_{j_1 m_1} |0\rangle$$

and as a consequence

$$[j\times j_1]_{IM} = -(-1)^{j+j_1-I}[j_1\times j]_{IM}.$$

When these results are combined with equations (5.37), (5.39), and (5.40) it follows that

$$\begin{aligned}\langle\psi_{I_f}\|\, T_\lambda^{(\xi)}\,\|\psi_{I_i}\rangle = \sqrt{(2I_i+1)}\big(&2\alpha\tilde{\alpha}\sqrt{(2j+1)}\langle\phi_j\|\, T_\lambda\,\|\phi_j\rangle W(jjI_f\lambda\,;\,I_ij)\\ &+\sqrt{2}\alpha\tilde{\beta}\sqrt{(2j_1+1)}\langle\phi_{j_1}\|\, T_\lambda\,\|\phi_j\rangle W(jjI_f\lambda\,;\,I_ij_1)\\ &+\sqrt{2}\beta\tilde{\alpha}\sqrt{(2j+1)}\rangle\phi_j\|\, T_\lambda\,\|\phi_{j_1}\rangle W(jj_1I_f\lambda\,;\,I_ij)\\ &+\beta\tilde{\beta}[\sqrt{(2j_1+1)}\langle\phi_{j_1}\|\, T_\lambda\,\|\phi_{j_1}\rangle W(jj_1I_f\lambda\,;\,I_ij_1)\\ &+(-1)^{I_i-I_f}\sqrt{(2j+1)}\langle\phi_j\|\, T_\lambda\,\|\phi_j\rangle W(j_1jI_f\lambda\,;\,I_ij)]\\ &+\sqrt{2}\beta\tilde{\gamma}(-1)^{j-j_1-I_i}\sqrt{(2j_1+1)}\langle\phi_{j_1}\|\, T_\lambda\,\|\phi_j\rangle\\ &\times W(j_1jI_f\lambda\,;\,I_ij_1)\\ &+\sqrt{2}\gamma\tilde{\beta}(-1)^{j-j_1-I_f}\sqrt{(2j+1)}\langle\phi_j\|\, T_\lambda\,\|\phi_{j_1}\rangle\\ &\times W(j_1j_1I_f\lambda\,;\,I_ij)\\ &+2\gamma\tilde{\gamma}\sqrt{(2j_1+1)}\langle\phi_{j_1}\|\, T_\lambda\,\|\phi_{j_1}\rangle W(j_1j_1I_f\lambda\,;\,I_ij_1)\big).\end{aligned} \tag{5.41}$$

This expression may now be used to calculate gamma-ray transition rates in any nucleus that can be described as two nucleons moving outside an inert core. In particular, we shall use it to predict the $E2$ lifetimes in ${}^{18}_{8}O_{10}$. In section 2, Chapter 1, we discussed the level structure of this nucleus on the assumption that ${}^{16}_{8}O_8$ was an inert core and that the low-lying states could be attributed to the motion of two-valence neutrons in the $d_{\frac{5}{2}}$ and $s_{\frac{1}{2}}$ single-particle orbits. Since the valence nucleons themselves have no electric charge, the entire effective $E2$ operator results from the core-polarization effect discussed in the preceding section. We assume a state-independent effective operator of the form

$$Q_{2\mu} = \tilde{e}\sum_i r_i^2 Y_{2\mu}(\theta_i, \phi_i)$$

and choose $\tilde{e}$ to fit experiment.

For the 2^+ and 0^+ levels in this nucleus the relevant wave functions are

$$\psi_{2M} = \alpha(d_{\frac{5}{2}}^2) + \beta(d_{\frac{5}{2}}s_{\frac{1}{2}}) \tag{5.42a}$$

$$\psi_{00} = \tilde{\alpha}(d_{\frac{5}{2}}^2) + \tilde{\gamma}(s_{\frac{1}{2}}^2). \tag{5.42b}$$

Thus from equation (5.41) and the definition of $B(E2)$ in equation (5.10) we find

$$B(E2; 2^+ \rightarrow 0^+) = \tfrac{1}{5}(2\alpha\tilde{\alpha}\langle\phi_{\frac{5}{2}}\| Q_2 \|\phi_{\frac{5}{2}}\rangle + \sqrt{2}\beta\tilde{\alpha}\langle\phi_{\frac{5}{2}}\| Q_2 \|\phi_{\frac{1}{2}}\rangle + \sqrt{2}\beta\tilde{\gamma}\langle\phi_{\frac{1}{2}}\| Q_2 \|\phi_{\frac{5}{2}}\rangle)^2. \quad (5.43)$$

If we assume harmonic-oscillator wave functions, the radial integrals can be evaluated from equation (1.11a). The angular factors are given by equation (A2.23) so that

$$\langle\phi_{\frac{5}{2}}\| \tilde{e}r^2 Y_2 \|\phi_{\frac{5}{2}}\rangle = -\sqrt{\frac{14}{4\pi}}\left(\frac{\tilde{e}\hbar}{m\omega}\right)$$

$$\langle\phi_{\frac{5}{2}}\| \tilde{e}r^2 Y_2 \|\phi_{\frac{1}{2}}\rangle = -\sqrt{\frac{10}{4\pi}}\left(\frac{\tilde{e}\hbar}{m\omega}\right)$$

$$\langle\phi_{\frac{1}{2}}\| \tilde{e}r^2 Y_2 \|\phi_{\frac{5}{2}}\rangle = -\sqrt{\frac{30}{4\pi}}\left(\frac{\tilde{e}\hbar}{m\omega}\right).$$

From this it is clear that only the product of the oscillator constant and the effective charge enter into the transition rate. With $\hbar\omega = 41/A^{\frac{1}{3}}$ and α, β, $\tilde{\alpha}$, and $\tilde{\beta}$ computed by use of the surface-delta interaction of Chapter 1, section 2.1 (equations (1.61) and (1.62)) it follows that the effective neutron charge needed to fit the lifetime (Berant *et al.* 1975) of the 1·98 MeV 2^+ level in $^{18}_{8}O_{10}$ is $\tilde{e} = 0{\cdot}83e$, where e is the proton charge. (Note that the surface-delta interaction does not reproduce the experimental excitation energy of the 2^+ level. However, in theoretical calculations of lifetimes the experimentally observed energy is always used in calculating the transition rate.)

In Table 5.3 we compare the known experimental and theoretical $E2$ rates in $^{18}_{8}O_{10}$ using a state-independent effective charge (i.e. a constant value of $\tilde{e}$). It is apparent that a single value of $\tilde{e}$ can give qualitative agreement with experiment for the yrast transitions. If we had chosen to fit the $4_1^+ \rightarrow 2_1^+$ lifetime the required value of $\tilde{e} = 0{\cdot}65e$. The fact that the effective charge needed for the $2_1^+ \rightarrow 0_1^+$ transition is larger than that for the other yrast transitions has also been found in the $0f_{\frac{7}{2}}$ nuclei (Brown *et al.* 1974). This is quite general and is a manifestation of the fact that the nucleus likes to take on a quadrupole deformation (see Chapter 7).

On the other hand, transitions between yrast and non-yrast levels are not well reproduced. As is apparent from Table 5.3 the transition probability for decay of the excited 0_3^+ state at 5·329 MeV to the first 2^+ level is underestimated by a factor of about 6·5. Similarly the 2_2^+ state, the excitation energy of which is fairly well reproduced by the $(d_{\frac{5}{2}}, s_{\frac{1}{2}})$ model (see Fig. 1.4), has a transition probability to the ground state which is underestimated by a factor of about 80.

TABLE 5.3

Mean lifetime in picoseconds (10^{-12} seconds) of various transitions in $^{18}_{8}O_{10}$.

Transition	Gamma-ray energy (MeV)	Mean life (picoseconds) Experiment	Theory
$2_1^+ \to 0_1^+$	1·982	3·58±0·18	3.6
$4_1^+ \to 2_1^+$	1·573	24·8±1·3	15·3
$0_3^+ \to 2_1^+$	3·347	0·37±0·11	2·4
$2_2^+ \to 0_1^+$	3·921	0·18±0·12	14·5

I_i^+ means that the ith state of that spin in the nucleus. The experimental lifetimes of the yrast levels are taken from Berant *et al.* (1975) and those for the non-yrast levels from Olness *et al.* (1973). The theoretical calculations are based on the wave functions determined from the surface-delta interaction (Chapter 1, section 2.1 and an effective charge of $0{\cdot}83e$.

To analyze this further, consider the form of the requisite wave functions (equations (1.61) and (1.62)). The 0^+ ground state has an eigenfunction, equation (5.42b), for which

$$\tilde{\alpha} = 0{\cdot}929 \qquad \tilde{\gamma} = 0{\cdot}371.$$

For the $2_1^+ \to 0_1^+$ transition the values of α and β in equation (5.42a) are

$$\alpha = 0{\cdot}764 \qquad \beta = 0{\cdot}645.$$

Substitution into equation (5.43) shows that all terms in $B(E2; 2_1^+ \to 0_1^+)$ interfere constructively. This is typical of yrast → yrast transitions and means that if the degree of configuration mixing is comparable in all yrast states a single and reasonable value of the effective charge ($\tilde{e} \cong e$) can be used to fit the data. A detailed analysis (Halbert *et al.* 1971) of all the known $E2$ transition rates in the $A = 17-22$ nuclei shows that a single value of the proton effective charge $e_p = 1{\cdot}5e$, and neutron charge $e_n = 0{\cdot}5e$, is consistent with the data on yrast transitions provided we are willing to say that theory and experiment are 'in agreement' when they differ by no more than about a factor of two.

However, for the $2_2^+ \to 0_1^+$ decay in $^{18}_{8}O_{10}$ the values of α and β to be used in equation (5.42a) are

$$\alpha = 0{\cdot}645 \qquad \beta = -0{\cdot}764.$$

Thus in $B(E2; 2_2^+ \to 0_1^+)$ destructive interference within the model space occurs and consequently the effects of neglected configurations cannot be mopped up so successfully with a reasonable effective charge. A study of the $A = 17$–22 data shows that in many cases the experimental $B(E2)$ between non-yrast and yrast levels is poorly reproduced by the restricted configuration shell-model calculation. In addition, the rates are generally quite sensitive to the assumed two-body interaction. Thus it is extremely important to study these types of transitions since they are much more sensitive to the assumed two-body interaction and are much better indicators of configuration impurity.

If the signs of all off-diagonal energy matrix elements within the model space are changed, there is no change in the calculated spectra. On the other hand, $B(E2)$ depends critically on these signs. For example, $B(E2; 2_1^+ \to 0_1^+)$, which for the signs used in the tabulated results has the value $1{\cdot}54\,(\tilde{e}\hbar/m\omega)^2$ becomes $0{\cdot}32\,(\tilde{e}\hbar/m\omega)^2$ when β and $\tilde{\gamma}$ have their signs changed. Since one would expect a simple short-range force to give at least the correct sign for energy matrix elements that have appreciable magnitude, this sign ambiguity in any energy fit is circumvented by requiring that the signs of empirically determined matrix elements agree with calculated ones; i.e. either those that come from the delta–function interaction of equation (A2.29) or from the realistic force calculations (Kuo 1967, Kuo and Brown 1968; for a review of calculations of this type see Barrett and Kirson 1973 and Ellis and Osnes 1977).

Finally a word of caution should be inserted for those whose calculate two-body matrix elements using the surface–delta interaction. If the matrix element is taken from equation (A2.29) with

$$\bar{R} = \int R_{j_1}(r)R_{j_2}(r)R_{j_3}(r)R_{j_4}(r)r^2\,\mathrm{d}r$$

a constant independent of the single-particle states, the implication is that all radial wave functions are positive at ∞. On the other hand, if $\bar{R}$ is given by equation (A2.30), the nuclear wave functions have been assumed to be positive at the origin. The matrix elements of $\langle r^L \rangle$ which enter into the transition rates must, of course, be calculated with wave functions obeying the same boundary conditions as those used in the energy calculation.

4. *E*1 transitions

According to equation (5.5) the operator governing $E1$ transitions is

$$\begin{aligned}(E1)_\nu &= \frac{e}{2}\sum_k r_k Y_{1\nu}(\theta_k, \phi_k)\{1-(\tau_k)_z\}\\ &= \frac{e}{2}\sqrt{\frac{3}{4\pi}}\sum_k (r_k)_\nu\{1-(\tau_k)_z\}\end{aligned} \tag{5.45}$$

where the second term of equation (5.5) has been omitted because it is small. In addition, we have inserted the proton projection operator $\{1-(\tau_k)_z\}/2$ because only protons have an electric charge, and in writing the second line use has been made of equation (A2.12). From the definition of the centre-of-mass coordinate $\tilde{R}$ given in equation (4.58) it follows that

$$(E1)_\nu = \frac{e}{2}\sqrt{\frac{3}{4\pi}}\left\{A\tilde{R}_\nu - \sum_k (r_k)_\nu (\tau_k)_z\right\}. \tag{5.45a}$$

The first term in this operator connects states with differing numbers of centre-of-mass quanta and as such gives a non-physical result. One can merely neglect this term (recognizing its spurious nature) or alternatively express the $E1$ operator in terms of coordinates relative to the centre-of-mass

$$\begin{aligned}(E1)_\nu &= \frac{e}{2}\sqrt{\frac{3}{4\pi}}\sum_k \{(r_k)_\nu - \tilde{R}_\nu\}\{1-(\tau_k)_z\} \\ &= -\frac{e}{2}\sqrt{\frac{3}{4\pi}}\sum_k \{(r_k)_\nu - \tilde{R}_\nu\}(\tau_k)_z.\end{aligned}$$

If N and Z are the number of neutrons and protons in the nucleus, it follows that

$$\sum_k (\tau_k)_z = N - Z$$

and consequently the $E1$ operator can be written as

$$(E1)_\nu = e\sqrt{\frac{3}{4\pi}}\left[-\frac{Z}{A}\sum_k (r_k)_\nu \left\{\frac{1+(\tau_k)_z}{2}\right\} + \frac{N}{A}\sum_k (r_k)_\nu \left\{\frac{1-(\tau_k)_z}{2}\right\}\right]. \tag{5.46}$$

In other words, this operator, which can be used when the spurious centre-of-mass motion has not been eliminated from the states, looks as if we deal with nucleons with an effective charge of $-Z/A$ for neutrons and N/A for protons. The effective charges in this case are the result of recoil corrections and have nothing to do with the core polarization effects discussed in section 2 of this chapter.

4.1. *Isospin effects*

Because the physical part of the $E1$ operator in equation (5.45a) is a tenor of rank 1 in both ordinary space and isospin space its matrix elements have the form

$$\begin{aligned}\langle \Psi_{I_f M_f; T_f T_z} | (E1)_\nu | \Psi_{I_i M_i; T_i T_z}\rangle &= \langle \Psi_{I_f; T_f} ||| (E1) ||| \Psi_{I_i; T_i}\rangle \\ &\quad \times (I_i 1 M_i \nu \mid I_f M_f)(T_i 1 T_z 0 \mid T_f T_z).\end{aligned} \tag{5.47}$$

From Table A1.2 one can write down the values of the isospin Clebsch–Gordan coefficients

$$
\begin{aligned}
(T_i 1 T_z 0 \mid T_f T_z) &= \left\{\frac{(T_i - T_z + 1)(T_i + T_z + 1)}{(T_i + 1)(2T_i + 1)}\right\}^{\frac{1}{2}} && \text{if} \quad T_f = T_i + 1 \\
&= \frac{T_z}{\sqrt{\{T_i(T_i+1)\}}} && \text{if} \quad T_f = T_i \qquad (5.48) \\
&= -\left\{\frac{(T_i - T_z)(T_i + T_z)}{T_i(2T_i + 1)}\right\}^{\frac{1}{2}} && \text{if} \quad T_f = T_i - 1
\end{aligned}
$$

and by inspection of these formulae certain isospin selection rules may be deduced (Warburton and Weneser 1972).

(a) $T=0 \rightarrow T=0$ $E1$ *transitions are forbidden.* One may find many examples of this selection rule. For instance, in $^{18}_{9}F_9$ the 1·08 MeV 0^- state has a mean lifetime for gamma decay to the 1^+ ground state of 27 ± 2 picoseconds (Endt and Van der Leun 1974a). From Table 5.2 it follows that

$$\frac{w_{\text{expt}}(0^- \rightarrow 1^+)}{w_W(0^- \rightarrow 1^+)} = 4{\cdot}3 \times 10^{-5}$$

where w_{expt} and w_W are the experimental and Weisskopf estimates for the transition probability/unit time. For nuclei with $A < 45$ the data on these isospin-forbidden transitions have been collected (Endt and Van der Leun 1974, 1974a), and it is found that on the average they are inhibited by about a factor of 10^4 from the Weisskopf estimate.

(b) $\Delta T = 0$ $E1$ *transitions in* $T_z = 0$ *nuclei are forbidden.* This selection rule implies that the same transition in neighbouring nuclei may differ markedly. An example would be provided by comparing the transition from the first excited 1^- state in $^{14}_{6}C_8$ (at 6·09 MeV) to the ground state and the analogous decay in $^{14}_{7}N_7$; i.e. 8·06 MeV $T=1$, $1^- \rightarrow$ 2·31 MeV $T=1$, 0^+. For the latter the transition is isospin-forbidden whereas there is no selection rule to inhibit the former. Unfortunately data such as these are difficult to obtain and only crude limits have so far been set on these lifetimes (Ajzenberg-Selove 1970).

(c) *Corresponding transitions in nuclei with the same* $|T_z|$ *have the same value of* $B(E1; I_i \rightarrow I_f)$. This follows because when $T_z \rightarrow -T_z$ the isospin Clebsch at most changes sign. This rule says, for example, that $B(E1)$ for the $\frac{1}{2}^- \rightarrow \frac{1}{2}^+$ decay in $^{19}_{9}F_{10}$ and $^{19}_{10}Ne_9$ should be the same. For the former, the mean lifetime of the 110 keV $\frac{1}{2}^-$ level for decay to the $\frac{1}{2}^+$ ground state is 849×10^{-12} s and for the latter the $\frac{1}{2}^-$, which lies at 280 keV, has a mean life of 61×10^{-12} s (Endt and Van der Leun 1974a). Thus from Table 5.1 it follows that

$$B(E1,\,{}^{19}_{9}\mathrm{F}_{10};\tfrac{1}{2}^{-}\rightarrow\tfrac{1}{2}^{+})=5{\cdot}6\times10^{-4}\,e^2\,\mathrm{fm}^2$$

$$B(E1,\,{}^{19}_{10}\mathrm{Ne}_{9};\tfrac{1}{2}^{-}\rightarrow\tfrac{1}{2}^{+})=4{\cdot}7\times10^{-4}\,e^2\,\mathrm{fm}^2.$$

The fact that these two numbers differ by so much could be construed to indicate a large difference in the radial wave functions of the particles and/or a substantial isospin mixing in the states. However, these decays both exhibit a very large inhibition compared to the Weisskopf estimate

$$B_{\mathrm{W}}(E1;\tfrac{1}{2}^{-}\rightarrow\tfrac{1}{2}^{+})=0{\cdot}45e^2\,\mathrm{fm}^2.$$

Consequently, although some difference in the structure of the states is indicated, it may not be large because the values of small matrix elements are usually easily changed. As yet no theoretical calculations concerning this difference have been carried out.

4.2. Deformation effects

The inhibition observed in the $\frac{1}{2}^{-}\rightarrow\frac{1}{2}^{+}$ ^{19}F and ^{19}Ne decays is typical of almost all $E1$ transitions between low-lying states. If one inspects the histograms of Endt and Van der Leun (1974, 1974a) shown in Fig. 5.1, one sees that the isospin-allowed $E1$ transitions are generally inhibited by a factor of about 10^3. In other words, transitions are inhibited even when they might be thought to go at approximately single-particle speed. We

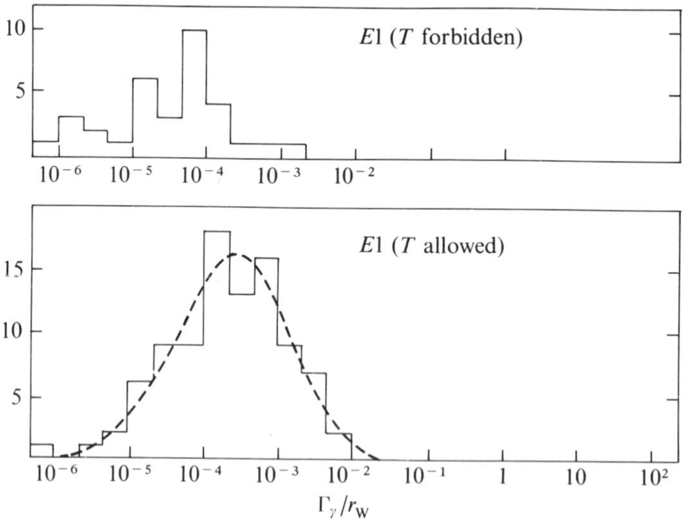

Fig. 5.1. Histograms of the strengths in Weisskopf units of the T-allowed and T-forbidden $E1$ transitions between bound states in the $A=21$–44 nuclei.

shall now show that that this inhibition is due to the fact that the nucleus likes to assume a quadrupole deformation.

From the classic work of Bohr and Mottelson (1953) one knows that many properties of the heavier nuclei can be understood in terms of collective rotational and vibrational motion. These nuclei have a permanent deformation, and nucleons no longer move in a spherical potential but instead move in a deformed one. For simplicity let us assume the single-particle potential can be described by a cylindrically symmetric harmonic oscillator. In this case the Hamiltonian governing the single-particle motion is

$$H=\frac{1}{2m}(p_x^2+p_y^2+p_z^2)+\tfrac{1}{2}m\omega^2(x^2+y^2)+\tfrac{1}{2}m\tilde{\omega}^2z^2. \tag{5.49}$$

When $\tilde{\omega}=\omega$ the potential becomes the spherical harmonic-oscillator potential discussed in Chapter 4 and Appendix 6. Clearly if $\tilde{\omega}\neq\omega$ the potential is deformed along the z axis. For $\tilde{\omega}<\omega$ particles are less constrained in the z than in the x and y directions; in other words the nucleus is elongated along the z direction. For $\tilde{\omega}>\omega$ the converse is true. Eigenvalues and eigenfunctions of this Hamiltonian, including the spin-orbit force, have been given by Nilsson (1955). We shall not reproduce his calculation here but merely give those details needed to demonstrate that $E1$ transitions are inhibited due to deformation (see Chapter 7 for more details).

The eigenvalues of equation (5.49) are found most simply by solving the oscillator potential in x, y, z space:

$$E=\hbar\omega(n_x+n_y+1)+\hbar\tilde{\omega}(n_z+\tfrac{1}{2}). \tag{5.50}$$

In Fig. 5.2 we show the level structure for $\tilde{\omega}<\omega$ where we have assumed that these two quantities are not so different that the major shell structure is completely washed out. That is, if n and l stand for the spherical oscillator quantum numbers in equation (4.11) we assume that levels with the same

$$N=2n+l=n_x+n_y+n_z \tag{5.51}$$

are still grouped in energy. Thus the lowest state for given N is the one with

$$n_z=N; \qquad n_x=n_y=0,$$

and because the nucleon has spin this state can accomodate two particles. The second state will be two-fold degenerate with

$$n_z=N-1, \qquad n_x=0, \qquad n_y=1$$

$$n_z=N-1, \qquad n_x=1, \qquad n_y=0$$

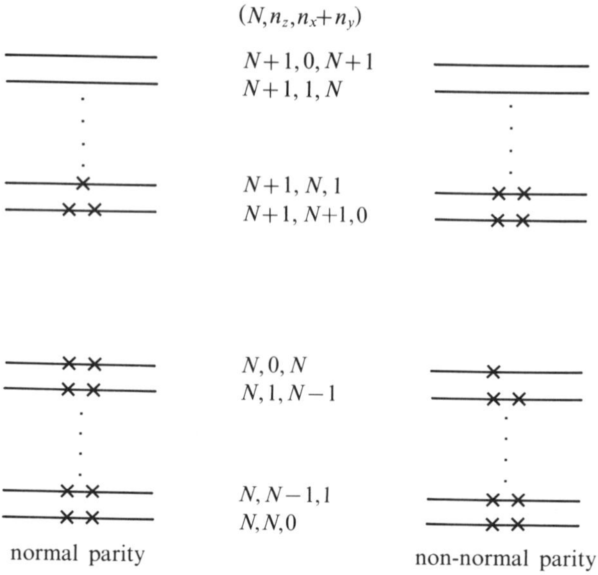

Fig. 5.2. Order of the single-particle levels in the deformed oscillator potential of equation (5.49). The energy is given by equation (5.50) with $n_x+n_y+n_z=N=2n+l$ and is shown for $\tilde{\omega}<\omega$. For simplicity we have assumed each state can accomodate at most two particles. Normal parity states are shown on the left and non-normal parity levels on the right.

so that this state can accommodate four neutrons. Finally the uppermost level in the shell will be $(N+1)$-fold degenerate and have $n_z=0$.

Let us consider the case shown in Fig. 5.2 in which the shell N is full and we are beginning to populate the shell $N+1$. Normal parity states would correspond to filling the lowest possible orbits as shown on the left-hand side of this diagram. Low-lying non-normal parity levels correspond to exciting a nucleon from the uppermost level in the N shell to the lowest available orbit in the shell $N+1$. For the non-normal parity state shown on the right-hand side of Fig. 5.2 to make an $E1$ transition to a normal parity one

$$(n_z=N;\, n_x+n_y=1)\longrightarrow(n_z=0;\, n_x+n_y=N).$$

Thus both n_z and n_x and/or n_y must change. However, in terms of x, y, and z coordinates, $(E1)_\nu$ has the form

$$\begin{aligned}(E1)_0&=-\frac{e}{2}\sqrt{\frac{3}{4\pi}}\sum_k z_k(\tau_k)_z\\(E1)_{\pm1}&=\pm\frac{e}{2}\sqrt{\frac{3}{8\pi}}\sum_k(x_k\pm iy_k)(\tau_k)_z,\end{aligned}\qquad(5.52)$$

and consequently these two states cannot be connected. In fact we would have to excite a particle up to the orbit $n_z = 1$; $n_x + n_y = N$ before an $E1$ transition could occur. Thus for deformed nuclei the $E1$ transition strength is concentrated in the high-lying opposite-parity states.

Although the nuclei we deal with in shell-model calculations are not strongly deformed the remnants of this selection rule, which was first enunciated by Strominger and Rasmussen (1957) still persist. In fact, as we shall show in Chapter 7, if one projects states of good angular momentum from a deformed oscillator well one gets eigenfunctions in excellent agreement with those that result from a detailed shell-model calculation. This implies that many of the effects of the residual two-body interaction used in the shell model can be simulated by allowing particles to move in a deformed Hartree–Fock potential. In shell-model language the inhibition results from the fact that the hole can polarize the valence nucleons. For example, consider the case of an odd number of particles in the state j moving outside a closed j_1 shell

$$\psi_{jm} = (j_1)_0^{2j_1+1}(j)_{jm}^n.$$

We are interested in the $E1$ transition between this state and the $(j_1)^{-1}$ hole state

$$\psi_{j_1m_1} = \alpha(j_1)_{j_1m_1}^{2j_1}(j)_{J=M=0}^{n+1} + \sum_{J'} \beta_{J'}[(j_1)_{j_1}^{2j_1} \times (j)_{J'}^{n+1}]_{j_1m_1}.$$

In zeroth approximation the hole state results from coupling the hole to the $J = 0$ state of the valence particles. However, because of the residual two-body force some admixtures with $J' \neq 0$ arise. The $E1$ matrix element therefore has contributions from both the α and $\beta_{J'}$ terms, and because the nucleus likes to assume a quadrupole deformation these terms interfere destructively. For $n = 1$, $\beta_{J'}$ is usually small, but for $n \gtrsim 3$ the destructive interference usually cuts down the matrix element appreciably. This leads to a fourth $E1$ rule:

(d) $E1$ *Decays between low-lying states of nuclei slightly removed from closed shells are expected to be substantially inhibited due to the fact the nucleus tends to have a quadrupole deformation.*

5. *M*2 transitions

In this section we review the properties of a second common type of decay connecting states of different parity. From equation (5.9) it follows that the operator governing the emission of $M2$ radiation is

$$(M2)_\nu = \sqrt{\tfrac{5}{2}}\mu_N \sum_k r_k \big(\mu_n[Y_1(k)\times\sigma_k]_{2\nu}\{1+(\tau_k)_z\}$$

$$+\{\mu_p[Y_1(k)\times\sigma_k]_{2\nu}+\tfrac{2}{3}[Y_1(k)\times\ell_k]_{2\nu}\}\{1-(\tau_k)_z\}\big)$$

$$=\sqrt{\tfrac{5}{2}}\mu_N \sum_k r_k \big((\mu_n+\mu_p-\tfrac{1}{3})[Y_1(k)\times\sigma_k]_{2\nu}+\tfrac{2}{3}[Y_1(k)\times \mathrm{j}_k]_{2\nu}$$

$$+(\mu_n-\mu_p+\tfrac{1}{3})[Y_1(k)\times\sigma_k]_{2\nu}(\tau_k)_z-\tfrac{2}{3}[Y_1(k)\times \mathrm{j}_k]_{2\nu}(\tau_k)_z\big) \quad (5.53)$$

where j_k and ℓ_k are the single-particle angular-momentum operators associated with the kth nucleon.

When the free-particle values of μ_p and μ_n are used one finds

$$\mu_p+\mu_n-\tfrac{1}{3}\cong 0{\cdot}55$$

$$\mu_n-\mu_p+\tfrac{1}{3}\cong -4{\cdot}37.$$

Thus the largest term in the operator is the one with the second of these factors. Since this term is proportional to τ_z it follows that its isospin structure is the same as that for the $E1$ operator given in equations (5.47) and (5.48). Thus one arrives at the result (Warburton 1958)

(a) *$M2$ transitions between $T=0$ states and $\Delta T=0$ decays in self-conjugate nuclei should be retarded relative to a normal $M2$.* This follows because only the terms independent of τ_z can contribute to these decays.

There are many instances where this retardation effect is observed. For example, in $^{14}_{7}N_7$ the 5·11 MeV 2^- $T=0$ state is observed to gamma decay to both the 1^+ $T=0$ ground and 2·31 MeV 0^+ $T=1$ states. from the known lifetimes (Endt and Van der Leun 1974a) it follows that

$$B(M2; 2^-T=0\rightarrow 1^+T=0)=0{\cdot}026\ (\mu_N\ \text{fermis})^2$$

and

$$B(M2; 2^-T=0\rightarrow 0^+T=1)=6{\cdot}84\ (\mu_N\ \text{fermis})^2$$

where μ_N is given by equation (5.4). The Weisskopf estimate for $B(M2)$ in this nucleus is

$$B_W(M2)=9{\cdot}47\ (\mu_N\ \text{fermis})^2.$$

The compilations of Endt and Van der Leun (1974, 1974a) show that on the average for $A\leq 45$ the isospin-allowed $M2$ decays have a strength of about $\frac{1}{3}$ of a Weisskopf unit whereas the isospin-retarded transitions have a strength of about $\frac{1}{25}$ of a Weisskopf unit.

Apart from the term depending on $\underline{\mathrm{j}}$, the $M2$ operator has a spatial structure identical to the $E1$ operator. In the preceding section we showed that matrix elements of $rY_{1\nu}(\theta,\phi)$ are generally small between low-lying states. Since the coefficient of the j term is small it follows that

$M2$ decays should exhibit the same sort of retardation as $E1$ transitions particularly when $[Y_1 \times \mathrm{j}]_{2\nu}$ cannot contribute to the decay. For the lighter nuclei $M2$ transitions involve changing a $d_{\frac{5}{2}}$ particle to a $p_{\frac{3}{2}}$ or $f_{\frac{7}{2}}$ and in these cases $[Y_1 \times \mathrm{j}]_{2\nu}$ can give a contribution. Thus, although some inhibition is observed, it is not as pronounced as in the case of $E1$. For nuclei near $A = 40$ the dominant term in the transition involves $d_{\frac{3}{2}} \rightarrow f_{\frac{7}{2}}$. In this case $[Y_1 \times \mathrm{j}]_{2\nu}$ cannot contribute. Thus for $A \geq 40$ one would expect $M2$ decays between low-lying states in nuclei a few particles away from closed shells to be inhibited and the data bear this out (Kurath and Lawson 1967). Consequently

(b) *Nuclei a few particles (or holes) away from a doubly-magic core are expected to have inhibited M2 decays between low-lying states.*

6. Magnetic dipole properties

So far we have discussed properties of operators that connect states of opposite parity. Because of this, there were no static moments (diagonal matrix elements), only transition rates. We now turn to those electromagnetic multipoles that can have both static and dynamic properties (see, for example, Yoshida and Zamick 1972, Horie and Sugimoto 1973). First we consider the magnetic dipole operator, which according to equation (5.9) has the form

$$(M1)_\nu = \sqrt{\left(\frac{3}{4\pi}\right)}\mu_{\mathrm{N}} \sum_k [\mu_{\mathrm{n}}(\sigma_k)_\nu\{1+(\tau_k)_z\}/2 + \{\mu_{\mathrm{p}}(\sigma_k)_\nu + (\ell_k)_\nu\}\{1-(\tau_k)_z\}/2] \qquad (5.54)$$

where μ_{N} is the nuclear magneton defined by equation (5.4). One often sees this operator expressed in terms of g factors. In that case it takes the form

$$(M1)_\nu = \sqrt{\left(\frac{3}{4\pi}\right)}\mu_{\mathrm{N}} \sum_k (\mathrm{g_s(n)}(\mathrm{s}_k)_\nu\{1+(\tau_k)_z\}/2 + \{\mathrm{g_s(p)}(\mathrm{s}_k)_\nu + \mathrm{g}_l(\ell_k)_\nu\}\{1-(\tau_k)_z\}/2) \qquad (5.55)$$

where

$$(\mathrm{s}_k)_\nu = \tfrac{1}{2}(\sigma_k)_\nu$$

is the spin operator of the kth nucleon and

$$\mathrm{g_s(n)} = 2\mu_{\mathrm{n}} = -3{\cdot}82$$

$$\mathrm{g_s(p)} = 2\mu_{\mathrm{p}} = 5{\cdot}58$$

$$\mathrm{g}_l = 1.$$

The $M1$ operator has the above structure and constants provided meson exchange between nucleons in the nucleus can be neglected and provided there is no core polarization. In actual fact both of these effects exist and modify the $M1$ operator by adding to it a term

$$(\delta M1)_\nu = \sqrt{\left(\frac{3}{4\pi}\right)}\mu_N \sum_k ((g_s^{ex}+g_s^{cp})(s_k)_\nu + (g_l^{ex}+g_l^{cp})(\ell_k)_\nu + (g_2^{ex}+g_2^{cp})[Y_2(k)\times s_k]_{1\nu}) \tag{5.56}$$

where g^{ex} is the addition due to meson exchange and g^{cp} is that brought about by core polarization.

Meson-exchange effects were originally studied by Miyazawa (1951) and more recently have been discussed by several authors (Chemtob 1969; Wahlborn and Blomqvist 1969; Arima and Huang-Lin 1972, 1972a; Shimizu *et al.* 1974). In general the exchange contributions are found to have isovector character and from Table 12 of Shimizu *et al.* (1974) it is clear that they depend on the configuration being considered. For light nuclei ($A \leq 40$)

$$g_l^{ex} \simeq -0{\cdot}2\tau_z$$
$$g_s^{ex} \simeq 0{\cdot}03\tau_z,$$

and for all cases they consider these quantities have the same sign. On the other hand, the sign of g_2^{ex} depends on the configuration involved and has the range of values

$$g_2^{ex} = \{-0{\cdot}03 - +0{\cdot}07\}\tau_z.$$

Thus with the exception of g_l^{ex} the meson-exchange effects give a small contribution to the magnetic properties of light nuclei.

For nuclei near ^{208}Pb empirical values of

$$\delta g_s = g_s^{ex} + g_s^{cp}$$
$$\delta g_l = g_l^{ex} + g_l^{cp}$$
$$\delta g_2 = g_2^{ex} + g_2^{cp}$$

have been determined by Maier *et al.* (1972) who made a fit to the observed magnetic moments under the assumption that these additional contributions are configuration independent. With this assumption they find

$$\delta g_s = 3{\cdot}43\tau_z$$
$$\delta g_l = +0{\cdot}015 - 0{\cdot}075\tau_z$$
$$\delta g_2 = -4{\cdot}55\tau_z.$$

Since aside from g_l^{ex} the meson-exchange effects are expected to be small, it follows that the most important effects that modify the magnetic-moment operator are those due to polarization of the core which was discussed in section 2 of this chapter. However, as stressed by Arima and Huang-Lin (1972, 1972a) these effects are expected to be strongly configuration dependent.

Consequently since the polarization effects are large and configuration dependent it is important to look for cases where the exact coefficients in the $M1$ operator are unimportant. That such cases may exist can be seen by noting that because of the Wigner–Eckhart theorem any tensor operator of rank one fulfills the condition

$$\langle\phi_{jm'}|\, T_{1\nu}\, |\phi_{jm}\rangle = (j1m\nu \mid jm')\langle\phi_j\|\, T_1\, \|\phi_j\rangle.$$

Consequently diagonal matrix elements of the $M1$ operator are all proportional to matrix elements of the single-particle angular-momentum operator j, so that

$$T_{1\nu} = \alpha(j)\mathrm{j}_\nu \tag{5.57}$$

with

$$\alpha(j) = \frac{\langle\phi_j\|\, T_1\, \|\phi_j\rangle}{\langle\phi_j\|\, \mathrm{j}\, \|\phi_j\rangle}. \tag{5.58}$$

Therefore if one takes ratios of $M1$ matrix elements between states of the configuration $(j)^n_J$, the detailed structure of the operator will cancel out and one will be left with ratios of matrix elements of the single-particle operator j. In many instances the same result will be found to be true when one considers ratios of $M1$ matrix elements within the configuration

$$\psi_{JM} = [(\pi j)^n_{I_P} \times (\nu j')^{n'}_{I_N}]_{JM}.$$

Throughout the remainder of our discussion of $M1$ properties we shall, wherever possible, exploit this result.

6.1. *Magnetic moments*

The magnetic moment of the state I is defined by the equation

$$\mu = gI = \langle\Psi_{IM=I}|\sqrt{\left(\frac{4\pi}{3}\right)}(M1)_0\,|\Psi_{IM=I}\rangle \tag{5.59}$$

where $(M1)_0$ is the z component of the operator given in equation (5.54) and the g factor defined by this equation is the constant of proportionality relating μ and I. A compilation of experimental values of the magnetic moments has been given by Fuller and Cohen (1968).

6.1.1. Single-particle estimate and the configuration $(j)^n$. We first consider the single-particle magnetic moments and g factors. By using equation

(A2.19) it follows that for the neutron

$$\mu/\mu_N = \mu_n \qquad \text{if} \quad j = l + \tfrac{1}{2} \tag{5.60a}$$

$$= \left(\frac{-j}{j+1}\right)\mu_n \quad \text{if} \quad j = l - \tfrac{1}{2}. \tag{5.60b}$$

Thus the neutron single-particle configuration with $j = l + \frac{1}{2}$ has a magnetic moment identical to that of the free neutron, $-1{\cdot}91\mu_N$, whereas the $j = l - \frac{1}{2}$ state has a moment with opposite sign to that of the free particle. The values given by equations (5.60) are known as the Schmidt or single-particle values (Schmidt 1937, Schüler 1937) and are shown as solid lines in Fig. 5.3.

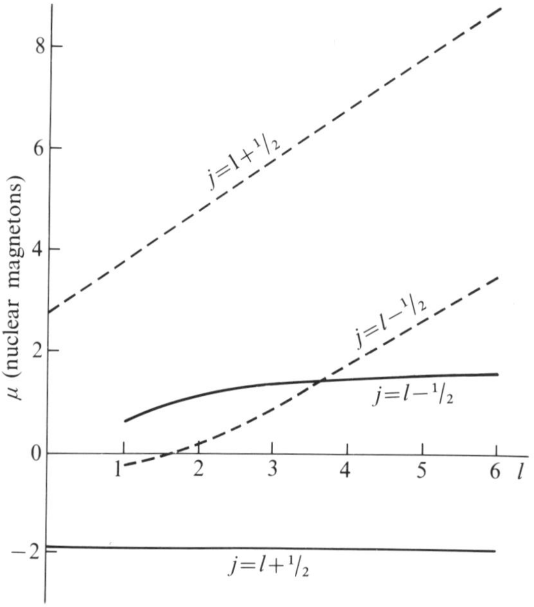

Fig. 5.3. Schmidt (single-particle) values for the magnetic moment of the neutron (solid line) and proton (dotted line) in units of nuclear magnetons. The values are calculated by use of equations (5.60) and (5.61) with $\mu_n = -1{\cdot}91$ and $\mu_p = 2{\cdot}79$.

Similar results can be derived for the proton by use of equations (A2.19) and (A2.20):

$$\mu/\mu_N = \mu_p + l \qquad \text{if} \quad j = l + \tfrac{1}{2} \tag{5.61a}$$

$$= \frac{j}{j+1}(l + 1 - \mu_p) \quad \text{if} \quad j = l - \tfrac{1}{2} \tag{5.61b}$$

and these are the Schmidt values for the proton magnetic moment shown as dashed lines in Fig. 5.3. The g factors $g(j)$ obtained by dividing these values by j are precisely the values of $\alpha(j)$ in equation (5.58).

Since the magnetic-moment operator cannot change a proton to a neutron it follows that when we deal with diagonal matrix elements of the configuration $[(\pi j)^n_{J_P}\times(\nu j_1)^{n'}_{J_N}]_{IM}$ the form given by equation (5.57) can be used for the magnetic-moment operator. Thus in this case

$$\begin{aligned}\mu_{\text{op}}=\sqrt{\left(\frac{4\pi}{3}\right)}(M1)_0&=g_p(j)\sum_p(\mathrm{j})_0+g_n(j_1)\sum_n(\mathrm{j}_1)_0\\&=g_p(j)(\mathrm{J_P})_0+g_n(j_1)(\mathrm{J_N})_0\\&=\tfrac{1}{2}\{g_p(j)+g_n(j_1)\}\mathrm{I}_0\\&\quad+\tfrac{1}{2}\{g_p(j)-g_n(j_1)\}\{(\mathrm{J_P})_0-(\mathrm{J_N})_0\}\end{aligned}\tag{5.62}$$

where the neutron and proton g factors are obtained from equations (5.60) and (5.61) by dividing by j. The sum over the proton and neutron single-particle operators leads to the total angular-momentum operators $\mathrm{J_P}$ and $\mathrm{J_N}$ for the protons and neutrons, respectively. In writing the last line of this equation use has been made of the fact that I, the total angular-momentum operator, is the sum of $\mathrm{J_P}$ and $\mathrm{J_N}$.

If we deal with the identical nucleon configuration, say $(\pi j)^n_I$, only the proton operator comes in and we arrive at the theorem

The g factor, μ/I, of any state of the identical nucleon configuration $(j)^n_I$ has the same value independent of n and I.

If core polarization and meson-exchange effects can be neglected this value is the Schmidt result given by equations (5.60) and (5.61) for neutrons and protons, respectively.

One can use this result to check the configuration purity and validity of the state-independent effective operator concept to describe the magnetic moments of the proton $(0f_{\frac{7}{2}})$ nuclei. In Table 5.3 we have tabulated the existing data and it is clear that the g factor is smaller than the free-particle value $g_p=(\mu_p+3)/3{\cdot}5=1{\cdot}654$. Furthermore, the effective g factor decreases as the shell is filled. From the structure of the $M1$ operator it is clear that only $j=l+\frac{1}{2}$ and $j=l-\frac{1}{2}$ states can be connected and consequently these moments are mainly sensitive to $0f_{\frac{5}{2}}$ admixtures into the predominantly $(\pi f_{\frac{7}{2}})^n$ configuration. Thus one must conclude that even though the spectrum of the $N=28$ isotones is well reproduced by the proton $(0f_{\frac{7}{2}})^n$ configuration (see section 3.2, Chapter 1) the $0f_{\frac{5}{2}}$ admixture into $(\pi f_{\frac{7}{2}})^n$ increases as Z increases.

TABLE 5.3

g values for the proton $f_{\frac{7}{2}}$ nuclei

Nucleus	Spin	μ/μ_N	g/μ_N
$^{51}_{23}V_{28}$	$\frac{7}{2}^-$	5·148	1·47
$^{53}_{25}Mn_{28}$	$\frac{7}{2}^-$	5·046	1·44
$^{54}_{26}Fe_{28}$	2^+	2·86	1·43
	6^+	8·22	1·37
$^{55}_{27}Co_{28}$	$\frac{7}{2}^-$	4·3	1·23

The values are taken from Horie and Sugimoto (1973). The Schmidt value of g is given by equation (5.61a) and has the value 1·654.

In Fig. 5.3 we have plotted the single-particle or Schmidt values for the magnetic moments. From Table 5.3 it is clear that the moments of the odd-A $0f_{\frac{7}{2}}$ nuclei lie between the two possible Schmidt values for $l=3$ (much closer, however, to the $l+\frac{1}{2}$ line). This result—lying between the Schmidt values—is generally found to be true and has been shown by Blin Stoyle (1953) and Arima and Horie (1954) to be a consequence of core polarization.

6.1.2. Weak-coupling multiplet. The magnetic moment of the more complicated configuration $[(\pi j)^n_{J_P}\times(\nu j_1)^{n'}_{J_N}]_{IM}$ requires the use of both terms in equation (5.62). The expectation value of $\underline{J}_P$ and $\underline{J}_N$ can be computed by use of the vector model; i.e.

$$\langle [J_P\times J_N]_{IM=I}|\,(J_P)_0\,|[J_P\times J_N]_{IM=I}\rangle = \frac{\langle \mathbf{J}_P\,.\,\mathbf{I}\rangle I}{I(I+1)}$$

$$= [I(I+1)+J_P(J_P+1)-J_N(J_N+1)]/2(I+1).$$

Thus one arrives at the result

The g factor of any state of the configuration

$$[(\pi j)^n_{J_P}\times(\nu j_1)^m_{J_N}]_{IM}$$

is

$$g = \tfrac{1}{2}\{g_p(j)+g_n(j_1)\} + \tfrac{1}{2}\{g_p(j)-g_n(j_1)\}\{J_P(J_P+1)-J_N(J_N+1)\}/I(I+1). \tag{5.63}$$

As an illustration of this result consider the moments of the low-lying states of $^{40}_{19}K_{21}$, the spectrum of which, as we saw in section 1.2 of Chapter 3, could be described by the configuration $[(\pi d_{\frac{3}{2}})^{-1}\times(\nu f_{\frac{7}{2}})]_{IM}$. The g factor of the $(\pi d_{\frac{3}{2}})^{-1}$ configuration can be obtained from the measured moment of $^{39}_{19}K_{20}$, $g=0{\cdot}39142\mu_N/1{\cdot}5=0{\cdot}261\mu_N$. The $\nu f_{\frac{7}{2}}$ g factor is determined from the moment of $^{41}_{20}Ca_{21}$, $g=-1{\cdot}595\mu_N/3{\cdot}5=-0{\cdot}456\mu_N$. Thus the g factor of any state of this multiplet is

$$g=\left(-0{\cdot}0974-\frac{4{\cdot}300}{I(I+1)}\right)\mu_N.$$

For the 4^- ground state the theoretical value is $\mu=-1{\cdot}25\mu_N$ compared to the experimental value of $-1{\cdot}298\mu_N$ (Endt and Van der Leun 1973). Recently the moment of the 29·6 keV 3^- state has been measured (Brandolini *et al.* 1974) and found to have the value $\mu=(-1{\cdot}29\pm0{\cdot}09)\mu_N$ whereas theory gives $\mu=-1{\cdot}368\mu_N$. Thus the moments of these two states are consistent with the assumption that a state-independent effective operator can be used to describe the magnetic moments of the states of the configuration $[(\pi d_{\frac{3}{2}})^{-1}\times(\nu f_{\frac{7}{2}})]_{IM}$.

6.1.3. The configuration $(\pi j)^n(\nu j)^{\pm n}$. When both neutrons and protons occupy the same single-particle orbit the eigenfunctions of the shell-model Hamiltonian can be written in the form

$$\psi_{IM}=\sum_{J_1J_2}A_I(J_1,J_2,n,n')[(\pi j)^n_{J_1}\times(\nu j)^{n'}_{J_2}]_{IM} \tag{5.64}$$

where the coefficients $A_I(J_1,J_2,n,n')$ are determined by diagonalizing the residual two-body interaction. If this potential is charge independent it can be written in the form

$$V=-\sum_J\sqrt{(2J+1)}E_J\big(\tfrac{1}{2}[[a_j^\dagger\times a_j^\dagger]_J\times[\tilde{a}_j\times\tilde{a}_j]_J]_{00}$$
$$+\tfrac{1}{2}[[b_j^\dagger\times b_j^\dagger]_J\times[\tilde{b}_j\times\tilde{b}_j]_J]_{00}+[[a_j^\dagger\times b_j^\dagger]_J\times[\tilde{a}_j\times\tilde{b}_j]_J]_{00}\big) \tag{5.65}$$

where $a^\dagger_{jm}$ and $b^\dagger_{jm}$ are creation operators for neutrons and protons, respectively, and $E_J=E_J(jj;jj)$ is the interaction energy in the state J.

It is clear that V is invariant to the simultaneous replacement of a proton by a neutron and a neutron by a proton. Consequently it follows that when $n'=n$ the eigenfunctions in equation (5.64) must either remain invariant i.e. have A_I's with property that

$$A_I(J_1,J_2,n,n)=(-1)^{J_1+J_2-I}A_I(J_2,J_1,n,n) \tag{5.66a}$$

or change sign

$$A_I(J_1,J_2,n,n)=-(1)^{J_1+J_2-I}A_I(J_2,J_1,n,n). \tag{5.66b}$$

Thus the states have a definite signature associated with them, positive in the first case and negative in the second.

When neutrons are replaced by neutron holes one may use equation (3.49) and equations (3.64)–(3.67) to show that V in equation (5.65) becomes

$$V' = V_0 - \sum_J \sqrt{(2J+1)}\{E_J(\tfrac{1}{2}[[a_j^\dagger \times a_j^\dagger]_J \times [\tilde{a}_j \times \tilde{a}_j]_J]_{00}$$
$$+\tfrac{1}{2}[[b_j^\dagger \times b_j^\dagger]_J \times [\tilde{b}_j \times \tilde{b}_j]_J]_{00}) + F_J[[a_j^\dagger \times b_j^\dagger]_J \times [\tilde{a}_j \times \tilde{b}_j]_J]_{00}\}$$

where

$$V_0 = \tfrac{1}{2}\sum\{1+(-1)^J\}(2J+1)E_J\left(1-\frac{2n'}{2j+1}\right)+\frac{n}{2j+1}\sum_J (2J+1)E_J$$

and F_J is given by equation (3.67),

$$F_J = -\sum_K (2K+1)W(jjjj;KJ)E_K.$$

Thus V' has exactly the same form as V except that F_J in the proton–neutron interaction replaces E_J and a constant I-independent term, V_0, is added to each diagonal matrix element of the Hamiltonian. Thus the potential is invariant to the simultaneous replacement of a neutron hole by a proton and a proton by a neutron hole. Consequently the A_I in

$$\psi_{IM} = \sum_{J_1J_2} A_I(J_1J_2n, 2j+1-n)[(\pi j)^n_{J_1} \times (\nu j)^{-n}_{J_2}]_{IM}$$

have exactly the symmetry properties of equation (5.66).

In general equation (5.63) can be used to evaluate the g factor of states described by ψ_{IM}, equation (5.64),

$$g = \tfrac{1}{2}\{g_p(j)+g_n(j)\}\sum_{J_1J_2} A_I^2(J_1, J_2, n, n')$$
$$+\tfrac{1}{2}\{g_p(j)-g_n(j)\}\sum_{J_1J_2} A_I^2(J_1, J_2, n, n')\left\{\frac{J_1(J_1+1)-J_2(J_2+1)}{I(I+1)}\right\}. \quad (5.67)$$

When $n' = n$ or $(2j+1-n)$ the symmetry property in equation (5.66) holds and the second term in equation (5.67) sums to zero. Furthermore, sincc

$$\sum_{J_1J_2} A_I^2(J_1, J_2, n, n') = 1$$

it follows (Lawson 1971) that

The g factor of any state of the configuration $[(\pi j)^n \times (\nu j)^{\pm n}]_{IM}$ *is independent of n and I and has the value* $g = \tfrac{1}{2}\{g_p(j)+g_n(j))\}$.

This result may be used to check the configuration purity of the low-lying states of the nucleus $^{48}_{23}V_{25}$. If the $(\pi f_{\frac{7}{2}})^3(\nu f_{\frac{7}{2}})^{-3}$ description of these states is correct it follows that all should have the same g factor $(2{\cdot}79+3-1{\cdot}91)/7=0{\cdot}554\mu_N$, if the Schmidt values are used. Experimentally (Weigt *et al.* 1968) the g factor of the 4^+ ground state is $0{\cdot}408\mu_N$, whereas that for the 306 keV 2^+ excited state is $0{\cdot}188\mu_N$ (Auerbach *et al.* 1967). Thus one must conclude that at least in these two states there are appreciable admixtures of configurations other than $f_{\frac{7}{2}}$.

6.2. *M1 transitions*

We now turn to properties of the $M1$ operator that are important in the calculation of transition rates. As before there are several simple rules that can be enunciated.

6.2.1. The identical-nucleon configuration. Within the identical nucleon configuration the operator governing the emission of $M1$ radiation can be obtained from equation (5.62) and for s.c.s. nuclei takes the form

$$(M1)_\nu = \sqrt{\left(\frac{3}{4\pi}\right)} g(j)\mathrm{J}_\nu$$
$$= \sqrt{\left(\frac{3}{4\pi}\right)} \mu_{\mathrm{op}}$$

where μ_{op} is the magnetic moment operator and $g(j)$ is $g_n(j)$ if we deal with the identical neutron configuration and $g_p(j)$ if we are considering protons. Since J_ν can at most change the z component of angular momentum of the state $(j)^n_I$ it follows that

> *$M1$ transactions between different states of the identical nucleon configuration, $(j)^n_I$, are strictly forbidden.*

An example of this selection rule is provided by the $M1$ transitions between low-lying states of ${}^{51}_{23}\mathrm{V}_{28}$. In section 3.2, Chapter 1, we showed that the spectrum of this nucleus could be described by the $(\pi f_{\frac{7}{2}})^3$ model. Thus the 320 keV transition between the $\frac{5}{2}^-$ excited and $\frac{7}{2}^-$ ground states should be severely inhibited. Experimentally the mean lifetime of the state is 290 picoseconds (Rao and Rapaport 1970) whereas the Weisskopf estimate is 10^{-12} s. Thus the transition is substantially retarded. The $M1$ operator has non-vanishing matrix elements only between the same single-particle states or spin-orbit partners, i.e. $j=l+\frac{1}{2}\rightarrow j=l\pm\frac{1}{2}$. Consequently the fact that this selection rule is closely obeyed tells us:

(a) there is no large $f_{\frac{5}{2}}$ contaminant in either state involved in the decay;
(b) there are no large components of the form $[(\pi f_{\frac{7}{2}})^2_J\times \pi j]_{IM}$ $(j\neq f_{\frac{5}{2}})$ in

both the initial and final states. This follows because

$$\langle[(\pi f_{\frac{7}{2}})^2_J\times\pi j]_{I_f}\| M1 \|[(\pi f_{\frac{7}{2}})^2_J\times\pi j]_{I_i}\rangle=\sqrt{\left(\frac{3}{4\pi}\right)}\sqrt{(2I_i+1)}\big(\sqrt{(2j+1)}$$

$$\times W(JjI_f1;\, I_ij)\langle\pi j\|\,\mu_{op}\,\|\pi j\rangle+(-1)^{I_i-I_f}\sqrt{(2J+1)}\,W(jJI_f1;\, I_iJ)$$

$$\times\langle(\pi f_{\frac{7}{2}})^2_J\|\,\mu_{op}\,\|(\pi f_{\frac{7}{2}})^2_J\rangle\big) \qquad (5.68)$$

where the matrix elements of μ_{op} will not, in general vanish.

6.2.2. Transitions within a weak-coupling multiplet. Equation (5.68) can be straightforwardly extended to describe transitions between states of the multiplet $[(\pi j_1)^n_{J_1}\times(\nu j)^m_J]_{IM}$. Thus

$$B(M1;\, I_i\to I_f)=\left(\frac{2I_f+1}{2I_i+1}\right)|\langle(\pi j_1)^n_{J_1}\times(\nu j)^m_J]_{I_f}\| M1 \|[(\pi j_1)^n_{J_1}\times(\nu j)^m_J]_{I_i}\rangle|^2$$

$$=\frac{3}{4\pi}(2I_f+1)\big((-1)^{I_i-I_f}\sqrt{(2J_1+1)}\,W(JJ_1I_f1;\, I_iJ_1)$$

$$\times\langle(\pi j_1)^n_{J_1}\|\,\mu_{op}\,\|(\pi j_1)^n_{J_1}\rangle$$

$$+\sqrt{(2J+1)}\,W(J_1JI_f1;\, I_iJ)\langle(\nu j)^m_J\|\,\mu_{op}\,\|(\nu j)^m_J\rangle\big)^2$$

$$=\frac{3}{4\pi}(2I_f+1)(2J+1)\,W^2(J_1JI_f1;\, I_iJ)\Bigg(\langle(\nu j)^m_J\|\,\mu_{op}\,\|(\nu j)^m_J\rangle$$

$$-\sqrt{\left\{\frac{J(J+1)}{J_1(J_1+1)}\right\}}\langle(\pi j_1)^n_{J_1}\|\,\mu_{op}\,\|(\pi j_1)^n_{J_1}\rangle\Bigg)^2. \qquad (5.69)$$

In writing the last line of this equation use has been made of the fact that for fixed J and J_1 there is only one way to make a state of given spin so that an $M1$ transition within the multiplet only takes place between states for which $|I_i-I_f|=1$. When this condition is satisfied it follows from equation (A4.24) that

$$\left(\frac{2J_1+1}{2J+1}\right)^{\frac{1}{2}}\frac{W(JJ_1I_f1;\, I_iJ_1)}{W(J_1JI_f1;\, I_iJ)}=\left\{\frac{J(J+1)}{J_1(J_1+1)}\right\}^{\frac{1}{2}}.$$

Thus we arrive at the rule

Provided the $M1$ operator is a sum of single-particle operators, ratios of $B(M1)$ values within the multiplet $[(\pi j_1)^n_{J_1}\times(\nu j)^m_J]_{IM}$ depend only on geometrical factors and do not depend on the detailed structure of the operator. Thus

$$\frac{B(M1;\, I_i\to I_f)}{B(M1;\, I_i'\to I_f')}=\frac{(2I_f+1)\,W^2(J_1JI_f1;\, I_iJ)}{(2I_f'+1)\,W^2(J_1JI_f'1;\, I_i'J)}.$$

From Table 5.1 it follows that this result can be rewritten in the form

$$(\mathrm{ME})^2 = \left(\langle(\nu j)^m_J \| \mu_{\mathrm{op}} \|(\nu j)^m_J\rangle - \sqrt{\left\{\frac{J(J+1)}{J_1(J_1+1)}\right\}}\langle(\pi j_1)^n_{J_1}\| \mu_{\mathrm{op}} \|(\pi j_1)^n_{J_1}\rangle\right)^2$$
$$= \frac{2\cdot 38\times 10^{-13}}{\tau E^3(2J+1)(2I_{\mathrm{f}}+1)W^2(J_1JI_{\mathrm{f}}1;\, I_{\mathrm{i}}J)} \tag{5.70}$$

where τ and E are the mean lifetime in seconds and the experimental energy in MeV of the transition $I_{\mathrm{i}} \rightarrow I_{\mathrm{f}}$.

In Table 5.4 we have collected the data concerning $M1$ transitions between states of the $[(\pi d_{\frac{3}{2}})^{3}_{\frac{3}{2}} \times (\nu f_{\frac{7}{2}})]_{IM}$ multiplet in ${}^{40}_{19}\mathrm{K}_{21}$ and the analogous $[\pi d_{\frac{3}{2}} \times \nu f_{\frac{7}{2}})]_{IM}$ states in ${}^{38}_{17}\mathrm{Cl}_{21}$ (Segel *et al.* 1970, Engelbertink and Olness 1972, Wedberg and Segel 1973). In Chapter 3, section 1.2, we showed that the energies of these low-lying states satisfy the particle-hole transformation to very high accuracy. Consequently one is tempted to say the states are quite pure. An inspection of Table 5.4 shows that this is not the case. In the first three lines of the table we have given the values of the left-hand side of equation (5.70) as deduced from the measured lifetimes and energies of the transitions. If the configurations were pure, all entries would be the same and, if the free-particle values (Schmidt limit) of the magnetic moments are appropriate, this quantity would be $6\cdot 25\mu_{\mathrm{N}}^2$. Since the single-particle $f_{\frac{7}{2}}$ nucleus ${}^{41}_{20}\mathrm{Ca}_{21}$ has $\mu/\mu_{\mathrm{N}} = -1\cdot 595$ (instead of $-1\cdot 91$) we have combined this with the measured moment of

TABLE 5.4

Values of $(\mathrm{ME})^2$ *in equation* (5.70) *deduced from the measured lifetimes and energies*

Transition	$\left(\frac{\mathrm{ME}}{\mu_{\mathrm{N}}}\right)^2$ ${}^{40}_{19}\mathrm{K}_{21}$	$\left(\frac{\mathrm{ME}}{\mu_{\mathrm{N}}}\right)^2$ ${}^{38}_{17}\mathrm{Cl}_{21}$
$3^- \rightleftarrows 2^-$	$6\cdot 6 \pm 0\cdot 7$	$18\cdot 4 \pm 1\cdot 7$
$4^- \rightleftarrows 3^-$	$8\cdot 3 \pm 0\cdot 1$	$4\cdot 4 \pm 1\cdot 1$
$5^- \rightleftarrows 4^-$	$3\cdot 3 \pm 0\cdot 4$	$15\cdot 7 \pm 4\cdot 2$
Free moments	$6\cdot 25$	$6\cdot 25$
Effective moments	$8\cdot 09$	$13\cdot 09$

For the pure configuration $[(\pi d_{\frac{3}{2}})^{\pm 1} \times (\nu f_{\frac{7}{2}})]_{IM}$ this quantity should be a constant. The Schmidt value is $6\cdot 25\mu_{\mathrm{N}}^2$ for both nuclei. If the experimental magnetic moment of ${}^{41}\mathrm{Ca}$ is used for the neutron and those of ${}^{37}\mathrm{Cl}$ and ${}^{39}\mathrm{K}$ for the $\pi d_{\frac{3}{2}}$ and $\pi d_{\frac{3}{2}}^{-1}$ configurations, the predicted values are those given in the last line.

${}^{39}_{19}K_{20}$, $\mu/\mu_N = 0{\cdot}391$, to give the effective moment value for ${}^{40}_{19}K_{21}$ listed on the last line of the table. In the same way the measured moment of ${}^{37}_{17}Cl_{20}$, $\mu/\mu_N = 0{\cdot}684$, leads to the effective-moment prediction for ${}^{38}_{17}Cl_{21}$. For both nuclei one of the three measured transition rates has an appreciably different value than that predicted by the use of effective operators.

In ${}^{40}_{19}K_{21}$ we have seen that the experimental magnetic moments of the 3^- and 4^- states can be well reproduced by the use of effective operators. However, from Table 5.4 it is clear that $(ME)^2$ governing the $5^- \to 4^-$ transition is at least a factor of 2 smaller than predicted. It has been shown (Becker and Warburton 1971, Kurath and Lawson 1972) that the computed transition rates are extremely sensitive to small components in the wave functions involving excitation of a proton out of the full $d_{\frac{5}{2}}$ proton state to the $d_{\frac{3}{2}}$ level. When this and the excitation of the neutron to the $f_{\frac{5}{2}}$ state are allowed, theory and experiment for ${}^{40}_{19}K_{21}$ can be brought into reasonable accord with only a small amount of configuration mixing.

On the other hand, for ${}^{38}_{17}Cl_{21}$ weak configuration mixing worsens the agreement between theory and experiment. In fact only when approximately 20% excitation out of the $s_{\frac{1}{2}}$ proton core state is allowed can experiment and theory be brought into agreement (Maripuu *et al.* 1973). This result clearly shows the need to study properties other than energies if one wishes to learn about shell-model configuration assignments.

6.2.3. The signature-selection rule. The fact that wave functions describing states of the $(\pi j)^n(\nu j)^{\pm n}$ system have a definite signature, equation (5.66), leads to the following selection rule

M1 transitions can only take place between states of the configuration $[(\pi j)^n \times (\nu j)^{\pm n}]_I$ that have opposite signature.

That this rule is true can be easily seen from equation (5.62). Since I_0 cannot contribute to transitions, only the second term is important. Thus the operator governing the transition changes sign when neutrons and protons are interchanged. Consequently, when applied to an initial state with say positive signature, the resulting function can only have a non-vanishing matrix element when the final state has negative signature.

${}^{48}_{22}Ti_{26}$ is a nucleus where this selection rule might be observed. In Fig. 5.4 we have shown the experimental spectrum below 3·25 MeV together with the observed (Bardin *et al.* 1973) $B(M1)$ values measured in Weisskopf units ($B_W(M1) = 1{\cdot}76\mu_N^2$). To the right of each level is the theoretical signature that arises when the $f_{\frac{7}{2}}$ shell-model calculation is

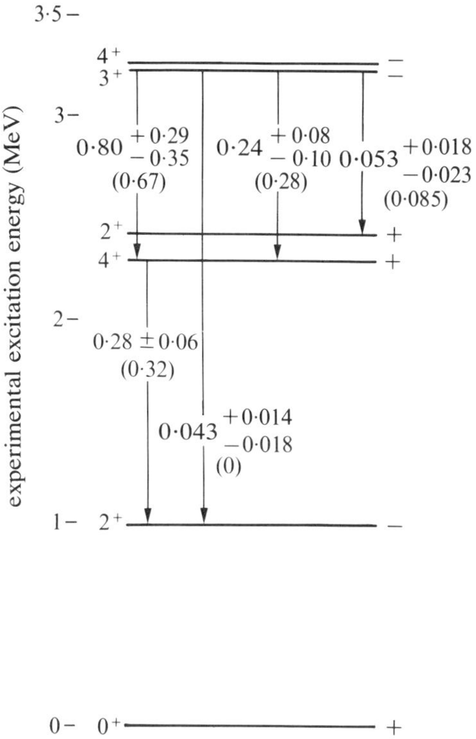

Fig. 5.4. $M1$ transitions observed in ${}^{48}_{22}\mathrm{Ti}_{26}$. The strengths are in Weisskopf units and the experimental data are taken from Bardin *et al.* (1973). The theoretical predictions (bracketed numbers) are based on the $f_{\frac{7}{2}}$ model using the ${}^{42}_{21}\mathrm{Sc}_{21}$ matrix elements for the residual two-body interaction. The plus or minus to the right of the levels is the theoretical signature.

carried out using for the residual interaction the ${}^{42}_{21}\mathrm{Sc}_{21}$ matrix elements (see Fig. 2.1). Although the signature of the 0^+ states is always positive, all other signatures can only be determined by carrying out the detailed shell-model calculation. Because the signature of the 3^+ and lowest 2^+ are the same, no $M1$ decay between them should be seen. Experimentally this decay probability is a factor of approximately ten smaller than that observed for most other $M1$s. According to the $f_{\frac{7}{2}}$ model the other $M1$s should be allowed, and to test the consistency of this picture we have computed $B(M1)$ for these transitions. The decay matrix element is proportional to

$$(g_p - g_n) = \tfrac{2}{7}(\mu_p + 3 - \mu_n).$$

Because I_0 of equation (5.62) cannot contribute to transition rates all $M1$

decays within $(\pi f_{\frac{7}{2}})^n(\nu f_{\frac{7}{2}})^m$ (both $n=m$ and $n\neq m$) are proportional to this combination, and it is found empirically that better agreement with experiment is obtained if instead of taking $\mu_p+3-\mu_n=7{\cdot}7$ one uses 4·5. This is the value employed in calculating the numbers quoted in Fig. 5.4.

Clearly one can understand the observed $M1$ properties in ${}^{48}_{22}Ti_{26}$ by using the $(\pi f_{\frac{7}{2}})^2(\nu f_{\frac{7}{2}})^{-2}$ model and a state-independent effective $M1$ operator. In this respect ${}^{48}_{22}Ti_{26}$ differs from its neighbour, the odd–odd nucleus ${}^{48}_{23}V_{25}$, where the state-independent operator approximation could not explain the observed magnetic moments. Thus it would appear that there is more configuration mixing in the odd–odd $f_{\frac{7}{2}}$ nuclei.

6.2.4. Isospin effects. We now turn to some isospin properties of the $M1$ operator which can be written in the form

$$(M1)_{1\nu}=\frac{1}{2}\sqrt{\left(\frac{3}{4\pi}\right)}\mu_N\Big((\mu_p+\mu_n-\tfrac{1}{2})\sum_i(\sigma_i)_\nu \\ +(\mu_n-\mu_p+\tfrac{1}{2})\sum_i(\tau_i)_z(\sigma_i)_\nu+\sum_i(\mathbf{j}_i)_\nu\{1-(\tau_i)_z\}\Big). \quad (5.71)$$

When the free-particle values of μ_p and μ_n are used, the coefficient of the first term is 0·38 whereas that of the second is −4·2. For transitions between $T=0$ states only the former can contribute (clearly terms proportional to τ_z vanish and $\sum_i(\mathbf{j}_i)_k=\mathbf{I}_k$, the total angular-momentum operator, cannot contribute to a transition rate). Since the coefficient of the contributing term is so small one is led to the rule (Morpurgo 1958)

> *$M1$ transitions between $T=0$ states in self-conjugate nuclei should be much slower (approximately $(0{\cdot}38/4{\cdot}2)^2$) than transitions in the same nucleus with $\Delta T=1$ or those in neighbouring $T\neq 0$ nuclei.*

An example of this rule is provided by the decay of the 3·95 MeV $I=1^+$ $T=0$ state om ${}^{14}_{7}N_7$. The decay to the $I=1^+$ $T=0$ ground state has a radiative width of $5{\cdot}8\times10^{-4}$ eV and that to the $I=0^+$ $T=1$ 2·31 MeV level has $\Gamma_\gamma=0{\cdot}14$ eV (Ajzenberg-Selove 1970) where Γ_γ is defined by equation (5.21). Thus

$$R_1=\frac{\Gamma_\gamma\{3{\cdot}95(1^+,0)\rightarrow \text{g.s.}(1^+,0)\}}{\Gamma_\gamma\{3{\cdot}95(1^+,0)\rightarrow 2{\cdot}31(0^+,1)\}}=4{\cdot}1\times10^{-3}.$$

Naively one would have anticipated that R_1 would be approximately the ratio of the cube of the transition energies

$$R_1=\left(\frac{3{\cdot}95}{3{\cdot}95-2{\cdot}31}\right)^3=14.$$

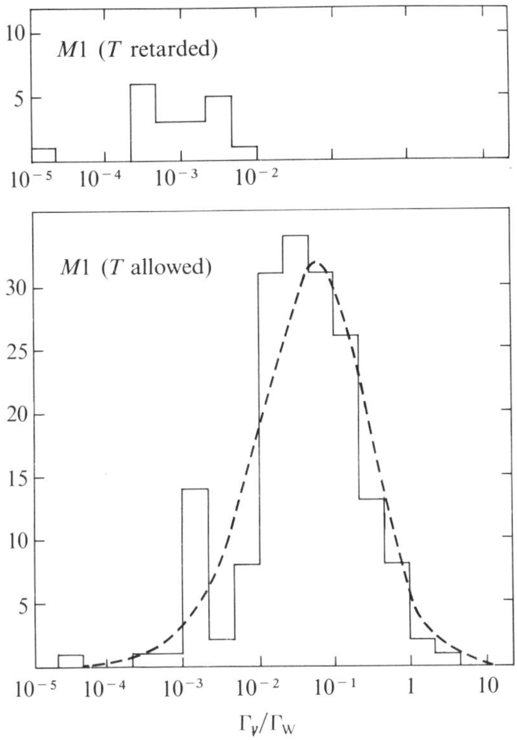

Fig. 5.5. Histograms of the strengths of T-allowed and T-retarded $M1$ transitions between bound states in the $A = 21$–44 nuclei.

Many examples of this inhibition, although not as marked as in this case, have been observed (Endt and Van der Leun 1974, 1974a), and to illustrate this we give in Fig. 5.5 histograms that compare $M1$ T-retarded and $M1$ T-allowed transitions.

6.2.5. Analogue to anti-analogue decays. When N is greater than Z, states of the nucleus (N, Z) can occur as excited states in the neighbouring nucleus $(N-1, Z+1)$ and these excitations are called analogue states. A particular type of analogue state is one in which a single particle with angular momentum j couples to a $J_c = 0$, $T_c = T_0$ core. For example, the simplest model for the $\frac{7}{2}^-$ ground state of ${}^{37}_{16}S_{21}$ corresponds to coupling an $f_{\frac{7}{2}}$ particle to the $(d_{\frac{3}{2}})^4_{J_c=0T_c=2}$ core. The analogue of this state in ${}^{37}_{17}Cl_{20}$ is the $T = \frac{5}{2}$, $I = \frac{7}{2}^-$ state observed at 9·09 MeV (Endt and Van der Leun 1973). Since ${}^{37}_{17}Cl_{20}$ has $T_z = \frac{3}{2}$, a state with $T = \frac{3}{2}$ of this configuration can also be realized. This state is called the anti-analogue state. Thus in general if we denote the $J_c = M_c = 0$ core wave function by $\Phi_{00;T_0T_z}$ the analogue state will have the form

$$\Psi_{jm;T_0+\frac{1}{2}T_0-\frac{1}{2}} = \sqrt{\left(\frac{2T_0}{2T_0+1}\right)}\Phi_{00;T_0T_0-1}\phi_{jm;\frac{1}{2}\frac{1}{2}}$$

$$+\frac{1}{\sqrt{(2T_0+1)}}\Phi_{00;T_0T_0}\phi_{jm;\frac{1}{2}-\frac{1}{2}}. \tag{5.72a}$$

where the coefficients in the wave function are just the isospin Clebsch–Gordan coefficients taken from Table A1.1. Similarly the anti-analogue eigenfunction is given by

$$\Psi_{jm;T_0-\frac{1}{2}T_0-\frac{1}{2}} = -\frac{1}{\sqrt{(2T_0+1)}}\Phi_{00;T_0T_0-1}\phi_{jm;\frac{1}{2}\frac{1}{2}}$$

$$+\sqrt{\left(\frac{2T_0}{2T_0+1}\right)}\Phi_{00;T_0T_0}\phi_{jm;\frac{1}{2}-\frac{1}{2}}. \tag{5.72b}$$

We shall now examine what can be learned about nuclear structure from a study of gamma-ray transitions between analogue and anti-analogue states.

Since the isospin of the initial and final states differ, analogue to anti-analogue transitions are a class of $M1$ decays that proceed via the τ_z part of the operator in equation (5.71). To calculate these transition rates we must, in general, find the effect of an operator of rank 1 in both space–spin and isospin acting on the state $[\Phi_{J_c;T_c}\times\chi_{J;T}]_{I_iM_i;T_iT_z}$. This may be done by a straightforward extension of equation (5.17):

$$(M1)_{1\nu;10}[\Phi_{J_c;T_c}\times\chi_{J;T}]_{I_iM_i;T_iT_z}$$

$$= \sum_{T'T_z'IM}(I_i1M_i\nu\mid IM)(T_i1T_z0\mid T'T_z')$$

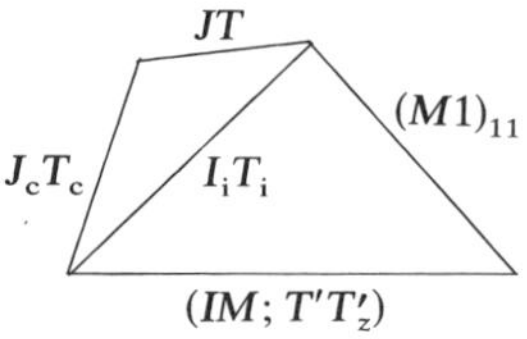

$$= \sum_{T'T_z'IM}\sqrt{\{(2I_i+1)(2T_i+1)\}}(T_i1T_z0\mid T'T_z')(I_i1M_i\nu\mid IM)$$

$$\times\Bigg(\sum_{J''T''}\sqrt{\{(2J''+1)(2T''+1)\}}W(J_cJI1;\,I_iJ'')W(T_cTT'1;\,T_iT'')$$

$$\times\langle\chi_{J'';T''}|||\,(M1)_{1;1}\,|||\chi_{J;T}\rangle$$

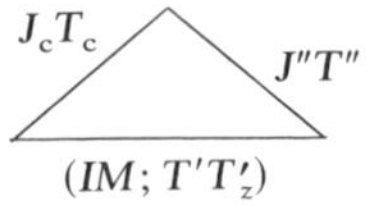

$$+\sum_{J_c''T_c''}(-1)^{\Delta J_c+\Delta T_c-I_i+I_f-T_i+T_f}\sqrt{\{(2J_c''+1)(2T_c''+1)\}}W(JJ_cI1;\,I_iJ_c'')$$

$$\times W(TT_cT'1;\,T_iT_c'')\langle\Phi_{J_c'';T_c''}|||\,(M1)_{1;1}\,|||\Phi_{J_c;T_c}\rangle\Bigg)$$

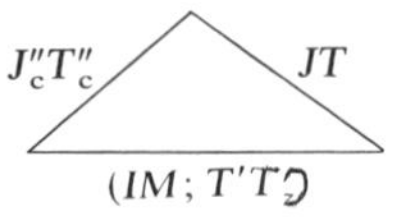

where use has been made of the fact that the sum of single-particle operators can be written as a part that operates on the nucleons making up $(J_c T_c)$ and a second part that operates on those in (J, T) and $\Delta J_c = J_c - J''_c$, $\Delta T_c = T_c - T''_c$. The overlap of this wave function with the final state gives the matrix element in question. Therefore

$$\begin{aligned}\langle[\Phi_{J_c';T_c'}\times\chi_{J';T'}]_{I_f M_f;T_f T_z}|\,(M1)_{1\nu;10}\,|[\Phi_{J_c;T_c}\times\chi_{J;T}]_{I_i M_i;T_i T_z}\rangle \\ = \sqrt{\{(2I_i+1)(2T_i+1)\}}(T_i 1 T_z 0\mid T_f T_z)(I_i 1 M_i \nu \mid I_f M_f) \\ \times\big(\sqrt{\{(2J'+1)(2T'+1)\}}\delta_{J_c J_c'}\delta_{T_c T_c'}W(J_c J I_f 1;\, I_i J')W(T_c T T_f 1;\, T_i T') \\ \times\langle\chi_{J';T'}|||\,(M1)_{1;1}\,|||\chi_{J;T}\rangle \\ +(-1)^{\Delta J_c+\Delta T_c-I_i+I_f-T_i+T_f}\sqrt{\{(2J_c'+1)(2T_c'+1)\}}\delta_{JJ'}\delta_{TT'} \\ \times W(J J_c I_f 1;\, I_i J_c')W(T T_c T_f 1;\, T_i T_c')\langle\Phi_{J_c';T_c'}|||\,(M1)_{1;1}\,|||\Phi_{J_c;T_c}\rangle\big).\end{aligned} \tag{5.73}$$

Transitions between analogue and anti-analogue states with the structure of equation (5.72) have $\chi_{JM;TT_z} = \phi_{jm;\frac{1}{2}\mu}$ and there is no contribution from the core states because $J_c = 0$. Furthermore, since the single-particle state is the same before and after the transition and because the initial and final isospins are different the effective part of the $M1$ operator is

$$(M1)_{1\nu;10} \doteq \frac{1}{2}\sqrt{\left(\frac{3}{4\pi}\right)}\{g_n(j)-g_p(j)\}\mu_N \mathrm{j}_\nu \tau_z$$

where $\doteq$ means the effective part of the operator. Thus

$$\begin{aligned}\langle\phi_{jj;\frac{1}{2}\frac{1}{2}}|\,(M1)_{10;10}\,|\phi_{jj;\frac{1}{2}\frac{1}{2}}\rangle &= \frac{1}{2}\sqrt{\left(\frac{3}{4\pi}\right)}\{g_n(j)-g_p(j)\}j\mu_N \\ &= \langle\phi_{j;\frac{1}{2}}|||\,(M1)_{1;1}\,|||\phi_{j;\frac{1}{2}}\rangle(j1j0\mid jj)(\tfrac{1}{2}1\tfrac{1}{2}0\mid\tfrac{1}{2}\tfrac{1}{2})\end{aligned}$$

so that

$$\langle\phi_{j;\frac{1}{2}}|||\,(M1)_{1;1}\,|||\phi_{j;\frac{1}{2}}\rangle = \frac{1}{2}\sqrt{\left(\frac{3}{4\pi}\right)}\{g_n(j)-g_p(j)\}\sqrt{\{3j(j+1)\}}\mu_N.$$

Consequently for transitions between the analogue and anti-analogue states of equations (5.72)

$$\begin{aligned}B(M1) &= \frac{9}{8\pi}(2T_i+1)(T_i 1 T_z 0\mid T_f T_z)^2 W^2(T_i T_f \tfrac{1}{2}\tfrac{1}{2};\, 1 T_0)j(j+1) \\ &\quad\times\{g_n(j)-g_p(j)\}^2\mu_N^2 \\ &= \frac{3}{2\pi}\frac{T_0}{(2T_0+1)^2}j(j+1)\{g_n(j)-g_p(j)\}^2\mu_N^2.\end{aligned}$$

In writing the last line of this equation we have made use of the fact that $T_i = T_0 + \frac{1}{2}$ and that $T_z = T_f = T_0 - \frac{1}{2}$.

By using equations (5.60) and (5.61) it follows that

$$B(M1) = \frac{3}{2\pi}\frac{T_0}{(2T_0+1)^2}\left(\frac{j+1}{j}\right)(\mu_p + l - \mu_n)^2\mu_N^2 \quad \text{if} \quad j = l + \tfrac{1}{2} \tag{5.74a}$$

$$= \frac{3}{2\pi}\frac{T_0}{(2T_0+1)^2}\left(\frac{j}{j+1}\right)(\mu_n + l + 1 - \mu_p)^2\mu_N^2 \quad \text{if} \quad j = l - \tfrac{1}{2}. \tag{5.74b}$$

In Table 5.5 we list the values of $B(M1)$ for the analogue to anti-analogue transitions described by equations (5.74). It is clear that when the single particle has $j = l + \frac{1}{2}$ the transition is strong, whereas when $j = l - \frac{1}{2}$ it is weak. Consequently we arrive at the rule (Maripuu 1969)

TABLE 5.5

Values of $B(M1)$ for analogue to anti-analogue state transitions

T_0	$s_{\frac{1}{2}}$	$p_{\frac{3}{2}}$	$p_{\frac{1}{2}}$	$d_{\frac{5}{2}}$	$d_{\frac{3}{2}}$	$f_{\frac{7}{2}}$	$f_{\frac{5}{2}}$	$g_{\frac{9}{2}}$
1	2·00	1·63	0·073	1·89	0·052	2·30	0·011	2·79
2	1·44	1·18	0·053	1·36	0·038	1·65	0·008	2·01
3	1·10	0·90	0·040	1·04	0·029	1·27	0·006	1·54
4	0·89	0·73	0·033	0·84	0·023	1·02	0·005	1·24
5	0·74	0·61	0·027	0·70	0·019	0·85	0·004	1·04

The values are given in Weisskopf units ($B_W(M1) = 1{\cdot}76\mu_N^2$) and are deduced from equations (5.74) using $\mu_p = 2{\cdot}79$ and $\mu_n = -1{\cdot}91$. T_0 is the isospin of the core which is assumed to have angular momentum zero.

If the analogue and anti-analogue states are described by the wave functions $[\Phi_{J_c=0;T_0} \times \phi_{j;\frac{1}{2}}]_{jm;TT_z}$ ($T = T_0 + \frac{1}{2}$ for the former and $T = T_0 - \frac{1}{2}$ for the latter) the $M1$ transition between the two states will be strong if $j = l + \frac{1}{2}$ and weak if $j = l - \frac{1}{2}$.

The data concerning these transitions have been collected by Klapdor (1971) and are shown in Fig. 5.6. It is clear that the $f_{\frac{7}{2}} \to f_{\frac{7}{2}}$ transitions observed at the end of the *sd* shell and the $g_{\frac{9}{2}} \to g_{\frac{9}{2}}$ decays near $A = 60$ have strengths of the order of one Weisskopf unit. This is somewhat smaller than the predictions given in Table 5.5, but, as already stated, configuration mixing leads to smaller effective values of μ_p and μ_n than the free-particle values. These transitions are about a factor of ten

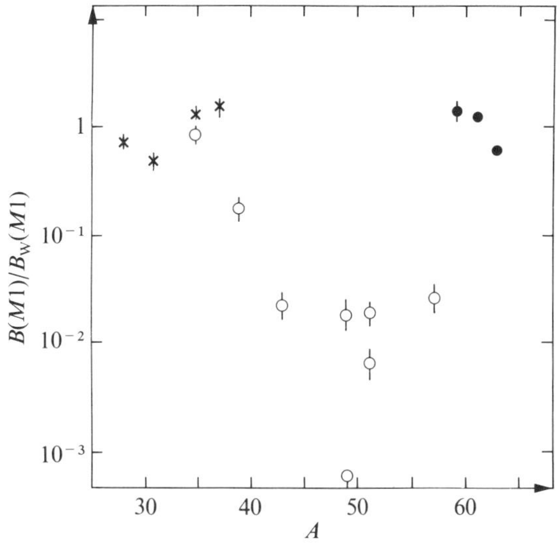

Fig. 5.6. Analogue to anti-analogue transition strengths in Weisskopf units. Crosses indicate $f_{\frac{7}{2}} \to f_{\frac{7}{2}}$ transitions, open circles $p_{\frac{3}{2}} \to p_{\frac{3}{2}}$ decays and filled circles $g_{\frac{9}{2}} \to g_{\frac{9}{2}}$ transitions. The figure is taken from the work of Klapdor (1971).

stronger than the average $M1$ T-allowed strengths observed for $21 \leq A \leq 44$ (see Fig. 5.5), and consequently the $M1$ decays between these analogue and anti-analogue states are consistent with the assumed structure for the states, equations 5.72.

On the other hand, most of the $p_{\frac{3}{2}} \to p_{\frac{3}{2}}$ transitions appear to be weak and we shall now examine what can be learned about nuclear structure from this result. In deriving equations (5.74) from equation (5.73) we assumed that $J_c = J'_c = 0$. Consequently only the extra core particle contributed to the transition rate. Although the dominant configuration in both the analogue and anti-analogue state may have $J_c = 0$, a $J_c = 1$ component in either state gives rise to a contribution to the decay matrix element linear in the admixture coefficient. To illustrate what effect this can have we consider the simple case in which the core is either $(j_1)^2$ with $T_c = 0$ or 1 or the configuration $(j_1)^{2j_1+1}$ with isospin $(2j_1+1)/2$ or $(2j_1-1)/2$. If one includes only $J_c = 0$ and 1 states the former always has the larger isospin and the latter has the smaller value. In this approximation the analogue state arises purely from coupling to $J_c = 0$ whereas the anti-analogue can be realized in two ways:

$$\Psi_{jm;T_0-\frac{1}{2}T_0-\frac{1}{2}} = \alpha[\Phi_{J_c=0;T_0} \times \phi_{j;\frac{1}{2}}]_{jm;T_0-\frac{1}{2}T_0-\frac{1}{2}} + \beta[\Phi_{J_c=1;T_0-1} \times \phi_{j;\frac{1}{2}}]_{jm;T_0-\frac{1}{2}T_0-\frac{1}{2}}. \quad (5.75)$$

Consequently $B(M1)$ governing the transition between these two states is

$$B(M1)=\frac{3}{2\pi}\frac{T_0}{(2T_0+1)^2}\mu_N^2\bigg(\alpha\{g_n(j)-g_p(j)\}\sqrt{\{j(j+1)\}} - \beta\sqrt{\left(\frac{2T_0-1}{T_0}\right)}\sqrt{\frac{4\pi}{3}}\langle\Phi_{J_c=1;T_0-1}|||(M1)_{1;1}|||\Phi_{J_c=0;T_0}\rangle\bigg)^2.$$

The core matrix element is simple to evaluate because when $n'=1$ or $2j$ there is only one coefficient $A_I(J_1=j_1, J_2=j_1,\ n=1, n'=2j)=1$ in the core wave function given by equation (5.64). Thus

$$\langle\Phi_{J_c=1;T_0-1}|||(M1)_{1;1}|||\Phi_{J_c;T_0}\rangle=\sqrt{\left(\frac{3}{4\pi}\right)}\sqrt{\left\{\frac{T_0(2T_0+1)}{3(2T_0-1)}\right\}}\{g_n(j_1)-g_p(j_1)\} \times\sqrt{\{j_1(j_1+1)\}}$$

so that

$$B(M1)=\frac{3}{2\pi}\frac{T_0}{(2T_0+1)^2}\mu_N^2\bigg(\alpha\{g_n(j)-g_p(j)\}\sqrt{\{j(j+1)\}} - \beta\sqrt{\left(\frac{2T_0+1}{3}\right)}\{g_n(j_1)-g_p(j_1)\}\sqrt{\{j_1(j_1+1)\}}\bigg)^2. \quad (5.76)$$

We now use equation (5.76) to interpret the weak transition between the $p_{\frac{3}{2}}$ analogue state in ${}^{49}_{21}Sc_{28}$ (the analogue of the ${}^{49}_{20}Ca_{29}$ ground state) observed at 11·56 MeV and the anti-analogue state at 3·08 MeV. Experimentally $B(M1)$ for this transition is (Gaarde *et al.* 1972)

$$B(M1)=2\times10^{-3}\ \mu_N^2$$

so that the transition is inhibited relative to the Weisskopf estimate by about a factor of 1000. The ${}^{48}_{20}Ca_{28}({}^3He, d){}^{49}_{21}Sc_{28}$ reaction (Erskine *et al.* 1966) leads to a value of $C^2\mathcal{S}=0{\cdot}6$ for $l=1$ transfer to the 3·08 MeV level. From equation (5.72b) it follows that the $T_0=4$ part of the anti-analogue state is

$$\alpha[\Phi_{J_c=0;T_0=4}\times\phi_{\frac{3}{2};\frac{1}{2}}]^{\frac{3}{2}}_{m;\frac{7}{2}\frac{7}{2}}=\alpha\left(\sqrt{\tfrac{8}{9}}\Phi_{00;44}\phi_{\frac{3}{2}m;\frac{1}{2}-\frac{1}{2}}-\frac{1}{\sqrt{9}}\Phi_{00;43}\phi_{\frac{3}{2}m;\frac{1}{2}\frac{1}{2}}\right),$$

and since this is the only component that will be populated in the stripping reaction it follows that

$$C^2\mathcal{S}=\tfrac{8}{9}\alpha^2=0{\cdot}6.$$

Thus in equation (5.75), $\alpha=\sqrt{0{\cdot}675}$ and $\beta=\pm\sqrt{0{\cdot}325}$. From equation (5.76) it follows that

$$\frac{B(M1)}{B_W(M_1)}=8{\cdot}9\times10^{-2}$$

when the free-particle values of g are used and the interference is assumed to be destructive. Detailed calculations show that the interference is destructive and in fact that the experimentally observed inhibition together with the correct spectroscopic factor can be obtained when the model space is extended to include the $J_c=2$ and 3 states of $(f_{\frac{7}{2}})^8$ (Maripuu 1970, Bloom *et al.* 1973).

The reason the destructive interference is so marked in this case is because the g factor of the $f_{\frac{7}{2}}$ core particles is large and consequently even a small admixture of $J_c=1$ can substantially change the calculated matrix element. On the other hand, if the g factor of the core nucleons is small, $J_c=1$ admixtures will barely affect the decay. For example, when the same values of α and β are assumed for the anti-analogue $\frac{7}{2}^-$ state in $^{37}_{17}Cl_{20}$ (observed at 3·10 MeV) and the admixed core state is assumed to be $(d_{\frac{3}{2}})^4_{J_c=1,T_0=1}$

$$\frac{B(M1)}{B_W(M1)}=0{\cdot}835$$

which is to be compared with the no-interference-value of 1·65 given in Table 5.5.

We can now understand the systematics shown in Fig. 5.6. For $35 \leq A \leq 40$ the core states are dominated by the configuration $(d_{\frac{3}{2}})^n$ which has a relatively small value of $\sqrt{\{j(j+1)\}}\{g_p(j)-g_n(j)\}=-1{\cdot}32$ compared to the $f_{\frac{7}{2}}$ value of 8·73. Consequently a $J_c \neq 0$ admixture into the anti-analogue state in this region will be difficult to detect and indeed the analogue to anti-analogue transitions do proceed at close to the single-particle speed. However, for a $p_{\frac{3}{2}}$ particle $\sqrt{\{j(j+1)\}}\{g_p(j)-g_n(j)\}=7{\cdot}36$ which is almost the same as for $f_{\frac{7}{2}}$ nucleons. Consequently from a study of the $p_{\frac{3}{2}}$ analogue to anti-analogue transitions in the region $40 \leq A \leq 60$ one learns that the states are far from pure. For the heavier nuclei, $A \geq 60$, the $g_{\frac{9}{2}}$ analogue to anti-analogue transition is again observed to be strong. In this case $\sqrt{\{j(j+1)\}}\{g_p(j)-g_n(j)\}=9{\cdot}62$ for $j=\frac{9}{2}$. Since the $p_{\frac{3}{2}}$ orbit has a similar value for this quantity, one must conclude that either the $J_c=1$ admixtures into the anti-analogue state are small or that the admixed core states are dominated by the $f_{\frac{5}{2}}$ orbit. In this latter case $\sqrt{\{j(j+1)\}}$ $\{g_p(j)-g_n(j)\}=-0{\cdot}59$, so that a large amount of this configuration can be admixed without changing the value of $B(M1)$.

6.2.6. l-Forbidden transitions. Transitions between single-particle states with $j=l+\frac{1}{2}$ and $j'=(l+2)-\frac{1}{2}$ are forbidden because the $M1$ operator of equation (5.54) cannot change the orbital angular momentum. Thus the observation of these transitions between single-particle neutron states gives evidence that core polarization (Arima *et al.* 1957) and meson

TABLE 5.6
l-Forbidden transitions between single-particle states

Nucleus	Transition	$\dfrac{B_{\mathrm{expt}}(M1)}{B_{\mathrm{W}}(M1)}$
${}^{39}_{19}\mathrm{K}_{20}$	$1s_{\frac{1}{2}} \to 0d_{\frac{3}{2}}$	$1{\cdot}5\times10^{-2}$
${}^{39}_{20}\mathrm{Ca}_{19}$	$1s_{\frac{1}{2}} \to 0d_{\frac{3}{2}}$	$8{\cdot}1\times10^{-3}$
${}^{57}_{28}\mathrm{Ni}_{29}$	$0f_{\frac{5}{2}} \to 1p_{\frac{3}{2}}$	$1{\cdot}1\times10^{-2}$
${}^{67}_{30}\mathrm{Zr}_{37}$	$1p_{\frac{3}{2}} \to 0f_{\frac{5}{2}}$	$2{\cdot}5\times10^{-3}$
${}^{87}_{37}\mathrm{Rb}_{50}$	$0f_{\frac{5}{2}} \to 1p_{\frac{3}{2}}$	$3{\cdot}2\times10^{-3}$
${}^{203}_{81}\mathrm{Tl}_{122}$	$3s_{\frac{1}{2}} \to 2d_{\frac{3}{2}}$	$1{\cdot}5\times10^{-3}$
${}^{209}_{82}\mathrm{Pb}_{127}$	$0i_{\frac{11}{2}} \to 1g_{\frac{9}{2}}$	$<2{\cdot}3\times10^{-2}$
${}^{209}_{83}\mathrm{Bi}_{126}$	$1f_{\frac{7}{2}} \to 0h_{\frac{9}{2}}$	$2{\cdot}4\times10^{-3}$

In the last column the ratio of the experimental $B(M1)$ to the Weisskopf estimate is tabulated ($B_{\mathrm{W}}(M1) = 1{\cdot}76\mu_{\mathrm{N}}$).

exchange (Arima and Huang Lin 1972) lead to a $[Y_2\times\sigma]_{1\nu}$ part of the effective $M1$ operator. For protons this addition to the basic operator could also originate in part from the $f(r)$ term considered in equation (5.7). In Table 5.6 we have tabulated $B(M1)/B_{\mathrm{W}}(M1)$ for those transitions that can be reasonably interpreted as being between single-particle or single-hole states. It is clear that these transitions are inhibited by about a factor of 100 compared to the Weisskopf estimate; that is, slowed down by an additional factor of 10 compared to the usual T-allowed $M1$ decays (see Fig. 5.5).

7. Electric quadrupole properties

It has already been shown (section 3 of this chapter) that to understand the observed $E2$ transition rates one must introduce an effective charge. Thus the electric quadrupole operator is

$$\begin{aligned}(E2)_{2\nu} &= \tfrac{1}{2}e_{\mathrm{p}}\sum_k \{1-(\tau_k)_z\}r_k^2 Y_{2\nu}(\theta_k,\phi_k)\\ &\quad+\tfrac{1}{2}e_{\mathrm{n}}\sum_k \{1+(\tau_k)_z\}r_k^2 Y_{2\nu}(\theta_k,\phi_k)\\ &=\tfrac{1}{2}(e_{\mathrm{p}}+e_{\mathrm{n}})\sum_k r_k^2 Y_{2\nu}(\theta_k,\phi_k)\\ &\quad+\tfrac{1}{2}(e_{\mathrm{n}}-e_{\mathrm{p}})\sum_k (\tau_k)_z r_k^2 Y_{2\nu}(\theta_k,\phi_k).\end{aligned} \tag{5.77}$$

If we define

$$e_{\rm p}/e = 1+\delta_{\rm p} \quad (5.78a)$$

$$e_{\rm n}/e = \delta_{\rm n}, \quad (5.78b)$$

the polarization charge $e\delta_{\rm p}$ is the change in the proton charge brought about by core polarization and $e\delta_{\rm n}$ is the effective electric charge for the neutrons.

We shall first discuss the expected and empirical properties of effective charges and then turn to special cases where the $E2$ matrix element can lead to information about nuclear structure.

7.1. *Effective charges*

The effective charge that is needed to explain the experimental $E2$ transition rates depends on the model space employed in the calculation. For example, in ${}^{42}_{20}\text{Ca}_{22}$ the gamma-ray transition between the first excited 2^+ state (observed at 1·524 MeV; see Fig. 2.1) and the ground state has

$$B(E2; 2^+ \to 0^+) = (81{\cdot}5 \pm 3{\cdot}0)e^2\,\text{fm}^4.$$

If both states involved are described as $(\nu f_{\frac{7}{2}})^2$ the effective neutron charge required to explain the transition is

$$e\delta_{\rm n} = (2{\cdot}08 \pm 0{\cdot}04)e.$$

If the entire $(0f, 1p)$ shell is used in the description of the states the required effective charge is reduced by about 20% and

$$e\delta_{\rm n} = (1{\cdot}73 \pm 0{\cdot}03)e.$$

Finally, if one diagonalizes the shell-model Hamiltonian using as basis states

$$(0f, 1p)^2 \quad \text{and} \quad (0f, 1p)^4(0d, 1s)^{-2}$$

the effective charge needed drops to

$$e\delta_{\rm n} = (0{\cdot}95 \pm 0.02)e.$$

These values (Brown *et al.* 1974) for the effective charge have all been calculated using harmonic oscillator wave functions to evaluate $\langle r^2 \rangle$ with $\hbar\omega = 41/A^{\frac{1}{3}}$ MeV. Since the oscillator spatially localizes the particles more than a square well or a Woods–Saxon potential, it is clear that $\langle r^2 \rangle$ will be somewhat larger for these latter two wells, and consequently the required value of $\delta_{\rm n}$ will be slightly smaller. Except for very loosely bound systems this is only about a 10% effect.

The polarization charges arise from two sources. First the valence particles themselves can be excited to orbits above the Fermi sea. For

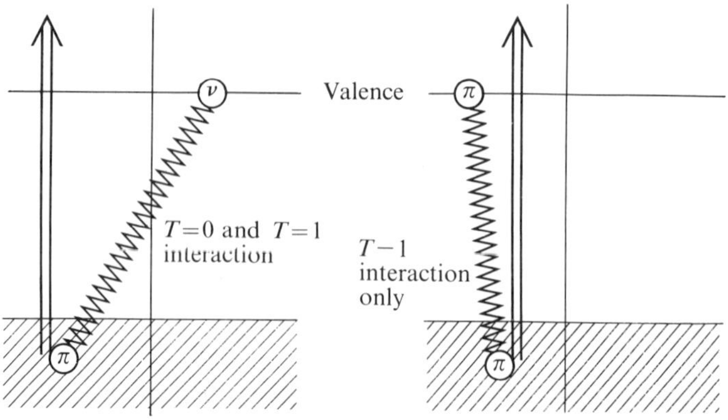

Fig. 5.7. Schematic illustration of the origin of polarization charge. Because a neutron can interact with a core proton through both the $T=0$ and $T=1$ potential, it is more effective in exciting protons out of the core.

example, if we deal with particles in an oscillator shell with energy $(N+\frac{3}{2})\hbar\omega$ these valence particles can be excited to states with energy $(N+2+\frac{3}{2})\hbar\omega$. Secondly, the valence nucleons can polarize the core and hence excite the core particles as shown in Fig. 5.7. Since only protons carry an electric charge it is only proton excitation that contributes to δ_p and δ_n. Two protons can only interact via the $T=1$ part of the nucleon–nucleon potential whereas the neutron–proton interaction involves both the $T=1$ and $T=0$ potential. Since the $T=0$ interaction is stronger than the $T=1$ (i.e. the deuteron is bound whereas the di-neutron is not) it follows that the neutron is more effective in producing proton core polarization (Federman and Zamick 1969). Consequently one would expect δ_n to be larger than δ_p. This expectation is borne out throughout the periodic table. For example, if one compares the lifetimes of the first excited 2^+ state in ${}^{42}_{20}Ca_{22}$ and ${}^{42}_{22}Ti_{20}$ one finds

$$B(E2;\,{}^{42}Ca,\,2^+\to 0^+)=81{\cdot}5e^2\,\mathrm{fm}^4$$

and

$$B(E2;\,{}^{42}\mathrm{Ti},\,2^+\to 0^+)=134e^2\,\mathrm{fm}^4.$$

On the basis of a pure $0f_{\frac{7}{2}}$-description of these states it follows that

$$\frac{B(E2;\,{}^{42}\mathrm{Ti},\,2^+\to 0^+)}{B(E2;\,{}^{42}\mathrm{Ca},\,2^+\to 0^+)}=\left(\frac{1+\delta_p}{\delta_n}\right)^2.$$

As already stated $\delta_n=2{\cdot}08$ to explain the ${}^{42}_{20}Ca_{22}$ data. Consequently to interpret ${}^{42}_{22}Ti_{20}$ one needs the somewhat smaller value $\delta_p=1{\cdot}67$.

The same effect has been observed in the lead region (Astner *et al.* 1972) and in the $(0d, 1s)$ shell (Elliott and Wilsdon 1968). In the former region

$$\delta_n - \delta_p \cong 0{\cdot}2,$$

whereas for the few transitions considered in the latter calculation the effect was much smaller

$$\delta_n - \delta_p \cong 0{\cdot}07.$$

If one assumes that the majority of the $2\hbar\omega$ electric quadrupole strength is contained in a state that can be characterized as a vibration of the nuclear surface (Bohr and Mottelson 1975), states with larger value of l will have larger values of δ_p and δ_n. This follows because the wave functions of these states tend to be more strongly localized near the nuclear surface and consequently can interact more strongly with a surface oscillation. There is some data near $^{208}_{82}Pb_{126}$ supporting this contention. For example, Astner *et al.* (1972), find

$$\delta_n(0i_{\frac{13}{2}}) \cong 0{\cdot}96$$

whereas

$$\delta_n(1g_{\frac{9}{2}}) \cong 0{\cdot}84$$

and

$$\delta_n(2p) \cong 0{\cdot}75.$$

Finally one can ask about the j dependence of the effective charge, namely the difference between $\delta(j = l + \frac{1}{2})$ and $\delta(j = l - \frac{1}{2})$. At present there is no experimental evidence concerning this point. However, the effect is expected to be small since in first-order perturbation theory it can be shown that this difference will vanish if the nucleon–nucleon interaction is velocity independent (Federman and Zamick 1969).

From the preceding arguments it is seen that one can understand qualitatively the observed trend of the effective charges. On the other hand, quantitative agreement between experiment and theory, when realistic two-body interactions are used to calculate δ_p and δ_n, has not yet been achieved (Barrett and Kirson 1973, Ellis and Osnes 1977). For example, it has been shown (Halbert *et al.* 1971) that transitions between the yrast levels in the $(0d, 1s)$ nuclei with $18 \leq A \leq 22$ can be understood (within about a factor of three) by use of a state-independent polarization charge of $\delta_p = \delta_n = 0{\cdot}5$. Theoretical values of these polarization charges have been calculated by Kirson (1974) and found to be

$$\delta_p \approx 0{\cdot}11 \text{———} 0{\cdot}25$$

$$\delta_n \approx 0{\cdot}30 \text{———} 0{\cdot}45.$$

Thus the calculated values are considerably smaller than needed to fit experiment.

It can be seen from the following simple arguments that the polarization charge is positive. In first-order perturbation theory the polarization charge is given by equations (5.28) and (5.29). For simplicity let us take the potential that induces the mixing to have Gaussian form, namely $V = \bar{V}_0 \exp\{-\beta^2(\underline{r}_1 - \underline{r}_2)^2\}$. Thus the $F_\lambda(r_k, r_n)$ of equation (5.28) are given by equation (4.4),

$$F_\lambda(r_1, r_2) = \bar{V}_0(2\lambda+1)i^\lambda e^{-\beta^2(r_1{}^2+r_2{}^2)} J_\lambda(-2i\beta^2 r_1 r_2)$$
$$\cong \frac{2^\lambda \bar{V}_0 \beta^{2\lambda} r_1^\lambda r_2^\lambda}{(2\lambda-1)!!}. \tag{5.79}$$

In writing the second line of this equation we have expanded F_λ in powers of $\beta^2 r_1 r_2$ and kept only the leading term. In this approximation the change in the $E2$ operator brought about by core polarization is

$$C_2(r) Y_{2\mu}(\theta, \phi) = \frac{16\pi}{15} \bar{V}_0 \beta^4 r^2 Y_{2\mu}(\theta, \phi)$$
$$\times \sum_q \left\{ \frac{1}{E_{I_i} - E_{I_f} - E_2(q)} + \frac{1}{E_{I_f} - E_{I_i} - E_2(q)} \right\}$$
$$\times |\langle \Phi_{2q} \| \sum_k r_k^2 Y_2(\theta_k, \phi_k) \| \Phi_0 \rangle|^2. \tag{5.80}$$

Provided we deal with low-lying nuclear states the energy of the core-excited state will be greater than $|E_{I_i} - E_{I_f}|$. Consequently the energy denominator in equation (5.80) will be negative. Since the nucleon–nucleon interaction is attractive $\bar{V}_0$ is negative. Thus the coefficient of $r^2 Y_2(\theta, \phi)$, which is to be interpreted as the polarization charge, is positive. Because all contributions to the polarization charge have the same sign it is usually found that quadrupole moments and $E2$ transition rates are strongly enhanced. In Fig. 5.8 a histogram of the $E2$ transition strengths for nuclei with $A \leq 45$ is shown (Endt and Van der Leun 1974). From this Figure it is apparent that for light nuclei the average $E2$ transition rate is enhanced over the Weisskopf estimate by a factor of between five and ten. For deformed nuclei this enhancement is even greater. For example, the quadrupole moment of the ground state of ${}^{175}_{71}\mathrm{Lu}_{104}$ is (Fuller and Cohen 1968)

$$|Q/e| = 560 \text{ fm}^2.$$

If one assumes this is due to a single particle in the $g_{\frac{7}{2}}$ orbit, the quadrupole moment (see equation (5.82)) is

$$|Q/e| = \tfrac{2}{3}\langle r^2 \rangle,$$

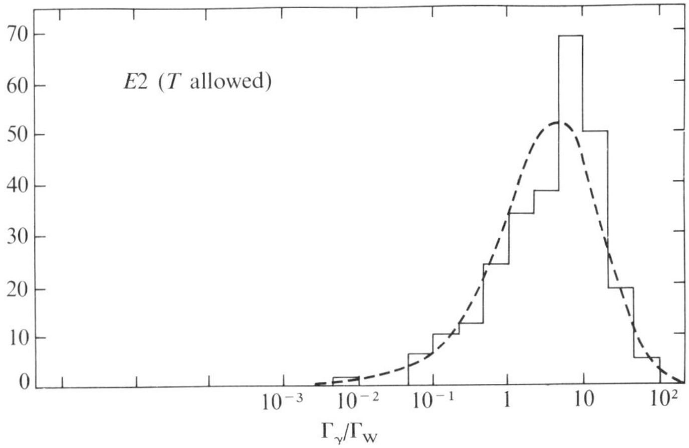

Fig. 5.8. $B(E2)/B_W(E2)$ for T-allowed $E2$ transitions in nuclei with $A \leq 45$.

and if one uses harmonic oscillator wave functions with $\hbar\omega = 41/A^{\frac{1}{3}}$ to evaluate $\langle r^2 \rangle$

$$|Q/e| = 20{\cdot}7 \text{ fm}^2.$$

Thus this quadrupole moment, which is proportional to the matrix element itself (and not the square as are the $B(E2)$ values shown in Fig. 5.8) is enhanced by a factor of 27.

7.2. *Quadrupole moments*

It can be verified that the polarization charge is indeed positive from a study of nuclear quadrupole moments. This moment is defined as

$$Q = \sqrt{\left(\frac{16\pi}{5}\right)} \langle \psi_{IM=I} | (E2)_0 | \psi_{IM=I} \rangle \tag{5.81}$$

where $(E2)_0$ is the z component of the $E2$ operator in equation (5.77). From equation (A2.23) one can evaluate this moment for the single-particle configuration

$$\begin{aligned} Q_{\text{sp}} &= 2(j2\tfrac{1}{2}0 \mid j\tfrac{1}{2})(j2j0 \mid jj) e\langle r^2 \rangle_j \\ &= -\frac{(2j-1)}{2(j+1)} e\langle r^2 \rangle_j. \end{aligned} \tag{5.82}$$

In writing the last line of this equation use has been made of the explicit form of the Clebsch–Gordan coefficients given in Table (A1.4), and $\langle r^2 \rangle_j$ is the expectation value of r^2 in the single-particle state j,

$$\langle r^2 \rangle_j = \int_0^\infty R_j^2 r^4 \, \mathrm{d}r.$$

Since $\langle r^2\rangle_j$ is positive it follows that the sign of the single-particle quadrupole moment is negative for positive e.

As already discussed, the ground state of the ${}^{17}_{8}O_9$ nucleus can be described as a single neutron in the $0d_{\frac{5}{2}}$ orbit. The quadrupole moment of this nucleus is (Fuller and Cohen 1968)

$$Q/e = -2{\cdot}6\ \mathrm{fm}^2.$$

Thus the polarization charge δ_n is positive as expected.

As discussed in Appendix 3 the electric quadrupole operator is a quasi-spin tensor of rank one. Thus the number dependence of its matrix elements have an extremely simple form given by equation (A3.31). When this result is combined with equation (5.82) one finds that

The quadrupole moment of the n-particle state $(j^n)_I$ with angular momentum $I = j$ and seniority one is

$$Q = -\frac{(2j+1-2n)}{2(j+1)}\, e\langle r^2\rangle_j. \tag{5.83}$$

From equation (5.83) it follows that the quadrupole moment of a single hole should be equal to but with opposite sign to that of a single particle. If ${}^{37}_{17}Cl_{20}$ is assumed to be a single $d_{\frac{3}{2}}$ proton outside an inert ${}^{36}_{16}S_{20}$ core its quadrupole moment should be equal and opposite to that of ${}^{39}_{19}K_{20}$ which is a single $d_{\frac{3}{2}}$ proton hole. Experimentally

$$Q({}^{37}_{17}Cl_{20})/e = -6{\cdot}2\ \mathrm{fm}^2$$

whereas

$$Q({}^{39}_{19}K_{20})/e = 4{\cdot}9\ \mathrm{fm}^2.$$

According to equation (5.83) the quadrupole moment of ${}^{39}_{19}K_{20}$ should be

$$\begin{aligned} Q({}^{39}_{19}K_{20})/e &= \tfrac{2}{5}\langle r^2\rangle \\ &= 4{\cdot}8\ \mathrm{fm}^2 \end{aligned}$$

when the radial integral is evaluated using harmonic oscillator wave functions with $\hbar\omega = 41/(39)^{\frac{1}{3}}$. Thus the quadrupole moment of the ${}^{39}_{19}K_{20}$ ground state is in excellent agreement with the single-particle estimate and indicates that $\delta_p \cong 0$. On the other hand, the ground state of ${}^{37}_{17}Cl_{20}$ is not as pure and requires $\delta_p \cong 0{\cdot}29$.

Quadrupole moments have been measured not only for nuclear ground states but also for excited states. The experimental values of these moments have been compiled by Fuller and Cohen (1968) and Christy and Hausser (1973).

7.3. $\Delta T \neq 0$ *transitions*

In section 7.1 we cited several examples that indicated that the neutron polarization charge was somewhat larger than that of the proton. Thus

$$e_p - e_n \lesssim e.$$

Since all T-changing $E2$ transitions are proportional to this quantity (see equation (5.77) it follows that

$\Delta T = 1$ $E2$ transitions are expected to have $B(E2)$ values less than, or of the order of, a Weisskopf unit.

Some data concerning this rule in light nuclei are tabulated in Table 5.7

TABLE 5.7
Isospin-retarded transitions

Nucleus	Energy Initial	Energy Final	Spin (isospin) Initial	Spin (isospin) Final	τ_{mean} (picoseconds)	$B(E2)/B_W(E2)$
$^{14}_{7}N_7$	7·03	2·31	$2^+(0)$	$0^+(1)$	1·0	0·17
$^{36}_{18}Ar_{18}$	6·61	0	$2^+(1)$	$0^+(0)$	0·12	0·08
$^{38}_{19}K_{19}$	2·41	0·46	$2^+(1)$	$1^+(0)$	13·1	0·29

and show that the expected effect indeed occurs. In addition to these $\Delta T = 1$ transitions one can also learn about the isovector part of the $E2$ operator by looking at analogous transitions that occur in mirror nuclei. Many cases of this are given in the tabulations of Endt and Van der Leun (1974, 1974a). For example, in both $^{25}_{13}Al_{12}$ and $^{25}_{12}Mg_{13}$ the first excited state is $\frac{1}{2}^+$ and the lifetime for decay to the $\frac{5}{2}^+$ ground state has been measured. In the former the excited state lies at 450 keV and has a mean lifetime of $3{\cdot}3\times10^{-9}$ s whereas for the latter the lifetime is $4{\cdot}86\times10^{-9}$ s and the level is at 590 keV. Thus

$$B(E2;\, ^{25}_{13}Al_{12};\, \tfrac{1}{2}^+ \rightarrow \tfrac{5}{2}^+) = 13{\cdot}35e^2\ \text{fm}^4$$

and

$$B(E2;\, ^{25}_{12}Mg_{13};\, \tfrac{1}{2}^+ \rightarrow \tfrac{5}{2}^+) = 2{\cdot}34e^2\ \text{fm}^4.$$

If one assumes the nuclear states have good isospin, the isoscalar part of the $E2$ operator has the same value for each transition. On the other hand, the isovector contribution will be proportional to the isospin Clebsch

$$(T1T_z0\,|\,TT_z) = T_z/\sqrt{\{T(T+1)\}}$$

and consequently changes sign for the two nuclei. Thus the isovector contribution to $B(E2)$ is

$$B(E2;\text{isovector}) = \left|\frac{\sqrt{\{B(E2;{}^{25}_{13}\mathrm{Al}_{12})\}} - \sqrt{\{B(E2;{}^{25}_{12}\mathrm{Mg}_{13})\}}}{2}\right|^2$$

$$= 1{\cdot}13e^2\ \mathrm{fm}^4.$$

From Tables 5.1 and 5.2 it follows that the Weisskopf estimate for an $E2$ transition in the $A = 25$ system is

$$B_{\mathrm{W}}(E2) = 4{\cdot}34e^2\ \mathrm{fm}^4$$

so that

$$\frac{B(E2;\text{isovector})}{B_{\mathrm{W}}(E2)} = 0{\cdot}26.$$

7.4. Properties of the half-filled shell

$\Delta T = 0$ electric quadrupole transitions between low-lying states are generally predicted and observed to be enhanced. On the other hand, it sometimes happens that a simple configuration assignment leads to the prediction of a cancellation in the $E2$ matrix element. Such predictions are likely to be very sensitive to configuration mixing, and consequently one may learn about configuration purity from a study of nuclei with these predicted properties.

In Appendix 3 we discussed the quasi-spin dependence of various operators and in particular found that the $E2$ operator is a quasi-spin vector. Because of this the number dependence of its matrix elements can be written down in a simple fashion. From equation (A3.31) it follows that if the initial and final states have the same seniority

$$\langle (j^n)_{I'v} \| E2 \| (j^n)_{Iv} \rangle = \frac{2j+1-2n}{2j+1-2v} \langle (j^v)_{I'v} \| E2 \| (j^v)_{Iv} \rangle. \qquad (5.84)$$

Thus it is sufficient to calculate transition rates in the v-particle state and use equation (5.84) to find the value in the more complicated n-particle configuration. It also follows from equation (5.84) that

> *$E2$ matrix elements between states of the same seniority vanish for the half-filled shell.*

This rule was also deduced in Chapter 3, section 2.3, by using the particle-hole conjugation operator.

An interesting example of this selection rule is in ${}^{136}_{54}\mathrm{Xe}_{82}$. The ground state of ${}^{133}_{51}\mathrm{Sb}_{82}$ corresponds to a single $0g_{\frac{7}{2}}$ proton moving outside the inert ${}^{132}_{50}Sn_{82}$ core. The first excited state lies at 963 keV and is interpreted as the $1d_{\frac{5}{2}}$ single-particle level. Thus in first approximation the yrast states with spin 0^+, 2^+, 4^+, and 6^+ that occur in the nuclei ${}^{134}_{52}Te_{82}$ and ${}^{136}_{54}Xe_{82}$ may be interpreted as arising from the configurations $(\pi g_{\frac{7}{2}})^2$ and $(\pi g_{\frac{7}{2}})^4$, respectively. If this is true one would expect the $\Delta v = 0$ transitions in ${}^{136}_{54}Xe_{82}$ to be severely inhibited. An isomeric state, which has been interpreted as the yrast 6^+ level, has been observed at 1·892 MeV (Carraz *et al.* 1970). The mean lifetime for its decay to the yrast 4^+ at 1·695 MeV is 4×10^{-6} s. That is

$$B(E2; 6^+ \rightarrow 4^+) = 0{\cdot}69 e^2 \text{ fm}^4$$

whereas the Weisskopf estimate for this transition strength is

$$B_W(E2; 6^+ \rightarrow 4^+) = 41{\cdot}5 e^2 \text{ fm}^4.$$

Thus the transition is severely inhibited in agreement with the prediction of the simple configuration assignment. Wildenthal and Larson (1971) have shown that this retardation remains $(B(E2; 6^+ \rightarrow 4^+) = 3{\cdot}5 e^2 \text{ fm}^4)$ even when one carries out a calculation which includes the $d_{\frac{5}{2}}$ orbit.

7.5. *Particle-hole multiplets*

From equation (5.83) it follows that the quadrupole moment of a single hole has the opposite sign to that of a single particle. This, of course, is exactly the result proved in Chapter 3, section 2.3; namely, if one deals with holes one must change the sign of the effective charge. When there is a single proton in the state j and a single neutron hole in the same state j, the reduced $E2$ matrix element is therefore given by

$$\begin{aligned}\langle[\pi j \times (\nu j)^{-1}]_{I_f} \| E2 \| [\pi j \times (\nu j)^{-1}]_{I_i}\rangle \\ = \sqrt{\left(\frac{5}{4\pi}\right)} \sqrt{\{(2I_i+1)(2j+1)\}} (j2\tfrac{1}{2}0 \mid j\tfrac{1}{2}) W(jjI_f2; I_ij) \\ \times \langle r^2 \rangle [-e_n + (-1)^{I_i - I_f} e_p]. \qquad (5.85)\end{aligned}$$

Thus one arrives at the rule

> *$\Delta I = 2$ transitions between states of the particle-hole multiplet $[\pi j \times (\nu j)^{-1}]_{IM}$ have matrix elements proportional to $(e_p - e_n)$ and consequently are expected to be inhibited compared with the Weisskopf estimate.*

This effect is illustrated by considering the $6^+ \rightarrow 8^+$ transition in ${}^{90}_{41}Nb_{49}$. If one assumes that ${}^{90}_{40}Zr_{50}$ is a closed core these two states should be the 6^+

and 8^+ members of the $[\pi g_{\frac{9}{2}} \times (\nu g_{\frac{9}{2}})^{-1}]$ configuration. The mean gamma-ray lifetime for decay of the 122 keV 6^+ state is 137×10^{-6} s (Holland *et al.* 1971). Thus

$$B(E2; 6^+ \rightarrow 8^+) = 0{\cdot}22e^2 \text{ fm}^4,$$

and compared to the Weisskopf estimate ($B_W(E2) = 23{\cdot}9e^2 \text{ fm}^4$), this corresponds to a transition inhibited by more than a factor of 100.

From equations (5.10) and (5.85) it follows that the theoretical value of $B(E2)$ based on the $[\pi g_{\frac{9}{2}} \times (\nu g_{\frac{9}{2}})^{-1}]$ configuration is

$$B(E2; 6^+ \rightarrow 8^+) = \frac{119}{4719\pi}(e_p - e_n)^2 \, |\langle r^2 \rangle_{g_{\frac{9}{2}}}|^2.$$

An analysis of the $38 \leq Z \leq 44$, $N = 50$ data has shown that if harmonic oscillator wave functions are used to evaluate $\langle r^2 \rangle$ the proton transition rates in this region are best fitted by an effective charge of $e_p = 1{\cdot}72e$ (Gloeckner *et al.* 1972). Thus in order to fit the observed $B(E2)$ in ${}^{90}_{41}Nb_{49}$ one must have

$$(1 - e_p/e_n) = \pm 0{\cdot}122,$$

i.e. $e_n = 1{\cdot}51e$ or $1{\cdot}93e$. The former value is quite close to that deduced from the quadrupole moment of ${}^{87}_{38}Sr_{49}$, namely $Q/e = 30 \text{ fm}^2$ (Fuller and Cohen 1968) which gives

$$e_n = 1{\cdot}68e.$$

7.6. *Signature rules*

A second consequence of the fact that the effective charge of a particle and hole have opposite signs concerns transitions between states of the configuration $[(\pi j)^n \times (\nu j)^{-n}]$. As shown in section 6.1.3, the wave functions describing these states, (5.64), have a definite signature. Thus

> *E2 matrix elements between states of the configuration* $[(\pi j)^n \times (\nu j)^{-n}]$ *that have the same signature will be proportional to* $(e_p - e_n)$ *and consequently will proceed at speeds comparable to or smaller than the single-particle estimate. Those involving states with opposite signature will be proportional to* $(e_p + e_n)$ *and hence be enhanced.*

In Fig. 5.4 we show the states of ${}^{48}_{22}Ti_{26}$ together with their predicted signatures when the $(\pi f_{\frac{7}{2}})^2 (\nu f_{\frac{7}{2}})^{-2}$ configuration spaced is used. According to the above rule the $E2$ transition $2_1^+ \rightarrow 0_1^+$ and $4_1^+ \rightarrow 2_1^+$ both involve a signature change and hence should be fast. Experimentally $B(E2; I_i \rightarrow I_f)/B_w(E2)$ is $14{\cdot}1 \pm 2{\cdot}2$ for the former and $8{\cdot}5 \pm 2{\cdot}5$ for the latter (Bardin *et al.* 1973). On the other hand, the second 2^+ to ground state transition involves no signature change and should be slow. Experimentally $B(E2; 2_2^+ \rightarrow 0_1^+)/B_W(E2) = 1{\cdot}34 \pm 0{\cdot}33$, and this is considerably slower than the preceding two examples. In addition, the second 4^+ to first 2^+ decay involves no signature change and experimentally $B(E2{:}\, 4_2^+ \rightarrow 2_1^+)/B_W(E2) \leq 0{\cdot}05$. Consequently the observed $E2$ behaviour of the low-lying states in ${}^{48}_{22}Ti_{26}$ can be understood qualitatively in terms of the $f_{\frac{7}{2}}$ model.

For nuclei described by $[(\pi j)^n \times (\nu j)^n]$ all states with the same isospin have the same signature. Since in this case we deal only with particles it follows that $\Delta T = 0$ transitions will be proportional to $(e_p + e_n)$. Consequently the signature rule in this case is the same as the isospin rule, i.e. $\Delta T = 1$ transitions are proportional to $(e_p - e_n)$.

7.7. *Weak-coupling multiplets*

In Chapter 2, section 3 the weak-coupling model was discussed and applied to the description of states that arose from the coupling of a single particle to the ground and excited states of the core. $B(E2)$ values governing transitions between the states $[j \times J_c]_{I_i M_i}$ and $[j \times J_c']_{I_f M_f}$ $(J_c \neq J_c')$ can be evaluated by the same methods as discussed in section 3.3 of Chapter 2 and one finds

$$B\{E2; (jJ_c)_{I_i} \rightarrow (jJ_c')_{I_f}\} = (2I_f+1)(2J_c+1)W^2(jJ_cI_f2; I_iJ_c')B(E2; J_c \rightarrow J_c'). \tag{5.86}$$

Consequently

The value of the E2 transition rate between the states $[j \times J_c]_{I_i M_i} \rightarrow [j \times J_c']_{I_f M_f}$ *is related by the geometrical factor in equation* (5.86) *to the E2 transition rate between the core states.*

One may use this rule in conjunction with other experiments to substantiate spin assignments and deduce the dominant structure of states in nuclei near closed shells. For example, in the ${}^{58}_{26}Fe_{32}({}^3He, d){}^{59}_{27}Co_{32}$ experiment the 1·099 MeV state is observed with $l = 1$ stripping. $C^2\mathcal{S} = 0{\cdot}11$ indicating the state is approximately 11% $p_{\frac{1}{2}}$ or $p_{\frac{3}{2}}$ coupled to the ${}^{58}_{26}Fe_{32}$ ground state (Vervier 1967). The observed mean lifetime for gamma decay to the $\frac{7}{2}^-$ ground state is $2{\cdot}9 \times 10^{-12}$ s. Since the Weisskopf estimate for the $M3$ transition $\frac{1}{2}^- \rightarrow \frac{7}{2}^-$ is $\tau_W = 2{\cdot}25 \times 10^{-4}$ s, the state is certainly $\frac{3}{2}^-$.

Thus, apart from a small $\pi p_{\frac{3}{2}}$ single-particle component, this state should be mainly

$$\Phi_{\frac{3}{2}M_i} = [(\pi f_{\frac{7}{2}})^{-1} \times \Psi_2({}^{60}_{28}\mathrm{Ni}_{32})]^{\frac{3}{2}}_{M_i}.$$

According to equation (5.86) the gamma-ray lifetime τ for decay of this level to the ${}^{59}_{27}\mathrm{Co}_{32}$ ground state

$$\Phi_{\frac{7}{2}M_f} = [(\pi f_{\frac{7}{2}})^{-1} \times \Psi_0({}^{60}_{28}\mathrm{Ni}_{32})]^{\frac{7}{2}}_{M_f}$$

should be

$$\tau = \tau_c \left(\frac{E_c}{E}\right)^5$$

where τ_c and E_c are the lifetime and energy of the 2^+ core state in ${}^{60}_{28}\mathrm{Ni}_{32}$. These quantities have the values $1{\cdot}15\times10^{-12}$ s and 1·333 MeV (Raman 1967) so that

$$\tau = 3{\cdot}02\times10^{-12}\ \mathrm{s}.$$

Consequently the structure of the 1·099 MeV $\frac{3}{2}^-$ state in ${}^{59}_{27}\mathrm{Co}_{32}$ is consistent with the wave function

$$\Psi_{\frac{3}{2}M} = \sqrt{0{\cdot}11}[\pi p_{\frac{3}{2}} \times \Psi_0({}^{58}_{26}\mathrm{Fe}_{32})]^{\frac{3}{2}}_{M} + \sqrt{0{\cdot}89}[(\pi f_{\frac{7}{2}})^{-1} \times \Psi_2({}^{60}_{28}\mathrm{Ni}_{32})]^{\frac{3}{2}}_{M}.$$

8. Collective states

In general the shell model predicts that the $E2$ transition strength between the yrast 2^+ state and the ground state is extremely strong. In fact most of the model space $0^+ \to 2^+$ strength is concentrated in this transition. It is also true that the strongest $E3$ transition between low-lying 3^- states and the ground state involves the yrast level. Because of this concentration of transition strength these states are often called collective. On the other hand, the lowest 1^- state has only a weak $E1$ branch to the ground state (see section 4 of this Chapter). We shall now show that these phenomena can be understood simply, at least for the two-particle system, if the nucleon–nucleon interaction is a separable potential; i.e. if its matrix elements can be written as

$$\langle \Psi_{IM;TT_z}(j_1j_2)|\, V\, |\Psi_{IM;TT_z}(j_3j_4)\rangle = -V_T D_I(j_1j_2)D_I(j_3j_4) - \mathcal{V}_T F_I(j_1j_2)F_I(j_3j_4). \tag{5.87}$$

From equation (A2.29) it follows that the surface-delta interaction satisfies this requirement since it can be written in the form of equation (5.87) with

$$D_I(j_1j_2) = \frac{(-1)^{j_1+l_1+\frac{1}{2}}\sqrt{\{(2j_1+1)(2j_2+1)\}}}{\sqrt{\{2(2I+1)(1+\delta_{j_1j_2})\}}}(j_1j_2\,\tfrac{1}{2}\,{-\tfrac{1}{2}}\,|\,I0) \tag{5.88a}$$

$$F_I(j_1j_2) = (-1)^{j_1+i_2}\left\{\frac{(2j_1+1)(2j_2+1)}{2(2I+1)(1+\delta_{j_1j_2})}\right\}^{\frac{1}{2}}(j_1j_2\tfrac{1}{2}\tfrac{1}{2}\,|\,I1) \tag{5.88b}$$

and

$$V_T = G_T\{1-(-1)^{1+T+l_3+l_4}\}$$

$$\mathscr{V}_T = G_T\{1+(-1)^T\}$$

where $G_T = V_T\bar{R}$. Since the surface-delta interaction gives suitable matrix elements for many shell-model calculations it follows that the valence nucleon–nucleon potential, at least approximately, satisfies equation (5.87).

To show that one eigenstate of the Hamiltonian contains most of the transition strength let us first consider even-parity $T=1$ states. In this case

$$\begin{aligned} V_1 &= 2G_1 \quad \text{when } I \text{ is even} \\ &= 0 \quad \text{when } I \text{ is odd} \end{aligned}$$

and

$$\mathscr{V}_1 = 0.$$

We expand the eigenfunction ψ_{IM} of the Hamiltonian

$$H = H_0 + V$$

in terms of the two-particle eigenfunctions Φ_{IM}, equation (1.42), which diagonalize H_0. Thus

$$\psi_{IM} = \sum_i C_I(i)\Phi_{IM}(i) \tag{5.89}$$

where

$$H_0\Phi_{IM}(i) = \tilde{\varepsilon}_i\Phi_{IM}(i)$$

$$\tilde{\varepsilon}_i = \varepsilon_{j_1} + \varepsilon_{j_2}$$

and the index i stands for (j_1j_2). Consequently the eigenvalue problem becomes

$$\sum_i (E_I - \tilde{\varepsilon}_i)C_I(i)\Phi_{IM}(i) = V\sum_i C_I(i)\Phi_{IM}(i).$$

If one multiplies this equation by $\Phi_{IM}(k)$ and makes use of equation (5.87) it follows that

$$(E_I - \tilde{\varepsilon}_k)C_I(k) = -2G_1 \sum_i C_I(i)D_I(i)D_I(k). \tag{5.90}$$

Thus

$$\sum_k C_I(k)D_I(k) = \left\{-2G_1 \sum_k \frac{D_I^2(k)}{(E_I - \tilde{\varepsilon}_k)}\right\} \sum_i C_I(i)D_I(i).$$

Since the summation over $C_I(i)D_I(i)$ is the same on each side of the equation, the energy eigenvalue is determined from the dispersion relationship

$$\sum_k \frac{D_I^2(k)}{E_I - \tilde{\varepsilon}_k} = -\frac{1}{2G_1}. \tag{5.91}$$

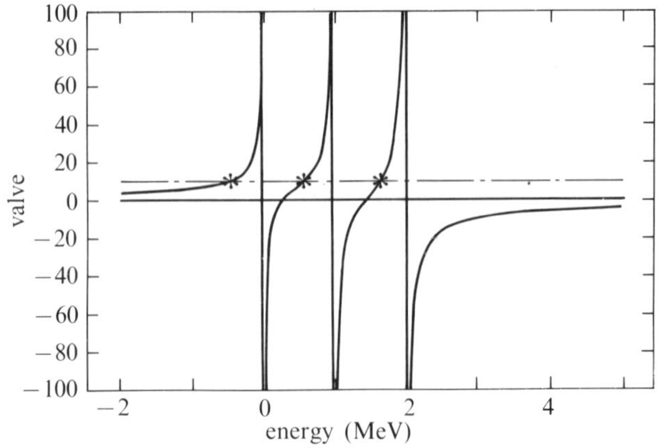

Fig. 5.9. Graph of $\sum_i [(2j+1)/(\tilde{\varepsilon}_i - E)] = 2/G_1$ (equation (5.91) when $I=0$). Three single-particle levels are assumed with $j=\frac{1}{2}$, $\frac{3}{2}$, and $\frac{5}{2}$ and energies $\tilde{\varepsilon}_i = 0$, 0·5, and 1 MeV, respectively. The right-hand side of the equation is constant as a function of E and is shown for $G_1>0$ by the dashed line. The intersections of the two curves (marked $*$) gives the three possible eigenvalues of E. The perpendicular lines are drawn for $E=0$, 1, and 2 MeV where the left-hand side of the equation becomes singular.

In Fig. 5.9 we have plotted the left- and right-hand sides of this equation as a function of E_I when $D_I(k)$ is given by equation (5.88a) and $I=0$. If there are n states $\Phi_{IM}(i)$ involved in the diagonalization all eigenvalues but one are trapped between the unperturbed energies $\tilde{\varepsilon}_1 - \tilde{\varepsilon}_2$, $\tilde{\varepsilon}_2 - \tilde{\varepsilon}_3, \ldots, \tilde{\varepsilon}_{n-1} - \tilde{\varepsilon}_n$. The remaining eigenvalue is pushed down in energy if G_1 is positive (attractive potential) and raised in energy if G_1 is negative (repulsive interaction). The larger is $|G_1|$, the further the eigenvalue is pushed.

From equation (5.90) it is clear that the expansion coefficients $C_I(k)$ are proportional to $D_I(k)/(E_I - \tilde{\varepsilon}_k)$. Since the eigenfunction must be normalized

$$C_I(k) = \left\{\sum_i \frac{D_I^2(i)}{(E_I - \tilde{\varepsilon}_i)^2}\right\}^{-\frac{1}{2}} \frac{D_I(k)}{E_I - \tilde{\varepsilon}_k}. \tag{5.92}$$

In general one gets non-vanishing values for all $C_I(k)$, the largest being the one whose unperturbed energy $\tilde{\varepsilon}_k$ is closest to E_I.

Let us now look at $I=0$. For this case $j_1 = j_2$, and from equation (A1.21) it follows that

$$D_0(j) = -\tfrac{1}{2}(-1)^l \sqrt{(2j+1)}.$$

Thus if we drop the trivial phase factor -1 and define ε_j as the energy of the single-particle state j

$$C_0(j)=(-1)^l\left\{\sum_{j_1}\frac{2j_1+1}{(E_0-2\varepsilon_{j_1})^2}\right\}^{-\frac{1}{2}}\frac{\sqrt{(2j+1)}}{E_0-2\varepsilon_j}.$$

Let us now look at the effect of the electric multipole operator $T^{(\mathscr{E})}_{\lambda\mu}$ on this state. In terms of annihilation and creation operators $T^{(\mathscr{E})}_{\lambda\mu}$ can be written as

$$T^{(\mathscr{E})}_{\lambda\mu}=\sqrt{\left(\frac{2\lambda+1}{4\pi}\right)}\sum_{j_1j_2m_1m_2}\langle r^\lambda\rangle_{j_1j_2}(-1)^{l_1+l_2+j_1-j_2}\sqrt{\left(\frac{2j_1+1}{2j_2+1}\right)}$$
$$\times(j_1\lambda\tfrac{1}{2}0\mid j_2\tfrac{1}{2})(j_1\lambda m_1\mu\mid j_2m_2)a^\dagger_{j_2m_2}a_{j_1m_1}$$

where the sums over j_1 and j_2 both run over all single-particle states. Thus

$$T^{(\mathscr{E})}_{\lambda\mu}\psi_{00}=\frac{1}{\sqrt{(4\pi)}}\left\{\sum_{j_1}\frac{(2j_1+1)}{(E_0-2\varepsilon_{j_1})^2}\right\}^{-\frac{1}{2}}\sum_{jj_2}(-1)^{l_2-\lambda}\frac{\sqrt{(2j+1)}}{(E_0-2\varepsilon_j)}\langle r^\lambda\rangle_{jj_2}$$
$$\times\sqrt{\{2(1+\delta_{jj_2})\}}(j\lambda\tfrac{1}{2}0\mid j_2\tfrac{1}{2})\Phi_{\lambda\mu}(jj_2) \qquad (5.93)$$

where

$$\langle r^\lambda\rangle_{jj_2}=\int R_{j_2}r^\lambda R_jr^2\,dr.$$

In equation (5.93) the sums over j and j_2 are each over all single-particle states included in the model space. Thus there is a contribution to $\psi_{\lambda\mu}$ arising from $[j\times j_2]_{\lambda\mu}$ and from $[j_2\times j]_{\lambda\mu}$. Since these two states correspond to the same physical situation only one of them should be included in the diagonalization of the Hamiltonian (i.e. in equation (5.89) the index i denotes an ordered pair $\{jj_1\}$ with say $j\geq j_1$). When this is realized and use is made of equation (A1.20) together with the fact that

$$(-1)^{l_2-\lambda}=(-1)^l$$

and

$$\Phi_{\lambda\mu}(jj_2)=-(-1)^{j+j_2-\lambda}\Phi_{\lambda\mu}(j_2j)$$

it follows that

$$T^{(\mathscr{E})}_{\lambda\mu}\psi_{00}=-\frac{1}{\sqrt{\pi}}\left\{\sum_{j_1}\frac{(2j_1+1)}{(E_0-2\varepsilon_{j_1})^2}\right\}^{-\frac{1}{2}}\sum_{j\geq j_2}\langle r^\lambda\rangle_{jj_2}D_\lambda(jj_2)$$
$$\times\left(\frac{1}{E_0-2\varepsilon_j}+\frac{1}{E_0-2\varepsilon_{j_2}}\right)\Phi_{\lambda\mu}(jj_2).$$

The resulting function is quite similar to the eigenfunction $\psi_{\lambda\mu}$ whose expansion coefficients are given by equation (5.92) when $I=\lambda$. There are two differences: first the radial matrix element $\langle r^\lambda\rangle_{jj_2}$ is contained in the summmation and second the energy denominators are different. However, for the case that the single-particle energies are degenerate and under the assumption that $\langle r^\lambda \rangle_{jj_2}$ is independent of (jj_2) it follows that

$$T^{(\mathscr{E})}_{\lambda\mu}\psi_{00} = -\frac{2}{\sqrt{\pi}}\left\{\sum_{j_1}\frac{(2j_1+1)}{(E_0-2\varepsilon)^2}\right\}^{-\frac{1}{2}}\left(\frac{E_\lambda-2\varepsilon}{E_0-2\varepsilon}\right)\langle r^\lambda\rangle \sum_{j\geq j_2}\frac{D_\lambda(jj_2)}{(E_\lambda-2\varepsilon)}\Phi_{\lambda\mu}(jj_2)$$

$$= -\frac{2}{\sqrt{\pi}}\left\{\sum_{j_1}\frac{(2j_1+1)}{(E_0-2\varepsilon)^2}\right\}^{-\frac{1}{2}}\left\{\sum_{j\geq j_2}\frac{D^2_\lambda(jj_2)}{(E_\lambda-2\varepsilon)^2}\right\}^{\frac{1}{2}}\langle r^\lambda\rangle\left(\frac{E_\lambda-2\varepsilon}{E_0-2\varepsilon}\right)\psi_{\lambda\mu} \tag{5.94}$$

where E_0 and E_λ are given by the dispersion relationship, equation (5.91) and $\psi_{\lambda\mu}$ is precisely the $(IM)=(\lambda\mu)$ yrast eigenfunction given by equations (5.89) and (5.92). (When the single particle states are degenerate and the interaction is attractive it is only the wave functions of the yrast levels that can be expressed with expansion coefficients given by equation (5.92) since it is only these states that have E_I different from ε.)

Thus in this approximation $T^{(\mathscr{E})}_{\lambda\mu}$ operating on the ground state wave function gives a constant times the yrast state with angular momentum $(\lambda\mu)$. Since the eigenfunctions of H are orthogonal it follows that all the $E\lambda$ strength is contained in the yrast $\rightarrow$ yrast transition. The fact that all single-particle energies are not degenerate, that the separable property of the interaction is not rigorously true, and that $\langle r^\lambda\rangle_{jj_2}$ does depend on (jj_2) means that the result is only approximate. However, shell-model calculations involving single closed-shell nuclei generally predict 80–90% of the transition strength to be concentrated in the yrast 2^+ state even when all the above restrictions are removed. The experimental observation of strong yrast to non-yrast transitions is a sensitive indicator that there are sizeable contributions to the wave functions coming from configurations outside the assumed model space. For example, the strong 0^+_2(3·63 MeV) $\rightarrow 2^+_1$ (1·98 MeV) $E2$ transition observed in ${}^{18}_{8}\mathrm{O}_{10}$ shows that a large component in these states comes from outside the $(0d, 1s)$ shell (Brown 1964, Engeland 1965).

We now turn to a discussion of the low-lying negative-parity states that arise in even–even nuclei. As discussed in section 4 of this chapter they arise from exciting a particle from one oscillator shell to the one next higher in energy. Consequently their position is determined by the particle–hole interaction discussed in detail in Chapter 3. From equation (3.78) and equations (A4.19)–(A4.22) it follows that for the surface-delta potential of equation (A2.29) the $T=0$ particle–hole interaction can be written in the form (Feshbach and Iachello 1974)

$$E_{JT=0}(j_1j_2^{-1}; j_3j_4^{-1})$$
$$=\tfrac{1}{2}(-1)^{j_1+j_2+j_3+j_4}\sum_I (2I+1)W(j_1j_2j_4j_3; JI)$$

$$\times(\{(3G_1-G_0)+(-1)^{I+l_3+l_4}(3G_1+G_0)\}D_I(j_1j_4)D_I(j_3j_2)$$
$$-2G_0F_I(j_1j_4)F_I(j_3j_2))$$
$$=-\left((-1)^{l_2+l_4}G_0+(-1)^{l_3+l_4+J}(3G_1+G_0)/2\right)D_J(j_1j_2)D_J(j_3j_4)$$
$$+(-1)^{l_2+l_4}\frac{(3G_1-G_0)}{2}F_J(j_1j_2)F_J(j_3j_4). \quad (5.95)$$

where

$$G_T=V_T\bar{R}$$

with $D_I(j_1j_2)$ and $F_I(j_1j_2)$ given by equations (5.88).

Let us now examine typical values of the numerical coefficients that occur in the $T=0$ particle-hole interaction. If we use $G_0=0{\cdot}9$ MeV and $G_1=0{\cdot}54$ MeV (values already used in discussing the ${}^{40}_{20}Ca_{20}$ particle-hole states in Chapter 3, section 3.1), we arrive at the conclusion that

$$E_{JT=0}(j_1j_2^{-1};j_3j_4^{-1})=\alpha D_J(j_1j_2)D_J(j_3j_4)+\beta F_J(j_1j_2)F_J(j_3j_4)$$

where

$$\alpha=-2{\cdot}16\text{ MeV},\quad \beta=0{\cdot}36\text{ MeV}\quad \text{if } J \text{ is odd}$$

and

$$\alpha=0{\cdot}36\text{ MeV},\quad \beta=0{\cdot}36\text{ MeV}\quad \text{if } J \text{ is even.}$$

Thus the odd-J, $T=0$ states experience a large attractive force whereas the even-J levels are subject to a small repulsive interaction.

The factor multiplying D_J for the odd J states is six times that multiplying F_J so that in first approximation the F_J term can be neglected in discussing these states. Thus the $T=0$ odd-J particle-hole states have eigenvalues given by equation (5.91) with $2G_1=2{\cdot}16$ MeV. Consequently the yrast odd-J negative-parity level will be lowered in energy from its unperturbed position. Moreover, by the same arguments used in arriving at equation (5.93) it follows that the transition between the yrast state and the ground state will exhaust most of the model space $E\lambda$ (λ odd) transition strength.

The yrast 1^- state deserves particular attention, however. Its structure, in the case that the single-particle levels are degenerate, is proportional to

$$\sum_i r_i Y_{1\mu}(\theta_i,\phi_i)\psi_{00}=\sqrt{\left(\frac{3}{4\pi}\right)}A\tilde{R}_\mu\psi_{00}$$

where $\underset{\sim}{R}$ is the centre-of-mass coordinate of the A-particle nucleus (see equation (4.58)) and ψ_{00} is the ground-state wave function. Thus the lowest 1^- state is, in first approximation, the spurious state. Since this state corresponds to a non-physical situation it must be removed from the calculation. Once this spuriosity has been taken out it follows that the yrast 1^- to ground-state transition will be weak.

When the surface-delta potential is used the $T=1$ particle-hole interaction is

$$E_{JT=1}(j_1j_2^{-1};j_3j_4^{-1})=\tfrac{1}{2}(-1)^{j_1+j_2+j_3+j_4}\sum_I(2I+1)W(j_1j_2j_4j_3;JI)$$
$$\times(\{(G_1+G_0)+(-1)^{I+l_3+l_4}(G_1-G_0)\}D_I(j_1j_4)D_I(j_3j_2)$$
$$+2G_0F_I(j_1j_4)F_I(j_3j_2))$$
$$=((-1)^{l_2+l_4}G_0-(-1)^{l_3+l_4+J}(G_1-G_0)/2)D_J(j_1j_2)D_J(j_3j_4)$$
$$+(-1)^{l_2+l_4}\frac{(G_1+G_0)}{2}F_J(j_1j_2)F_J(j_3j_4). \tag{5.96}$$

For $G_0=0{\cdot}9$ MeV and $G_1=0{\cdot}54$ MeV

$$E_{JT=1}(j_1j_2^{-1};j_3j_4^{-1})=\alpha_1D_J(j_1j_2)D_J(j_3j_4)+\beta_1F_J(j_1j_2)F_J(j_3j_4)$$

where

$$\alpha_1=1{\cdot}08\text{ MeV},\qquad \beta_1=0{\cdot}72\text{ MeV}\qquad \text{for } J \text{ odd}$$

and

$$\alpha_1=0{\cdot}72\text{ MeV},\qquad \beta_1=0{\cdot}72\text{ MeV}\qquad \text{for } J \text{ even.}$$

Thus the $T=1$ particle-hole interaction is repulsive and is stronger for the odd-J states than for the even-J ones.

The procedure used to solve the eigenvalue problem when the interaction is a sum of separable potentials is similar to that already discussed. We express the solution as a sum over the eigenfunctions of H_0 in equation (5.89) so that the analogue of equation (5.90) is

$$(E_I-\tilde{\varepsilon}_k)C_I(k)=\alpha\sum_i C_I(i)D_I(i)D_I(k)+\beta\sum C_I(i)F_I(i)F_I(k).$$

Consequently

$$\sum_k C_I(k)D_I(k)=\left\{\alpha\sum_k\frac{D_I^2(k)}{E_I-\tilde{\varepsilon}_k}\right\}\sum_i C_I(i)D_I(i)$$
$$+\left\{\beta\sum_k\frac{D_I(k)F_I(k)}{E_I-\tilde{\varepsilon}_k}\right\}\sum_i C_I(i)F_I(i) \tag{5.97a}$$

$$\sum_k C_I(k)F_I(k)=\left(\alpha\sum_k\frac{D_I(k)F_I(k)}{E_I-\tilde{\varepsilon}_k}\right\}\sum_i C_I(i)D_I(i)$$
$$+\left\{\beta\sum_k\frac{F_I^2(k)}{E_I-\tilde{\varepsilon}_k}\right\}\sum_i C_I(i)F_I(i). \tag{5.97b}$$

Equations (5.97) constitute a pair of coupled homogeneous equations for the unknowns $\sum_i C_I(i)D_I(i)$ and $\sum_i C_I(i)F_I(i)$, and the condition for their solution is that the determinant of the coefficients of the unknowns vanish. Thus

$$\text{determinant}\,(Z)=0$$

where

$$Z_{11}=\left(\alpha\sum_k \frac{D_I^2(k)}{E_I-\tilde{\varepsilon}_k}\right\}-1$$

$$Z_{22}=\left(\beta\sum_k \frac{F_I^2(k)}{E_I-\tilde{\varepsilon}_k}\right\}-1$$

$$Z_{12}=\beta\sum_k \frac{D_I(k)F_I(k)}{E_I-\tilde{\varepsilon}_k}$$

$$Z_{21}=\alpha\sum_k \frac{D_I(k)F_I(k)}{E_I-\tilde{\varepsilon}_k}$$

from which the eigenvalue E_I may be determined.

Equations (5.97) uncouple if $\sum_i [D_I(i)F_I(i)/(E_I-\tilde{\varepsilon}_i)]$ is much smaller than $\sum_i [D_I^2(i)/(E_I-\tilde{\varepsilon}_i)]$ and $\sum_i [F_I^2(i)/(E_I-\tilde{\varepsilon}_i)]$. That this is likely to be the case follows from the fact that for the eigenvalue which is most shifted, all terms in the latter two sums have the same sign. On the other hand, for the delta function interaction

$$D_I(i)F_I(i)=(-1)^{l_1+\frac{1}{2}-j_2}\frac{(2j_1+1)(2j_2+1)}{2(2I+1)(1+\delta_{j_1j_2})}(j_1j_2\tfrac{1}{2}-\tfrac{1}{2}\mid I0)(j_1j_2\tfrac{1}{2}\tfrac{1}{2}\mid I1)$$

come in with random signs and are likely to sum to a small value. When this approximation is made, it is clear that two states are pushed up in energy and their eigenvalues are given by

$$\sum_k \frac{D_I^2(k)}{E_I-\tilde{\varepsilon}_k}=\frac{1}{\alpha} \tag{5.99a}$$

$$\sum_k \frac{F_I^2(k)}{E_I-\tilde{\varepsilon}_k}=\frac{1}{\beta}. \tag{5.99b}$$

In particular for $J=1$, $T=1$ it is clear that since α is positive one state is shifted to high excitation energy, and in the limit of degenerate single-particle levels its eigenfunction looks like the electric dipole operator applied to the ground-state wave function. Thus in the absence of the coupling term all the electric dipole strength would be contained in this one level which is known as the 'giant dipole' state (Brown and Bolsterli 1959).

In addition, since β is also positive, one state of equations (5.99b) is also pushed up in energy. Furthermore, since β is comparable to α the coupling term $\sum_k [D_I(k)F_I(k)/(E_I-\tilde{\varepsilon}_k)]$ may substantially mix the two pushed up states. Consequently one would expect that the majority of the dipole strength would be concentrated in two states, and this is indeed what has been found to happen in the calculations that have been carried out for ${}^{16}_{8}O_8$ and ${}^{40}_{20}Ca_{20}$ (Gloeckner and Lawson 1975).

6

QUASI-PARTICLES

One of the difficulties in carrying out nuclear-structure calculations is the rapidity with which the computation becomes too large to handle even with a high-speed computer. To see this consider the number of ways an $I=0$ state can be made when six neutrons occupy the single-particle orbits $0f_{\frac{5}{2}}$, $1p_{\frac{3}{2}}$, and $1p_{\frac{1}{2}}$—a reasonable model for the ${}^{62}_{28}\mathrm{Ni}_{34}$ ground state. From the procedures outlined in Chapter 1, section 1.3 it follows that this angular momentum can be realized in the 14 ways listed below:

$$
\begin{array}{rl}
(f_{\frac{5}{2}})^6_0 & (f_{\frac{5}{2}})^3_{\frac{3}{2}}(p_{\frac{3}{2}})^2_2(p_{\frac{1}{2}}) \\
(f_{\frac{5}{2}})^4_0(p_{\frac{3}{2}})^2_0 & (f_{\frac{5}{2}})^3_{\frac{3}{2}}(p_{\frac{3}{2}})(p_{\frac{1}{2}})^2_0 \\
(f_{\frac{5}{2}})^4_0(p_{\frac{1}{2}})^2_0 & (f_{\frac{5}{2}})^2_0(p_{\frac{3}{2}})^4_0 \\
(f_{\frac{5}{2}})^4_2(p_{\frac{3}{2}})^2_2 & (f_{\frac{5}{2}})^2_0(p_{\frac{3}{2}})^2_0(p_{\frac{1}{2}})^2_0 \\
(f_{\frac{5}{2}})^4_2(p_{\frac{3}{2}}p_{\frac{1}{2}})_2 & (f_{\frac{5}{2}})^2_2(p_{\frac{3}{2}})^3_{\frac{3}{2}}(p_{\frac{1}{2}}) \\
(f_{\frac{5}{2}})^3_{\frac{5}{2}}(p_{\frac{3}{2}})^2_2(p_{\frac{1}{2}}) & (f_{\frac{5}{2}})^2_2(p_{\frac{3}{2}})^2_2(p_{\frac{1}{2}})^2_0 \\
(f_{\frac{5}{2}})^3_{\frac{3}{2}}(p_{\frac{3}{2}})^3_{\frac{3}{2}} & (p_{\frac{3}{2}})^4_0(p_{\frac{1}{2}})^2_0.
\end{array}
$$

To set up and diagonalize a 14×14 matrix is a rather modest task particularly if one has a computer. However, if one tries to get a somewhat better approximation to the wave function by including say the $0g_{\frac{9}{2}}$ orbit (still maintaining an inert $N=Z=28$ core) the dimensionality of the Hamiltonian matrix is 88×88.

The situation becomes even worse when we deal with $I\neq0$ states. For example, in Chapter 1, section 1.3, we showed there were 33 ways to make $I=2$ when six particles occupy the $0f_{\frac{5}{2}}$, $1p_{\frac{3}{2}}$, and $1p_{\frac{1}{2}}$ orbits. With the addition of the $0g_{\frac{9}{2}}$ orbit this number becomes 301.

Clearly some approximation scheme must be found to keep the problem within bounds. For s.c.s. nuclei we showed in Chapter 1, section 5.2, that seniority is a fairly good quantum number. Thus a reasonable approximation for these nuclei would be a seniority-zero truncation for the ground state and seniority-two for low-lying excited states of even–even nuclei. This would seem to be a good procedure until one counts the number of ways one can make a seniority-zero $I=0$ state when the number of single-particle levels is only slightly greater than that already considered. For example, the tin isotopes with $Z=50$ have neutrons in the $1d_{\frac{5}{2}}$, $1d_{\frac{3}{2}}$, $2s_{\frac{1}{2}}$, $0g_{\frac{7}{2}}$ and $0h_{\frac{11}{2}}$ orbits. To describe the ground state of ${}^{116}_{50}Sn_{66}$, even in the seniority-zero approximation, one must diagonalize a matrix the dimensionality of which is greater than 100×100! Thus a different approximation procedure is clearly called for.

As shown in Chapter 1 the reason seniority is a good quantum number is that the residual $T=1$ force is much stronger in $I=0$ states than in all others. For example, from the binding-energy analysis carried out for the calcium isotopes in Chapter 1 (see equation (1.135)) one finds that

$$E_{I=0}(\tfrac{7}{2}\tfrac{7}{2};\tfrac{7}{2}\tfrac{7}{2}) = -3\text{ MeV}$$

whereas

$$\frac{\sum_{I\neq0}(2I+1)E_I(\tfrac{7}{2}\tfrac{7}{2};\tfrac{7}{2}\tfrac{7}{2})}{\sum_{I\neq0}(2I+1)} = -0{\cdot}13\text{ MeV}.$$

For this reason an interaction which has non-vanishing matrix elements only in $I=0$ states is often introduced. This potential, which is called the pairing force, is defined by the equation

$$V_p = -G\sum_{jj'} S_+(j)S_-(j') \tag{6.1}$$

where $S_+(j)$ and $S_-(j)$ are the zero-coupled pair-creation and pair-annihilation operators defined in Chapter 1, section 5,

$$S_+(j) = \sum_{jm>0}(-1)^{j-m}a^\dagger_{jm}a^\dagger_{j-m} \tag{6.2a}$$

$$S_-(j) = \sum_{jm>0}(-1)^{j-m}a_{j-m}a_{jm} \tag{6.2b}$$

and G is a constant.

As discussed in Chapter 1 the operator $S_+(j)$ creates a pair of particles in the single-particle orbit j coupled to angular momentum zero and $S_-(j)$ destroys a zero-coupled pair. Consequently

$$\langle(j)^2_I|\,V_p\,|(j')^2_I\rangle = -\frac{G}{2}\sqrt{\{(2j+1)(2j'+1)\}}\,\delta_{I,0}. \tag{6.3}$$

The reason the pairing-force matrix elements are taken to have this j dependence follows from the fact that the residual two-body interaction is short range. In the limit that this potential can be approximated by a delta function its matrix elements in the $I=0$ state are

$$\langle(j)^2_0|\,-4\pi V_0\delta(\underline{r}_1-\underline{r}_2)\,|(j')^2_0\rangle$$
$$= -\left\{\frac{V_0}{2}(-1)^{l+l'}\int R^2_{njl}(r)R^2_{n'j'l'}(r)r^2\,\mathrm{d}r\right\}\sqrt{\{(2j+1)(2j'+1)\}}$$

where use has been made of equation (A2.29). Apart from the phase factor $(-1)^{l+l'}$, which we shall discuss later, it follows that the matrix elements of the pairing force have the same j dependence as the delta-function interaction provided the radial integral does not have too great a state dependence.

Since this part of the force is most important, it is imperative that an adequate way is found to take it into account. The Bardeen–Cooper–Schrieffer (BCS) theory (1957), which is an approximation to a seniority-zero shell-model calculation, provides a relatively simple and accurate method for considering this part of the interaction. Furthermore, as we shall see, the BCS wave function can be thought of as the vacuum state for entities called quasi-particles which are really seniority excitations. Once this is realized a simple approximation procedure can be shown to exist for the treatment of low-lying nuclear states. Because the approximation is based on a seniority truncation, it is best suited to s.c.s. nuclei. However, it can also be applied to cases where neutrons and protons are filling different shells provided there are not too many valence nucleons, i.e. provided the nuclei do not have a permanent deformation.

1. BCS approximation

The Bardeen–Cooper–Schrieffer theory, originally used in the theory of superconductivity, can be exploited to provide an approximation to a seniority-zero calculation for the ground-state energy of an even-A s.c.s. nucleus (Bohr *et al.* 1958, Soloviev 1958, Belyaev 1959). In this theory a trial wave function

$$\psi = \prod_{jm>0} \{u_j + (-1)^{j-m} v_j a^\dagger_{jm} a^\dagger_{j-m}\} |0\rangle, \tag{6.4}$$

which does not conserve particle number, is introduced and a variational calculation is carried out with it. The product $\prod_{jm>0}$ is over all single-particle orbits j that one wishes to consider in the calculation and $|0\rangle$ is the vacuum state. For example in the nickel calculation just discussed the j values would be $1p_{\frac{1}{2}}$, $1p_{\frac{3}{2}}$, $0f_{\frac{5}{2}}$, and $0g_{\frac{9}{2}}$ and $|0\rangle$ would be the inert $N = Z = 28$ core. The quantities u_j and v_j are to be determined from the condition that the energy be an extremum.

Let us start out by examining some properties of this proposed trial function.

(*a*) *All particles are in zero-coupled pair states.* To see this, consider the m products for one j. To simplify the discussion and notation take $j = \frac{3}{2}$

and suppress the j index on all quantities. Thus

$$\prod_{m>0} \{u+(-1)^{\frac{3}{2}-m} v a_m^\dagger a_{-m}^\dagger\} |0\rangle = (u+va_{\frac{3}{2}}^\dagger a_{-\frac{3}{2}}^\dagger)(u-va_{\frac{1}{2}}^\dagger a_{-\frac{1}{2}}^\dagger)|0\rangle$$
$$= \{u^2+uv(a_{\frac{3}{2}}^\dagger a_{-\frac{3}{2}}^\dagger - a_{\frac{1}{2}}^\dagger a_{-\frac{1}{2}}^\dagger) - v^2 a_{\frac{3}{2}}^\dagger a_{-\frac{3}{2}}^\dagger a_{\frac{1}{2}}^\dagger a_{-\frac{1}{2}}^\dagger\}|0\rangle$$
$$= \{u^2+uvS_+(\tfrac{3}{2}) + \frac{v^2}{2!} S_+^2(\tfrac{3}{2})\}|0\rangle$$

where $S_+(\frac{3}{2})$ is given by equation (6.2a) and use has been made of the fact that we deal with fermions so that each state can accommodate only one particle. From this equation it is apparent that the BCS wave function has seniority-zero and consequently describes a 0^+ state in an even–even nucleus. Furthermore, the state it describes has a finite probability, in the above case u^4, of having no particles, $2u^2v^2$ probability that there are two particles, and v^4 probability that there are four nucleons.

For an arbitrary value of j the product on m can be written as

$$\prod_m \{u_j+(-1)^{j-m} v_j a_{jm}^\dagger a_{j-m}^\dagger\}|0\rangle = \sum_{n=0}^{\Omega_j} \frac{u_j^{\Omega_j-n} v_j^n}{n!} S_+^n(j)|0\rangle$$
$$= u_j^{\Omega_j} \exp\left\{\frac{v_j}{u_j} S_+(j)\right\}|0\rangle$$

where Ω_j is the pair degeneracy of the orbit j

$$\Omega_j = (2j+1)/2. \tag{6.5}$$

Consequently an alternative way of writing the BCS wave function is

$$\psi = \prod_j u_j^{\Omega_j} \exp\left(\frac{v_j}{u_j} S_+(j)\right)|0\rangle$$
$$= \left(\prod_j u_j^{\Omega_j}\right) \exp\left(\sum_j \frac{v_j}{u_j} S_+(j)\right)|0\rangle. \tag{6.6}$$

(*b*) *Probability interpretation for u_j and v_j.* Before looking at the probability interpretation consider the restriction imposed on the variational coefficients by normalization. For simplicity return to the $j=\frac{3}{2}$ example. Thus

$$\langle\psi|\psi\rangle = \langle 0|(u-va_{-\frac{1}{2}}a_{\frac{1}{2}})(u+va_{-\frac{3}{2}}a_{\frac{3}{2}})(u+va_{\frac{3}{2}}^\dagger a_{-\frac{3}{2}}^\dagger)(u-va_{\frac{1}{2}}^\dagger a_{-\frac{1}{2}}^\dagger)|0\rangle$$
$$= \langle 0|(u-va_{-\frac{1}{2}}a_{\frac{1}{2}})(u-va_{\frac{1}{2}}^\dagger a_{-\frac{1}{2}}^\dagger)|0\rangle$$
$$\times \langle 0|(u+va_{-\frac{3}{2}}a_{\frac{3}{2}})(u+va_{\frac{3}{2}}^\dagger a_{-\frac{3}{2}}^\dagger)|0\rangle$$
$$= (u^2+v^2)^2. \tag{6.7}$$

Therefore the condition that the BCS wave function is normalized is that

$$u_j^2+v_j^2 = 1. \tag{6.8}$$

The physical interpretation of u_j and v_j is provided by the matrix elements of the number operator

$$\mathrm{N}_{j'} = \sum_{m'} a^{\dagger}_{j'm'} a_{j'm'}. \tag{6.9}$$

The expectation value of $a^{\dagger}_{j'm'} a_{j'm'}$—a particular term in the number operator sum—may be computes as follows: The term obviously affects only that part of the product wave function with $j = j'$ and $m = m'$ so

$$\begin{aligned} a^{\dagger}_{j'm'} a_{j'm'} \psi &= \left[\prod_{jm>0}{}' \{u_j + (-1)^{j-m} v_j a^{\dagger}_{jm} a^{\dagger}_{j-m}\} \right] a^{\dagger}_{j'm'} a_{j'm'} \\ &\quad \times \{u_{j'} + (-1)^{j'-m'} v_{j'} a^{\dagger}_{j'm'} a^{\dagger}_{j'-m'}\} |0\rangle \\ &= (-1)^{j'-m'} v_{j'} a^{\dagger}_{j'm'} a^{\dagger}_{j'-m'} \\ &\quad \times \prod_{jm>0}{}' \{u_j + (-1)^{j-m} v_j a^{\dagger}_{jm} a^{\dagger}_{j-m}\} |0\rangle, \end{aligned}$$

where the prime on the product indicates that the term with $j = j'$, $m = m'$ is missing. As seen from equation (6.7) each individual term in the expression for $\langle \psi \mid \psi \rangle$ leads to $u_j^2 + v_j^2 = 1$. Consequently

$$\langle \psi | \, a^{\dagger}_{j'm'} a_{j'm'} \, | \psi \rangle = v_{j'}^2. \tag{6.10}$$

In other words, $v_{j'}^2$ is the probability that the state $(j'm')$ is occupied. Since $u_{j'}^2 = 1 - v_{j'}^2$, $u_{j'}^2$ is the probability that the state $(j'm')$ is empty.

1.1. Variational calculation for the ground-state energy

To use the BCS wave function in a variational calculation one must remember that it does not describe a state with a definite number of particles. Therefore the variational calculation must be carried out subject to the constraint that the expectation value of the number operator is equal to the number of valence nucleons in the nucleus under investigation. To illustrate how to proceed we consider the calculation of the ground-state energy associated with N particles whose motion is governed by the Hamiltonian

$$H = \sum_{jm} \varepsilon_j a^{\dagger}_{jm} a_{jm} - G \sum_{jj'} \mathrm{S}_+(j) \mathrm{S}_-(j') \tag{6.11}$$

where ε_j are single-particle energies and the residual two-body interaction is taken to be the pairing force.

To minimize H subject to the number constraint means we must calculate the expectation value of

$$\mathcal{H} = H - \lambda \sum_j \mathrm{N}_j \tag{6.12}$$

using the BCS wave function and then vary v_j (u_j) until the result is an extremum. If n single-particle orbits are considered there will be n equations

$$\frac{\partial}{\partial v_j}\langle\psi|\,\mathcal{H}\,|\psi\rangle=0. \tag{6.13}$$

These together with the expectation value of the number operator

$$\langle\psi|\sum_j \mathrm{N}_j\,|\psi\rangle=\sum_j(2j+1)v_j^2=N \tag{6.14}$$

provide $(n+1)$ equations for the determination of n values of v_j and the constraint constant λ which is often called the chemical potential.

The $(\varepsilon_j-\lambda)$ part of this expectation value is easily seen to be

$$\mathcal{H}_1=\langle\psi|\sum_{jm}(\varepsilon_j-\lambda)a_{jm}^{\dagger}a_{jm}\,|\psi\rangle=\sum_j(2j+1)(\varepsilon_j-\lambda)v_j^2 \tag{6.15a}$$

when use is made of equation (6.10). To calculate the contribution from the $\mathrm{S}_+(j_2)\mathrm{S}_-(j_1)$ term note that

$$a_{j_1-m_1}a_{j_1m_1}\prod_{jm}\{u_j+(-1)^{j-m}v_ja_{jm}^{\dagger}a_{j-m}^{\dagger}\}\,|0\rangle$$
$$=(-1)^{j_1-m_1}v_{j_1}\prod_{jm}{}'\{u_j+(-1)^{j-m}v_ja_{jm}^{\dagger}a_{j-m}^{\dagger}\}\,|0\rangle$$

where the prime on $\prod$ means that one term in the product (the term $\{u_{j_1}+(-1)^{j_1-m_1}v_{j_1}a_{j_1m_1}^{\dagger}a_{j_1-m_1}^{\dagger}\}$) is missing. Thus

$$\mathcal{H}_2=\langle\psi|-G\sum_{j_1j_2}\mathrm{S}_+(j_2)\mathrm{S}_-(j_1)\,|\psi\rangle$$
$$=-G\sum_{j_1j_2}\sum_{m_1m_2>0}v_{j_1}v_{j_2}\langle 0|\prod_{j'm'}{}''\{u_{j'}+(-1)^{j'-m'}v_{j'}a_{j'-m'}a_{j'm'}\}$$
$$\times\prod_{jm}{}'(u_j+(-1)^{j-m}v_ja_{jm}^{\dagger}a_{j-m}^{\dagger})\,|0\rangle$$

and $\prod''$ means the term with (j_2m_2) is missing in the product. The same collapse of these products that occurred in equation (6.7) also comes about here. In other words, every term in $\prod''$ can be paired with an analogous term in $\prod'$ except for the $\{u_{j_1}+(-1)^{j_1-m_1}v_{j_1}a_{j_1-m_1}a_{j_1m_1}\}$ in the former and the term $\{u_{j_2}+(-1)^{j_2-m_2}v_{j_2}a_{j_2m_2}^{\dagger}a_{j_2-m_2}^{\dagger}\}$ in the latter. Thus

$$\mathcal{H}_2=-G\sum_{j_1j_2}\sum_{m_1m_2>0}v_{j_1}v_{j_2}\langle 0|\,\{u_{j_1}+(-1)^{j_1-m_1}v_{j_1}a_{j_1-m_1}a_{j_1m_1}\}$$
$$\times\{u_{j_2}+(-1)^{j_2-m_2}v_{j_2}a_{j_2m_2}^{\dagger}a_{j_2-m_2}^{\dagger}\}\,|0\rangle$$
$$=-G\sum_{j_1j_2}\sum_{m_1m_2>0}u_{j_1}v_{j_1}u_{j_2}v_{j_2}-G\sum_{j_1}\sum_{m_1>0}v_{j_1}^4. \tag{6.15b}$$

In writing the last line of this equation use has been made of the fact that $|0\rangle$ is the vacuum state for particles in the single-particle levels under consideration. When equations (6.15a) and (6.15b) are combined with the definition of Ω_j in equation (6.5), one sees that

$$\langle\psi|\mathscr{H}|\psi\rangle=2\sum_j\Omega_j(\varepsilon_j-\lambda)v_j^2-\Delta^2/G-G\sum_j\Omega_jv_j^4 \tag{6.16}$$

where

$$\Delta=G\sum_j\Omega_ju_jv_j. \tag{6.17}$$

To minimize $\langle\psi|\mathscr{H}|\psi\rangle$ we make use of equation (6.8) from which it follows that

$$\frac{\partial u_j}{\partial v_j}=-v_j/u_j$$

so that

$$\begin{aligned}\frac{\partial\Delta}{\partial v_j}&=G\Omega_j\left(u_j+\frac{\partial u_j}{\partial v_j}v_j\right)\\&=G\Omega_j\left(\frac{u_j^2-v_j^2}{u_j}\right)\\&=G\Omega_j\left(\frac{1-2v_j^2}{u_j}\right).\end{aligned}$$

Thus

$$\frac{\partial\langle\psi|\mathscr{H}|\psi\rangle}{\partial v_j}=4\Omega_j(\varepsilon_j-\lambda)v_j-2\Delta\Omega_j\left(\frac{1-2v_j^2}{u_j}\right)-4G\Omega_jv_j^3.$$

If there are n active single-particle orbits j, there are n equations of the above form and each of these must be zero for $\langle\psi|\mathscr{H}|\psi\rangle$ to be an extremum. Consequently in addition to the constraint equation (6.14), v_j and λ must be chosen to satisfy the n equations

$$2(\varepsilon_j-\lambda)u_jv_j-\Delta(1-2v_j^2)-2Gv_j^3u_j=0. \tag{6.18}$$

The easiest way to solve these equations is first to neglect the v_j^3 term in equation (6.18). As we shall see later, once a solution in this approximation has been obtained, the one including v_j^3 can be obtained by a simple modification of λ. With this approximation the square of equation (6.18) becomes

$$4(\varepsilon_j-\lambda)^2v_j^2(1-v_j^2)=\Delta^2(1-4v_j^2+4v_j^4).$$

From this it follows that

$$v_j^2=\frac{1}{2}\left[1\pm\frac{(\varepsilon_j-\lambda)}{\sqrt{\{(\varepsilon_j-\lambda)^2+\Delta^2\}}}\right]$$

and the sign ambiguity arises because we squared equation (6.18) to find the solution simply. If one substitutes the above value of v_j^2 in the original equation and neglects the v_j^3 term, only the minus sign is appropriate. Thus

$$v_j^2=\frac{1}{2}\left[1-\frac{(\varepsilon_j-\lambda)}{\surd\{(\varepsilon_j-\lambda)^2+\Delta^2\}}\right] \tag{6.19a}$$

$$u_j^2=\frac{1}{2}\left[1+\frac{(\varepsilon_j-\lambda)}{\surd\{(\varepsilon_j-\lambda)^2+\Delta^2\}}\right]. \tag{6.19b}$$

These values of u_j and v_j may now be substituted in equations (6.14) and (6.17) to solve for λ and Δ; i.e. λ and Δ must satisfy the equations

$$N=\sum_j \Omega_j\left[1-\frac{(\varepsilon_j-\lambda)}{\surd\{(\varepsilon_j-\lambda)^2+\Delta^2\}}\right] \tag{6.20a}$$

$$\Delta=\frac{G}{2}\sum_j \Omega_j\left\{1-\frac{(\varepsilon_j-\lambda)^2}{(\varepsilon_j-\lambda)^2+\Delta^2}\right\}^{\frac{1}{2}},$$

and the second of these equations can be rewritten as

$$\frac{G}{2}\sum_j \frac{\Omega_j}{\surd\{(\varepsilon_j-\lambda)^2+\Delta^2\}}=1. \tag{6.20b}$$

Thus the procedure is to first solve equations (6.20) for λ and Δ and then use these values to compute u_j^2 and v_j^2. It is clear that the effect of the neglected term $-2Gv_j^3u_j$ in equation (6.18) is to modify the equation by replacing λ by $(\lambda+Gv_j^2)$. Consequently the effect of the v_j^3 term may simply be taken into account iteratively by a change in λ.

Once u_j and v_j have been determined one may use the BCS wave function to compute the expectation value of H. This, of course, is given by neglecting the λv_j^2 term in equation (6.16) so that E_0, the ground-state energy of the nucleus, is

$$E_0=2\sum_j \Omega_j\varepsilon_j v_j^2-\Delta^2/G-G\sum_j \Omega_j v_j^4. \tag{6.21}$$

As stated before v_j^2 is the probability that the state (jm) is occupied and u_j^2 is the probability that it is empty. In a shell-model calculation in which there is no residual interaction v_j^2 will be identically zero for certain states and have a finite value for others. For example, from the single-particle level scheme shown in Fig. 1.1 it follows that $v_{\frac{5}{2}}=v_{\frac{7}{2}}=1$ and $v_{\frac{11}{2}}=v_{\frac{1}{2}}=v_{\frac{3}{2}}=0$ for ${}^{114}_{50}\mathrm{Sn}_{64}$. However, once the residual interaction is taken into account this probability is smeared out as shown in Fig. 6.1 and only those states with $\varepsilon_j \gg \lambda$ will have $v_j \to 0$.

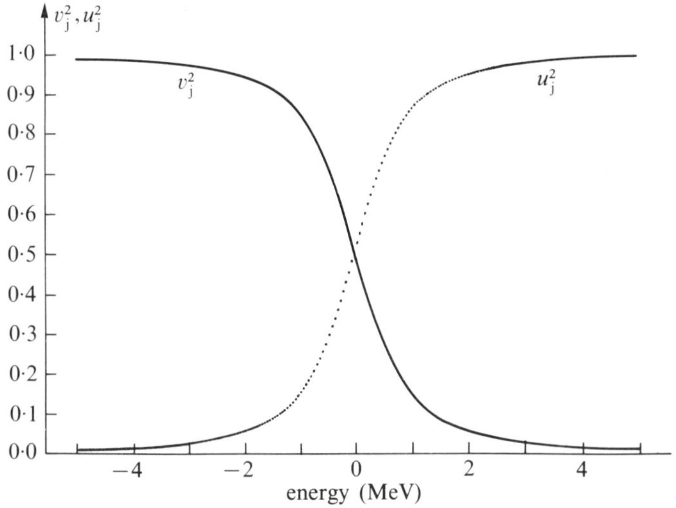

Fig. 6.1. v_j^2 and u_j^2 defined by equations (6.19) are plotted as a function of $(\varepsilon_j - \lambda)$. The curve is drawn for $\Delta = 1$ MeV. For smaller values of Δ the curves become steeper.

1.2. The degenerate model; comparison of BCS and exact solutions

The BCS procedure for finding the energy is much simpler than diagonalizing the shell-model energy matrix. For example, in the $^{116}_{50}Sn_{66}$ calculation one would have to diagonalize a matrix larger than 100×100 even if one restricted the calculation to seniority zero. On the other hand, to carry out the BCS calculation one has to solve six equations for the six unknowns λ and v_j for $j = 1d_{\frac{3}{2}}$, $1d_{\frac{5}{2}}$, $2s_{\frac{1}{2}}$, $0g_{\frac{7}{2}}$, and $0h_{\frac{11}{2}}$. However, the fact that this calculation is so much easier only has importance if the approximate energy is close to the exact one. For the case that the single-particle levels are degenerate one can simply solve the pairing-force problem and compare the results with the BCS calculation. We shall now do this.

To solve the degenerate case exactly one must find the eigenvalues and eigenvectors of

$$H_{\mathrm{d}} = \varepsilon \sum_j \mathrm{N}_j - G\mathrm{S}_+\mathrm{S}_-$$

where

$$\mathrm{S}_\pm = \sum_j \mathrm{S}_\pm(j).$$

Since the pairing force only gives a contribution to the energy when a zero-coupled pair exists, it is clear that the lowest eigenvalue of the N-particle system has a wave function in which all nucleons are in zero-coupled pair states. Thus if we define

$$\Omega = \sum_j \Omega_j$$

it follows from equation (1.120) that

$$\psi_{\mathrm{d}} = \left\{\frac{2^N(2\Omega - N)!!}{N!!(2\Omega!!)}\right\}^{\frac{1}{2}} (\mathrm{S}_+)^{N/2}\,|0\rangle.$$

Consequently

$$\begin{aligned}(E_0)_{\text{exact}} &= \langle\psi_{\mathrm{d}}|\,\mathrm{H}_{\mathrm{d}}\,|\psi_{\mathrm{d}}\rangle \\ &= N\varepsilon - \frac{GN}{4}(2\Omega - N + 2).\end{aligned} \tag{6.22}$$

In evaluating this matrix element use has been made of equation (1.117)

$$[\mathrm{S}_+, \mathrm{S}_-] = \mathrm{N} - \Omega$$

where

$$\mathrm{N} = \sum_j \mathrm{N}_j$$

is the total number operator for the system.

To compute this energy using the BCS approximation, we note that because $\varepsilon_j = \varepsilon$, equation (6.19a) implies v_j is the same for all the degenerate levels. This, of course, is what one would expect since there is no way to distinguish between the states. From equation (6.14) it follows that

$$v^2 = \frac{N}{2\Omega}$$

and

$$\begin{aligned}u^2 &= 1 - v^2 \\ &= 1 - \frac{N}{2\Omega}.\end{aligned}$$

Thus Δ^2 in equation (6.17) becomes

$$\Delta^2 = \frac{G^2 N\Omega}{2}\left(1 - \frac{N}{2\Omega}\right).$$

When these values are inserted into the BCS expression for the energy given by equation (6.21) one sees that

$$(E_0)_{\text{BCS}} = \varepsilon N - \frac{GN}{4}\left(2\Omega - N + \frac{N}{\Omega}\right). \tag{6.23}$$

Since the single-particle energy εN is trivial to obtain correctly, the relevant ratio to look at in assessing the validity of the BCS theory is

$$\frac{\{(E_0)_{\text{exact}}-\varepsilon N\}-\{(E_0)_{\text{BCS}}-\varepsilon N\}}{(E_0)_{\text{exact}}-\varepsilon N}=\frac{2-N/\Omega}{2\Omega-N+2}.$$

In the case of ${}^{116}_{50}\text{Sn}_{66}$, where the 16 valence nucleons occupy the $1d_{\frac{3}{2}}$, $1d_{\frac{5}{2}}$, $2s_{\frac{1}{2}}$, $0g_{\frac{7}{2}}$, and $0h_{\frac{11}{2}}$ orbits and $\Omega=16$, this ratio has the value $\frac{1}{18}$; in other words the BCS approximation comes within 6% of the 'true' energy.

By use of equation (6.6) it is simple to project the N-particle part of the BCS wave function. Clearly this is just

$$\psi_{\text{proj}}=\prod_j u_j^{\Omega_j}\left\{\sum_j \frac{v_j}{u_j} S_+(j)\right\}^{N/2}|0\rangle=\left(1-\frac{N}{2\Omega}\right)^{(2\Omega-N)/4}\left(\frac{N}{2\Omega}\right)^{N/4} S_+^{N/2}|0\rangle.$$

In other words, apart from a normalization factor, the N-particle part of the BCS wave function is identical to the exact eigenfunction. Thus in this case the entire error in the calculated energy is due to number non-conservation in the wave function. These conclusions are not appreciably altered when the single-particle levels are not degenerate (Kerman *et al.* 1961). For typical values of G it is found that the ground-state energy as calculated with the BCS eigenfunction is usually within about 500 keV of the exact result. Moreover, if the wave function obtained by projecting and normalizing that part of the trial wave function with the correct number of particles is used to calculate $\langle\psi|H|\psi\rangle$, results in excellent agreement with the exact calculations are obtained. Thus the BCS wave function gives an excellent approximation when one deals with a pairing force.

Finally, in the degenerate model or in a system with only one active single-particle orbit, one can easily write down the eigenvalues of the pairing-force Hamiltonian for N-particle states with seniority v different from zero. In Chapter 1, section 5.2 we showed that the N-particle seniority-v state $\psi_{IM\alpha}$ is related to the v-particle seniority-v state $\Phi_{IM\alpha}$ by the relationship

$$\psi_{IM\alpha}=\left\{\frac{2^{N-v}(2\Omega-N-v)!!}{(N-v)!!(2\Omega-2v)!!}\right\}^{\frac{1}{2}}(S_+)^{(N-v)/2}\Phi_{IM\alpha}.$$

By definition $\Phi_{IM\alpha}$ has no zero-coupled pairs in its make-up so that

$$S_-\Phi_{IM\alpha}=0.$$

Thus from the arguments used to deduce equation (6.22) one easily shows that

$$H\psi_{IM\alpha}=E_v\psi_{IM\alpha}$$

where

$$E_v=\varepsilon N-\frac{G(N-v)}{4}(2\Omega-N-v+2). \tag{6.24}$$

Consequently, when the pairing force alone is used, all N-particle states with the same seniority are degenerate. This, of course, is not observed experimentally (see, for example, the spectrum of ${}^{42}_{20}Ca_{22}$ shown in Fig. 2.1; the 2^+, 4^+, and 6^+ $v=2$ states lie at 1·524, 2·751, and 3·191 MeV, respectively). Consequently it is clear that the pairing force alone is not the complete story. In section 2 of this chapter we shall show how one uses the BCS approximation to treat a more general interaction.

1.3. Lead isotopes with a pairing force

In this section we shall use the BCS theory to calculate the binding energies of the lead isotopes on the assumption that the residual two-body interaction is a pairing force. In addition, various properties of the BCS wave function will be examined so that one may understand why the theory is so successful.

We shall consider the lead isotopes with $N<126$ and consequently deal with neutron holes in the doubly-magic ${}^{208}_{82}Pb_{126}$ core. A suitable model space and single-particle energies can be determined from examination of the spectrum of the single-hole nucleus ${}^{207}_{82}Pb_{125}$. Below 2·4 MeV the only states observed in this nuleus are the $2p_{\frac{1}{2}}^{-1}$ ground state and the $1f_{\frac{5}{2}}^{-1}$, $2p_{\frac{3}{2}}^{-1}$, $0i_{\frac{13}{2}}^{-1}$, and $1f_{\frac{7}{2}}^{-1}$ levels at excitation energies of 0·57, 0·90, 1·634 and 2·35 MeV, respectively (Schmorak and Auble 1971). A model space composed of these five orbits together with their experimental single-hole energies will be used. The strength of the pairing force will be taken to be that suggested by Kisslinger and Sorensen (1960), i.e. $G=0{\cdot}111$ MeV.

Having decided on the model space, the single-hole energies, and the value of G, it remains to solve equations (6.20a) and (6.20b) for λ and Δ. In Table 6.1 the values obtained for these quantities together with the values of v_j^2 calculated when they are used in equation (6.19a) are listed. Throughout we have assumed that $\varepsilon_{p\frac{1}{2}}=0$; consequently the energy calculated from equation (6.21) must be compared to the experimental energy

$$B_N = \mathrm{BE}({}^{208-N}_{82}\mathrm{Pb}_{126-N}) - \mathrm{BE}({}^{208}_{82}\mathrm{Pb}_{126}) \\ - N\{\mathrm{BE}({}^{207}_{82}\mathrm{Pb}_{125}) - \mathrm{BE}({}^{208}_{82}\mathrm{Pb}_{126})\}. \qquad (6.25)$$

In columns 9, 10, and 11 of Table 6.1 the BCS value for this quantity, the results of an exact seniority-zero solution (Kerman *et al.* 1961), and the experimental value obtained from the binding-energy tables are listed. From a comparison of columns 9 and 10 we see that the BCS theory provides an excellent approximation to the exact calculation. The lack of agreement between theory and experiment merely points to the fact already discussed at the end of the last section; namely there are parts of the residual two-body interaction other than the pairing force.

TABLE 6.1

Values of λ, Δ, v_j^2, and B_N for the lead isotopes

Nucleus	λ (MeV)	Δ (MeV)	$v_{\frac{1}{2}}^2$	$v_{\frac{3}{2}}^2$	$v_{\frac{5}{2}}^2$	$v_{\frac{7}{2}}^2$	$v_{\frac{13}{2}}^2$	B_N (MeV) BCS	Exact	Experiment
$^{206}_{82}Pb_{124}$	0·111	0·256	0·699	0·024	0·063	0·003	0·007	−0·128	−0·315	−0·627
$^{204}_{82}Pb_{122}$	0·337	0·421	0·812	0·100	0·258	0·011	0·024	0·303	0·098	−0·547
$^{202}_{82}Pb_{120}$	0·515	0·517	0·853	0·201	0·447	0·019	0·046	1·096	0·801	−0·096
$^{200}_{82}Pb_{118}$	0·695	0·581	0·884	0·334	0·605	0·028	0·075	2·234	1·892	1·018
$^{198}_{82}Pb_{116}$	0·878	0·628	0·907	0·482	0·720	0·040	0·115	3·701	3·441	2·762

The single-hole levels included in the model space were $2p_{\frac{1}{2}}^{-1}$, $1f_{\frac{5}{2}}^{-1}$, $2p_{\frac{3}{2}}^{-1}$, $0i_{\frac{13}{2}}^{-1}$, and $1f_{\frac{7}{2}}^{-1}$ with excitation energies of 0, 0·57, 0·90, 1·634, and 2·35 MeV, respectively. The pairing-force strength G of equation (6.1) was taken to be 0·111 MeV. The BCS approximation to the binding energy, B_N, is calculated using these values of λ, Δ, and v_j in conjunction with equation (6.21), the column labelled exact gives the results of a seniority zero shell model calculation and the experimental value is given by equation (6.25) with the binding energies taken from the compilation of Wapstra and Gove 1971.

In this case, since we deal with holes in the 126 shell,

$$N_j = (2j+1)v_j^2$$

is the number of holes in the single-particle orbit j. Thus for example, in $^{206}_{82}Pb_{124}$ the $2p_{\frac{1}{2}}$ orbit has 1·398 holes whereas in $^{198}_{82}Pb_{116}$ the level has $N_{\frac{1}{2}} = 1{\cdot}814$.

An important property of the BCS wave function is its number distribution, i.e. the probability that the wave function contains a part with $2n$ particles (where $2n$ may or may not be equal to the number of particles N in the physical system under consideration). The $2n$-particle part of the BCS wave function may be obtained from equation (6.6)

$$\psi_{2n} = \left(\prod_j u_j^{\Omega_j}\right)\frac{1}{n!}\left\{\sum_j \frac{v_j}{u_j} S_+(j)\right\}^n |0\rangle, \tag{6.26}$$

and the probability that there are $2n$ particles is $\langle\psi_{2n} \mid \psi_{2n}\rangle$. The binomial theorem together with equation (1.120) may be used to calculate this quantity.

For $^{202}_{82}Pb_{120}$ $(N=6)$ and $^{198}_{82}Pb_{116}$ $(N=10)$ the number distribution in the BCS state is given numerically in Table 6.2 and shown graphically in Fig. 6.2. The probability peaks at the desired number of particles; however, the peak only corresponds to about 30% of the wave function. Thus the probability that in the BCS wave function one has the desired number of particles is only about 30% even though the v_j^2 were chosen so that the expectation value of the number operator is the correct number of particles. Despite this fact energies calculated with this function come extremely close to the results of the exact calculation, particularly if one projects out of the BCS function that part with the desired number of particles, renormalizes the wave function, and then uses this normalized projected eigenfunction to calculate the energy (Kerman *et al.* 1961).

TABLE 6.2

Probability, P_{2n}, that the BCS wave function has a part corresponding to 2n particles

Number $2n$	P_{2n} $N=6$	$N=10$
0	0·01057	0·00020
2	0·09668	0·00404
4	0·25300	0·02963
6	0·31256	0·10899
8	0·21361	0·22597
10	0·08702	0·27828
12	0·02231	0·21060
14	0·00377	0·10139
16	0·00044	0·03248
18	0·00004	0·00718
20	—	0·00112
22	—	0·00013
24	—	0·00001

This probability is listed for $N=6$ and 10. $2n$-states with probability $<10^{-5}$ have been neglected.

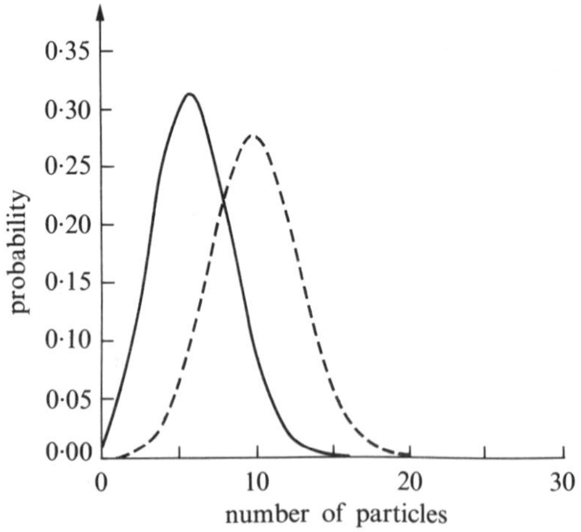

Fig. 6.2. Probability distribution as a function of the number of particles in the BCS wave function. The solid curve is for $^{202}_{82}Pb_{120}$ ($N=6$) and the dashed curve for $^{198}_{82}Pb_{116}$ ($N=10$).

The reason that even the unprojected BCS wave function is so successful is because its number distribution is approximately symmetric about the desired number of particles. Consequently, if the quantity one wants to calculate is constant as a function of N or varies more or less linearly with N, any error introduced by non-conservation of the number of particles will tend to cancel. From Table 6.1 it is clear that B_N, calculated by either the BCS or the exact theory, does not oscillate as a function of N but instead increases with increasing N. Consequently the error due to number non-conservation tends to cancel. As we shall see in section 3 of this chapter one does not always have such near linearity of the binding energy and consequently one must often resort to projection.

2. Quasi-particle calculations

Since the BCS wave function gives the lowest energy state of an even-A nucleus, it can be thought of as the vacuum state for entities created by an operator $\alpha^{\dagger}_{jm}$ which we shall now construct. If the BCS wave function is to be the vacuum state for these entities, then

$$\alpha_{jm}\psi = 0.$$

By inspection it is clear that an operator with this property can be constructed by taking a linear combination of a_{jm} and $a^{\dagger}_{j-m}$. Thus if we can make the ansatz

$$\alpha_{jm} = Aa_{jm} + Ba^{\dagger}_{j-m}$$

it follows that

$$\alpha_{j_1m_1}\psi = \left[\prod_{jm}{}' \{u_j + (-1)^{j-m} v_j a^{\dagger}_{jm} a^{\dagger}_{j-m}\}\right](Aa_{j_1m_1} + Ba^{\dagger}_{j_1-m_1})$$
$$\times\{u_{j_1} + (-1)^{j_1-m_1} v_{j_1} a^{\dagger}_{j_1m_1} a^{\dagger}_{j_1-m_1}\}\,|0\rangle$$
$$= \left[\prod_{jm}{}' \{u_j + (-1)^{j-m} v_j a^{\dagger}_{jm} a^{\dagger}_{j-m}\}\right]\{Bu_{j_1} + (-1)^{j_1-m_1} Av_{j_1}\} a^{\dagger}_{j_1-m_1}\,|0\rangle.$$

Consequently the operator

$$\alpha_{jm} = u_j a_{jm} - (-1)^{j-m} v_j a^{\dagger}_{j-m} \tag{6.27a}$$

has the property that when it operates on the BCS wave function it gives zero. This operator and its Hermitian adjoint

$$\alpha^{\dagger}_{jm} = u_j a^{\dagger}_{jm} - (-1)^{j-m} v_j a_{j-m} \tag{6.27b}$$

satisfy the anticommutation relationships

$$\{\alpha_{jm}, \alpha_{j'm'}\} = \{\alpha^{\dagger}_{jm}, \alpha^{\dagger}_{j'm'}\} = 0 \tag{6.28a}$$

and

$$\{\alpha_{jm}, \alpha^{\dagger}_{j'm'}\} = (u_j^2 + v_j^2)\delta_{jj'}\delta_{mm'} = \delta_{jj'}\delta_{mm'}. \tag{6.28b}$$

Thus the entities created and destroyed by these operators act like fermions and are called quai-particles. The BCS wave function is the quasi-particle vacuum.

From the properties of u_j and v_j shown in Fig. 6.1 (page 345) it follows that for states high above the Fermi surface ($\varepsilon_j \gg \lambda$)

$$v_j \to 0$$
$$u_j \to 1$$
$$\alpha^{\dagger}_{jm} \to a^{\dagger}_{jm}$$

so that the quasi-particle becomes an ordinary particle. For states far below the Fermi surface ($\varepsilon_j \ll \lambda$)

$$v_j \to 1$$
$$u_j \to 0$$
$$\alpha^{\dagger}_{jm} \to -(-1)^{j-m} a_{j-m}$$

and the quasi-particle is a hole in the Fermi sea. For situations midway between these two extremes the quasi-particle is partly particle and partly hole.

It is simple to see that one quasi-particle states correspond to seniority-one eigenfunctions. This follows because

$$\begin{aligned}
\psi_{j_1m_1} = \alpha^{\dagger}_{j_1m_1}\psi &= \prod_{jm}{}' \{u_j + (-1)^{j-m} v_j a^{\dagger}_{jm} a^{\dagger}_{j-m}\} \\
&\quad \times \{u_{j_1} a^{\dagger}_{j_1m_1} - (-1)^{j_1-m_1} v_{j_1} a_{j_1-m_1}\} \\
&\quad \times \{u_{j_1} + (-1)^{j_1-m_1} v_{j_1} a^{\dagger}_{j_1m_1} a^{\dagger}_{j_1-m_1}\} |0\rangle \\
&= \prod_{jm}{}' \{u_j + (-1)^{j-m} v_j a^{\dagger}_{jm} a^{\dagger}_{j-m}\} a^{\dagger}_{j_1m_1} |0\rangle \\
&= \frac{1}{u_{j_1}} \prod_{jm} \{u_j + (-1)^{j-m} v_j a^{\dagger}_{jm} a^{\dagger}_{j-m}\} a^{\dagger}_{j_1m_1} |0\rangle
\end{aligned} \tag{6.29}$$

where the prime on $\prod$ indicates the term (j_1m_1) is absent in the product, and in writing the last line of this equation use has been made of the fact that a state can accomodate only one fermion. Thus the one quasi-particle state has one unpaired particle. Consequently the state has seniority one and can be used to approximate a seniority-one state in an odd-A nucleus.

In a similar way one can show that two quasi-particle states are either seniority-zero when $I=0$ or seniority-two when $I\neq 0$. These two quasi-particle states can be used to approximate the low-lying states in even–even nuclei. Returning to the example of the tin isotopes, for ${}^{116}_{50}\mathrm{Sn}_{66}$ there are more than 1000 ways of making $I=2^+$ by putting 16 particles into the valence orbits. Consequently the normal shell-model approach would require the diagonalization of an energy matrix of this size. On the other hand, if one assumes the lowest 2^+ state is some linear combination of $I=2^+$ two-quasi-particle states, there are only nine ways the state can arise, namely

$$(1d_{\frac{5}{2}})^2;\,(1d_{\frac{3}{2}})^2;\,(0g_{\frac{7}{2}})^2;\,(0h_{\frac{11}{2}})^2;\,(1d_{\frac{5}{2}},\,1d_{\frac{3}{2}});$$
$$(1d_{\frac{5}{2}},\,2s_{\frac{1}{2}});\,(1d_{\frac{5}{2}},\,0g_{\frac{7}{2}});\,(1d_{\frac{3}{2}},\,2s_{\frac{1}{2}});\,(1d_{\frac{3}{2}},\,0g_{\frac{7}{2}}).$$

We shall now examine how shell-model calculations are carried out in terms of quasi-particles (Baranger 1960) and show that to a good approximation the first 2^+ state in an even–even nucleus can be described in terms of two quasi-particle eigenfunctions.

2.1. *Energy calculations with quasi-particles*

In this section we outline the procedure for calculating the energies of states in s.c.s. nuclei when a general two-body residual interaction is used (Baranger 1960). A convenient way to proceed is to write $\mathcal{H}$, equation (6.12), in terms of quasi-particle operators. To do this we must know the inverse of equations (6.27); that is how to express $a^\dagger_{jm}$ and a_{jm} in terms of $\alpha^\dagger_{jm}$ and α_{jm}. Since $u_j^2+v_j^2=1$, it follows that

$$a^\dagger_{jm}=u_j\alpha^\dagger_{jm}+(-1)^{j-m}v_j\alpha_{j-m} \tag{6.30a}$$

$$a_{jm}=u_j\alpha_{jm}+(-1)^{j-m}v_j\alpha^\dagger_{j-m}. \tag{6.30b}$$

Substitution of these results into the number-operator part of $\mathcal{H}$ gives

$$\sum_{jm}(\varepsilon_j-\lambda)a^\dagger_{jm}a_{jm}=\sum_{jm}(\varepsilon_j-\lambda)\{u_j\alpha^\dagger_{jm}+(-1)^{j-m}v_j\alpha_{j-m}\}\{u_j\alpha_{jm}+(-1)^{j-m}v_j\alpha^\dagger_{j-m}\}$$
$$=\sum_j(2j+1)(\varepsilon_j-\lambda)v_j^2+\sum_{jm}(\varepsilon_j-\lambda)(u_j^2-v_j^2)\alpha^\dagger_{jm}\alpha_{jm}$$
$$+\sum_{jm}(-1)^{j-m}(\varepsilon_j-\lambda)u_jv_j(\alpha^\dagger_{jm}\alpha^\dagger_{j-m}+\alpha_{j-m}\alpha_{jm})$$

where use has been made of equations (6.28) and the fact that m is a dummy summation index. Thus when the quasi-particle operators are written in normal order (i.e. destruction operators always on the extreme right) the particle number operator becomes the sum of three terms: a c-

number, the number operator for quasi-particles, and a term dependent on $(\alpha^\dagger_{jm}\alpha^\dagger_{j-m}+\alpha_{j-m}\alpha_{jm})$.

According to equation (1.47) the general two-body interaction V can be written as

$$V=\tfrac{1}{4}\sum_{JM}\sum_{j_1j_2j_3j_4}(1+\delta_{j_1j_2})(1+\delta_{j_3j_4})E_J(j_1j_2;\,j_3j_4)A^\dagger_{JM}(j_1j_2)A_{JM}(j_3j_4)$$

where $A^\dagger_{JM}(j_1j_2)$ and $A_{JM}(j_3j_4)$ are given by equations (1.42) and (1.43), respectively. One may now write V in terms of quasi-particle operators, and after a straightforward but tedious calculation one can show that the total Hamiltonian $\mathcal{H}$ becomes

$$\mathcal{H}=\mathcal{H}_0+\mathcal{H}_1+\mathcal{H}_2+\mathcal{H}_{\text{int}}$$

where

$$\mathcal{H}_0=\sum_j(2j+1)(\eta_j-\tfrac{1}{2}\mu_j)v_j^2-\tfrac{1}{2}\sum_j(2j+1)\Delta_ju_jv_j \tag{6.31}$$

$$\mathcal{H}_1=\sum_{jm}\{\eta_j(u_j^2-v_j^2)+2u_jv_j\Delta_j\}\alpha^\dagger_{jm}\alpha_{jm} \tag{6.32}$$

$$\mathcal{H}_2=\sum_{jm}(-1)^{j-m}\{\eta_ju_jv_j-\tfrac{1}{2}(u_j^2-v_j^2)\Delta_j\}(\alpha^\dagger_{jm}\alpha^\dagger_{j-m}+\alpha_{j-m}\alpha_{jm}) \tag{6.33}$$

with

$$\Delta_j=-\sum_{j'}\left(\frac{2j'+1}{2j+1}\right)^{\frac{1}{2}}u_{j'}v_{j'}E_0(jj;\,j'j') \tag{6.34}$$

$$\mu_j=\frac{1}{2j+1}\sum_{j'J}(2J+1)v_{j'}^2(1+\delta_{jj'})E_J(jj';\,jj') \tag{6.35}$$

$$\eta_j=\varepsilon_j-\lambda+\mu_j. \tag{6.36}$$

The remaining term $\mathcal{H}_{\text{int}}$, which we shall discuss later, is the product of four quasi-particle operators (with destruction operators always standing on the extreme right) and corresponds to the quasi-particle-quasi-particle residual interaction.

The BCS procedure of minimizing $\mathcal{H}$ as a function of v_j is equivalent to setting the coefficient of $(\alpha^\dagger_{jm}\alpha^\dagger_{j-m}+\alpha_{j-m}\alpha_{jm})$ in equation (6.33) equal to zero, i.e. to require that

$$\eta_ju_jv_j-\tfrac{1}{2}(u_j^2-v_j^2)\Delta_j=0 \tag{6.37}$$

for each value of j. That this condition is equivalent to equation (6.18) (the BCS prescription) in the case that V is the pairing force may be seen simply by noting that for the pairing force

$$E_J(j_1j_2;\,j_3j_4)=-\frac{G}{2}\sqrt{\{(2j_1+1)(2j_3+1)\}}\,\delta_{J0}\delta_{j_1j_2}\delta_{j_3j_4}.$$

From this it follows that Δ_j, equation (6.34), is independent of j and has exactly the value of Δ given by equation (6.17). Also μ_j in equation (6.35) is $-Gv_j^2$ and $\eta_j = \varepsilon_j - \lambda - Gv_j^2$ so that equation (6.37) becomes identical to equation (6.18). Furthermore, when these conditions are fulfilled $\mathscr{H}_0 + \lambda \sum_j (2j+1)v_j^2$ is precisely the BCS energy given by equation (6.21).

When v_j satisfies equation (6.37) the transformation to quasi-particles is known as the Bogoliubov–Valatin transformation (Bogoliubov 1958, Valatin 1958), and the BCS wave function is the quasi-particle vacuum state. By squaring equation (6.37) one obtains expressions for u_j and v_j, namely

$$v_j^2 = \frac{1}{2}\left\{1 - \frac{\eta_j}{\sqrt{(\eta_j^2 + \Delta_j^2)}}\right\} \tag{6.38a}$$

$$u_j^2 = \frac{1}{2}\left\{1 + \frac{\eta_j}{\sqrt{(\eta_j^2 + \Delta_j^2)}}\right\}, \tag{6.38b}$$

and the number of particles constraint can then be written as

$$N = \sum (2j+1)v_j^2 = \sum_j \Omega_j\left\{1 - \frac{\eta_j}{\sqrt{(\eta_j^2 + \Delta_j^2)}}\right\}. \tag{6.39}$$

Consequently for the general potential one must choose λ and v_j to satisfy equations (6.37) and (6.39).

When v_j satisfies equation (6.37) it follows that $\mathscr{H}$ may be written as

$$\mathscr{H} = \mathscr{H}_0 + \sum_{jm} E_j \alpha_{jm}^\dagger \alpha_{jm} + \mathscr{H}_{\text{int}} \tag{6.40}$$

where

$$E_j = \sqrt{(\eta_j^2 + \Delta_j^2)}. \tag{6.41}$$

Clearly $\mathscr{H}_0$ is the energy of the quasi-particle vacuum. If one interprets E_j as the energy of a quasi-particle in the state (j, m), it follows that the second term is just the energy associated with a system of non-interacting quasi-particles. From equation (6.41) it is clear that $E_j \geq \Delta_j$. Thus a two quasi-particle state (j_1, j_2) will, in the absence of $\mathscr{H}_{\text{int}}$, lie at an energy $(E_{j_1} + E_{j_2}) \geq (\Delta_{j_1} + \Delta_{j_2})$ above the vacuum. For this reason Δ_j is often called the gap energy and $2\Delta_j$ gives the energy it takes to break a zero-coupled pair, in other words to go from the seniority-zero to a seniority-two state of the j^n configuration. The interpretation of equation (6.40) is now clear. Instead of a system of nucleons with single-particle energies ε_j interacting by means of the residual two-body potential V, the quasi-particle transformation given in equations (6.27) introduces a system of quasi-particles with energies E_j that interact via $\mathscr{H}_{\text{int}}$, the form of which we shall now discuss.

$\mathcal{H}_{\text{int}}$ is that part of the Hamiltonian which has products of four quasi-particle operators. By substitution of equations (6.30) into the expression for V, one can show (Kuo *et al.* 1966, Pal *et al.* 1967) by straightforward recoupling techniques together with equations (1.46) that

$$\mathcal{H}_{\text{int}} = \mathcal{H}_{04} + \mathcal{H}_{13} + \mathcal{H}_{22} + \mathcal{H}_{31} + \mathcal{H}_{40} \tag{6.42}$$

where

$$\mathcal{H}_{04} = -\tfrac{1}{4}\sum_{Jj_i}(1+\delta_{j_1j_2})(1+\delta_{j_3j_4})\sqrt{(2J+1)}E_J(j_1j_2;\, j_3j_4) \times v_{j_1}v_{j_2}u_{j_3}u_{j_4}[\tilde{\mathcal{A}}_J(j_1j_2)\times\tilde{\mathcal{A}}_J(j_3j_4)]_{00} \tag{6.43a}$$

$$\mathcal{H}_{13} = -\tfrac{1}{2}\sum_{Jj_i}(1+\delta_{j_1j_2})(1+\delta_{j_3j_4})\sqrt{(2J+1)}E_J(j_1j_2;\, j_3j_4) \times (u_{j_1}v_{j_2}u_{j_3}u_{j_4} - v_{j_1}u_{j_2}v_{j_3}v_{j_4})[\beta_J(j_1j_2)\times\tilde{\mathcal{A}}_J(j_3j_4)]_{00} \tag{6.43b}$$

$$\begin{aligned}\mathcal{H}_{22} = \tfrac{1}{4}\sum_{j_iJ}\sqrt{(2J+1)}\big((1+\delta_{j_1j_2})(1+\delta_{j_3j_4})E_J(j_1j_2;\, j_3j_4) \\ \times (u_{j_1}u_{j_2}u_{j_3}u_{j_4} + v_{j_1}v_{j_2}v_{j_3}v_{j_4}) \\ + 4u_{j_1}v_{j_2}u_{j_3}v_{j_4}\sqrt{\{(1+\delta_{j_1j_4})(1+\delta_{j_2j_3})(1+\delta_{j_1j_2})(1+\delta_{j_3j_4})\}} \\ \times E_J(j_1j_2^{-1};\, j_3j_4^{-1})\big)[\mathcal{A}_J^{\dagger}(j_1j_2)\times\tilde{\mathcal{A}}_J(j_3j_4)]_{00}\end{aligned} \tag{6.43c}$$

$$\mathcal{H}_{31} = \mathcal{H}_{13}^{\dagger} \tag{6.43d}$$

$$\mathcal{H}_{40} = \mathcal{H}_{04}^{\dagger}. \tag{6.43e}$$

In writing these equations the following notation has been used

$$\mathcal{A}_{JM}^{\dagger}(j_1j_2) = \frac{1}{\sqrt{(1+\delta_{j_1j_2})}}\sum_{m_1m_2}(j_1j_2m_1m_2 \mid JM)\alpha_{j_1m_1}^{\dagger}\alpha_{j_2m_2}^{\dagger} \tag{6.44a}$$

$$\tilde{\mathcal{A}}_{JM}(j_1j_2) = \frac{-1}{\sqrt{(1+\delta_{j_1j_2})}}\sum_{m_1m_2}(j_1j_2m_1m_2 \mid JM)\tilde{\alpha}_{j_1m_1}\tilde{\alpha}_{j_2m_2} \tag{6.44b}$$

$$\beta_{JM}(j_1j_2) = \frac{1}{\sqrt{(1+\delta_{j_1j_2})}}\sum_{m_1m_2}(j_1j_2m_1m_2 \mid JM)\alpha_{j_1m_1}^{\dagger}\tilde{\alpha}_{j_2m_2} \tag{6.44c}$$

with

$$\tilde{\alpha}_{jm} = (-1)^{j+m}\alpha_{j-m}$$

and $E_J(j_1j_2^{-1};\, j_3j_4^{-1})$, the particle-hole interaction, given by equation (3.68)

$$E_J(j_1j_2^{-1};\, j_3j_4^{-1}) = -(-1)^{j_1+j_2+j_3+j_4}\sum_K(2K+1)W(j_4j_3j_1j_2;\, JK) \times E_K(j_1j_4;\, j_3j_2).$$

Consequently the residual quasi-particle- quasi-particle interaction has a number-conserving part (i.e. a seniority-conserving part) which is given by $\mathscr{H}_{22}$, and in addition there are parts that change the number of quasi-particles by two ($\mathscr{H}_{13}$ and $\mathscr{H}_{31}$) and four ($\mathscr{H}_{04}$ and $\mathscr{H}_{40}$), that is mix a seniority-v state with $v \pm 2$ and $v \pm 4$ states.

Thus in the quasi-particle approximation the steps to be followed in carrying out a shell-model calculation are:

(a) The residual two-body interaction and the single-particle energies ε_j are chosen as in the usual shell-model calculation.

(b) There are $(n+1)$ parameters which determine the quasi-particle transformation, namely the n values of v_j, if there are n single-particle levels, and the chemical potential λ. Since equation (6.37) must hold for each value of j it provides n conditions on these quantities. In addition, the expectation value of the number operator given by equation (6.39) must be equal to the number of nucleons in the system. When solved these $(n+1)$ equations yield values for the parameters and hence the ground-state energy, $\mathscr{H}_0 + \lambda \sum_j (2j+1)v_j^2$, can be computed.

(c) For an even-A nucleus the low-lying excited states are two quasi-particle configurations with wave functions

$$\psi_{IM} = \mathscr{A}^{\dagger}_{IM}(j_1 j_2)\psi_{\text{BCS}} \tag{6.45}$$

with ψ_{BCS} given by equation (6.4). The energy of this state is

$$E_I = \mathscr{H}_0 + \lambda \sum_j (2j+1)v_j^2 + E_{j_1} + E_{j_2} + \langle \psi_{IM} | \mathscr{H}_{22} | \psi_{IM} \rangle \tag{6.46}$$

where E_j is computed from equation (6.41) using the values of u_j, v_j, and λ determined in (b). The same values of these quantities are used in the expression for $\mathscr{H}_{22}$ in equation (6.43c). If there are several ways to realize the angular momentum I, one must set up all possible two quasi-particle states with this spin and diagonalize $\mathscr{H}$ using these eigenfunctions as basis states. This procedure, which is known as the Tamm–Dancoff approximation, provides an enormous simplification since the dimensionality of the Hamiltonian matrix is now identical to that which one would encounter for two particles outside the closed shell. Thus, for example, in ${}^{62}_{28}\text{Ni}_{34}$ where the $1p_{\frac{3}{2}}$, $1p_{\frac{1}{2}}$, $0f_{\frac{5}{2}}$ model space leads to 33 ways of producing a 2^+ level, the two quasi-particle approximation involves diagonalizing only a 5×5 matrix!

One can also construct four quasi-particle states which will correspond to seniority-zero, -two or -four eigenfunctions. These describe other possible states in the even-A nucleus which could be included in the calculation. This would of course, lead to a larger matrix and obviously should be a better approximation to the true situation. If this procedure is carried out then in addition to $\mathcal{H}_{22}$ one must also include $\mathcal{H}_{31}$ and $\mathcal{H}_{13}$ given in equations (6.43b) and (6.43d) (Pal *et al.* 1967). In section 3 of this chapter we shall discuss this matter further.

(d) There is one place where the two quasi-particle approximation runs into difficulty, and that is in the consideration of the excited 0^+ states. From equation (6.6) it is clear that the BCS and two quasi-particle eigenfunctions have contributions corresponding to various numbers of particles. Certainly at a minimum one would expect that if a two quasi-particle eigenfunction is to describe a system with N nucleons it should at least contain an N-particle part. We shall now show that the two quasi-particle state vector

$$\psi_{00}=\sum_j \sqrt{(2j+1)}u_jv_j[\alpha_j^\dagger\times\alpha_j^\dagger]_{00}\psi_{\rm BCS} \tag{6.47}$$

has no part corresponding to N particles (where N is the expectation value of the number operator) and consequently should be excluded from the calculation. To see this we note that in terms of quasi-particles the number operator $\sum_j \mathrm{N}_j$ becomes

$$\begin{aligned}\sum_j \mathrm{N}_j &= \sum_{jm}(u_j\alpha_{jm}^\dagger+(-1)^{j-m}v_j\alpha_{j-m})(u_j\alpha_{jm}+(-1)^{j-m}v_j\alpha_{j-m}^\dagger)\\ &=\sum_{jm}\{v_j^2+(u_j^2-v_j^2)\alpha_{jm}^\dagger\alpha_{jm}\\ &\quad+(-1)^{j-m}u_jv_j\,(\alpha_{jm}^\dagger\alpha_{j-m}^\dagger+\alpha_{j-m}\alpha_{jm})\}.\end{aligned}$$

Since the BCS wave function is the quasi-particle vacuum and since $\sum_j (2j+1)v_j^2$ is chosen to be equal to N in equation (6.39), it follows that

$$\left(\sum_j \mathrm{N}_j-N\right)\psi_{\rm BCS}=\sum_j \sqrt{(2j+1)}u_jv_j[\alpha_j^\dagger\times\alpha_j^\dagger]_{00}\psi_{\rm BCS}. \tag{6.48}$$

If we denote by ψ_n a normalized n-particle wave function and by C_n^2 the probability that ψ_n is contained in $\psi_{\rm BCS}$ it follows that

$$\begin{aligned}\left(\sum_j \mathrm{N}_j-N\right)\psi_{\rm BCS}&=\left\{\sum_j \mathrm{N}_j-N\right)\sum_n C_n\psi_n\\ &=\sum_n C_n(n-N)\psi_n.\end{aligned}$$

Thus $(\sum_j \mathrm{N}_j-N)\psi_{\rm BCS}$ has no component with $n=N$. Consequently from equation (6.48) it follows that the wave function given by equation (6.47) has no N-particle part and hence should be excluded when calculating the properties of excited $I=0$ two quasi-particle states. Furthermore, it has been shown (Kuo *et al.* 1966) that, unless this state is excluded from the calculation, the first excited 0^+ level lies at much too low an excitation energy.

(e) In a similar way the lowest levels of an odd-A nucleus should be the one quasi-particle states whose energies are given by

$$\tilde{E}_{j_1} = \mathcal{H}_0 + \lambda \sum_j (2j+1)v_j^2 + E_{j_1} \tag{6.49}$$

where $\mathcal{H}_0$ and E_{j_1} are calculated using u_j, v_j, and λ determined from equations (6.37) and (6.39). The one quasi-particle approximation within a given shell leads to only one state of spin j_1. Consequently $\mathcal{H}_{22}$, which involves at least two quasi-particles, does not enter into the calculation and the energy of the state is given directly by equation (6.49).

Strictly speaking equation (6.39) is not the expectation value of the number operator when we deal with one quasi-particle eigenfunctions. For a one quasi-particle state with angular momentum (j_1m_1)

$$\psi_{j_1m_1} = \alpha^\dagger_{j_1m_1}\psi_{\text{BCS}} \tag{6.50}$$

and

$$\begin{aligned} N &= \langle\psi_{j_1m_1}|\sum_j \text{N}_j\,|\psi_{j_1m_1}\rangle \\ &= \sum_j (2j+1)v_j^2 + (u_{j_1}^2 - v_{j_1}^2). \end{aligned} \tag{6.51}$$

If this relationship is used instead of equation (6.39) the calculation becomes much more laborious since these are then different values of λ and Δ_j for each angular momentum j_1. One can argue that for particles near the Fermi surface, $\varepsilon_{j_1} \cong \lambda - \mu_{j_1}$, and hence the term $u_{j_1}^2 - v_{j_1}^2$ is small since according to equation (6.38) the difference is proportional to $(\varepsilon_{j_1} - \lambda + \mu_{j_1})$. In the particular case of the pairing force it has been shown (Kerman *et al.* 1961) that the difference in energies and wave functions brought about by using equation (6.51) instead of equation (6.39) is too small to be of any practical importance. Thus to the accuracy with which these calculations can be believed one can use equation (6.39) for the number-operator constraint.

One can also include three quasi-particle states in the diagonalization of the Hamiltonian, and such calculations have been carried out for the tin isotopes by Kuo *et al.* (1966). (We shall discuss this in more detail in section 3 of this chapter.) In actual fact the one quasi-particle calculations predict that all the spectroscopic strength for single-nucleon direct transfer should be concentrated in a single state. This follows because the BCS wave function is the quasi-particle vacuum, and hence the effect of the single-particle creation operator $a^{\dagger}_{jm}$ on ψ_{BCS} is to produce exactly the one quasi-particle state given by equation (6.50). Since the spectroscopic strength for given (jl) is usually spread over several states, it is clear that at a minimum one must generally do a $1+3$ quasi-particle calculation to approximate states in odd-A nuclei.

(f) In section 1.2 of this chapter we showed that it is possible to improve the accuracy of calculated energies by determining u_j, v_j, and λ from the variational calculation, then projecting out of the BCS wave function that part which corresponds to N (the desired number of particles), and using this projected-normalized eigenfunction to calculate energies. Clearly in the more general case we are now discussing a better approximation to the 'true' energies and eigenfunctions would be obtained if the original Hamiltonian,

$$H=\sum_{jm}\varepsilon_j a^{\dagger}_{jm}a_{jm}+V$$

were diagonalized using projected normalized wave functions. If C_I denotes the appropriate normalization constant then according to equation (6.6) a better approximation to the 'true' ground-state energy would be obtained if one calculates

$$\tilde{E}_0=\langle\tilde{\psi}_{00}|\, H\, |\tilde{\psi}_{00}\rangle$$

with

$$\tilde{\psi}_{00}=C_0\left\{\sum_j \frac{v_j}{u_j}\,\mathrm{S}_+(j)\right\}^{N/2}|0\rangle. \tag{6.51a}$$

In the two quasi-particle approximation the energies and eigenfunctions of the low-lying excited states of s.c.s. even-A nuclei would be given by diagonalizing H using as basis states all the projected normalized two quasi-particle wave functions

$$\tilde{\psi}_{IM}=C_2\left\{\sum_j \frac{v_j}{u_j}\,\mathrm{S}_+(j)\right\}^{(N-2)/2}A^{\dagger}_{IM}(j_1j_2)\,|0\rangle. \tag{6.51b}$$

Similar considerations would, of course, hold true for an odd-A nucleus, so that a better approximation to the energy of a one quasi-particle state would be given by

$$\tilde{E}'_j=\langle\tilde{\psi}_{jm}|\, H\, |\tilde{\psi}_{jm}\rangle$$

with

$$\tilde{\psi}_{jm}=C_1\left\{\sum_j \frac{v_j}{u_j}\,\mathrm{S}_+(j)\right\}^{(N-1)/2}a^{\dagger}_{jm}\,|0\rangle. \tag{6.51c}$$

2.2. M4 transitions in the lead nuclei

In order to calculate gamma-ray transition rates using this formalism one must first write the transition-rate operator in terms of quasi-particles. From equations (6.30) it follows that any single-particle operator T_{LM} can be written as

$$\begin{aligned}T_{LM} &= \sum_{jj'}\sum_{mm'} \langle\phi_{j'}\|\, T_L \,\|\phi_j\rangle(jLmM \mid j'm')a^{\dagger}_{j'm'}a_{jm}\\ &= \delta_{L0}\sum_{j}(2j+1)v_j^2\langle\phi_j\|\, T_L \,\|\phi_j\rangle + \sum_{jj'}\sum_{mm'}\langle\phi_{j'}\|\, T_L \,\|\phi_j\rangle\\ &\quad\times(jLmM \mid j'm')\big((u_ju_{j'}-(-1)^Lv_jv_{j'})\alpha^{\dagger}_{j'm'}\alpha_{jm}\\ &\quad+(-1)^{j-m}u_{j'}v_j\alpha^{\dagger}_{j'm'}\alpha^{\dagger}_{j-m}+(-1)^{j'-m'}u_jv_{j'}\alpha_{j'-m'}\alpha_{jm}\big)\end{aligned} \tag{6.52}$$

where $\langle\phi_{j'}\|\, T_L \,\|\phi_j\rangle$ is the single-particle reduced matrix element and use has been made of equation (5.24).

Thus for L different from zero the transition-rate operator contains a part which conserves the number of quasi-particles (conserves seniority) and a second part which changes the number of quasi-particles by two. Built into the term that conserves the number of quasi-particles are the seniority selection rules for the configuration j^n discussed in Chapter 3, section 2.3 and Appendix 3 i.e. for $j=j'$ the coefficient of $\alpha^{\dagger}_{jm'}\alpha_{jm}$ becomes $u_j^2+v_j^2=1$ for L odd and $(u_j^2-v_j^2)$ for L even. Since all the dependence on particle number is contained in u_j and v_j it follows that the matrix elements of odd-L operators between states with the same number of quasi-particles (same seniority) are independent of N. For the half-filled shell $u_j^2=v_j^2=\frac{1}{2}$ and hence for L even, the matrix element between states with the same number of quasi-particles vanishes.

We now use equation (6.52) to calculate the transition rate between one quasi-particle states in an odd-A s.c.s. nucleus. From equation (6.50) it follows that

$$\langle\psi_{j_f}\|\, T_L \,\|\psi_{j_i}\rangle = \{u_{j_f}u_{j_i}-(-1)^Lv_{j_f}v_{j_i}\}\langle\phi_{j_f}\|\, T_L \,\|\phi_{j_i}\rangle. \tag{6.53}$$

Thus the reduced matrix element is precisely the single-particle one modified by a multiplicative factor that depends on the occupation probabilities of the states involved in the transition. Because of this factor it is important to understand the signs associated with the various v_j. When the pairing force in equation (6.1) is used the assumption is made that all matrix elements $\langle(j)^2_0|\, V\,|(j_1)^2_0\rangle$ are negative, and because of this all v_j can be chosen positive. On the other hand, when one uses the phase conventions employed in this book (namely ϕ_{jm} given by equation (1.4)) the experimental data on gamma decay favours a positive sign for $\langle(j)^2_0|\, V\,|(j_1)^2_0\rangle$ if j and j_1 have opposite parity (Gloeckner and Serduke

1974). This change of sign is consistent with that given by a short-range attractive potential. For example, the surface-delta interaction, the matrix elements of which are given by equation (A2.29), has $J=0$ matrix elements

$$V'_{\mathrm{p}} = -G\sum_{jj'}(-1)^{l+l'}\mathrm{S}_{+}(j)\mathrm{S}_{-}(j') \tag{6.54}$$

where $G = V_0\bar{R}$. If V'_{p} is used the mathematical machinery of section 1.1 of this chapter goes through unchanged provided u_jv_j is replaced by $(-1)^l u_jv_j$ where l is the orbital angular momentum associated with the orbit j. Since the overall sign of the wave function is unimportant it is sufficient to merely change the sign of v_j corresponding to the one opposite-parity orbit that appears in the description of the state. For example, in the lead calculation described in section 1.3 of this chapter, the normal parity orbits are the $2p_{\frac{1}{2}}$, $2p_{\frac{3}{2}}$, $1f_{\frac{5}{2}}$, and $1f_{\frac{7}{2}}$ levels and the non-normal parity level is the $0i_{\frac{13}{2}}$. Thus when the phase conventions of this book are used $v_{\frac{13}{2}}$ must be negative in order that one have overall sign consistency. Of course, if the $E_J(j_1j_2;j_3j_4)$ used in the explicit calculation of v_j are matrix elements of a conventional potential calculated using our ϕ_{jm}, this sign change will automatically be taken into account.

We now use equation (6.53) to calculate the $M4$ transition rates between the $0i_{\frac{13}{2}}$ and $1f_{\frac{5}{2}}$ one quasi-hole states in the lead isotopes. In $^{207}_{82}\mathrm{Pb}_{125}$ the lifetime of the 1·633 MeV $\frac{13}{2}^{+}$ state for decay to the 570 keV $\frac{5}{2}^{-}$ level is 1·32 s. From Table 5.1 it follows that in this nucleus

$$B(M4; \tfrac{13}{2}^{+} \to \tfrac{5}{2}^{-}) = 2{\cdot}33 \times 10^5 \mu_{\mathrm{N}}^2\ \mathrm{fm}^6.$$

These states correspond to a single neutron hole in the 126 shell and consequently $u_{\frac{5}{2}} = u_{\frac{13}{2}} = 1$ (the quasi-particle calculations for the lead isotopes deal with holes instead of particles). Thus this value for $B(M4)$ should correspond to the single-particle value. Since the $M4$ operator has the form given by equation (5.9) it follows that for a neutron

$$\begin{aligned}\langle\phi_{1f_{\frac{5}{2}}}\| T_4^{(\mathcal{M})} \|\phi_{0i_{\frac{13}{2}}}\rangle &= 6\mu_{\mathrm{n}}\mu_{\mathrm{N}}\langle\phi_{1f_{\frac{5}{2}}}\| r^3[Y_3\times\sigma]_4 \|\phi_{0i_{\frac{13}{2}}}\rangle\\ &= 30\sqrt{\left(\frac{21}{\pi}\right)}\mu_{\mathrm{n}}\mu_{\mathrm{N}}\left(\frac{\hbar}{m\omega}\right)^{\frac{3}{2}}\\ &= -2{\cdot}17\times10^3\mu_{\mathrm{N}}\ \mathrm{fm}^3\end{aligned}$$

where use has been made of equation (A2.24) to evaluate the angle-spin part of the matrix element and harmonic-oscillator wave functions to evaluate the radial integral.

From the experimental value of $B(M4)$ needed to fit the data it follows that

$$\langle\phi_{1f_{\frac{5}{2}}}\| T_4^{(\mathcal{M})} \|\phi_{0i_{\frac{13}{2}}}\rangle_{\mathrm{expt.}} = \pm737\ \mu_{\mathrm{N}}\ \mathrm{fm}^3.$$

Thus the single-particle matrix element is a factor of 2·95 larger in absolute value than the experimentally required one. (This quenching is almost identical to that required in the $N=50$ region (Serduke *et al.* 1976) where $\langle\phi_{1p_{\frac{1}{2}}}\| T_4^{(\mathcal{M})} \|\phi_{0g_{\frac{9}{2}}}\rangle$ must be reduced by a factor of 2·87 to fit experiment.) Thus we have the situation discussed in section 2 of Chapter 5; i.e. core polarization modifies the matrix element and requires us to introduce an effective operator if a truncated model space is used. It has been shown that configuration mixing does indeed tend to decrease the single-particle $M4$ matrix elements (Gupta and Lawson 1959).

In column 5 of Table 6.3 we list the predicted values of $B(M4)$ based

TABLE 6.3

M4 transition rates in the lead isotopes

Nucleus	E_γ (MeV)	Experimental τ_m(s)	$B(M4; \frac{13}{2}^+ \to \frac{5}{2}^-)$ Experimental $(\mu_N^2\,\mathrm{fm}^6)$	Quasi-particle $(\mu_N^2\,\mathrm{fm}^6)$	$u_{\frac{5}{2}}$	$v_{\frac{5}{2}}$	$u_{\frac{13}{2}}$	$v_{\frac{13}{2}}$
$^{207}_{82}Pb_{125}$	1·063	1·32	$2{\cdot}33\times10^5$	$2{\cdot}33\times10^5$	1	0	1	0
$^{205}_{82}Pb_{123}$	1·014	1·58	$2{\cdot}97\times10^5$	$2{\cdot}14\times10^5$	0·92	0·40	0·99	−0·12
$^{203}_{82}Pb_{121}$	0·825	12·2	$2{\cdot}46\times10^5$	$1{\cdot}91\times10^5$	0·81	0·59	0·98	−0·19
$^{201}_{82}Pb_{119}$	0·629	154·	$2{\cdot}24\times10^5$	$1{\cdot}62\times10^5$	0·68	0·73	0·97	−0·24
$^{199}_{82}Pb_{117}$	0·424	4965·	$2{\cdot}42\times10^5$	$1{\cdot}48\times10^5$	0·57	0·82	0·95	−0·30

The experimental data were taken from the 'Nuclear Data' sheets and have been corrected for internal conversion. The effective single-particle $M4$ matrix element was chosen to fit ^{207}Pb. The values of u and v were calculated using a pairing force with $G=0{\cdot}111$ MeV and single-hole energies $\varepsilon_{p_{\frac{1}{2}}}=0$, $\varepsilon_{f_{\frac{5}{2}}}=0{\cdot}57$ MeV, $\varepsilon_{p_{\frac{3}{2}}}=0{\cdot}90$ MeV, $\varepsilon_{i_{\frac{13}{2}}}=1{\cdot}634$ MeV, and $\varepsilon_{f_{\frac{7}{2}}}=2{\cdot}35$ MeV (Kisslinger and Sorensen 1960).

on the experimental value of the single-particle matrix element. The u_j and v_j needed in equation (6.53) are taken from the paper of Kisslinger and Sorensen (1960) who studied these isotopes using a pairing force with $G=0{\cdot}111$ MeV and single-particle energies taken from the experimental data on ^{207}Pb. The theoretical values of $B(M4)$ decrease slightly with decreasing A ($B(M4;\,^{199}\text{Pb})/B(M4;\,^{207}\text{Pb})=0{\cdot}634$) whereas the experimental numbers are essentially constant as a function of A. Thus the effective operator needed to fit experiment changes slightly from nucleus to nucleus and must be increased by about 20% in going from $A=207$ to $A=199$, and this situation is the same as found in the $N=50$ region.

Although the quasi-particle value for $B(M4)$ does decrease slightly with A, the decrease is much less marked than if the non-interacting shell model had been used to calculate the transition rates. Since the $1f_{\frac{5}{2}}$ level is the first excited state in ^{207}Pb it follows that for the non-interacting shell model the transition in the nucleus ${}^{205-n}_{82}\mathrm{Pb}_{123-n}$ should correspond to $(2p_{\frac{1}{2}})^{-2}(1f_{\frac{5}{2}})^{-n}(0i_{\frac{13}{2}})^{-1} \rightarrow (2p_{\frac{1}{2}})^{-2}(1f_{\frac{5}{2}})^{-(n+1)}$ ($n \geq 0$ and even). By using the seniority-one c.f.p's in equation (A5.55) it is simple to show that

$$B(M4;\,{}^{205-n}_{82}\mathrm{Pb}_{123-n}) = \left(\frac{6-n}{6}\right) B(M4:\,{}^{207}_{82}Pb_{125}).$$

On the basis of this model the rate in ${}^{205}\mathrm{Pb}$ is slightly larger than the quasi-particle result, but all others are smaller ($B(M4;\,{}^{203}_{82}\mathrm{Pb}_{121}) = 1{\cdot}55 \times 10^5 \mu_{\mathrm{N}}^2\,\mathrm{fm}^6$ and $B(M4;\,{}^{201}_{82}\mathrm{Pb}_{199}) = 0{\cdot}78 \times 10^5 \mu_{\mathrm{N}}^2\,\mathrm{fm}^6$). Consequently a collectivity similar to that given by quasi-particle theory leads to a more nearly nucleus independent effective operator than does the non-interacting model.

2.3. Pick-up and stripping on the tin isotopes

The cross-section for single-nucleon direct-transfer reactions discussed in Chapter 2, section 5 is proportional to the spectroscopic factor $\mathscr{S}_{jl}$ which is defined by equation (2.115) to be

$$\mathscr{S}_{jl}(\mathrm{A} \rightleftarrows \mathrm{B}) = |\langle \psi_{I_{\mathrm{B}};T_{\mathrm{B}}} |\!|\!|\, a^{\dagger}_{j;\frac{1}{2}} \,|\!|\!| \psi_{I_{\mathrm{A}};T_{\mathrm{A}}} \rangle|^2 \tag{6.55}$$

where the triple bars indicate that the matrix element is reduced in both ordinary and isospin space. In this section we show how to calculate this factor when quasi-particle wave functions are used (Yoshida 1961) and consequently how to extract u_j and v_j from experiment. Since these latter quantities change in going from one nucleus to another, care must be taken in the calculation of $\mathscr{S}_{jl}$. We therefore begin our discussion by showing how the quasi-particle wave functions change when the number of particles goes from N to $N \pm 1$.

We denote the values for the N-particle system by u_j and v_j and those for the neighbouring $N \pm 1$ nucleus by $\bar{u}_j$ and $\bar{v}_j$. δu_j and δv_j are defined by the equations

$$\delta u_j = \bar{u}_j - u_j \tag{6.56a}$$

$$\delta v_j = \bar{v}_j - v_j, \tag{6.56b}$$

and these changes are assumed to be small. Since

$$u_j^2 + v_j^2 = \bar{u}_j^2 + \bar{v}_j^2 = 1$$

it follows that

$$u_j \delta u_j = -v_j \delta v_j. \tag{6.57}$$

Thus the change in the BCS wave function ψ in equation (6.4) induced by this change in u_j and v_j is

$$\begin{aligned}\bar{\psi} - \psi &= \sum_{jm>0} \{\delta u_j + (-1)^{j-m} \delta v_j a^{\dagger}_{jm} a^{\dagger}_{j-m}\} \Pi' \\ &= \sum_{jm>0} \frac{\delta u_j}{v_j} \{v_j - (-1)^{j-m} u_j a^{\dagger}_{jm} a^{\dagger}_{j-m}\} \Pi'\end{aligned}$$

where
$$\Pi' = \prod'_{j_1 m_1 > 0} \{u_{j_1} + (-1)^{j_1 - m} v_{j_1} a^\dagger_{j_1 m_1} a^\dagger_{j_1 - m_1}\} |0\rangle$$

with the prime indicating the term with $(j_1 m_1) = (jm)$ is excluded from the product. If one uses the definition of the quasi-particle creation operator given by equation (6.27b) one sees that

$$(-1)^{j-m} \alpha^\dagger_{j-m} \alpha^\dagger_{jm} \psi = \{v_j - (-1)^{j-m} u_j a^\dagger_{jm} a^\dagger_{j-m}\} \Pi'$$

so that
$$\bar{\psi} = (1 + \tfrac{1}{2} \sum_{jm} (-1)^{j-m} \frac{\delta u_j}{v_j} \alpha^\dagger_{j-m} \alpha^\dagger_{jm}) \psi \tag{6.58}$$

where the factor $\frac{1}{2}$ arises because the sum is over all m.

The quasi-particle creation operator for the $N \pm 1$ particle system is

$$\begin{aligned} \bar{\alpha}^\dagger_{jm} &= \bar{u}_j a^\dagger_{jm} - (-1)^{j-m} \bar{v}_j a_{j-m} \\ &= \alpha^\dagger_{jm} + \frac{\delta u_j}{v_j} \{v_j a^\dagger_{jm} + (-1)^{j-m} u_j a_{j-m}\} \\ &= \alpha^\dagger_{jm} + (-1)^{j-m} \frac{\delta u_j}{v_j} \alpha_{j-m}. \end{aligned}$$

Consequently, since $\alpha_{jm} \psi = 0$, it follows tnat to order δu_j

$$\bar{\alpha}^\dagger_{jm} \bar{\psi} = \alpha^\dagger_{jm} \bar{\psi}. \tag{6.59}$$

In order to illustrate how one calculates spectroscopic factors we consider in detail the (d, t) reaction on an odd-A target. We assume only neutrons are active and hence the isospin dependence of equation (6.55) trivially gives unity. *In the following the target wave function will always be expressed in terms of u_j and v_j and N will refer to the number of valence nucleons in the target.* The target ground state is assumed to be the one quasi-particle state $\alpha^\dagger_{jm} \psi$, and we first consider the spectroscopic factor connecting this to the 0^+ ground state $\bar{\psi}$ in the neighbouring even–even nucleus.

By using equation (6.30a) the nucleon creation operator $a^\dagger_{jm}$ can be expressed in terms of quasi-particle operators. When this is combined with equation (6.58) one sees that

$$\begin{aligned} \langle \alpha^\dagger_{jm} \psi | \, a^\dagger_{jm} \, | \bar{\psi} \rangle &= (0j0m \mid jm) \{\mathscr{S}_{jl}(j \to 0)\}^{\frac{1}{2}} \\ &= \langle \psi | \, \alpha_{jm} (u_j \alpha^\dagger_{jm} + (-1)^{j-m} v_j \alpha_{j-m}) \mid \bar{\psi} \rangle \\ &= u_j + \tfrac{1}{2} v_j (-1)^{j-m} \langle \psi | \, \alpha_{jm} \alpha_{j-m} \sum_{j'm'} (-1)^{j'-m'} \frac{\delta u_{j'}}{v_{j'}} \alpha^\dagger_{j'-m'} \alpha^\dagger_{j'm'} \, | \psi \rangle \\ &= u_j + \delta u_j \\ &= \bar{u}_j. \end{aligned}$$

Since the Clebsch–Gordan coefficient is unity

$$\mathcal{S}_{jl}(j \to 0) = \bar{u}_j^2. \tag{6.60}$$

Thus the pickup reaction on an odd-A target proceeding to the ground state of the final nucleus is determined by $\bar{u}_j$ of the final even–even nucleus.

To see that this corresponds to the result deduced in Chapter 2, section 5.1, for the configuration j^N we consider the single-level model discussed in section 1.2 of this chapter. Because the 'barred' or final nucleus contains only $N-1$ particles

$$\bar{v}_j^2 = (N-1)/(2j+1).$$

Since $\bar{u}_j^2 + \bar{v}_j^2 = 1$ it follows that

$$\begin{aligned}\mathcal{S}_{jl}(j \to 0) &= 1 - \frac{(N-1)}{2j+1}\\ &= \frac{2j+2-N}{2j+1}\end{aligned}$$

in agreement with the result given in Table 2.7.

The $J \neq 0$ states are described by the two quasi-particle wave functions of equation (6.45), and the spectroscopic factor governing their population is

$$\begin{aligned}\langle \alpha^\dagger_{jm}\psi|\, a^\dagger_{j_1m_1}\, |\mathcal{A}^\dagger_{JM}(jj_1)\bar{\psi}\rangle &= (Jj_1Mm_1 \mid jm)[\mathcal{S}_{j_1l_1}\{j \to (jj_1)_J\}]^{\frac{1}{2}}\\ &= \frac{1}{\sqrt{(1+\delta_{jj_1})}} \sum_{\bar{m}\bar{m}_1} (jj_1\bar{m}\bar{m}_1 \mid JM)\langle \psi|\, \alpha_{jm}\{u_{j_1}\alpha^\dagger_{j_1m_1}\\ &\quad + (-1)^{j_1-m_1}v_{j_1}\alpha_{j_1-m_1}\}\alpha^\dagger_{j\bar{m}}\alpha^\dagger_{j_1\bar{m}_1}\,|\bar{\psi}\rangle\\ &= (-1)^{j_1+m_1}v_{j_1}\sqrt{(1+\delta_{jj_1})}(jj_1m - m_1 \mid JM)\\ &= (-1)^{j+j_1-J}\left(\frac{(1+\delta_{jj_1})(2J+1)}{2j+1}\right)^{\frac{1}{2}} v_{j_1}(Jj_1Mm_1 \mid jm)\end{aligned}$$

where equation (6.59) was used in writing the second line of this equation, and in the final expression the symmetry properties of the Clebsch–Gordan coefficients given by equations (A1.18)–(A1.20) were used. When $\bar{\psi}$ is converted to ψ in the above equation, the term proportional to δu_j has four quasi-particle creation operators, and since the matrix element contains at most two destruction operators it follows that there is no contribution proportional to δu_j. *Consequently for $J \neq 0$ the spectroscopic factor depends on the occupation numbers of the target.*

If the state J arises from diagonalizing the quasi-particle Hamiltonian in the Tamm–Dancoff approximation (see the discussion associated with equations (6.45) and (6.46)) it follows that

$$\bar{\psi}_{JM} = \sum_{\{jj_1\}} \beta_J(jj_1)\mathscr{A}^{\dagger}_{JM}(jj_1)\bar{\psi} \tag{6.61}$$

and

$$\mathscr{S}_{j_1l_1}\{j \to (jj_1)_J\} = \frac{(1+\delta_{jj_1})(2J+1)}{2j+1}\, v_{j_1}^2 \beta_J^2(jj_1). \tag{6.62}$$

In order to see that this result reduces to that deduced in Chapter 2 for the transition

$$(j)_j^N \xrightarrow{(d,t)} (j)_J^{N-1}$$

one must extract correctly the number dependence of v_j^2. In section 2.1 of this chapter we pointed out that the expectation value of the number operator in a one-quasi-particle state is different from that for the BCS wave function (see equation (6.51)). At that time it was argued that as long as the state j_1 is near the Fermi surface, $u_{j_1}^2 \cong v_{j_1}^2 \cong \frac{1}{2}$ and the extra term is unimportant. However, for the single-level model this is not the case and consequently equation (6.51) itself must be used to determine v_j. When this is done it follows that

$$v_j^2 = (N-1)/(2j-1)$$

and

$$\mathscr{S}_{jl}\{j \to (j^2)_J\} = \frac{2(N-1)(2J+1)}{(2j-1)(2j+1)}$$

in agreement with the result of Table 2.7.

Similar analyses can be carried out for the other cases of interest, and in Table 6.4 we summarise the results for the pick-up and stripping spectroscopic factors. Clearly these reactions provide a useful tool for determining the values of u_j and v_j that characterize the nuclear states. For example, if one wishes to find these quantities for the even–even s.c.s. nucleus with N valence nucleons one can proceed as follows:

(a) the value of v_j^2 for the even–even nucleus can be determined from either the pick-up reaction $N(d,t)\{N-1\}$ or the stripping process $\{N-1\}(d,p)N$ proceeding to the ground state of the nucleus N.
(b) u_j^2 for the even–even nucleus can be obtained from the stripping cross-section $N(d,p)\{N+1\}$ or the pick-up process $\{N+1\}(d,t)N$ in which the ground state of the nucleus N is populated.

If one measures both u_j and v_j one has a check on the extraction of these quantities since $u_j^2 + v_j^2$ should equal one.

TABLE 6.4
Spectroscopic factors for stripping and pick-up

N even	N odd
Stripping	
$\mathcal{S}_{jl}(0 \to j) = u_j^2$	$\mathcal{S}_{jl}(j \to 0) = (2j+1)\bar{v}_j^2$ $\mathcal{S}_{j_1 l_1}\{j \to (jj_1)_J\} = (1+\delta_{jj_1})u_{j_1}^2\beta_J^2(jj_1)$
Pick-up	
$\mathcal{S}_{jl}(0 \to j) = (2j+1)v_j^2$	$\mathcal{S}_{jl}(j \to 0) = \bar{u}_j^2$ $\mathcal{S}_{j_1 l_1}\{j \to (jj_1)_J\} = \dfrac{(1+\delta_{jj_1})(2J+1)}{(2j+1)} v_{j_1}^2\beta_J^2(jj_1)$

u_j and v_j refer to the target whereas $\bar{u}_j$ and $\bar{v}_j$ are for the final nucleus. The quantity $\beta_J^2(jj_1)$ is the probability that in the two quasi-particle wave function of equation (6.61) the configuration $\mathscr{A}^{\dagger}_{JM}(jj_1)\psi$ will be realized.

Schneid *et al.* (1967) have carried out measurements on the tin isotopes ($Z = 50$) and have measured both u_j^2 and v_j^2. Because of the wide range of stable tin nuclei one can examine the filling of the $1d_{\frac{5}{2}}$, $0g_{\frac{7}{2}}$, $2s_{\frac{1}{2}}$, $1d_{\frac{3}{2}}$, and $0h_{\frac{11}{2}}$ orbits as one proceeds from $N = 50$ to 82. In Fig. 6.3 we show the experimental values of v_j^2 obtained in these studies. For comparison, in the same Figure, we give the values predicted when a pairing force with $G = 0{\cdot}187$ MeV is used in conjunction with the single-particle energies 0, 0·22, 1·90, 2·20, and 2·80 MeV for the $1d_{\frac{5}{2}}$, $0g_{\frac{7}{2}}$, $2s_{\frac{1}{2}}$, $1d_{\frac{3}{2}}$, and $0h_{\frac{11}{2}}$ orbits, respectively (Kisslinger and Sorensen 1960). As can be seen the experimental values of v_j^2 follow quite closely the predictions of the quasi-particle model, particularly when one considers the uncertainty (~20%) in extracting spectroscopic factors using the DWBA approximation.

As already stated both u_j^2 and v_j^2 were measured and the sum of the two should be unity. In Table 6.5 we show that this expectation is closely borne out. A further consistency check on the data would be the number of particles constraint in equation (6.14). However, since it was difficult to obtain reliable absolute cross-sections for the $\frac{11}{2}^-$ states, $v_{\frac{11}{2}}^2$ was chosen by Schneid *et al.* so that equation (6.14) was satisfied identically and moreover $u_{\frac{11}{2}}^2$ was set equal to $(1 - v_{\frac{11}{2}}^2)$.

If the entire spectroscopic strength for a given j-transfer were concentrated in a single state of the final odd-A nucleus, this would be the only state with that j populated in the stripping or pick-up reaction. In

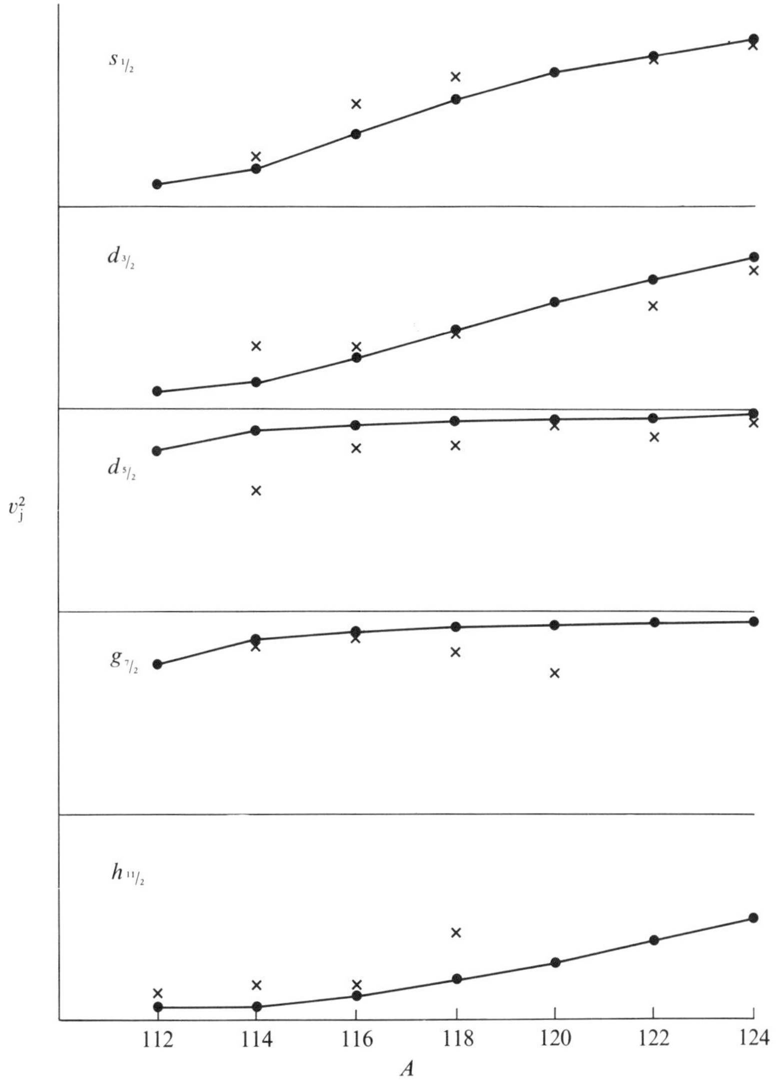

Fig. 6.3. Experimental values (x) of v_j^2 for the nucleus ${}^{A}_{50}\mathrm{Sn}_{A-50}$ as determined by Schneid *et al.* (1967). The dots (connected by solid lines) are the theoretical predictions of Kisslinger and Sorensen (1960) calculated by use of the pairing interaction of equation (6.1) with $G = 0{\cdot}187$ MeV. In the theoretical calculation $N = Z = 50$ was taken to be a closed shell and the neutrons were restricted to the $1d_{\frac{5}{2}}$, $0g_{\frac{7}{2}}$, $2s_{\frac{1}{2}}$, $1d_{\frac{3}{2}}$, and $0h_{\frac{11}{2}}$ single-particle states with energies of 0, 0·22, 1·90, 2·20, and 2·80 MeV, respectively.

TABLE 6.5

Experimental values of u_j^2 and v_j^2 for various tin isotopes taken from the work of Schneid et al. (1967)

j		$^{114}_{50}Sn_{64}$ Experiment	$^{114}_{50}Sn_{64}$ Theory	$^{116}_{50}Sn_{66}$ Experiment	$^{116}_{50}Sn_{66}$ Theory	$^{118}_{50}Sn_{68}$ Experiment	$^{118}_{50}Sn_{68}$ Theory
$\frac{1}{2}$	u^2	0·73		0·49		0·45	
	v^2	0·26	0·19	0·52	0·37	0·64	0·53
$\frac{3}{2}$	u^2	0·72		0·64		0·60	
	v^2	0·31	0·13	0·32	0·25	0·38	0·39
$\frac{5}{2}$	u^2	0·31		0·18		0·21	
	v^2	0·60	0·90	0·81	0·93	0·82	0·94
$\frac{7}{2}$	u^2	0·19		0·13		0·14	
	v^2	0·86	0·87	0·88	0·92	0·81	0·93
$\frac{11}{2}$	u^2	0·83		0·85		0·70	
	v^2	0·17	0·06	0·15	0·11	0·30	0·19

The theoretical values are from Kisslinger and Sorensen (1960) who solved the BCS problem using the $1d_{\frac{5}{2}}$, $0g_{\frac{7}{2}}$, $2s_{\frac{1}{2}}$, $1d_{\frac{3}{2}}$, and $0h_{\frac{11}{2}}$ model space with single-particle energies 0, 0·22, 1·90, 2·20, and 2·80 MeV, respectively. The strength of the pairing force, G, was taken to be 0·187 MeV.

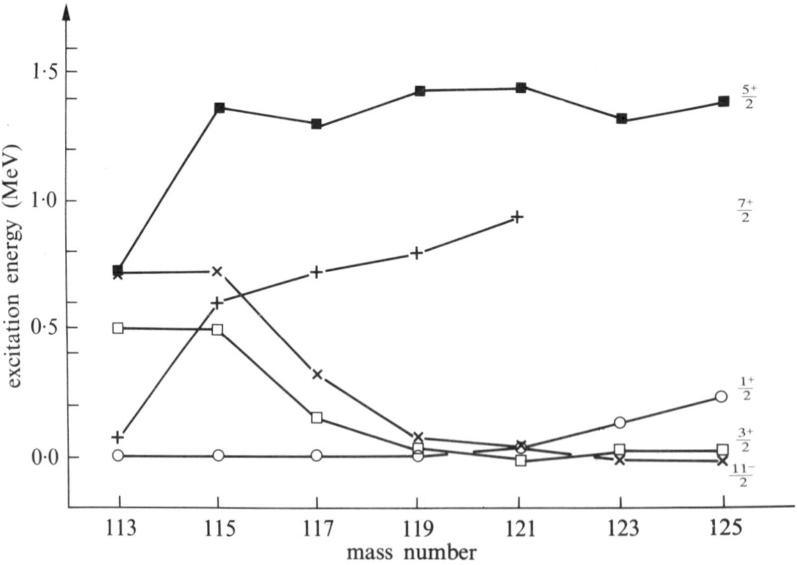

Fig. 6.4. Experimental excitation energies of the one quasi-particle states in the tin isotopes taken from Schneid *et al.* (1967).

practice, however, the strength is fragmented and one must use equation (2.124) to find the energy of the pure one quasi-particle state. In Fig. 6.4 the excitation energies of the one quasi-particle states computed from this equation are shown. The observed systematics can be easily understood by noting that when v_j in equation (6.38a) is small, $(\varepsilon_j - \lambda + \mu_j)$ is large and positive, and when $v_j \to 1$, $(\varepsilon_j - \lambda + \mu_j)$ is large and negative. The quasi-particle energy

$$E_j = \sqrt{\{(\varepsilon_j - \lambda + \mu_j)^2 + \Delta_j^2\}}$$

will be smallest when $(\varepsilon_j - \lambda + \mu_j) = 0$, i.e. when $v_j = \frac{1}{2}$. Thus one would expect that when v_j goes through 0·5 the state with spin j would lie lowest in energy. Comparison of Figs. 6.3 and 6.4 shows that this is indeed the case so that the quasi-particle description gives a consistent picture of the experimental data. The fact that in the mass region studied the $j = \frac{5}{2}$ and $\frac{7}{2}$ states are never the ground states merely reflects the fact that throughout the region both $v_{\frac{5}{2}}^2$ and $v_{\frac{7}{2}}^2$ are close to unity.

2.4. *Beta decay of the indium isotopes*

As a final application of quasi-particle methods we examine the allowed beta decay of the indium ($Z = 49$) isotopes to the s.c.s. tin nuclei. In order to do this we first briefly review the theory of allowed beta decay. For a more detailed discussion the reader is referred to the treatments of Konopinski (1966) and Wu and Moszkowski (1966).

In beta decay the energy of the outgoing electron or positron is of the order of a few MeV. Consequently the wavelength $\lambda = \hbar/p$ of the emitted particle will be of the order of 100 fm i.e. large compared to nuclear dimensions. Thus, as in gamma emission, decays in which the electron carries off orbital angular momentum l will be inhibited relative to those for which $l = 0$ by a factor of about $(R/\lambda)^{2l}$, where R is the nuclear radius. Even for the largest nuclei this factor is $\sim(\frac{1}{10})^{2l}$, and hence transitions in which the electron-antineutrino (or positron-neutrino) carry no orbital angular momentum are preferred and are called allowed decays. Since the electron and antineutrono each have intrinsic spin $\frac{1}{2}$ this implies that allowed decays proceed between two nuclear states that satisfy the selection rules

$$T_i - T_f = 0, \pm 1$$

$$I_i - I_f = 0, \pm 1$$

no parity change.

There are two single-particle operators involving the nucleon coordinates but not involving the orbital angular momentum that can lead to this type of transition. The first is called the Fermi operator (Fermi, (1934))

$$F = \sum_k (\tau_k)_\pm = \mathrm{T}_\pm \tag{6.63}$$

where $(\tau_k)_\pm$ is the nucleon isospin operator of equation (2.15)

$$\tau_+\pi = \nu \qquad \tau_+\nu = 0$$

$$\tau_-\pi = 0 \qquad \tau_-\nu = \pi$$

with π and ν standing for the proton and neutron isospin eigenfunctions, respectively. The sum over k in the above equation is over all nucleons in the nucleus and hence leads to the step-up or step-down operator for the total isospin. Consequently only transitions in which

$$T_i = T_f$$

$$I_i = I_f$$

$$\text{no parity change,}$$

can be induced by this operator.

The second way allowed decay can take place is through the mediation of the Gamow–Teller operator (Gamow and Teller (1936))

$$(GT)_\mu = \sum_k (\tau_k)_\pm (\sigma_k)_\mu \tag{6.64}$$

where $(\sigma_k)_\mu$ is the μth component of the Pauli spin operator for particle number k. Clearly this operator has nonvanishing matrix elements for

$$T_i - T_f = 0, \pm 1$$

$$I_i - I_f = 0, \pm 1; \qquad I = 0 \to I = 0 \text{ forbidden}$$

$$\text{no parity change.}$$

Thus if one observes the beta decay of randomly oriented nuclei the transition probability/unit time is proportional to the sum of the squares of the nuclear matrix elements of these two operators.

$$M(\beta; I_i \to I_f, T_i \to T_f, T_z \to T_z \pm 1) = \delta_{I_i I_f}\delta_{T_i T_f}(T_i \mp T_z)(T_i \pm T_z + 1)$$
$$+ \left(\frac{G_A}{G_V}\right)^2 \left(\frac{2I_f+1}{2I_i+1}\right) \left| \langle \psi_{I_f;T_f T_z \pm 1} \| \sum_k (\tau_k)_\pm \sigma_k \| \psi_{I_i;T_i T_z} \rangle \right|^2 \tag{6.65}$$

where (G_A/G_V) is the ratio of the Gamow–Teller to Fermi coupling constants.

As in the theory of gamma decay, the lifetime of a nuclear state depends critically on the amount of energy available to the emitted particles. In analogy with the reduced transition probability in gamma decay (i.e. $B(\xi L; I_i \to I_f)$) the comparative half-life or ft value for beta decay is introduced. In this definition t is the measured half-life of the transition in seconds, and all the energy and Z dependence are incorporated into the statistical factor f which can be obtained from the tabulation of Gove and Martin (1971), or if greater accuracy is desired from the work of Wilkinson and Macefield (1974). Thus it is customary to quote either the ft value or $\log_{10} ft$ governing the decay where

$$ft = \frac{C_0}{M(\beta; I_i \to I_f, T_i \to T_f, T_z \to T_z \pm 1)}.$$

For $0^+ \to 0^+$ $\Delta T = 0$ transitions, the Gamow–Teller operator does not contribute and hence an analysis of these decays will yield the value of C_0. This has recently been carried out by Raman *et al.* (1975) who conclude that

$$C_0 = 6176{\cdot}8 \text{ s}.$$

The ratio of Gamow–Teller to Fermi coupling constant can be taken from the work of Krohn and Ringo (1975) who obtain

$$G_A/G_V = 1{\cdot}258.$$

Thus the final formula governing the allowed beta decay of randomly oriented nuclei is

$$ft = \frac{6177}{(T_i \mp T_z)(T_i \pm T_z + 1)\delta_{T_iT_f}\delta_{I_iI_f} + 1{\cdot}583 B_\sigma(I_i \to I_f)} \tag{6.66}$$

where

$$B_\sigma(I_i \to I_f) = \frac{2I_f + 1}{2I_i + 1} |\langle \psi_{I_f; T_fT_z \pm 1} \| \sum_k (\tau_k)_\pm \sigma_k \| \psi_{I_i; T_iT_z} \rangle|^2 \tag{6.67}$$

and ft is measured in seconds.

We now use these results to discuss the decay of the $Z = 49$ (indium) nuclei. The ground state of an odd-A indium nucleus is described as a $\pi g_{\frac{9}{2}}$ hole coupled to the BCS ground state of the even–even tin core and is denoted by

$$\psi_{\frac{9}{2}m} = (\pi g_{\frac{9}{2}})^{9}_{\frac{9}{2}m} \bar{\psi}. \tag{6.68}$$

The allowed beta decay of this nucleus populates the one quasi-particle $g_{\frac{7}{2}}$ state of the neighbouring tin isotope:

$$\psi_{\frac{7}{2}m'} = (\pi g_{\frac{9}{2}})^{10}_{0} \alpha^{\dagger}_{\frac{7}{2}m'} \psi. \tag{6.69}$$

For the odd–odd indium isotopes we restrict our discussion to decay of the 1^+ state. The reason for considering only this decay is that the structure of the 1^+ eigenfunction should be well described as a $g_{\frac{9}{2}}$

proton–hole coupling to a $g_{\frac{7}{2}}$ neutron quasi-particle. That this is so follows because the low-lying proton–hole states are either $\pi g_{\frac{9}{2}}$ or $\pi p_{\frac{1}{2}}$ whereas the low-lying neutron configurations are the one quasi-particle $d_{\frac{5}{2}}$, $g_{\frac{7}{2}}$, $d_{\frac{3}{2}}$, $s_{\frac{1}{2}}$, and $h_{\frac{11}{2}}$ states. Consequently

$$\psi_{1M} = [(\pi g_{\frac{9}{2}})^{9}_{\frac{9}{2}} \times \bar{\alpha}^{\dagger}_{\frac{7}{2}}]_{1M}\bar{\psi} \tag{6.70}$$

is the only way to realize a 1^+ eigenfunction. We shall consider the decay of this state to either the BCS ground state or to the two quasi-particle 2^+ level in the neighbouring tin nucleus. Since the beta-decay operator cannot change orbital angular momentum it follows that the only part of the 2^+ eigenfunction that can be populated in this decay is

$$\psi_{2M} = \frac{1}{\sqrt{2}}[\alpha^{\dagger}_{\frac{7}{2}} \times \alpha^{\dagger}_{\frac{7}{2}}]_{2M}\psi. \tag{6.71}$$

The isospin of the initial and final nuclei are different so that only the Gamow–Teller operator given by equation (6.64) contributes to the decay. Furthermore, the decay always involves the creation of a $g_{\frac{9}{2}}$ proton and the destruction of a $g_{\frac{7}{2}}$ quasi-particle (neutron). If we denote the proton creation operator by $b^{\dagger}_{\frac{9}{2}m} = a^{\dagger}_{\frac{9}{2}m;\frac{1}{2}-\frac{1}{2}}$ and the neutron operator by $a_{\frac{7}{2}m} = a_{\frac{7}{2}m;\frac{1}{2}\frac{1}{2}}$ the Gamow–Teller operator becomes

$$\begin{aligned}(GT)_{\mu} &= \sum_{m_1 m_2} \langle g_{\frac{9}{2}}\| \sigma \| g_{\frac{7}{2}}\rangle (\tfrac{7}{2} 1 m_1 \mu \mid \tfrac{9}{2} m_2) b^{\dagger}_{\frac{9}{2}m_2} a_{\frac{7}{2}m_1} \\ &= \sqrt{(\tfrac{10}{3})}\langle g_{\frac{9}{2}}\| \sigma \| g_{\frac{7}{2}}\rangle [b^{\dagger}_{\frac{9}{2}} \times \tilde{a}_{\frac{7}{2}}]_{1\mu}\end{aligned} \tag{6.72}$$

where $\tilde{a}_{jm}$ is the modified Hermitian adjoint operator of equation (A2.9),

$$\tilde{a}_{jm} = (-1)^{j+m} a_{j-m}, \tag{6.73}$$

and in writing the second line of equation (6.72) use has been made of the properties of the Clebsch–Gordan coefficients given by equations (A1.18)–(A1.20).

To calculate the effect of the Gamow–Teller operator we note that the indium wave functions have the diagramatic form

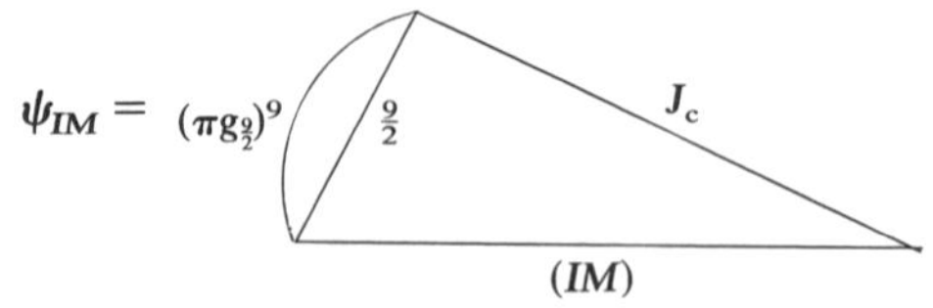

where J_c is the neutron core wave function $\bar{\psi}$ (the BCS eigenfunction) if we deal with an odd-A isotrope or $\bar{\alpha}^{\dagger}_{\frac{7}{2}m}\bar{\psi}$ for an even-A indium. Thus from equation (5.17) and the definition of the $9j$ coefficient in equation

(A4.8) it follows that

$$(GT)_\mu \psi_{IM} = \sqrt{(\tfrac{10}{3})}\langle g_{\frac{9}{2}} \| \sigma \| g_{\frac{7}{2}} \rangle \sum_{I''M''} (I1M\mu \mid I''M'')$$

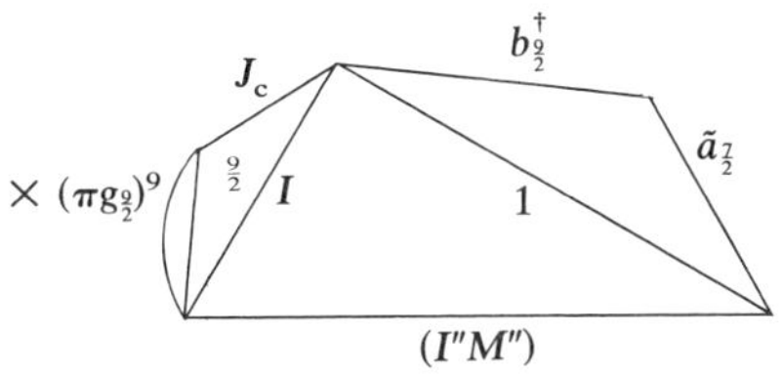

$$= \langle g_{\frac{9}{2}} \| \sigma \| g_{\frac{7}{2}} \rangle \sum_{J_c' I'' M''} \sqrt{\{10(2I+1)(2J_c'+1)\}}(I1M\mu \mid I''M'')$$

$$\times \begin{Bmatrix} \frac{9}{2} & J_c & I \\ \frac{9}{2} & \frac{7}{2} & 1 \\ 0 & J_c' & I'' \end{Bmatrix}$$

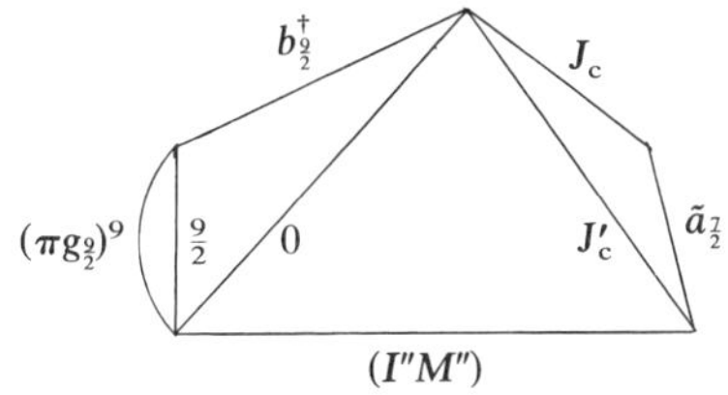

$$= (-1)^{J_c - J_c' - \frac{7}{2}} \sqrt{(2I+1)} \sum_{J_c' I'' M''} \delta_{J_c' I''} W(J_c I \tfrac{7}{2} 1; \tfrac{9}{2} J_c') \langle g_{\frac{9}{2}} \| \sigma \| g_{\frac{7}{2}} \rangle$$

$$\times \langle (\pi g_{\frac{9}{2}})^{10}_0 \| b^\dagger_{\frac{9}{2}} \| (\pi g_{\frac{9}{2}})_{\frac{9}{2}} \rangle \langle J_c' \| \tilde{a}_{\frac{7}{2}} \| J_c \rangle (I1M\mu \mid I''M'')$$

$$\times \qquad (6.74)$$

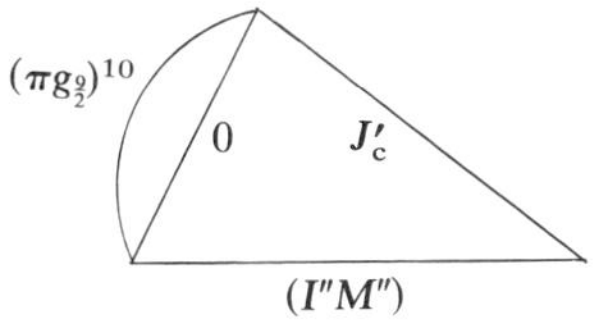

and in writing the last line of this equation use has been made of the simple form of the $9j$ coefficient when one angular momentum is zero (equation (A4.33)) together with the definition of the reduced matrix element.

The proton reduced matrix element is just $\sqrt{10}$ times the full-shell c.f.p. which is unity. Furthermore we are interested in the transition probability from the indium state $I = I_i$ to a tin state with angular momentum I_f. Thus the term with $I'' = I_f$ is picked out of the sum and from its definition in equation (A2.7) it follows that the coefficient of $(I_i 1 M_i \mu \mid I_f M_f)$ in the last line of equation (6.74) is the reduced matrix element governing the beta decay. Consequently $B_\sigma(I_i \to I_f)$ given by equation (6.67) is

$$B_\sigma(I_i \to I_f) = 10(2I_f+1)\langle g_{\frac{9}{2}} \| \sigma \| g_{\frac{7}{2}} \rangle^2 W^2(J_c I_i \tfrac{7}{2} 1; \tfrac{9}{2} J_c')$$
$$\times \langle J_c' \| \tilde{a}_{\frac{7}{2}} \| J_c \rangle^2 \delta_{J_c' I_f} \qquad (6.75)$$

The quasi-particle reduced matrix element in equation (6.75) is evaluated by the methods outlined in the last Section and in the following discussion u_j and v_j are the values associated with the tin nuclei and $\bar{u}_j$ and $\bar{v}_j$ those with indium. To illustrate the method of calculation we consider the $\frac{9}{2}^+ \to \frac{7}{2}^+$ decay. In this case J_c, the neutron state associated with the odd-A indium isotope, is the BCS wave function $\bar{\psi}$. J'_c stands for the one quasi-particle state $\alpha^\dagger_{\frac{7}{2}m}\psi$. Thus

$$\langle \alpha^\dagger_{\frac{7}{2}m}\psi|\, \tilde{a}_{\frac{7}{2}m'}\, |\bar{\psi}\rangle = (0\tfrac{7}{2}0m' \mid \tfrac{7}{2}m)\langle J'_c \|\, \tilde{a}_{\frac{7}{2}}\, \| J_c\rangle. \tag{6.76}$$

In order to evaluate this matrix element we express all operators and wave functions in terms of u_j and v_j. Thus from equation (6.30b)

$$\tilde{a}_{jm} = v_j \alpha^\dagger_{jm} + (-1)^{j+m} u_j \alpha_{j-m}.$$

When this is combined with the definition of $\bar{\psi}$, equation (6.58), one sees that

$$\begin{aligned}
\langle \alpha^\dagger_{\frac{7}{2}m}\psi|\, \tilde{a}_{\frac{7}{2}m'}\, |\bar{\psi}\rangle &= \langle \psi |\, \alpha_{\frac{7}{2}m}\{v_{\frac{7}{2}}\alpha^\dagger_{\frac{7}{2}m'} + (-1)^{\frac{7}{2}+m'} u_{\frac{7}{2}}\alpha_{\frac{7}{2}-m'}\} \\
&\quad \times \left\{1 + \tfrac{1}{2}\sum_{\bar{m}} (-1)^{\frac{7}{2}-\bar{m}} \frac{\delta u_{\frac{7}{2}}}{v_{\frac{7}{2}}} \alpha^\dagger_{\frac{7}{2}-\bar{m}}\alpha^\dagger_{\frac{7}{2}\bar{m}}\right\} |\psi\rangle \\
&= v_{\frac{7}{2}} - \frac{u_{\frac{7}{2}}\delta u_{\frac{7}{2}}}{v_{\frac{7}{2}}} \\
&= \bar{v}_{\frac{7}{2}}.
\end{aligned}$$

Since the Clebsch–Gordan coefficient in equation (6.76) is unity, one sees that in this case

$$\langle J'_c \|\, \tilde{a}_{\frac{7}{2}}\, \| J_c\rangle = \bar{v}_{\frac{7}{2}}$$

and that

$$B_\sigma(\tfrac{9}{2}^+ \to \tfrac{7}{2}^+) = \langle g_{\frac{9}{2}} \|\, \sigma\, \| g_{\frac{7}{2}}\rangle^2 \bar{v}^2_{\frac{7}{2}}. \tag{6.77a}$$

From the definition of B_σ in equation (6.67) and the reduced matrix element in equation (A2.10), it follows that the inverse transition $\frac{7}{2}^+ \to \frac{9}{2}^+$ has

$$B_\sigma(\tfrac{7}{2}^+ \to \tfrac{9}{2}^+) = \tfrac{10}{8}\langle g_{\frac{9}{2}} \|\, \sigma\, \| g_{\frac{7}{2}}\rangle^2 \bar{v}^2_{\frac{7}{2}} \tag{6.77b}$$

where $\bar{v}$ is the value of v_j associated with the indium nucleus.

By the same procedure one can show that for the decay of the odd–odd indium isotopes

$$B_\sigma(I_i = 1^+ \to I_f = 0^+) = \tfrac{10}{3} u^2_{\frac{7}{2}} \langle g_{\frac{9}{2}} \|\, \sigma\, \| g_{\frac{7}{2}}\rangle^2 \tag{6.78}$$

and

$$\begin{aligned}
B_\sigma(I_i = 1^+ \to I_f = 2^+) &= 100\, W^2(11\tfrac{7}{2}\tfrac{7}{2};\, 2\tfrac{9}{2})\bar{v}^2_{\frac{7}{2}}\beta^2_2(\tfrac{7}{2}\tfrac{7}{2})\langle g_{\frac{9}{2}} \|\, \sigma\, \| g_{\frac{7}{2}}\rangle^2 \\
&= \tfrac{7}{36}\bar{v}^2_{\frac{7}{2}}\beta^2_2(\tfrac{7}{2}\tfrac{7}{2})\langle g_{\frac{9}{2}} \|\, \sigma\, \| g_{\frac{7}{2}}\rangle^2
\end{aligned} \tag{6.79}$$

where $\beta_2^2(\frac{7}{2}\frac{7}{2})$ is the probability that the two quasi-particle state ψ_{2M} of equation (6.71) will occur in the yrast 2^+ wave function.

The value of $\bar{v}_{\frac{7}{2}}^2$ extracted from the transition ${}^{99+N}_{49}\mathrm{In}_{50+N}(\frac{9}{2}^+) \rightleftarrows {}^{99+N}_{50}\mathrm{Sn}_{50+N-1}(\frac{7}{2}^+)$ is the $v_{\frac{7}{2}}^2$ associated with the even–even ${}^{100+N}_{50}\mathrm{Sn}_{50+N}$ nucleus. According to equation (6.78) $u_{\frac{7}{2}}^2$ for this nucleus can be extracted from the decay ${}^{100+N}_{49}\mathrm{In}_{50+N+1}(1^+) \rightleftarrows {}^{100+N}_{50}\mathrm{Sn}_{50+N}(0^+)$. If one assumes that the reduced matrix element is the same for these transitions it follows from equation (6.66) and the fact that $u_{\frac{7}{2}}^2 + v_{\frac{7}{2}}^2 = 1$ that this quantity is given by the equation

$$\langle g_{\frac{9}{2}} \| \sigma \| g_{\frac{7}{2}} \rangle^2 = \frac{6177}{1{\cdot}583}\left\{\frac{3}{10}(ft)_{\mathrm{e}}^{-1} + \frac{8}{2I_{\mathrm{f}}+1}(ft)_{\mathrm{o}}^{-1}\right\} \tag{6.80}$$

where $(ft)_{\mathrm{e}}$ and $(ft)_{\mathrm{o}}$ are the ft values for the decay of the even-A and odd-A members of the pair, respectively. Once the effective reduced matrix element has been obtained the value of $v_{\frac{7}{2}}^2$ can be computed and compared with that obtained from an analysis of the stripping data.

There are three pairs of indium isotopes for which this data is available $\{{}^{117}\mathrm{In}, {}^{118}\mathrm{In}\}$, $\{{}^{113}\mathrm{In}, {}^{114}\mathrm{In}\}$, and $\{{}^{111}\mathrm{In}, {}^{112}\mathrm{In}\}$. In Table 6.6 the values of

TABLE 6.6

Values of the reduced matrix element needed to fit the In ⇄ Sn *decays*

N	12	14	18
$\log_{10}(ft)_{\mathrm{o}}$	4·8	4·6	4·5
$\log_{10}(ft)_{\mathrm{e}}$	4·1	4·5	4·9
$\langle g_{\frac{9}{2}} \| \sigma \| g_{\frac{7}{2}} \rangle^2$	0·142	0·115	0·138
$v_{\frac{7}{2}}^2$	0·35	0·68	0·89

$(ft)_{\mathrm{o}}$ is the comparative half life for the decay ${}^{99+N}_{49}\mathrm{In}_{50+N} \rightleftarrows {}^{99+N}_{50}\mathrm{Sn}_{50+N-1}$. For $N = 12$ and 14 the transition is $\frac{7}{2}^+ \to \frac{9}{2}^+$ whereas for $N = 18$ it is $\frac{9}{2}^+ \to \frac{7}{2}^+$. $(ft)_{\mathrm{e}}$ is the comparative half life for the ${}^{100+N}_{49}\mathrm{In}_{50+N+1}(1^+) \to {}^{100+N}_{50}\mathrm{Sn}_{50+N}(0^+)$ transition. $8v_{\frac{7}{2}}^2$ is the number of $g_{\frac{7}{2}}$ neutrons in the various isotopes.

$\log_{10} ft$ governing the decays are tabulated together with the value of the single-particle reduced matrix element deduced from each pair. From the table it is clear that this quantity is constant to within about 25% with an average value of

$$\langle g_{\frac{9}{2}} \| \sigma \| g_{\frac{7}{2}} \rangle^2 = 0{\cdot}132. \tag{6.81}$$

The values of $v_{\frac{7}{2}}^2$ that arise from this analysis are compared in Fig. 6.5 with those that were deduced from the single-nucleon transfer data of

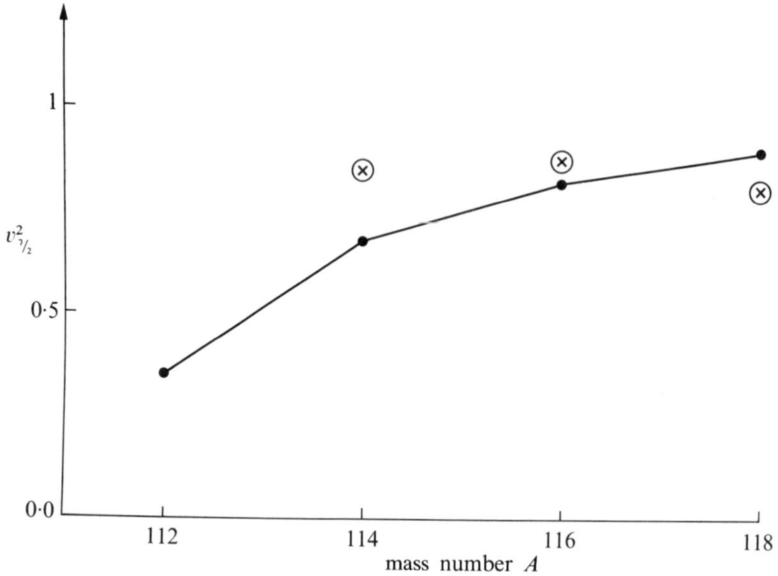

Fig. 6.5. The dots (connected by a solid line) are the values of $v^2_{\frac{7}{2}}$ for the tin nucleus ${}^{A}_{50}\mathrm{Sn}_{A-50}$ determined from the beta decay of the indium nuclei. For comparison the values obtained from the one-nucleon transfer experiments of Schneid *et al.* (1967) are shown by ⊗ in the figure.

Schneid *et al.* (1967), and it is apparent that the two methods give quite similar results. In this figure the value for $A=116$ was obtained from the ${}^{116}\mathrm{In}(1^+)\rightarrow{}^{116}\mathrm{Sn}(0^+)$ transition which has $\log_{10} ft=4{\cdot}7$. This experimental value when combined with the average value of the single-particle matrix element in equation (6.81) gives $u^2_{\frac{7}{2}}=0{\cdot}18$ from which it follows that $v^2_{\frac{7}{2}}=0{\cdot}82$.

From equation (6.79) one can deduce the probability that the 2^+ state in a given tin nucleus has the configuration $(1/\sqrt{2})[\alpha^\dagger_{\frac{7}{2}}\times\alpha^\dagger_{\frac{7}{2}}]_{2M}\psi$. For example, the decay ${}^{114}_{49}\mathrm{In}_{65}(1^+)\rightarrow{}^{114}_{50}\mathrm{Sn}_{64}(2^+)$ has $\log_{10} ft=5.5$. If one interpolates linearly between the $N=14$ and 16 systems one concludes that $v^2_{\frac{7}{2}}$ for $N=15$ (upon which this decay depends) is $0{\cdot}75$. Consequently

$$\beta^2_2(\tfrac{7}{2}\tfrac{7}{2})=0{\cdot}64.$$

In other words in ${}^{114}\mathrm{Sn}$ the probability that the 2^+ state is composed of two $g_{\frac{7}{2}}$ quasi-particles is about 64%.

Finally we compare the value of the effective matrix element with the single-particle estimate. According to equation (A2.19)

$$\langle\phi_{j=l+\frac{1}{2}}\|\,\sigma\,\|\phi_{j=l-\frac{1}{2}}\rangle=-\left(\frac{4l}{2l+1}\right)^{\frac{1}{2}} \qquad (6.82)$$

so that

$$\langle g_{\frac{9}{2}} \| \sigma \| g_{\frac{7}{2}} \rangle = -1{\cdot}333.$$

This is to be compared with the average effective value $-0{\cdot}363$. Thus to fit the $\pi g_{\frac{9}{2}} \rightleftarrows \nu g_{\frac{7}{2}}$ beta-decay data one must quench the operator by approximately a factor of four. However, despite this large quenching, the effective operator does not change appreciably in going from nucleus to nucleus.

This rather large reduction in the value of the Gamow–Teller operator seems to be standard for this mass region. For example, the same sort of analysis as just conducted can be carried out for the decay of the antimony isotopes ($Z = 51$) to the tin nuclei where the decay involves $\pi d_{\frac{5}{2}} \rightleftarrows \nu d_{\frac{3}{2}}$. In this case if one assumes the 1^+ state in the odd–odd Sb nucleus can be described as $[\pi d_{\frac{5}{2}} \times \alpha^\dagger_{\frac{3}{2}}]_{1M}\psi$, the analogue of equation (6.80) is

$$\langle d_{\frac{5}{2}} \| \sigma \| d_{\frac{3}{2}} \rangle^2 = \frac{6177}{1{\cdot}583} \left\{ \frac{1}{2} (ft)_e^{-1} + \frac{4}{2I_f + 1} (ft)_o^{-1} \right\} \tag{6.83}$$

where $(ft)_e$ is the comparative half-life ${}^{100+N}_{51}\mathrm{Sb}_{50+N-1}(1^+) \rightarrow {}^{100+N}_{50}\mathrm{Sn}_{50+N}(0^+)$ and $(ft)_o$ that for the decay ${}^{101+N}_{51}\mathrm{Sb}_{50+N}(\frac{5}{2}^+) \rightleftarrows {}^{101+N}_{50}\mathrm{Sn}_{50+N+1}(\frac{3}{2}^+)$. There are two pairs of transitions from which we can extract a value for the effective matrix element. The decay ${}^{120}_{51}\mathrm{Sb}_{69}(1^+) \rightarrow {}^{120}_{50}\mathrm{Sn}_{70}(0^+)$ has $(ft)_e = 10^{4\cdot5}$ and ${}^{121}_{50}\mathrm{Sn}_{71}(\frac{3}{2}^+) \rightarrow {}^{121}_{51}\mathrm{Sb}_{70}(\frac{5}{2}^+)$ has $(ft)_o = 10^5$. When these values are inserted into equation (6.83) one concludes that

$$\langle d_{\frac{5}{2}} \| \sigma \| d_{\frac{3}{2}} \rangle^2 = 0{\cdot}0877.$$

A second set of decays yielding the same information is ${}^{118}_{51}\mathrm{Sb}_{67}(1^+) \rightarrow {}^{118}_{50}\mathrm{Sn}_{68}(0^+)$ with $(ft)_e = 10^{4\cdot8}$ and ${}^{119}_{51}\mathrm{Sb}_{68}(\frac{5}{2}^+) \rightarrow {}^{119}_{50}\mathrm{Sn}_{69}(\frac{3}{2}^+)$ where $(ft)_o = 10^5$. These values lead to

$$\langle d_{\frac{5}{2}} \| \sigma \| d_{\frac{3}{2}} \rangle^2 = 0{\cdot}0699.$$

Thus the average value of the reduced matrix element is $-0{\cdot}281$ compared to the single-particle estimate $\langle d_{\frac{5}{2}} \| \sigma \| d_{\frac{3}{2}} \rangle = -1{\cdot}264$. Nevertheless once this large reduction factor is incorporated, the qualitative features of the decay of the odd-A Sb isotopes have been shown to be in agreement with quasi-particle theory (Silverberg and Winther (1963).

In section 2 of Chapter 5 we showed how small admixtures in the nuclear wave functions can lead to effective electro-magnetic operators that are substantially different than those for free nucleons and the same effect can occur for beta decay. As an example consider the decay ${}^{111}_{49}\mathrm{In}_{62}(\frac{9}{2}^+) \rightarrow {}^{111}_{50}\mathrm{Sn}_{61}(\frac{7}{2}^+)$, and for simplicity consider the truncated model space consisting of only the $g_{\frac{9}{2}}$, $g_{\frac{7}{2}}$, and $d_{\frac{5}{2}}$ single-particle levels. Furthermore, as in the case of the one quasi-particle approximation, pure seniority-one eigenfunctions will be used. Thus the eigenfunction of the initial state is

$$\psi_{\frac{9}{2}m} = (\pi g_{\frac{9}{2}})^{9}_{\frac{9}{2}m}(\nu g_{\frac{9}{2}})^{10}_{0}\{\alpha(\nu d_{\frac{5}{2}})^{6}_{0}(\nu g_{\frac{7}{2}})^{6}_{0} + \beta(\nu d_{\frac{5}{2}})^{4}_{0}(\nu g_{\frac{7}{2}})^{8}_{0}\}$$

and that of the final state is

$$\psi_{\frac{7}{2}m'} = (\pi g_{\frac{9}{2}})^{10}_{0}(\nu g_{\frac{9}{2}})^{10}_{0}\{\bar{\alpha}(d_{\frac{5}{2}})^{6}_{0}(\nu g_{\frac{7}{2}})^{5}_{\frac{7}{2}m'} + \bar{\beta}(\nu d_{\frac{5}{2}})^{4}_{0}(\nu g_{\frac{7}{2}})^{7}_{\frac{7}{2}m'}\}.$$

Since the Gamow–Teller operator cannot change the orbital angular momentum, the allowable types of mixing that can contribute linearly in the admixture coefficients are the same as those that modify the magnetic dipole operator. Consequently, in addition to contributions from $(\nu g_{\frac{7}{2}})^{6}_{K}$ $(K \neq 0)$ the admixtures in the final state that can contribute linearly to the transition rate are

$$\bar{\chi}_1 = [[(\pi g_{\frac{9}{2}})^{9}_{\frac{9}{2}} \times (\nu g_{\frac{9}{2}})^{10}_{0}]_{\frac{9}{2}} \times [[(\pi d_{\frac{5}{2}}) \times (\nu d_{\frac{5}{2}})^{5}_{\frac{5}{2}}]_1 \times (\nu g_{\frac{7}{2}})^{6}_{0}]_1]_{\frac{7}{2}m'}$$

$$\bar{\chi}_2 = [[(\pi g_{\frac{9}{2}})^{9}_{\frac{9}{2}} \times (\nu g_{\frac{9}{2}})^{10}_{0}]_{\frac{9}{2}} \times [(\nu d_{\frac{5}{2}})^{6}_{0} \times [\pi g_{\frac{7}{2}} \times (\nu g_{\frac{7}{2}})^{5}_{\frac{7}{2}}]_1]_1]_{\frac{7}{2}m'}$$

$$\bar{\chi}_3 = [[(\pi g_{\frac{9}{2}})^{9}_{\frac{9}{2}} \times (\nu g_{\frac{9}{2}})^{10}_{0}]_{\frac{9}{2}} \times [[(\pi d_{\frac{5}{2}}) \times (\nu d_{\frac{5}{2}})^{3}_{\frac{5}{2}}]_1 \times (\nu g_{\frac{7}{2}})^{8}_{0}]_1]_{\frac{7}{2}m'}$$

$$\bar{\chi}_4 = [[(\pi g_{\frac{9}{2}})^{9}_{\frac{9}{2}} \times (\nu g_{\frac{9}{2}})^{10}_{0}]_{\frac{9}{2}} \times [[(\nu d_{\frac{5}{2}})^{4}_{0} \times [(\pi g_{\frac{7}{2}}) \times (\nu g_{\frac{7}{2}})^{7}_{\frac{7}{2}}]_1]_1]_{\frac{7}{2}m'}.$$

Thus each component in the neutron wave function leads to two contributions. In general, any neutron configuration that consists of particles in m different single-particle orbits will give rise to at least m different admixtures that can contribute linearly to the beta decay. Since the BCS wave function has many components, it is clear that there will be many admixed components in the wave function that can contribute linearly in the admixture coefficients. Fujita and Ikeda (1965) have shown that these contributions interfere destructively and consequently lead to a small effective value of the Gamow–Teller matrix element.

Since the foregoing argument is based on the fact that both spin–orbit partners have not been included in the model space, it follows that when both are included the Gamow–Teller matrix element should be close to the single-particle value. Wilkinson (1973) has shown that this is indeed the case: for example, when the Cohen–Kurath (1965) wave functions for the $0p$ shell are used, the effective Gamow–Teller matrix element is approximately state independent and has about 90% of the single-particle value. In the same way, at the beginning of the $(0d, 1s)$ shell the Lanford–Wildenthal (1973) wave functions (which include $0d_{\frac{5}{2}}$, $1s_{\frac{1}{2}}$, and $0d_{\frac{3}{2}}$) can explain the $\log ft$ values with a state-independent operator provided the operator has approximately 90% of its single-particle value.

3. Some limitations of quasi-particle calculations

In this section we shall compare an exact calculation of the nickel isotopes (on the assumption that $^{56}_{28}Ni_{28}$ is an inert core) to the quasi-particle results. The model space used is $1p_{\frac{3}{2}}$, $1p_{\frac{1}{2}}$, and $0f_{\frac{5}{2}}$, and the energies of the single-particle states are taken to be 0, 1·08, and 0·78 MeV, respectively. The residual two-body potential is the best-fit interaction of Cohen *et al.* (1967). In Table 6.7 we give the energies of the low-lying states of $^{62}_{28}Ni_{34}$

TABLE 6.7

Energies of the low-lying states of $^{62}_{28}Ni_{34}$ as given by the full ($1p_{\frac{3}{2}}$, $1p_{\frac{1}{2}}$, $0f_{\frac{5}{2}}$) shell-model calculation and as computed using various truncation schemes

State	Exact	Lowest seniority	Tamm–Dancoff approximation	Projected Tamm–Dancoff
0_1^+	(0·45)	(0·46)	(1·93)	(0·57)
0_2^+	2·01	2·13	2·39	1·98
2_1^+	1·53	1·53	1·07	1·46
2_2^+	2·25	2·36	1·86	2·27
4_1^+	2·20	2·28	1·83	2·23

All energies are in MeV and except for that of the 0_1^+ level are excitation energies. The ground state energy is the total energy of the nucleus relative to $^{56}_{28}Ni_{28}$ with the $p_{\frac{3}{2}}$ single-particle energy set equal to zero (see equation (6.25)).

computed using various approximation techniques (Macfarlane 1966). The lowest-seniority approximation means that the energy was computed using only seniority-zero eigenfunctions for the 0^+ levels and seniority-two for the others. The Tamm–Dancoff approximation is the usual two quasi-particle calculation discussed in section 2.1 (see equation (6.45) and the discussion following it) and the projected Tamm-Dancoff approximation is the one described at the end of section 2.1 (see equation (6.51b)).

It is clear (as pointed out in Chapter 1, section 5.5) that seniority truncation does not appreciably affect the computed energies. On the other hand, the use of the BCS and Tamm–Dancoff approximations lead to results in rather bad agreement with the exact calculations, particularly for the binding energy. The reason for this can be seen from the following considerations. In Fig. 6.6 we plot the experimental values of B_N (see equation (6.25)) for the nickel isotopes. Clearly B_N is a very non-linear function of the number of nucleons. As already discussed, the BCS and quasi-particle wave functions do not have a fixed number of particles; instead they have a distribution which is more or less symmetric about the desired number of particles (see Fig. 6.2). That the BCS theory can have no success in calculating such a non-linear B_N may be seen from the following numerical example. Suppose we have an interaction that does fit the nickel data of Fig. 6.6 (the best-fit interaction used in this calculation comes quite close when the conventional shell-model calculation is used). If we now perform a BCS calculation for the ground-state energy of ${}^{62}_{28}\text{Ni}_{34}$, there will be certain probabilities $C_0^2, C_2^2, \ldots$ that the wave function will have a zero, two,... particle parts. For the sake of numerics assume $C_2^2 = C_{10}^2 = 0{\cdot}1$, $C_4^2 = C_8^2 = 0{\cdot}15$, and $C_6^2 = 0{\cdot}5$. With this wave function one would find that

$$\begin{aligned} B_{N=6}(\text{theory}) &= 0{\cdot}1 \times (9{\cdot}802 - 1{\cdot}936) + 0{\cdot}15 \times (4{\cdot}362 - 1{\cdot}789) \\ &\quad + 0{\cdot}5 \times 0{\cdot}328 = 1{\cdot}337 \text{ MeV} \end{aligned}$$

where the numbers $-1{\cdot}936, -1{\cdot}789, \ldots, 9{\cdot}802$ MeV are the values of B_N taken from Fig. 6.6 for $N = 2, 4, \ldots 10$. In other words, one would miss

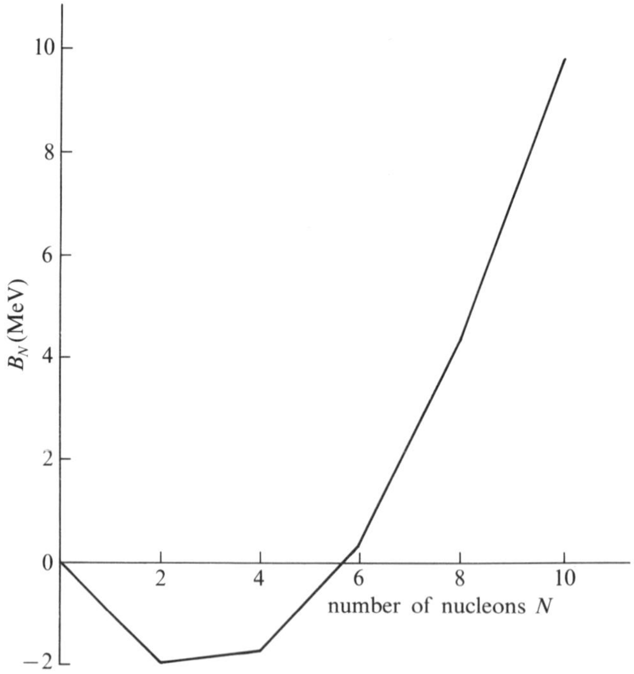

Fig. 6.6. Graph of the binding energy B_N (equation (6.25)) for the nickel isotopes. N is the number of neutrons outside the ${}^{56}_{28}\text{Ni}_{28}$ core.

the 'true' value of B_N by more than 1 MeV! Thus we arrive at the first limitation of quasi-particle calculations.

Limitation. Quasi-particle calculations can never explain discontinuities in any physical effect. Since they are averages of the physical quantity one wishes to describe over several neighbouring nuclei, at best they can reproduce only an average effect. Therefore one should only use this type of calculation to attempt to explain quantities that are more or less constant as a function of N or which vary more or less linearly with N.

In section 1.3 of this Chapter we saw that the BCS calculation reproduced fairly well the results of an exact shell-model study for the binding energies of the lead isotopes. Thus the degree of non-linearity exhibited in column 10, Table 6.1, can be tolerated in a BCS calculation whereas that shown in Fig. 6.6 cannot be tolerated. On the other hand, it is apparent from the last column in Table 6.7 that the projected Tamm–Dancoff calculation, which uses the wave function of equation (6.51b), can quite successfully reproduce the results of an exact calculation.

A second limitation on two quasi-particle calculations is evident from an inspection of Table 1.3. In this tabulation the seniority decomposition of the lowest states in ${}^{62}_{28}\mathrm{Ni}_{34}$ is given, and it is apparent that any attempt to describe the 1^+ and 3^+ states as two quasi-particle excitations would be nonsensical since the 'exact' eigenfunctions have less than 50% probability of being seniority-two.

In addition, one must be very careful when one attempts to describe low-lying states of the odd-A nuclei in terms of one quasi-particle states. For example, as seen in Chapter 1, section 3.2 the lowest $\frac{5}{2}^-$ state in ${}^{51}_{23}\mathrm{V}_{28}$ at 320 keV excitation energy is the seniority-three $(f_{\frac{7}{2}})^3_{I=\frac{5}{2}}$ state and this is certainly not a one quasi-particle level. In the odd nickel isotopes the reason this effect is not so important in, for example, the lowest $\frac{3}{2}^-$ state is because the single-particle $f_{\frac{5}{2}}$ state lies 780 keV above the $p_{\frac{3}{2}}$ state. If the situation had been reversed, as it is in the lead nuclei, the lowest $\frac{3}{2}^-$ state would undoubtedly have been dominated by the seniority-three configuration $(f_{\frac{5}{2}})^3_{\frac{3}{2}}$. Because the state $(j)^3_{I=j-1}$ lies close to the state $(j)^3_{I=j}$ it follows that a seniority-one description of the low-lying states of odd-A nuclei must be carefully scrutinized.

Thus we arrive at the second limitation on quasi-particle calculations:

Limitation. In even–even nuclei, two quasi-particle calculations should not be expected to reproduce the energies of the even-parity odd angular-momentum states (i.e. the $1^+, 3^+, 5^+, \ldots$ levels). In addition, one must be very careful when one attempts to describe the lowest state of spin j in an odd-A nucleus as a one quasi-particle state, particularly if the single-particle state $j_1 = j+1$ lies below the orbit j. Clearly a one quasi-particle description for states with angular momenta j_2 (where j_2 is an angular momentum different from that of one of the single-particle orbits considered) is impossible.

In order to assess the accuracy of the two quasi-particle approximation for computing gamma-ray transition rates we compare in Table 6.8 the values of

$$Q(E2) = \text{const} \times (2I_i + 1) B(E2; I_i \to I_f)$$

The values of $Q(E2) = const \times (2I_i + 1) \times B(E2; I_i \to I_f)$ for various transitions in $^{62}_{28}Ni_{34}$

Transition	Exact	Lowest seniority	Projected Tamm–Dancoff
$2_1^+ \to 0_1^+$	180·8	172·8	196·7
$2_2^+ \to 0_1^+$	6·5	1·0	0·9
$0_2^+ \to 2_1^+$	1·7	4·3	7·6
$2_2^+ \to 0_2^+$	86·3	46·7	26·0
$4_1^+ \to 2_1^+$	232·8	128·1	76·3
$4_1^+ \to 2_2^+$	6·7	2·4	0·5
$2_2^+ \to 2_1^+$	132·5	49·6	42·6

for various transitions in $^{62}_{28}Ni_{34}$. It is clear that the $B(E2)$ from the first 2^+ state to the ground state is well reproduced by both the seniority-truncated and projected Tamm–Dancoff calculations. On the other hand, the predicted $E2$ transition rates from other states cannot be trusted to better than a factor of two when a lowest-seniority truncation is carried out. Consequently even the projected Tamm–Dancoff approximation must have this same limitation. This illustrates clearly the importance of small admixtures (~10%) of higher-seniority components in these states (see Table 1.3). Because even the lowest state of given spin in an odd-A nuclei has a 5–10% admixture of seniority higher than one in its wave function (see Table 1.4), one should not rely on one quasi-particle wave functions to give predictions to better than a factor of about 2.

Thus we arrive at a third limitation on the quasi-particle calculations:

Limitation. If one considers only zero and two quasi-particle states in

> the description of the properties of an even-A nucleus, then only the $E2$ properties of the lowest 2^+ states can be predicted with some degree of certainty. The other transition rates should not be trusted to better than a factor of two. The electro-magnetic transition rates predicted by use of a one quasi-particle wave function for the odd-A nuclei should also be considered uncertain by about the same factor.

However, with the truncated model space used in shell-model calculations, theory and experiment seldom agree to better than a factor of two (see Chapter 5) even when the exact calculation is carried out, so that in general this is not a very serious limitation. The exception is when we deal with transitions that are predicted to be weak, and then one should be very sceptical about the quasi-particle predictions.

Finally, we turn to the validity of the BCS approximation for other than s.c.s. nuclei. In this scheme, the fact that the $T=1$ $J=0$ interaction is strongest is exploited. However, when both neutrons and protons fill the same orbit this is not always the case. For example, the ground state of both ${}^{6}_{3}\mathrm{Li}_3$ and ${}^{18}_{9}\mathrm{F}_9$ is $J=1$ $T=0$. As soon as one gets to the $f_{\frac{7}{2}}$ shell, the $J=0$ interaction is the strongest, but not by much—both the $J=1$ and $J=7$ states of ${}^{42}_{21}\mathrm{Sc}_{21}$, as shown in Fig. 2.1 have matrix elements only 600 keV less attractive than that for the $J=0$ state. Thus the BCS and TDA (as discussed here) are not likely to provide a very good description of nuclei in which neutrons and protons fill the same single-particle orbits.

> *Limitation.* When neutrons and protons are filling the same orbits, the $T=0$ interaction in states with $J=1$ and $2j$ is comparable to that in the $J=0$ state. Since the BCS approximation assumes that the $J=0$ interaction is by far the most important, it is not likely to give results in agreement with experiment in this case.

On the other hand, in those cases in which there are only a few neutrons in one major shell and a few protons or proton holes in a second major shell, the weak-coupling model has validity and the quasi-particle approximation can be used in the following way. The BCS and quasi-particle calculations are performed separately for neutrons and protons (i.e. using the assumption that the neutron–proton interaction is zero). If this interaction is indeed weak, one can then choose a small subset of the neutron and proton quasi-particle states, couple them together, and use them as basis states for diagonalizing the Hamiltonian. Extensive calculations along these lines have been carried out by Kisslinger and Sorensen

(1963) and Yoshida (1962). In this type of calculation, a simple form called the quadrupole force is often taken for the neutron–proton interaction

$$
\begin{aligned}
V_Q &= -V_0 \sum_{\mu,i,k} (-1)^{\mu} r_i^2 Y_{2\mu}(\theta_i, \phi_i) r_k^2 Y_{2-\mu}(\theta_k, \phi_k) \\
&= -\sqrt{5}\, V_0 \sum_{i,k} r_i^2 r_k^2 [Y_2(\theta_i, \phi_i) \times Y_2(\theta_k, \phi_k)]_{00}
\end{aligned}
\tag{6.84}
$$

where V_0 is the strength of the interaction.

7

POOR MAN'S HARTREE–FOCK

When neutrons and protons occupy the same single-particle orbit j the residual interaction in two-particle states with $I=1$ and $2j$ is comparable to that in the $I=0$ configuration (see Fig. 2.1). Thus the quasi-particle approximation, which singles out the $I=0$ interaction as being the most important, can no longer be used. It is even more important to find an approximation procedure for the neutron-proton problem, since in this case the dimensionality of the shell-model calculation is much larger than for s.c.s. nuclei. For example, if one wanted to study the properties of the $I=3^+$ $T=0$ state in ${}^{28}_{14}\mathrm{Si}_{14}$ under the assumption that the 12 valence nucleons outside the ${}^{16}_{8}\mathrm{O}_8$ core were restricted to the $(0d_{\frac{5}{2}}, 1s_{\frac{1}{2}}, 0d_{\frac{3}{2}})$ shell, one would have to construct and diagonalize in an isospin formalism a $3,711\times 3,711$ matrix (Harvey and Sebe 1968). Alternatively, if one did not use isospin eigenfunctions but instead used the neutron-proton formalism discussed in Chapter 2, (sections 2.3 and 2.4), the dimensionality of the matrix would be $15,385\times 15,385$. Clearly it is imperative that an approximation scheme be found.

In order to see what type of approximation might be appropriate we examine the form of the nuclear Hamiltonian when the nucleon-nucleon interaction is spin independent

$$H=\sum_k \frac{p_k^2}{2m}+\sum_{k<i} V(\beta\,|\underline{r}_k-\underline{r}_i|)$$

$$=\sum_k \frac{p_k^2}{2m}+4\pi\sum_{k<i}\sum_l \frac{F_l(\beta;\,r_k,\,r_i)}{2l+1}\sum_m (-1)^m Y_{lm}(\theta_k,\phi_k)Y_{l-m}(\theta_i,\phi_i) \quad (7.1)$$

where $1/\beta$ is the range of the interaction. In writing the last line of this equation a Slater decomposition of the potential has been carried out and the functions $F_l(\beta;\,r_k,\,r_i)$, which depend on the scalar variables r_k and r_i, are given by equation (4.2). A possible way to treat this many-body problem is via the Hartree–Fock theory (Ripka 1966, 1968)—that is look for the single-particle potential that best reproduces the interaction of the kth nucleon with all others in the nucleus. The Hartree–Fock calculation is an iterative one in which one first guesses the single-particle eigenfunctions, inserts them into the Hartree–Fock equations and calculates the single-particle potential, recalculates the eigenfunctions, and continues in this vein until self-consistency is attained. This procedure is time consuming and requires a high-speed computer for its execution. In this Chapter we shall show that an excellent approximation to the Hartree–Fock eigenfunctions can often be obtained by considering the single-particle motion to take place in a deformed harmonic oscillator potential.

To see what type of deformation is likely to be important, we note that experimentally nuclei with several neutrons and protons outside closed shells have quadrupole moments and $E2$ transition rates greatly enhanced over the single-particle values (Bohr and Mottelson 1975). Since this is the case, many nucleons must contribute collectively to the observed properties. Because the exclusion principle prohibits more than two protons and two neutrons in contact, this means that the long-range part of the effective interaction mediates the collectivity; in other words, that part of equation (7.1) corresponding to small values of β is important. To understand the implications of this, we assume that the long-range part of the interaction has the Gaussian form given by equations (4.3) and (4.4) and expand about $\beta=0$. To order β^4 the $l=0$ multipole gives (Harvey 1968)

$$\begin{aligned}\sum_{k<i} F_0(\beta; r_k, r_i) &= \bar{V}_0 \sum_{k<i} e^{-\beta^2(r_k^2+r_i^2)} j_0(-2i\beta^2 r_k r_i)\\ &\simeq \bar{V}_0 \sum_{k<i} \{1-\beta^2(r_k^2+r_i^2)+\tfrac{1}{2}\beta^4(r_k^4+r_i^4)+\tfrac{5}{3}\beta^4 r_k^2 r_i^2\}\\ &= \frac{A(A-1)}{2}\bar{V}_0 - \bar{V}_0\beta^2(A-1)\sum_k r_k^2\left(1-\frac{\beta^2}{2}r_k^2\right)\\ &\quad +\tfrac{5}{3}\bar{V}_0\beta^4 \sum_{k<i} r_k^2 r_i^2. \qquad (7.2a)\end{aligned}$$

The first term, which is proportional to the number of pairs in the nucleus, gives the energy when the interaction has infinite range. The second is the oscillator potential flattened out because of the $(\beta^2/2)r^2$ term, for large r. This shows that the average spherical potential arises from the long-range part of the residual interaction and can be approximated by the harmonic oscillator. Finally, the last term is a constant within the configuration $(j_1^m j_2^n)_I$ (independent of I) and does not vary appreciably within an oscillator shell when the configuration is changed to $(j_1^{m'} j_2^{n'})_I$ provided $m'+n'=m+n$. Thus the main effect of the $l=0$ part of the long-range interaction is to lead to the harmonic oscillator ordering of the single-particle states.

Turning now to the $l=1$ part of the Slater decomposition we see that

$$\frac{4\pi}{3}\sum_{k<i} F_1(\beta;\, r_k, r_i)\sum_m (-1)^m Y_{1m}(\theta_k, \phi_k)Y_{1-m}(\theta_i, \phi_i)$$

$$=4\pi i\bar{V}_0 \sum_{k<i} \mathrm{e}^{-\beta^2(r_k{}^2+r_i{}^2)} j_1(-2i\beta^2 r_k r_i)\sum_m (-1)^m Y_{1m}(\theta_k, \phi_k)Y_{1-m}(\theta_i, \phi_i)$$

$$\cong \frac{8\pi}{3}\bar{V}_0\beta^2 \sum_{k<i} r_k r_i[1-\beta^2(r_k^2+r_i^2)]\sum_m (-1)^m Y_{1m}(\theta_k, \phi_k)Y_{1-m}(\theta_i, \phi_i). \quad (7.2b)$$

Because of parity considerations this term will have zero matrix elements in a shell-model calculation involving only one active oscillator shell.

Finally, we consider the $l=2$ term in the Slater decomposition of equation (7.1):

$$\frac{4\pi}{5}\sum_{k<i} F_2(\beta;\, r_k, r_i)\sum (-1)^m Y_{2m}(\theta_k, \phi_k)Y_{2-m}(\theta_i, \phi_i)$$

$$=-4\pi\bar{V}_0 \sum_{k<i} \mathrm{e}^{-\beta^2(r_k{}^2+r_i{}^2)} j_2(-2i\beta^2 r_k r_i)\sum_m (-1)^m Y_{2m}(\theta_k, \phi_k)Y_{2-m}(\theta_i, \phi_i)$$

$$\cong \frac{16\pi\bar{V}_0}{15}\beta^4 \sum_{k<i}\sum_m (-1)^m \{r_k^2 Y_{2m}(\theta_k, \phi_k\}\{r_i^2 Y_{2-m}(\theta_i, \phi_i)\}.$$

This potential, which was introduced in equation (2.94), is called the quadrupole-quadrupole interaction or P_2 force, and to order β^4 is the main term responsible for splitting the oscillator degeneracy.

Thus in the long-range limit a spin-independent residual interaction gives rise to a shell-model Hamiltonian that can be approximated by

$$H_1 \cong \sum_k \frac{p_k^2}{2m} - \bar{V}_0\beta^2(A-1)\sum_k r_k^2$$
$$+\frac{16\pi\bar{V}_0}{15}\beta^4 \sum_{k<i}\sum_m (-1)^m \{r_k^2 Y_{2m}(\theta_k, \phi_k)\}\{r_i^2 Y_{2-m}(\theta_i, \phi_i)\}. \quad (7.3)$$

Consequently, since $\bar{V}_0<0$, in this approximation particles move independently in an oscillator potential under the influence of an attractive quadrupole-quadrupole effective residual interaction. Elliott (1958) has shown that within an oscillator shell this Hamiltonian can be diagonalized exactly by use of the SU_3 coupling scheme and leads to a series of rotational bands with energies proportional to $L(L+1)$, where L is the orbital angular momentum. However, it is known that the j–j coupling shell model requires the existence of a strong one-body spin-orbit interaction, and consequently the bands, which characterize the solutions of H_1, have to be mixed in order to describe the physical situation. An extensive literature exists in which various truncations of the basis states of H_1, within an oscillator shell, are used to diagonalize the Hamiltonian including the spin-orbit interaction (Harvey 1968).

We shall now exploit the fact that the most important part of the long-range interaction, over and above the central potential, is the quadrupole interaction, and we shall assume that an adequate Hartree–Fock potential is given by the Nilsson model in which the single-particle Hamiltonian is assumed to have the form (Nilsson 1955)

$$H=\frac{p^2}{2m}+V(r)-\frac{4}{3}\left(\frac{\pi}{5}\right)^{\frac{1}{2}}\delta m\omega^2 r^2 Y_{20}(\theta,\phi), \tag{7.4}$$

where δ is a dimensionless quantity that measures the nuclear deformation and $V(\mathrm{r})$ is the spherical shell-model potential which has a strong $\underset{\sim}{\sigma}\cdot\underset{\sim}{\ell}$ component. We shall first demonstrate that when there are only a few particles outside a closed shell, the eigenfunctions of this Hamiltonian are almost identical to those emerging from Hartree–Fock calculations. Once this has been established, we shall show that many-particle nuclear wave functions that have a high degree of overlap with those that result from direct diagonalization of the shell-model Hamiltonian, can be obtained by filling the lowest possible states (consistent with the exclusion principle) of the Nilsson Hamiltonian and then projecting from this resulting intrinsic state eigenfunctions with well defined angular momentum. We shall then show that many observed nuclear properties can be understood as arising from the fact that the nucleus likes to take on a quadrupole deformation.

1. Nilsson eigenfunctions

Since the Nilsson Hamiltonian in equation (7.4) is not rotationally invariant, its eigenfunctions are not those of j^2. On the other hand, since the Hamiltonian is axially symmetric, μ, the eigenvalue of j_z, is a good quantum number. In this Section we shall diagonalize H using as basis states all j values within a given major shell. That is, we assume that the N or $Z=2$, 8, 20 etc. spherical shells are inert and that the deformed potential acts only between nucleons within a major shell. (This assumption corresponds to that made in a restricted Hartree–Fock calculation.) Thus, for example, to find the $\mu=\frac{1}{2}$ eigenfunctions in the $(0d, 1s)$ shell requires the diagonalization of a 3×3 matrix.

In general, we denote the possible eigenvalues for given μ by $\tilde{\varepsilon}_\mu(\alpha)$

and the eigenfunction associated with this energy by

$$\chi_\mu = \sum_j c_j(\mu, \alpha)\phi_{j\mu} \tag{7.5}$$

where $c_j(\mu, \alpha)$ are expansion coefficients and have the property that

$$\sum_j c_j(\mu, \alpha)c_j(\mu, \alpha') = \delta_{\alpha\alpha'}.$$

$\phi_{j\mu}$ are the single-particle eigenfunctions of equation (1.4) which satisfy the spherical shell-model Schrödinger equation

$$\left\{\frac{p^2}{2m} + V(r)\right\}\phi_{j\mu} = \varepsilon_j\phi_{j\mu}.$$

For simplicity we shall assume the radial wave functions are the harmonic oscillator functions given by equation (4.11). However, we shall not take oscillator energies for the ε_j but instead will use the experimental values seen when there is only one nucleon outside a closed shell.

The expansion coefficients $c_j(\mu, \alpha)$ of equation (7.5) are calculated by diagonalizing the Nilsson Hamiltonian. From equation (A2.23) one may easily show that

$$\begin{aligned}\langle\phi_{j'\mu}|\, H\, |\phi_{j\mu}\rangle &= \varepsilon_j\delta_{jj'}\delta_{nn'}\delta_{ll'} - (-1)^{j-j'}\tfrac{2}{3}\delta m\omega^2\left(\frac{2j+1}{2j'+1}\right)^{\frac{1}{2}}\\ &\quad\times(j2\tfrac{1}{2}0\mid j'\tfrac{1}{2})(j2\mu 0\mid j'\mu)\langle R_{n'l'}|\, r^2\, |R_{nl}\rangle\\ &= \varepsilon_j\delta_{jj'}\delta_{nn'}\delta_{ll'} - (-1)^{j-j'}\tfrac{2}{3}\bar{E}\left(\frac{2j+1}{2j'+1}\right)^{\frac{1}{2}}\\ &\quad\times(j2\mu 0\mid j'\mu)(j2\tfrac{1}{2}0\mid j'\tfrac{1}{2})\\ &\quad\times[(2n+l+\tfrac{3}{2})\delta_{ll'} - 2\surd\{(n+1)(n+l+\tfrac{1}{2})\}\delta_{l',l-2}\\ &\quad - 2\surd\{(n'+1)(n'+l'+\tfrac{1}{2})\}\delta_{l'-2,l}]\end{aligned} \tag{7.6}$$

where

$$\bar{E} = \delta\hbar\omega \tag{7.7}$$

is the deformation energy. In writing the second line of this equation use has been made of equation (1.11a) to evaluate the matrix elements of r^2 within a major oscillator shell.

Because of the assumption of axial symmetry the energy eigenvalues $\tilde{\varepsilon}_\mu(\alpha)$ and $\tilde{\varepsilon}_{-\mu}(\alpha)$ are the same and the expansion coefficients $c_j(\mu, \alpha)$ and $c_j(-\mu, \alpha)$ are related. To see that these assertions are true we note that when μ is replaced by $-\mu$ the only change in equation (7.6) is that

$$(j2\mu 0\mid j'\mu) \rightarrow (j2-\mu 0\mid j'-\mu).$$

However, equation (A1.19) tells us that

$$(j2-\mu 0\,|\,j'-\mu)=(-1)^{j-j'}(j2\mu 0\,|\,j'\mu).$$

Thus when $\mu\to-\mu$ the most that happens is that off-diagonal matrix elements of the Nilsson Hamiltonian change sign. This change of sign would not have occurred if the phase of the basis states (which is completely arbitrary and cannot change the predicted value of any observable quantity) for $-\mu$ had been chosen to be $(-1)^{j-\bar{j}}\phi_{j-\mu}$, where $\bar{j}$ is any half-integral number. When this latter phase convention is used, the matrix elements of H proportional to $(j2-\mu 0\,|\,j'-\mu)$ become

$$(-1)^{j+j'-2\bar{j}}(j2-\mu 0\,|\,j'-\mu)=(-1)^{2(j-\bar{j})}(j2\mu 0\,|\,j'\mu').$$

Thus for any half-integral value of $\bar{j}$ the Hamiltonian matrices for μ and $-\mu$ become identical. Consequently

$$\tilde{\varepsilon}_{\mu}(\alpha)=\tilde{\varepsilon}_{-\mu}(\alpha) \tag{7.8}$$

and because of the exclusion principle each Nilsson orbit is four-fold degenerate and can accommodate both a neutron and a proton with $+\mu$ and $-\mu$. Furthermore, in the eigenfunction expansion in equation (7.5)

$$c_j(-\mu,\alpha)=(-1)^{j-\bar{j}}c_j(\mu,\alpha). \tag{7.9}$$

Throughout this discussion we shall always take $\bar{j}$ equal to μ. This corresponds to defining $\phi_{j-\mu}$ to be the wave function obtained from $\phi_{j\mu}$ by a rotation of 180° about the y axis.

For the $0p$ shell the solutions of the Nilsson Hamiltonian can be written down explicitly. For $\mu=\frac{3}{2}$ the matrix is (1×1), so that

$$\begin{aligned}\tilde{\varepsilon}_{\frac{3}{2}}&=\varepsilon_{\frac{3}{2}}-\tfrac{5}{3}\bar{E}(\tfrac{3}{2}2\tfrac{1}{2}0\,|\,\tfrac{3}{2}\tfrac{1}{2})(\tfrac{3}{2}2\tfrac{3}{2}0\,|\,\tfrac{3}{2}\tfrac{3}{2})\\&=\varepsilon_{\frac{3}{2}}+\tfrac{1}{3}\bar{E}\end{aligned}$$

where $\varepsilon_{\frac{3}{2}}$ is the spherical $p_{\frac{3}{2}}$ single-particle energy and use has been made of Table A1.4 to evaluate the Clebsch–Gordan coefficients. Clearly the $\mu=\frac{3}{2}$ eigenfunction for all values of $\bar{E}$ is

$$\chi_{\frac{3}{2}\frac{3}{2}}=\phi_{\frac{3}{2}\frac{3}{2}}$$

The $\mu=\frac{1}{2}$ energy matrix in this shell is (2×2) and has the form

$$H=\begin{bmatrix}\varepsilon_{\frac{3}{2}}-\frac{1}{3}\bar{E} & \dfrac{\sqrt{2}\bar{E}}{3}\\ \dfrac{\sqrt{2}\bar{E}}{3} & \varepsilon_{\frac{1}{2}}\end{bmatrix}$$

which has eigenvalues

$$\tilde{\varepsilon}_{\frac{1}{2}} = \varepsilon_{\frac{3}{2}} + \tfrac{1}{2}[(\Delta - \bar{E}/3) \mp \sqrt{\{(\Delta + \bar{E}/3)^2 + 8\bar{E}^2/9\}}]$$

where

$$\Delta = \varepsilon_{\frac{1}{2}} - \varepsilon_{\frac{3}{2}}.$$

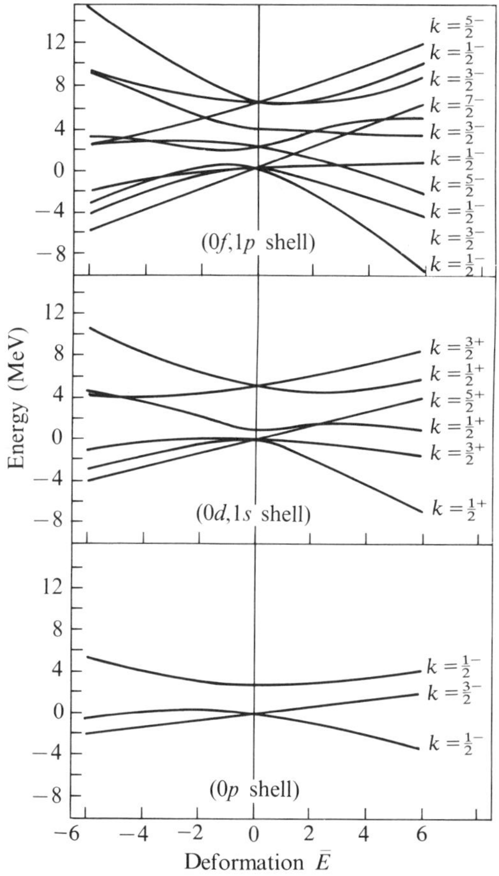

Fig. 7.1. Nilsson diagram for the $0p$, $(0d, 1s)$, and $(0f, 1p)$ shells. Numerical values for the expansion coefficients in the wave functions (equation (7.5)) are given in Table 7.1.

In Fig. 7.1 we have plotted for $\Delta = 2{\cdot}6$ MeV (Lauritsen and Ajzenberg-Selove 1966) the dependence of these energies on the deformation parameter $\bar{E}$, and in Table 7.1 the eigenfunctions for the states are given. For positive (prolate) deformation ($\bar{E} > 0$) the lowest Nilsson state has $\mu = \frac{1}{2}$, whereas for negative (oblate) deformation the $\mu = \frac{3}{2}$ orbit lies

TABLE 7.1

Eigenvalues and eigenfunctions of the Nilsson Hamiltonian as a function of $\bar{E}=\delta\hbar\omega$

$\bar{E}$(MeV)	$\to -\infty$	−6	−4	−2	0	2	4	6	$\to +\infty$
$\mu=\frac{3}{2}^-$				0*p* shell					
$\tilde{\varepsilon}_{\frac32}=\frac13\bar{E}$		$\chi_{\frac32}=\phi_{\frac32\frac32}$							
$\mu=\frac{1}{2}^-$									
$\tilde{\varepsilon}_{\frac12}(1)$ in MeV	$\frac13\bar{E}$	−0·544	−0·022	0·283	0	−0·919	−2·091	−3·346	$-\frac23\bar{E}$
$c_{\frac32}$	0·577	0·743	0·812	0·926	1·000	0·966	0·928	0·903	0·816
$c_{\frac12}$	0·816	0·669	0·584	0·377	0	−0·259	−0·373	−0·430	−0·577
$\tilde{\varepsilon}_{\frac12}(2)$ in MeV	$-\frac23\bar{E}$	5·144	3·956	2·984	2·6	2·853	3·358	3·946	$\frac13\bar{E}$
$c_{\frac32}$	−0·816	−0·669	−0·584	−0·377	0	0·259	0·373	0·430	0·577
$c_{\frac12}$	0·577	0·743	0·812	0·926	1·000	0·966	0·928	0·903	0·816
$\mu=\frac{5}{2}^+$				(0*d*, 1*s*) shell					
$\tilde{\varepsilon}_{\frac52}=\frac23\bar{E}$		$\chi_{\frac52}=\phi_{\frac52\frac52}$							
$\mu=\frac{3}{2}^+$									
$\tilde{\varepsilon}_{\frac32}(1)$ in MeV	$\frac23\bar{E}$	−0·971	−0·214	0·108	0	−0·367	−0·861	−1·419	$-\frac13\bar{E}$
$c_{\frac52}$	0·447	0·805	0·906	0·981	1·000	0·992	0·980	0·968	0·894
$c_{\frac32}$	0·894	0·594	0·423	0·194	0	−0·124	−0·201	−0·250	−0·447
$\tilde{\varepsilon}_{\frac32}(2)$ in MeV	$-\frac13\bar{E}$	4·052	3·960	4·305	5·08	6·114	7·275	8·499	$\frac23\bar{E}$
$c_{\frac52}$	−0·894	−0·594	−0·423	−0·194	0	0·124	0·201	0·250	0·447
$c_{\frac32}$	0·447	0·805	0·906	0·981	1·000	0·992	0·980	0·968	0·894
$\mu=\frac{1}{2}^+$									
$\tilde{\varepsilon}_{\frac12}(1)$ in MeV	$\frac23\bar{E}$	−3·001	−1·746	−0·563	0	−1·974	−4·368	−6·879	$-\frac43\bar{E}$
$c_{\frac52}$	0·447	0·540	0·573	0·652	1·000	0·855	0·796	0·762	0·632
$c_{\frac32}$	−0·365	−0·219	−0·177	−0·104	0	−0·143	−0·243	−0·305	−0·516
$c_{\frac12}$	0·816	0·813	0·800	0·751	0	−0·499	−0·554	−0·572	−0·577
$\tilde{\varepsilon}_{\frac12}(2)$ in MeV	$-\frac13\bar{E}$	4·458	3·461	2·157	0·871	1·386	1·443	1·019	$-\frac13\bar{E}$
$c_{\frac52}$	−0·632	−0·746	−0·769	−0·747	0	0·518	0·580	0·586	0·632
$c_{\frac32}$	−0·775	−0·571	−0·453	−0·253	0	0·287	0·570	0·701	0·775
$c_{\frac12}$	0	0·342	0·451	0·614	1·000	0·806	0·582	0·407	0
$\tilde{\varepsilon}_{\frac12}(3)$ in MeV	$-\frac43\bar{E}$	10·494	8·236	6·357	5·08	4·539	4·876	5·811	$\frac23\bar{E}$
$c_{\frac52}$	−0·632	−0·390	−0·283	−0·126	0	−0·028	−0·175	−0·276	−0·447
$c_{\frac32}$	0·516	0·791	0·874	0·962	1·000	0·947	0·785	0·645	0·365
$c_{\frac12}$	0·577	0·471	0·395	0·243	0	−0·319	−0·595	−0·712	−0·816
$\mu=\frac{7}{2}^-$				(0*f*, 1*p*) shell					
$\tilde{\varepsilon}_{\frac72}=\bar{E}$		$\chi_{\frac72}=\phi_{\frac72\frac72}$							
$\mu=\frac{5}{2}^-$									
$\tilde{\varepsilon}_{\frac52}(1)$ in MeV	$\bar{E}$	−2·124	−1·047	−0·381	0	0·224	0·367	0·463	0
$c_{\frac72}$	0·378	0·856	0·947	0·991	1·000	0·996	0·989	0·983	0·926
$c_{\frac52}$	0·926	0·517	0·322	0·134	0	−0·087	−0·145	−0·185	−0·378
$\tilde{\varepsilon}_{\frac52}(2)$ in MeV	0	2·624	3·547	4·881	6·5	8·276	10·133	12·037	$\bar{E}$
$c_{\frac72}$	−0·926	−0·517	−0·322	−0·134	0	0·087	0·145	0·185	0·378
$c_{\frac52}$	0·378	0·856	0·947	0·91	1·000	0·996	0·989	0·983	0·926
$\mu=\frac{3}{2}^-$									
$\tilde{\varepsilon}_{\frac32}(1)$ in MeV	$\bar{E}$	−4·166	−2·249	−0·480	0	−1·337	−3·027	−4·839	$-\bar{E}$
$c_{\frac72}$	0·378	0·490	0·545	0·704	1·000	0·950	0·911	0·885	0·756
$c_{\frac52}$	−0·239	−0·121	−0·089	−0·031	0	−0·102	−0·183	−0·239	−0·478
$c_{\frac32}$	0·894	0·863	0·834	0·709	0	−0·296	−0·369	−0·399	−0·447
$\tilde{\varepsilon}_{\frac32}(2)$ in MeV	0	3·376	2·818	2·028	2·1	3·539	4·576	4·889	0
$c_{\frac72}$	−0·535	−0·769	−0·791	−0·699	0	0·312	0·401	0·414	0·535
$c_{\frac52}$	−0·845	−0·526	−0·383	−0·201	0	0·236	0·603	0·795	0·845
$c_{\frac32}$	0	0·363	0·476	0·686	1·000	0·920	0·690	0·443	0
$\tilde{\varepsilon}_{\frac32}(3)$ in MeV	$-\bar{E}$	9·390	8·031	7·052	6·5	6·398	7·051	8·550	$\bar{E}$
$c_{\frac72}$	−0·756	−0·410	−0·277	−0·121	0	0·024	−0·096	−0·211	−0·378
$c_{\frac52}$	0·478	0·842	0·919	0·979	1·000	0·966	0·777	0·558	0·239
$c_{\frac32}$	0·447	0·351	0·279	0·163	0	−0·256	−0·623	−0·803	−0·894

TABLE 7.1 *(Continued)*

$\bar{E}$(MeV)	$\rightarrow -\infty$	−6	−4	−2	0	2	4	6	$\rightarrow +\infty$
$\mu = \frac{1}{2}^-$									
$\tilde{\varepsilon}_{\frac{1}{2}}(1)$ in MeV	$\bar{E}$	−3·074	−1·310	0·066	0	−2·579	−6·113	−9·888	$-2\bar{E}$
$c_{\frac{7}{2}}$	0·293	0·472	0·565	0·770	1·000	0·835	0·715	0·653	0·478
$c_{\frac{5}{2}}$	0·338	0·200	0·146	0·056	0	−0·080	−0·167	−0·223	−0·414
$c_{\frac{3}{2}}$	0·516	0·616	0·633	0·572	0	−0·521	−0·623	−0·648	−0·632
$c_{\frac{1}{2}}$	0·730	0·598	0·509	0·278	0	0·158	0·269	0·323	0·447
$\tilde{\varepsilon}_{\frac{1}{2}}(2)$ in MeV	0	2·624	2·620	2·653	2·1	0·913	−0·644	−2·485	$-\bar{E}$
$c_{\frac{7}{2}}$	−0·586	−0·640	−0·629	−0·537	0	0·521	0·600	0·604	0·586
$c_{\frac{5}{2}}$	0·507	0·386	0·335	0·204	0	0·236	0·452	0·551	0·676
$c_{\frac{3}{2}}$	−0·516	−0·238	−0·078	0·340	1·000	0·647	0·316	0·140	−0·258
$c_{\frac{1}{2}}$	0·365	0·620	0·698	0·745	0	−0·504	−0·580	−0·559	−0·365
$\tilde{\varepsilon}_{\frac{1}{2}}(3)$ in MeV	$-\bar{E}$	9·401	7·260	5·054	3·9	3·696	3·284	3·086	0
$c_{\frac{7}{2}}$	0·586	0·544	0·501	0·338	0	0·166	0·377	0·426	0·586
$c_{\frac{5}{2}}$	0·676	0·607	0·540	0·396	0	−0·593	−0·664	−0·634	−0·507
$c_{\frac{3}{2}}$	−0·258	−0·578	−0·666	−0·718	0	0·534	0·643	0·646	0·516
$c_{\frac{1}{2}}$	−0·365	−0·037	0·118	0·462	1·000	0·580	0·182	−0·004	−0·365
$\tilde{\varepsilon}_{\frac{1}{2}}(4)$ in MeV	$-2\bar{E}$	15·549	11·930	8·727	6·5	6·470	7·973	9·787	$\bar{E}$
$c_{\frac{7}{2}}$	−0·478	−0·267	−0·188	−0·075	0	0·056	0·126	0·168	0·293
$c_{\frac{5}{2}}$	0·414	0·665	0·758	0·894	1·000	0·766	0·572	0·495	0·338
$c_{\frac{3}{2}}$	0·632	0·480	0·387	0·205	0	0·159	0·313	0·379	0·516
$c_{\frac{1}{2}}$	−0·447	−0·506	−0·490	−0·392	0	0·620	0·747	0·764	0·730

The wave-function expansion coefficients c_j are defined by equation (7.5). The single-particle $p_{\frac{3}{2}}$ energy is set equal to zero in the $0p$ shell, and in the $(0d, 1s)$ and $(0f, 1p)$ configurations the $0d_{\frac{5}{2}}$ and $0f_{\frac{7}{2}}$ energies are taken to be zero. The single-particle energies of the other levels in a major shell are given by the tabulated energies for $\bar{E} = 0$.

lowest. From Fig. 7.1 and from Table 7.1 it is clear that $\tilde{\varepsilon}_{\frac{1}{2}}(1)$ is lower than $\tilde{\varepsilon}_{\frac{3}{2}}$ and consequently prolate deformation is preferred at the beginning of the $0p$ shell.

Hartree–Fock calculations have been carried out for ${}^{8}_{4}Be_4$ by Bassichis *et al.* (1967) using a residual interaction that reproduces the low-energy nucleon–nucleon scattering and bound-state data. These authors find that the wave function of the lowest $\frac{1}{2}^-$ state is*

$$\chi_{\frac{1}{2}} = 0{\cdot}859\phi_{\frac{3}{2}\frac{1}{2}} - 0{\cdot}512\phi_{\frac{1}{2}\frac{1}{2}} \tag{7.10}$$

Thus, as seen from Table 7.1, the phasing of the Hartree–Fock wave function is the same as given by the Nilsson calculation for prolate deformation. Since there are only two components in $\chi_{\frac{1}{2}}$ it is trivial to find a value of $\bar{E}$ that exactly reproduces the Hartree–Fock calculation. The required value satisfies the equation

$$\frac{0{\cdot}859}{2}\left[\left\{\left(\Delta+\frac{\bar{E}}{3}\right)^2+\frac{8\bar{E}^2}{9}\right\}^{\frac{1}{2}}-\left(\Delta+\frac{\bar{E}}{3}\right)\right]=\frac{\sqrt{2}\bar{E}}{3}(0{\cdot}512)$$

* In actual fact these authors carry out an extended Hartree–Fock calculation that includes the $(0f, 1p)$ shell. The above numbers represent the renormalized values to be used in the $0p$ shell alone.

which has the solution

$$\Delta/\bar{E}=0{\cdot}1766. \tag{7.11}$$

This gives $\bar{E}=14{\cdot}72$ MeV if $\Delta=2{\cdot}6$ MeV. Although this value of $\bar{E}$ is considerably larger than required for reproducing the Hartree–Fock calculations in the $(0d, 1s)$ or $(0f, 1p)$ shells, we shall show in the next section a large $\bar{E}$ is also needed to reproduce the shell-model wave functions in the $0p$ shell.

In addition to ${}^8_4\mathrm{Be}_4$, Bassichis *et al.* (1967) also calculate the wave function of the occupied orbits in ${}^{12}_{6}\mathrm{C}_6$. They predict these orbits to be $\frac{3}{2}^-$ and the $\frac{1}{2}^-$ state whose eigenfunction is

$$\chi'_{\frac{1}{2}}=0{\cdot}718\phi_{\frac{3}{2}\frac{1}{2}}+0{\cdot}696\phi_{\frac{1}{2}\frac{1}{2}}. \tag{7.12}$$

Clearly this is what one would expect on the basis of the deformed single-particle model. From Fig. 7.1 it is apparent that the Nilsson state with lowest energy in ${}^{12}_{6}\mathrm{C}_6$ occurs for negative values of $\bar{E}$ and would correspond to filled $\frac{3}{2}^-$ and $\frac{1}{2}^-$ orbits. The phasing of the Hartree–Fock wave function in equation (7.12) is exactly that predicted for the lowest $\frac{1}{2}^-$ Nilsson orbit with oblate deformation (see Table 7.1). The value of $\Delta/\bar{E}$ that exactly reproduces the Hartree–Fock eigenfunction is

$$\Delta/\bar{E}=-0{\cdot}3627$$

or $\bar{E}=-7{\cdot}17$ MeV when $\Delta=2{\cdot}6$ MeV.

In Fig. 7.1 and Table 7.1 we have also given the results for the $(0d, 1s)$ and $(0f, 1p)$ shells. As before, it is clear that if there are only a few particles in the shell, positive deformation is favoured whereas for only a few holes a negative value of $\bar{E}$ is preferred. There have been several Hartree–Fock calculations for the early $(0d, 1s)$ shell. Bassichis *et al.* (1967) find that the occupied orbit in ${}^{20}_{10}\mathrm{Ne}_{10}$ has a wave function

$$\chi_{\frac{1}{2}}=0{\cdot}862\phi_{\frac{5}{2}\frac{1}{2}}-0{\cdot}311\phi_{\frac{3}{2}\frac{1}{2}}-0{\cdot}401\phi_{\frac{1}{2}\frac{1}{2}}. \tag{7.13a}$$

The same nucleus has been studied by Bar–Touv and Kelson (1965) using as the residual nucleon–nucleon potential the Rosenfeld–Yukawa interaction

$$V(r)=\frac{V_0}{3}\,\underline{\tau}_1\cdot\underline{\tau}_2(0{\cdot}3+0{\cdot}7\underline{\sigma}_1\cdot\underline{\sigma}_2)\Big(\exp(-r/r_0)\Big)/(r/r_0).$$

With $V_0=50$ MeV and r_0 equal to the pi-meson Compton wave-length, $1{\cdot}37$ fm, they find

$$\chi'_{\frac{1}{2}}=0{\cdot}819\phi_{\frac{5}{2}\frac{1}{2}}-0{\cdot}381\phi_{\frac{3}{2}\frac{1}{2}}-0{\cdot}429\phi_{\frac{1}{2}\frac{1}{2}}. \tag{7.13b}$$

Thus both calculations lead to similar wave functions for the occupied state. Moreover, from Table 7.1 it is apparent that these wave functions have precisely the phasing predicted by the deformed single-particle model when $\bar{E}>0$. Although not identical, these wave functions are quite similar to Nilsson eigenfunctions. For $\bar{E}=4$ MeV

$$\int \chi^*_{\frac{1}{2}}(\mathrm{HF})\chi_{\frac{1}{2}}(\mathrm{Nilsson})\,\mathrm{d}\tau = 0{\cdot}984$$

when $\chi_{\frac{1}{2}}(\mathrm{HF})$ is the Bassichis, Kerman, Svenne wave function and

$$\int \chi'^*_{\frac{1}{2}}(\mathrm{HF})\chi_{\frac{1}{2}}(\mathrm{Nilsson})\,\mathrm{d}\tau = 0{\cdot}982$$

for the Bar–Touv Kelson state.

At the beginning of the $(0f, 1p)$ shell the similarity between the Hartree–Fock and Nilsson states persists. When the residual two-body matrix elements of Kuo and Brown (1968) are used to describe the interaction of the nucleons outside the $N=Z=20$ closed shell, Sharma and Bhatt (1972) find that the occupied state in ${}^{44}_{22}\mathrm{Ti}_{22}$ is

$$\chi_{\frac{1}{2}} = 0{\cdot}83\phi_{\frac{7}{2}\frac{1}{2}} - 0{\cdot}24\phi_{\frac{5}{2}\frac{1}{2}} - 0{\cdot}44\phi_{\frac{3}{2}\frac{1}{2}} + 0{\cdot}23\phi_{\frac{1}{2}\frac{1}{2}}. \tag{7.14}$$

For $\bar{E}=2$ MeV this has an overlap of $0{\cdot}978$ with the lowest $\frac{1}{2}^-$ Nilsson state in the $(0f, 1p)$ shell.

For four or less particles or holes, the simple Nilsson model reproduces the Hartree–Fock wave functions to a high degree of accuracy. However, for more particles or holes in a shell this is not always the case. For example, Bar-Touv and Kelson (1965) have shown that for ${}^{24}_{12}\mathrm{Mg}_{12}$ the Hartree–Fock solution no longer has axial symmetry. In other words, the wave functions are not eigenfunctions of j_z and this means that more than one Nilsson state must be used to reproduce the Hartree–Fock calculations. This result is also true in the $(0f, 1p)$ shell (Sandhya Devi *et al.* 1970) and, as we shall see in the next Section, 'mixed-μ' states are sometimes needed in order to obtain agreement with the shell-model wave functions.

A hint that this might happen comes from inspection of the Nilsson diagram. The lowest Nilsson state for $\bar{E}>0$ or the highest for $\bar{E}<0$ is well isolated from all others, and consequently only a strong component in the residual nucleon–nucleon force (over and above that which has approximately been taken into account by the introduction of a deformed potential) would lead to mixing between the various single-particle states. On the other hand, the remaining orbits are not so well separated in energy and, in particular when the deformation is large, degeneracies result (see Table 7.1 and Fig. 7.1). Thus even a weak component in the residual interaction can mix these orbits.

2. Projected wave functions

In the preceding Section we showed that the Nilsson eigenfunctions and the Hartree–Fock wave functions are quite similar. However, neither of these are angular-momentum eigenfunctions and hence, as they stand, are not appropriate for the description of nuclear states with given angular momentum. We now show that when an appropriate choice of the deformation energy $\bar{E}$ is made the angular-momentum eigenfunctions $\psi_{IK}(x, \bar{E})$ obtained by the following procedure are almost identical to the wave functions of the yrast levels that arise from diagonalization of the shell-model Hamiltonian:

(a) Construct an intrinsic state χ_K by filling the lowest possible Nilsson orbits. Because of the exclusion principle each level can accommodate at most two identical nucleons, one with $+\mu$ and the other with $-\mu$. The subscript K on χ_K is the sum of the z components of angular momenta.

(b) Carry out an angular-momentum decomposition of this intrinsic state

$$\chi_K(x, \bar{E}) = \sum_I C_I(K, \bar{E})\psi_{IK}(x, \bar{E})$$

$$= \sum_{I\gamma} \tilde{C}_I(K, \gamma, \bar{E})\tilde{\Phi}_{IK}(x, \gamma) \tag{7.15}$$

where the $\tilde{\Phi}_{IK}(x, \gamma)$ are normalized eigenfunctions with angular momentum I and z component K, the index γ serves to enumerate the various ways one can arrive at a state with (I, K) within the model space considered, and $|\tilde{C}_I(K, \gamma, \bar{E})|^2$ is the probability that the state $\tilde{\Phi}_{IK}(x, \gamma)$ is contained in $\chi_K(x, \bar{E})$. The wave function $\psi_{IK}(x, \bar{E})$ is the normalized linear combination of the $\tilde{\Phi}_{IK}(x, \gamma)$ defined by

$$\psi_{IK}(x, \bar{E}) = \left\{\sum_\gamma |\tilde{C}_I(K, \gamma, \bar{E})|^2\right\}^{-\frac{1}{2}} \sum_\gamma \tilde{C}_I(K, \gamma, \bar{E})\tilde{\Phi}_{IK}(x, \gamma) \tag{7.16a}$$

with

$$|C_I(K, \bar{E})| = \left\{\sum_\gamma |\tilde{C}_I(K, \gamma, \bar{E})|^2\right\}^{\frac{1}{2}}, \tag{7.16b}$$

and in all these equations the coordinate x is a short-hand abbreviation for the space-spin coordinates of the N nucleons making up the wave function. Provided the intrinsic states χ_K are well separated in energy, Kurath and Pičman (1959) have shown that in the $0p$ shell the $\psi_{IK}(x, \bar{E})$ of equation (7.16a) are almost identical to those of the yrast levels emerging from shell-model calculations. The same has also been shown to be true at the beginning of the $(0d, 1s)$ shell (Redlich 1958) and in the $0f_{\frac{7}{2}}$ region (Lawson 1961, McCullen *et al.* 1964).

In order to carry out the above angular-momentum decomposition it is convenient to note that if equations (7.15) and (7.16) hold for the coordinate system x, they are also true when the coordinates of χ_K and $\tilde{\Phi}_{IK}$ are taken with respect to a rotated set of axes. If the coordinates x referred to the rotated axes are designated by x' it follows from equation (A7.7) that

$$\tilde{\Phi}_{IM'}(x') = \sum_M D^{I*}_{MM'}(R)\tilde{\Phi}_{IM}(x) \tag{7.17}$$

where R is the rotation that links the primed and unprimed coordinate systems and the properties of the rotation matrix $D^I_{MM'}(R)$ are discussed in Appendix 7. Thus

$$\chi_K(x', \bar{E}) = \sum_{I\gamma} \tilde{C}_I(K, \gamma, \bar{E})\tilde{\Phi}_{IK}(x', \gamma)$$

$$= \sum_{I\gamma M'} \tilde{C}_I(K, \gamma, \bar{E}) D^{I*}_{M'K}(R)\tilde{\Phi}_{IM'}(x, \gamma).$$

If one multiplies each side of this equation by $(2I+1)D^I_{MK}(R)$ and integrates over the Euler angles specifying the rotation, it follows from equation (A7.21) that

$$(2I+1)\int D^I_{MK}(R)\chi_K(x', \bar{E})\,\mathrm{d}R = \sum_\gamma \tilde{C}_I(K, \gamma, \bar{E})\tilde{\Phi}_{IM}(x, \gamma). \tag{7.18}$$

This method for obtaining the wave function is often referred to as the Elliott generating procedure (Elliott 1958) or the Hill–Wheeler integral (Hill and Wheeler 1953).

We shall now give several examples of the usefulness of this generating procedure and show what modifications must be made when there are several χ_K with almost the same energy.

2.1. *Wave functions for* ${}^7_3\mathrm{Li}_4$

To good approximation ${}^4_2\mathrm{He}_2$ forms an inert closed core. Consequently the structure of the low-lying states of ${}^7_3\mathrm{Li}_4$ should be ascribed to the motion of one proton and two neutrons confined to the $0p$ shell. Since this is the beginning of the shell the $\mu = \frac{1}{2}$ state with positive (prolate) deformation lies lowest (see Fig. 7.1). Thus the appropriate generator χ_K is

$$\chi_{K=\frac{1}{2}} = b^\dagger_{\frac{1}{2}} a^\dagger_{\frac{1}{2}} a^\dagger_{-\frac{1}{2}} |0\rangle \tag{7.19}$$

where $b^\dagger_{\frac{1}{2}}$ is the creation operator for a proton in the lower $\frac{1}{2}^-$ level and $a^\dagger_{\pm\frac{1}{2}}$ is the neutron creation operator for the same orbit. Because the isospin step-up operator t_+ changes a proton to a neutron

$$t_+ b_\mu^\dagger = a_\mu^\dagger.$$

it follows that

$$\begin{aligned} T_+\chi_{K=\frac{1}{2}} &= \sum_i (t_i)_+ \chi_{K=\frac{1}{2}} \\ &= a_{\frac{1}{2}}^\dagger a_{\frac{1}{2}}^\dagger a_{-\frac{1}{2}}^\dagger |0\rangle \\ &\equiv 0 \end{aligned}$$

so that the intrinsic state given by equation (7.19) has $T = T_z = \frac{1}{2}$.
From equation (7.9) it follows that

$$\begin{aligned} \chi_{\frac{1}{2}} &= (\alpha b_{\frac{3}{2}\frac{1}{2}}^\dagger + \beta b_{\frac{1}{2}\frac{1}{2}}^\dagger)(\alpha a_{\frac{3}{2}\frac{1}{2}}^\dagger + \beta a_{\frac{1}{2}\frac{1}{2}}^\dagger)(-\alpha a_{\frac{3}{2}-\frac{1}{2}}^\dagger + \beta a_{\frac{1}{2}-\frac{1}{2}}^\dagger)\,|0\rangle \\ &= (\alpha b_{\frac{3}{2}\frac{1}{2}}^\dagger + \beta b_{\frac{1}{2}\frac{1}{2}}^\dagger)\{-\alpha^2 a_{\frac{3}{2}\frac{1}{2}}^\dagger a_{\frac{3}{2}-\frac{1}{2}}^\dagger + \beta^2 a_{\frac{1}{2}\frac{1}{2}}^\dagger a_{\frac{1}{2}-\frac{1}{2}}^\dagger + \alpha\beta(a_{\frac{3}{2}\frac{1}{2}}^\dagger a_{\frac{1}{2}-\frac{1}{2}}^\dagger \\ &\quad + a_{\frac{3}{2}-\frac{1}{2}}^\dagger a_{\frac{1}{2}\frac{1}{2}}^\dagger)\}\,|0\rangle \\ &= (\alpha b_{\frac{3}{2}\frac{1}{2}}^\dagger + \beta b_{\frac{1}{2}\frac{1}{2}}^\dagger)[-\alpha^2 \sum_J \sqrt{\{1+(-1)^J\}}(\tfrac{3}{2}\tfrac{3}{2}\tfrac{1}{2}-\tfrac{1}{2}\,|\,J0)A_{J0}^\dagger(\tfrac{3}{2}\tfrac{3}{2}) \\ &\quad + \beta^2 A_{00}^\dagger(\tfrac{1}{2}\tfrac{1}{2}) + \alpha\beta \sum_J \{1+(-1)^J\}(\tfrac{3}{2}\tfrac{1}{2}\tfrac{1}{2}-\tfrac{1}{2}\,|\,J0) \\ &\quad \times A_{J0}^\dagger(\tfrac{3}{2}\tfrac{1}{2})]\,|0\rangle \end{aligned} \tag{7.20}$$

where α and β, the expansion coefficients of equation (7.5), are functions of $\bar{E}$, $A_{J0}^\dagger(j_1 j_2)\,|0\rangle$ is the two neutron state given by equation (1.42) that arises from coupling j_1 and j_2 to angular momentum J with z component zero, and use has been made of the fact that

$$a_{j_1 m_1}^\dagger a_{j_2 m_2}^\dagger\,|0\rangle = \sum_J \sqrt{\{1+(-1)^J \delta_{j_1 j_2}\}}(j_1 j_2 m_1 m_2\,|\,JM) A_{JM}^\dagger(j_1 j_2)\,|0\rangle.$$

To proceed further we make use of the projection integral given in equation (7.18). Since $b_{\frac{3}{2}\frac{1}{2}}^\dagger$ creates a proton in the state $j = \frac{3}{2}$ $m = \frac{1}{2}$ its spatial representation is given by equation (1.4). Furthermore, we define the spatial representation of $A_{J0}^\dagger(j_1 j_2)\,|0\rangle$ to be $\Phi_{J0}(\bar{x}; j_1 j_2)$. Thus if we denote the α^3 term in equation (7.20) by $\tilde{\chi}_{\frac{1}{2}}(x', \bar{E})$ it follows that

$$\begin{aligned} \tilde{\chi}_{\frac{1}{2}}(x', \bar{E}) &= -\alpha^3 \phi_{\frac{3}{2}\frac{1}{2}}(\bar{x}') \sum_J \sqrt{\{1+(-1)^J\}}(\tfrac{3}{2}\tfrac{3}{2}\tfrac{1}{2}-\tfrac{1}{2}\,|\,J0)\Phi_{J0}(\bar{x}'; \tfrac{3}{2}^2) \\ &= -\alpha^3 \sum_{JM'm} \sqrt{\{1+(-1)^J\}}(\tfrac{3}{2}\tfrac{3}{2}\tfrac{1}{2}-\tfrac{1}{2}\,|\,J0) D_{m\frac{1}{2}}^{\frac{3}{2}*}(R) D_{M'0}^{J*}(R) \\ &\quad \times \phi_{\frac{3}{2}m}(\bar{x})\Phi_{JM'}(\bar{x}; \tfrac{3}{2}^2) \end{aligned}$$

where $\tilde{x}$ is the coordinate of the proton in the unprimed system and $\bar{x}$ stands for those of the two neutrons. Consequently by use of equation (A7.22), which gives the value of the integral of the product of three D functions, one sees that

$$(2I+1)\int \tilde{\chi}_{\frac{1}{2}}(x',\bar{E})D^{I}_{M\frac{1}{2}}(R)\,\mathrm{d}R=-\alpha^3\sum_J\sqrt{\{1+(-1)^J\}}$$

$$\times(\tfrac{3}{2}\tfrac{3}{2}\tfrac{1}{2}-\tfrac{1}{2}\mid J0)(\tfrac{3}{2}J\tfrac{1}{2}0\mid I\tfrac{1}{2})(\tfrac{3}{2}JmM'\mid IM)\phi_{\frac{3}{2}m}(\tilde{x})\Phi_{JM'}(\bar{x};\tfrac{3}{2}^2)$$

$$=-\alpha^3\sum_J\sqrt{\{1+(-1)^J\}}(\tfrac{3}{2}\tfrac{3}{2}\tfrac{1}{2}-\tfrac{1}{2}\mid J0)(\tfrac{3}{2}J\tfrac{1}{2}0\mid I\tfrac{1}{2})$$

$$\times[\phi_{\frac{3}{2}}(\tilde{x})\times\Phi_J(\bar{x};\tfrac{3}{2}^2)]_{IM}.$$

If we compare this result with equation (7.15) and denote by $\gamma=1$ the wave function when $J=0$ and $\gamma=2$ when $J=2$ we see that

$$\tilde{\Phi}_{IM}(x;1)=[\phi_{\frac{3}{2}}(\tilde{x})\times\Phi_0(\bar{x};\tfrac{3}{2}^2)]_{IM}$$

$$\tilde{C}_I(\tfrac{1}{2},1,\bar{E})=\frac{\alpha^3}{\sqrt{2}}\,\delta_{I,\frac{3}{2}}$$

$$\tilde{\Phi}_{IM}(x;2)=[\phi_{\frac{3}{2}}(\tilde{x})\times\Phi_2(\bar{x};\tfrac{3}{2}^2)]_{IM}$$

$$\tilde{C}_I(\tfrac{1}{2},2,\bar{E})=-\frac{\alpha^3}{\sqrt{2}}(\tfrac{3}{2}2\tfrac{1}{2}0\mid I\tfrac{1}{2}).$$

Consequently the normalized projected wave function with angular momentum (IM) that arises from $\tilde{\chi}_{\frac{1}{2}}(x',\bar{E})$ is

$$\psi_{IM}(x,\bar{E})=\frac{1}{\sqrt{\{\delta_{I,\frac{3}{2}}+(\frac{3}{2}2\frac{1}{2}0\mid I\frac{1}{2})^2\}}}\big([\phi_{\frac{3}{2}}(\tilde{x})\times\Phi_0(\bar{x};\tfrac{3}{2}^2)]_{IM}$$

$$-(\tfrac{3}{2}2\tfrac{1}{2}0\mid I\tfrac{1}{2})[\phi_{\frac{3}{2}}(\tilde{x})\times\Phi_2(\bar{x};\tfrac{3}{2}^2)]_{IM}\big)$$

with

$$C_I(\tfrac{1}{2},\bar{E})=\frac{\alpha^3}{\sqrt{2}}\left(\delta_{I,\frac{3}{2}}+(\tfrac{3}{2}2\tfrac{1}{2}0\mid I\tfrac{1}{2})^2\right)^{\frac{1}{2}}.$$

Since only α and β are functions of $\bar{E}$, the dependence of the wave function on this quantity cancels out in this case. In general this will not be true.

The same procedure may now be followed for the entire intrinsic state. Since at most the two neutrons can couple to $J=2$ it follows that the only angular momenta that can be generated from the intrinsic state given by equation (7.20) are $I=\frac{1}{2},\frac{3}{2},\frac{5}{2}$, and $\frac{7}{2}$. If one inserts the appropriate values for the Clebsch–Gordan coefficients, one arrives at the following generated wave functions.

$$\psi_{\frac{1}{2}m} = \frac{1}{\sqrt{\{\alpha^6+9\alpha^4\beta^2+10\beta^6\}}}\Big(\alpha^3[\phi_{\frac{3}{2}}(\tilde{x})\times\Phi_2(\bar{x};\tfrac{3}{2}^{2})]_{\frac{1}{2}m}$$
$$-2\alpha^2\beta[\phi_{\frac{3}{2}}(\tilde{x})\times\Phi_2(\bar{x};\tfrac{31}{22})]_{\frac{1}{2}m}$$
$$+\sqrt{5}\alpha^2\beta[\phi_{\frac{1}{2}}(\tilde{x})\times\Phi_0(\bar{x};\tfrac{3}{2}^{2})]_{\frac{1}{2}m}$$
$$+\sqrt{10}\beta^3[\phi_{\frac{1}{2}}(\tilde{x})\times\Phi_0(\bar{x};\tfrac{1}{2}^{2})]_{\frac{1}{2}m}\Big) \tag{7.21a}$$

$$\psi_{\frac{3}{2}m} = \frac{1}{\sqrt{\{6(\alpha^6+\alpha^4\beta^2+3\alpha^2\beta^4)\}}}\Big(\sqrt{5}\alpha^3[\phi_{\frac{3}{2}}(\tilde{x})\times\Phi_0(\bar{x};\tfrac{3}{2}^{2})]_{\frac{3}{2}m}$$
$$+\alpha^3[\phi_{\frac{3}{2}}(\tilde{x})\times\Phi_2(\bar{x};\tfrac{3}{2}^{2})]_{\frac{3}{2}m}-2\alpha^2\beta[\phi_{\frac{3}{2}}(\tilde{x})\times\Phi_2(\bar{x};\tfrac{31}{22})]_{\frac{3}{2}m}$$
$$+\sqrt{10}\alpha\beta^2[\phi_{\frac{3}{2}}(\tilde{x})\times\Phi_0(\bar{x};\tfrac{1}{2}^{2})]_{\frac{3}{2}m}$$
$$-\sqrt{2}\alpha^2\beta[\phi_{\frac{1}{2}}(\tilde{x})\times\Phi_2(\bar{x};\tfrac{3}{2}^{2})]_{\frac{3}{2}m}$$
$$+2\sqrt{2}\alpha\beta^2[\phi_{\frac{1}{2}}(\tilde{x})\times\Phi_2(\bar{x};\tfrac{31}{22})]_{\frac{3}{2}m}\Big) \tag{7.21b}$$

$$\psi_{\frac{5}{2}m} = \frac{1}{\sqrt{(\alpha^6+11\alpha^4\beta^2+28\alpha^2\beta^4)}}\Big(\alpha^3[\phi_{\frac{3}{2}}(\tilde{x})\times\Phi_2(\bar{x};\tfrac{3}{2}^{2})]_{\frac{5}{2}m}$$
$$-2\alpha^2\beta[\phi_{\frac{3}{2}}(\tilde{x})\times\Phi_2(\bar{x};\tfrac{31}{22})]_{\frac{5}{2}m}+\sqrt{7}\alpha^2\beta[\phi_{\frac{1}{2}}(\tilde{x})\times\Phi_2(\bar{x};\tfrac{3}{2}^{2})]_{\frac{5}{2}m}$$
$$-2\sqrt{7}\alpha\beta^2[\phi_{\frac{1}{2}}(\tilde{x})\times\Phi_2(\bar{x};\tfrac{31}{22})]_{\frac{5}{2}m}\Big) \tag{7.21c}$$

$$\psi_{\frac{7}{2}m} = \frac{1}{\sqrt{(\alpha^6+4\alpha^4\beta^2)}}\Big(\alpha^3[\phi_{\frac{3}{2}}(\tilde{x})\times\Phi_2(\bar{x};\tfrac{3}{2}^{2})]_{\frac{7}{2}m}$$
$$-2\alpha^2\beta[\phi_{\frac{3}{2}}(\tilde{x})\times\Phi_2(\bar{x};\tfrac{31}{22})]_{\frac{7}{2}m}\Big) \tag{7.21d}$$

with

$$C_{I=\frac{1}{2}}(K=\tfrac{1}{2},\bar{E})=\{(\alpha^6+9\alpha^4\beta^2+10\beta^6)/10\}^{\frac{1}{2}}, \tag{7.22a}$$
$$C_{I=\frac{3}{2}}(K=\tfrac{1}{2},\bar{E})=\{3(\alpha^6+\alpha^4\beta^2+3\alpha^2\beta^4)/5\}^{\frac{1}{2}}, \tag{7.22b}$$
$$C_{I=\frac{5}{2}}(K=\tfrac{1}{2},\bar{E})=-\{3(\alpha^6+11\alpha^4\beta^2+28\alpha^2\beta^4)/70\}^{\frac{1}{2}}, \tag{7.22c}$$
$$C_{I=\frac{7}{2}}(K=\tfrac{1}{2},\bar{E})=-\{9(\alpha^6+4\alpha^4\beta^2)/35\}^{\frac{1}{2}} \tag{7.22d}$$

where the negative sign for $C_{I=\frac{5}{2}}(\frac{1}{2},\bar{E})$ and $C_{I=\frac{7}{2}}(\frac{1}{2},\bar{E})$ merely reflects the fact that we have chosen the coefficient of $[\phi_{\frac{3}{2}}(\tilde{x})\times\Phi_2(\bar{x};\frac{3}{2}^{2})]_{IM}$ to be positive in all cases. Furthermore, since $|C_I(K,\bar{E})|^2$ represents the probability that the state (IM) is contained in χ_K, it follows that

$$\sum_I |C_I(K,\bar{E})|^2=1$$

which provides a check on the algebra.

To demonstrate that the wave functions of equation (7.21) have a high degree of overlap with the shell-model eigenfunctions one must set up and diagonalize the Hamiltonian matrices for this nucleus. Once the residual nucleon–nucleon interaction and the single-particle energies have been chosen this may be done by the methods discussed in Chapter 2. Instead of the matrix elements of Cohen and Kurath (1965), which were chosen to fit the data for $A \geq 8$, we have taken the values obtained by Kumar (1973) in his fit to the $A=6$ and 7 systems. These are listed in Table 7.2, and in Fig. 7.2 we show that his calculated spectrum of ${}^7_3\mathrm{Li}_4$ is

TABLE 7.2

0p-shell two-body interaction matrix elements for use in the $A=6$ and 7 systems

$E_{JT}(j_1j_2; j_3j_4)$	Value of matrix element (MeV)	$E_{JT}(j_1j_2; j_3j_4)$	Value of matrix element (MeV)
$E_{30}(\frac{3}{2}\frac{3}{2}; \frac{3}{2}\frac{3}{2})$	−5·400	$E_{21}(\frac{3}{2}\frac{3}{2}; \frac{3}{2}\frac{3}{2})$	−1·020
$E_{20}(\frac{3}{2}\frac{1}{2}; \frac{3}{2}\frac{1}{2})$	−7·070	$E_{21}(\frac{3}{2}\frac{1}{2}; \frac{3}{2}\frac{1}{2})$	−2·310
$E_{10}(\frac{3}{2}\frac{3}{2}; \frac{3}{2}\frac{3}{2})$	−2·823	$E_{21}(\frac{3}{2}\frac{3}{2}; \frac{3}{2}\frac{1}{2})$	−1·824
$E_{10}(\frac{3}{2}\frac{1}{2}; \frac{3}{2}\frac{1}{2})$	−5·424	$E_{11}(\frac{3}{2}\frac{1}{2}; \frac{3}{2}\frac{1}{2})$	−0·360
$E_{10}(\frac{1}{2}\frac{1}{2}; \frac{1}{2}\frac{1}{2})$	−2·703	$E_{01}(\frac{3}{2}\frac{3}{2}; \frac{3}{2}\frac{3}{2})$	−2·850
$E_{10}(\frac{3}{2}\frac{3}{2}; \frac{3}{2}\frac{1}{2})$	4·950	$E_{01}(\frac{1}{2}\frac{1}{2}; \frac{1}{2}\frac{1}{2})$	0·140
$E_{10}(\frac{3}{2}\frac{3}{2}; \frac{1}{2}\frac{1}{2})$	3·229	$E_{01}(\frac{3}{2}\frac{3}{2}; \frac{1}{2}\frac{1}{2})$	−4·228
$E_{10}(\frac{3}{2}\frac{1}{2}; \frac{1}{2}\frac{1}{2})$	−0·053		

The $p_{\frac{1}{2}}$-$p_{\frac{3}{2}}$ single-particle splitting to be used with these matrix elements is 3·99 MeV. The values are taken from Kumar (1973).

in good agreement with experiment (Ajzenberg-Selove and Lauritsen 1974).

The shell-model wave functions can be written in the form

$$\psi_{IM}(x) = \sum_{\gamma} d_I(\gamma)\psi_{IM}(x, \gamma)$$

where once more γ denotes the various possible ways of attaining the spin. In Table 7.3 we list the values of $d_I(\gamma)$ for yrast levels and compare them with those predicted by the generating procedure in equation (7.21). In the comparison the coefficients α and β are taken from Table 7.1 with $\bar{E}=6$ MeV. Clearly for all components the phasing is correct and for the $I=\frac{1}{2}^-, \frac{3}{2}^-$, and $\frac{7}{2}^-$ states the generated and shell-model wave functions are almost identical. To be quantitative about this similarity we define the

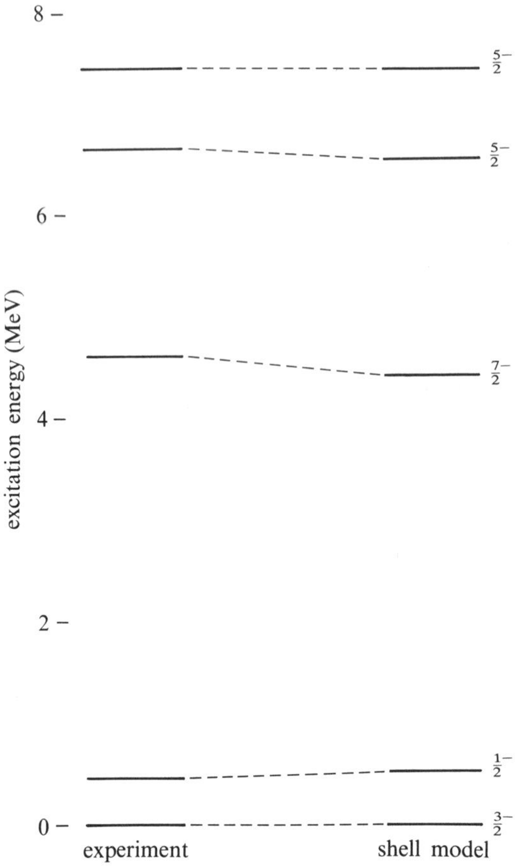

Fig. 7.2. The levels of ${}^{7}_{3}\mathrm{Li}_{4}$ below 8 MeV excitation energy. The experimental data are taken from the compilation of Ajzenberg-Selove and Lauritsen (1974). The shell-model spectrum was calculated using the matrix elements of Table 7.2.

overlap of the wave functions as

$$\text{overlap} = \int \psi^{*}_{IM}(\text{s.m.})\psi_{IM}(\text{gen})\, d\tau. \tag{7.23}$$

In the worst case of these three, when $I = \frac{3}{2}$, the generated wave function and the shell-model eigenfunction (which is obtained by constructing and diagonalizing an 8×8 matrix) have an overlap of 0·991.

The numerical values of the expansion coefficients that characterize the generated wave functions depend on the value of $\bar{E}$ chosen. However, the relative phasing of the various components is the same for all values of

TABLE 7.3

The shell-model yrast wave functions of $^{7}_{3}Li_4$ calculated from the matrix elements of Table 7.2 compared with those generated from the intrinsic state in equation (7.19) with coefficients α and β taken from Table 7.1 when $\bar{E}=6$ MeV

	Coefficients in wave function expansion							
	$I=\frac{1}{2}$		$I=\frac{3}{2}$		$I=\frac{5}{2}$		$I=\frac{7}{2}$	
Basis states	Shell model	Generated	Shell model	Generated	Shell model	Generated	Shell model	Generated
$[\phi_{\frac{3}{2}}(\tilde{x})\times\Phi_0(\bar{x};\frac{3}{2}^2)]_{IM}$			0·704	0·777				
$[\phi_{\frac{3}{2}}(\tilde{x})\times\Phi_0(\bar{x};\frac{1}{2}^2)]_{IM}$			0·306	0·249				
$[\phi_{\frac{1}{2}}(\tilde{x})\times\Phi_0(\bar{x};\frac{3}{2}^2)]_{IM}$	−0·582	−0·599						
$[\phi_{\frac{1}{2}}(\tilde{x})\times\Phi_0(\bar{x};\frac{1}{2}^2)]_{IM}$	−0·204	−0·192						
$[\phi_{\frac{3}{2}}(\tilde{x})\times\Phi_1(\bar{x};\frac{3}{2}\frac{1}{2})]_{IM}$	−0·016	0·000	0·023	0·000	−0·362	0·000		
$[\phi_{\frac{1}{2}}(\tilde{x})\times\Phi_1(\bar{x};\frac{3}{2}\frac{1}{2})]_{IM}$	−0·076	0·000	−0·013	0·000				
$[\phi_{\frac{3}{2}}(\tilde{x})\times\Phi_2(\bar{x};\frac{3}{2}^2)]_{IM}$	0·597	0·563	0·315	0·347	0·760	0·450	0·780	0·724
$[\phi_{\frac{1}{2}}(\tilde{x})\times\Phi_2(\bar{x};\frac{3}{2}^2)]_{IM}$			0·296	0·234	−0·358	−0·567		
$[\phi_{\frac{3}{2}}(\tilde{x})\times\Phi_2(\bar{x};\frac{3}{2}\frac{1}{2})]_{IM}$	0·508	0·536	0·379	0·331	0·134	0·429	0·625	0·690
$[\phi_{\frac{1}{2}}(\tilde{x})\times\Phi_2(\bar{x};\frac{3}{2}\frac{1}{2})]_{IM}$			0·284	0·223	−0·381	−0·540		
Overlap	0·996		0·991		0·808		0·996	

The overlap integral is defined by equation (7.23) and its square is the probability that the generated wave function is contained in the shell-model eigenfunction.

$\bar{E}>0$. Thus although the exact magnitude of a matrix element computed with these eigenfunctions will depend on the precise value of $\bar{E}$ used, the fact that the predicted interference between the various terms is constructive or destructive will not. Consequently one is often able to say whether or not a transition (or moment) is likely to be enhanced or retarded simply by noting that the expected deformation has a certain sign.

The one place where the shell-model and generated wave functions are in bad agreement is the $\frac{5}{2}^-$ state. In this case, as shown in Fig. 7.2, both experimentally and in the shell-model calculation two states of spin $\frac{5}{2}^-$ lie close in energy. Since it is well known that many properties of the $(0p)$, $(0d, 1s)$, and $(0f, 1p)$ shell nuclei can be explained not only on the basis of the shell model but also in terms of the rotational model (Clegg 1961, Paul 1957, Bhatt 1962, Malik and Scholz 1966, 1967), a second intrinsic state that can lead to an $I=\frac{5}{2}^-$ level in ${}^7_3\mathrm{Li}_4$ must lie close in energy to the one obtained from the χ_K of equation (7.19). Under such conditions the Coriolis coupling will cause the rotational bands built on these two intrinsic states to mix (Kerman 1956), and the use of a single intrinsic state to generate the yrast $\frac{5}{2}^-$ wave function is likely to be a bad approximation. On the other hand, the conclusion to be drawn from the calculations just presented is that provided the yrast state with spin I is reasonably separated in energy from other states of the same spin an excellent approximation to its wave function can be obtained through projection from the lowest Nilsson orbital.

2.2. *Wave function for the ${}^6_3\mathrm{Li}_3$ ground state; band mixing*

To show how one gets around the difficulty that arises when there are two almost degenerate intrinsic states that can give rise to the same angular momentum, we consider the yrast 1^+ $T=0$ level of ${}^6_3\mathrm{Li}_3$. ${}^4_2\mathrm{He}_2$ is again assumed to form an inert closed core. The single-neutron and proton outside the core are confined to the $0p$ shell and it is assumed that the Kumar interaction of Table 7.2 is operative. When the $p_{\frac{1}{2}}$, $p_{\frac{3}{2}}$ single-particle energy difference is taken to be 3·99 MeV, the yrast 1^+ level has the structure

$$\psi_{1M}(\text{s.m.})=0{\cdot}765(p_{\frac{3}{2}}^2)_{1M}-0{\cdot}614(p_{\frac{3}{2}}p_{\frac{1}{2}})_{1M}-0{\cdot}194(p_{\frac{1}{2}}^2)_{1M}. \quad (7.24)$$

As far as the Nilsson model is concerned, the extra core particles go into the lowest $\mu=\pm\frac{1}{2}$ state with positive deformation. Since the valence particles are a neutron and a proton both may have $\mu=+\frac{1}{2}$. This gives rise to the $K=1$ generator

$$\chi_1=a_{\frac{1}{2}}^{\dagger}b_{\frac{1}{2}}^{\dagger}\,|0\rangle \quad (7.25a)$$

which obviously has $T=0$ since $T_+\chi_1$ vanishes. There are two other $T=0$

intrinsic states that are degenerate in energy with χ_1 and these are

$$\chi_0 = 2^{-\frac{1}{2}}\{a^\dagger_{\frac{1}{2}}b^\dagger_{-\frac{1}{2}} - b^\dagger_{\frac{1}{2}}a^\dagger_{-\frac{1}{2}}\}\,|0\rangle \tag{7.25b}$$

$$\chi_{-1} = a^\dagger_{-\frac{1}{2}}b^\dagger_{-\frac{1}{2}}\,|0\rangle. \tag{7.25c}$$

One may simply find the angular-momentum decomposition of these states by use of the inverse of equation (2.22)

$$a^\dagger_{j_1m_1;\frac{1}{2}\mu_1}a^\dagger_{j_2m_2;\frac{1}{2}\mu_2}\,|0\rangle = \sum_{JT}\sqrt{\{1-(-1)^{J+T}\delta_{j_1j_2}\}}(j_1j_2m_1m_2\mid JM)$$
$$\times(\tfrac{1}{2}\tfrac{1}{2}\mu_1\mu_2\mid TT_z)A^\dagger_{JM;TT_z}(j_1j_2)\,|0\rangle. \tag{7.26}$$

When use is made of the fact that $a^\dagger_{jm} = a^\dagger_{jm;\frac{1}{2}\frac{1}{2}}$ and $b^\dagger_{jm} = a^\dagger_{jm;\frac{1}{2}-\frac{1}{2}}$ one finds, by the same procedure used in the preceding Section, that

$$\chi_1 = \sqrt{(\tfrac{3}{5})}\alpha^2(p^2_{\frac{3}{2}})_{31} + \sqrt{(\tfrac{3}{2})}\alpha\beta(p_{\frac{3}{2}}p_{\frac{1}{2}})_{21} - \left(\frac{4\alpha^4+5\alpha^2\beta^2+10\beta^4}{10}\right)^{\frac{1}{2}}\psi_{11}, \tag{7.27a}$$

$$\chi_0 = \frac{-3}{\sqrt{10}}\alpha^2(p^2_{\frac{3}{2}})_{30} + \left(\frac{\alpha^4+20\alpha^2\beta^2+10\beta^4}{10}\right)^{\frac{1}{2}}\tilde{\psi}_{10}, \tag{7.27b}$$

$$\chi_{-1} = \sqrt{(\tfrac{3}{5})}\alpha^2(p^2_{\frac{3}{2}})_{3-1} - \sqrt{(\tfrac{3}{2})}\alpha\beta(p_{\frac{3}{2}}p_{\frac{1}{2}})_{2-1} - \left(\frac{4\alpha^4+5\alpha^2\beta^2+10\beta^4}{10}\right)^{\frac{1}{2}}\psi_{1-1} \tag{7.27c}$$

where

$$\psi_{1\pm1} = \frac{1}{\sqrt{(4\alpha^4+5\alpha^2\beta^2+10\beta^4)}}\{2\alpha^2(p^2_{\frac{3}{2}})_{1\pm1} + \sqrt{5}\alpha\beta(p_{\frac{3}{2}}p_{\frac{1}{2}})_{1\pm1} - \sqrt{10}\beta^2(p^2_{\frac{1}{2}})_{1\pm1}\} \tag{7.28a}$$

and

$$\tilde{\psi}_{10} = \frac{1}{\sqrt{(\alpha^4+20\alpha^2\beta^2+10\beta^4)}}\{a^2(p^2_{\frac{3}{2}})_{10} + 2\sqrt{5}\alpha\beta(p_{\frac{3}{2}}p_{\frac{1}{2}})_{10} + \sqrt{10}\beta^2(p^2_{\frac{1}{2}})_{10}\}. \tag{7.28b}$$

In writing these equations we have denoted the two-particle wave functions by $(j_1j_2)_{IK}$ and the isospin quantum numbers, which are always $T = T_z = 0$, have been suppressed.

By use of the generating procedure in equation (7.18) one obtains precisely the wave functions of equation (7.28) except that K is replaced by M. In Table 7.4 we list the coefficients of the two generated 1^+ states

TABLE 7.4

The yrast 1^+ wave function of ${}^6_3Li_3$ generated from the $K=1$ and $K=0$ intrinsic states of equations (7.25)

K	$\bar{E}$ (MeV)	Configurations			Overlap
		$(p^2_{\frac{3}{2}})_{1M}$	$(p_{\frac{3}{2}}p_{\frac{1}{2}})_{1M}$	$(p^2_{\frac{1}{2}})_{1M}$	
	2	0·952	−0·285	−0·108	0·924
	4	0·888	−0·399	−0·227	0·968
1	6	0·842	−0·448	−0·302	0·978
	∞	0·667	−0·527	−0·527	0·936
	2	0·634	−0·760	0·144	0·924
0	4	0·472	−0·848	0·241	0·835
	6	0·407	−0·866	0·292	0·786
	∞	0·272	−0·861	0·430	0·653
	2	0·904	−0·426	−0·044	0·962
Mixed	4	0·812	−0·578	−0·086	0·993
	6	0·759	−0·642	−0·112	0·997
	∞	0·609	−0·770	−0·192	0·976

The overlap with the shell-model eigenfunction given by equation (7.24) is presented in the last column. The designation 'mixed' gives the wave function obtained from the lowest 1^+ eigenvector of the rotational Hamiltonian (see equation (7.33)).

for various values of $\bar{E}$. (The same 1^+ state is obtained from χ_1 and χ_{-1}.) In addition, in the last column we give the overlap (equation (7.23)) with the shell-model wave function. When either a pure $K=0$ or pure $K=1$ intrinsic state is used the best overlap is 0·978; this is considerably worse than for the $I=\frac{1}{2}^-$, $\frac{3}{2}^-$, and $\frac{7}{2}^-$ states in ${}^7_3Li_4$. However, from inspection of Table 7.4 it is clear that one would do much better if the generated wave function was calculated from a linear combination of the two intrinsic states. For example, the 1^+ state obtained from

$$\chi=\sqrt{0{\cdot}91}\chi_{K=1}-0{\cdot}3\chi_{K=0}$$

has an overlap with the shell-model eigenfunction of 0·999 when $\bar{E}=$ 6 MeV. We shall now examine a possible mechanism for mixing the intrinsic states.

It is well known that many properties of light nuclei can be explained on the basis of either the shell model or the rotational model. In the latter model the energy of the system is the sum of the rotational kinetic energy of the core plus the energy of the extra-core particles (Bohr and Mottelson 1975, Davidson 1968). Thus

$$H=\frac{\hbar^2}{2\mathscr{I}}(\mathrm{R}_x^2+\mathrm{R}_y^2)+\frac{\hbar^2}{2\mathscr{I}_z}\mathrm{R}_z^2+H_{\text{intr}},$$

where the z axis has been chosen to be the axis of symmetry and $\mathscr{I}$ is the moment of inertia of the core whose angular momentum is $\underset{\sim}{\mathrm{R}}$. H_{intr}, which we shall take as the Nilsson Hamiltonian of equation (7.4), describes the motion of the extra core particles in the nonspherical potential well which, on the average, describes their interaction with the core. For this Hamiltonian neither the angular momentum of the core nor that of the particles is a constant of the motion. However, their sum

$$\begin{aligned}\underset{\sim}{\mathrm{I}}&=\underset{\sim}{\mathrm{R}}+\sum_i \underset{\sim}{\mathrm{j}}_i\\&=\underset{\sim}{\mathrm{R}}+\underset{\sim}{\mathrm{J}}\end{aligned}$$

is conserved. Expressed in terms of $\underset{\sim}{\mathrm{I}}$ and $\underset{\sim}{\mathrm{J}}$ the Hamiltonian becomes

$$\begin{aligned}H&=\frac{\hbar^2}{2\mathscr{I}}(\mathrm{I}^2+\mathrm{J}^2-2\underset{\sim}{\mathrm{I}}\cdot\underset{\sim}{\mathrm{J}})+\left(\frac{\hbar^2}{2\mathscr{I}_z}-\frac{\hbar^2}{2\mathscr{I}}\right)(\mathrm{I}_z-\mathrm{J}_z)^2+H_{\text{intr}}\\&=\frac{\hbar^2}{2\mathscr{I}}(\mathrm{I}^2-2\mathrm{I}_z\mathrm{J}_z)+\left(\frac{\hbar^2}{2\mathscr{I}_z}-\frac{\hbar^2}{2\mathscr{I}}\right)(\mathrm{I}_z-\mathrm{J}_z)^2+H_{\text{intr}}+\frac{\hbar^2}{2\mathscr{I}}\mathrm{J}^2+\frac{\hbar^2}{\mathscr{I}}(\mathrm{I}_1\mathrm{J}_{-1}+\mathrm{I}_{-1}\mathrm{J}_1)\end{aligned}\tag{7.29}$$

where the components of $\underset{\sim}{\mathrm{I}}$ and $\underset{\sim}{\mathrm{J}}$ are referred to the body fixed axes of the rotor.

In the usual rotational model, eigenfunctions of the form

$$\psi_{IMK}=\sqrt{(2I+1)}D^I_{MK}(R)\chi_K(x')$$

are used to diagonalize H and when the various intrinsic states $\chi_K(x')$ are well separated in energy the physical states of the system are described by a linear combination of ψ_{IMK} and ψ_{IM-K} only. However, when several intrinsic states with different K have approximately the same energy one must diagonalize H using all the ψ_{IMK} which have comparable energy. With this in mind we investigate the hypothesis that the shell-model eigenfunction ψ_{IM} should be generated from the state that diagonalizes the rotational Hamiltonian. That is

$$\psi_{IM}=N_I\sqrt{(2I+1)}\int \mathrm{d}R\sum_K \alpha_{IK}\psi_{IMK}\tag{7.30}$$

where N_I is a normalization constant and the α_{IK} are determined from the diagonalization of the Hamiltonian given by equation (7.29).

Let us now examine the results obtained for ${}^6_3\mathrm{Li}_3$ by use of this prescription. In this case the three degenerate states are

$$\psi_{IM1} = \sqrt{(2I+1)}D^I_{M1}(R)\chi_1(x')$$
$$\psi_{IM0} = \sqrt{(2I+1)}D^I_{M0}(R)\chi_0(x')$$
$$\psi_{IM-1} = \sqrt{(2I+1)}D^I_{M-1}(R)\chi_{-1}(x')$$

where χ_1, χ_0, and χ_{-1} are given by equations (7.27). Since $I_z = J_z = K$ it follows that the diagonal matrix elements of H have the form

$$\langle\psi_{IMK}|\, H\, |\psi_{IMK}\rangle = 2\tilde{\varepsilon}_{\frac{1}{2}}(1) + \frac{\hbar^2}{2\mathscr{I}}\{I(I+1) - 2K^2 + \sum_J J(J+1)\,|C_J(K, \bar{E})|^2\} \tag{7.31a}$$

where $\tilde{\varepsilon}_{\frac{1}{2}}(1)$ is the Nilsson energy (given in Table 7.1) of the first $\mu = \frac{1}{2}^-$ orbit in the $0p$ shell and the $C_J(K, \bar{E})$ are the angular-momentum decomposition coefficients given by equation (7.15). For the generators used in the ${}^6_3\mathrm{Li}_3$ calculation these may be obtained directly from equations (7.27). Since the K value for each of the intrinsic states is different only the Coriolis interaction $(\mathrm{I}_1\mathrm{J}_{-1} + \mathrm{I}_{-1}\mathrm{J}_1)$ contributes to the off-diagonal matrix elements of H. From equations (A7.30) it follows that

$$\begin{aligned}\langle\psi_{IMK'}|\, H\, |\psi_{IMK}\rangle = -\frac{\hbar^2}{2\mathscr{I}}\sum_{J\gamma} (&\sqrt{\{(I+K+1)(I-K)(J+K+1)(J-K)\}}\\ &\times \tilde{C}_J(K, \gamma, \bar{E})\tilde{C}_J(K', \gamma, \bar{E})\delta_{K',K+1}\\ &+\sqrt{\{(I-K+1)(I+K)(J-K+1)(J+K)\}}\\ &\times \tilde{C}_J(K, \gamma, \bar{E})\tilde{C}_J(K', \gamma, \bar{E})\delta_{K',K-1})\end{aligned} \tag{7.31b}$$

where $\tilde{C}_J(K, \gamma, \bar{\varepsilon})$ is defined in equation (7.15). From equations (7.27) and (7.28) it is clear that χ_1 and χ_{-1} have, except for the sign of $C_2(K, \bar{E})$ and $\tilde{C}_2(K, \gamma, \bar{E})$, the same angular-momentum decompositions, i.e.

$$C_J(1, \bar{E}) = (-1)^{1-J}C_J(-1, \bar{E})$$
$$\tilde{C}_J(1, \gamma, \bar{E}) = (-1)^{1-J}\tilde{C}_J(-1, \gamma, \bar{E}). \tag{7.32}$$

When this result is combined with equations (7.31b) one sees that

$$\langle\psi_{IM+}|\, H\, |\psi_{IM-}\rangle = \langle\psi_{IM-}|\, H\, |\psi_{IM0}\rangle \equiv 0$$

where

$$\psi_{IM\pm} = \sqrt{\left(\frac{2I+1}{2}\right)}\left\{D^I_{M1}(R)\chi_1(x') \pm D^I_{M-1}(R)\chi_{-1}(x')\right\}.$$

Thus the 3×3 Hamiltonian matrix partitions. Moreover, from equation (7.32) it follows that no 1^+ state can be generated from ψ_{IM-}. Consequently only ψ_{IM0} and ψ_{IM+} need be considered, and the Hamiltonian matrix obtained with these two basis states has the form

$$\langle H\rangle=\begin{bmatrix} 2\bar{\varepsilon}_{\frac{1}{2}}(1)+\dfrac{\hbar^2}{2\mathscr{I}}\{I(I+1)+11\alpha^4 & \dfrac{\hbar^2}{\mathscr{I}}\sqrt{\{I(I+1)\}}(2\alpha^4+\alpha^2\beta^2-\beta^4)\\ \quad +4\alpha^2\beta^2+2\beta^4\} & \\ \dfrac{\hbar^2}{\mathscr{I}}\sqrt{\{I(I+1)\}}(2\alpha^4+\alpha^2\beta^2-\beta^4) & 2\bar{\varepsilon}_{\frac{1}{2}}(1)+\dfrac{\hbar^2}{2\mathscr{I}}\{I(I+1)-2\\ & \quad +8\alpha^4+10\alpha^2\beta^2+2\beta^4\}\end{bmatrix} \tag{7.33}$$

For $\bar{E}\to\infty$, $\alpha=\sqrt{\frac{2}{3}}$ and $\beta=-1/\sqrt{3}$. Thus for $I=1$ the lower eigenvalue of H is

$$\varepsilon=-\tfrac{4}{3}\bar{E}+\frac{2\hbar^2}{\mathscr{I}}$$

and the corresponding eigenvector is

$$\sum_K \alpha_{1K}\psi_{1MK}=3^{-\frac{1}{2}}(-\psi_{IM0}+\sqrt{2}\psi_{IM+})$$
$$=-D^1_{M0}(R)\chi_0(x')+D^1_{M1}(R)\chi_1(x')+D^1_{M-1}(R)\chi_{-1}(x').$$

Thus from equation (7.30) one sees that the generated 1^+ wave function has the form

$$\psi_{1M}=\frac{1}{3\sqrt{3}}\{\sqrt{10}(p^2_{\frac{3}{2}})_{1M}-4(p_{\frac{3}{2}}p_{\frac{1}{2}})_{1M}-(p^2_{\frac{1}{2}})_{1M}\}$$

which has an overlap of 0·976 with the shell-model eigenfunction of equation (7.24). In Table 7.4 we give the generated wave function for other values of $\bar{E}$. Clearly for $\bar{E}\cong 6$ MeV this wave function is an excellent approximation to the shell-model eigenfunction. *Thus whenever two or more intrinsic states lie close in energy the wave function used to approximate the shell-model eigenfunction should be generated from the combination of states that diagonalizes the rotational Hamiltonian of equation* (7.29).

2.3. Intruder states.

In Chapter 1, section 2.1 we discussed the structure of ${}^{18}_{8}O_{10}$ based on the assumption that ${}^{16}_{8}O_8$ was an inert core and that the two valence neutrons were restricted to the $(0d, 1s)$ shell. Experimentally one observes one more state of each of the spins 0^+, 2^+, and 4^+ than can be accounted for on the basis of this model. In section 2.2 of Chapter 1 we showed that these extra states are to be expected if one considers excitation out of the $0p$ shell (i.e. if one considers states whose structure is

$$[(0p)^{-2}_{J_p}\times(sd)^4_{J_d}]_{IM'}).$$

Such low-lying states are often called intruder states and must be taken into account in a shell-model calculation if one is to explain the experimental data.

In order to treat these intruder states there are two ways one may proceed:

(a) The Hamiltonian matrix describing the nucleus may be written in the form

$$H = H_p + H_{sd} + H'$$

where H_p operates only within the $0p$ shell, H_{sd} has non-vanishing matrix elements only between particles in the $(0d, 1s)$ shell, and H' describes the interaction between the two shells. As discussed in Chapter 2, section 3, one may first diagonalize H_p and H_{sd} separately. If one denotes the p-shell eigenfunctions by $\Phi_{J_pM_p\alpha}$ and the four-particle (sd) wave functions by $\Phi_{J_dM_d\beta}$, one may then diagonalize the total Hamiltonian using the basis states

$$\psi_{IM\gamma} = (0s^4_{\frac{1}{2}})(0p^8_{\frac{3}{2}})(0p^4_{\frac{1}{2}})\Phi_{IM\gamma}\{(sd)^2\}$$

and

$$\psi_{IM} = [\Phi_{J_p\alpha} \times \Phi_{J_d\beta}]_{IM}$$

where the $\psi_{IM\gamma}\{(sd)^2\}$ are wave functions describing the motion of two $(0d, 1s)$ shell particles outside a ${}^{16}_{8}O_8$ core. Of course, if one uses all possible ψ_{IM} in the diagonalization the problem becomes enormous. As a consequence one takes only a few of these, the choice being dictated by energy considerations. This way of examining the properties of ${}^{18}_{8}O_{10}$ has been studied in detail by Ellis and Engeland (1970) and Engeland and Ellis (1972).

(b) In the low-lying spectrum of ${}^{18}_{8}O_{10}$ there seems to be evidence for only one intruder 0^+ level and one extra model-space 2^+ state. Therefore, instead of explicitly carrying out the diagonalization described above, one may assumc that a suitable description of the core-excited states of these spins can be obtained by the generating procedure

$$\psi_{IM} = (2I+1)N_I \int D^I_{MK=0}(R)\chi_{K=0}(x')\, dR \tag{7.34}$$

where $\chi_{K=0}(x')$ is the intrinsic state shown in Fig. 7.3 and N_I is a normalization constant chosen so that

$$\langle \psi_{IM} \mid \psi_{IM} \rangle = 1.$$

According to equation (7.13) the structure of the $(ds)^4$ states is well

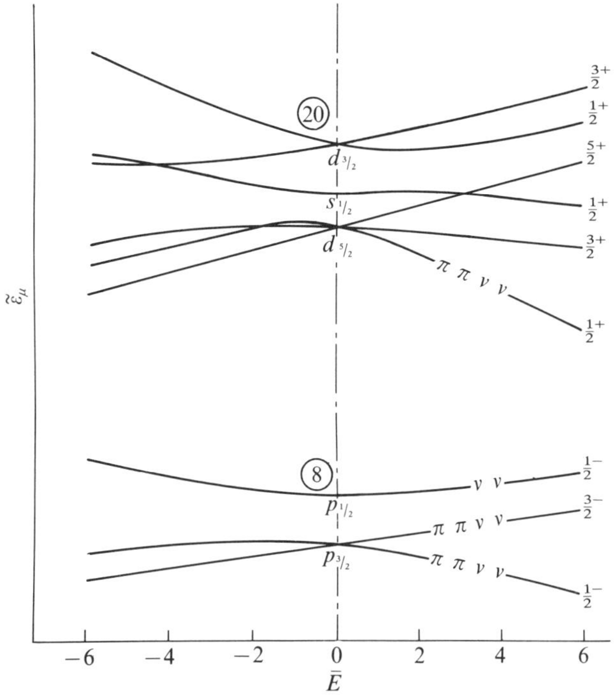

Fig. 7.3. The intrinsic state to be used in generating the intruder states in $^{18}_{8}O_{10}$.

reproduced by the Nilsson model when $\bar{E} \cong 4$ MeV, and one might therefore choose this value to determine the generated wave functions.

By this procedure it is possible to pick out one particular ψ_{IM}, and consequently the number of basis states to be used in the diagonalization of H is increased by only one for each I. For $^{18}_{8}O_{10}$ this scheme was originally used by Brown (1964) and Engeland (1965) and more recently has been exploited to find not only the wave functions that best reproduce all the static and dynamic properties of the low-lying states but also to find the Hamiltonian that gives rise to these wave functions (Lawson *et al.* 1976).

This procedure for constructing the intruder states has also been used near $N = Z = 20$ by Flowers and Skouras (1969) who studied the properties of $^{42}_{20}Ca_{22}$ and $^{42}_{21}Sc_{21}$ and by Johnstone (1968) in a study of the low-lying positive-parity states in $^{43}_{21}Sc_{22}$.

2.4. Beta decay involving the odd-A calcium isotopes

The beta decays connecting the odd-A calcium and Sc isotopes show the interesting trend

$$ {}^{43}_{21}\mathrm{Sc}_{22}(\tfrac{7}{2}^-) \to {}^{43}_{20}\mathrm{Ca}_{23}(\tfrac{7}{2}^-) \quad \log ft = 5{\cdot}05 \quad \text{(Endt and Van der Leun 1973)} $$

$$ {}^{45}_{20}\mathrm{Ca}_{25}(\tfrac{7}{2}^-) \to {}^{45}_{21}\mathrm{Sc}_{24}(\tfrac{7}{2}^-) \quad \log ft = 6{\cdot}0 \quad \text{(Lewis 1970)} $$

$$ {}^{47}_{20}\mathrm{Ca}_{27}(\tfrac{7}{2}^-) \to {}^{47}_{21}\mathrm{Sc}_{26}(\tfrac{7}{2}^-) \quad \log ft = 8{\cdot}5 \quad \text{(Lewis 1970a).} $$

All of these transitions have $\Delta I = 0$, $\Delta T = 1$ and consequently might be expected to proceed at about the same rate. However, the $A = 45$ decay is a factor of ten slower than that for $A = 43$, and the $A = 47$ beta transition is inhibited by about a factor of 3000 compared to the $A = 43$ rate.

If one assumes seniority-one eigenfunctions of $(0f_{\frac{7}{2}})^n$ are applicable for the states involved in the decays, one can write down a simple expression (Grayson and Nordheim 1956) for B_σ (equation (6.67)), the reduced matrix element governing the decay,

$$ B_\sigma = \frac{(4j+4-n)^2-(2T)^2}{16j(j+1)(2T)^2}(T+T_z)(T+T_z-1) \tag{7.35} $$

where T and T_z refer to the isospin of the calcium nuclei. When this value for B_σ is inserted in equation (6.66) one finds that $\log ft = 3{\cdot}83$, $3{\cdot}93$, and $4{\cdot}20$ for the $A = 43$, 45, and 47 decays, respectively. Thus pure seniority-one $f_{\frac{7}{2}}$ eigenfunctions fail to explain the large variations. However, we shall now show that this can be simply understood by using $f_{\frac{7}{2}}$ wave functions that arise when the nucleus has a quadrupole deformation.

To see this we consider the $f_{\frac{7}{2}}$ eigenfunctions predicted by the Nilsson model. We assume the $f_{\frac{7}{2}}$ level is well isolated from all others and because of this the energy matrix of equation (7.6) is only (1×1) with

$$ \tilde{\varepsilon}_\mu = \varepsilon_{\frac{7}{2}} + (\tfrac{15}{7})^{\frac{1}{2}}(\tfrac{7}{2}2\mu 0 \mid \tfrac{7}{2}\mu)\bar{E} \tag{7.36a} $$

and

$$ \chi_\mu = \phi_{\frac{7}{2}\mu} \tag{7.36b} $$

for all values of $\bar{E}$. By filling the Nilsson orbits within this truncated space and then using the generating procedure given in equation (7.18) one obtains states of mixed seniority.

The magnitude of the beta-decay matrix element does not depend on whether we consider the transition Ca → Sc or Sc → Ca. Since it is convenient to consider the latter we have illustrated in Fig. 7.4 the intrinsic states to be used in conjunction with equation (7.18) to generate wave functions for the low-lying levels in ${}^{43}_{21}\mathrm{Sc}_{23}$, ${}^{45}_{21}\mathrm{Sc}_{24}$, and ${}^{47}_{21}\mathrm{Sc}_{26}$. The Gamow-Teller operator that induces the transition is given by equation

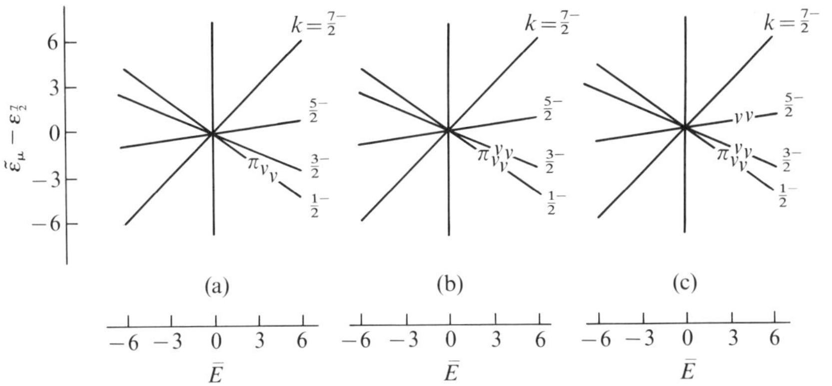

Fig. 7.4. The Nilsson diagram for the pure $0f_{\frac{7}{2}}$ shell. The energies are linear functions of $\bar{E}$ given by equation (7.36a) and the wave functions are those of equation (7.36b). Diagram (a) shows the intrinsic state from which the yrast eigenfunctions of ${}^{43}_{21}Sc_{22}$ should be generated and (b) and (c) those appropriate for ${}^{45}_{21}Sc_{24}$ and ${}^{47}_{21}Sc_{26}$.

(6.64). Consequently

$$\Big\{\sum_i (\tau_i)_+(\sigma_i)_\mu\Big\}\psi_{IM}(x) = (2I+1)N_I \sum_i (\tau_i)_+(\sigma_i)_\mu \int D^I_{MK}(R)\chi_K(x')\,\mathrm{d}R$$

$$= (2I+1)N_I \sum_\nu \int D^1_{\mu\nu}(R)D^I_{MK}(R)\Big\{\sum_i (\tau_i)_+(\sigma'_i)_\nu \chi_K(x')\Big\}\,\mathrm{d}R \quad (7.37)$$

where N_I is the normalization constant the value of which we shall presently calculate. In writing the second line of this equation use has been made of the inverse of equation (7.17), namely

$$\Phi_{JM}(x) = \sum_{M'} D^J_{MM'}(R)\Phi_{JM'}(x'), \quad (7.38)$$

to express the $\underset{\sim}{\sigma}$ operator in the rotated (primed) coordinate system. Once this transformation has been made, $\tau_+\sigma'_\nu$ can be allowed to operate prior to carrying out the integration over dR. From equation (A2.19) it follows that the operator $\tau_+\sigma_\nu$ within the $f_{\frac{7}{2}}$ model space has the form

$$\sum_i (\tau_i)_+(\sigma_i)_\nu = \sum_{mm'} \langle\phi_{\frac{7}{2}}\|\,\sigma\,\|\phi_{\frac{7}{2}}\rangle(\tfrac{7}{2}1m\nu\mid\tfrac{7}{2}m')a^\dagger_{m'}b_m$$

$$= \frac{3}{\sqrt{7}}\sum_{mm'}(\tfrac{7}{2}1m\nu\mid\tfrac{7}{2}m')a^\dagger_{m'}b_m.$$

In the above equation we have suppressed the subscript $\frac{7}{2}$ on the creation and destruction operators and retained only the m values. Throughout the remainder of this section we shall replace $a^\dagger_{\frac{7}{2}m}$ by $a^\dagger_m$. Thus if one assumes positive deformation for ${}^{43}_{21}\mathrm{Sc}_{22}$ one finds

$$\sum_{i\nu} (\tau_i)_+(\sigma_i)_\nu \chi_{K=\frac{1}{2}} = \frac{3}{\sqrt{7}} \sum_{mm'\nu} (\tfrac{7}{2}1m\nu \mid \tfrac{7}{2}m')a^\dagger_{m'}b_m b^\dagger_{\frac{1}{2}} a^\dagger_{\frac{1}{2}} a^\dagger_{-\frac{1}{2}} |0\rangle$$

$$= \frac{3}{\sqrt{7}} (\tfrac{7}{2}1\tfrac{1}{2}1 \mid \tfrac{7}{2}\tfrac{3}{2}) a^\dagger_{\frac{3}{2}} a^\dagger_{\frac{1}{2}} a^\dagger_{-\frac{1}{2}} |0\rangle. \tag{7.39}$$

The reason that only the $\nu = 1$ term can contribute has its origin in the exclusion principle—the neutron orbits with $\mu = +\frac{1}{2}$ and $\mu = -\frac{1}{2}$ are both occupied and consequently when a $\mu = \frac{1}{2}$ proton is changed to a neutron it must be promoted to the $\mu = \frac{3}{2}$ state.

This result immediately tells us why the ${}^{45}_{20}\mathrm{Ca}_{25} \rightarrow {}^{45}_{21}\mathrm{Sc}_{24}$ and ${}^{47}_{20}\mathrm{Ca}_{27} \rightarrow {}^{47}_{21}\mathrm{Sc}_{26}$ transitions are inhibited. In the former the $\mu = \pm\frac{3}{2}$ orbits are occupied and consequently only an operator that can yield a neutron in a state with $|\mu| \geq \frac{5}{2}$ can have a non-vanishing matrix element. Since $\underline{\sigma}$ is a vector operator it can change μ by at most one unit and hence the operation demanded by equation (7.37) yields a vanishing result. Thus on the basis of the pure $f_{\frac{7}{2}}$ model the ${}^{45}_{20}\mathrm{Ca}_{25} \rightarrow {}^{45}_{21}\mathrm{Sc}_{24}$ beta decay has one degree of μ forbiddenness and consequently would be expected to be inhibited compared to the ${}^{43}_{21}\mathrm{Sc}_{22} \rightarrow {}^{43}_{20}\mathrm{Ca}_{23}$ decay. For $A = 47$ the $\mu = \pm\frac{3}{2}$ and $\pm\frac{5}{2}$ neutron orbitals are occupied so that the transition has two degrees of forbiddeness and should be even more inhibited than the $A = 45$ decay (Lawson 1961). Thus the observed trend in the beta decays can be understood without further calculation and the inhibition is due to the fact that the nucleus likes to assume a quadrupole deformation.

We now turn to the explicit evaluation of log ft for the ${}^{43}_{21}\mathrm{Sc}_{22} \rightarrow {}^{43}_{20}\mathrm{Ca}_{23}$ transition. From equations (7.37) and (7.39) it follows that

$$\sum_i (\tau_i)_+(\sigma_i)_\mu \psi_{IM}(x) = \frac{3}{\sqrt{7}} (\tfrac{7}{2}1\tfrac{1}{2}1 \mid \tfrac{7}{2}\tfrac{3}{2})(2I+1)N_I \int \mathrm{d}R D^1_{\mu 1}(R) D^I_{M\frac{1}{2}}(R) \chi_{\frac{3}{2}}(x')$$

$$= \frac{3}{\sqrt{7}} (\tfrac{7}{2}1\tfrac{1}{2}1 \mid \tfrac{7}{2}\tfrac{3}{2})(2I+1)N_I \left\{ \sum_{I''M''} (I1M\mu \mid I''M'')(I1\tfrac{1}{2}1 \mid I''\tfrac{3}{2}) \right.$$

$$\left. \times \int \mathrm{d}R D^{I''}_{M''\frac{3}{2}}(R) \chi_{\frac{3}{2}}(x') \right\} \tag{7.40}$$

where use has been made of equation (A7.13) to express the product of two D functions as one, and in terms of creation operators the generator $\chi_{\frac{3}{2}}(x')$ is

$$\chi_{\frac{3}{2}} = a^\dagger_{\frac{3}{2}} a^\dagger_{\frac{1}{2}} a^\dagger_{-\frac{1}{2}} |0\rangle. \tag{7.41}$$

For three neutrons the lowest intrinsic state with positive deformation is $\chi_{\frac{3}{2}}$ given by equation (7.41). Consequently the low-lying states of ${}^{40}_{20}\mathrm{Ca}_{23}$

should be obtained from this generator

$$\psi_{I'M'}(x)=(2I'+1)N_{I'}\int \mathrm{d}RD^{I'}_{M'\frac{3}{2}}(R)\chi_{\frac{3}{2}}(x'). \qquad (7.42)$$

If one calculates the matrix element $\langle\psi_{I'M'}(x)|\sum_i(\tau_i)_+(\sigma_i)_\mu|\psi_{IM}(x)\rangle$, the term with $I''=I'$, $M''=M'$ is picked out of the sum in equation (7.40) and one finds

$$\langle\psi_{I'M'}(x)|\sum_i(\tau_i)_+(\sigma_i)_\mu|\psi_{IM}(x)\rangle=\left\{\frac{3}{\sqrt{7}}\left(\frac{N_I}{N_{I'}}\right)\left(\frac{2I+1}{2I'+1}\right)\right.$$
$$\left.\times(\tfrac{7}{2}1\tfrac{1}{2}1\,|\tfrac{7}{2}\tfrac{3}{2})(I1\tfrac{1}{2}1\,|\,I'\tfrac{3}{2})\right\}(I1M\mu\,|\,I'M'). \qquad (7.43)$$

Thus the calculation of the matrix element governing the beta decay involves calculating only the normalization constants N_I and $N_{I'}$ associated with the initial and final states.

To calculate the normalization constant N_I for the initial state we note that the intrinsic state used to generate the $^{43}_{21}Sc_{22}$ wave function is

$$\chi_{K=\frac{1}{2}}=b^\dagger_{\frac{1}{2}}a^\dagger_{\frac{1}{2}}a^\dagger_{-\frac{1}{2}}|0\rangle. \qquad (7.44)$$

Thus since

$$a^\dagger_{\frac{1}{2}}a^\dagger_{-\frac{1}{2}}|0\rangle=\sqrt{2}\sum_J(\tfrac{7}{2}\tfrac{7}{2}\tfrac{1}{2}-\tfrac{1}{2}\,|\,J0)A^\dagger_{J0}(\tfrac{7}{2}\tfrac{7}{2})|0\rangle$$

where $A^\dagger_{J0}(\frac{7}{2}\frac{7}{2})$ is the creation operator for a pair of $f_{\frac{7}{2}}$ neutrons coupled to even J and $M=0$, it follows that the wave function with angular momentum I generated from this intrinsic state is

$$\psi_{IM}(x)=\sqrt{2}(2I+1)N_I\sum_J(\tfrac{7}{2}\tfrac{7}{2}\tfrac{1}{2}-\tfrac{1}{2}\,|\,J0)\int \mathrm{d}RD^I_{M\frac{1}{2}}(R)\phi_{\frac{7}{2}\frac{1}{2}}(\bar{x}')\,\Phi_{J0}(\bar{x}';\tfrac{7}{2}^2)$$
$$=\sqrt{2}(2I+1)N_I\sum_{JM'm}(\tfrac{7}{2}\tfrac{7}{2}\tfrac{1}{2}-\tfrac{1}{2}\,|\,J0)\int \mathrm{d}RD^I_{M\frac{1}{2}}(R)D^{\frac{7}{2}*}_{m\frac{1}{2}}(R)$$
$$\times D^{J*}_{M'0}(R)\phi_{\frac{7}{2}m}(\bar{x})\Phi_{JM'}(\bar{x};\tfrac{7}{2}^2)$$
$$=\sqrt{2}N_I\sum_J(\tfrac{7}{2}\tfrac{7}{2}\tfrac{1}{2}-\tfrac{1}{2}\,|\,J0)(\tfrac{7}{2}J\tfrac{1}{2}0\,|\,I\tfrac{1}{2})[\phi_{\frac{7}{2}}(\bar{x})\times\Phi_J(\bar{x};\tfrac{7}{2}^2)]_{IM}. \qquad (7.45)$$

For $\psi_{IM}(x)$ to be normalized

$$2(N_I)^2\sum_J(\tfrac{7}{2}\tfrac{7}{2}\tfrac{1}{2}-\tfrac{1}{2}\,|\,J0)^2(\tfrac{7}{2}J\tfrac{1}{2}0\,|\,I\tfrac{1}{2})^2=1 \qquad (7.46)$$

so that when $I=\frac{7}{2}$

$$N_{I=\frac{7}{2}}=(\tfrac{91}{33})^{\frac{1}{2}}. \qquad (7.47)$$

To find the normalization coefficient associated with the ${}^{43}_{20}\mathrm{Ca}_{23}$ ground state one must know the angular-momentum decomposition of the three-particle generator given by equation (7.41). To find this decomposition we note that a three particle state (JM) can be written as

$$\psi_{JM}(x)=\sum_k \beta_{Jk}\chi_M(x;k) \tag{7.48}$$

where $\chi_M(x;k)$ are three particle Slater determinants with z-component of angular momentum M and k is an index that labels the various possible determinants. The coefficients β_{Jk} can be obtained by the methods outlined in Chapter 1, section (1.2). This equation can be inverted to find the angular momentum decomposition of $\chi_M(x;k)$ and since the transformation is unitary it follows that

$$\chi_M(x,k)=\sum_J \beta_{Jk}\psi_{JM}(x). \tag{7.49}$$

If one projects from equation (7.49) the state with angular momentum I it follows that

$$N_I=1/\beta_{Ik}.$$

For the $(\nu f_{\frac{7}{2}})^3$ configuration there is only one state with $I=\frac{7}{2}$ and this is the seniority-one level the wave function of which is given by equation (1.125):

$$\psi_{\frac{7}{2}\frac{3}{2}}=\frac{1}{\sqrt{3}}\{a^\dagger_{\frac{3}{2}}a^\dagger_{\frac{7}{2}}a^\dagger_{-\frac{7}{2}}-a^\dagger_{\frac{3}{2}}a^\dagger_{\frac{5}{2}}a^\dagger_{-\frac{5}{2}}-a^\dagger_{\frac{3}{2}}a^\dagger_{\frac{1}{2}}a^\dagger_{-\frac{1}{2}}\}.$$

The last term in this equation corresponds to the generator given by equation (7.41) and consequently when $I'=\frac{7}{2}$

$$N_{I'=\frac{7}{2}}=-\sqrt{3}. \tag{7.50}$$

When the values given by equations (7.47) and (7.50) for N_I and $N_{I'}$ are used in equation (7.43) one finds that

$$\langle\psi_{\frac{7}{2}}({}^{43}_{20}\mathrm{Ca}_{23})\|\sum_i(\tau_i)_+\sigma_i\|\psi_{\frac{7}{2}}({}^{43}_{21}\mathrm{Sc}_{22})\rangle=-\frac{10\sqrt{13}}{21\sqrt{11}}$$

where the values of the Clebsch–Gordan coefficients have been obtained from Table A2.2. From equations (6.66) and (6.67), which relate the reduced matrix element to the ft value, one finds that

$$\log ft({}^{43}_{21}\mathrm{Sc}_{22}\rightarrow{}^{43}_{20}\mathrm{Ca}_{23})=4{\cdot}16.$$

Thus, as already seen in Chapter 6, use of unquenched values of the $\underline{\sigma}$ operator lead to beta-decay lifetimes much shorter than observed experimentally. If one wishes to obtain the experimental lifetime for the ${}^{43}_{21}\mathrm{Sc}_{22}$

decay using generated wave functions one must take

$$\langle \phi_{\frac{7}{2}} \| \sigma \| \phi_{\frac{7}{2}} \rangle = 0{\cdot}408,$$

in other words 36% of the single-particle estimate, $3/\sqrt{7}$. This value is somewhat smaller than that found by Alburger and Wilkinson (1973) in their study of the ${}^{41}_{21}\mathrm{Sc}_{20} \rightarrow {}^{41}_{20}\mathrm{Ca}_{21}$ decay which required $\langle \phi_{\frac{7}{2}} \| \sigma \| \phi_{\frac{7}{2}} \rangle = 0{\cdot}863$ to give agreement with experiment. Therefore, as might be expected, the mixing of neglected configurations, which as discussed in Chapter 6 leads to a smaller $\underset{\sim}{\sigma}$ matrix element, is larger when there are three particles outside the closed shell than it is when there is only one.

APPENDIX 1

CLEBSCH–GORDAN COEFFICIENTS

In this Appendix we consider some properties of the angular-momentum operators that can be used to find the explicit form of the single-particle j–j coupled wave functions. We then discuss the role of the Clebsch–Gordan coefficients in coupling two angular momenta to a resultant total spin.

1. Angular-momentum operators

The components j_x, j_y, and j_z of the single-particle angular-momentum operator $\underline{\mathrm{j}}$ satisfy the commutation relationships

$$\begin{aligned}[\mathrm{j}_x, \mathrm{j}_y] &= \mathrm{j}_x \mathrm{j}_y - \mathrm{j}_y \mathrm{j}_x \\ &= i\mathrm{j}_z \\ [\mathrm{j}_y, \mathrm{j}_z] &= \mathrm{j}_y \mathrm{j}_z - \mathrm{j}_z \mathrm{j}_y \\ &= i\mathrm{j}_x \\ [\mathrm{j}_z, \mathrm{j}_x] &= \mathrm{j}_z \mathrm{j}_x - \mathrm{j}_x \mathrm{j}_z \\ &= i\mathrm{j}_y\end{aligned}$$

or in vector notation

$$\underline{\mathrm{j}} \times \underline{\mathrm{j}} = i\underline{\mathrm{j}} \tag{A1.1}$$

where $\times$ stands for the cross-product encountered in ordinary vector analysis.

ϕ_{jm}, the single-particle angular-momentum eigenfunction, has the property that

$$\underline{\mathrm{j}}^2 \phi_{jm} = j(j+1)\phi_{jm} \tag{A1.2a}$$

$$\mathrm{j}_z \phi_{jm} = m\phi_{jm} \tag{A1.2b}$$

where $j(j+1)$ is the square of the angular momentum measured in units of $\hbar$ and m is the z component measured in the same units. Furthermore, since

$$\mathrm{j}_+ = \mathrm{j}_x + i\mathrm{j}_y \tag{A1.3a}$$

$$\mathrm{j}_- = \mathrm{j}_x - i\mathrm{j}_y \tag{A1.3b}$$

commute with j^2 they do not change the angular momentum of the state but merely change the m value. The effect of these operators on ϕ_{jm} may be deduced by use of the commutation relationship given in equation (A1.1)

$$\begin{aligned}\mathrm{j}_z(\mathrm{j}_\pm \phi_{jm}) &= m(\mathrm{j}_\pm \phi_{jm}) + [\mathrm{j}_z, \mathrm{j}_\pm]\phi_{jm} \\ &= (m \pm 1)(\mathrm{j}_\pm \phi_{jm}).\end{aligned}$$

Consequently j_+ and j_- are referred to as the step-up and step-down operators, respectively.

From the above discussion it is clear that

$$j_+\phi_{jm} = \alpha_+\phi_{jm+1}$$

where α_+ is a numerical factor that can be evaluated by taking the absolute square of the equation, integrating over all space, and using the fact that ϕ_{jm} is normalized. Thus

$$\begin{aligned}|\alpha_+|^2 &= \langle\phi_{jm}|\, j_-j_+ \,|\phi_{jm}\rangle \\ &= \langle\phi_{jm}|\,(j^2 - j_z^2 - j_z)\,|\phi_{jm}\rangle \\ &= (j-m)(j+m+1).\end{aligned}$$

A similar calculation can be carried out to determine α_- defined by

$$j_-\phi_{jm} = \alpha_-\phi_{jm-1},$$

and one finds

$$|\alpha_-|^2 = (j+m)(j-m+1).$$

We shall follow the Condon and Shortley (1951) phase convention and take $\alpha_\pm$ to be real and positive so that

$$j_\pm\phi_{jm} = \surd\{(j\mp m)(j\pm m+1)\}\phi_{jm\pm1}. \tag{A1.4}$$

As we shall show in Appendix 2 it is convenient to define components of $\underline{j}$ that satisfy the commutation relationships

$$[j_z, j_\mu] = \mu j_\mu \qquad [j_\pm, j_\mu] = \surd\{(1\mp\mu)(2\pm\mu)\}j_{\mu\pm1}.$$

Thus j_μ, the spherical components of the operator $\underline{j}$, have the form

$$\begin{aligned} j_{+1} &= -\frac{1}{\surd 2}(j_x + ij_y) \\ &= -\frac{1}{\surd 2}j_+ \end{aligned} \tag{A1.5a}$$

$$j_0 = j_z \tag{A1.5b}$$

$$\begin{aligned} j_{-1} &= \frac{1}{\surd 2}(j_x - ij_y) \\ &= \frac{1}{\surd 2}j_-. \end{aligned} \tag{A1.5c}$$

Finally, in a system consisting of several particles the operator

$$J_\mu = \sum_i (j_i)_\mu \tag{A1.6}$$

is the μth component of the total angular momentum if $(j_i)_\mu$ is the μth component of the angular momentum of particle number i.

2. Single-particle eigenfunctions

Because the spin-orbit interaction in nuclei is strong it is convenient to work in the j–j coupling representation. Since the orbital angular-momentum operator $\underset{\sim}{\ell}$ obeys the commutation rule given in equation (A1.1), it is clear that if $V(r)$ and $f(r)$ are functions of the scalar variable r, $\underset{\sim}{\ell}^2$ commutes with the Hamiltonian

$$H_0 = \frac{p^2}{2m} + V(r) + f(r)\underset{\sim}{\sigma} \cdot \underset{\sim}{\ell}$$

where $\underset{\sim}{\sigma}$ stands for the Pauli spin operators

$$\sigma_x = \begin{pmatrix} 0 & 1 \\ 1 & 0 \end{pmatrix}$$
$$\sigma_y = \begin{pmatrix} 0 & -i \\ i & 0 \end{pmatrix} \tag{A1.7}$$
$$\sigma_z = \begin{pmatrix} 1 & 0 \\ 0 & -1 \end{pmatrix}.$$

Thus the angular part of the single-particle eigenfunction is the spherical harmonic

$$Y_{lm}(\theta, \phi) = \frac{(-1)^{l+m}}{(2l)!!} \left\{\frac{(2l+1)(l-m)!}{4\pi(l+m)!}\right\}^{\frac{1}{2}} \times (\sin\theta)^m e^{im\phi} \frac{d^{l+m}(\sin\theta)^{2l}}{\{d(\cos\theta)\}^{l+m}}. \tag{A1.8}$$

and these functions satisfy equations (A1.2) with $\underset{\sim}{j}$ replaced by $\underset{\sim}{\ell}$.

In terms of the Pauli matrices, the nucleon spin operator satisfying equation (A1.1) is

$$s_\mu = \tfrac{1}{2}\sigma_\mu. \tag{A1.9}$$

The eigenvectors of the spin operators s^2 and s_z which describe a state with angular momentum $\frac{1}{2}$ and z component $+\frac{1}{2}$ or $-\frac{1}{2}$ are

$$\chi_{\frac{1}{2}} = \begin{pmatrix} 1 \\ 0 \end{pmatrix}$$
$$\chi_{-\frac{1}{2}} = \begin{pmatrix} 0 \\ 1 \end{pmatrix}. \tag{A1.10}$$

Since the nucleon has both an orbital and a spin angular momentum its total $\underset{\sim}{j}$ is the sum of the two

$$j_\mu = \ell_\mu + s_\mu.$$

Clearly since H_0 is a scalar, j^2 commutes with H_0 and so does j_z. However, although both ℓ^2 and s^2 commute with the Hamiltonian, ℓ_z and s_z do not. Consequently the eigenfunctions of H_0 have definite values for l, $s=\frac{1}{2}$, j, and m (the z component of the total angular momentum) but are not eigenfunctions of

ℓ_z and s_z. The possible values of j are $j = l + \frac{1}{2}$ or $j = l - \frac{1}{2}$, and we now deduce the appropriate linear combinations of $Y_{lm-\frac{1}{2}}\chi_{\frac{1}{2}}$ and $Y_{lm+\frac{1}{2}}\chi_{-\frac{1}{2}}$ (both of which are eigenfunctions of j_z with eigenvalue m) needed to describe the two possible j states for given l. To do this we write

$$\phi_{jm} = R_j(r)(\alpha_m Y_{lm-\frac{1}{2}}\chi_{\frac{1}{2}} + \beta_m Y_{lm+\frac{1}{2}}\chi_{-\frac{1}{2}})$$

where $R_j(r)$ is the radial wave function associated with a particle in the state j. From equation (A1.4) it follows that

$$\begin{aligned}
j_+\phi_{jm} &= \sqrt{\{(j-m)(j+m+1)\}}\phi_{jm+1} \\
&= \sqrt{\{(j-m)(j+m+1)\}}R_j(r) \\
&\quad \times(\alpha_{m+1} Y_{lm+\frac{1}{2}}\chi_{\frac{1}{2}} + \beta_{m+1} Y_{lm+\frac{3}{2}}\chi_{-\frac{1}{2}}) \\
&= R_j(r)([\sqrt{\{(l-m+\tfrac{1}{2})(l+m+\tfrac{1}{2})\}}\alpha_m + \beta_m]Y_{lm+\frac{1}{2}}\chi_{\frac{1}{2}} \\
&\quad + \sqrt{\{(l-m-\tfrac{1}{2})(l+m+\tfrac{3}{2})\}}\beta_m Y_{lm+\frac{3}{2}}\chi_{-\frac{1}{2}})
\end{aligned}$$

where we have made use of the fact that $\underline{j}$ does not affect the radial wave function $R_j(r)$.

If ϕ_{jm} is to be an eigenfunction of j^2 and j_z then the coefficients of $Y_{lm+\frac{1}{2}}\chi_{\frac{1}{2}}$ and $Y_{lm+\frac{3}{2}}\chi_{-\frac{1}{2}}$ must be the same for the two different ways of writing $j_+\phi_{jm}$. By equating coefficients of $Y_{lm+\frac{3}{2}}\chi_{-\frac{1}{2}}$ we see that when $j = l + \frac{1}{2}$

$$\frac{\beta_{m+1}}{\beta_m} = \left(\frac{l-m-\frac{1}{2}}{l-m+\frac{1}{2}}\right)^{\frac{1}{2}}$$

so that

$$\beta_m = A\sqrt{(l-m+\tfrac{1}{2})}$$

where A is an arbitrary constant independent of m. Similarly, equality of the coefficients of $Y_{lm+\frac{1}{2}}\chi_{\frac{1}{2}}$ leads to

$$\alpha_{m+1}\sqrt{(l+m+\tfrac{3}{2})} - \alpha_m\sqrt{(l+m+\tfrac{1}{2})} = A$$

which is satisfied if

$$\alpha_m = A\sqrt{(l+m+\tfrac{1}{2})}.$$

Thus the normalized $j = l + \frac{1}{2}$ wave function with z component of angular momentum m is

$$\phi_{(j=l+\frac{1}{2})m} = R_j(r)\left\{\left(\frac{l+m+\frac{1}{2}}{2l+1}\right)^{\frac{1}{2}} Y_{lm-\frac{1}{2}}\chi_{\frac{1}{2}} + \left(\frac{l-m+\frac{1}{2}}{2l+1}\right)^{\frac{1}{2}} Y_{lm+\frac{1}{2}}\chi_{-\frac{1}{2}}\right\} \quad \text{(A1.11a)}$$

An analogous calculation can be carried out for $j = l - \frac{1}{2}$ and yields

$$\phi_{(j=l-\frac{1}{2})m} = R_j(r)\left\{-\left(\frac{l-m+\frac{1}{2}}{2l+1}\right)^{\frac{1}{2}} Y_{lm-\frac{1}{2}}\chi_{\frac{1}{2}} + \left(\frac{l+m+\frac{1}{2}}{2l+1}\right)^{\frac{1}{2}} Y_{lm+\frac{1}{2}}\chi_{-\frac{1}{2}}\right\} \quad \text{(A1.11b)}$$

The coefficients in the wave functions generated by this procedure are known as Clebsch–Gordan coefficients and some of their properties will now be discussed.

3. Clebsch–Gordan coefficients

In analogy with equations (A1.11) it follows that in order to obtain a two-particle state with definite angular momentum (IM) when nucleons occupy the single-particle orbits j_1 and j_2, one must take a linear combination of the various ϕ_{jm}. The coefficients $(j_1j_2m_1m_2 \mid IM)$ in the expansion

$$\psi_{IM}(1,2) = \sum_{m_1m_2} (j_1j_2m_1m_2 \mid IM)\phi_{j_1m_1}(1)\phi_{j_2m_2}(2) \tag{A1.12}$$

are called Clebsch–Gordan coefficients and are chosen so that

$$\mathrm{J}^2\psi_{IM}(1,2) = I(I+1)\psi_{IM}(1,2)$$
$$\mathrm{J}_z\psi_{IM}(1,2) = M\psi_{IM}(1,2)$$

with the total angular-momentum operator defined by equation (A1.6)

$$\mathrm{J}_\mu = (\mathrm{j}_1)_\mu + (\mathrm{j}_2)_\mu.$$

Since the total angular-momentum operator is the sum of the operators for the individual nucleons, these coefficients vanish unless

$$m_1 + m_2 = M$$

and

$$|j_1 - j_2| \leq I \leq j_1 + j_2.$$

The physical interpretation of the Clebsch–Gordan coefficients follows immediately from the fact that the ϕ_{jm} are orthonormal. Thus

$$(j_1j_2m_1m_2 \mid IM) = \langle \phi_{j_2m_2}(2)\phi_{j_1m_1}(1) \mid \psi_{IM}(1,2)\rangle$$

so that $(j_1j_2m_1m_2 \mid IM)^2$ is the probability that in the state $\psi_{IM}(1,2)$ particle number 1 will be in the state (j_1m_1) and particle number 2 in (j_2m_2).

Since both ϕ_{jm} and ψ_{IM} are orthonormal states

$$\sum_{m_1m_2} (j_1j_2m_1m_2 \mid IM)(j_1j_2m_1m_2 \mid I'M') = \delta_{II'}\delta_{MM'}.$$

Moreover, because the ψ_{IM} are a complete set of states,

$$\begin{aligned}\sum_{IM} (j_1j_2m_1m_2 \mid IM)(j_1j_2m_1'm_2' \mid IM) &= \sum_{IM} \langle \phi_{j_2m_2}(2)\phi_{j_1m_1}(1) \mid \psi_{IM}(1,2)\rangle \\ &\quad \times \langle \psi_{IM}(1,2) \mid \phi_{j_1m_1'}(1)\phi_{j_2m_2'}(2)\rangle \\ &= \langle \phi_{j_1m_1}(1) \mid \phi_{j_1m_1'}(1)\rangle\langle \phi_{j_2m_2}(2) \mid \phi_{j_2m_2'}(2)\rangle\end{aligned}$$

so that

$$\sum_{IM} (j_1j_2m_1m_2 \mid IM)(j_1j_2m_1'm_2' \mid IM) = \delta_{m_1m_1'}\delta_{m_2m_2'}.$$

As a consequence of this last equation, equation (A1.12) may be turned inside-out to yield

$$\phi_{j_1m_1}(1)\phi_{j_2m_2}(2) = \sum_{IM} (j_1j_2m_1m_2 \mid IM)\psi_{IM}(1,2). \tag{A1.13}$$

The Clebsch–Gordan coefficients defined by equations (A1.12) and (A1.13) are related to the Wigner $3j$ coefficient (Wigner 1959) by the equation

$$(j_1j_2m_1m_2 \mid j_3m_3) = (-1)^{j_1-j_2+m_3}\sqrt{(2j_3+1)}\begin{pmatrix} j_1 & j_2 & j_3 \\ m_1 & m_2 & -m_3 \end{pmatrix}. \qquad \text{(A1.14)}$$

An extensive numerical tabulation of the Wigner $3j$ coefficient has been given by Rotenberg *et al.* (1959).

Racah (1942) has shown that the Clebsch–Gordan coefficient can be written in the form of a finite series

$$(j_1j_2m_1m_2 \mid j_3m_3) = \left\{\frac{(2j_3+1)(j_1+j_2-j_3)!(j_1-j_2+j_3)!(-j_1+j_2+j_3)!}{(j_1+j_2+j_3+1)!}\right\}^{\frac{1}{2}}$$
$$\times\sqrt{\{(j_1+m_1)!(j_1-m_1)!(j_2+m_2)!(j_2-m_2)!(j_3+m_3)!(j_3-m_3)!\}}$$
$$\times\sum_k \frac{(-1)^k}{k!(j_1+j_2-j_3-k)!(j_1-m_1-k)!(j_2+m_2-k)!(j_3-j_2+m_1+k)!(j_3-j_1-m_2+k)!} \qquad \text{(A1.15)}$$

where

$$k! = k(k-1)(k-2)\ldots(1); \qquad 0! = 1$$

and the factorial of a negative number is meaningless. From this equation three fundamental symmetry properties of the coefficients can be deduced:

(a) If one writes the analogue of equation (A1.15) for the coefficient $(j_2j_1m_2m_1 \mid j_3m_3)$ one sees that the only difference between this coefficient and $(j_1j_2m_1m_2 \mid j_3m_3)$ is in the summation over k. If one replaces k by

$$k = j_1+j_2-j_3-z$$

then the summation over z is identical to that in equation (A1.15) except for multiplication by the factor $(-1)^{j_1+j_2-j_3}$. Thus

$$(j_1j_2m_1m_2 \mid j_3m_3) = (-1)^{j_1+j_2-j_3}(j_2j_1m_2m_1 \mid j_3m_3). \qquad \text{(A1.16a)}$$

(b) A similar calculation leads to the conclusion that

$$(j_1j_2m_1m_2 \mid j_3m_3) = (-1)^{j_1+j_2-j_3}(j_1j_2-m_1-m_2 \mid j_3-m_3). \qquad \text{(A1.16b)}$$

(c) The Clebsch–Gordan coefficient $(j_1j_3m_1-m_3 \mid j_2-m_2)$ can be shown to be related to the coefficient $(j_1j_2m_1m_2 \mid j_3m_3)$ if one replaces k in the summation by

$$k = j_1-m_1-z$$

and makes use of the fact that

$$m_1+m_2 = m_3$$

When this is done one finds that

$$(j_1j_2m_1m_2 \mid j_3m_3) = (-1)^{j_1-m_1}\left(\frac{2j_3+1}{2j_2+1}\right)^{\frac{1}{2}}(j_1j_3m_1-m_3 \mid j_2-m_2). \qquad \text{(A1.16c)}$$

For the case that $j_2=\frac{1}{2}$, 1, $\frac{3}{2}$, and 2 the summation over k can be carried out simply and in section 5 of this Appendix we give the results for these values of j_2 (Condon and Shortley 1951).

From Table A1.1, given in Section 5 of this Appendix, it follows that the single-particle wave functions of equation (A1.11) are

$$\phi_{jm}=R_j(r)\sum_{k\mu}(l\tfrac{1}{2}k\mu\mid jm)Y_{lk}\chi_\mu. \tag{A1.17}$$

Consequently *the coupling scheme used in writing these single-particle wave functions is l plus s coupled to j.*

By use of the two-particle generalization of the step-up operator in equation (A1.3a) some useful relationships between the Clebsch–Gordan coefficients can be derived. If this operator is applied to equation (A1.13) one finds that

$$\begin{aligned}\sqrt{\{(j_1-m_1)(j_1+m_1+1)\}}\phi_{j_1m_1+1}\phi_{j_2m_2}+\sqrt{\{(j_2-m_2)(j_2+m_2+1)\}}\phi_{j_1m_1}\phi_{j_2m_2+1}\\=\sum_{IM}\sqrt{\{(I-M)(I+M+1)\}}(j_1j_2m_1m_2\mid IM)\psi_{IM+1}.\end{aligned}$$

(For notational convenience we have dropped the particle number in this expression.) By writing $\phi_{j_1m_1+1}\phi_{j_2m_2}$ and $\phi_{j_1m_1}\phi_{j_2m_2+1}$ in terms of $\psi_{I'M+1}$ and equating coefficients on both sides of the equation one finds

$$\begin{aligned}\sqrt{\{(I-M)(I+M+1)\}}&(j_1j_2m_1m_2\mid IM)\\&=\sqrt{\{(j_1-m_1)(j_1+m_1+1)\}}(j_1j_2m_1+1m_2\mid IM+1)\\&\quad+\sqrt{\{(j_2-m_2)(j_2+m_2+1)\}}(j_1j_2m_1m_2+1\mid IM+1).\end{aligned}$$

Alternatively, if one applies the step-up operator to equation (A1.12) a second recursion relationship is obtained by equating coefficients of $\phi_{j_1m_1}\phi_{j_2m_2}$:

$$\begin{aligned}\sqrt{\{(I-M)(I+M+1)\}}&(j_1j_2m_1m_2\mid IM+1)\\&=\sqrt{\{(j_1+m_1)(j_1-m_1+1)\}}(j_1j_2m_1-1m_2\mid IM)\\&\quad+\sqrt{\{(j_2+m_2)(j_2-m_2+1)\}}(j_1j_2m_1m_2-1\mid IM).\end{aligned}$$

When use is made of equation (A1.16b) one sees that in the special case $m_1=m_2=\frac{1}{2}$ the above equation becomes

$$(j_1j_2\tfrac{1}{2}\tfrac{1}{2}\mid I1)=(-1)^{j_1+j_2-I}\frac{(j_1j_2\frac{1}{2}-\frac{1}{2}\mid I0)\eta_I(j_1j_2)}{\sqrt{\{I(I+1)\}}}$$

where

$$\eta_I(j_1j_2)=\tfrac{1}{2}\{(2j_1+1)+(-1)^{j_1+j_2-I}(2j_2+1)\}.$$

4. Useful formulae involving Clebsch–Gordan coefficients

An extensive numerical tabulation of the Wigner $3j$ coefficient in equation (A1.14) has been given by Rotenberg *et al.* (1959). In addition to numerical values, it is often useful to know some analytic properties of these coefficients and these are listed below.

$$(j_1j_2m_1m_2 \mid IM) = (-1)^{j_1+j_2-I}(j_2j_1m_2m_1 \mid IM) \tag{A1.18}$$

$$(j_1j_2m_1m_2 \mid IM) = (-1)^{j_1+j_2-I}(j_1j_2 - m_1 - m_2 \mid I - M) \tag{A1.19}$$

$$(j_1j_2m_1m_2 \mid IM) = (-1)^{j_1-m_1}\left(\frac{2I+1}{2j_2+1}\right)^{\frac{1}{2}}(j_1Im_1 - M \mid j_2 - m_2) \tag{A1.20}$$

$$(jjm - m \mid 00) = (-1)^{j-m}(2j+1)^{-\frac{1}{2}} \tag{A1.21}$$

$$(j_1j_2\tfrac{1}{2}\tfrac{1}{2} \mid I1) = \frac{(j_1j_2\tfrac{1}{2} - \tfrac{1}{2} \mid I0)}{2\sqrt{\{I(I+1)\}}}\{(2j_2+1) + (-1)^{j_1+j_2-I}(2j_1+1)\} \tag{A1.22}$$

$$\sum_{m_1m_2}(j_1j_2m_1m_2 \mid IM)(j_1j_2m_1m_2 \mid I'M') = \delta_{II'}\delta_{MM'} \tag{A1.23}$$

$$\sum_{IM}(j_1j_2m_1m_2 \mid IM)(j_1j_2m_1'm_2' \mid IM) = \delta_{m_1m_1'}\delta_{m_2m_2'} \tag{A1.24}$$

$$\begin{aligned}\sqrt{\{(j_1-m_1)(j_1+m_1+1)\}}&(j_1j_2m_1+1m_2 \mid IM+1)\\ &+\sqrt{\{(j_2-m_2)(j_2+m_2+1)\}}(j_1j_2m_1m_2+1 \mid IM+1)\\ &=\sqrt{\{(I-M)(I+M+1)\}}(j_1j_2m_1m_2 \mid IM)\end{aligned} \tag{A1.25}$$

$$\begin{aligned}\sqrt{\{(j_1+m_1)(j_1-m_1+1)\}}&(j_1j_2m_1-1m_2 \mid IM)\\ &+\sqrt{\{(j_2+m_2)(j_2-m_2+1)\}}(j_1j_2m_1m_2-1 \mid IM)\\ &=\sqrt{\{(I-M)(I+M+1)\}}(j_1j_2m_1m_2 \mid IM+1)\end{aligned} \tag{A1.26}$$

$$\begin{aligned}(l_1l_200 \mid L0) &= \left\{\frac{1+(-1)^J}{2}\right\}(-1)^{\frac{1}{2}(L-l_1-l_2)}\left\{\frac{(2L+1)(J-2l_1)!(J-2l_2)!(J-2L)!}{(J+1)!}\right\}^{\frac{1}{2}}\\ &\times\frac{(\frac{1}{2}J)!}{(\frac{1}{2}J-l_1)!(\frac{1}{2}J-l_2)!(\frac{1}{2}J-L)!} \quad \text{where} \quad (J = l_1+l_2+L)\end{aligned} \tag{A1.27}$$

$$(j_1j_2\tfrac{1}{2} - \tfrac{1}{2} \mid L0) = -\frac{1}{2}\left\{\frac{(K+2)(K-2L+1)}{(j_1+\frac{1}{2})(j_2+\frac{1}{2})}\right\}^{\frac{1}{2}}(j_1+\tfrac{1}{2}j_2+\tfrac{1}{2}00 \mid L0)$$

$$\text{for} \quad K = j_1+j_2+L \quad \text{odd} \tag{A1.28}$$

5. Analytic expressions for Clebsch–Gordan coefficients

In this Section analytic expressions for the Clebsch–Gordan coefficients are given when one angular momentum is $\frac{1}{2}$, 1, $\frac{3}{2}$, and 2 (Condon and Shortley 1951). Similar expressions when $j_2 = \frac{5}{2}$ can be found in the literature (Melvin and Swamy 1957).

TABLE A1.1
$(j\tfrac{1}{2}mm' \mid IM)$

I	$m'=+\frac{1}{2}$	$m'=-\frac{1}{2}$
$j+\frac{1}{2}$	$\left(\frac{j+M+\frac{1}{2}}{2j+1}\right)^{\frac{1}{2}}$	$\left(\frac{j-M+\frac{1}{2}}{2j+1}\right)^{\frac{1}{2}}$
$j-\frac{1}{2}$	$-\left(\frac{j-M+\frac{1}{2}}{2j+1}\right)^{\frac{1}{2}}$	$\left(\frac{j+M+\frac{1}{2}}{2j+1}\right)^{\frac{1}{2}}$

TABLE A1.2
$(j1mm' \mid IM)$

I	$m'=+1$	$m'=0$	$m'=-1$
$j+1$	$\left\{\frac{(j+M)(j+M+1)}{(2j+1)(2j+2)}\right\}^{\frac{1}{2}}$	$\left\{\frac{(j-M+1)(j+M+1)}{(2j+1)(j+1)}\right\}^{\frac{1}{2}}$	$\left\{\frac{(j-M)(j-M+1)}{(2j+1)(2j+2)}\right\}^{\frac{1}{2}}$
j	$-\left\{\frac{(j+M)(j-M+1)}{2j(j+1)}\right\}^{\frac{1}{2}}$	$\frac{M}{\sqrt{\{j(j+1)\}}}$	$\left\{\frac{(j-M)(j+M+1)}{2j(j+1)}\right\}^{\frac{1}{2}}$
$j-1$	$\left\{\frac{(j-M)(j-M+1)}{2j(2j+1)}\right\}^{\frac{1}{2}}$	$-\left\{\frac{(j-M)(j+M)}{j(2j+1)}\right\}^{\frac{1}{2}}$	$\left\{\frac{(j+M+1)(j+M)}{2j(2j+1)}\right\}^{\frac{1}{2}}$

TABLE A1.3
$(j\tfrac{3}{2}mm' \mid IM)$

I	$m'=\frac{3}{2}$	$m'=\frac{1}{2}$
$j+\frac{3}{2}$	$\left\{\frac{(j+M-\frac{1}{2})(j+M+\frac{1}{2})(j+M+\frac{3}{2})}{(2j+1)(2j+2)(2j+3)}\right\}^{\frac{1}{2}}$	$\left\{\frac{3(j+M+\frac{1}{2})(j+M+\frac{3}{2})(j-M+\frac{3}{2})}{(2j+1)(2j+2)(2j+3)}\right\}^{\frac{1}{2}}$
$j+\frac{1}{2}$	$-\left\{\frac{3(j+M-\frac{1}{2})(j+M+\frac{1}{2})(j-M+\frac{3}{2})}{2j(2j+1)(2j+3)}\right\}^{\frac{1}{2}}$	$-(j-3M+\frac{3}{2})\left\{\frac{j+M+\frac{1}{2}}{2j(2j+1)(2j+3)}\right\}^{\frac{1}{2}}$
$j-\frac{1}{2}$	$\left\{\frac{3(j+M-\frac{1}{2})(j-M+\frac{1}{2})(j-M+\frac{3}{2})}{(2j-1)(2j+1)(2j+2)}\right\}^{\frac{1}{2}}$	$-(j+3M-\frac{1}{2})\left\{\frac{j-M+\frac{1}{2}}{(2j-1)(2j+1)(2j+2)}\right\}^{\frac{1}{2}}$
$j-\frac{3}{2}$	$-\left\{\frac{(j-M-\frac{1}{2})(j-M+\frac{1}{2})(j-M+\frac{3}{2})}{2j(2j-1)(2j+1)}\right\}^{\frac{1}{2}}$	$\left\{\frac{3(j+M-\frac{1}{2})(j-M-\frac{1}{2})(j-M+\frac{1}{2})}{2j(2j-1)(2j+1)}\right\}^{\frac{1}{2}}$
I	$m'=-\frac{1}{2}$	$m'=-\frac{3}{2}$
$j+\frac{3}{2}$	$\left\{\frac{3(j+M+\frac{3}{2})(j-M+\frac{1}{2})(j-M+\frac{3}{2})}{(2j+1)(2j+2)(2j+3)}\right\}^{\frac{1}{2}}$	$\left\{\frac{(j-M-\frac{1}{2})(j-M+\frac{1}{2})(j-M+\frac{3}{2})}{(2j+1)(2j+2)(2j+3)}\right\}^{\frac{1}{2}}$
$j+\frac{1}{2}$	$(j+3M+\frac{3}{2})\left\{\frac{j-M+\frac{1}{2}}{2j(2j+1)(2j+3)}\right\}^{\frac{1}{2}}$	$\left\{\frac{3(j+M+\frac{3}{2})(j-M-\frac{1}{2})(j-M+\frac{1}{2})}{2j(2j+1)(2j+3)}\right\}^{\frac{1}{2}}$
$j-\frac{1}{2}$	$-(j-3M-\frac{1}{2})\left\{\frac{j+M+\frac{1}{2}}{(2j-1)(2j+1)(2j+2)}\right\}^{\frac{1}{2}}$	$\left\{\frac{3(j+M+\frac{1}{2})(j+M+\frac{3}{2})(j-M-\frac{1}{2})}{(2j-1)(2j+1)(2j+2)}\right\}^{\frac{1}{2}}$
$j-\frac{3}{2}$	$-\left\{\frac{3(j+M-\frac{1}{2})(j+M+\frac{1}{2})(j-M-\frac{1}{2})}{2j(2j-1)(2j+1)}\right\}^{\frac{1}{2}}$	$\left\{\frac{(j+M-\frac{1}{2})(j+M+\frac{1}{2})(j+M+\frac{3}{2})}{2j(2j-1)(2j+1)}\right\}^{\frac{1}{2}}$

TABLE A1.4
$(j2mm' \mid IM)$

I	$m'=2$	$m'=1$	$m'=0$
$j+2$	$\left\{\frac{(j+M-1)(j+M)(j+M+1)(j+M+2)}{(2j+1)(2j+2)(2j+3)(2j+4)}\right\}^{\frac{1}{2}}$	$\left\{\frac{(j-M+2)(j+M+2)(j+M+1)(j+M)}{(2j+1)(j+1)(2j+3)(j+2)}\right\}^{\frac{1}{2}}$	$\left\{\frac{3(j-M+2)(j-M+1)(j+M+2)(j+M+1)}{(2j+1)(2j+2)(2j+3)(j+2)}\right\}^{\frac{1}{2}}$
$j+1$	$-\left\{\frac{(j+M-1)(j+M)(j+M+1)(j-M+2)}{2j(j+1)(j+2)(2j+1)}\right\}^{\frac{1}{2}}$	$-(j-2M+2)\left\{\frac{(j+M+1)(j+M)}{2j(2j+1)(j+1)(j+2)}\right\}^{\frac{1}{2}}$	$M\left\{\frac{3(j-M+1)(j+M+1)}{j(2j+1)(j+1)(j+2)}\right\}^{\frac{1}{2}}$
j	$\left\{\frac{3(j+M-1)(j+M)(j-M+1)(j-M+2)}{(2j-1)2j(j+1)(2j+3)}\right\}^{\frac{1}{2}}$	$(1-2M)\left\{\frac{3(j-M+1)(j+M)}{(2j-1)(2j+2)(2j+3)j}\right\}^{\frac{1}{2}}$	$\frac{3M^2-j(j+1)}{\{(2j-1)j(j+1)(2j+3)\}^{\frac{1}{2}}}$
$j-1$	$-\left\{\frac{(j+M-1)(j-M)(j-M+1)(j-M+2)}{2(j-1)(j+1)(2j+1)j}\right\}^{\frac{1}{2}}$	$(j+2M-1)\left\{\frac{(j-M+1)(j-M)}{(j-1)j(2j+1)(2j+2)}\right\}^{\frac{1}{2}}$	$-M\left\{\frac{3(j-M)(j+M)}{(j-1)j(2j+1)(j+1)}\right\}^{\frac{1}{2}}$
$j-2$	$\left\{\frac{(j-M-1)(j-M)(j-M+1)(j-M+2)}{(2j-2)(2j-1)2j(2j+1)}\right\}^{\frac{1}{2}}$	$-\left\{\frac{(j-M+1)(j-M)(j-M-1)(j+M-1)}{(j-1)(2j-1)j(2j+1)}\right\}^{\frac{1}{2}}$	$\left\{\frac{3(j-M)(j-M-1)(j+M)(j+M-1)}{(2j-2)(2j-1)j(2j+1)}\right\}^{\frac{1}{2}}$

I	$m'=-1$	$m'=-2$
$j+2$	$\left\{\frac{(j-M+2)(j-M+1)(j-M)(j+M+2)}{(2j+1)(j+1)(2j+3)(j+2)}\right\}^{\frac{1}{2}}$	$\left\{\frac{(j-M-1)(j-M)(j-M+1)(j-M+2)}{(2j+1)(2j+2)(2j+3)(2j+4)}\right\}^{\frac{1}{2}}$
$j+1$	$(j+2M+2)\left\{\frac{(j-M+1)(j-M)}{j(2j+1)(2j+2)(j+2)}\right\}^{\frac{1}{2}}$	$\left\{\frac{(j-M-1)(j-M)(j-M+1)(j+M+2)}{j(2j+1)(j+1)(2j+4)}\right\}^{\frac{1}{2}}$
j	$(2M+1)\left\{\frac{3(j-M)(j+M+1)}{(2j-1)j(2j+2)(2j+3)}\right\}^{\frac{1}{2}}$	$\left\{\frac{3(j-M-1)(j-M)(j+M+1)(j+M+2)}{(2j-1)j(2j+2)(2j+3)}\right\}^{\frac{1}{2}}$
$j-1$	$-(j-2M-1)\left\{\frac{(j+M+1)(j+M)}{(j-1)j(2j+1)(2j+2)}\right\}^{\frac{1}{2}}$	$\left\{\frac{(j-M-1)(j+M)(j+M+1)(j+M+2)}{(j-1)j(2j+1)(2j+2)}\right\}^{\frac{1}{2}}$
$j-2$	$-\left\{\frac{(j-M-1)(j+M+1)(j+M)(j+M-1)}{(j-1)(2j-1)j(2j+1)}\right\}^{\frac{1}{2}}$	$\left\{\frac{(j+M-1)(j+M)(j+M+1)(j+M+2)}{(2j-2)(2j-1)2j(2j+1)}\right\}^{\frac{1}{2}}$

APPENDIX 2

REDUCED MATRIX ELEMENTS

In this Appendix we evaluate the matrix elements of the most commonly encountered single-particle operators. To do this we first define what is meant by an irreducible tensor operator and then show that the dependence of matrix elements of these operators on the z components of angular momentum are given by a Clebsch–Gordan coefficient. This leads to a great simplification in that once a matrix element has been evaluated for one set of m values its value for all others is known immediately.

Throughout this discussion we shall often exploit the fact that the spherical harmonic $Y_{lm}(\theta, \phi)$ given by equation (A1.8) has a particularly simple form when $\theta = 0°$, namely

$$Y_{lm}(\theta = 0°, \phi) = \left(\frac{2l+1}{4\pi}\right)^{\frac{1}{2}} \delta_{m,0} \tag{A2.1}$$

1. The Wigner–Eckart theorem

An irreducible tensor operator of rank λ is an operator with $(2\lambda + 1)$ components $Q_{\lambda\mu}$ that satisfy the commutation relationships

$$[\mathrm{J}_z, Q_{\lambda\mu}] = \mu Q_{\lambda\mu} \tag{A2.2}$$

$$[\mathrm{J}_{\pm}, Q_{\lambda\mu}] = \surd\{(\lambda \mp \mu)(\lambda \pm \mu + 1)\} Q_{\lambda\mu\pm 1}. \tag{A2.3}$$

Clearly the angular-momentum wave functions ψ_{IM} discussed in Appendix 1 are irreducible tensors as are the spherical harmonics. Furthermore, the product of two irreducible tensor operators

$$Q_{\lambda_3\mu_3} = \sum_{\mu_1\mu_2} (\lambda_1\lambda_2\mu_1\mu_2 \mid \lambda_3\mu_3) Q_{\lambda_1\mu_1} Q_{\lambda_2\mu_2} \tag{A2.4}$$

can easily be shown to be a tensor operator of rank λ_3. Since $\mu_1 + \mu_2 = \mu_3$ it is evident that $Q_{\lambda_3\mu_3}$ satisfies equation (A2.2). To see that equation (A2.3) is obeyed we note that $Q_{\lambda_1\mu_1}$ and $Q_{\lambda_2\mu_2}$ each satisfy equation (A2.3) so that

$$\begin{aligned}
[\mathrm{J}_{\pm}, Q_{\lambda_3\mu_3}] &= \sum_{\mu_1\mu_2} (\lambda_1\lambda_2\mu_1\mu_2 \mid \lambda_3\mu_3)[\surd\{(\lambda_1 \mp \mu_1)(\lambda_1 \pm \mu_1 + 1)\} \\
&\quad \times Q_{\lambda_1\mu_1\pm 1} Q_{\lambda_2\mu_2} + \surd\{(\lambda_2 \mp \mu_2)(\lambda_2 \pm \mu_2 + 1)\} Q_{\lambda_1\mu_1} Q_{\lambda_2\mu_2\pm 1}] \\
&= \sum_{\mu_1\mu_2} [\surd\{(\lambda_1 \pm \mu_1)(\lambda_1 \mp \mu_1 + 1)\}(\lambda_1\lambda_2\mu_1 \mp 1\mu_2 \mid \lambda_3\mu_3) \\
&\quad + \surd\{(\lambda_2 \pm \mu_2)(\lambda_2 \mp \mu_2 + 1)\}(\lambda_1\lambda_2\mu_1\mu_2 \mp 1 \mid \lambda_3\mu_3)] Q_{\lambda_1\mu_1} Q_{\lambda_2\mu_2} \\
&= \surd\{(\lambda_3 \mp \mu_3)(\lambda_3 \pm \mu_3 + 1)\} Q_{\lambda_3\mu_3\pm 1}.
\end{aligned}$$

The last line in this equation is identical to equation (A2.3) and comes about when the recursion relationships between the Clebsch-Gordan coefficients given by equations (A1.25) and (A1.26) are used.

We now consider the evaluation of the matrix element $\langle\psi_{I'M'}|\, Q_{\lambda\mu}\, |\psi_{IM}\rangle$ where ψ_{IM}, $Q_{\lambda\mu}$, and $\psi_{I'M'}$ are irreducible tensor operators of rank I, λ, and I', respectively. From equation (A2.4) it follows that

$$Q_{\lambda\mu}\psi_{IM} = \sum_{I''M''} (I\lambda M\mu \mid I''M'')\Phi_{I''M''}. \tag{A2.5}$$

Since $\Phi_{I''M''}$ is the product of two irreducible tensor operators it follows that it is itself an irreducible tensor operator of rank I''. Because states with different values of I and M are orthogonal, it is clear that

$$\langle\psi_{I'M'}|\, Q_{\lambda\mu}\, |\psi_{IM}\rangle = (I\lambda M\mu \mid I'M')\langle\psi_{I'M'} \mid \Phi_{I'M'}\rangle.$$

One may easily show that $\langle\psi_{I'M'} \mid \Phi_{I'M'}\rangle$ is independent of M'. This follows from equation (A1.4) which tells us that

$$\begin{aligned}\langle\psi_{I'M'+1} \mid \Phi_{I'M'+1}\rangle &= \{(I'-M')(I'+M'+1)\}^{-\frac{1}{2}}\langle \mathrm{J}_{+}\psi_{I'M'} \mid \Phi_{I'M'+1}\rangle \\ &= \{(I'-M')(I'+M'+1)\}^{-\frac{1}{2}}\langle\psi_{I'M'} \mid \mathrm{J}_{-}\Phi_{I'M'+1}\rangle \\ &= \langle\psi_{I'M'} \mid \Phi_{I'M'}\rangle \\ &\equiv \langle\psi_{I'}\|\, Q_{\lambda}\, \|\psi_{I}\rangle.\end{aligned}$$

Thus we arrive at the Wigner–Eckart theorem (Eckart 1930, Wigner 1959) which states that if ψ_{IM}, $Q_{\lambda\mu}$, and $\psi_{I'M'}$ are irreducible tensors of rank I, λ, and I' then

$$\langle\psi_{I'M'}|\, Q_{\lambda\mu}\, |\psi_{IM}\rangle = (I\lambda M\mu \mid I'M')\langle\psi_{I'}\|\, Q_{\lambda}\, \|\psi_{I}\rangle \tag{A2.6}$$

where the reduced or double barred matrix element $\langle\psi_{I'}\|\, Q_{\lambda}\, \|\psi_{I}\rangle$ is independent of the z components of angular momentum. Provided ψ_{IM} form a complete orthonormal set, equation (A2.5) can also be written as

$$Q_{\lambda\mu}\psi_{IM} = \sum_{I'M'} (I\lambda M\mu \mid I'M')\langle\psi_{I'}\|\, Q_{\lambda}\, \|\psi_{I}\rangle\psi_{I'M'}, \tag{A2.7}$$

or if use is made of equation (A1.23) as

$$\langle\psi_{I'}\|\, Q_{\lambda}\, \|\psi_{I}\rangle\psi_{I'M'} = \sum_{M\mu} (I\lambda M\mu \mid I'M')Q_{\lambda\mu}\psi_{IM}. \tag{A2.8}$$

Finally before proceeding with the evaluation of the reduced matrix elements we examine the properties of the Hermitian adjoint $Q^{\dagger}_{\lambda\mu}$ of the irreducible tensor operator $Q_{\lambda\mu}$. The Hermitian adjoints of equations (A2.2) and (A2.3) are

$$[\mathrm{J}_z, Q^{\dagger}_{\lambda\mu}] = -\mu Q^{\dagger}_{\lambda\mu} \qquad [\mathrm{J}_{\pm}, Q^{\dagger}_{\lambda\mu}] = -\sqrt{\{(\lambda \pm \mu)(\lambda \mp \mu + 1)\}}Q^{\dagger}_{\lambda\mu\mp1}.$$

Thus $Q^{\dagger}_{\lambda\mu}$ is not an irreducible tensor operator. However, the modified Hermitian adjoint

$$\tilde{Q}_{\lambda\mu} = (-1)^{\lambda+\mu}Q^{\dagger}_{\lambda-\mu} \tag{A2.9}$$

does satisfy equations (A2.2) and (A2.3) and consequently is an irreducible tensor operator of rank λ.

The relationship between the reduced matrix elements of $Q_{\lambda\mu}$ and $\tilde{Q}_{\lambda\mu}$ is often needed and may be deduced directly by taking the adjoint of both sides of equation (A2.6). Since the Clebsch–Gordan coefficients are real

$$\begin{aligned}\langle\psi_{I'M'}|\, Q_{\lambda\mu}\,|\psi_{IM}\rangle^{\dagger} &= (I\lambda M\mu \mid I'M')\langle\psi_{I'}\|\, Q_{\lambda}\,\|\psi_{I}\rangle^{*}\\ &= \langle\psi_{IM}|\, Q^{\dagger}_{\lambda\mu}\,|\psi_{I'M'}\rangle\\ &= (-1)^{-\lambda+\mu}(I'\lambda M'-\mu \mid IM)\langle\psi_{I}\|\,\tilde{Q}_{\lambda}\,\|\psi_{I'}\rangle\end{aligned}$$

where $\langle\psi_{I'}\|\, Q_{\lambda}\,\|\psi_{I}\rangle^{*}$ is the complex conjugate of the reduced matrix element. When use is made of the symmetry properties of the Clebsch–Gordan coefficients given by equations (A1.18)–(A1.20) it follows that

$$\sqrt{(2I'+1)}\langle\psi_{I'}\|\, Q_{\lambda}\,\|\psi_{I}\rangle^{*} = (-1)^{I+\lambda-I'}\sqrt{(2I+1)}\langle\psi_{I}\|\,\tilde{Q}_{\lambda}\,\|\psi_{I'}\rangle. \qquad \text{(A2.10)}$$

2. $\langle \mathbf{Y}_{l'}\|\, \mathbf{Y}_{\lambda}\, \|\mathbf{Y}_{l}\rangle$

If Y_{lm}, $Y_{\lambda\mu}$, and $Y_{l'm'}$ are associated with the ψ_{IM}, $Q_{\lambda\mu}$, and $\psi_{I'M'}$ of equation (A2.8) it follows that

$$\langle Y_{l'}\|\, Y_{\lambda}\,\|Y_{l}\rangle Y_{l'm'} = \sum_{m\mu}(l\lambda m\mu \mid l'm')Y_{\lambda\mu}Y_{lm}.$$

Since this equation is true for all values of θ and ϕ it must be true for $\theta = 0°$. When use is made of equation (A2.1) one sees that

$$\langle Y_{l'}\|\, Y_{\lambda}\,\|Y_{l}\rangle = \left\{\frac{(2l+1)(2\lambda+1)}{4\pi(2l'+1)}\right\}^{\frac{1}{2}}(l\lambda 00 \mid l'0). \qquad \text{(A2.11)}$$

3. $\langle \boldsymbol{R}_{l'}\boldsymbol{Y}_{l'm}|\, \boldsymbol{r}_{\mu}\, |\boldsymbol{R}_{l}\boldsymbol{Y}_{lm}\rangle$

In this matrix element R_l and $R_{l'}$ are the radial wave functions that multiply Y_{lm} and $Y_{l'm'}$, respectively. The coordinate vector $\underline{r}$ satisfies the commutation relations in equations (A2.2) and (A2.3) for an irreducible tensor operator of rank one provided we define

$$r_0 = z \qquad r_{\pm1} = \mp(x \pm iy)/\sqrt{2}.$$

When this is combined with the definition of the spherical harmonics in equation (A1.8) one sees that

$$r_{\mu} = \left(\frac{4\pi}{3}\right)^{\frac{1}{2}} rY_{1\mu} \qquad \text{(A2.12)}$$

where r is the scalar radial coordinate. Thus from equation (A2.11) we obtain the result

$$\langle R_{l'}Y_{l'm'}|\, r_{\mu}\, |R_{l}Y_{lm}\rangle = \left(\frac{2l+1}{2l'+1}\right)^{\frac{1}{2}}(l100 \mid l'0)\,(l1m\mu \mid l'm')\int R_{l'}rR_{l}r^{2}\,\mathrm{d}r. \qquad \text{(A2.13)}$$

The reduced matrix element is, of course, just this matrix element divided by $(l1m\mu \mid l'm')$.

4. $\langle \boldsymbol{R}_{l'} \boldsymbol{Y}_{l'm'} | \boldsymbol{\nabla}_\mu | \boldsymbol{R}_l \boldsymbol{Y}_{lm} \rangle$

To evaluate this matrix element we make use of the fact that

$$\nabla_\mu = \tfrac{1}{2}[\nabla^2, r_\mu]$$

where

$$\nabla^2 = \frac{\partial^2}{\partial x^2} + \frac{\partial^2}{\partial y^2} + \frac{\partial^2}{\partial z^2}$$

$$= \frac{\partial^2}{\partial r^2} + \frac{2}{r}\frac{\partial}{\partial r} - \frac{\ell^2}{r^2}$$

and ℓ^2 is the orbital angular-momentum operator. From equation (A2.12) it follows that

$$\frac{1}{2}\left[\left(\frac{\partial^2}{\partial r^2} + \frac{2}{r}\frac{\partial}{\partial r}\right), r_\mu\right] = \left(\frac{4\pi}{3}\right)^{\frac{1}{2}} Y_{1\mu}\left(\frac{\partial}{\partial r} + \frac{1}{r}\right).$$

Consequently

$$\langle R_{l'} Y_{l'm'} | \nabla_\mu | R_l Y_{lm} \rangle = \left(\frac{4\pi}{3}\right)^{\frac{1}{2}} \langle R_{l'} Y_{l'm'} | \left(\frac{\partial}{\partial r} + \frac{1}{r}\right) Y_{1\mu} - \frac{\ell^2}{2r} Y_{1\mu} + Y_{1\mu}\frac{\ell^2}{2r} | R_l Y_{lm} \rangle.$$

If we allow ℓ^2 to operate on Y_{lm} when it stands to the right of $Y_{1\mu}$ and on $Y_{l'm'}$ when it stands on the left, we may use equation (A2.13) to write down the result

$$\langle R_{l'} Y_{l'm'} | \nabla_\mu | R_l Y_{lm} \rangle = \left(\frac{2l+1}{2l'+1}\right)^{\frac{1}{2}} (l100 \mid l'0)(l1m\mu \mid l'm')$$

$$\times \int R_{l'} \left\{\frac{\mathrm{d}}{\mathrm{d}r} + \frac{1}{r} + \frac{l(l+1) - l'(l'+1)}{2r}\right\} R_l r^2 \,\mathrm{d}r. \qquad \text{(A2.14)}$$

5. $\langle \boldsymbol{\phi}_{j'} \| \boldsymbol{\sigma} \| \boldsymbol{\phi}_j \rangle$

Since $\underset{\sim}{\sigma}$ is a vector operator it has the properties of an irreducible tensor of rank-one with

$$\sigma_0 = \sigma_z \qquad \sigma_\pm = \mp 2^{-\frac{1}{2}}(\sigma_x \pm i\sigma_y). \qquad \text{(A2.15)}$$

Although $\underset{\sim}{\sigma}$ cannot change the orbital angular momentum of a state it can change the j value and consequently the operator has non-vanishing matrix elements between spin-orbit partners. When $m = l - \frac{1}{2}$ the single-particle wave functions ϕ_{jm} in equation (A1.11) have the form

$$\phi_{j=l+\frac{1}{2}, m=l-\frac{1}{2}} = (2l+1)^{-\frac{1}{2}}\{\sqrt{(2l)}\, Y_{ll-1}\chi_{\frac{1}{2}} + Y_{ll}\chi_{-\frac{1}{2}}\} R_{j=l+\frac{1}{2}}(r) \qquad \text{(A2.16)}$$

$$\phi_{j=l-\frac{1}{2}, m=l-\frac{1}{2}} = (2l+1)^{-\frac{1}{2}}\{-Y_{ll-1}\chi_{\frac{1}{2}} + \sqrt{(2l)}\, Y_{ll}\chi_{-\frac{1}{2}}\} R_{j=l-\frac{1}{2}}(r). \qquad \text{(A2.17)}$$

Since $\sigma_z\chi_{\pm\frac{1}{2}}=\pm\chi_{\pm\frac{1}{2}}$

$$\langle\phi_{j'l-\frac{1}{2}}|\,\sigma_z\,|\phi_{jl-\frac{1}{2}}\rangle=\frac{\delta_{ll'}}{(2l+1)}\left((-1)^{l+\frac{1}{2}-j}(2l-1)\delta_{jj'}-2\sqrt{(2l)}\delta_{j,l\pm\frac{1}{2}}\delta_{j',l\mp\frac{1}{2}}\right)\int R_{j'}R_jr^2\,\mathrm{d}r$$

$$=\langle\phi_{j'}\|\,\sigma\,\|\phi_j\rangle(j1l-\tfrac{1}{2}0\,|\,j'l-\tfrac{1}{2})\int R_{j'}R_jr^2\,\mathrm{d}r.$$

The value of the Clebsch–Gordan coefficient can be obtained from Table A1.2

$$(j1l-\tfrac{1}{2}0\,|\,j'l-\tfrac{1}{2})=\frac{(2l-1)}{2\sqrt{\{j(j+1)\}}}\,\delta_{jj'}+(-1)^{l+\frac{1}{2}-j'}\left\{\frac{2l}{(l+\frac{1}{2})(2j+1)}\right\}^{\frac{1}{2}}\delta_{j,l\pm\frac{1}{2}}\delta_{j',l\mp\frac{1}{2}}. \quad \text{(A2.18)}$$

Consequently

$$\langle\phi_{j'}\|\,\sigma\,\|\phi_j\rangle=\left[(-1)^{l+\frac{1}{2}-j}\frac{2\sqrt{\{j(j+1)\}}}{2l+1}\,\delta_{ll'}\delta_{jj'}\right.$$
$$\left.-(-1)^{l+\frac{1}{2}-j'}\left\{\frac{2(2j+1)}{2l+1}\right\}^{\frac{1}{2}}\delta_{ll'}\delta_{j,l\pm\frac{1}{2}}\delta_{j',l\mp\frac{1}{2}}\right]\int R_{j'}R_jr^2\,\mathrm{d}r. \quad \text{(A2.19)}$$

6. $\langle\boldsymbol{\phi}_{j'}\|\,\boldsymbol{\ell}\,\|\boldsymbol{\phi}_j\rangle$

The same procedure may be used to evaluate the matrix element of the orbital angular momentum operator. Since $\ell_zY_{lm}=mY_{lm}$

$$\langle\phi_{j'l-\frac{1}{2}}|\,\ell_z\,|\phi_{jl-\frac{1}{2}}\rangle=\frac{\delta_{ll'}}{2l+1}\left(l(2l-1)\delta_{jj'}\delta_{j,l+\frac{1}{2}}+(l+1)(2l-1)\delta_{jj'}\delta_{j,l-\frac{1}{2}}+\sqrt{(2l)}\delta_{j,l\pm\frac{1}{2}}\delta_{j',l\mp\frac{1}{2}}\right)$$
$$\times\int R_{j'}R_jr^2\,\mathrm{d}r.$$

By use of equation (A2.18) one finds that

$$\langle\phi_{j'}\|\,\ell\,\|\phi_j\rangle=\delta_{ll'}\left\{\frac{2\sqrt{\{j(j+1)\}}}{(2l+1)}[\delta_{jj'}\{l\delta_{j,l+\frac{1}{2}}+(l+1)\delta_{j,l-\frac{1}{2}}\}]\right.$$
$$\left.+(-1)^{l+\frac{1}{2}-j'}\left\{\frac{(2j+1)}{2(2l+1)}\right\}^{\frac{1}{2}}\delta_{j,l\pm\frac{1}{2}}\delta_{j',l\mp\frac{1}{2}}\right\}\int R_{j'}R_jr^2\,\mathrm{d}r \quad \text{(A2.20)}$$

7. $\langle\boldsymbol{\phi}_{j'}\|\,\mathbf{Y}_{l''}\,\|\boldsymbol{\phi}_j\rangle$

By use of equation (A1.24) for the single-particle wave functions one finds that for $\theta=0°$

$$\phi_{jm}(\theta=0°)=\left\{\frac{(2l+1)}{4\pi}\right\}^{\frac{1}{2}}(l\tfrac{1}{2}0m\,|\,jm)\chi_mR_j(r). \quad \text{(A2.21)}$$

When this result is combined with equation (A2.8) one sees that

$$\sqrt{(2l'+1)}(l'\tfrac{1}{2}0m \mid j'm)\langle\phi_{j'}\| Y_{l''} \|\phi_j\rangle\chi_m R_{j'}(r)$$
$$=\left\{\frac{(2l+1)(2l''+1)}{4\pi}\right\}^{\frac{1}{2}}(jl''m0 \mid j'm)(l\tfrac{1}{2}0m \mid jm)\chi_m R_j(r)$$

where m is either $+\frac{1}{2}$ or $-\frac{1}{2}$.

From Table A1.1 it follows that

$$\sqrt{(2l+1)}(l\tfrac{1}{2}0\tfrac{1}{2} \mid j\tfrac{1}{2})=(-1)^{l+\frac{1}{2}-j}\left(\frac{2j+1}{2}\right)^{\frac{1}{2}}. \tag{A2.22}$$

Since $\int R_{j'}^2 r^2\,\mathrm{d}r=1$

$$\langle\phi_{j'}\| Y_{l''} \|\phi_j\rangle=(-1)^{l+l'+j-j'}\left\{\frac{(2l''+1)(2j+1)}{4\pi(2j'+1)}\right\}^{\frac{1}{2}}(jl''\tfrac{1}{2}0 \mid j'\tfrac{1}{2})\left\{\frac{1+(-1)^{l+l'+l''}}{2}\right\}\int R_{j'}R_j r^2\,\mathrm{d}r. \tag{A2.23}$$

The factor $[1+(-1)^{l+l'+l''}]/2$ arises because the original equation must hold true for both $m=+\frac{1}{2}$ and $-\frac{1}{2}$ and in the above we considered $m=+\frac{1}{2}$ only. If we carry out the calculation for $m=-\frac{1}{2}$ we obtain the same result except for a phase factor $(-1)^{l+l'+l''}$. This tells us that the matrix element vanishes unless $(l+l'+l'')$ is even, that is unless parity is conserved.

8. $\langle\boldsymbol{\phi}_{j'}\| [\boldsymbol{\sigma}\times \mathbf{Y}_{l''}]_L \|\boldsymbol{\phi}_j\rangle$

The notation $[\sigma\times Y_{l''}]_L$ implies the vector coupling of the σ operator and the spherical harmonic $Y_{l''m''}$ to total angular momentum $L=l''$, $l''\pm 1$. When equations (A2.8) and (A2.21) are combined one finds

$$\sqrt{(2l'+1)}(l'\tfrac{1}{2}0m' \mid j'm')\langle\phi_{j'}\| [\sigma\times Y_{l''}]_L \|\phi_j\rangle\chi_{m'}R_{j'}(r)$$
$$=\left\{\frac{(2l''+1)(2l+1)}{4\pi}\right\}^{\frac{1}{2}}\sum_{m\nu}(jLm\nu \mid j'm')(1l''\nu 0 \mid L\nu)(l\tfrac{1}{2}0m \mid jm)\sigma_\nu\chi_m R_j(r)$$

where the equation must hold for $m'=\frac{1}{2}$ or $-\frac{1}{2}$.

When use is made of equation (A2.22) and the symmetry property of the Clebsch–Gordan coefficients given in equation (A1.19), it follows that for $m'=\frac{1}{2}$

$$\langle\phi_{j'}\| [\sigma\times Y_{l''}]_L \|\phi_j\rangle=(-1)^{l+l'+j-j'}\left\{\frac{(2l''+1)(2j+1)}{4\pi(2j'+1)}\right\}^{\frac{1}{2}}$$
$$\times\sum_{m\nu}\langle\chi_{\frac{1}{2}}| \sigma_\nu |\chi_m\rangle(jLm\nu \mid j'\tfrac{1}{2})(1l''\nu 0 \mid L\nu)$$
$$\times\{\delta_{m,\frac{1}{2}}+(-1)^{l+\frac{1}{2}-j}\,\delta_{m,-\frac{1}{2}}\}\int R_{j'}R_j r^2\,\mathrm{d}r.$$

From the definition of the σ operator in equation (A2.15) one easily shows that

$$\langle\chi_{\frac{1}{2}}| \sigma_\nu |\chi_m\rangle=1 \quad \text{if} \quad m=\tfrac{1}{2} \text{ and } \nu=0$$
$$=-\sqrt{2} \quad \text{if} \quad m=-\tfrac{1}{2} \text{ and } \nu=1.$$

If one uses the properties of the Clebsch–Gordan coefficients given by equations (A1.19), (A1.20), and (A1.22) together with Table A1.2, which gives analytic expressions for these coefficients when one angular momentum is unity, one can carry out the $(m\nu)$-sum and finally arrive at the expression

$$\langle \phi_{j'} \| [\sigma \times Y_{l''}]_L \, |\phi_j\rangle = (-1)^{l+l'+j'-\frac{1}{2}} \left\{ \frac{(2j+1)}{4\pi(2L+1)} \right\}^{\frac{1}{2}} (jj'\tfrac{1}{2}-\tfrac{1}{2} \mid L0)$$
$$\times \left\{ \frac{1+(-1)^{l+l'+l''}}{2} \right\} \int R_{j'} R_j r^2 \, dr$$
$$\times \begin{pmatrix} \sqrt{(l''+1)} & 1/\sqrt{(l''+1)} \\ 0 & \sqrt{\{(2l''+1)/l''(l''+1)\}} \\ -\sqrt{l''} & 1/\sqrt{l''} \end{pmatrix} \begin{pmatrix} 1 \\ (-1)^{l+\frac{1}{2}-j} \, \eta_L(jj') \end{pmatrix} \tag{A2.24}$$

where

$$\eta_L(jj') = \tfrac{1}{2}\{(2j+1) + (-1)^{j+j'-L}(2j'+1)\}.$$

The (3×2) and (2×1) arrays in the equation are matrices and matrix multiplication is implied. The resulting matrix A_{mn} is a (3×1) array with A_{11}, A_{21}, and A_{31} the values of the reduced matrix elements when $L = l''+1$, l'', and $l''-1$, respectively. The factor $[1+(-1)^{l+l'+l''}]/2$ arises in the same way that it did in equation (A2.23).

9. The delta-function potential

The residual nucleon-nucleon interaction is often approximated by a delta-function potential

$$V = -4\pi V_0 \delta(\underline{r}_1 - \underline{r}_2) \tag{A2.25}$$

where V_0 is the strength of the interaction. Matrix elements of this operator are to be calculated between normalized two-particle states that are either antisymmetric to interchange of the space-spin coordinates of the particles ($T = 1$ states) or symmetric to this interchange ($T = 0$ states). Because of the symmetry property in equation (A1.18) of the Clebsch–Gordan coefficients, the normalized two-particle wave function is

$$\psi_{IM}(j_1 j_2) = \{2(1+\delta_{j_1 j_2})\}^{-\frac{1}{2}} \sum_{m_1 m_2} (j_1 j_2 m_1 m_2 \mid IM)$$
$$\times \{\phi_{j_1 m_1}(1)\phi_{j_2 m_2}(2)$$
$$+ (-1)^T \phi_{j_2 m_2}(1)\phi_{j_1 m_1}(2)\}. \tag{A2.26}$$

Matrix elements of V will be denoted by

$$E_{IT}(j_1 j_2; j_3 j_4) = \langle \psi_{IM}(j_1 j_2) | -4\pi V_0 \delta(\underline{r}_1 - \underline{r}_2) \, | \psi_{IM}(j_3 j_4)\rangle$$

and since V is a scalar operator E_{IT} is independent of M.

Since we deal with a delta-function potential, the only time two nucleons can interact is when their spatial coordinates are the same. We denote the wave function of equation (A2.26) when $\underline{r}_1 = \underline{r}_2$ by $\Phi_{IM}(j_1 j_2)$. Thus when the form of ϕ_{jm}

given in equation (A1.17) is used

$$\Phi_{IM}(j_1j_2)=\{2(1+\delta_{j_1j_2})\}^{-\frac{1}{2}}R_{j_1}(r_1)R_{j_2}(r_1)\sum_i(j_1j_2m_1m_2\mid IM)$$
$$\times(l_1\tfrac{1}{2}k_1\mu_1\mid j_1m_1)(l_2\tfrac{1}{2}k_2\mu_2\mid j_2m_2)Y_{l_1k_1}(\theta_1,\phi_1)Y_{l_2k_2}(\theta_1,\phi_1)$$
$$\times\{\chi_{\mu_1}(1)\chi_{\mu_2}(2)+(-1)^T\chi_{\mu_1}(2)\chi_{\mu_2}(1) \tag{A2.27}$$

where the sum on i stands for a sum over m_1, m_2, k_1, k_2, μ_1, and μ_2.

Although we shall ultimately carry out the calculation without explicitly making an L–S decomposition, it is convenient first to imagine such a decomposition has been made. In this case

$$\Phi_{IM}(j_1j_2)=\sum_{LS}\sum_{M_LM_S}\alpha_{LS}(j_1j_2IT)(LSM_LM_S\mid IM)Y_{LM_L}(\theta_1,\phi_1)\chi_{SM_S}(1,2)R_{j_1}(r_1)R_{j_2}(r_1)$$

where $\chi_{SM_S}(1,2)$ is the two particle spin eigenfunction

$$\chi_{SM_S}=\sum_{\mu_1\mu_2}(\tfrac{1}{2}\tfrac{1}{2}\mu_1\mu_2\mid SM_S)\chi_{\mu_1}(1)\chi_{\mu_2}(2)$$

and

$$\alpha_{LS}(j_1j_2IT)=\left(\frac{(2l_1+1)(2l_2+1)}{4\pi(2L+1)}\right)^{\frac{1}{2}}\frac{\{1-(-1)^{S+T}\}}{\sqrt{\{2(1+\delta_{j_1j_2}\delta_{l_1l_2})\}}}(l_1l_200\mid L0)\gamma_{LS}^{(I)}(j_1l_1;j_2l_2)$$

where $\gamma_{LS}^{(I)}(j_1l_1;j_2l_2)$ is defined by equation (4.15) and equation (A2.11) has been used to express the product of two spherical harmonics in terms of a single one. As we have already remarked, the E_{IT} are independent of M. Thus we may sum on M provided we divide by $(2I+1)$:

$$E_{IT}(j_1j_2;j_3j_4)=\frac{1}{(2I+1)}\sum_M\langle\Phi_{IM}(j_1j_2)|-4\pi V_0\delta(\underline{r}_1-\underline{r}_2)|\Phi_{IM}(j_3j_4)\rangle$$
$$=\frac{-4\pi V_0\bar{R}}{(2I+1)}\sum_{LS}\alpha_{LS}(j_1j_2IT)\alpha_{LS}(j_3j_4IT)\sum_{MM_LM_S}(LSM_LM_S\mid IM)^2$$
$$\times\int Y^*_{LM_L}(\theta_1,\phi_1)Y_{LM_L}(\theta_1,\phi_1)\,d\Omega_1$$

where

$$\bar{R}=\bar{R}(j_1j_2j_3j_4)=\int R_{j_1}(r)R_{j_2}(r)R_{j_3}(r)R_{j_4}(r)r^2\,dr. \tag{A2.28}$$

From equations (A1.20) and (A1.23)

$$\sum_{MM_S}(LSM_LM_S\mid IM)^2=(2I+1)/(2L+1).$$

Thus the integration over the spherical harmonics takes the form

$$\sum_{M_L}\int Y^*_{LM_L}Y_{LM_L}\,d\Omega=\sum_{M_L}\int(-1)^{M_L}Y_{L-M_L}Y_{LM_L}\,d\Omega=\sqrt{(2L+1)}(-1)^L\int[Y_L\times Y_L]_{00}\,d\Omega$$

where use has been made of the fact that $Y^*_{LM}=(-1)^M Y_{L-M}$ and that the Clebsch–Gordan coefficient $(LLM-M\,|\,00)=(-1)^{L-M}/\surd(2L+1)$. Thus the spatial integral is over a scalar integrand and is identical to 4π times the integrand evaluated at any convenient point. Obviously, in view of equation (A2.1), $\theta=0°$ is a good choice. In addition, it is clearly desirable to make this simplification at as early a stage as possible, namely when equation (A2.27) is being used in the evaluation. Thus

$$E_{IT}(j_1j_2;\,j_3j_4)=\frac{1}{(2I+1)}\sum_M\langle\Phi_{IM}(j_1j_2)|\,-4\pi V_0\delta(\underline{r}_1-\underline{r}_2)\,|\Phi_{IM}(j_3j_4)\rangle$$

$$=\frac{4\pi}{(2I+1)}\langle\Phi_{IM}(j_1j_2)|\,-4\pi V_0\delta(\underline{r}_1-\underline{r}_2)\,|\Phi_{IM}(j_3j_4)\rangle_{\theta=0°}.$$

When equation (A2.21) is used it follows that

$$E_{IT}(j_1j_2;\,j_3j_4)=\frac{-V_0\bar{R}}{(2I+1)}\left\{\frac{(2l_1+1)(2l_2+1)(2l_3+1)(2l_4+1)}{(1+\delta_{j_1j_2}\delta_{l_1l_2})(1+\delta_{j_3j_4}\delta_{l_3l_4})}\right\}^{\frac{1}{2}}$$

$$\times\sum_{\mu_i}(j_1j_2\mu_1\mu_2\,|\,IM)(j_3j_4\mu_3\mu_4\,|\,IM)(l_1\tfrac{1}{2}0\mu_1\,|\,j_1\mu_1)$$

$$\times(l_2\tfrac{1}{2}0\mu_2\,|\,j_2\mu_2)(l_3\tfrac{1}{2}0\mu_3\,|\,j_3\mu_3)(l_4\tfrac{1}{2}0\mu_4\,|\,j_4\mu_4)$$

$$\times\{\delta_{\mu_1\mu_3}\delta_{\mu_2\mu_4}+(-1)^T\delta_{\mu_1\mu_4}\delta_{\mu_2\mu_3}\}.$$

The sum over μ_i $(i=1-4)$ can be carried out and when the property of the Clebsch–Gordan coefficients given by equation (A2.22) is used one finds

$$E_{IT}(j_1j_2;\,j_3j_4)=-(-1)^{j_1+j_2+j_3+j_4}\frac{V_0\bar{R}}{4(2I+1)}\left\{\frac{(2j_1+1)(2j_2+1)(2j_3+1)(2j_4+1)}{(1+\delta_{j_1j_2}\delta_{l_1l_2})(1+\delta_{j_3j_4}\delta_{l_3l_4})}\right\}^{\frac{1}{2}}$$

$$\times\{1+(-1)^{l_1+l_2+l_3+l_4}\}[\{1+(-1)^T\}(j_1j_2\tfrac{1}{2}\tfrac{1}{2}\,|\,I1)(j_3j_4\tfrac{1}{2}\tfrac{1}{2}\,|\,I1)$$

$$+(-1)^{l_2+l_4+j_2-j_4}\{1-(-1)^{I+T+l_3+l_4}\}(j_1j_2\tfrac{1}{2}-\tfrac{1}{2}\,|\,I0)(j_3j_4\tfrac{1}{2}-\tfrac{1}{2}\,|\,I0)].\quad\text{(A2.29)}$$

Because of their simple form harmonic-oscillator wave functions are often used in the evaluation of $\bar{R}$ (equation (A2.28)). Since the oscillator has infinite walls, the radial motion of the particles is much more confined in this potential than in one that goes to zero as $r\to\infty$. This is particularly true for states with small values of the orbital angular momentum. For high l values the oscillator approximation is better since the particles are naturally constrained by the centrifugal barrier $l(l+1)/r^2$. To illustrate this, we consider the radial integrals that would arise in a study of ${}^{18}_{8}O_{10}$ where the nucleons are filling the $(0d, 1s)$ shell. When oscillator wave functions are used

$$\frac{\bar{R}\{(1s^4)\}}{\bar{R}\{(0d^4)\}}=\frac{205}{84},$$

whereas the ratio is close to unity if the nucleons have the same binding energy and eigenfunctions of a finite well are used.

To compensate for this effect, Green and Moszkowski (1965) and Arvieu and

Moszkowski (1966) introduced the surface-delta interaction. This interaction is identical to that of equation (A2.25) except that the magnitude of $\bar{R}$ is taken independent of j_1, j_2, j_3, and j_4:

$$\bar{R} = (-1)^{n_1+n_2+n_3+n_4} \tag{A2.30}$$

where n_i is the number of radial nodes in the eigenfunction the angular momentum of which is j_i. The phase factor in this expression is introduced so that the sign of $\bar{R}$ will be the same as would be obtained by use of equation (A2.28) when radial wave functions that are positive at the origin are used.

APPENDIX 3

QUASI-SPIN AND NUMBER DEPENDENCE OF MATRIX ELEMENTS

In Chapter 1, section 5, we introduced the seniority quantum number v, the number of unpaired nucleons in the wave function describing the state (Racah 1943, 1949). In this Appendix we shall show the relationship between seniority and quasi-spin and exploit this relationship to deduce the number dependence of matrix elements when we deal with the identical nucleon configuration j^n. In order to simplify the notation we shall, wherever possible, suppress the index j and write

$$a_{jm}^{\dagger} = a_m^{\dagger}$$

1. Relationship between quasi-spin and seniority

The S_+ and S_- operators

$$S_+ = \sum_{m>0} (-1)^{j-m} a_m^{\dagger} a_{-m}^{\dagger} \tag{A3.1a}$$

$$S_- = \sum_{m>0} (-1)^{j-m} a_{-m} a_m \tag{A3.1b}$$

which were introduced in Chapter 1, section 5, create and destroy pairs of particles coupled to angular momentum zero. These operators together with

$$S_z = \tfrac{1}{2}(N - \Omega), \tag{A3.1c}$$

where

$$N = \sum_m a_m^{\dagger} a_m \tag{A3.2}$$

is the number operator and

$$\Omega = (2j+1)/2 \tag{A3.3}$$

is the pair degeneracy of the state j, obey the usual angular-momentum commutation relationships (Kerman 1961)

$$[S_k, S_l] = S_k S_l - S_l S_k$$
$$= i S_m \quad (k, l, m \text{ cyclic}). \tag{A3.4}$$

$\underset{\sim}{S}$ is called the quasi-spin operator and we shall now deduce the relationship between quasi-spin and seniority.

By definition, a v-particle state with seniority v has no zero-coupled pairs in its make-up. Thus if

$$(j^v)_{IMv} = \Phi_{IMv} |0\rangle$$

is a v-particle wave function with seniority v

$$S_- \Phi_{IMv} |0\rangle = 0.$$

(For convenience we have suppressed any other quantum numbers that may be needed to give a complete characterization of the state.) Because $\underset{\sim}{S}$ obeys the commutation rules in equation (A3.4),

$$S^2 = \tfrac{1}{2}(S_+S_- + S_-S_+) + S_z^2$$
$$= S_+S_- + S_z(S_z - 1).$$

Thus

$$S_z \Phi_{IMv} |0\rangle = -\tfrac{1}{2}(\Omega - v)\Phi_{IMv} |0\rangle \quad \text{(A3.5a)}$$

and

$$S^2\Phi_{IMv} |0\rangle = S_z(S_z - 1)\Phi_{IMv} |0\rangle$$
$$= \{\tfrac{1}{2}(\Omega - v)\}\{\tfrac{1}{2}(\Omega - v) + 1\}\Phi_{IMv} |0\rangle. \quad \text{(A3.5b)}$$

Thus a v-particle state with seniority v is an eigenfunction of the quasi-spin operators S^2 and S_z with eigenvalues $S = \tfrac{1}{2}(\Omega - v)$ and $S_z = -\tfrac{1}{2}(\Omega - v)$, and these eigenvalues are independent of any other quantum numbers of the state.

Since S_+ commutes with S^2 it follows that the state

$$(j^n)_{IMv} = \Phi_{IMv} S_+^{(n-v)/2} |0\rangle$$

has the same quasi-spin as $(j^v)_{IMv}$ and has $S_z = -\tfrac{1}{2}(\Omega - n)$. Therefore there is a one-to-one correspondence between the quasi-spin of a state and its seniority

$$S = \tfrac{1}{2}(\Omega - v) \quad \text{(A3.6a)}$$

and the z component of quasi-spin gives the number of particles in the state

$$S_z = -\tfrac{1}{2}(\Omega - n). \quad \text{(A3.6b)}$$

Consequently one can classify states either by their seniority and number of particles or by their quasi-spin and z component of quasi-spin. Thus

$$(j^n)_{IMv} \equiv \phi_{IM;SS_z}. \quad \text{(A3.7)}$$

Since $\underset{\sim}{S}$ satisfies the same commutation relationships as the ordinary angular momentum $\underset{\sim}{J}$, it may be taken as the generator of infinitesimal rotations in quasi-spin space. In Appendix 2, section 1, we discussed the properties that an operator must satisfy to be a tensor of rank k in ordinary space. Thus if R_{kq} is a tensor of rank k with z component q in quasi-spin space, it will satisfy the commutation relationships

$$[S_z, R_{kq}] = qR_{kq} \qquad [S_\pm, R_{kq}] = \sqrt{\{(k \mp q)(k \pm q + 1)\}} R_{kq\pm 1}. \quad \text{(A3.8)}$$

The advantage of looking at the tensor character of operators in quasi-spin space is that one can use the Wigner–Eckart theorem in this space to extract the S_z dependence of the matrix element

$$\langle \phi_{I'M';S'S'_z} | R_{kq} | \phi_{IM;SS_z} \rangle = (SkS_zq \mid S'S'_z) \langle \phi_{I'M';S'} \| R_k \| \phi_{IM;S} \rangle \qquad (A3.9)$$

where $\langle \| \| \rangle$ is used to denote the reduced matrix element in quasi-spin space. Since the reduced matrix element does not depend on S_z or S'_z, the entire number dependence is contained in the Clebsch–Gordan coefficient. In the following sections we shall use equation (A3.9) to find the number dependence of matrix elements for the identical particle system (Arima and Kawarda 1964, Lawson and MacFarlane 1965, Arima and Ichimura 1966, Ichimura 1968). Relationships for the n dependence of matrix elements can also be deduced when neutrons and protons occupy the orbit j (Hecht 1967), but these will not be discussed here since, as was shown in Chapter 2, section 1.4, seniority is not a good quantum number for the (n, p) system.

2. Number dependence of the fractional-parentage coefficients

The operators $a_m^\dagger$ and $(-1)^{j-m} a_{-m}$ satisfy equations (A3.8) with $k = \frac{1}{2}$ and $q = \pm\frac{1}{2}$. Thus these two operators form the components of a quasi-spin tensor of rank $\frac{1}{2}$

$$\begin{aligned} R_{\frac{1}{2}\frac{1}{2}} &= a_m^\dagger \\ R_{\frac{1}{2}-\frac{1}{2}} &= (-1)^{j-m} a_{-m} \\ &= -\tilde{a}_m \end{aligned} \qquad (A3.10)$$

where $\tilde{a}_m$ is the modified Hermitian adjoint operator defined by equation (A2.9). Consequently

$$\begin{aligned} \langle (j^{n+1})_{I'M'v'} | a_m^\dagger | (j^n)_{IMv} \rangle &= (S\tfrac{1}{2}S_z\tfrac{1}{2} \mid S'S'_z) \langle \phi_{I'M';S'} \| R_{\frac{1}{2}} \| \phi_{IM;S} \rangle = \frac{1}{\sqrt{\{2(\Omega - v + 1)\}}} \\ &\times \{\sqrt{(n - v + 2)}\, \delta_{S',S+\frac{1}{2}} - \sqrt{(2\Omega - n - v)}\, \delta_{S',S-\frac{1}{2}}\} \\ &\times \langle \phi_{I'M';S'} \| R_{\frac{1}{2}} \| \phi_{IM;S} \rangle \end{aligned}$$

where $S = \frac{1}{2}(\Omega - v)$, $S' = \frac{1}{2}(\Omega - v')$, $S_z = \frac{1}{2}(n - \Omega)$, $S'_z = \frac{1}{2}(n + 1 - \Omega)$, and use has been made of Table A1.1 to evaluate the quasi-spin Clebsch–Gordan coefficients.

One may evaluate the reduced matrix element for any convenient value of the z-component of quasi-spin, and clearly the most convenient value is when the number of particles is a minimum; i.e. when $S_z = -S$, $S'_z = -S'$. Since

$$\langle \phi_{I'M';S'} \| R_{\frac{1}{2}} \| \phi_{IM;S} \rangle = \frac{\langle \phi_{I'M';S'-S'} | R_{\frac{1}{2}q} | \phi_{IM;S-S} \rangle}{(S\frac{1}{2} - Sq \mid S' - S')}$$

one sees that

$$\langle (j^{n+1})_{I'M'v'}|\, a_m^\dagger\, |(j^n)_{IMv}\rangle = \left\{\frac{n-v+2}{2(\Omega-v+1)}\right\}^{\frac{1}{2}} \langle \phi_{I'M';S'-S'}|\, R_{\frac{1}{2}-\frac{1}{2}}\, |\phi_{IM;S-S}\rangle \delta_{S',S+\frac{1}{2}}$$

$$+\left\{\frac{(2\Omega-n-v)}{2(\Omega-v)}\right\}^{\frac{1}{2}} \langle \phi_{I'M';S'-S'}|\, R_{\frac{1}{2}\frac{1}{2}}\, |\phi_{IM;S-S}\rangle \delta_{S',S-\frac{1}{2}}$$

$$= -\left\{\frac{n-v+2}{2(\Omega-v+1)}\right\}^{\frac{1}{2}} \langle (j^{v-1})_{I'M'v'}|\, \tilde{a}_m\, |(j^v)_{IMv}\rangle \delta_{v',v-1}$$

$$+\left\{\frac{(2\Omega-n-v)}{2(\Omega-v)}\right\}^{\frac{1}{2}} \langle (j^{v+1})_{I'M'v'}|\, a_m^\dagger\, |(j^v)_{IMv}\rangle \delta_{v',v+1}. \tag{A3.11}$$

Equation (A3.11) can equally well be written as

$$\langle (j^{n+1})_{I'v'}\|\, a_j^\dagger\, \|(j^n)_{Iv}\rangle (IjMm \mid I'M')$$

$$= \left\{ -\left\{\frac{n-v+2}{2(\Omega-v+1)}\right\}^{\frac{1}{2}} \langle (j^{v-1})_{I'v'}\|\, \tilde{a}_j\, \|(j^v)_{Iv}\rangle \delta_{v',v-1} \right.$$

$$\left. + \left\{\frac{(2\Omega-n-v)}{2(\Omega-v)}\right\}^{\frac{1}{2}} \langle (j^{v+1})_{I'v'}\|\, a_j^\dagger\, \|(j^v)_{Iv}\rangle \delta_{v',v+1} \right\} (IjMm \mid I'M')$$

where the reduced matrix element is now the conventional one and according to equation (A5.6) is related to the one-nucleon c.f.p. Furthermore, because the c.f.p. are real, it follows from equation (A2.10) that

$$\langle (j^{v-1})_{I'v'}\|\, \tilde{a}_j\, \|(j^v)_{Iv}\rangle = (-1)^{I-j-I'} \left(\frac{2I+1}{2I'+1}\right)^{\frac{1}{2}}$$

$$\times \langle (j^v)_{Iv}\|\, a_j^\dagger\, \|(j^{v-1})_{I'v'}\rangle.$$

Therefore

$$\langle j^n Iv, j|\}j^{n+1}I'v'\rangle = \left\{\frac{(v+1)(2\Omega-n-v)}{2(n+1)(\Omega-v)}\right\}^{\frac{1}{2}} \langle j^v Iv, j|\}j^{v+1}I'v'\rangle \delta_{v',v+1}$$

$$+(-1)^{I-j-I'} \left\{\frac{v(n-v+2)(2I+1)}{2(n+1)(\Omega-v+1)(2I'+1)}\right\}^{\frac{1}{2}} \langle j^{v-1}I'v', j|\}j^v Iv\rangle \delta_{v',v-1}. \tag{A3.12}$$

Thus if one wishes to find c.f.p. for the $(n+1)$-particle senority-v state one must know the parentage coefficients involving the v- and $(v\pm1)$-particle wave functions. In general, the construction of multi-particle wave functions is tedious but can, if necessary, be carried out by the methods outlined in Chapter 1, section 1.2. For the special case of one and two particles, the parentage coefficients are trivial:

$$\langle j^0 I=0, v=0, j|\}j^1 I'=j, v'=1\rangle = 1 \tag{A3.13a}$$

$$\langle j^1 I=j, v=1, j|\}j^2 I'v'\rangle = \frac{1+(-1)^{I'}}{2}. \tag{A3.13b}$$

When these relationships are used one can write down explicit expressions for the seniority-zero and seniority-one fractional-parentage coefficients

$$\langle j^n Iv, j| \} j^{n+1}00\rangle = \delta_{I,j}\delta_{v,1} \tag{A3.14}$$

$$\langle j^n Iv, j| \} j^{n+1}j1\rangle = \left\{\frac{2j+1-n}{(n+1)(2j+1)}\right\}^{\frac{1}{2}}\delta_{I,0}$$

$$= -\left\{\frac{2n(2I+1)}{(n+1)(2j-1)(2j+1)}\right\}^{\frac{1}{2}}\left\{\frac{1+(-1)^I}{2}\right\}\delta_{v,2} \quad (I\neq 0) \tag{A3.15}$$

Although it is necessary to know the three-particle states to obtain all the parentage coefficients for the $v'=2$ system, one can use equations (A3.12) and (A3.13b) to write down the one involving $I=j$, $v=1$:

$$\langle j^n j1, j| \} j^{n+1}I'v'=2\rangle = \left\{\frac{2(2j-n)}{(n+1)(2j-1)}\right\}^{\frac{1}{2}}\left\{\frac{1+(-1)^{I'}}{2}\right\}. \tag{A3.16}$$

By using equations (A3.14)–(A3.16) one obtains the spectroscopic factors for single-nucleon transfer (given in Table 2.7). When the target is in the state $(j^n)_{IMv}$ with $v=0$ or 1.

3. Quasi-spin tensors of rank zero and one

Since $a^\dagger_m$ and $-\tilde{a}_m$ are the components of a quasi-spin tensor of rank $\frac{1}{2}$, they can be used as building blocks in the construction of tensors of rank 0 and 1:

$$R_{kq} = \sum_{\mu\mu'} (\tfrac{1}{2}\tfrac{1}{2}\mu\mu' \mid kq) R_{\frac{1}{2}\mu} R_{\frac{1}{2}\mu'}.$$

Since the R_{kq} are bilinear products of creation and destruction operators, it is convenient to write them for a definite angular momentum. Thus we define

$$R_{IM;kq} = \frac{-\sqrt{\Omega}}{2}\sum_{\mu\mu'} (\tfrac{1}{2}\tfrac{1}{2}\mu\mu' \mid kq)[R_{\frac{1}{2}\mu} \times R_{\frac{1}{2}\mu'}]_{IM} \tag{A3.17}$$

where $[\times]$ means the coupling in ordinary angular-momentum space to (I, M), and the normalization constant has been chosen, as we shall see, so that $R_{00;1q} = S_q$.

We first consider the form of $R_{IM;k0}$. From the definition of $R_{\frac{1}{2}\mu}$ in equation (A3.10) one sees that

$$R_{IM;k0} = \frac{\sqrt{\Omega}}{2\sqrt{2}}\{[a^\dagger_j \times \tilde{a}_j]_{IM} - (-1)^k[\tilde{a}_j \times a^\dagger_j]_{IM}\}.$$

By use of the anti-commutation relationship that $a^\dagger_{jm}$ and a_{jm} satisfy (equation (1.31)) it follows that

$$[\tilde{a}_j \times a^\dagger_j]_{IM} = \sum_{mm'} (jjmm' \mid IM)\tilde{a}_{jm}a^\dagger_{jm'}$$

$$= \sum_{mm'} (jjmm' \mid IM)\{(-1)^{j+m}\delta_{m',-m} - a^\dagger_{jm'}\tilde{a}_{jm}\}$$

$$= (-1)^I[a^\dagger_j \times \tilde{a}_j]_{IM} - \sqrt{(2\Omega)}\delta_{I0},$$

where we have made use of equations (A1.18), (A1.21), and (A1.23) in writing the last line of this equation. Consequently

$$R_{IM;k0}=\left(\frac{\Omega}{2}\right)^{\frac{1}{2}}\left[\left\{\frac{1-(-1)^{I+k}}{2}\right\}[a_j^\dagger\times\tilde{a}_j]_{IM}+(-1)^k\left(\frac{\Omega}{2}\right)^{\frac{1}{2}}\delta_{I0}\right]. \qquad \text{(A3.18)}$$

Therefore for $k=0$, $R_{IM;k0}$ vanishes identically for all even non-zero values of I and reduces to a numerical constant when $I=0$. Consequently only for odd values of I can one construct a non-trivial quasi-spin scalar operator

$$R_{IM;00}=\left(\frac{\Omega}{2}\right)^{\frac{1}{2}}U_{IM}(jj) \qquad (I \text{ odd}) \qquad \text{(A3.19)}$$

where

$$U_{IM}(jj)=[a_j^\dagger\times\tilde{a}_j]_{IM} \qquad \text{(A3.20)}$$

On the other hand, when $k=1$ equations (A3.17) and (A3.18) lead to the conclusion that the operator which is a vector in quasi-spin space has components

$$R_{IM;11}=-\left(\frac{\Omega}{2}\right)^{\frac{1}{2}}A^\dagger_{IM}(jj)$$

$$R_{IM;10}=\left(\frac{\Omega}{2}\right)^{\frac{1}{2}}\left\{U_{IM}(jj)-\left(\frac{\Omega}{2}\right)^{\frac{1}{2}}\delta_{I0}\right\}$$

$$R_{IM;1-1}=\left(\frac{\Omega}{2}\right)^{\frac{1}{2}}\tilde{A}_{IM}(jj) \qquad (I \text{ even}) \qquad \text{(A3.21)}$$

where

$$A^\dagger_{IM}(jj)=\frac{1}{\sqrt{2}}[a_j^\dagger\times a_j^\dagger]_{IM}$$

$$\tilde{A}_{IM}(jj)=\frac{-1}{\sqrt{2}}[\tilde{a}_j\times\tilde{a}_j]_{IM}. \qquad \text{(A3.22)}$$

The fact that only even values of I are possible in equation (A3.18) when $k=1$ is merely a restatement of the fact that the Pauli principle does not allow two identical particles in the same j state to couple to odd values of I. Furthermore, because of the special form of the Clebsch–Gordan coefficient when $I=0$, (equation (A1.21)),

$$R_{00;11}=-\mathrm{S}_+/\sqrt{2}.$$

Thus since the $+1$ spherical component of the quasi-spin operator S_1 is $-\mathrm{S}_+/\sqrt{2}$, it follows that with the normalization of equation (A3.17) $R_{00;1q}=\mathrm{S}_q$.

3.1. Number dependence of the two-particle parentage coefficients

By the same arguments that were used in section 2 of this Appendix one can deduce the number dependence of the double-parentage coefficients defined by equation (A5.14). From equation (A3.6a) it follows that $S'=S$ means $v'=v$ and

$S'=S\pm1$ implies $v'=v\mp2$. When this is used in conjunction with the fact that

$$\langle\phi_{I'M';S'S_z'}|\,R_{JM'';1q}\,|\phi_{IM;SS_z}\rangle=\frac{(S1S_zq\mid S'S_z')}{(S1-Sq'\mid S'-S')}\times\langle\phi_{I'M';S'-S'}|\,R_{JM'';1q'}\,|\phi_{IM;S-S}\rangle \qquad \text{(A3.23)}$$

one finds

$$-\left(\frac{\Omega}{2}\right)^{\frac{1}{2}}\langle(j^{n+2})_{I'M'v'}|\,A^{\dagger}_{JM''}(jj)\,|(j^n)_{IMv}\rangle$$

$$=\langle\phi_{I'M';S'S_z'}|\,R_{JM'';11}\,|\phi_{IM;SS_z}\rangle$$

$$=\left\{\frac{(n-v+2)(n-v+4)}{4(\Omega-v+1)(\Omega-v+2)}\right\}^{\frac{1}{2}}\langle\phi_{I'M';S'-S'}|\,R_{JM'';1-1}\,|\phi_{IM;S-S}\rangle\delta_{S',S+1}$$

$$+\left\{\frac{(2\Omega-n-v)(n-v+2)}{2(\Omega-v)(\Omega-v)}\right\}^{\frac{1}{2}}\langle\phi_{I'M';S'-S'}|\,R_{JM'';10}\,|\phi_{IM;S-S}\rangle\delta_{S',S}$$

$$+\left\{\frac{(2\Omega-n-v)(2\Omega-n-v-2)}{4(\Omega-v)(\Omega-v-1)}\right\}^{\frac{1}{2}}\langle\phi_{I'M';S'-S'}|\,R_{JM'';11}\,|\phi_{IM;S-S}\rangle\delta_{S',S-1}$$

$$=\left(\frac{\Omega}{2}\right)^{\frac{1}{2}}\left[\left\{\frac{(n-v+2)(n-v+4)}{4(\Omega-v+1)(\Omega-v+2)}\right\}^{\frac{1}{2}}\langle(j^{v-2})_{I'M'v'}|\,\tilde{A}_{JM''}(jj)\,|(j^v)_{IMv}\rangle\delta_{v',v-2}\right.$$

$$+\left\{\frac{(2\Omega-n-v)(n-v+2)}{2(\Omega-v)(\Omega-v)}\right\}^{\frac{1}{2}}\langle(j^v)_{I'M'v'}|\,U_{JM''}(jj)-\left(\frac{\Omega}{2}\right)^{\frac{1}{2}}\delta_{J0}\,|(j^v)_{IMv}\rangle\delta_{v',v}$$

$$\left.-\left\{\frac{(2\Omega-n-v)(2\Omega-n-v-2)}{4(\Omega-v)(\Omega-v-1)}\right\}^{\frac{1}{2}}\langle(j^{v+2})_{I'M'v'}|\,A^{\dagger}_{JM''}(jj)\,|(j^v)_{IMv}\rangle\delta_{v',v+2}\right].$$

Since the ordinary reduced matrix element of $A^{\dagger}_{JM}(jj)$ is related to the two-particle c.f.p. by equation (A5.14)

$$\langle(j^{n+2})_{I'v'}\|\,A^{\dagger}_J(jj)\,\|(j^n)_{Iv}\rangle=\left\{\frac{(n+2)(n+1)}{2}\right\}^{\frac{1}{2}}\langle j^nIv,j^2J|\}j^{n+2}I'v'\rangle \qquad \text{(A3.24)}$$

it follows that

$$\left\{\frac{(n+2)(n+1)}{2}\right\}^{\frac{1}{2}}\langle j^nIv,j^2J|\}j^{n+2}I'v'\rangle=-\left\{\frac{(n-v+2)(n-v+4)}{4(\Omega-v+1)(\Omega-v+2)}\right\}^{\frac{1}{2}}$$

$$\times\langle(j^{v-2})_{I'v'}\|\,\tilde{A}_J(jj)\,\|(j^v)_{Iv}\rangle\delta_{v',v-2}$$

$$-\left\{\frac{(2\Omega-n-v)(n-v+2)}{2(\Omega-v)(\Omega-v)}\right\}^{\frac{1}{2}}\langle(j^v)_{I'v'}\|\,U_J(jj)-\left(\frac{\Omega}{2}\right)^{\frac{1}{2}}\delta_{J0}\,\|(j^v)_{Iv}\rangle\delta_{v',v}$$

$$+\left\{\frac{(v+2)(v+1)(2\Omega-n-v)(2\Omega-n-v-2)}{8(\Omega-v)(\Omega-v-1)}\right\}^{\frac{1}{2}}\langle j^vIv,j^2J\mid\}j^{v+2}I'v'\rangle\delta_{v',v+2}.$$

To write this expression in its final form we make use of equation (A2.10) which tells us that

$$\langle (j^{v-2})_{I'v'} \| \tilde{A}_J(jj) \| (j^v)_{Iv} \rangle = (-1)^{I-J-I'} \left(\frac{2I+1}{2I'+1}\right)^{\frac{1}{2}} \langle (j^v)_{Iv} \| A_J^\dagger(jj) \| (j^{v-2})_{I'v'} \rangle$$

$$= (-1)^{I-I'} \left\{\frac{v(v-1)(2I+1)}{2(2I'+1)}\right\}^{\frac{1}{2}} \langle j^{v-2} I'v', j^2 J | \} j^v I v \rangle. \tag{A3.25}$$

Thus we arrive at the final expression for the recursion relationship between the double-parentage coefficients:

$$\begin{aligned}
&\langle j^n I v, j^2 J | \} j^{n+2} I'v' \rangle \\
&= -(-1)^{I-I'} \left\{\frac{(2I+1)v(v-1)(n-v+2)(n-v+4)}{4(2I'+1)(n+2)(n+1)(\Omega-v+1)(\Omega-v+2)}\right\}^{\frac{1}{2}} \\
&\quad \times \langle j^{v-2} I'v', j^2 J | \} j^v I v \rangle \delta_{v',v-2} \\
&\quad - \left\{\frac{(2\Omega-n-v)(n-v+2)}{(n+2)(n+1)(\Omega-v)(\Omega-v)}\right\}^{\frac{1}{2}} \langle (j^v)_{I'v'} \| U_J(jj) - \left(\frac{\Omega}{2}\right)^{\frac{1}{2}} \delta_{J0} \| (j^v)_{Iv} \rangle \delta_{v',v} \\
&\quad + \left\{\frac{(v+2)(v+1)(2\Omega-n-v)(2\Omega-n-v-2)}{4(n+2)(n+1)(\Omega-v)(\Omega-v-1)}\right\}^{\frac{1}{2}} \langle j^v I v, j^2 J | \} j^{v+2} I'v' \rangle \delta_{v',v+2}.
\end{aligned} \tag{A3.26}$$

Consequently in order to find the two-particle c.f.p. for the $(n+2)$-particle state with seniority v' one must know the double-parentage coefficients for the $(v'+2)$- and $(v'-2)$-particle configurations together with the reduced matrix elements of the operator $U_{JM}(jj)$ (equation (A3.20)) in the v'-particle configuration.

The n dependence of the spectroscopic factors for two-nucleon transfer when the target is in the state $(j^n)_{I=M=v=0}$ are given in Chapter 4, section 4 and can be deduced from equation (A3.26) when one notes that

$$\langle j^0 I=0 v=0, j^2 J | \} j^2 I'v' \rangle = \left\{\frac{1+(-1)^{I'}}{2}\right\} \delta_{I',J}$$

and

$$\langle (j^0)_{I=v=0} \| U_J(jj) - \left(\frac{\Omega}{2}\right)^{\frac{1}{2}} \delta_{J0} \| (j^0)_{I'=v'=0} \rangle = -\left(\frac{\Omega}{2}\right)^{\frac{1}{2}} \delta_{J,0}.$$

When these results are inserted into equation (A3.26) one finds

$$\langle j^n I=0 v=0, j^2 J=0 | \} j^{n+2} I'=0 v'=0 \rangle = \left\{\frac{2j+1-n}{(n+1)(2j+1)}\right\}^{\frac{1}{2}} \tag{A3.27}$$

$$\langle j^n I v=2, j^2 I | \} j^{n+2} I'=0 v'=0 \rangle = -\left\{\frac{2n(2I+1)}{(n+1)(2j+1)(2j-1)}\right\}^{\frac{1}{2}} \left\{\frac{1+(-1)^I}{2}\right\} \tag{A3.28}$$

$$\langle j^n I=0 v=0, j^2 I' | \} j^{n+2} I'v'=2 \rangle = \left\{\frac{2(2j+1-n)(2j-1-n)}{(n+2)(n+1)(2j+1)(2j-1)}\right\}^{\frac{1}{2}} \left\{\frac{1+(-1)^{I'}}{2}\right\}. \tag{A3.29}$$

3.2. *Number dependence of gamma-decay matrix elements*

As we have seen in Chapter 5, nuclear gamma-decay is governed by matrix elements of the operator

$$
\begin{aligned}
T^{(\xi)}_{LM} &= \sum_{jj'mm'} \langle \phi_{j'} \| T^{(\xi)}_{L} \| \phi_j \rangle (jLmM \mid j'm') a^{\dagger}_{j'm'} a_{jm} \\
&= \sum_{jj'} \left(\frac{2j'+1}{2L+1} \right)^{\frac{1}{2}} \langle \phi_{j'} \| T^{(\xi)}_{L} \| \phi_j \rangle [a^{\dagger}_{j'} \times \tilde{a}_j]_{LM} \\
&= \sum_{jj'} \left(\frac{2j'+1}{2L+1} \right)^{\frac{1}{2}} \langle \phi_{j'} \| T^{(\xi)}_{L} \| \phi_j \rangle U_{LM}(j'j) \qquad \text{(A3.30)}
\end{aligned}
$$

where use has been made of equations (A1.19) and (A1.20) and the explicit form of the operator $T^{(\xi)}_{LM}$ is given by equations (5.5) and (5.9) for electric and magnetic radiation respectively. Because of parity, transitions between states of the configuration j^n can only proceed by the emission of $M1$, $M3$, $M5, \ldots$, and $E2$, $E4$, $E6, \ldots$ radiation. Thus magnetic multipole moments and radiation are restricted to be magnetic dipole, magnetic octupole, etc; in other words L can only be odd. However, as we have seen $U_{LM}(jj)$ is a quasi-spin scalar when L is odd. Therefore we arrive at the following selection rule governing the configuration j^n:

The matrix elements of the magnetic multipole operator of any order connecting the states $(j^n)_{I'M'v'}$ *and* $(j^n)_{IMv}$

(*i*) *vanish unless* $v = v'$

(*ii*) *are independent of* n.

The magnetic dipole relationships given in Chapter 5, sections 5.1a and 5.2a are just special cases of this rule.

As far as electric multipole moments and transitions are concerned, they too are governed by matrix elements of $U_{LM}(jj)$. However, in this case since L is even ($L > 0$) the operator is a quasi-spin vector. One may use equation (A3.23) to write down explicit expressions for the number dependence of these operators:

$$
\begin{aligned}
\langle (j^n)_{I'M'v'} | T^{(\mathscr{E})}_{LM} | (j^n)_{IMv} \rangle &= \left\{ \frac{2(2j+1)}{\Omega(2L+1)} \right\}^{\frac{1}{2}} \langle \phi_j \| T^{(\mathscr{E})}_{L} \| \phi_j \rangle \, \langle \phi_{I'M';S'S_z} | R_{LM;10} | \phi_{IM;SS_z} \rangle \\
&= \left\{ \frac{2(2j+1)}{\Omega(2L+1)} \right\}^{\frac{1}{2}} \langle \phi_j \| T^{(\mathscr{E})}_{L} \| \phi_j \rangle \left[\left(\frac{\Omega - n}{\Omega - v} \right) \langle \phi_{I'M';S'-S'} | R_{LM;10} | \phi_{IM;S-S} \rangle \delta_{S',S} \right. \\
&\quad + \left\{ \frac{(2\Omega - n - v + 2)(n - v + 2)}{2(\Omega - v + 1)(\Omega - v + 2)} \right\}^{\frac{1}{2}} \langle \phi_{I'M';S'-S'} | R_{LM;1-1} | \phi_{IM;S-S} \rangle \delta_{S',S+1} \\
&\quad \left. - \left\{ \frac{(n - v)(2\Omega - n - v)}{2(\Omega - v)(\Omega - v - 1)} \right\}^{\frac{1}{2}} \langle \phi_{I'M';S'-S'} | R_{LM;11} | \phi_{IM;S-S} \rangle \delta_{S',S-1} \right] \\
&= \left(\frac{2j+1}{2L+1} \right)^{\frac{1}{2}} \langle \phi_j \| T^{(\mathscr{E})}_{L} \| \phi_j \rangle \left[\left(\frac{\Omega - n}{\Omega - v} \right) \langle (j^v)_{I'M'v'} | U_{LM}(jj) | (j^v)_{IMv} \rangle \delta_{v',v} \right.
\end{aligned}
$$

$$+\left\{\frac{(2\Omega-n-v+2)(n-v+2)}{2(\Omega-v+1)(\Omega-v+2)}\right\}^{\frac{1}{2}}\langle(j^{v-2})_{I'M'v'}|\,\tilde{A}_{LM}(jj)\,|(j^{v})_{IMv}\rangle\,\delta_{v',v-2}$$

$$+\left\{\frac{(n-v)(2\Omega-n-v)}{2(\Omega-v)(\Omega-v-1)}\right\}^{\frac{1}{2}}\langle(j^{v+2})_{I'M'v'}|\,A^{\dagger}_{LM}(jj)\,|(j^{v})_{IMv}\rangle\delta_{v',v+2}\Bigg].$$

When use is made of equations (A3.24) and (A3.25) the reduced matrix element governing the electric multipole moments or transition rates within the configuration j^n takes the form

$$\langle(j^n)_{I'v'}\|\,T_L^{(\mathscr{E})}\,\|(j^n)_{Iv}\rangle$$

$$=\left(\frac{2j+1}{2L+1}\right)^{\frac{1}{2}}\langle\phi_j\|\,T_L^{(\mathscr{E})}\,\|\phi_j\rangle\Bigg[\left(\frac{\Omega-n}{\Omega-v}\right)\langle(j^v)_{I'v'}\|\,U_L(jj)\,\|(j^v)_{Iv}\rangle\delta_{v',v}$$

$$+(-1)^{I-I'}\left\{\frac{(2I+1)v(v-1)(2\Omega-n-v+2)(n-v+2)}{4(2I'+1)(\Omega-v+1)(\Omega-v+2)}\right\}^{\frac{1}{2}}$$

$$\times\langle j^{v-2}I'v',j^2L|\}j^vIv\rangle\delta_{v',v-2}$$

$$+\left\{\frac{(v+2)(v+1)(n-v)(2\Omega-n-v)}{4(\Omega-v)(\Omega-v-1)}\right\}^{\frac{1}{2}}\langle j^vIv,j^2L|\}j^{v+2}I'v'\rangle\delta_{v',v+2}\Bigg] \quad \text{(A3.31)}$$

When the seniorities of the initial and final states are the same, only the first term in equation (A3.31) contributes and from this we obtain the particle-hole rules deduced in Chapter 3, section 2.3:

(i) *for the half-filled shell all static electric multipole moments vanish.*
(ii) *for the half-filled shell there are no electric multipole transitions between states of the same seniority.*
(iii) *the reduced matrix element for the n-particle system is equal in magnitude to that for the n-hole configuration. The signs are opposite when $v=v'$.*

In addition, when $v=1$ or 2 the reduced matrix element of $U_{LM}(jj)$ can be evaluated simply. Thus for $v=v'=1$

$$\langle(j^n)_{I'=jv'=1}\|\,T_L^{(\mathscr{E})}\,\|(j^n)_{I'=jv'=1}\rangle=\left(\frac{\Omega-n}{\Omega-1}\right)\langle\phi_j\|\,T_L^{(\mathscr{E})}\,\|\phi_j\rangle. \quad \text{(A3.32)}$$

When $v=v'=2$ the two-particle reduced matrix element is given by equation (5.39). Thus

$$\langle(j^n)_{I'v'=2}\|\,T_L^{(\mathscr{E})}\,\|(j^n)_{Iv=2}\rangle\doteq\left(\frac{\Omega-n}{\Omega-2}\right)2\sqrt{\{(2j+1)(2I+1)\}}$$

$$\times W(jjI'L;Ij)\langle\phi_j\|\,T_L^{(\mathscr{E})}\,\|\phi_j\rangle. \quad \text{(A3.33)}$$

Finally, if one makes use of the fact that the two-particle to zero-particle double-parentage coefficient is unity, a simple expression can be written for the $v=2\to v'=0$ transition

$$\langle(j^n)_{I'=0v'=0}\|\,T_L^{(\mathscr{E})}\,\|(j^n)_{Iv=2}\rangle=\left\{\frac{2n(2j+1-n)}{2j-1}\right\}^{\frac{1}{2}}\langle\phi_j\|\,T_L^{(\mathscr{E})}\,\|\phi_j\rangle\delta_{I,L}. \quad \text{(A3.34)}$$

4. Two-body interaction in terms of quasi-spin

According to equation (1.48) the two-body interaction for the configuration j^n can be written as

$$V=\sum_{JM}(-1)^{J-M}E_J\,A^{\dagger}_{JM}(jj)\tilde{A}_{J-M}(jj)$$
$$=-\frac{2}{\Omega}\sum_{JM}(-1)^{J-M}E_J\,R_{JM;11}R_{J-M;1-1} \qquad \text{(A3.35)}$$

where E_J stands for $E_J(jj;jj)$ and use has been made of the definition of the quasi-spin tensors of rank 1 given in equation (A3.21). Clearly, $R_{JM;11}$ and $R_{J-M;1-1}$ can couple, in quasi-spin space, to $s=0$, 1, or 2. We define these tensors to be

$$V_{JM;ss_z}=\sum_{\mu\mu'}(11\mu\mu'\,|\,ss_z)R_{JM;1\mu}R_{J-M;1\mu'} \qquad \text{(A3.36a)}$$

and from equation (A1.24) one sees that

$$R_{JM;1\mu}R_{J-M;1\mu'}=\sum_{ss_z}(11\mu\mu'\,|\,ss_z)V_{JM;ss_z}. \qquad \text{(A3.36b)}$$

This last equation can be used to express V in terms of these quasi-spin tensors:

$$V=-\frac{2}{\Omega}\sum_{JMs}(-1)^{J-M}E_J\,(111-1\,|\,s0)V_{JM;s0}$$
$$=-\frac{\sqrt{2}}{\Omega\sqrt{3}}\sum_{JM}(-1)^{J-M}E_J(V_{JM;20}+\sqrt{3}V_{JM;10}+\sqrt{2}V_{JM;00}) \qquad \text{(A3.37)}$$

where the explicit values of the Clebsch–Gordan coefficients have been taken from Table A1.2.

We first examine the form of $V_{JM;10}$. From equation (A3.36a) it follows that

$$\sum_M(-1)^{J-M}V_{JM;10}=\frac{1}{\sqrt{2}}\sum_M(-1)^{J-M}(R_{JM;11}R_{J-M;1-1}-R_{JM;1-1}R_{J-M;11})$$
$$=-\frac{\Omega}{2\sqrt{2}}\sum_M(-1)^{J-M}\{A^{\dagger}_{JM}(jj)\tilde{A}_{J-M}(jj)-\tilde{A}_{JM}(jj)A^{\dagger}_{J-M}(jj)\}$$
$$=-\frac{\Omega}{4\sqrt{2}}\sum_{Mm_i}(jjm_1m_2\,|\,JM)(jjm_3m_4\,|\,JM)(a^{\dagger}_{m_1}a^{\dagger}_{m_2}a_{m_4}a_{m_3}-a_{m_4}a_{m_3}a^{\dagger}_{m_1}a^{\dagger}_{m_2})$$
$$=-\frac{\Omega}{4\sqrt{2}}\sum_{Mm_i}(jjm_1m_2\,|\,JM)(jjm_3m_4\,|\,JM)(\delta_{m_2m_4}a^{\dagger}_{m_1}a_{m_3}-\delta_{m_2m_3}a^{\dagger}_{m_1}a_{m_4}$$
$$+\delta_{m_1m_4}a_{m_3}a^{\dagger}_{m_2}-\delta_{m_1m_3}a_{m_4}a^{\dagger}_{m_2})$$

where use has been made of the anticommutation relationships in equations (1.28), (1.30), and (1.31) that the creation and destruction operators satisfy. From the symmetry and summation properties of the Clebsch–Gordan coefficients given in equations (A1.18), (A1.20), and (A1.24), one can show that

$$\sum_M (-1)^{J-M} V_{JM;10} = -\frac{\Omega}{2\sqrt{2}}\left(\frac{2J+1}{2j+1}\right)\sum_m (a_m^\dagger a_m - a_m a_m^\dagger)$$

$$= -\frac{(2J+1)}{2\sqrt{2}}(\mathrm{N}_j - \Omega)$$

$$= -\frac{(2J+1)}{\sqrt{2}}\mathrm{S}_z \qquad \text{(A3.38)}$$

where N_j is the number operator for the orbit j and S_z is the z component of the quasi-spin operator given by equation (A3.1c).

In equation (3.68) the particle-hole interaction was shown to be

$$F_K(jj;jj) = E_K(jj^{-1};jj^{-1})$$

$$= -\sum_J (2J+1)W(jjjj;JK)E_J(jj;jj). \qquad \text{(A3.39)}$$

Therefore, from the form of the Racah coefficient when $K=0$ given in equation (A4.14), it follows that

$$\sum_J (2J+1)E_J = +2\Omega F_0. \qquad \text{(A3.40)}$$

When these results are incorporated into equation (A3.37) V may be written as

$$V = -\frac{\sqrt{2}}{\Omega\sqrt{3}}\sum_{JM}(-1)^{J-M}E_J(V_{JM;20} + \sqrt{2}V_{JM;00}) + 2F_0\mathrm{S}_z. \qquad \text{(A3.41)}$$

We shall now use this expression to deduce the number dependence of the matrix elements of V.

4.1. *Number dependence of matrix elements off-diagonal in seniority*

In equation (A3.41) $V_{JM;00}$ is a quasi-spin scalar and consequently has vanishing matrix elements between states with different seniority. Similarly

$$\langle \phi_{IM;S'S_z}|\,\mathrm{S}_z\,|\phi_{IM;SS_z}\rangle = S_z\delta_{S',S}.$$

Thus all matrix elements off-diagonal in seniority are governed by the quasi-spin tensor of rank 2, $V_{JM;20}$. Since $S'=S+2$ implies that $v'=v-4$ and $S'=S+1$ means that $v'=v-2$ one sees that

$$\langle (j^n)_{IMv'}|\,V\,|(j^n)_{IMv}\rangle = \langle \phi_{IM;S'S_z}|\,V\,|\phi_{IM;SS_z}\rangle$$

$$= \frac{(S2S_z0\,|\,S'S_z)}{(S2-S0\,|\,S'-S)}\langle \phi_{IM;S'-S}|\,V\,|\phi_{IM;S-S}\rangle$$

$$= \left\{\frac{(2\Omega-n-v+4)(2\Omega-n-v+2)(n-v+4)(n-v+2)}{32(\Omega-v+2)(\Omega-v+1)}\right\}^{\frac{1}{2}}$$

$$\times \langle (j^v)_{IMv'}|\,V\,|(j^v)_{IMv}\rangle \delta_{v',v-4}$$

$$+\left(\frac{\Omega-n}{\Omega-v}\right)\left\{\frac{(2\Omega-n-v+2)(n-v+2)}{4(\Omega-v+1)}\right\}^{\frac{1}{2}}$$

$$\times \langle (j^v)_{IMv'}|\,V\,|(j^v)_{IMv}\rangle \delta_{v',v-2} \qquad \text{(A3.42)}$$

where Table A1.4 has been used to find the values of the Clebsch–Gordan coefficients.

Equation (A3.42) shows that for the half-filled shell the only non-vanishing matrix elements off-diagonal in seniority are those connecting states with $|v-v'|=4$. This is precisely the result (equation (3.50)) obtained by use of the particle-hole conjugation operator and provides an alternative explanation as to why all two-body interactions conserve seniority within the $f_{\frac{7}{2}}$ and $g_{\frac{7}{2}}$ single-particle orbits. The origin of the selection rule in this case is the fact that when $n=(2j+1)/2$, $S_z=0$ and the "parity-Clebsch" $(S200\,|\,S'0)$ vanishes unless $S+2+S'$ is even.

4.2. Seniority conservation with the delta-function interaction

In order to demonstrate that the delta-function potential conserves seniority within the identical nucleon configuration j^n, we examine explicitly the form of the seniority-mixing interaction $V_{JM;20}$. From equation (A3.36a) it follows that

$$\sum_M (-1)^{J-M} V_{JM;20} = \frac{1}{\sqrt{6}} \sum_M (-1)^{J-M} (R_{JM;11} R_{J-M;1-1} + 2R_{JM;10} R_{J-M;10} + R_{JM;1-1} R_{J-M;11}).$$

A calculation similar to that used in deducing equation (A3.38) leads to the conclusion that

$$\sum_M (-1)^{J-M} (R_{JM;11} R_{J-M;1-1} + R_{JM;1-1} R_{J-M;11}) = 2\sum_M (-1)^{J-M} R_{JM;11} R_{J-M;1-1} + (2J+1) S_z.$$

Therefore the seniority breaking part of V in equation (A3.41) takes the form

$$\begin{aligned} -\frac{\sqrt{2}}{\Omega\sqrt{3}} \sum_{JM} (-1)^{J-M} E_J V_{JM;20} &= -\frac{2}{3\Omega} \sum_{JM} (-1)^{J-M} E_J (R_{JM;11} R_{J-M;1-1} \\ &\quad + R_{JM;10} R_{J-M;10}) - \tfrac{2}{3} F_0 S_z \\ &= \tfrac{1}{3} \sum_{JM} (-1)^{J-M} E_J A^{\dagger}_{JM}(jj) \tilde{A}_{J-M}(jj) \\ &\quad - \frac{2}{3\Omega} \sum_{JM} (-1)^{J-M} E_J\, R_{JM;10} R_{J-M;10} - \tfrac{2}{3} F_0 S_z \qquad \text{(A3.43)} \end{aligned}$$

where F_0 is given by equation (A3.40).

The second term in this equation can be written in terms of $A^{\dagger}_{JM}(jj)$ and $\tilde{A}_{JM}(jj)$. From the definition of $R_{JM;10}$ in equation (A3.21) one sees that

$$\begin{aligned} \sum_{JM} (-1)^{J-M} E_J R_{JM;10} R_{J-M;10} &= \frac{\Omega}{2} \sum_{JM} (-1)^{J-M} E_J \left\{ U_{JM}(jj) - \left(\frac{\Omega}{2}\right)^{\frac{1}{2}} \delta_{J0} \right\} \left\{ U_{J-M}(jj) - \left(\frac{\Omega}{2}\right)^{\frac{1}{2}} \delta_{J0} \right\} \\ &= \frac{\Omega}{2} \sum_{JM} (-1)^{J-M} E_J \left\{ U_{JM}(jj) U_{J-M}(jj) - 2\left(\frac{\Omega}{2}\right)^{\frac{1}{2}} U_{JM}(jj) \delta_{J0} + \frac{\Omega}{2} \delta_{J0} \right\}. \end{aligned}$$

Furthermore,

$$U_{00}(jj)=\sum_{mm'}(jjmm'\mid 00)a^{\dagger}_{jm}\tilde{a}_{jm'}$$
$$=\frac{1}{\sqrt{(2\Omega)}}\mathrm{N}_j.$$

Thus when use is made of the fact that $(-1)^{J-M}=\sqrt{(2J+1)}(JJM-M\mid 00)$, one sees that

$$\sum_{JM}(-1)^{J-M}E_J R_{JM;10}R_{J-M;10}=\frac{\Omega}{2}\sum_J\sqrt{(2J+1)}E_J[U_J(jj)\times U_J(jj)]_{00}-\Omega E_0\mathrm{S}_z-\left(\frac{\Omega}{2}\right)^2E_0. \tag{A3.44}$$

The first term in this equation can be written as

$$\frac{\Omega}{2}\sum_J\sqrt{(2J+1)}E_J[U_J(jj)\times U_J(jj)]_{00}=\frac{\Omega}{2}\sum_J\sqrt{(2J+1)}E_J$$

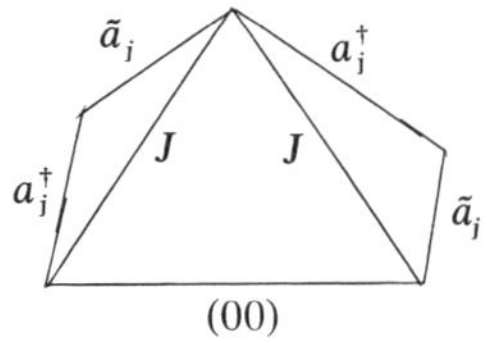

$$=\frac{\Omega}{2}\left\{-\sum_{JK}(2J+1)(2K+1)\sqrt{(2J+1)}E_J\begin{Bmatrix} j & j & J\\ j & j & J\\ K & K & 0\end{Bmatrix}\right.$$

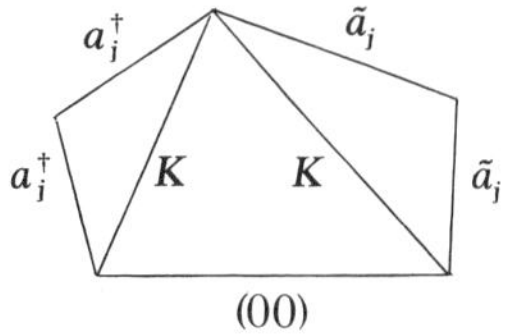

$$\left.+\sum_{JMm_i}(-1)^{J-M+j+m_2}E_J(jjm_1m_2\mid JM)(jjm_3m_4\mid J-M)a^{\dagger}_{m_1}\tilde{a}_{m_4}\delta_{m_3,-m_2}\right\} \tag{A3.45}$$

where we have used the definition of the $9j$ coefficient given in equation (A4.8). The minus sign in the first term arises because we anticommute $a^{\dagger}_{jm}$ and $\tilde{a}_{jm'}$ in order to get the two $a^{\dagger}_j$ together. In addition, this anticommutation leads to the second term of the equation. Since the $9j$ coefficient has one argument equal to zero, equation (A4.33) can be used to simplify the first term. Moreover, from the definition of $A^{\dagger}_{KM}(jj)$ and $\tilde{A}_{K-M}(jj)$ in equations (A3.22), it follows that

$$-\frac{\Omega}{2}\sum_{JK}(2J+1)(2K+1)\sqrt{(2J+1)}E_J\begin{Bmatrix} j & j & J\\ j & j & J\\ K & K & 0\end{Bmatrix}$$

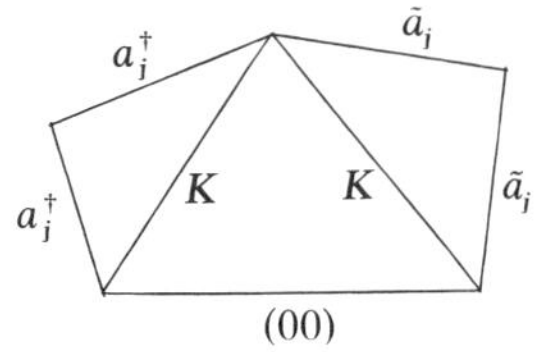

$$=\frac{\Omega}{2}\sum_{JK}(2J+1)\sqrt{(2K+1)}W(jjjj;JK)E_J$$

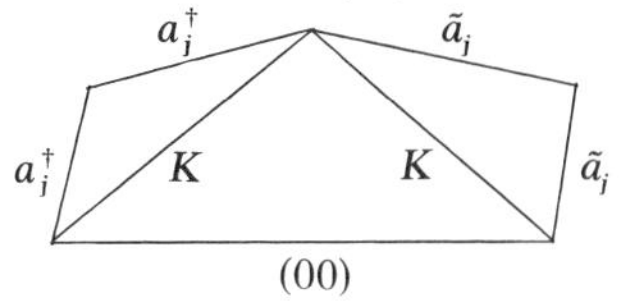

$$=\Omega\sum_{KM}(-1)^{K-M}F_K A^{\dagger}_{KM}(jj)\tilde{A}_{K-M}(jj) \tag{A3.46}$$

where F_K is given by equation (A3.39).

The summation in the second term of equation (A3.45) can easily be carried out by use of equations (A1.18)–(A1.20) and equation (A1.23):

$$\frac{\Omega}{2}\sum_{JMm_i}(-1)^{J-M+j+m_2}E_J(jjm_1m_2\,|\,JM)(jj-m_2m_4\,|\,J-M)\alpha^{\dagger}_{m_1}\tilde{a}_{m_4}$$

$$=\frac{\Omega}{2}\sum_J\frac{2J+1}{2j+1}E_J\sum_m(-1)^{j-m}a^{\dagger}_m\tilde{a}_{-m}$$

$$=\frac{\Omega}{2}F_0\mathrm{N}_j. \tag{A3.47}$$

When the results of equations (A3.45)–(A3.47) are inserted into equation (A3.44) it follows that

$$\sum_{JM}(-1)^{J-M}E_JR_{JM;10}R_{J-M;10}=\Omega\sum_{JM}(-1)^{J-M}F_JA^{\dagger}_{JM}(jj)\tilde{A}_{J-M}(jj)$$
$$+\frac{\Omega}{2}F_0\mathrm{N}_j-\Omega E_0\mathrm{S}_z-\left(\frac{\Omega}{2}\right)^2E_0 \tag{A3.48}$$

and, as a consequence, the seniority mixing part of V in (A3.43) takes the form

$$-\frac{\sqrt{2}}{\Omega\sqrt{3}}\sum_{JM}(-1)^{J-M}E_JV_{JM;20}$$
$$=\tfrac{1}{3}\sum_{JM}(-1)^{J-M}(E_J-2F_J)A^{\dagger}_{JM}(jj)\tilde{A}_{J-M}(jj)+\tfrac{1}{3}(E_0-2F_0)\left(\mathrm{N}_j-\frac{\Omega}{2}\right). \tag{A3.49}$$

Since the operator $(\mathrm{N}_j-\Omega/2)$ has only diagonal matrix elements, the first term in equation (A3.49) is the only one that can lead to seniority mixing. We shall now show that for the delta-function interaction

$$E_J-2F_J\equiv 0.$$

To see this we make use of equation (A2.29) which gives the matrix elements of the delta-function potential. Since we deal with identical particles, $T=1$ and

$$E_J=\frac{-V_0\bar{R}(2j+1)^2}{4}\left[\frac{\{1+(-1)^J\}(jj\tfrac{1}{2}-\tfrac{1}{2}\,|\,J0)^2}{(2J+1)}\right].$$

From the definition of F_J given in equation (A3.39) it follows that

$$2F_J=\frac{V_0\bar{R}(2j+1)^2}{2}\sum_K\{1+(-1)^K\}W(jjjj;JK)(jj\tfrac{1}{2}-\tfrac{1}{2}\,|\,K0)^2$$

$$=\frac{V_0\bar{R}(2j+1)^2}{2(2J+1)}\{(jj\tfrac{1}{2}\tfrac{1}{2}\,|\,J1)^2-(-1)^J(jj\tfrac{1}{2}-\tfrac{1}{2}\,|\,J0)^2\}$$

where use has been made of equations (A4.19) and (A4.20) to carry out the summation over K. Consequently

$$E_J-2F_J=\frac{-V_0\bar{R}(2j+1)^2}{4(2J+1)}[\{1-(-1)^J\}(jj\tfrac{1}{2}-\tfrac{1}{2}\,|\,J0)^2+2(jj\tfrac{1}{2}\tfrac{1}{2}\,|\,J1)^2]. \tag{A3.50}$$

Because of the Pauli principle, $A^{\dagger}_{JM}(jj)$ and $\tilde{A}_{J-M}(jj)$ only exist for even values of J, and when this is true the first term in equation (A3.50) vanishes. Furthermore, because of equation (A1.18),

$$(jj\tfrac{1}{2}\tfrac{1}{2} \mid J1) = (-1)^{1+J}(jj\tfrac{1}{2}\tfrac{1}{2} \mid J1)$$

so that

$$(jj\tfrac{1}{2}\tfrac{1}{2} \mid J1) = 0 \quad \text{for even values of } J.$$

Consequently for the delta-function interaction $(E_J - 2F_J) \equiv 0$ and the matrix elements off-diagonal in seniority vanish.

4.3. Number dependence of matrix elements diagonal in seniority

To simplify the notation we rewrite V of equation (A3.41) as

$$V = V_{20} + V_{00} + 2F_0 S_z$$

where V_{20} is a quasi-spin tensor of rank two and V_{00} is the quasi-spin scalar part of the interaction. Furthermore, since S_z has only diagonal matrix elements we must introduce a quantum number α in case seniority and angular momentum are not sufficient to completely specify the states of j^n. Thus

$$\begin{aligned}
\langle V\rangle_{nv\alpha'\alpha} &= \langle (j^n)_{IMv\alpha'} | \, V \, | (j^n)_{IMv\alpha}\rangle \\
&= \langle \phi_{IM\alpha';SS_z} | \, V \, | \phi_{IM\alpha;SS_z}\rangle \\
&= \left\{\frac{(S2S_z0 \mid SS_z)}{(S2-S0 \mid S-S)}\right\}\langle V_{20}\rangle_{vv\alpha'\alpha} + \langle V_{00}\rangle_{vv\alpha'\alpha} + F_0(n-\Omega)\delta_{\alpha',\alpha} \\
&= \left\{\frac{3S_z^2 - S(S+1)}{3S^2 - S(S+1)}\right\}\langle V_{20}\rangle_{vv\alpha'\alpha} + \langle V_{00}\rangle_{vv\alpha'\alpha} + F_0(n-\Omega)\delta_{\alpha',\alpha}
\end{aligned} \tag{A3.51}$$

where we have made use of the fact that the matrix element of the quasi-spin scalar V_{00} is independent of n and the explicit form of the Clebsch–Gordan coefficient given in Table A1.4 has been used.

Thus to calculate matrix elements of v one must know the values of $\langle V_{20}\rangle_{vv\alpha'\alpha}$ and $\langle V_{00}\rangle_{vv\alpha'\alpha}$. Clearly if one knew the value of $\langle V\rangle_{nv\alpha'\alpha}$ for two different values of n one could deduce values for these quantities. For example, when $n = v$, $S_z = -S$ and equation (A3.51) takes the form

$$\langle V\rangle_{vv\alpha'\alpha} = \langle V_{20}\rangle_{vv\alpha'\alpha} + \langle V_{00}\rangle_{vv\alpha'\alpha} + F_0(v-\Omega)\delta_{\alpha',\alpha}, \tag{A3.52a}$$

and when $n = v+2$, $S_z = -S+1$ so that

$$\langle V\rangle_{v+2,v\alpha'\alpha} = \left\{\frac{3(S-1)^2 - S(S+1)}{3S^2 - S(S+1)}\right\}\langle V_{20}\rangle_{vv\alpha'\alpha} + \langle V_{00}\rangle_{vv\alpha'\alpha} + F_0(v+2-\Omega)\delta_{\alpha'\alpha}. \tag{A3.52b}$$

If one subtracts equation (A3.52a) from equation (A3.52b) one can obtain an expression for $\langle V_{20}\rangle_{v v\alpha'\alpha}$ in terms of the matrix elements of V. Thus

$$\langle V_{20}\rangle_{vv\alpha'\alpha}=\left\{\frac{S(S+1)-3S^2}{3(2S-1)}\right\}\{\langle V\rangle_{v+2,v\alpha'\alpha}-\langle V\rangle_{vv\alpha'\alpha}-2F_0\delta_{\alpha',\alpha}\} \quad \text{(A3.53a)}$$

and

$$\langle V_{00}\rangle_{vv\alpha'\alpha}=-\left\{\frac{S(S+1)-3S^2}{3(2S-1)}\right\}\{\langle V\rangle_{v+2,v\alpha'\alpha}-\langle V\rangle_{vv\alpha'\alpha}-2F_0\delta_{\alpha',\alpha}\}$$
$$+\langle V\rangle_{vv\alpha'\alpha}-F_0(v-\Omega)\delta_{\alpha',\alpha}. \quad \text{(A3.53b)}$$

By substituting equations (A3.53) into equation (A3.51) one arrives at the relationship

$$\langle V\rangle_{nv\alpha'\alpha}=\left\{\frac{(S-S_z)(S+S_z)}{(2S-1)}\right\}\{\langle V\rangle_{v+2,v\alpha'\alpha}-\langle V\rangle_{vv\alpha'\alpha}-2F_0\delta_{\alpha',\alpha}\}$$
$$+\langle V\rangle_{vv\alpha'\alpha}-F_0(v-n)\delta_{\alpha',\alpha}$$
$$=\left\{\frac{(2\Omega-n-v)(n-v)}{4(\Omega-v-1)}\right\}\{\langle V\rangle_{v+2,v\alpha'\alpha}-\langle V\rangle_{vv\alpha'\alpha}-2F_0\delta_{\alpha',\alpha}\}$$
$$+\langle V\rangle_{vv\alpha'\alpha}-F_0(v-n)\delta_{\alpha',\alpha}. \quad \text{(A3.54)}$$

From equation (A3.54) it follows that

$$\langle V\rangle_{2j+1-n,v\alpha'\alpha}-\langle V\rangle_{nv\alpha'\alpha}=(2j+1-2n)F_0\delta_{\alpha'\alpha}.$$

Thus aside from an additive I-independent term, the energy of the n-particle state $(j^n)_{IMv\alpha}$ is identical to the energy of the n-hole configuration $(j^{-n})_{IMv\alpha}$—a result that was demonstrated in Chapter 3, section 2.2, by use of the particle-hole conjugation operator.

In order to simplify equation (A3.54), consider the evaluation of $\langle V\rangle_{v+2,v\alpha'\alpha}$. In equation (1.124) we showed that if $\Phi_{IMv\alpha}\,|0\rangle$ is a v-particle seniority-v eigenfunction, then the normalized $(v+2)$-particle state with the same $(IMv\alpha)$ quantum numbers is

$$(j^{v+2})_{IMv\alpha}=\frac{1}{\sqrt{(\Omega-v)}}\,\mathrm{S}_+\Phi_{IMv\alpha}\,|0\rangle. \quad \text{(A3.55)}$$

Thus

$$\langle V\rangle_{v+2,v\alpha'\alpha}=\frac{1}{\Omega-v}\langle 0|\,\Phi^\dagger_{IMv\alpha}\mathrm{S}_-V\mathrm{S}_+\Phi_{IMv\alpha}\,|0\rangle$$
$$=\frac{1}{\Omega-v}\langle \mathrm{S}_-V\mathrm{S}_+\rangle_{vv\alpha'\alpha}. \quad \text{(A3.56)}$$

In other words, the matrix element of V in the $(v+2)$-particle seniority-v state is the same as the matrix element of the modified operator $(\mathrm{S}_-V\mathrm{S}_+)/(\Omega-v)$ evaluated in the v-particle seniority-v state. The fact that $\underset{\sim}{\mathrm{S}}$ is the generator of infinitesimal rotations in quasi-spin space can be used to simplify the form of

S_-VS_+. From equations (A3.35) and (A3.8) one can rewrite this operator as

$$\begin{aligned}\frac{S_-VS_+}{\Omega-v}&=\frac{-2}{\Omega(\Omega-v)}\sum_{JM}(-1)^{J-M}E_JS_-R_{JM;11}R_{J-M;1-1}S_+\\&=\frac{-2}{\Omega(\Omega-v)}\sum_{JM}(-1)^{J-M}E_J(R_{JM;11}S_-+\sqrt{2}R_{JM;10})(S_+R_{J-M;1-1}\\&\quad-\sqrt{2}R_{J-M;10})\\&=\frac{-2}{\Omega(\Omega-v)}\sum_{JM}(-1)^{J-M}E_J(R_{JM;11}S_-S_+R_{J-M;1-1}\\&\quad-\sqrt{2}R_{JM;11}S_-R_{J-M;10}+\sqrt{2}R_{JM;10}S_+R_{J-M;1-1}-2R_{JM;10}R_{J-M;10}).\end{aligned}\tag{A3.57}$$

By use of equations (A3.4) and (A3.8) one can simplify the first term in this expression

$$\begin{aligned}R_{JM;11}S_-S_+R_{J-M;1-1}&=-2R_{JM;11}S_zR_{J-M;1-1}\\&\quad+R_{JM;11}S_+S_-R_{J-M;1-1}\\&=-2R_{JM;11}R_{J-M;1-1}S_z+2R_{JM;11}R_{J-M;1-1}\\&\quad+R_{JM;11}S_+R_{J-M;1-1}S_-.\end{aligned}$$

However, since

$$S_-(j^v)_{IMv\alpha}=0\tag{A3.58}$$

$$\langle R_{JM;11}S_-S_+R_{J-M;1-1}\rangle_{vv\alpha'\alpha}=(\Omega-v+2)\langle R_{JM;11}R_{J-M;1-1}\rangle_{vv\alpha'\alpha}.\tag{A3.59}$$

The second term in equation (A3.57) can be rewritten, by use of equation (A3.8), as

$$\begin{aligned}-\sqrt{2}\langle R_{JM;11}S_-R_{J-M;10}\rangle_{vv\alpha'\alpha}&=-\sqrt{2}\langle R_{JM;11}\{R_{J-M;10}S_-+\sqrt{2}R_{J-M;1-1}\}\rangle_{vv\alpha'\alpha}\\&=-2\langle R_{JM;11}R_{J-M;1-1}\rangle_{vv\alpha'\alpha}\end{aligned}\tag{A3.60}$$

where use has been made of equation (A3.58) in writing the final line of this equation. In the same way, when we commute $R_{JM;10}$ with S_+ the third term in equation (A3.57) becomes

$$\begin{aligned}\sqrt{2}\langle R_{JM;10}S_+R_{J-M;1-1}\rangle_{vv\alpha'\alpha}&=-2\langle R_{JM;11}R_{J-M;1-1}\rangle_{vv\alpha'\alpha}\\&\quad+\sqrt{2}\langle 0|\,\Phi^\dagger_{IMv\alpha}S_+R_{JM;10}R_{J-M;1-1}\Phi_{IMv\alpha}\,|0\rangle.\end{aligned}\tag{A3.61}$$

However, since

$$\Phi^\dagger_{IMv\alpha}S_+\,|0\rangle=(S_-\Phi_{IMv\alpha})^\dagger\,|0\rangle=0$$

only the first term in this equation has a non-vanishing value.

Finally, $R_{JM;10}R_{J-M;10}$ has the value given by equation (A3.48). When these results (equations (A3.48) and equations (A3.59)–(A3.61)) are used one finds

$$\begin{aligned}\langle V\rangle_{v+2,v\alpha'\alpha} &= \frac{1}{\Omega-v}\langle S_- V S_+\rangle_{vv\alpha'\alpha}\\ &= \frac{\Omega-v-2}{\Omega-v}\langle V\rangle_{vv\alpha'\alpha}\\ &\quad+\frac{4}{\Omega-v}\langle\sum_{JM}(-1)^{J-M}F_J A^{\dagger}_{JM}(jj)\tilde{A}_{J-M}(jj)\rangle_{vv\alpha'\alpha}\\ &\quad-\frac{2v}{\Omega-v}F_0\delta_{\alpha'\alpha}+\frac{\Omega-2v}{\Omega-v}E_0\delta_{\alpha'\alpha}.\end{aligned}$$

When this result is substituted into equation (A3.54) one obtains an expression for the matrix element of V in the n-particle system in terms of the matrix element evaluated in the v-particle configuration

$$\begin{aligned}\langle V\rangle_{nv\alpha'\alpha} &= \langle V\rangle_{vv\alpha'\alpha}-F_0(v-n)\delta_{\alpha'\alpha}\\ &\quad+\frac{(2\Omega-n-v)(n-v)}{4(\Omega-v-1)(\Omega-v)}((\Omega-2v)(E_0-2F_0)\delta_{\alpha'\alpha}-2\langle\bar{V}\rangle_{vv\alpha'\alpha})\end{aligned}\tag{A3.62}$$

where

$$\bar{V}=\sum_{JM}(-1)^{J-M}(E_J-2F_J)A^{\dagger}_{JM}(jj)\tilde{A}_{J-M}(jj)\tag{A3.63}$$

with F_J given by equation (A3.39).

For the seniority-zero and -one configurations, equation (A3.62) may be evaluated simply. Since V and $\bar{V}$ both have the two-particle destruction operator $\tilde{A}_{J-M}(jj)$ standing to the extreme right, the matrix elements of both these operators vanish in the zero- and one-particle configurations. Thus the energy of the $I=0$ seniority-zero state is

$$\langle V\rangle_{n,I=0,v=0}=nF_0+\frac{n(2j+1-n)}{2(2j-1)}(E_0-2F_0)$$

and by using the definition of F_0 in equation (A3.40) this can be shown to be identical to the result deduced in equation (1.127). In the same way the energy of the $I=j$, $v=1$ state is

$$\langle V\rangle_{n,I=j,v=1}=(n-1)F_0+\frac{(n-1)(2j-n)}{2(2j-1)}(E_0-2F_0)$$

which is identical to the result given by equation (1.129).

APPENDIX 4

RACAH AND 9J COEFFICIENTS

In this appendix we discuss properties of the Racah and $9j$ coefficients and tabulate some useful formulae involving them.

1. Racah coefficients

The Racah coefficient (Racah 1942) arises in calculations involving the coupling of three angular momenta j_1, j_2, and j_3 to total spin $(j_4 m_4)$. There are several ways this coupling can be carried out, and the resultant wave functions, as we shall show, are not orthogonal. For example, j_1 and j_2 can be coupled to J and then J and j_3 can be compounded to give $(j_4 \bar{m}_4)$. We denote the state vector obtained with this order of coupling by $\Psi_{j_4 \bar{m}_4}$

$$\Psi_{j_4\bar{m}_4} = [[\phi_{j_1}(1)\times\phi_{j_2}(2)]_J \times \phi_{j_3}(3)]_{j_4\bar{m}_4} =$$

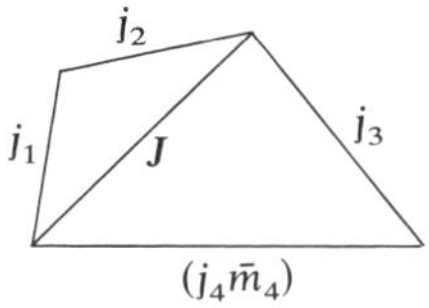

A second way $(j_4 \bar{m}_4)$ can be arrived at is by coupling j_2 and j_3 to spin K and then j_1 and K to the final result

$$\Phi_{j_4\bar{m}_4} = [\phi_{j_1}(1)\times[\phi_{j_2}(2)\times\phi_{j_3}(3)]_K]_{j_4\bar{m}_4} =$$

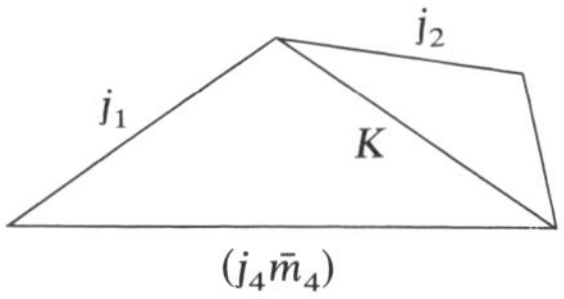

Provided the $\phi_{j_i m_i}$ are orthonormal wave functions it follows that

$$\begin{aligned}
\langle \Psi_{j_4\bar{m}_4} \mid \Phi_{j_4\bar{m}_4}\rangle
&= \sum_{m_i M m_i' M'} (j_1 j_2 m_1 m_2 \mid JM)(J j_3 M m_3 \mid j_4 \bar{m}_4)(j_1 K m_1' M' \mid j_4 \bar{m}_4)(j_2 j_3 m_2' m_3' \mid KM') \\
&\quad \times \langle \phi_{j_1 m_1} \mid \phi_{j_1 m_1'}\rangle \langle \phi_{j_2 m_2} \mid \phi_{j_2 m_2'}\rangle \langle \phi_{j_3 m_3} \mid \phi_{j_3 m_3'}\rangle \\
&= \sum_{m_i M M'} (j_1 j_2 m_1 m_2 \mid JM)(J j_3 M m_3 \mid j_4 \bar{m}_4)(j_1 K m_1 M' \mid j_4 \bar{m}_4)(j_2 j_3 m_2 m_3 \mid KM') \\
&= \sqrt{\{(2J+1)(2K+1)\}}\, W(j_1 j_2 j_4 j_3; JK) \qquad \text{(A4.1)}
\end{aligned}$$

where $W(j_1 j_2 j_4 j_3; JK)$ is the Racah coefficient (Racah 1942).

In writing equation (A4.1) we have assumed that W is independent of $\bar{m}_4$. That this is true can be shown by evaluating the overlap integral for the case $(\bar{m}_4+1)$. Provided the $\phi_{j_i m_i}$ are irreducible tensors (see Appendix 2, section 1) it follows that $\Psi_{j_4\bar{m}_4}$ and $\Phi_{j_4\bar{m}_4}$, which are products of these operators, are irreducible tensors of rank j_4. Thus

$$\mathrm{J}_+\Psi_{j_4\bar{m}_4}=\sqrt{\{(j_4-\bar{m}_4)(j_4+\bar{m}_4+1)\}}\Psi_{j_4\bar{m}_4+1}$$

$$\mathrm{J}_+\Phi_{j_4\bar{m}_4}=\sqrt{\{(j_4-\bar{m}_4)(j_4+\bar{m}_4+1)\}}\Phi_{j_4\bar{m}_4+1}.$$

Consequently

$$\begin{aligned}\langle\Psi_{j_4\bar{m}_4+1}\mid\Phi_{j_4\bar{m}_4+1}\rangle&=\frac{\langle \mathrm{J}_+\Psi_{j_4\bar{m}_4}\mid \mathrm{J}_+\Phi_{j_4\bar{m}_4}\rangle}{(j_4-\bar{m}_4)(j_4+\bar{m}_4+1)}\\&=\frac{\langle\Psi_{j_4\bar{m}_4}\mid \mathrm{J}_-\mathrm{J}_+\Phi_{j_4\bar{m}_4}\rangle}{(j_4-\bar{m}_4)(j_4+\bar{m}_4+1)}\\&=\frac{\langle\Psi_{j_4\bar{m}_4}\mid(\mathrm{J}^2-\mathrm{J}_z^2-\mathrm{J}_z)\Phi_{j_4\bar{m}_4}\rangle}{(j_4-\bar{m}_4)(j_4+\bar{m}_4+1)}\\&=\langle\Psi_{j_4\bar{m}_4}\mid\Phi_{j_4\bar{m}_4}\rangle.\end{aligned}$$

Because the Clebsch–Gordan coefficient $(j_a j_b m_a m_b \mid LM)$ vanishes unless $|j_a-j_b|\leq L\leq(j_a+j_b)$, it follows that the Racah coefficient has non-zero values only if

$$|j_1-j_2|\leq J\leq j_1+j_2;\qquad |j_4-j_3|\leq J\leq j_4+j_3$$

$$|j_1-j_4|\leq K\leq j_1+j_4;\qquad |j_2-j_3|\leq K\leq j_2+j_3.$$

Furthermore, the symmetry properties of the Clebsch–Gordan coefficients given in equations (A1.18)–(A1.20) can be used to show that

$$\begin{aligned}W(abcd;ef)&=W(badc;ef)=W(cdab;ef)=W(acbd;fe)\\&=(-1)^{e+f-a-d}W(ebcf;ad)\\&=(-1)^{e+f-b-c}W(aefd;bc).\end{aligned}\qquad\text{(A4.2)}$$

The Clebsch–Gordan coefficients take on an extremely simple form when one angular momentum is zero

$$(j_1j_2m_1m_2\mid 00)=(-1)^{j_1-m_1}(2j_1+1)^{-\frac{1}{2}}\delta_{j_1j_2}\delta_{m_1m_2}$$

$$(0j_30m_3\mid j_4m_4)=\delta_{j_3j_4}\delta_{m_3m_4}.$$

When these results plus the symmetry properties (equations (A1.19) and (A1.20)) are used in equation (A4.1) one finds that

$$W(j_1j_2j_4j_3;0K)=(-1)^{j_1+j_4-K}\{(2j_1+1)(2j_4+1)\}^{-\frac{1}{2}}\delta_{j_1j_2}\delta_{j_3j_4}.$$

From the orthogonality property of the Clebsch–Gordan coefficients

$$\sum_{m_1m_2}(j_1j_2m_1m_2\mid JM)(j_1j_2m_1m_2\mid J'M')=\delta_{JJ'}\delta_{MM'}$$

one sees that

$$\langle[[\phi_{j_1}(1)\times\phi_{j_2}(2)]_J\times\phi_{j_3}(3)]_{j_4m_4}\,|\,[[\phi_{j_1}(1)\times\phi_{j_2}(2)]_{J'}\times\phi_{j_3}(3)]_{j_4m_4}\rangle=\delta_{JJ'}$$

and

$$\langle[\phi_{j_1}(1)\times[\phi_{j_2}(2)\times\phi_{j_3}(3)]_K]_{j_4m_4}\,|\,[\phi_{j_1}(1)\times[\phi_{j_2}(2)\times\phi_{j_3}(3)]_{K'}]_{j_4m_4}\rangle=\delta_{KK'}.$$

Consequently

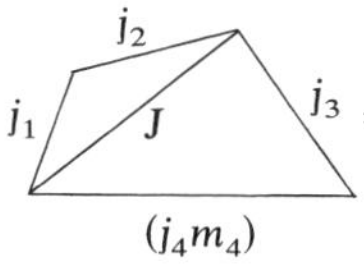

$$=\sum_K\sqrt{\{(2J+1)(2K+1)\}}W(j_1j_2j_4j_3;JK)$$

$\times$

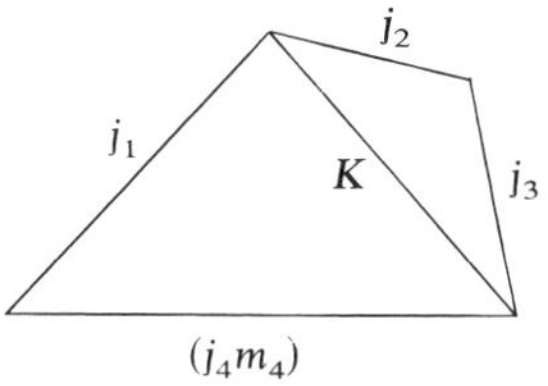

(A4.3)

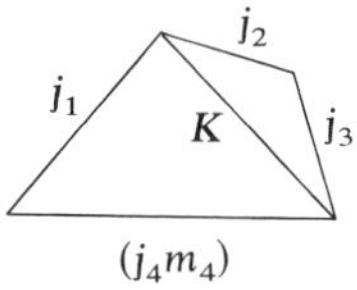

$$=\sum_J\sqrt{\{(2J+1)(2K+1)\}}W(j_1j_2j_4j_3;JK)$$

$\times$

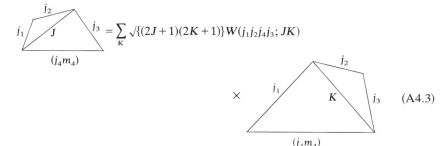

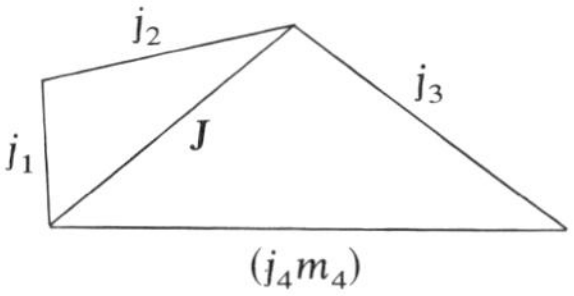

(A4.4)

and

$$\sum_K(2J+1)(2K+1)W(j_1j_2j_4j_3;JK)W(j_1j_2j_4j_3;J'K)=\delta_{JJ'}$$

$$\sum_J(2J+1)(2K+1)W(j_1j_2j_4j_3;JK)W(j_1j_2j_4j_3;JK')=\delta_{KK'}.$$

Since the wave functions $\phi_{j_im_i}$ in equation (A4.3) are orthonormal it follows that

$$\sum_M(j_1j_2m_1m_2\,|\,JM)(Jj_3Mm_3\,|\,j_4m_4)=\sum_{KM'}\sqrt{\{(2J+1)(2K+1)\}}$$
$$\times W(j_1j_2j_4j_3;JK)(j_1Km_1M'\,|\,j_4m_4)(j_2j_3m_2m_3\,|\,KM'). \quad \text{(A4.5)}$$

Multiplication of each side of this equation by $(j_2j_3m_2m_3\,|\,K''M'')$ and summation over m_2 and m_3 leads to the relationship

$$\sum_{m_2m_3M}(j_1j_2m_1m_2\,|\,JM)(Jj_3Mm_3\,|\,j_4m_4)(j_2j_3m_2m_3\,|\,KM')$$
$$=\sqrt{\{(2J+1)(2K+1)\}}W(j_1j_2j_4j_3;JK)(j_1Km_1M'\,|\,j_4m_4).$$

Equation (A4.5) can be written in a slightly different form if one exploits the symmetry properties of the Clebsch–Gordan coefficients given in equations (A1.18) and (A1.20):

$$(2J+1)\sum_{KM'} W(j_1j_2j_4j_3;JK)(j_1j_4m_1-m_4\mid K-M')(j_2j_3m_2m_3\mid KM')$$
$$=(-1)^{j_4-J-j_1+m_1-m_3}\sum_{M}(j_1j_2m_1m_2\mid JM)(j_3j_4m_3-m_4\mid J-M). \quad \text{(A4.6)}$$

In Appendix 2, section 9, we showed that the matrix elements of the delta-function potential are proportional to Clebsch–Gordan coefficients in which $|m_i|=\frac{1}{2}$ (see equation (A2.29)). Since equation (A4.6) holds for any value of m_i, one may obtain several relationships which are quite useful when a short-range potential is considered. In particular, if $m_1=m_2=m_4=\frac{1}{2}$ and $m_3=-\frac{1}{2}$ it follows that

$$(2J+1)\sum_{K} W(j_1j_2j_4j_3;JK)(j_1j_4\tfrac{1}{2}-\tfrac{1}{2}\mid K0)(j_2j_3\tfrac{1}{2}-\tfrac{1}{2}\mid K0)$$
$$=(-1)^{j_3-j_1}(j_1j_2\tfrac{1}{2}\tfrac{1}{2}\mid J1)(j_3j_4\tfrac{1}{2}\tfrac{1}{2}\mid J1).$$

Other relationships that may be written when $|m_i|=\frac{1}{2}$ are given in section 3 of this Appendix.

In the following discussion it will be useful to remember that whether j is integral or half integral $(-1)^{4j}=1$ so that $(-1)^{2j}=(-1)^{-2j}$. Furthermore, if j_a and j_b couple to a resultant j_c then $(j_a+j_b+j_c)$ is an integer and $(-1)^{2(j_a+j_b+j_c)}=1$.

With these results in mind we shall now deduce two additional summation relationships involving Racah coefficients that are often encountered. To derive these, one first does the recoupling the straightforward way and then by a more round about method. For example, equation (A4.3) is the simple way to change from $[[j_1\times j_2]_J\times j_3]_{j_4m_4}$ to $[j_1\times[j_2\times j_3]_K]_{j_4m_4}$. Alternatively, one can do this transformation in two steps. First, the order of the (j_1j_2) coupling is changed and then j_1 and j_3 are coupled to a dummy angular momentum L. Thus

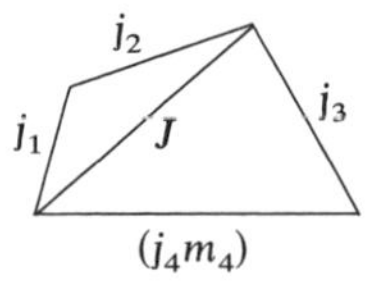

$=(-1)^{j_1+j_2-J}$
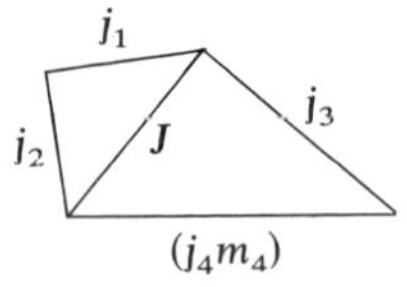

$$=(-1)^{j_1+j_2-J}\sum_{L}\sqrt{\{(2J+1)(2L+1)\}}\,W(j_2j_1j_4j_3;JL)$$
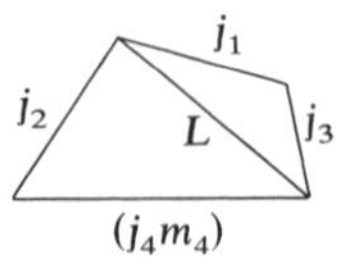

In order to obtain a diagram with the same form as that on the right-hand side of equation (A4.3), the order of j_1 and j_3 is changed and then the recoupling given

by equation (A4.4) is used. Thus

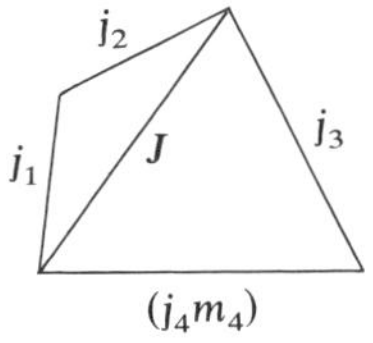

$$= (-1)^{j_2-j_3+L-J}\sum_{LK}(2L+1)\surd\{(2J+1)(2K+1)\}$$

$$\times W(j_2j_1j_4j_3;\, JL)W(j_2j_3j_4j_1;\, KL)$$

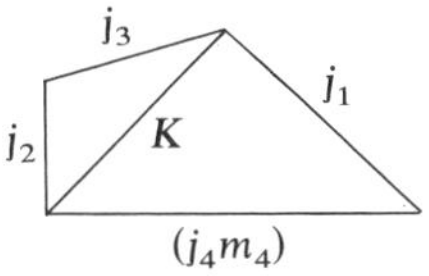

$$= \sum_{LK}(-1)^{j_2-j_3+L-J+K+j_1-j_4}(2L+1)\surd\{(2J+1)(2K+1)\}$$

$$\times W(j_2j_1j_4j_3;\, JL)W(j_2j_3j_4j_1;\, KL)$$

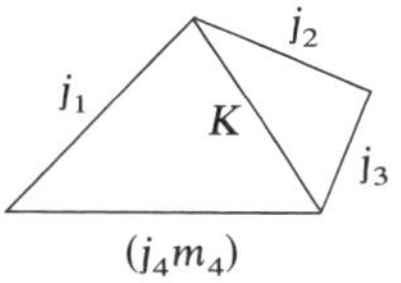

Since the 'simple' and 'hard' way of doing the recoupling must be identical it follows, once the symmetry properties of equation (A4.2) are used, that

$$\sum_{L}(-1)^{j_1+j_3-L}(2L+1)W(j_2j_4j_1j_3;\, LJ)W(j_2j_4j_3j_1;\, LK) = W(j_1JKj_3;\, j_2j_4).$$

The second summation relationship is one in which the sum of the product of three Racah coefficients can be expressed as two. Since one way of deriving this sum rule is similar to the method one uses to evaluate matrix elements of the residual two-body spin–orbit interaction or the tensor force, we formulate the derivation along those lines. In L–S coupling the total orbital angular momentum λ and the total spin S are compounded to give I. One can express λ in terms of the relative angular momentum l and the centre-of-mass angular momentum L of the two interacting nucleons. Consequently

$$\Psi_{IM}(lL\lambda S) =$$

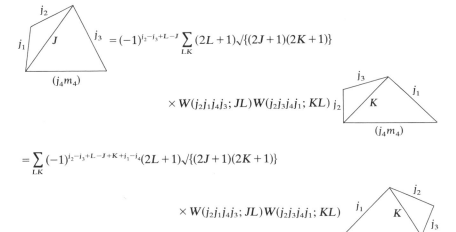

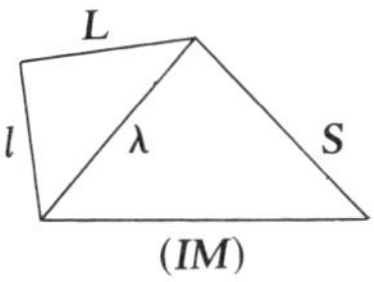

We consider the effect of a potential V on $\Psi_{IM}(lL\lambda S)$ where V is the product of two tensor operators X_k and Z_k of rank k. The former we assume acts only on the spin part of the wave function and the latter on the relative angular-momentum part. Since V must be a scalar it can be written as

$$V = [X_k \times Z_k]_{00}.$$

To illustrate what is meant by this equation, we consider the two-body spin–orbit force

$$V = \tfrac{1}{2}V_{so}(r)(\underset{\sim}{\sigma}_1 + \underset{\sim}{\sigma}_2)\,.\,\underset{\sim}{\ell}$$
$$= V_{so}(r)\underset{\sim}{S}\,.\,\underset{\sim}{\ell}$$
$$= -\sqrt{3}\,V_{so}(r)[\underset{\sim}{S} \times \underset{\sim}{\ell}]_{00}$$

where

$$\underset{\sim}{S} = \tfrac{1}{2}(\underset{\sim}{\sigma}_1 + \underset{\sim}{\sigma}_2)$$

is the total spin operator. Thus in this case

$$X_k = X_{k=1}$$
$$= \underset{\sim}{S}$$

and

$$Z_k = Z_{k=1}$$
$$= -\sqrt{3}\,V_{so}(r)\underset{\sim}{\ell}.$$

In terms of diagrams the operation $V\Psi_{IM}(lL\lambda S)$ takes the form

$V\Psi_{IM}(lL\lambda S) = X_k$ 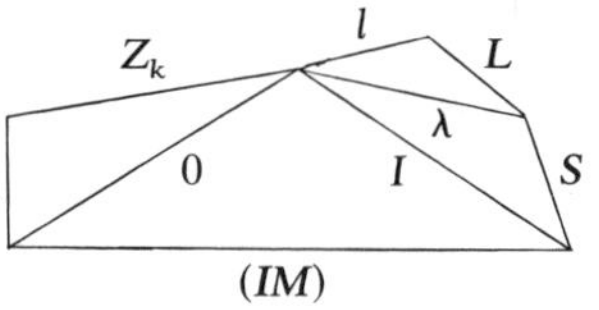

From the definition of the reduced matrix element in equation (A2.8) it follows that

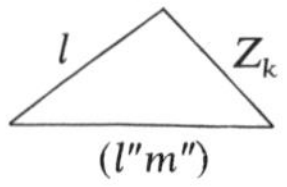

$= (-1)^{l+k-l''}$ 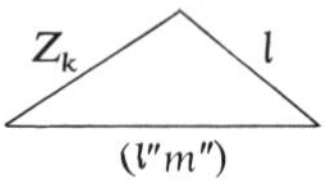

$= \langle l'' \| Z_k \| l \rangle$ 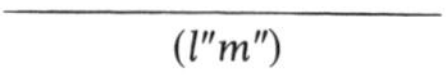 .

When this is combined with the fact that

$$W(0bcd\,;\,ef) = \{(2b+1)(2c+1)\}^{-\frac{1}{2}}\delta_{be}\delta_{cf}$$

one finds that

$V\Psi_{IM}(lL\lambda S) =$ 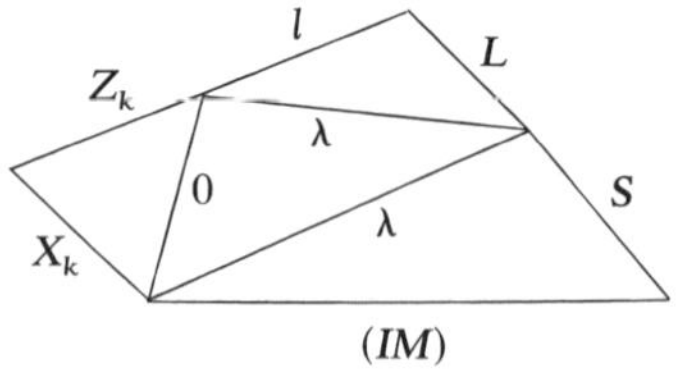

$=$ 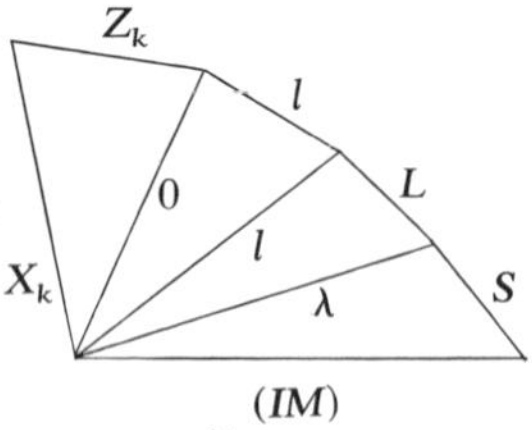

$$= \sum_{l''} \left\{\frac{(2l''+1)}{(2k+1)(2l+1)}\right\}^{\frac{1}{2}} \langle l'' \| Z_k \| l \rangle$$ 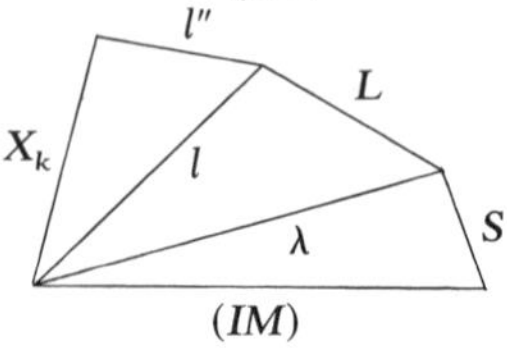 .

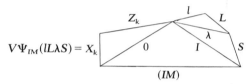

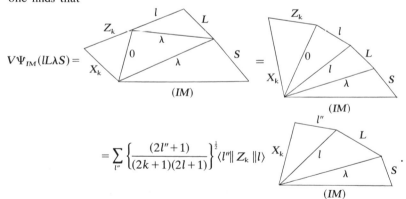

The final step in the calculation involves bringing X_k and S into the same triangle and this can be done by making two recouplings:

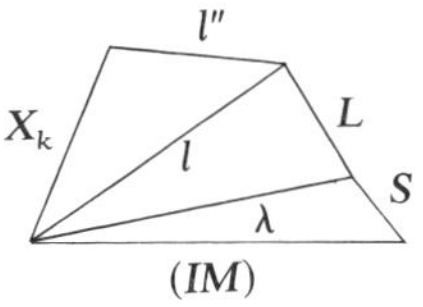

$$= \sum_{\lambda''} (-1)^{k+\lambda''-\lambda} \sqrt{\{(2l+1)(2\lambda''+1)\}} W(kl''\lambda L; l\lambda'')$$

$\times$ 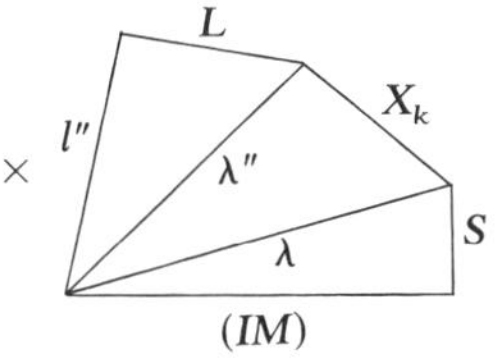

$$= \sum_{\lambda'' S''} (-1)^{2k+\lambda''-\lambda+S-S''} \sqrt{\{(2l+1)(2\lambda''+1)(2\lambda+1)(2S''+1)\}} \langle S'' \| X_k \| S \rangle$$

$$\times W(kl''\lambda L; l\lambda'') W(\lambda'' k I S; \lambda S'')$$

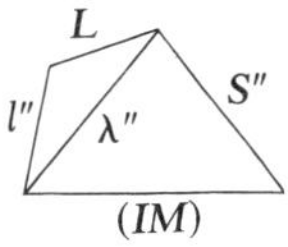

Since typically we want to calculate matrix elements of V between $\Psi_{IM}(lL\lambda S)$ and $\Psi_{IM}(l'L\lambda' S')$, $l''=l'$, $\lambda''=\lambda'$ and $S''=S'$. Furthermore, since (k, S, S') form a triangle $(-1)^{2k+S-S'} = (-1)^{S'-S}$ so that

$$\langle \Psi_{IM}(l'L\lambda' S') | V | \Psi_{IM}(lL\lambda S) \rangle = (-1)^{S'+\lambda'-S-\lambda} \langle l' \| Z_k \| l \rangle \langle S' \| X_k \| S \rangle$$

$$\times \left\{ \frac{(2l'+1)(2\lambda+1)(2S'+1)(2\lambda'+1)}{(2k+1)} \right\}^{\frac{1}{2}} W(kl'\lambda L; l\lambda') W(\lambda' k I S; \lambda S') \quad \text{(A4.6)}$$

An alternative way of evaluating the matrix element is to first put l and S into the same triangle

$$\Psi_{IM}(lL\lambda S) = (-1)^{l+L-\lambda} \sum_K \sqrt{\{(2\lambda+1)(2K+1)\}} W(LlIS; \lambda K)$$

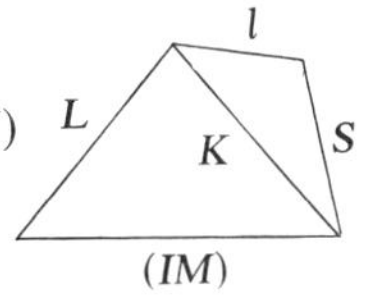

$$= \sum_K (-1)^{K-I+\lambda-l} \sqrt{\{(2\lambda+1)(2K+1)\}} W(LlIS; \lambda K)$$

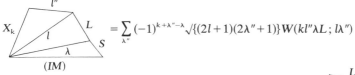

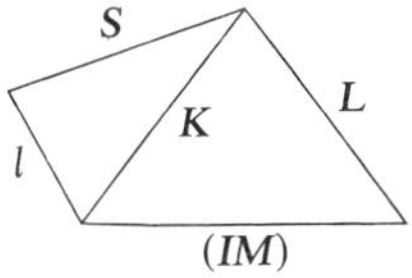

$$= \sum_K (-1)^{K-I+\lambda-l} \sqrt{\{(2\lambda+1)(2K+1)\}} W(LlIS; \lambda K) \tilde{\Psi}_{IM}(lL\lambda S) \quad \text{(A4.7)}$$

If one now applies V to $\tilde{\Psi}_{IM}(lL\lambda S)$ one obtains

$V\tilde{\Psi}_{IM}(lL\lambda S)=$ 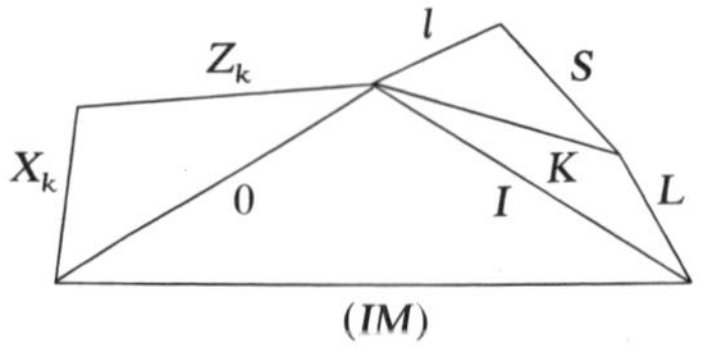

$=$ 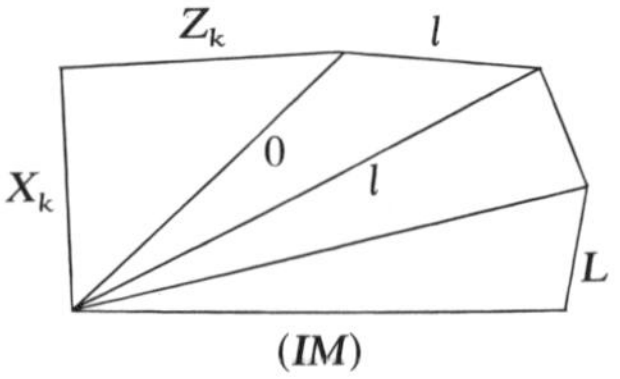

$$=\sum_{l''}\langle l''\|\, Z_k\,\|l\rangle\left\{\frac{(2l''+1)}{(2k+1)(2l+1)}\right\}^{\frac{1}{2}}$$ 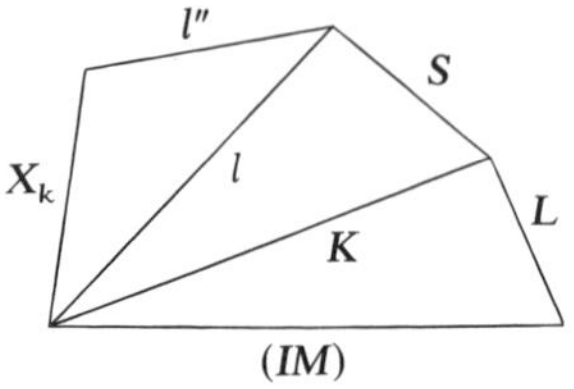

$$=\sum_{l''S''}(-1)^{l-l''+S-S''}\langle l''\|\, Z_k\,\|l\rangle\langle S''\|\, X_k\,\|S\rangle\left\{\frac{(2l''+1)(2S''+1)}{2k+1}\right\}^{\frac{1}{2}}$$

$$\times\, W(l''kKS;\, lS'')$$ 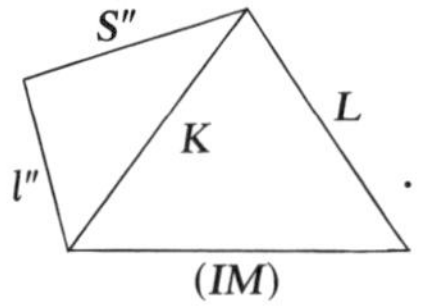

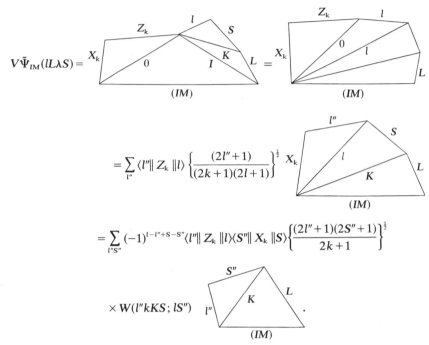

.

The final step in evaluating the matrix element in this way is to interchange l'' and S'' and recouple l'' and L to λ''. Once this is carried out and the result inserted into equation (A4.7) one finds

$$\begin{aligned}\langle\Psi_{IM}(l'L\lambda'S)|\, V\,|\Psi_{IM}(lL\lambda S)\rangle &\\ =(-1)^{S'+\lambda'-S-\lambda}&\langle l'\|\, Z_k\,\|l\rangle\,\langle S'\|\, X_k\,\|S\rangle\\ \times&\left\{\frac{(2l'+1)(2\lambda+1)(2S'+1)(2\lambda'+1)}{(2k+1)}\right\}^{\frac{1}{2}}\\ \times&\sum_K(2K+1)\,W(LlIS\colon\lambda K)\,W(l'kKS;\, lS')\,W(S'l'IL;\, K\lambda').\end{aligned}$$

Comparison of this way of deriving the matrix element with equation (A4.6) leads to the three-to-two-summation relationship. By use of the symmetry properties in equation (A4.2) this may be written in the form

$$\sum_K(-1)^{I+L-K}(2K+1)\,W(lLSI;\,\lambda K)\,W(l'LS'I;\,\lambda'K)\,W(ll'SS'\colon kK)$$
$$=(-1)^{\lambda+\lambda'-k}\,W(ll'\lambda\lambda';\, kL)\,W(\lambda\lambda'SS';\, kI).$$

(Although we formulated this problem as if we were evaluating the matrix elements of a two-body potential, in no place have we actually assumed that any of the angular momenta involved have integer values.)

Finally there are two special relationships involving Racah coefficients that are often useful. In Appendix 2, section 7, we showed that between single-particle states the reduced matrix element of the spherical harmonic $Y_{l''m''}$ could be written as

$$\langle \phi_{j'l'} \| Y_{l''} \| \phi_{jl} \rangle = (-1)^{l+l'+j-j'} \left\{ \frac{(2l''+1)(2j+1)}{4\pi(2j'+1)} \right\}^{\frac{1}{2}} \left\{ \frac{1+(-1)^{l+l'+l''}}{2} \right\}$$

$$\times (jl''\tfrac{1}{2}0 \mid j'\tfrac{1}{2}) \int R_{j'} R_j r^2 \, dr.$$

Alternatively, the value of this matrix element can be deduced by decomposing ϕ_{jlm} and $\phi_{j'l'm'}$ into their angular and spin parts. The angular-spin part of $[Y_{l''} \times \phi_{jl}]_{j_1 m_1}$ is

$$[Y_{l''} \times \phi_{jl}]_{j_1 m_1} =$$

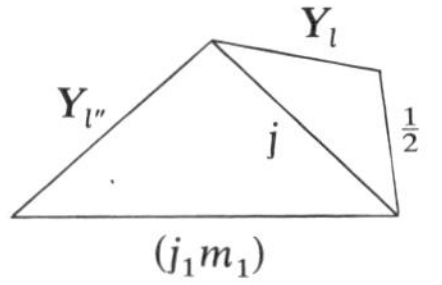

$$= \sum_{l_1} (-1)^{l+l''-l_1} \langle Y_{l_1} \| Y_{l''} \| Y_l \rangle \sqrt{\{(2l_1+1)(2j+1)\}}$$

$$\times W(l''lj_1\tfrac{1}{2};\, l_1 j)$$

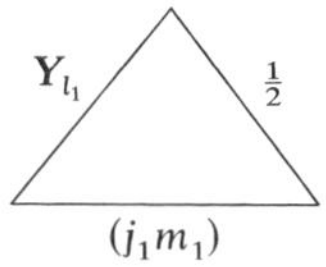

It follows, once use is made of equation (A2.11) for the reduced matrix element $\langle Y_{l_1} \| Y_{l''} \| Y_l \rangle$, that

$$\langle \phi_{j'l'} \| Y_{l''} \| \phi_{jl} \rangle = \left\{ \frac{(2l+1)(2j+1)(2l''+1)}{4\pi} \right\}^{\frac{1}{2}} (ll''00 \mid l'0) \, W(l''lj'\tfrac{1}{2};\, l'j) \int R_{j'} R_j r^2 \, dr.$$

Comparison of these two ways of deriving the reduced matrix element leads to the relationship

$$(jl''\tfrac{1}{2}0 \mid j'\tfrac{1}{2}) \left\{ \frac{1+(-1)^{l+l'+l''}}{2} \right\} = (-1)^{l'+\frac{1}{2}-j'} \sqrt{\{(2l+1)(2j'+1)\}} (ll''00 \mid l'0) \, W(l'j'lj;\, \tfrac{1}{2}l'').$$

The second result involves the Racah coefficient in which one angular momentum is unity. To deduce this relationship we note that the total angular-momentum operator

$$\underline{\mathrm{I}} = \underline{\mathrm{J}}_1 + \underline{\mathrm{J}}_2$$

cannot change the angular momentum of the state

$$\Psi_{IM}(J_1 J_2) =$$

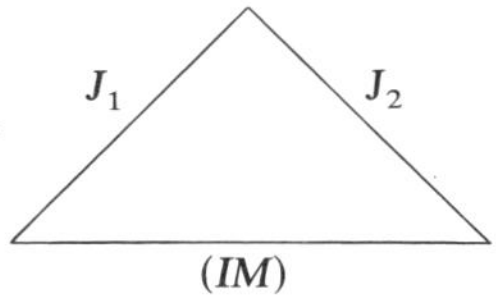

Thus the matrix element $\langle\Psi_{I'M'}(J_1J_2)|\,(\mathbf{J}_1)_\mu+(\mathbf{J}_2)_\mu\,|\Psi_{IM}(J_1J_2)\rangle$ vanishes unless $I=I'$. Since $\mathbf{J}_1$ and $\mathbf{J}_2$ are tensor operators of rank one, it follows that

$$\{(\mathbf{J}_1)_\mu+(\mathbf{J}_2)_\mu\}\Psi_{IM}(J_1J_2)=\sum_{I''M''}(1I\mu M\mid I''M'')$$

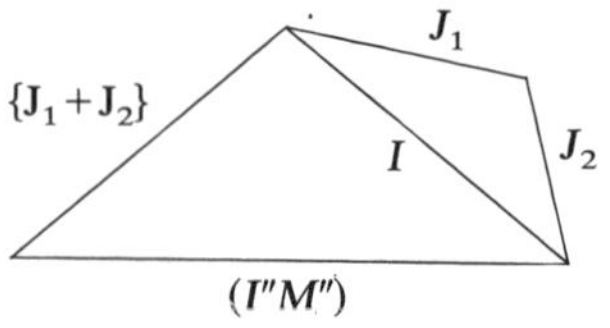

$$=\sum_{I''M''}(1I\mu M\mid I''M'')\surd(2I+1)\Big\{\surd(2J_1+1)W(1J_1I''J_2;\,J_1I)$$

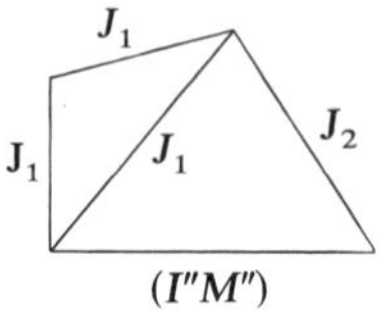

$$+(-1)^{J_1+J_2-I}\surd(2J_2+1)W(1J_2I''J_1;\,J_2I)$$

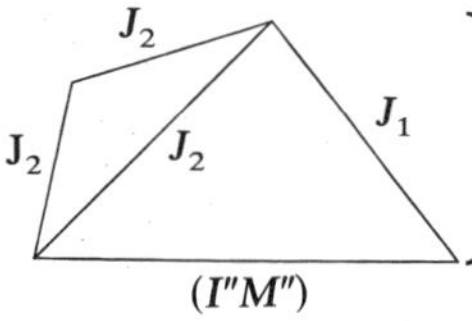

$\Big\}$

where we have made use of the fact that $\mathbf{J}_i$ is the total angular-momentum operator for the nucleons in group i and hence cannot change their angular momentum. Furthermore,

$$\begin{aligned}\langle\Phi_{J_1M}|\,(\mathbf{J}_1)_z\,|\Phi_{J_1M}\rangle &=\langle\Phi_{J_1}\|\,\mathbf{J}_1\,\|\Phi_{J_1}\rangle(J_11M0\mid J_1M)\\ &=M;\end{aligned}$$

and from Table A1.2 it follows that

$$(J_11M0\mid J_1M)=M/\surd\{J_1(J_1+1)\}.$$

Thus one obtains the result

$$\langle\Phi_{J_1}\|\,\mathbf{J}_1\,\|\Phi_{J_1}\rangle=\surd\{J_1(J_1+1)\}.$$

Consequently, since the matrix element $\langle\Psi_{I'M'}(J_1J_2)|\,(\mathbf{J}_1)_\mu+(\mathbf{J}_2)_\mu\,|\Psi_{IM}(J_1J_2)\rangle$ must vanish for $I'=I\pm1$

$$\begin{aligned}&\surd\{J_1(J_1+1)(2J_1+1)\}W(1J_1I'J_2;\,J_1I)\\ &\qquad=\surd\{J_2(J_2+1)(2J_2+1)\}W(1J_2I'J_1;\,J_2I),\qquad I'=I\pm1.\end{aligned}$$

Various other properties of the Racah coefficients are given by Biedenharn *et al.* (1952) and by Rotenberg *et al.* (1959). In the latter reference, numerical values are given for the 6j-coefficient $\begin{Bmatrix} j_1 & j_2 & J\\ j_3 & j_4 & K\end{Bmatrix}$. The relationship between this coefficient and $W(j_1j_2j_4j_3;\,JK)$ is

$$W(j_1j_2j_4j_3;\,JK)=(-1)^{j_1+j_2+j_3+j_4}\begin{Bmatrix} j_1 & j_2 & J\\ j_3 & j_4 & K\end{Bmatrix}.$$

2. 9j coefficients

When four angular momenta are coupled together there are, once again, several ways the coupling can be carried out. For example, j_1 and j_2 may be coupled to J_{12}; j_3 and j_4 to J_{34}; and then J_{12} and J_{34} are compounded to give the resultant (IM)

$$\chi_{IM} = [[\phi_{j_1} \times \phi_{j_2}]_{J_{12}} \times [\phi_{j_3} \times \phi_{j_4}]_{J_{34}}]_{IM} =$$

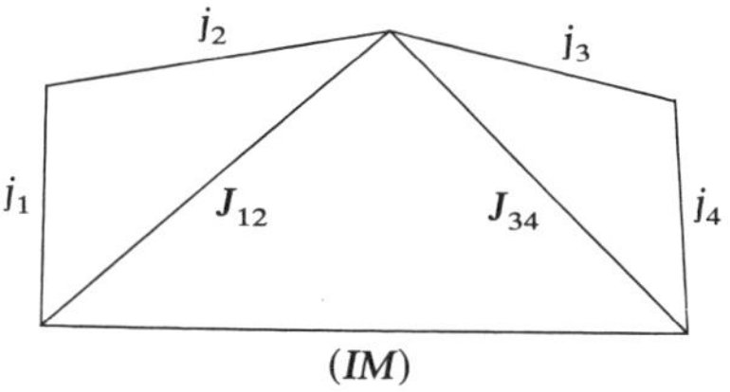

An alternative possibility is to couple j_1 and j_3 to J_{13}; j_2 and j_4 to J_{24}; and finally to arrive at (IM) by adding J_{13} and J_{24}

$$\xi_{IM} = [[\phi_{j_1} \times \phi_{j_3}]_{J_{13}} \times [\phi_{j_2} \times \phi_{j_4}]_{J_{24}}]_{IM} =$$

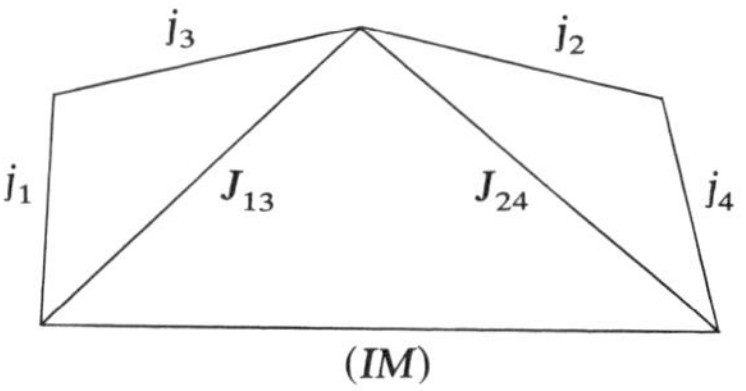

Provided the ϕ_{jm} are orthonormal wave functions, χ_{IM} and ξ_{IM} are normalized. However, they are not orthogonal and their overlap is

$$\begin{aligned}\langle \chi_{IM} \mid \xi_{IM} \rangle &= \sum_{m_i M_{ij}} (j_1 j_2 m_1 m_2 \mid J_{12} M_{12})(j_3 j_4 m_3 m_4 \mid J_{34} M_{34})(J_{12} J_{34} M_{12} M_{34} \mid IM) \\ &\quad \times (j_1 j_3 m_1 m_3 \mid J_{13} M_{13})(j_2 j_4 m_2 m_4 \mid J_{24} M_{24})(J_{13} J_{24} M_{13} M_{24} \mid IM) \\ &= \sqrt{\{(2J_{12}+1)(2J_{34}+1)(2J_{13}+1)(2J_{24}+1)\}} \begin{Bmatrix} j_1 & j_2 & J_{12} \\ j_3 & j_4 & J_{34} \\ J_{13} & J_{24} & I \end{Bmatrix}\end{aligned}$$

where the curly bracket is called the 9j coefficient. By the same argument used in connection with the Racah coefficient, one may show that the 9j coefficient defined by the above equation is independent of M.

From the symmetry properties of the Clebsch–Gordan coefficients, equations (A1.18)–(A1.20), it follows that

$$\begin{Bmatrix} j_1 & j_2 & J_{12} \\ j_3 & j_4 & J_{34} \\ J_{13} & J_{24} & I \end{Bmatrix} = \begin{Bmatrix} j_1 & j_3 & J_{13} \\ j_2 & j_4 & J_{24} \\ J_{12} & J_{34} & I \end{Bmatrix}$$

$$= \begin{Bmatrix} I & J_{34} & J_{12} \\ J_{24} & j_4 & j_2 \\ J_{13} & j_3 & j_1 \end{Bmatrix},$$

and that a permutation of rows or columns multiplies the $9j$ coefficient by $+1$ if the permutation is even and $(-1)^{\tilde{J}}$ if the permutation is odd, where $\tilde{J}$ is the sum of the angular momenta occurring in the symbol.

In the same manner as equations (A4.3) and (A4.4) were arrived at one can show that

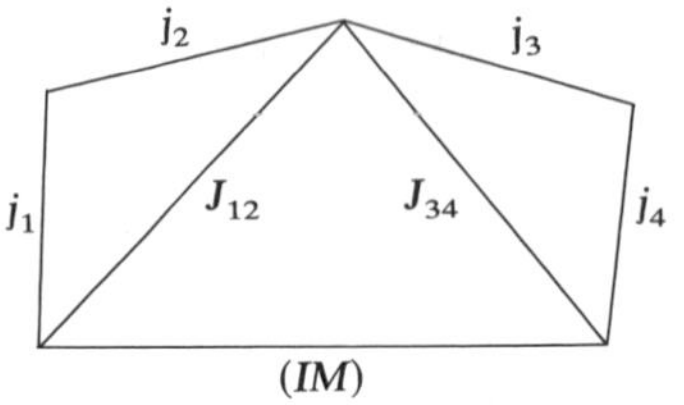

$$= \sum_{J_{13}J_{24}} \sqrt{\{(2J_{12}+1)(2J_{34}+1)(2J_{13}+1)(2J_{24}+1)\}} \begin{Bmatrix} j_1 & j_2 & J_{12} \\ j_3 & j_4 & J_{34} \\ J_{13} & J_{24} & I \end{Bmatrix}$$

$\times$ 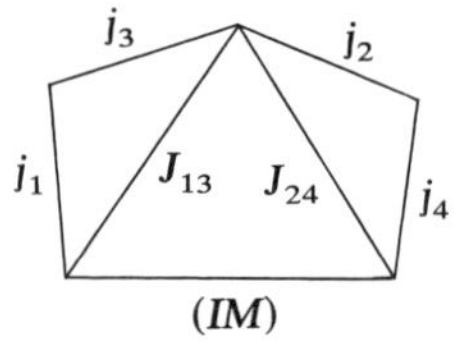

(A4.8)

and that

$$(2J_{12}+1)(2J_{34}+1) \sum_{J_{13}J_{24}} (2J_{13}+1)(2J_{24}+1) \begin{Bmatrix} j_1 & j_2 & J_{12} \\ j_3 & j_4 & J_{34} \\ J_{13} & J_{24} & I \end{Bmatrix} \begin{Bmatrix} j_1 & j_2 & J'_{12} \\ j_3 & j_4 & J'_{34} \\ J_{13} & J_{24} & I \end{Bmatrix} = \delta_{J_{12}J'_{12}} \delta_{J_{34}J'_{34}}.$$

It is clear that the j–j to L–S transformation coefficients alluded to in Appendix 2, section 9, and used in Chapter 4, section 1 are proportional to this coefficient

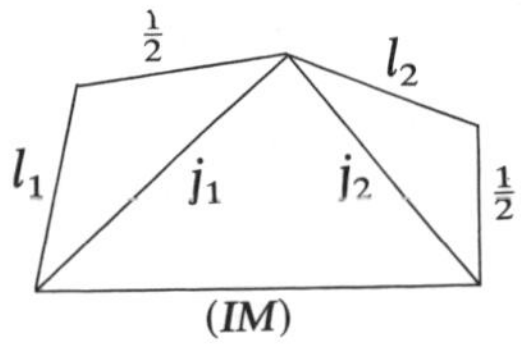

$$= \sum_{LS} \sqrt{\{(2j_1+1)(2j_2+1)(2L+1)(2S+1)\}} \begin{Bmatrix} l_1 & \frac{1}{2} & j_1 \\ l_2 & \frac{1}{2} & j_2 \\ L & S & I \end{Bmatrix}$$

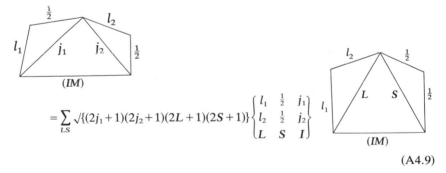

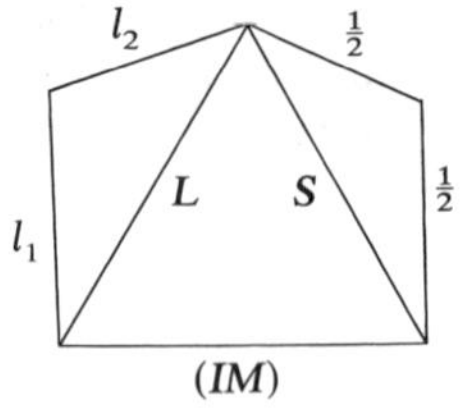

(A4.9)

The $9j$ coefficients may be expressed in terms of a sum over Racah coefficients. By use of equations (A4.3) and (A4.4) plus the symmetry property in equation (A1.18) of the Clebsch–Gordan coefficients one may write the left-hand side of

equation (A4.8) as

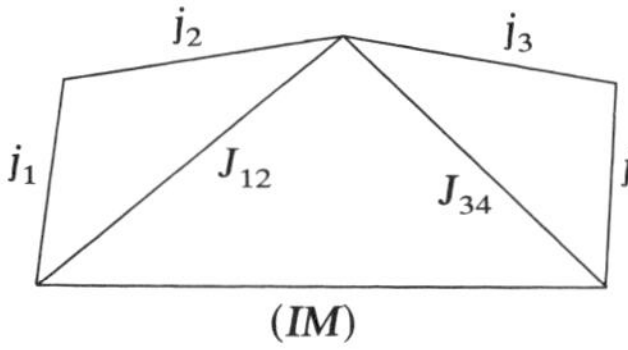

$$= \sum_K (-1)^{j_3+j_4-J_{34}} \sqrt{\{(2J_{34}+1)(2K+1)\}}$$

$$\times W(J_{12}j_4Ij_3;\, KJ_{34})$$

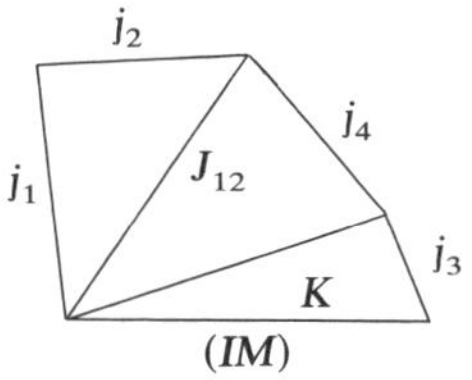

$$= (-1)^{j_3+j_4-J_{34}} \sum_{KJ'} \sqrt{\{(2J_{12}+1)(2J_{34}+1)(2J'+1)(2K+1)\}}$$

$$\times W(J_{12}j_4Ij_3;\, KJ_{34})W(j_1j_2Kj_4;\, J_{12}J')$$

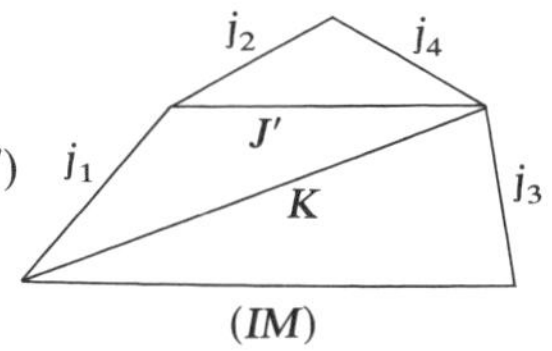

$$= \sum_{KJ'J''} (-1)^{j_3+j_4-J_{34}+j_1+J'-K}(2K+1)$$

$$\times \sqrt{\{(2J_{12}+1)(2J_{34}+1)(2J'+1)(2J''+1)\}}W(J_{12}j_4Ij_3;\, KJ_{34})$$

$$\times W(j_1j_2Kj_4;\, J_{12}J')W(J'j_1Ij_3;\, KJ'')$$

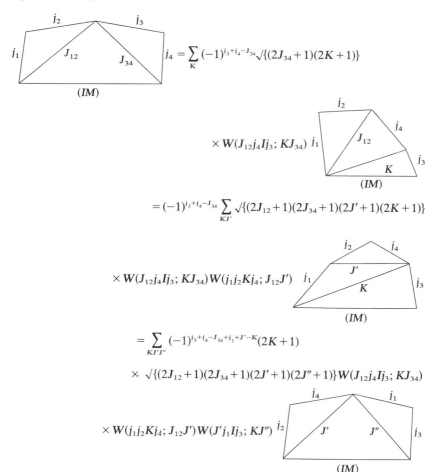

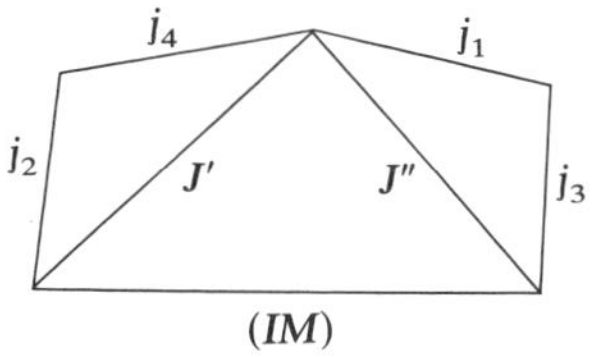

If one interchanges J' and J'' in the triangle (J', J'', I) this diagram becomes identical to the one on the right-hand side of equation (A4.8) provided $J' = J_{24}$ and $J'' = J_{13}$. When the symmetry properties of the Racah coefficients in equations (A4.2) are used one finds

$$\sum_K (2K+1)W(IJ_{12}j_3j_4;\, J_{34}K)W(j_1J_{12}J_{24}j_4;\, j_2K)W(J_{24}Ij_1j_3;\, J_{13}K)$$

$$= (-1)^{\bar{J}} \begin{Bmatrix} j_1 & j_2 & J_{12} \\ j_3 & j_4 & J_{34} \\ J_{13} & J_{24} & I \end{Bmatrix}$$

where

$$\bar{J} = j_1 + j_2 + j_3 + j_4 + J_{12} + J_{13} + J_{24} + J_{34} + I.$$

Since the Racah coefficient has a particularly simple form when one angular momentum is zero, one easily shows that

$$\begin{Bmatrix} j_1 & j_2 & J_{12} \\ j_3 & j_4 & J_{34} \\ J_{13} & J_{24} & 0 \end{Bmatrix} = (-1)^{j_1+j_4-J_{12}-J_{13}} \frac{W(j_1j_2j_3j_4;\, J_{12}J_{13})}{\sqrt{\{(2J_{12}+1)(2J_{13}+1)\}}}\,\delta_{J_{12}J_{34}}\delta_{J_{13}J_{24}}.$$

Other properties of these coefficients are given by Jahn and Hope (1954), Arima *et al.* (1954*a*) and Rotenberg *et al.* (1959). Tables of numerical values are given by Kennedy *et al.* (1954) and Smith and Stephenson (1957) and some useful formulae are collected in the next section of this appendix.

3. Some formulae involving Racah and 9*j* coefficients

$$\sum_M (j_1j_2m_1m_2\,|\,JM)(Jj_3Mm_3\,|\,j_4m_4) = \sum_{KM'} \sqrt{\{(2J+1)(2K+1)\}} \times W(j_1j_2j_4j_3;\, JK)(j_1Km_1M'\,|\,j_4m_4)(j_2j_3m_2m_3\,|\,KM') \qquad \text{(A4.10)}$$

$$\sum_{m_2m_3M} (j_1j_2m_1m_2\,|\,JM)(Jj_3Mm_3\,|\,j_4m_4)(j_2j_3m_2m_3\,|\,KM') = \sqrt{\{(2J+1)(2K+1)\}}\,W(j_1j_2j_4j_3;\, JK)(j_1Km_1M'\,|\,j_4m_4) \qquad \text{(A4.11)}$$

$$W(j_1j_2j_3j_4;\, JK) = W(j_2j_1j_4j_3;\, JK) = W(j_3j_4j_1j_2;\, JK) = W(j_1j_3j_2j_4;\, KJ) \qquad \text{(A4.12)}$$

$$W(j_1j_2j_3j_4;\, JK) = (-1)^{j_2+j_3-J-K} W(j_1JKj_4;\, j_2j_3) = (-1)^{j_1+j_4-J-K} W(Jj_2j_3K;\, j_1j_4) \qquad \text{(A4.13)}$$

$$W(j_1j_2j_3j_4;\, 0K) = (-1)^{j_1+j_3-K}\{(2j_1+1)(2j_3+1)\}^{-\frac{1}{2}}\delta_{j_1j_2}\delta_{j_3j_4} \qquad \text{(A4.14)}$$

$$\sum_K (2K+1)(2J+1)\,W(j_1j_2j_3j_4;\, JK)\,W(j_1j_2j_3j_4;\, J'K) = \delta_{JJ'} \qquad \text{(A4.15)}$$

$$\sum_J (2K+1)(2J+1)\,W(j_1j_2j_3j_4;\, JK)\,W(j_1j_2j_3j_4;\, JK') = \delta_{KK'} \qquad \text{(A4.16)}$$

$$\sum_L (-1)^{j_1+j_3-L}(2L+1)\,W(j_2j_4j_1j_3;\, LJ)\,W(j_2j_4j_3j_1;\, LK) = W(j_1JKj_3;\, j_2j_4) \qquad \text{(A4.17)}$$

$$\sum_K (-1)^{I+L-K}(2K+1)\,W(lLSI;\, \lambda K)\,W(l'LS'I;\, \lambda' K)\,W(ll'SS';\, kK) = (-1)^{\lambda+\lambda'-k}\,W(ll'\lambda\lambda';\, kL)\,W(\lambda\lambda' SS';\, kI) \qquad \text{(A4.18)}$$

$$(2J+1)\sum_K W(j_1j_2j_4j_3;\, JK)(j_1j_4\tfrac{1}{2}-\tfrac{1}{2}\,|\,K0)(j_2j_3\tfrac{1}{2}-\tfrac{1}{2}\,|\,K0) = (-1)^{j_3-j_1}(j_1j_2\tfrac{1}{2}\tfrac{1}{2}\,|\,J1)(j_3j_4\tfrac{1}{2}\tfrac{1}{2}\,|\,J1). \qquad \text{(A4.19)}$$

$$(2J+1)\sum_K(-1)^K W(j_1j_2j_4j_3;JK)(j_1j_4\tfrac{1}{2}-\tfrac{1}{2}\mid K0)(j_2j_3\tfrac{1}{2}-\tfrac{1}{2}\mid K0)$$
$$=-(-1)^{j_1+j_2+j_3+j_4-J}(j_1j_2\tfrac{1}{2}-\tfrac{1}{2}\mid J0)(j_3j_4\tfrac{1}{2}-\tfrac{1}{2}\mid J0) \quad \text{(A4.20)}$$

$$(2J+1)\sum_K(-1)^K W(j_1j_2j_4j_3;JK)(j_1j_4\tfrac{1}{2}\tfrac{1}{2}\mid K1)(j_2j_3\tfrac{1}{2}\tfrac{1}{2}\mid K1)$$
$$=(-1)^{j_1+j_2}(j_1j_2\tfrac{1}{2}-\tfrac{1}{2}\mid J0)(j_3j_4\tfrac{1}{2}-\tfrac{1}{2}\mid J0). \quad \text{(A4.21)}$$

$$(2J+1)\sum_K W(j_1j_2j_4j_3;JK)(j_4j_1\tfrac{1}{2}\tfrac{1}{2}\mid K1)(j_2j_3\tfrac{1}{2}\tfrac{1}{2}\mid K1)$$
$$=(-1)^{j_2-j_4}(j_1j_2\tfrac{1}{2}-\tfrac{1}{2}\mid J0)(j_3j_4\tfrac{1}{2}-\tfrac{1}{2}\mid J0). \quad \text{(A4.22)}$$

$$(-1)^{l'+\frac{1}{2}-j'}\sqrt{\{(2l+1)(2j'+1)\}}(ll''00\mid l'0)W(l'j'lj;\tfrac{1}{2}l'')$$
$$=(jl''\tfrac{1}{2}0\mid j'\tfrac{1}{2})\left\{\frac{1+(-1)^{l+l'+l''}}{2}\right\} \quad \text{(A4.23)}$$

$$\sqrt{\{J_1(J_1+1)(2J_1+1)\}}W(1J_1I'J_2;J_1I)$$
$$=\sqrt{\{J_2(J_2+1)(2J_2+1)\}}W(1J_2I'J_1;J_2I), \qquad I'=I\pm 1 \quad \text{(A4.24)}$$

$$W(abcd;a+b,f)$$
$$=\left\{\frac{2a!2b!(a+b+c+d+1)!(a+b+c-d)!(a+b+d-c)!(c+f-a)!(d+f-b)!}{(2a+2b+1)!(c+d-a-b)!(a+c-f)!(a+f-c)!\times(a+c+f+1)!(b+d-f)!(b+f-d)!(b+f+d+1)!}\right\}^{\frac{1}{2}}$$
(A4.25)

$$\sqrt{\{(2J_{12}+1)(2J_{13}+1)(2J_{24}+1)(2J_{34}+1)\}}\begin{Bmatrix} j_1 & j_2 & J_{12}\\ j_3 & j_4 & J_{34}\\ J_{13} & J_{24} & I\end{Bmatrix}$$
$$=\sum_{m_iM_{ij}}(j_1j_2m_1m_2\mid J_{12}M_{12})(j_3j_4m_3m_4\mid J_{34}M_{34})(J_{12}J_{34}M_{12}M_{34}\mid IM)$$
$$\times(j_1j_3m_1m_3\mid J_{13}M_{13})(j_2j_4m_2m_4\mid J_{24}M_{24})(J_{13}J_{24}M_{13}M_{24}\mid IM)$$
(A4.26)

$$\begin{Bmatrix} j_1 & j_2 & J_{12}\\ j_3 & j_4 & J_{34}\\ J_{13} & J_{24} & I\end{Bmatrix}=\begin{Bmatrix} j_1 & j_3 & J_{13}\\ j_2 & j_4 & J_{24}\\ J_{12} & J_{34} & I\end{Bmatrix}$$
$$=\begin{Bmatrix} I & J_{34} & J_{12}\\ J_{24} & j_4 & j_2\\ J_{13} & j_3 & j_1\end{Bmatrix} \quad \text{(A4.27)}$$

$$\mathcal{P}\begin{Bmatrix} j_1 & j_2 & J_{12}\\ j_3 & j_4 & J_{34}\\ J_{13} & J_{24} & I\end{Bmatrix}=\begin{Bmatrix} j_1 & j_2 & J_{12}\\ j_3 & j_4 & J_{34}\\ J_{13} & J_{24} & I\end{Bmatrix}$$ ($\mathcal{P}$ an even permutation of the rows or columns) (A4.28)

$$\mathcal{P}\begin{Bmatrix} j_1 & j_2 & J_{12}\\ j_3 & j_4 & J_{34}\\ J_{13} & J_{24} & I\end{Bmatrix}=(-1)^{\tilde{J}}\begin{Bmatrix} j_1 & j_2 & J_{12}\\ j_3 & j_4 & J_{34}\\ J_{13} & J_{24} & I\end{Bmatrix}$$ ($\mathcal{P}$ an odd permutation of the rows or columns $\tilde{J}=j_1+j_2+j_3+j_4+J_{12}+J_{13}+J_{24}+J_{34}+I$) (A4.29)

$$(2J_{12}+1)(2J_{34}+1)\sum_{J_{13}J_{24}}(2J_{13}+1)(2J_{24}+1)$$
$$\times\begin{Bmatrix} j_1 & j_2 & J_{12}\\ j_3 & j_4 & J_{34}\\ J_{13} & J_{24} & I\end{Bmatrix}\begin{Bmatrix} j_1 & j_2 & J'_{12}\\ j_3 & j_4 & J'_{34}\\ J_{13} & J_{24} & I\end{Bmatrix}=\delta_{J_{12}J'_{12}}\,\delta_{J_{34}J'_{34}} \qquad \text{(A4.30)}$$

$$(2J_{13}+1)(2J_{24}+1)\sum_{J_{12}J_{34}}(2J_{12}+1)(2J_{34}+1)$$
$$\times\begin{Bmatrix} j_1 & j_2 & J_{12}\\ j_3 & j_4 & J_{34}\\ J_{13} & J_{24} & I\end{Bmatrix}\begin{Bmatrix} j_1 & j_2 & J_{12}\\ j_3 & j_4 & J_{34}\\ J'_{13} & J'_{24} & I\end{Bmatrix}=\delta_{J_{13}J'_{13}}\,\delta_{J_{24}J'_{24}} \qquad \text{(A4.31)}$$

$$(-1)^{\bar{J}}\begin{Bmatrix} j_1 & j_2 & J_{12}\\ j_3 & j_4 & J_{34}\\ J_{13} & J_{24} & I\end{Bmatrix}=\sum_K (2K+1)W(IJ_{12}j_3j_4;\,J_{34}K)W(j_1J_{12}J_{24}j_4;\,j_2K)$$
$$\times W(J_{24}Ij_1j_3;\,J_{13}K) \qquad (\bar{J}=j_1+j_2+j_3+j_4+J_{12}+J_{13}+J_{24}+J_{34}+I) \qquad \text{(A4.32)}$$

$$\begin{Bmatrix} j_1 & j_2 & J_{12}\\ j_3 & j_4 & J_{34}\\ J_{13} & J_{24} & 0\end{Bmatrix}=(-1)^{j_1+j_4-J_{12}-J_{13}}\frac{W(j_1j_2j_3j_4;\,J_{12}J_{13})}{\sqrt{\{(2J_{12}+1)(2J_{13}+1)\}}}\,\delta_{J_{12}J_{34}}\delta_{J_{13}J_{24}} \qquad \text{(A4.33)}$$

4. Analytic expressions for Racah coefficients

Analytic expressions for the Racah coefficients with one angular momentum equal to $\frac{1}{2}$ or 1 are given. Values can also be found in the literature when one J value in the coefficient is $\frac{3}{2}$ or 2 (Biedenharn *et al.* 1952).

TABLE A4.1
$W(l_1 J_1 l_2 J_2; \frac{1}{2}, L)$

	$l_1 = J_1 + \frac{1}{2}$	$l_1 = J_1 - \frac{1}{2}$
$l_2 = J_2 + \frac{1}{2}$	$(-1)^{J_1+J_2-L}\left\{\frac{(J_1+J_2+L+2)(J_1+J_2-L+1)}{(2J_1+1)(2J_1+2)(2J_2+1)(2J_2+2)}\right\}^{\frac{1}{2}}$	$(-1)^{J_1+J_2-L}\left\{\frac{(L-J_1+J_2+1)(L+J_1-J_2)}{(2J_1)(2J_1+1)(2J_2+1)(2J_2+2)}\right\}^{\frac{1}{2}}$
$l_2 = J_2 - \frac{1}{2}$	$(-1)^{J_1+J_2-L}\left\{\frac{(L+J_1-J_2+1)(L-J_1+J_2)}{(2J_1+1)(2J_1+2)(2J_2)(2J_2+1)}\right\}^{\frac{1}{2}}$	$(-1)^{J_1+J_2-L-1}\left\{\frac{(J_1+J_2+L+1)(J_1+J_2-L)}{2J_1(2J_1+1)(2J_2)(2J_2+1)}\right\}^{\frac{1}{2}}$

TABLE A4.2
$W(l_1J_1l_2J_2;1,L)$

$l_2=J_2+1$	
$l_1=J_1+1$	$(-1)^{J_1+J_2-L}\left\{\dfrac{(L+J_1+J_2+3)(L+J_1+J_2+2)(-L+J_1+J_2+2)(-L+J_1+J_2+1)}{4(2J_1+3)(J_1+1)(2J_1+1)(2J_2+3)(J_2+1)(2J_2+1)}\right\}^{\frac{1}{2}}$
$l_1=J_1$	$(-1)^{J_1+J_2-L}\left\{\dfrac{(L+J_1+J_2+2)(-L+J_1+J_2+1)(L-J_1+J_2+1)(L+J_1-J_2)}{4J_1(2J_1+1)(J_1+1)(2J_2+1)(J_2+1)(2J_2+3)}\right\}^{\frac{1}{2}}$
$l_1=J_1-1$	$(-1)^{J_1+J_2-L}\left\{\dfrac{(L+J_1-J_2)(L+J_1-J_2-1)(L-J_1+J_2+2)(L-J_1+J_2+1)}{4(2J_1+1)(2J_1-1)(J_1)(J_2+1)(2J_2+1)(2J_2+3)}\right\}^{\frac{1}{2}}$
$l_2=J_2$	
$l_1=J_1+1$	$(-1)^{J_1+J_2-L}\left\{\dfrac{(L+J_1+J_2+2)(L+J_1-J_2+1)(J_1+J_2-L+1)(L-J_1+J_2)}{4(2J_1+1)(J_1+1)(2J_1+3)(J_2)(J_2+1)(2J_2+1)}\right\}^{\frac{1}{2}}$
$l_1=J_1$	$(-1)^{J_1+J_2-L-1}\left\{\dfrac{J_1(J_1+1)+J_2(J_2+1)-L(L+1)}{\sqrt{\{4J_1(J_1+1)(2J_1+1)(J_2)(J_2+1)(2J_2+1)\}}}\right\}$
$l_1=J_1-1$	$(-1)^{J_1+J_2-L-1}\left\{\dfrac{(L+J_1+J_2+1)(-L+J_1+J_2)(L+J_1-J_2)(L-J_1+J_2+1)}{4(2J_1+1)(J_1)(2J_1-1)(J_2)(2J_2+1)(J_2+1)}\right\}^{\frac{1}{2}}$
$l_2=J_2-1$	
$l_1=J_1+1$	$(-1)^{J_1+J_2-L}\left\{\dfrac{(L-J_1+J_2)(L-J_1+J_2-1)(L+J_1-J_2+2)(L+J_1-J_2+1)}{4(2J_1+1)(J_1+1)(2J_1+3)(2J_2-1)(J_2)(2J_2+1)}\right\}^{\frac{1}{2}}$
$l_1=J_1$	$(-1)^{J_1+J_2-L-1}\left\{\dfrac{(L+J_1+J_2+1)(L+J_1-J_2+1)(L+J_2-J_1)(J_1+J_2-L)}{4J_1(2J_1+1)(J_1+1)(J_2)(2J_2+1)(2J_2-1)}\right\}^{\frac{1}{2}}$
$l_1=J_1-1$	$(-1)^{J_1+J_2-L}\left\{\dfrac{(L+J_1+J_2+1)(L+J_1+J_2)(-L+J_1+J_2)(-L+J_1+J_2-1)}{4(2J_1+1)(J_1)(2J_1-1)(2J_2+1)(J_2)(2J_2-1)}\right\}^{\frac{1}{2}}$

APPENDIX 5

FRACTIONAL-PARENTAGE COEFFICIENTS

When one deals with the configuration j^n it is often convenient to single out particle number n for special treatment. To do this requires the introduction of the concept of parentage coefficients. Let $\Psi_{IM\alpha}(1, \ldots, n)$ be an antisymmetric state of the configuration j^n with angular momentum (IM), α stands for any other quantum numbers needed to completely specify the state, and denote by ϕ_{jm} the single-particle eigenfunction for the state j. If the states $\Phi_{J\bar{M}'\beta}(1, \ldots, n-1)$ are the complete set of antisymmetric wave functions of the configuration (j^{n-1}) one may write

$$\Psi_{IM\alpha}(1, \ldots, n) = \sum_{J\beta} \langle j^{n-1}J\beta, j| \}j^n I\alpha\rangle [\Phi_{J\beta}(1, \ldots, n-1) \times \phi_j(n)]_{IM} \quad \text{(A5.1)}$$

provided the coefficients $\langle j^{n-1}J\beta, j| \}j^n I\alpha\rangle$, known as the fractional-parentage coefficients, are chosen so that the n-particle wave function is antisymmetric with respect to interchange of any two particles (Racah 1943).

In a similar way, if one wants to single out particles number n and $n-1$ for special treatment the concept of a double-parentage coefficient must be introduced. Consequently if $\psi_{JM'\beta}(1, \ldots, n-2)$ are the complete set of antisymmetric states of (j^{n-2}) and $\phi_{KM''}(n-1, n)$ denotes all the two-particle states $(j^2)_{KM''}$, one may write

$$\Psi_{IM\alpha}(1, \ldots, n) = \sum_{J\beta K} \langle j^{n-2}J\beta, j^2K| \}j^n I\alpha\rangle [\psi_{J\beta}(1, \ldots, n-2) \times \phi_K(n-1, n)]_{IM} \quad \text{(A5.2)}$$

with the double-parentage coefficient $\langle j^{n-2}J\beta, j^2K| \}j^n I\alpha\rangle$ chosen so that the state vector is antisymmetric with respect to interchange of any two particles.

We shall now discuss some properties of these parentage coefficients and tabulate the values encountered in this book.

1. Single-particle parentage coefficients

Because of the orthonormality of the $\Phi_{J\beta}(1, \ldots, n-1)$ the fractional-parentage coefficients of equation (A5.1) are given by

$$\langle j^{n-1}J\beta, j| \}j^n I\alpha\rangle = \langle [\Phi_{J\beta}(1, \ldots, n-1) \times \phi_j(n)]_{IM} \mid \Psi_{IM\alpha}(1, \ldots, n)\rangle. \quad \text{(A5.3)}$$

One may easily show that the coefficients defined by this equation are independent of M. To do this we evaluate equation (A5.3) for a different value of M, namely $(M+1)$. Since

$$\Psi_{IM+1}=\frac{\mathrm{J}_+}{\sqrt{\{(I-M)(I+M+1)\}}}\Psi_{IM}$$

$$\begin{aligned}\langle[\Phi_{J\beta}(1,\ldots,n-1)\times\phi_j(n)]_{IM+1}\mid\Psi_{IM+1\alpha}(1,\ldots,n)\rangle\\ &=\frac{\langle \mathrm{J}_+[\Phi_{J\beta}(1,\ldots,n-1)\times\phi_j(n)]_{IM}\mid \mathrm{J}_+\Psi_{IM\alpha}\rangle}{(I-M)(I+M+1)}\\ &=\frac{\langle[\Phi_{J\beta}(1,\ldots,n-1)\times\phi_j(n)]_{IM}\mid \mathrm{J}_-\mathrm{J}_+\Psi_{IM\alpha}\rangle}{(I-M)(I+M+1)}.\end{aligned}$$

But from the definition of the $\mathrm{J}_\pm$ operator in equation (A1.4) and the commutation relationship it satisfies in equation (A1.1), one sees that

$$\begin{aligned}\mathrm{J}_-\mathrm{J}_+\Psi_{IM\alpha}&=(\mathrm{J}^2-\mathrm{J}_z^2-\mathrm{J}_z)\Psi_{IM\alpha}\\ &=(I-M)(I+M+1)\Psi_{IM\alpha}.\end{aligned}$$

Thus the right-hand side of equation (A5.3) is independent of M and consequently so are the c.f.p.

In addition, if $\Psi_{IM\alpha}(1,\ldots,n)$ and $\Psi_{IM\alpha'}(1,\ldots,n)$ are orthonormal wave functions of j^n, it follows from equation (A5.1) that

$$\sum_{J\beta}\langle j^{n-1}J\beta,j|\}j^nI\alpha\rangle\langle j^{n-1}J\beta,j|\}j^nI\alpha'\rangle=\delta_{\alpha'\alpha} \tag{A5.4}$$

where we have made use of the fact that in the representation we have chosen the parentage coefficients are real.

The c.f.p. can be related to the reduced matrix element of the nucleon-creation operator. To find this relationship we first note that since $a^\dagger_{jm}$ is an irreducible tensor operator of rank j (see Appendix 2, section 1) it satisfies the Wigner–Eckart theorem

$$\langle\Psi_{IM}|\,a^\dagger_{jm}\,|\Phi_{JM_1}\rangle=\langle\Psi_I\|\,a^\dagger_j\,\|\Phi_J\rangle(JjM_1m\mid IM).$$

Consequently

$$\begin{aligned}\langle\Psi_I\|\,a^\dagger_j\,\|\Phi_J\rangle^\dagger(JjM_1m\mid IM)&=\langle\Psi_{IM}|\,a^\dagger_{jm}\,|\Phi_{JM_1}\rangle^\dagger\\ &=\langle\Phi_{JM_1}|\,a_{jm}\,|\Psi_{IM}\rangle.\end{aligned}$$

With the aid of this relationship we examine the effect of the number operator

$$\mathrm{N}_j=\sum_m a^\dagger_{jm}a_{jm} \tag{A5.5}$$

on $\Psi_{IM\alpha}$. Since the $\Phi_{JM'\beta}$ form a complete set of $(n-1)$-particle states

$$\sum_{JM'\beta}|\Phi_{JM'\beta}\rangle\langle\Phi_{JM'\beta}|=1$$

so that

$$\begin{aligned}
\mathrm{N}_j \Psi_{IM\alpha} &= \sum_{mM_1J\beta} a^\dagger_{jm} \Phi_{JM_1\beta} \langle \Phi_{JM_1\beta} | \, a_{jm} \, | \Psi_{IM} \rangle \\
&= \sum_{mM_1J\beta} \langle \Psi_{I\alpha} \| \, a^\dagger_j \, \| \Phi_{J\beta} \rangle^\dagger (JjM_1m \mid IM) a^\dagger_{jm} \Phi_{JM_1\beta} \\
&= (-1)^{n-1} \sum_{J\beta} \langle \Psi_{I\alpha} \| \, a^\dagger_j \, \| \Phi_{J\beta} \rangle^\dagger [\Phi_{J\beta} \times a^\dagger_j]_{IM}
\end{aligned}$$

where the phase factor in the last line arises because we have commuted $a^\dagger_{jm}$ through the $(n-1)$-particle wave function $\Phi_{JM_1\beta}$.

A comparison of this result with equation (A5.1) shows that the c.f.p. is proportional to the reduced matrix element of $a^\dagger_{jm}$. To find the exact coefficient in this proportionality we note that

$$\begin{aligned}
n &= \langle \Psi_{IM\alpha} | \, \mathrm{N}_j \, | \Psi_{IM\alpha} \rangle \\
&= \sum_{J\beta} |\langle \Psi_{I\alpha} \| \, a^\dagger_j \, \| \Phi_{J\beta} \rangle|^2 \sum_{M_1m} \{(JjM_1m \mid IM)\}^2 \\
&= \sum_{J\beta} |\langle \Psi_{I\alpha} \| \, a^\dagger_j \, \| \Phi_{J\beta} \rangle|^2.
\end{aligned}$$

Because the c.f.p. are normalized (equation (A5.4)) and because with our choice of representation they are real, one sees that

$$\langle j^{n-1} J\beta, j | \} j^n I\alpha \rangle = \frac{(-1)^{n-1}}{\sqrt{n}} \langle \Psi_{I\alpha} \| \, a^\dagger_j \, \| \Phi_{J\beta} \rangle. \tag{A5.6}$$

For the one- and two-particle wave functions, the one-particle c.f.p. can be evaluated simply. To do this we note that the one-particle state ψ_{jm} is

$$\psi_{jm} = a^\dagger_{jm} \, |0\rangle.$$

Thus if we denote the vacuum by

$$\Phi_{00} = |0\rangle$$

$$\begin{aligned}
\langle \psi_{jm} | \, a^\dagger_{jm} \, |0\rangle &= (0j0m \mid jm) \langle \psi_j \| \, a^\dagger_j \, \| \Phi_0 \rangle \\
&= \langle 0 | \, a_{jm} a^\dagger_{jm} \, |0\rangle \\
&= 1.
\end{aligned}$$

Since $(0j0m \mid jm) = 1$ it follows from equation (A5.6) that

$$\langle j^0 0, j | \} j^1 j \rangle = 1. \tag{A5.7}$$

Similarly the normalized two-particle wave function can be written as

$$\Psi_{IM} = 2^{-\frac{1}{2}} \sum_{m_i} (jjm_1m_2 \mid IM) a^\dagger_{jm_1} a^\dagger_{jm_2} \, |0\rangle$$

provided I is even. Consequently

$$\langle\Psi_{IM}|\, a^\dagger_{jm}\, |\psi_{jm'}\rangle = (jjm'm \mid IM)\langle\Psi_I\|\, a^\dagger_j\, \|\psi_j\rangle$$
$$= 2^{-\frac{1}{2}} \sum_{m_1m_2} (jjm_1m_2 \mid IM)\langle 0|\, a_{jm_2}a_{jm_1}a^\dagger_{jm}a^\dagger_{jm'}\, |0\rangle$$
$$= 2^{-\frac{1}{2}} \sum_{m_1m_2} (jjm_1m_2 \mid IM)(\delta_{m,m_1}\delta_{m',m_2} - \delta_{m,m_2}\delta_{m',m_1})$$
$$= 2^{-\frac{1}{2}}(jjmm' \mid IM)(1+(-1)^I).$$

Therefore

$$\langle\psi_I\|\, a^\dagger_j\, \|\psi_j\rangle = -2^{-\frac{1}{2}}(1+(-1)^I)$$

and by use of equation (A5.6) one sees that

$$\langle j^1 j, j| \} j^2 I\rangle = \left\{\frac{1+(-1)^I}{2}\right\} \tag{A5.8}$$

For $n > 2$ extensive tabulations of the one-particle c.f.p. exist (Bayman and Lande 1966) and in section 6 of this Appendix we give the explicit values of those encountered in this book.

2. Double-parentage coefficients

In analogy with equation (A5.3) the double-parentage coefficient is defined by the overlap integral

$$\langle j^{n-2}J\beta, j^2K| \} j^n I\alpha\rangle = \langle[\psi_{J\beta}(1,\ldots,n-2)\times\phi_K(n-1,n)]_{IM} \mid \Psi_{IM\alpha}\rangle \tag{A5.9}$$

and by the same reasoning as used in the previous section one can show that these coefficients are independent of M. Furthermore, because of the orthonormality of the wave functions involved

$$\sum_{J\beta K} \langle j^{n-2}J\beta, j^2K| \} j^n I\alpha'\rangle \langle j^{n-2}J\beta, j^2K| \} j^n I\alpha\rangle = \delta_{\alpha'\alpha}, \tag{A5.10}$$

and in writing this equation we have made use of the fact that in the representation we use the double-parentage coefficients are real.

As might be expected, the double-parentage coefficients are related to the reduced matrix element of the two-particle creation operator

$$A^\dagger_{KM}(jj) = 2^{-\frac{1}{2}} \sum_{m_i} (jjm_1m_2 \mid KM)a^\dagger_{jm_1}a^\dagger_{jm_2}. \tag{A5.11}$$

To see this we note that since

$$A_{KM}(jj) = 2^{-\frac{1}{2}} \sum_{m_i} (jjm_1m_2 \mid KM)a_{jm_2}a_{jm_1} \tag{A5.12}$$

it follows that

$$\sum_{K\mu} A^{\dagger}_{K\mu}(jj)A_{K\mu}(jj)=\frac{1}{2}\left\{\sum_{K\mu}\sum_{m_i}(jjm_1m_2\mid K\mu)(jjm_3m_4\mid K\mu)\,a^{\dagger}_{jm_1}a^{\dagger}_{jm_2}a_{jm_4}a_{jm_3}\right\}$$
$$=\frac{1}{2}\left(\sum_{m_1m_2} a^{\dagger}_{jm_1}a^{\dagger}_{jm_2}a_{jm_2}a_{jm_1}\right)$$
$$=\tfrac{1}{2}(\mathrm{N}_j^2-\mathrm{N}_j) \tag{A5.13}$$

where N_j is the number operator defined by equation (A5.5), and in deducing this expression use has been made of the anticommutation properties of $a^{\dagger}_{jm}$ and a_{jm} together with the summation property of the Clebsch–Gordan coefficients in equation (A1.24). One may now use exactly the same arguments employed in the previous section (except that $\sum_{K\mu} A^{\dagger}_{K\mu}(jj)A_{K\mu}(jj)$ replaces N_j) to show that

$$\langle j^{n-2}J\beta, j^2K|\}j^nI\alpha\rangle=\left\{\frac{2}{n(n-1)}\right\}^{\frac{1}{2}}\langle\Psi_{I\alpha}\|\,A^{\dagger}_K(jj)\,\|\psi_{J\beta}\rangle. \tag{A5.14}$$

Once the one-nucleon c.f.p. are known, the double-parentage coefficients may be calculated simply. In particular, for three-particle wave functions the one- and two-particle parentage coefficients are identical except for a phase factor. In this case

$$\Psi_{IM\alpha}(1,2,3)=\sum_J\langle j^2J, j|\}j^3I\alpha\rangle[\psi_J(1,2)\times\phi_j(3)]_{IM}$$
$$=\sum_J\langle j^1j, j^2J|\}j^3I\alpha\rangle[\phi_j(3)\times\psi_J(1,2)]_{IM}$$

so that when use is made of equation (A1.18) one sees that

$$\langle j^1j, j^2J|\}j^3I\alpha\rangle=(-1)^{J+j-I}\langle j^2J, j|\}j^3I\alpha\rangle. \tag{A5.15}$$

For $n>3$ the recoupling rules discussed in Appendix 4 may be used to find the relationship between the one- and two-nucleon parentage coefficients. Thus

$$=\sum_{J\beta}\langle j^{n-1}J\beta, j|\}j^nI\alpha\rangle$$

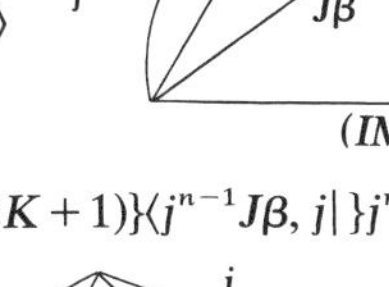

$$=\sum_{JJ'}\sum_{\beta\beta'}\langle j^{n-1}J\beta, j|\}j^nI\alpha\rangle\times\langle j^{n-2}J'\beta', j|\}j^{n-1}J\beta\rangle$$

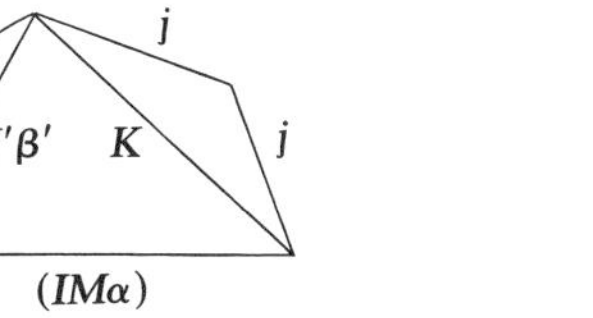

$$=\sum_{JJ'K}\sum_{\beta\beta'}\sqrt{\{(2J+1)(2K+1)\}}\langle j^{n-1}J\beta, j|\}j^nI\alpha\rangle\langle j^{n-2}J'\beta', j|\}j^{n-1}J\beta\rangle$$

$$\times\, W(J'jIj;JK) \tag{A5.16}$$

On the other hand, the diagrammatic representation of the n-particle wave function in terms of a double-parentage decomposition is

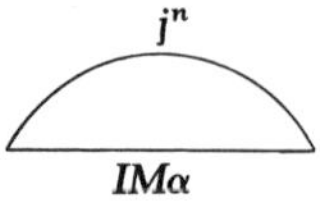

$$\sum_{J''\beta''L} \langle j^{n-2}J''\beta'', j^2L|\}j^nI\alpha\rangle$$

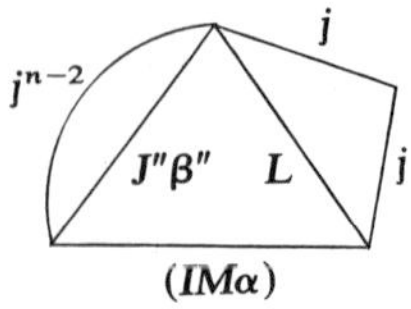

(A5.17)

Comparison of equations (A5.16) and (A5.17) shows that

$$\langle j^{n-2}J''\beta'', j^2L|\}j^nI\alpha\rangle = \sum_{J\beta} \sqrt{\{(2J+1)(2L+1)\}}\langle j^{n-1}J\beta, j|\}j^nI\alpha\rangle \times\langle j^{n-2}J''\beta'', j|\}j^{n-1}J\beta\rangle W(J''jIj; JL). \tag{A5.18}$$

For the case of two particles one may use the $1\rightarrow 0$ and $2\rightarrow 1$ one-nucleon parentage coefficients given by equations (A5.7) and (A5.8) together with the simple form of the Racah coefficient when one angular momentum is zero (equation A4.33) to show that

$$\langle j^0 0, j^2L|\}j^2I\rangle = \left\{\frac{1+(-1)^I}{2}\right\}\delta_{IL}. \tag{A5.19}$$

It is important that in any calculation that simultaneously uses one- and two-nucleon parentage coefficients that the phases be consistent. Thus once the one-nucleon c.f.p.'s are given, the phase of the two-nucleon parentage coefficients must satisfy equation (A5.18).

3. Particle–hole relationship

In all but the tabulation of Glaudemans *et al.* (1964) the fractional-parentage coefficients are only given up to the half-filled shell. By use of the particle–hole conjugation operator introduced in Chapter 3, section 2, one may deduce the c.f.p. when the shell is more than half filled. We shall use wavefunctions that obey the phase convention of equation (1.124) so that according to equation (3.45a) we must take

$$\Psi_{IMv}(j^{2j+1-n}) = (-1)^{\tilde{v}}\Gamma\Psi_{IMv}(j^n) \tag{A5.20}$$

where

$$\tilde{v} = (n-v)/2 \tag{A5.20a}$$

with v the seniority of the state. For simplicity in notation we assume seniority alone is enough to characterize the states.

As shown in Chapter 3, the particle–hole conjugation operator, Γ, has the properties

$$\begin{aligned}\Gamma^\dagger a^\dagger_{jm}\Gamma &= (-1)^{j-m}a_{j-m}\\ &= -\tilde{a}_{jm}\end{aligned} \tag{A5.21a}$$

$$\Gamma^\dagger a_{jm}\Gamma = (-1)^{j-m}a^\dagger_{j-m}. \tag{A5.21b}$$

We now use these relationships to evaluate the parentage coefficients

$$\langle j^{-(n+1)}Jv', j|\}j^{-n}Iv\rangle \quad \text{and} \quad \langle j^{-(n+2)}Jv', j^2K|\}j^{-n}Iv\rangle.$$

If Ψ_{IMv} and $\Phi_{JM'v'}$ represent the n- and $(n+1)$-particle wave functions, respectively, it follows from the definition of the fractional-parentage coefficient in equation (A5.6) that

$$\begin{aligned}\langle j^{-(n+1)}Jv', j|\}j^{-n}Iv\rangle &= \frac{(-1)^{\Delta v}}{\sqrt{(2j+1-n)}} \frac{\langle \Psi_{IMv}|\,\Gamma^{\dagger} a^{\dagger}_{jm}\Gamma\,|\Phi_{JM'v'}\rangle}{(JjM'm\mid IM)}\\ &= \frac{(-1)^{\Delta v+1}}{\sqrt{(2j+1-n)}} \frac{\langle \Psi_{IMv}|\,\tilde{a}_{jm}\,|\Phi_{JM'v'}\rangle}{(JjM'm\mid IM)}\\ &= \frac{(-1)^{\Delta v+1}}{\sqrt{(2j+1-n)}}\langle \Psi_{Iv}\|\,\tilde{a}_j\,\|\Phi_{Jv'}\rangle.\end{aligned} \tag{A5.22}$$

where

$$\Delta v = (v+v'+1)/2 \tag{A5.23}$$

From equation (A2.10) and the definition of the one-nucleon c.f.p. in equation (A5.6) one sees that

$$\begin{aligned}\langle \Psi_{Iv}\|\,\tilde{a}_j\,\|\Phi_{Jv'}\rangle &= (-1)^{I+j-J}\left(\frac{2J+1}{2I+1}\right)^{\frac{1}{2}}\langle \Phi_{Jv'}\|\,a^{\dagger}_j\,\|\Psi_{Iv}\rangle\\ &= (-1)^{I+j-J+n}\left\{\frac{(n+1)(2J+1)}{(2I+1)}\right\}^{\frac{1}{2}}\langle j^nIv, j|\}j^{n+1}Jv'\rangle.\end{aligned}$$

When this result is substituted into equation (A5.22) one finds that

$$\begin{aligned}\langle j^{-(n+1)}Jv', j|\}j^{-n}Iv\rangle = (-1)^{I-j-J+\Delta v+n}&\left\{\frac{(n+1)(2J+1)}{(2j+1-n)(2I+1)}\right\}^{\frac{1}{2}}\\ &\times\langle j^nIv, j|\}j^{n+1}Jv'\rangle.\end{aligned} \tag{A5.24}$$

We shall frequently need the parentage coefficients corresponding to the full-shell $(n=0)$ and the one-hole $(n=1)$ configurations. By use of equation (A5.7) one sees that when $n=0$

$$\langle j^{-1}j, j|\}j^{-0}0\rangle = 1. \tag{A5.25}$$

Furthermore, from equation (A5.8) it follows that when $n=1$

$$\langle j^{-2}jv\, j|\}j^{-1}j\rangle = (-1)^{v/2}\left\{\frac{2J+1}{4j(2j+1)}\right\}^{\frac{1}{2}}\{1+(-1)^J\}. \tag{A5.26}$$

A relationship similar to equation (A5.24) can be deduced for the two-particle parentage coefficients. From equations (A5.14) and (A5.20) one sees that

$$\langle j^{-(n+2)}Jv', j^2K|\}j^{-n}Iv\rangle = (-1)^{\Delta v-n+\frac{1}{2}}\left\{\frac{2}{(2j+1-n)(2j-n)}\right\}^{\frac{1}{2}}\frac{\langle \Psi_{IMv}|\,\Gamma^{\dagger}A^{\dagger}_{KM''}\Gamma\,|\psi_{JM'v'}\rangle}{(JKM'M''\mid IM)}$$

where Ψ_{IMv} and $\psi_{JM'v'}$ are the wave functions describing $(j^n)_{IMv}$ and $(j^{n+2})_{JM'v'}$, respectively and Δv is given by equation (A5.23).

By use of equation (A5.21a) it follows that

$$\begin{aligned}\Gamma^\dagger A^\dagger_{KM}\Gamma &= 2^{-\frac{1}{2}}[\tilde{a}_j \times \tilde{a}_j]_{KM}\\ &= -\tilde{A}_{KM}\end{aligned}$$

where the modified adjoint operator $\tilde{A}_{KM}$ is defined by equation (A2.9). By use of equation (A2.10) together with the fact that in the representation we have chosen the double-parentage coefficients are real, one sees that

$$\begin{aligned}(-1)^{\Delta v-n+\frac{1}{2}}\frac{\langle\Psi_{IMv}|\,\Gamma^\dagger A^\dagger_{KM''}\Gamma\,|\psi_{JM'v'}\rangle}{(JKM'M''\mid IM)} &= (-1)^{\Delta v-n-\frac{1}{2}}\langle\Psi_{Iv}\,\|\,\tilde{A}_K\,\|\psi_{Jv'}\rangle\\ &= (-1)^{\Delta v-n-\frac{1}{2}+I+K-J}\left(\frac{2J+1}{2I+1}\right)^{\frac{1}{2}}\langle\psi_{Jv'}\|\,A^\dagger_K\,\|\Psi_{Iv}\rangle\\ &= (-1)^{\Delta v-n-\frac{1}{2}+I+K-J}\left\{\frac{(n+2)(n+1)(2J+1)}{2(2I+1)}\right\}^{\frac{1}{2}}\\ &\quad\times\langle j^nIv, j^2K|\}j^{n+2}Jv'\rangle.\end{aligned}$$

Thus

$$\begin{aligned}\langle j^{-2}Jv, j^2K|\}j^{-0}0\rangle &= (-1)^{\Delta v-n-\frac{1}{2}+I+K-J}\left\{\frac{(n+2)(n+1)(2J+1)}{(2j+1-n)(2j-n)(2I+1)}\right\}^{\frac{1}{2}}\\ &\quad\times\langle j^nIv, j^2K|\}j^{n+2}Jv'\rangle. \qquad (A5.27)\end{aligned}$$

For the full shell ($n=0$) use of equation (A5.19) leads to the conclusion that

$$\langle j^{-2}Jv, j^2K|\}j^{-0}0\rangle = (-1)^{v/2}\left\{\frac{2J+1}{4j(2j+1)}\right\}^{\frac{1}{2}}\{1+(-1)^J\}\delta_{JK}. \qquad (A5.28)$$

4. Isospin parentage coefficients

One may easily extend the definition of the c.f.p. to include the isospin quantum number. Thus the analogue of equation (A5.1) is

$$\begin{aligned}\Psi_{IM;TT_z\alpha}(1,2,\ldots,n) &= \sum_{JT'\beta}\langle j^{n-1}JT'\beta, j\tfrac{1}{2}|\}j^nIT\alpha\rangle\\ &\quad\times[\Phi_{J;T'\beta}(1,\ldots,n-1)\times\phi_{j;\frac{1}{2}}(n)]_{IM;TT_z} \qquad (A5.29)\end{aligned}$$

where $[\times]$ implies coupling in both spin and isospin space. Since $\Phi_{JM';T'T_z'\beta}$ and $\Psi_{IM;TT_z\alpha}$ are a complete orthornomal set of functions for j^{n-1} and j^n, respectively, it follows that

$$\sum_{JT'\beta}\langle j^{n-1}JT'\beta, j\tfrac{1}{2}|\}j^nIT\alpha\rangle\langle j^{n-1}JT'\beta, j\tfrac{1}{2}|\}j^nIT\alpha'\rangle = \delta_{\alpha'\alpha}. \qquad (A5.30)$$

By exactly the same arguments that were used in section 2 of this appendix one may show that

$$\langle j^{n-1}JT'\beta, j\tfrac{1}{2}|\}j^nIT\alpha\rangle = \frac{(-1)^{n-1}}{\sqrt{n}}\langle\Psi_{I;T\alpha}|||\,a^\dagger_{j;\frac{1}{2}}\,|||\Phi_{J;T'\beta}\rangle \qquad (A5.31)$$

where $a^{\dagger}_{jm;\frac{1}{2}\mu}$ creates a neutron (proton) if $\mu=+\frac{1}{2}$ $(-\frac{1}{2})$ and the triple-barred matrix element is the matrix element of $a^{\dagger}_{jm;\frac{1}{2}\mu}$ reduced in both ordinary and isospin space.

The same definitions go through for the two-particle parentage coefficients so that

$$\Psi_{IM;TT_z\alpha}(1,\ldots,n)=\sum_{JT'\beta KT''}\langle j^{n-2}JT'\beta, j^2KT''|\}j^nIT\alpha\rangle \times[\psi_{J;T'\beta}(1,\ldots,n-2)\times\phi_{K;T''}(n-1,n)]_{IM;TT_z} \quad \text{(A5.32)}$$

with

$$\sum_{JT'\beta KT''}\langle j^{n-2}JT'\beta, j^2KT''|\}j^nIT\alpha\rangle\langle j^{n-2}JT'\beta, j^2KT''|\}j^nIT\alpha'\rangle=\delta_{\alpha'\alpha} \quad \text{(A5.33)}$$

and

$$\langle j^{n-2}JT'\beta, j^2KT''|\}j^nIT\alpha\rangle=\left\{\frac{2}{n(n-1)}\right\}^{\frac{1}{2}}\langle\Psi_{I;T\alpha}|||A^{\dagger}_{K;T''}|||\psi_{J;T'\beta}\rangle \quad \text{(A5.34)}$$

where

$$A^{\dagger}_{KM;TT_z}=2^{-\frac{1}{2}}\sum_{m_i\mu_i}(jjm_1m_2\mid KM)(\tfrac{1}{2}\tfrac{1}{2}\mu_1\mu_2\mid TT_z)a^{\dagger}_{jm_1;\frac{1}{2}\mu_1}a^{\dagger}_{jm_2;\frac{1}{2}\mu_2}.$$

For $n=1$ and 2 the same arguments that were used to deduce equations (A5.7) and (A5.8) may be used to show that

$$\langle j^0 00, j\tfrac{1}{2}|\}j^1 j\tfrac{1}{2}\rangle=1 \quad \text{(A5.35)}$$

$$\langle j^1 j\tfrac{1}{2}, j\tfrac{1}{2}|\}j^2 IT\rangle=\frac{1-(-1)^{I+T}}{2} \quad \text{(A5.36)}$$

and

$$\langle j^0 00, j^2KT'|\}j^2IT\rangle=\frac{1-(-1)^{I+T}}{2}\delta_{IK}\delta_{TT'}. \quad \text{(A5.37)}$$

For $n>2$ the one-nucleon isospin c.f.p.s have been tabulated by several authors. Glaudemans *et al.* (1964) give these coefficients for $j=\frac{1}{2}$ and $\frac{3}{2}$; Towner and Hardy (1969a) tabulate them for $j=\frac{3}{2}$ and $\frac{5}{2}$; Hubbard (1971) gives them for $j=\frac{7}{2}$ $(n\leq 5)$; and Shlomo (1972) gives them for $j=\frac{1}{2}$, $\frac{3}{2}$, $\frac{5}{2}$, and $\frac{7}{2}$ $(n\leq 4$ when $j=\frac{7}{2})$. For seniority-zero and -one states analytic expressions for the coefficients are given by Grayson and Nordheim (1956) and DeShalit and Talmi (1963).

For seniority-zero and -one wave functions Towner and Hardy (1969) give analytic expressions for the two-particle parentage coefficients, and of course when $n=3$ these coefficients are simply related to the one-nucleon c.f.p.s

$$\langle j^1 j\tfrac{1}{2}, j^2JT'|\}j^3IT\alpha\rangle=(-1)^{J+j-I+T'+\frac{1}{2}-T}\langle j^2JT', j\tfrac{1}{2}|\}j^3IT\alpha\rangle. \quad \text{(A5.38)}$$

In general, once the one-nucleon c.f.p. are known the two-particle parentage coefficients can be calculated by a straightforward generalization of equation (A5.18), namely

$$\langle j^{n-2}J'T'\beta', j^2LT''|\}j^nIT\alpha\rangle=\sum_{JT_1\beta}\sqrt{\{(2J+1)(2L+1)(2T_1+1)(2T''+1)\}} \times\langle j^{n-1}JT_1\beta, j\tfrac{1}{2}|\}j^nIT\alpha\rangle\langle j^{n-2}J'T'\beta', j\tfrac{1}{2}|\}j^{n-1}JT_1\beta\rangle \times W(J'jIj;JL)W(T'\tfrac{1}{2}T\tfrac{1}{2};T_1T''). \quad \text{(A5.39)}$$

Finally, by the arguments used in section 3 of this appendix one may deduce a particle–hole relationship for these coefficients. For all values of n except the half filled shell $n=2j+1$, (which we shall not discuss) we may define the hole-state wave function as done in Chapter 3, section 3

$$\Gamma_t \Psi_{IM;TT_z\alpha}(n\text{-particles}) = \Psi_{IM;TT_z\alpha}(n\text{-holes})$$

and

$$\begin{aligned}\Gamma_t^{\dagger} a^{\dagger}_{jm;\frac{1}{2}\mu} \Gamma_t &= -(-1)^{j+m+\frac{1}{2}+\mu} a_{j-m;\frac{1}{2}-\mu}\\ &= -\tilde{a}_{jm;\frac{1}{2}\mu}\end{aligned}$$

where $\tilde{a}_{jm;\frac{1}{2}\mu}$ is the modified Hermitian adjoint operator of equation (A2.9) suitably generalized to include isospin. Thus if $\Psi_{IM;TT_z\alpha}$ and $\Phi_{JM';T'T_z'\beta}$ are the wave functions describing the configurations $(j^n)_{IM;TT_z\alpha}$ and $(j^{n+1})_{JM';T'T_z'\beta}$, respectively, it follows from equation (A5.31) that

$$\begin{aligned}\langle j^{-(n+1)} JT'\beta, j\tfrac{1}{2}|\}j^{-n} IT\alpha\rangle &= \frac{(-1)^{4j+1-n}}{\sqrt{(4j+2-n)}}\\ &\quad\times\frac{\langle \Psi_{IM;TT_z\alpha}|\, \Gamma_t^{\dagger} a^{\dagger}_{jm;\frac{1}{2}\mu} \Gamma_t\, |\Phi_{JM;T'T_z'\beta}\rangle}{(JjM'm\mid IM)(T'\tfrac{1}{2}T_z'\mu\mid TT_z)}\\ &= \frac{(-1)^n}{\sqrt{(4j+2-n)}}\langle \Psi_{I;T\alpha}|||\, \tilde{a}_{j;\frac{1}{2}}\, |||\Phi_{J;T'\beta}\rangle\end{aligned}$$

where use has been made of the fact that when isospin is included, the single-particle level j can accommodate $(4j+2)$ particles and that $(-1)^{4j}=1$.

The generalization of equation (A2.10) to include isospin implies that

$$\begin{aligned}\langle \Psi_{I;T\alpha}|||\, \tilde{a}_{j;\frac{1}{2}}\, |||\Phi_{J;T'\beta}\rangle &= (-1)^{I+j-J+T+\frac{1}{2}-T'}\left\{\frac{(2J+1)(2T'+1)}{(2I+1)(2T+1)}\right\}^{\frac{1}{2}}\\ &\quad\times\langle \Phi_{J;T'\beta}|||\, a^{\dagger}_{j;\frac{1}{2}}\, |||\Psi_{I;T\alpha}\rangle\\ &= (-1)^{I+j-J+T+\frac{1}{2}-T'+n}\left\{\frac{(n+1)(2J+1)(2T'+1)}{(2I+1)(2T+1)}\right\}^{\frac{1}{2}}\\ &\quad\times\langle j^n IT\alpha, j\tfrac{1}{2}|\}j^{n+1} JT'\beta\rangle.\end{aligned}$$

Thus except at the centre of the shell

$$\begin{aligned}\langle j^{-(n+1)} JT'\beta, j\tfrac{1}{2}|\}j^{-n}\, IT_\alpha\rangle &= (-1)^{I+j-J+T+\frac{1}{2}-T'}\left\{\frac{(n+1)(2J+1)(2T'+1)}{(4j+2-n)(2I+1)(2T+1)}\right\}^{\frac{1}{2}}\\ &\quad\times\langle j^n IT\alpha, j\tfrac{1}{2}|\}j^{n+1} JT'\beta\rangle. \qquad\text{(A5.40)}\end{aligned}$$

By exactly the same arguments it follows that

$$\begin{aligned}&\langle j^{-(n+2)} JT'\beta, j^2 KT''|\}j^{-n} IT\alpha\rangle\\ &\quad= (-1)^{I+K-J+T+T''-T'+1}\left\{\frac{(n+2)(n+1)(2J+1)(2T'+1)}{(4j+2-n)(4j+1-n)(2I+1)(2T+1)}\right\}^{\frac{1}{2}}\\ &\quad\times\langle j^n IT\alpha, j^2 KT''|\}j^{n+2} JT'\beta\rangle. \qquad\text{(A5.41)}\end{aligned}$$

These particle–hole relationships may now be used to write down the parentage coefficients for the full-shell and one-hole states. Thus from equations (A5.35)–(A5.37), (A5.40), and (A5.41) it follows that

$$\langle j^{-1}j\tfrac{1}{2}, j\tfrac{1}{2}|\}j^{-0}00\rangle = 1 \tag{A5.42}$$

$$\langle j^{-2}JT', j\tfrac{1}{2}|\}j^{-1}j\tfrac{1}{2}\rangle = -\left\{\frac{(2J+1)(2T'+1)}{(2j+1)(4j+1)}\right\}^{\frac{1}{2}}\left\{\frac{1-(-1)^{J+T'}}{2}\right\} \tag{A5.43}$$

$$\langle j^{-2}JT', j^2KT''|\}j^{-0}00\rangle = -\left\{\frac{(2J+1)(2T'+1)}{(2j+1)(4j+1)}\right\}^{\frac{1}{2}}\left\{\frac{1-(-1)^{J+T'}}{2}\right\}\delta_{JK}\delta_{T'T''}. \tag{A5.44}$$

5. Some useful formulae

In this section we collect some of the useful formulae involving the one- and two-nucleon fractional-parentage coefficients. The analytic expressions for the seniority-zero-, and -one c.f.p.s were derived in Appendix 3. For the zero one- and two-nucleon configurations the states are uniquely specified by their angular momentum and isospin so that in these cases the auxiliary quantum numbers α and β have been suppressed. Whenever α and/or β are set equal to 0, 1, or 2 they refer to the seniority of the state. The phases of the wavefunctions are as given in equation (1.124).

$$\sum_{J\beta}\langle j^{n-1}J\beta, j|\}j^nI\alpha'\rangle\langle j^{n-1}J\beta, j|\}j^nI\alpha\rangle = \delta_{\alpha'\alpha} \tag{A5.45}$$

$$\sum_{J\beta K}\langle j^{n-2}J\beta, j^2K|\}j^nI\alpha'\rangle\langle j^{n-2}J\beta, j^2K|\}j^nI\alpha\rangle = \delta_{\alpha'\alpha} \tag{A5.46}$$

$$\langle j^1j, j^2K|\}j^3I\alpha\rangle = (-1)^{K+j-I}\langle j^2K, j|\}j^3I\alpha\rangle \tag{A5.47}$$

$$\langle j^00, j\,|\}j^1j\rangle = 1 \tag{A5.48}$$

$$\langle j^1j, j|\}j^2I\rangle = \left\{\frac{1+(-1)^I}{2}\right\} \tag{A5.49}$$

$$\langle j^00, j^2K|\}j^2I\rangle = \left\{\frac{1+(-1)^I}{2}\right\}\delta_{IK} \tag{A5.50}$$

$$\langle j^{-1}j, j|\}j^{-0}0\rangle = 1 \tag{A5.51}$$

$$\langle j^{-2}Jv, j|\}j^{-1}j\rangle = (-1)^{v/2}\left\{\frac{2J+1}{4j(2j+1)}\right\}^{\frac{1}{2}}\{1+(-1)^J\} \tag{A5.52}$$

$$\langle j^{-2}Jv, j^2K|\}j^{-0}0\rangle = (-1)^{v/2}\left\{\frac{2J+1}{4j(2j+1)}\right\}^{\frac{1}{2}}\{1+(-1)^J\}\delta_{JK}. \tag{A5.53}$$

$$\langle j^nIv, j|\}j^{n+1}00\rangle = \delta_{I,j}\delta_{v,1} \tag{A5.54}$$

$$\begin{aligned}\langle j^nIv, j|\}j^{n+1}j1\rangle &= \left\{\frac{2j+1-n}{(n+1)(2j+1)}\right\}^{\frac{1}{2}}\delta_{I,0}\delta_{v,0}\\ &= -\left\{\frac{2n(2I+1)}{(n+1)(2j-1)(2j+1)}\right\}^{\frac{1}{2}}\left\{\frac{1+(-1)^I}{2}\right\}\delta_{v,2}\end{aligned} \tag{A5.55}$$

$$\langle j^n j1, j| \} j^{n+1} I2\rangle = \left\{\frac{2(2j-n)}{(n+1)(2j-1)}\right\}^{\frac{1}{2}} \left\{\frac{1+(-1)^I}{2}\right\} \quad \text{(A5.56)}$$

$$\langle j^n 00, j^2 0| \} j^{n+2} 00\rangle = \left\{\frac{2j+1-n}{(n+1)(2j+1)}\right\}^{\frac{1}{2}} \quad \text{(A5.57)}$$

$$\langle j^n I2, j^2 I| \} j^{n+2} 00\rangle = -\left\{\frac{2n(2I+1)}{(n+1)(2j+1)(2j-1)}\right\}^{\frac{1}{2}} \left\{\frac{1+(-1)^I}{2}\right\} \quad \text{(A5.58)}$$

$$\langle j^n 00, j^2 I| \} j^{n+2} I2\rangle = \left\{\frac{2(2j+1-n)(2j-1-n)}{(n+2)(n+1)(2j+1)(2j-1)}\right\}^{\frac{1}{2}} \left\{\frac{1+(-1)^I}{2}\right\}. \quad \text{(A5.59)}$$

When the isospin quantum number is included, the following formulae hold true:

$$\sum_{J\beta T'} \langle j^{n-1} JT'\beta, j\tfrac{1}{2}| \} j^n IT\alpha'\rangle \langle j^{n-1} JT'\beta, j\tfrac{1}{2}| \} j^n IT\alpha\rangle = \delta_{\alpha'\alpha} \quad \text{(A5.60)}$$

$$\sum_{J\beta T' KT''} \langle j^{n-2} JT'\beta, j^2 KT''| \} j^n IT\alpha'\rangle \langle j^{n-2} JT'\beta, j^2 KT''| \} j^n IT\alpha\rangle = \delta_{\alpha'\alpha} \quad \text{(A5.61)}$$

$$\langle j^1 j\tfrac{1}{2}, j^2 KT'| \} j^3 IT\alpha\rangle = (-1)^{K+j-I+T'+\frac{1}{2}-T} \langle j^2 KT', j\tfrac{1}{2}| \} j^3 IT\alpha\rangle \quad \text{(A5.62)}$$

$$\langle j^0 00, j\tfrac{1}{2}| \} j^1 j\tfrac{1}{2}\rangle = 1 \quad \text{(A5.63)}$$

$$\langle j^1 j\tfrac{1}{2}, j\tfrac{1}{2}| \} j^2 IT\rangle = \left\{\frac{1-(-1)^{I+T}}{2}\right\} \quad \text{(A5.64)}$$

$$\langle j^0 00, j^2 KT'| \} j^2 IT\rangle = \left\{\frac{1-(-1)^{I+T}}{2}\right\} \delta_{IK}\delta_{TT'} \quad \text{(A5.65)}$$

$$\langle j^{-1} j\tfrac{1}{2}, j\tfrac{1}{2}| \} j^{-0} 00\rangle = 1 \quad \text{(A5.66)}$$

$$\langle j^{-2} JT', j\tfrac{1}{2}| \} j^{-1} j\tfrac{1}{2}\rangle = -\left\{\frac{(2J+1)(2T'+1)}{(2j+1)(4j+1)}\right\}^{\frac{1}{2}} \left\{\frac{1-(-1)^{J+T'}}{2}\right\} \quad \text{(A5.67)}$$

$$\langle j^{-2} JT', j^2 KT''| \} j^{-0} 00\rangle = -\left\{\frac{(2J+1)(2T'+1)}{(2j+1)(4j+1)}\right\}^{\frac{1}{2}} \left\{\frac{1-(-1)^{J+T'}}{2}\right\} \delta_{JK}\delta_{T'T''}. \quad \text{(A5.68)}$$

6. Tables of one- and two-nucleon parentage coefficients

Tables A5.1–A5.5 give the identical nucleon one- and two-particle parentage coefficients encountered in this book. Our phase convention is that the seniority-zero, -one and -two eigenfunctions are given by (see equation (1.124))

$$\Psi_{000} = \left\{\frac{2^n(2j+1-n)!!}{n!!(2j+1)!!}\right\}^{\frac{1}{2}} S_+^{n/2} |0\rangle \quad \text{(A5.69)}$$

$$\Psi_{jm1} = \left\{\frac{2^{n-1}(2j-n)!!}{(n-1)!!(2j-1)!!}\right\}^{\frac{1}{2}} a_{jm}^\dagger S_+^{(n-1)/2} |0\rangle \quad \text{(A5.70)}$$

$$\Psi_{IM2} = \left\{\frac{2^{n-2}(2j-1-n)!!}{(n-2)!!(2j-3)!!}\right\}^{\frac{1}{2}} A_{IM}^\dagger S_+^{(n-2)/2} |0\rangle \quad \text{(A5.71)}$$

where

$$S_+ = \sum_{m>0} (-1)^{j-m} a^\dagger_{jm} a^\dagger_{j-m}$$

and $A^\dagger_{IM}$ is given by equation (A5.11). For seniorities higher than two the wave functions have the phase given by Bayman and Lande (1966) where a more extensive tabulation can be found.

When the shell is more than half full the parentage coefficients should be calculated by use of equations (A5.24) and (A5.27).

TABLE A5.1

$\langle \tfrac{5}{2}^2 J, \tfrac{5}{2} | \} \tfrac{5}{2}^3 I \rangle$

I	$J=0$	$J=2$	$J=4$
$\frac{3}{2}$		$-\sqrt{\frac{5}{7}}$	$\sqrt{\frac{2}{7}}$
$\frac{5}{2}$	$\frac{\sqrt{2}}{3}$	$-\frac{1}{3}\sqrt{\frac{5}{2}}$	$-\frac{1}{\sqrt{2}}$
$\frac{9}{2}$		$-\sqrt{\frac{3}{14}}$	$\sqrt{\frac{11}{14}}$

TABLE A5.2

$\langle \tfrac{7}{2}^2 J, \tfrac{7}{2} | \} \tfrac{7}{2}^3 I \rangle$

I	$J=0$	$J=2$	$J=4$	$J=6$
$\frac{3}{2}$		$-\sqrt{\frac{3}{14}}$	$\sqrt{\frac{11}{14}}$	
$\frac{5}{2}$		$-\frac{1}{3}\sqrt{\frac{11}{2}}$	$-\sqrt{\frac{2}{33}}$	$\frac{1}{3}\sqrt{\frac{65}{22}}$
$\frac{7}{2}$	$\frac{1}{2}$	$-\frac{\sqrt{5}}{6}$	$-\frac{1}{2}$	$-\frac{\sqrt{13}}{6}$
$\frac{9}{2}$		$\frac{1}{3}\sqrt{\frac{13}{14}}$	$-5\sqrt{\frac{2}{77}}$	$\frac{7}{3\sqrt{22}}$
$\frac{11}{2}$		$-\frac{1}{3}\sqrt{\frac{5}{2}}$	$\sqrt{\frac{13}{66}}$	$\frac{2}{3}\sqrt{\frac{13}{11}}$
$\frac{15}{2}$			$-\sqrt{\frac{5}{22}}$	$\sqrt{\frac{17}{22}}$

TABLE A5.3

$\langle \frac{7}{2}^3 J, \frac{7}{2} | \} \frac{7}{2}^4 I v \rangle$

v	I	$J=\frac{3}{2}$	$J=\frac{5}{2}$	$J=\frac{7}{2}$	$J=\frac{9}{2}$	$J=\frac{11}{2}$	$J=\frac{15}{2}$
0	0			1			
2	2	$-\frac{3}{2\sqrt{35}}$	$\frac{1}{2}\sqrt{\frac{11}{10}}$	$\frac{1}{\sqrt{3}}$	$-\frac{1}{2}\sqrt{\frac{13}{42}}$	$-\frac{1}{2}$	
2	4	$\frac{1}{2}\sqrt{\frac{11}{21}}$	$\frac{1}{\sqrt{66}}$	$\frac{1}{\sqrt{3}}$	$5\sqrt{\frac{5}{462}}$	$\frac{1}{2}\sqrt{\frac{13}{33}}$	$-\sqrt{\frac{5}{33}}$
2	6		$-\frac{1}{2}\sqrt{\frac{5}{22}}$	$\frac{1}{\sqrt{3}}$	$-\frac{7}{2}\sqrt{\frac{5}{858}}$	$\sqrt{\frac{2}{11}}$	$\sqrt{\frac{51}{143}}$
4	2	$-\sqrt{\frac{11}{35}}$	$-\frac{1}{2\sqrt{10}}$		$\frac{3}{2}\sqrt{\frac{39}{154}}$	$\frac{-1}{\sqrt{11}}$	
4	4	$-\frac{1}{2}\sqrt{\frac{13}{105}}$	$-\sqrt{\frac{13}{30}}$		$\sqrt{\frac{3}{182}}$	$\frac{1}{2}\sqrt{\frac{5}{3}}$	$\frac{2}{\sqrt{39}}$
4	5	$\frac{1}{2}\sqrt{\frac{3}{5}}$	$\frac{-7}{2\sqrt{110}}$		$\frac{3}{2}\sqrt{\frac{3}{22}}$	$-\frac{1}{2}\sqrt{\frac{65}{77}}$	$\sqrt{\frac{17}{77}}$
4	8				$-2\sqrt{\frac{5}{143}}$	$-3\sqrt{\frac{2}{77}}$	$\sqrt{\frac{57}{91}}$

TABLE A5.4

$\langle \frac{9}{2}^2 J, \frac{9}{2} | \} \frac{9}{2}^3 I v \rangle$

v	I	$J=0$	$J=2$	$J=4$	$J=6$	$J=8$
1	$\frac{9}{2}$	$\frac{2}{\sqrt{15}}$	$-\frac{1}{2\sqrt{3}}$	$-\frac{\sqrt{3}}{2\sqrt{5}}$	$-\frac{\sqrt{13}}{2\sqrt{15}}$	$-\frac{\sqrt{17}}{2\sqrt{15}}$
3	$\frac{3}{2}$			$-\frac{2\sqrt{2}}{\sqrt{11}}$	$\sqrt{\frac{3}{11}}$	
3	$\frac{5}{2}$		$-\frac{\sqrt{5}}{3\sqrt{2}}$	$\sqrt{\frac{13}{66}}$	$\frac{2\sqrt{13}}{3\sqrt{11}}$	
3	$\frac{7}{2}$		$\frac{2\sqrt{13}}{3\sqrt{11}}$	$\frac{-2\sqrt{5}}{\sqrt{143}}$	$-\frac{1}{3\sqrt{55}}$	$\sqrt{\frac{238}{715}}$
3	$\frac{9}{2}$		$\frac{\sqrt{13}}{6\sqrt{11}}$	$\frac{-7\sqrt{5}}{2\sqrt{143}}$	$\frac{31}{6\sqrt{55}}$	$-\frac{3\sqrt{17}}{2\sqrt{715}}$
3	$\frac{11}{2}$		$\frac{\sqrt{17}}{3\sqrt{11}}$	$-\sqrt{\frac{170}{429}}$	$-\frac{2\sqrt{14}}{3\sqrt{55}}$	$\frac{6\sqrt{19}}{\sqrt{2145}}$
3	$\frac{13}{2}$		$-\sqrt{\frac{10}{33}}$	$\frac{3}{\sqrt{143}}$	$\sqrt{\frac{17}{66}}$	$\sqrt{\frac{323}{858}}$
3	$\frac{15}{2}$			$\sqrt{\frac{19}{143}}$	$-\sqrt{\frac{7}{11}}$	$\sqrt{\frac{3}{13}}$
3	$\frac{17}{2}$			$-\frac{5\sqrt{5}}{\sqrt{429}}$	$\sqrt{\frac{19}{110}}$	$\sqrt{\frac{209}{390}}$
3	$\frac{21}{2}$				$-\sqrt{\frac{7}{30}}$	$\sqrt{\frac{23}{30}}$

TABLE A5.5

$$\langle (\tfrac{7}{2})^2 J, (\tfrac{7}{2})^2 J' | \} (\tfrac{7}{2})^4 I v \rangle$$

(I, v)	$J \backslash J'$	0	2	4	6	(I, v)	$J \backslash J'$	0	2	4	6
(0, 0)	0	$\frac{1}{2}$				(2, 4)	0				
	2		$-\frac{\sqrt{5}}{6}$				2		$\frac{\sqrt{11}}{7\sqrt{2}}$	$\frac{2\sqrt{2}}{7\sqrt{3}}$	
	4			$-\frac{1}{2}$			4		$\frac{2\sqrt{2}}{7\sqrt{3}}$	$-\frac{51}{77}$	$-\frac{2\sqrt{13}}{11\sqrt{3}}$
	6				$-\frac{\sqrt{13}}{6}$		6			$-\frac{2\sqrt{13}}{11\sqrt{3}}$	$\frac{\sqrt{13}}{11\sqrt{2}}$
(2, 2)	0		$\frac{1}{2\sqrt{3}}$			(4, 4)	0				
	2	$\frac{1}{2\sqrt{3}}$	$\frac{4\sqrt{2}}{21}$	$-\frac{\sqrt{22}}{7\sqrt{3}}$			2		$-\frac{5\sqrt{13}}{21\sqrt{6}}$	$-\frac{2\sqrt{13}}{21}$	$\frac{2}{3\sqrt{3}}$
	4		$-\frac{\sqrt{22}}{7\sqrt{3}}$	$\frac{1}{7\sqrt{11}}$	$-\frac{\sqrt{13}}{2\sqrt{33}}$		4		$-\frac{2\sqrt{13}}{21}$	$\frac{5\sqrt{5}}{7\sqrt{66}}$	$\frac{1}{3\sqrt{11}}$
	6			$-\frac{\sqrt{13}}{2\sqrt{33}}$	$-\frac{\sqrt{26}}{3\sqrt{11}}$		6		$\frac{2}{3\sqrt{3}}$	$\frac{1}{3\sqrt{11}}$	$\frac{\sqrt{85}}{3\sqrt{33}}$
(4, 2)	0			$\frac{1}{2\sqrt{3}}$		(5, 4)	0				
	2		$-\frac{\sqrt{110}}{21\sqrt{3}}$	$\frac{\sqrt{5}}{21\sqrt{11}}$	$-\frac{\sqrt{65}}{6\sqrt{33}}$		2			$\frac{\sqrt{7}}{\sqrt{66}}$	$\frac{\sqrt{13}}{\sqrt{154}}$
	4	$\frac{1}{2\sqrt{3}}$	$\frac{\sqrt{5}}{21\sqrt{11}}$	$\frac{8\sqrt{26}}{77\sqrt{3}}$	$-\frac{2\sqrt{65}}{33}$		4		$-\frac{\sqrt{7}}{\sqrt{66}}$		$-\frac{\sqrt{13}}{\sqrt{42}}$
	6		$-\frac{\sqrt{65}}{6\sqrt{33}}$	$-\frac{2\sqrt{65}}{33}$	$\frac{\sqrt{221}}{33\sqrt{3}}$		6		$-\frac{\sqrt{13}}{\sqrt{154}}$	$\frac{\sqrt{13}}{\sqrt{42}}$	
(6, 2)	0				$\frac{1}{2\sqrt{3}}$	(8, 4)	0				
	2			$-\frac{\sqrt{5}}{2\sqrt{33}}$	$-\frac{\sqrt{10}}{3\sqrt{11}}$		2				$\frac{\sqrt{5}}{3\sqrt{7}}$
	4		$-\frac{\sqrt{5}}{2\sqrt{33}}$	$-\frac{2\sqrt{5}}{11}$	$\frac{\sqrt{17}}{11\sqrt{3}}$		4			$-\frac{\sqrt{35}}{11\sqrt{2}}$	$-\frac{\sqrt{114}}{11\sqrt{7}}$
	6	$\frac{1}{2\sqrt{3}}$	$-\frac{\sqrt{10}}{3\sqrt{11}}$	$\frac{\sqrt{17}}{11\sqrt{3}}$	$\frac{\sqrt{323}}{33}$		6		$\frac{\sqrt{5}}{3\sqrt{7}}$	$-\frac{\sqrt{114}}{11\sqrt{7}}$	$\frac{7\sqrt{19}}{33\sqrt{2}}$

APPENDIX 6

RELATIVE CENTRE-OF-MASS TRANSFORMATION COEFFICIENTS

The Schrödinger equation governing the motion of a single nucleon in a harmonic-oscillator potential is

$$H_1\phi_{nl} = \varepsilon_{nl}\phi_{nl}$$

where

$$H_1 = \frac{p_1^2}{2m} + \tfrac{1}{2}m\omega^2 r_1^2, \tag{A6.1}$$

$$\varepsilon_{nl} = (2n + l + \tfrac{3}{2})\hbar\omega$$

and

$$\phi_{nl} = R_{nl}(r_1)Y_{lm}(\theta_1, \phi_1)$$

with

$$R_{nl}(r_1) = \left[\frac{2^{l-n+2}(2l+2n+1)!!\alpha^{2l+3}}{\sqrt{\pi}n!\{(2l+1)!!\}^2}\right]^{\frac{1}{2}}(\exp -\tfrac{1}{2}\alpha^2 r_1^2)r_1^l$$

$$\times \sum_{k=0}^{n} \frac{(-1)^k 2^k n!(2l+1)!!(\alpha^2 r_1^2)^k}{k!(n-k)!(2l+2k+1)!!} \tag{A6.2}$$

$$\alpha^2 = m\omega/\hbar.$$

Two equal mass non-interacting particles moving in a common oscillator potential and coupling their angular momenta to $(\lambda\mu)$ will satisfy the wave equation

$$H\psi_{\lambda\mu} = (H_1 + H_2)\psi_{\lambda\mu} = E\psi_{\lambda\mu}$$

with

$$E = (2n_1 + l_1 + 2n_2 + l_2 + 3)\hbar\omega \tag{A6.3}$$

and

$$\psi_{\lambda\mu}(\underline{r}_1, \underline{r}_2) = R_{n_1l_1}(r_1)R_{n_2l_2}(r_2)[Y_{l_1}(\theta_1, \phi_1) \times Y_{l_2}(\theta_2, \phi_2)]_{\lambda\mu} \tag{A6.4}$$

Instead of writing the Hamiltonian in terms of the variables $\underline{r}_1$ and $\underline{r}_2$ one can introduce the relative and centre-of-mass coordinates of the two particles

$$\underline{r} = \frac{1}{\sqrt{2}}(\underline{r}_1 - \underline{r}_2)$$

$$\underline{R} = \frac{1}{\sqrt{2}}(\underline{r}_1 + \underline{r}_2) \tag{A6.5}$$

and their canonically conjugate momenta

$$\underset{\sim}{p}=\frac{1}{\sqrt{2}}(\underset{\sim}{p}_1-\underset{\sim}{p}_2)$$

$$\underset{\sim}{P}=\frac{1}{\sqrt{2}}(\underset{\sim}{p}_1+\underset{\sim}{p}_2) \tag{A6.6}$$

and rewrite H as (Talmi 1952)

$$\begin{aligned} H &= \frac{p_1^2}{2m}+\tfrac{1}{2}m\omega^2 r_1^2+\frac{p_2^2}{2m}+\tfrac{1}{2}m\omega^2 r_2^2 \\ &= \frac{p^2}{2m}+\tfrac{1}{2}m\omega^2 r^2+\frac{P^2}{2m}+\tfrac{1}{2}m\omega^2 R^2. \end{aligned} \tag{A6.7}$$

Since H is separable in both $(\underset{\sim}{r}_1, \underset{\sim}{r}_2)$ and $(\underset{\sim}{r}, \underset{\sim}{R})$ space, the eigenfunction in equation (A6.4) can also be separated in $(\underset{\sim}{r}, \underset{\sim}{R})$ space

$$\begin{aligned} R_{n_1 l_1}(r_1)R_{n_2 l_2}(r_2)[Y_{l_1}(\theta_1, \phi_1)\times Y_{l_2}(\theta_2, \phi_2)]_{\lambda\mu} &= \sum_{nlNL} M_\lambda(nlNL; n_1 l_1 n_2 l_2)R_{nl}(r)R_{NL}(R) \\ &\quad \times[Y_l(\theta, \phi)\times Y_L(\Theta, \Phi)]_{\lambda\mu}. \end{aligned} \tag{A6.8}$$

The coefficient M_λ is often called the Moshinsky coefficient and its square gives the probability that the two-particle system will be found in a state of relative motion characterized by the oscillator function $R_{nl}(r)Y_{lm}(\theta, \phi)$ and centre-of-mass motion $R_{NL}(R)Y_{LM}(\Theta, \Phi)$. Furthermore, since the wave-function on the right-hand side of equation (A6.8) must describe a state with the same energy as that on the left-hand side the summation over $(nlNL)$ is finite and

$$2n_1+l_1+2n_2+l_2=2n+l+2N+L. \tag{A6.9}$$

Since

$$\exp\left\{-\frac{\alpha^2}{2}(r_1^2+r_2^2)\right\}=\exp\left\{-\frac{\alpha^2}{2}(r^2+R^2)\right\}$$

the transformation of the exponential is straightforward. In addition, a term on the right-hand side of equation (A6.8) of the form $r_1^p r_2^q$ will transform, via equation (A6.5), into

$$r_1^p r_2^q=\sum_{st} c_{st} r^s R^t$$

with

$$s+t=p+q.$$

Thus equation (A6.8) holds not only for the oscillator wave-functions but also for each individual power of $r_1^p r_2^q$. In particular, the maximum power of $r_1 r_2$ is

$$r_1^{2n_1+l_1} r_2^{2n_2+l_2}$$

and because of equation (A6.9) the coefficient of this power of $r_1 r_2$ must be equal to the coefficient of $r^{2n+l}R^{2N+L}$ on the right-hand side. From the definition of the oscillator functions in equation (A6.2), the coefficient of this power of $r_1 r_2$ is

$$(-1)^{n_1+n_2}\left\{\frac{2^{2n_1+l_1+2n_2+l_2+4}\alpha^{2(2n_1+l_1+2n_2+l_2+3)}}{\pi(2n_1)!!(2l_1+2n_1+1)!!(2n_2)!!(2l_2+2n_2+1)!!}\right\}^{\frac{1}{2}}.$$

Thus in addition to satisfying equation (A6.8) the transformation coefficients also satisfy the simpler equation

$$A_{n_1l_1}A_{n_2l_2}r^{2n_1+l_1}r_2^{2n_2+l_2}[Y_{l_1}(\theta_1,\phi_1)\times Y_{l_2}(\theta_2,\phi_2)]_{\lambda\mu}$$

$$=\sum_{nlNL} M_\lambda(nlNL;n_1l_1n_2l_2)A_{nl}A_{NL}r^{2n+l}R^{2N+L}[Y_l(\theta,\phi)\times Y_L(\Theta,\Phi)]_{\lambda\mu} \quad \text{(A6.10)}$$

where each of the A_{nl} in equation (A6.10) are given by

$$A_{nl}=(-1)^n\left\{\frac{1}{(2n)!!(2l+2n+1)!!}\right\}^{\frac{1}{2}} \quad \text{(A6.11)}$$

1. Transformation brackets when $n_1=n_2=0$

To find the transformation coefficients when $n_1=n_2=0$ is a straightforward exercise in recoupling theory once the expansion of $r_1^l Y_{lm}(\theta_1,\phi_1)$ in terms of relative and centre-of-mass coordinates is known. To find this we make a Taylor series expansion of the function about the point $\underline{r}_1=\underline{R}/\sqrt{2}$. According to equation (A2.14)

$$\nabla_\mu r^l Y_{lm}=(2l+1)\left(\frac{2l+1}{2l-1}\right)^{\frac{1}{2}}(l100\mid l-1,0)(l1m\mu\mid l-1,\bar{m})r^{l-1}Y_{l-1,\bar{m}} \quad \text{(A6.12)}$$

and consequently we may write

$$r_1^l Y_{lm}(\theta_1,\phi_1)=\sum_k G(l,k)r^k R^{l-k}[Y_k(\theta,\phi)\times Y_{l-k}(\Theta,\Phi)]_{lm}. \quad \text{(A6.13)}$$

The coefficient $G(l,k)$ may be found simply by considering the particular case $m=l$. According to equation (A1.8)

$$Y_{ll}(\theta,\phi)=\frac{(-1)^l}{2^l l!}\left\{\frac{(2l+1)!}{4\pi}\right\}^{\frac{1}{2}}(\sin\theta\, e^{i\phi})^l$$

so that

$$r_1^l Y_{ll}(\theta_1,\phi_1)=\frac{(-1)^l}{2^l l!}\left\{\frac{(2l+1)!}{4\pi}\right\}^{\frac{1}{2}}(x_1+iy_1)^l.$$

The vector relationships in equation (A6.5) become scalar relationships in rectangular coordinates. Consequently x_1 may be replaced by $(X+x)/\sqrt{2}$ and y_1 by $(Y+y)/\sqrt{2}$. The binomial expansion then yields

$$r_1^l Y_{ll}(\theta_1,\phi_1)=\frac{(-1)^l}{2^l l!}\left\{\frac{(2l+1)!}{4\pi}\right\}^{\frac{1}{2}}\left(\frac{1}{\sqrt{2}}\right)^l\sum_k\frac{l!}{k!(l-k)!}(X+iY)^{l-k}(x+iy)^k$$

$$=\left(\frac{1}{\sqrt{2}}\right)^l\sum_k\left\{\frac{4\pi(2l+1)!}{(2k+1)!(2l-2k+1)!}\right\}^{\frac{1}{2}}r^k R^{l-k}Y_{kk}(\theta,\phi)Y_{l-kl-k}(\Theta,\Phi).$$

By noting that when $m=l$ the Clebsch in equation (A6.13) is unity one obtains the value of $G(l, k)$. Thus

$$r_1^l Y_{lm}(\theta_1, \phi_1) = \sum_k \left\{\frac{4\pi(2l+1)!}{2^l(2k+1)!(2l-2k+1)!}\right\}^{\frac{1}{2}} r^k R^{l-k}$$
$$\times [Y_k(\theta,\phi)\times Y_{l-k}(\Theta,\Phi)]_{lm}. \quad \text{(A6.14)}$$

A similar expression holds for $r_2^l Y_{lm}(\theta_2, \phi_2)$, the only change being that $r \to -r$. By inserting these expansions into equation (A6.10) with $n_1 = n_2 = 0$ we arrive at the relationship

$$4\pi A_{0l_1} A_{0l_2} \sum_{k_1k_2} (-1)^{k_2} \left\{\frac{(2l_1+1)!(2l_2+1)!}{2^{l_1+l_2}(2k_1+1)!(2k_2+1)!(2l_1-2k_1+1)!(2l_2-2k_2+1)!}\right\}^{\frac{1}{2}}$$
$$\times r^{k_1+k_2} R^{l_1+l_2-k_1-k_2} [[Y_{k_1}(\theta,\phi)\times Y_{l_1-k_1}(\Theta,\Phi)]_{l_1} \times [Y_{k_2}(\theta,\phi)\times Y_{l_2-k_2}(\Theta,\Phi)]_{l_2}]_{\lambda\mu}$$
$$= \sum_{nlNL} M_\lambda(nlNL; 0l_1 0l_2) A_{nl} A_{NL} r^{2N+l} R^{2N+L} [Y_l(\theta,\phi)\times Y_L(\Theta,\Phi)]_{\lambda\mu}. \quad \text{(A6.15)}$$

Since Y_{k_1} and Y_{k_2} are both functions of the relative coordinates and the other two spherical harmonics depend on the centre-of-mass angles, a recoupling is obviously called for. According to equation (A4.8) this will lead to a $9j$ coefficient; i.e.

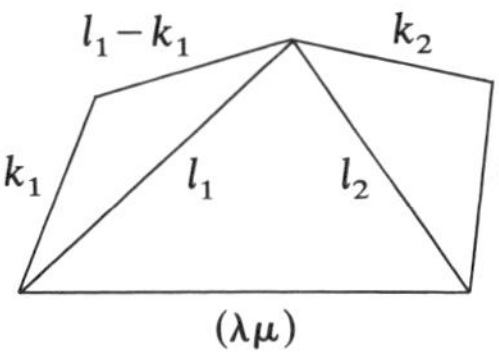

$$= \sum_{\tilde{l}\tilde{L}} \sqrt{\{(2l_1+1)(2l_2+1)(2\tilde{l}+1)(2\tilde{L}+1)\}}$$

$$\times \begin{Bmatrix} k_1 & k_2 & \tilde{l} \\ l_1-k_1 & l_2-k_2 & \tilde{L} \\ l_1 & l_2 & \lambda \end{Bmatrix}$$

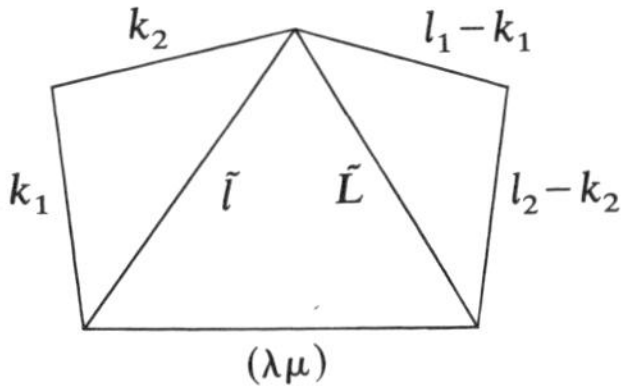

$$= \frac{1}{4\pi} \sum_{\tilde{l}\tilde{L}} \sqrt{\{(2l_1+1)(2l_2+1)(2k_1+1)(2k_2+1))(2l_1-2k_1+1)(2l_2-2k_2+1)\}}$$
$$\times (k_1k_2 00 \mid \tilde{l}0)(l_1-k_1 l_2-k_2 00 \mid \tilde{L}0) \begin{Bmatrix} k_1 & k_2 & \tilde{l} \\ l_1-k_1 & l_2-k_2 & \tilde{L} \\ l_1 & l_2 & \lambda \end{Bmatrix}$$
$$\times [Y_{\tilde{l}}(\theta,\phi)\times Y_{\tilde{L}}(\Theta,\Phi)]_{\lambda\mu}.$$

In writing the last line of this equation we have made use of the reduced matrix element (equation (A2.11)) that comes into the coupling of two spherical harmonics with the same argument to a single Y_{lm}.

When this expression is incorporated in equation (A6.15) one sees that $\bar{l}=l$ and $\bar{L}=L$. By equating powers of r and R on each side of the equation, one arrives at the expression for the $n_1=n_2=0$ transformation coefficient (Moshinsky 1959)

$$M_\lambda(nlNL;0l_10l_2)=\frac{A_{0l_1}A_{0l_2}}{A_{nl}A_{NL}}\times\sum_{\substack{k_1k_2\\ \{k_1+k_2=2n+l\}}}(-1)^{k_2}\left\{\frac{(2l_1+1)(2l_2+1)(2l_1+1)!(2l_2+1)!}{2^{l_1+l_2}(2k_1)!(2k_2)!(2l_1-2k_1)!(2l_2-2k_2)!}\right\}^{\frac{1}{2}}$$
$$\times(k_1k_200\mid l0)(l_1-k_1l_2-k_200\mid L0)\begin{Bmatrix}k_1 & k_2 & l\\ l_1-k_1 & l_2-k_2 & L\\ l_1 & l_2 & \lambda\end{Bmatrix}. \qquad (A6.16)$$

2. Transformation brackets for the general case

For n_1 and n_2 different from zero there are many ways the transformation coefficients can be obtained (Moshinsky 1959, Arima and Teresawa 1960, Lawson and Goeppert–Mayer 1960, Smirnov 1961, Kumar 1966, Baranger and Davies 1966, Bakri 1967, Chasman and Wahlborn 1967, Talman and Lande 1971). One can carry out the transformation straightforwardly by using Sack's formula (Sack 1964) for expressing $r_1^{2n+l}Y_{lm}(\theta_1,\phi_1)$ in terms of relative and centre-of-mass coordinates

$$r_1^{2n_1+l_1}Y_{l_1m_1}(\theta_1,\phi_1)=\sqrt{\pi}\sum_{k_1k_2\nu}\left\{\frac{(2k_1+1)(2k_2+1)}{2l_1+1}\right\}^{\frac{1}{2}}(k_1k_200\mid l_10)$$
$$\times f_\nu(n_1l_1k_1k_2)r^{k_1+2\nu}R^{2n_1+l_1-k_1-2\nu}[Y_{k_1}(\theta,\phi)\times Y_{k_2}(\Theta,\Phi)]_{l_1m_1} \qquad (A6.17)$$

where

$$f_\nu(n_1l_1k_1k_2)=\frac{\sqrt{\pi}n_1!\Gamma(n_1+l_1+\frac{3}{2})(\sqrt{2})^{-2n_1-l_1}}{\nu!\left(\frac{2n_1+l_1-k_1-k_2}{2}-\nu\right)!\Gamma\left(\frac{2n_1+l_1+3+k_2-k_1}{2}-\nu\right)\Gamma(k_1+\frac{3}{2}+\nu)} \qquad (A6.18)$$

and Γ is the gamma function

$$\Gamma(z+1)=z\Gamma(z),\qquad \Gamma(\tfrac{1}{2})=\sqrt{\pi},\qquad \Gamma(0)=1.$$

Once this expansion is known one can carry out the same calculation which led to equation (A6.16). Since equation (A6.17), the analogue of equation (A6.13), contains three summations it is clear that the expression for the transformation coefficient will now involve the sum over six dummy indices (instead of two) with one constraint. Thus as shown by Bakri (1967)

$$M_\lambda(nlNL; n_1l_1n_2l_2) = \frac{1}{4}\frac{A_{n_1l_1}A_{n_2l_2}}{A_{nl}A_{NL}} \times \sum_{k_1k_2}\sum_{k_3k_4}\sum_{\substack{\nu_1\nu_2 \\ k_1+k_3+2\nu_1+2\nu_2=2n+l}} (-1)^{k_3}(2k_1+1)(2k_2+1)$$

$$\times(2k_3+1)(2k_4+1)(k_1k_200 \mid l_10)(k_3k_400 \mid l_20)(k_1k_300 \mid l0)(k_2k_400 \mid L0)$$

$$\times\begin{Bmatrix} k_1 & k_3 & l \\ k_2 & k_4 & L \\ l_1 & l_2 & \lambda \end{Bmatrix} f_{\nu_1}(n_1l_1k_1k_2)f_{\nu_2}(n_2l_2k_3k_4). \qquad \text{(A6.19)}$$

When $n = l = 0$ (or $N = L = 0$) a simple expression can be given for the transformation coefficient. This can be deduced by taking in equation (A6.10) the limit that $\underset{\sim}{r}_1 \to \underset{\sim}{r}_2$. In this limit $\underset{\sim}{r} \to 0$ and $\underset{\sim}{R} \to \sqrt{2}\underset{\sim}{r}_1$ so that only the $n = l = 0$ part of equation (A6.10) remains. Thus $L = \lambda$ and N is fixed by energy conservation (equation (A6.9)). Consequently

$$M_\lambda(00N\lambda; n_1l_1n_2l_2) = (-1)^{n_1+n_2-N}(l_1l_200 \mid \lambda 0)$$

$$\times\left\{\frac{(2l_1+1)(2l_2+1)(2N)!!(2N+2\lambda+1)!!}{2^{2N+\lambda}(2\lambda+1)(2n_1)!!(2n_2)!!(2n_1+2l_1+1)!!(2n_2+2l_2+1)!!}\right\}^{\frac{1}{2}}. \qquad \text{(A6.20)}$$

The foregoing expressions for the transformation coefficients show that they are real and independent of μ. Moreover, from their defining equation (A6.8) it follows that they are precisely the overlap integral

$$M_\lambda(nlNL; n_1l_1n_2l_2) = \langle\psi_{\lambda\mu}(nlNL) \mid \psi_{\lambda\mu}(n_1l_1n_2l_2)\rangle$$

where $\psi_{\lambda\mu}(n_1l_1n_2l_2)$ is the two-particle wave function given by equation (A6.4) and $\psi_{\lambda\mu}(nlNL)$ is the analogous two-particle relative centre-of-mass eigenfunction. Since the $\psi_{\lambda\mu}(n_1l_1n_2l_2)$ form a complete set when summed over $(n_1l_1n_2l_2)$ and a similar statement holds for $\psi_{\lambda\mu}(nlNL)$ it follows that

$$\sum_{nlNL} M_\lambda(nlNL; n_1l_1n_2l_2)M_\lambda(nlNL; n_3l_3n_4l_4) = \delta_{n_1n_3}\delta_{n_2n_4}\delta_{l_1l_3}\delta_{l_2l_4} \qquad \text{(A6.21a)}$$

$$\sum_{n_1l_1n_2l_2} M_\lambda(nlNL; n_1l_1n_2l_2)M_\lambda(n'l'N'L'; n_1l_1n_2l_2) = \delta_{nn'}\delta_{NN'}\delta_{ll'}\delta_{LL'} \qquad \text{(A6.21b)}$$

3. Symmetry properties of the transformation brackets

In order to use the Brody Moshinsky (1960) tabulation of the transformation coefficients there are certain symmetry properties that must be known. We now derive these symmetries.

$$\text{(a)} \quad M_\lambda(nlNL; n_1l_1n_2l_2) = (-1)^{L-\lambda}M_\lambda(nlNL; n_2l_2n_1l_1). \qquad \text{(A6.22)}$$

Consider the transformation Q which has the property that

$$Q\underline{r}_1 = \underline{r}_2$$
$$Q\underline{r}_2 = \underline{r}_1.$$

Under this transformation

$$Q\underline{r} = -\underline{r}$$
$$Q\underline{R} = \underline{R}.$$

The effect of Q on a two-particle wave function is

$$Q[\phi_{n_1l_1}(\underline{r}_1)\times\phi_{n_2l_2}(\underline{r}_2)]_{\lambda\mu} = [\phi_{n_1l_1}(\underline{r}_2)\times\phi_{n_2l_2}(\underline{r}_1)]_{\lambda\mu}$$
$$= (-1)^{l_1+l_2-\lambda}[\phi_{n_2l_2}(\underline{r}_1)\times\phi_{n_1l_1}(\underline{r}_2)]_{\lambda\mu}$$
$$= (-1)^{l_1+l_2-\lambda}\sum_{nlNL} M_\lambda(nlNL; n_2l_2n_1l_1)[\phi_{nl}(\underline{r})\times\phi_{NL}(\underline{R})]_{\lambda\mu}.$$

Since

$$\phi_{nl}(-\underline{r}) = (-1)^l\phi_{nl}(\underline{r}) \tag{A6.23}$$

Q applied to both sides of equation (A6.8) leads to the result

$$(-1)^{l_1+l_2-\lambda}M_\lambda(nlNL; n_2l_2n_1l_1) = (-1)^l M_\lambda(nlNL; n_1l_1n_2l_2).$$

The n are integers, consequently equation (A6.9) implies that

$$(-1)^{l_1+l_2} = (-1)^{l+L} \tag{A6.24}$$

so that equation (A6.22) follows.

$$\text{(b)}\quad M_\lambda(nlNL; n_1l_1n_2l_2) = (-1)^{l_1-\lambda}M_\lambda(NLnl; n_1l_1n_2l_2). \tag{A6.25}$$

To derive this property we consider the effect on the wave-function of an operator Q which has the property that

$$Q\underline{r}_1 = \underline{r}_1$$
$$Q\underline{r}_2 = -\underline{r}_2$$

so that

$$Q\underline{r} = \underline{R}$$
$$Q\underline{R} = \underline{r}.$$

By applying this operator to each side of equation (A6.8) the same arguments as used in the preceding derivation give

$$(-1)^{l_2}\sum_{nlNL} M_\lambda(nlNL; n_1l_1n_2l_2)[\phi_{nl}(\underline{r})\times\phi_{NL}(\underline{R})]_{\lambda\mu}$$
$$= \sum_{n'l'N'L'} M_\lambda(n'l'N'L'; n_1l_1n_2l_2)(-1)^{l'+L'-\lambda}[\phi_{N'L'}(\underline{r})\times\phi_{n'l'}(\underline{R})]_{\lambda\mu}.$$

By equating coefficients on each side of this equation and using equation (A6.24) the desired relationship in equation (A6.25) follows.

By taking the product of the two preceding transformations it follows that

$$\text{(c)}\ M_\lambda(nlNL; n_1l_1n_2l_2) = (-1)^{L-l_2} M_\lambda(NLnl; n_2l_2n_1l_1). \tag{A6.26}$$

Alternatively, this result can be deduced in the same way as before by considering the transformation

$$Q\underline{r}_1 = \underline{r}_2$$

$$Q\underline{r}_2 = -\underline{r}_1.$$

APPENDIX 7

THE ROTATION MATRIX

In this Appendix we shall discuss some of the properties of the rotation matrix $D^I_{MM'}(R)$ which will be useful in the angular-momentum decomposition of Nilsson wave functions.

Let (x, y, z) be the coordinates of a fixed point in space measured with respect to a particular frame of reference. The coordinates of this same point measured with respect to a new set of axes obtained by rotating the original set about the origin will be (x', y', z'). If $\mathscr{R}$ is the operator function of the parameters specifying the rotation

$$x' = \mathscr{R}x$$

$$y' = \mathscr{R}y$$

$$z' = \mathscr{R}z.$$

For example, if α is the angle of rotation about the $0z$ axis (see Fig. A7.1) it follows that the point (x, y, z) has in the rotated coordinate system the specification

$$x_1 = x \cos \alpha + y \sin \alpha$$

$$y_1 = -x \sin \alpha + y \cos \alpha$$

$$z_1 = z.$$

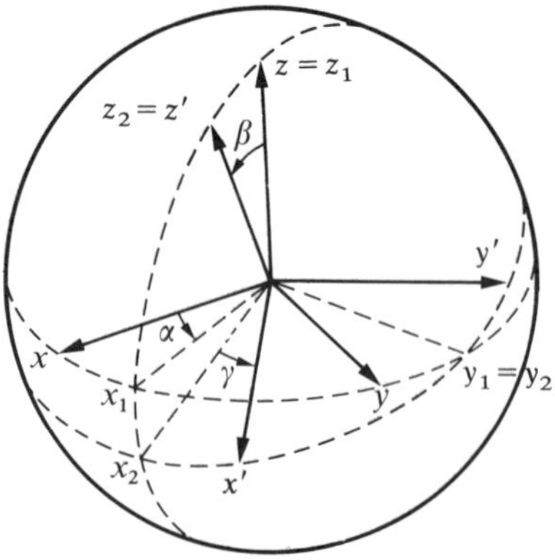

Fig. A7.1. Euler angles specifying the relationship between the initial reference frame S and the final frame S′.

Under such a rotation a function $\psi(x, y, z)$ will become $\psi(x_1, y_1, z_1)$, i.e.

$$\begin{aligned}\mathcal{R}\psi(x, y, z) &= \psi(x_1, y_1, z_1)\\ &= \psi(x \cos\alpha + y \sin\alpha, -x \sin\alpha + y \cos\alpha, z).\end{aligned} \tag{A7.1}$$

If we had used spherical coordinates (r, θ, ϕ) to label the point in the original coordinate system the effect of the rotation would have been simpler, namely to replace ϕ by $\phi - \alpha$

$$\mathcal{R}\psi(r, \theta, \phi) = \psi(r, \theta, \phi - \alpha).$$

A Taylor's series expansion of ψ about $\alpha = 0$ yields

$$\begin{aligned}\psi(r, \theta, \phi - \alpha) &= \psi(r, \theta, \phi) - \alpha\left(\frac{\partial\psi}{\partial\phi}\right)_{\alpha=0} + \frac{\alpha^2}{2!}\left(\frac{\partial^2\psi}{\partial\phi^2}\right)_{\alpha=0} + \dots\\ &= e^{-\alpha(\partial/\partial\phi)}\psi(r, \theta, \phi).\end{aligned}$$

Since $\partial/\partial\phi = iL_z$, where L_z is the z component of orbital angular momentum measured in units of $\hbar$, it follows that

$$\mathcal{R}\psi(x, y, z) = e^{-i\alpha L_z}\psi(x, y, z). \tag{A7.2}$$

This argument can easily be extended to the case that ψ depends on both the spin and spatial coordinates of the particles, and as might be expected the only change is that L_z is replaced by $J_z = L_z + S_z$.

For a rotation through an angle α about an arbitrary axis, equation (A7.2) becomes

$$\mathcal{R}\psi = e^{-i\alpha(\underline{J}\cdot\underline{n})}\psi$$

where $\underline{n}$ is the unit vector along the axis of rotation. Thus to first order in α the change in ψ induced by the rotation is

$$\begin{aligned}\Delta\psi &= \mathcal{R}\psi - \psi\\ &= -i\alpha(\underline{J}\cdot\underline{n})\psi.\end{aligned}$$

For this reason the angular-momentum operator is called the generator of infinitesimal rotations.

The rotation $\mathcal{R}$ that transforms ψ into $\mathcal{R}\psi$ also induces a transformation on an arbitrary operator Q that acts on ψ. To find this transformation we allow $\mathcal{R}$ to operate on the product $Q\psi = \tilde{\psi}$. Thus

$$\mathcal{R}\tilde{\psi} = \mathcal{R}Q\mathcal{R}^{-1}(\mathcal{R}\psi)$$

so that under the rotation $\mathcal{R}$ the operator Q becomes

$$Q \to \mathcal{R}Q\mathcal{R}^{-1} \tag{A7.3}$$

where $\mathcal{R}^{-1}$ is the inverse of $\mathcal{R}$.

To specify a general rotation it is convenient to introduce Euler angles (α, β, γ). We assume the initial reference frame S is rotated to the final frame S' in three steps illustrated in Fig. A7.1:

(i) $S \to S_1$ by a rotation through an angle α about $0z$
(ii) $S_1 \to S_2$ by a rotation through an angle β about $0y_1$
(iii) $S_2 \to S'$ by a rotation through an angle γ about $0z_2$.

With these rotations it follows that $\mathcal{R}$ has the form

$$\mathcal{R}(\alpha, \beta, \gamma) = \exp(-i\gamma \mathrm{J}_{z_2})\exp(-i\beta \mathrm{J}_{y_1})\exp(-i\alpha \mathrm{J}_z)$$

where J_{y_1} and J_{z_2} are rotations about the y axis of S_1 and the z axis of S_2.

By use of equation (A7.3) one may express $\mathcal{R}(\alpha, \beta, \gamma)$ in terms of rotations about the original coordinate axes. We first note that the rotation β transforms the $0z_1$ axis of $0z_2$. Thus from equation (A7.3) it follows that

$$\exp(-i\gamma \mathrm{J}_{z_2}) = \exp(-i\beta \mathrm{J}_{y_1})\exp(-i\gamma \mathrm{J}_{z_1})\exp(i\beta \mathrm{J}_{y_1}).$$

Since the $0z_1$ and $0z$ axes are identical, $\mathcal{R}(\alpha, \beta, \gamma)$ becomes

$$\begin{aligned}\mathcal{R}(\alpha, \beta, \gamma) &= \exp(-i\beta \mathrm{J}_{y_1})\exp(-i\gamma \mathrm{J}_z)\exp(-i\alpha \mathrm{J}_z)\\ &= \exp(-i\beta \mathrm{J}_{y_1})\exp(-i\alpha \mathrm{J}_z)\exp(-i\gamma \mathrm{J}_z).\end{aligned}$$

Applying equation (A7.3) once more to take into account the fact that the rotation α transforms $0y$ to $0y_1$ we find that

$$\mathcal{R}(\alpha, \beta, \gamma) = \mathrm{e}^{-i\alpha \mathrm{J}_z}\mathrm{e}^{-i\beta \mathrm{J}_y}\mathrm{e}^{-i\gamma \mathrm{J}_z}. \tag{A7.4}$$

For the particular case that ψ is an angular-momentum eigenfunction $\psi_{JM'}(\underline{r})$ it follows that

$$\begin{aligned}\mathcal{R}(\alpha, \beta, \gamma)\psi_{JM'}(\underline{r}) &= \psi_{JM'}(\underline{r}')\\ &= \sum_M \langle \psi_{JM}(\underline{r})|\, \mathcal{R}(\alpha, \beta, \gamma)\, |\psi_{JM'}(\underline{r})\rangle \psi_{JM}(\underline{r}).\end{aligned} \tag{A7.5}$$

In writing the second line of this equation use has been made of the fact that under a rotation the total angular momentum J is invariant so that $\psi_{JM'}(\underline{r}')$ can be expressed as a sum over M (and not J) in the original coordinate system. In terms of the rotation matrix defined by the equation

$$D^{J*}_{MM'}(R) = \langle \psi_{JM}(\underline{r})|\, \mathcal{R}(\alpha, \beta, \gamma)\, |\psi_{JM'}(\underline{r})\rangle \tag{A7.6}$$

$\psi_{JM'}(\underline{r}')$ becomes

$$\psi_{JM'}(\underline{r}') = \sum_M D^{J*}_{MM'}(R)\psi_{JM}(\underline{r}) \tag{A7.7}$$

where we have used the shorthand notation R to stand for the arguments (α, β, γ) of the D function. This definition of $D^J_{MM'}(R)$ agrees with that of Bohr and Mottelson (1975) and is the one generally used in the description of heavy deformed nuclei.

If one makes use of the completeness relationship

$$\sum_{JM} |\psi_{JM}(\underline{r})\rangle\langle\psi_{JM}(\underline{r})| = 1$$

together with the fact that

$$\mathcal{R}^*(\alpha, \beta, \gamma)\mathcal{R}(\alpha, \beta, \gamma) = 1$$

(which follows immediately from the definition of $\mathcal{R}(\alpha, \beta, \gamma)$ given in equation (A7.4)) one easily shows that

$$\begin{aligned}\sum_M D^J_{MM_1'}(R)D^{J*}_{MM_2'}(R) &= \sum_M \langle\psi_{JM_1'}(\underline{r})|\,\mathcal{R}^*(\alpha, \beta, \gamma)\,|\psi_{JM}(\underline{r})\rangle \\ &\quad\times\langle\psi_{JM}(\underline{r})|\,\mathcal{R}(\alpha, \beta, \gamma)\,|\psi_{JM_2'}(\underline{r})\rangle \\ &= \delta_{M_1'M_2'}\end{aligned} \tag{A7.8}$$

and also that

$$\sum_{M'} D^J_{M_1M'}(R)D^{J*}_{M_2M'}(R) = \delta_{M_1M_2}. \tag{A7.9}$$

By use of this last equation one can turn equation (A7.7) inside out and write

$$\psi_{JM}(\underline{r}) = \sum_{M'} D^J_{MM'}(R)\psi_{JM'}(\underline{r}') \tag{A7.10}$$

If one applies equation (A7.10) to the particular case that $\psi_{JM}(\underline{r})$ is the spherical harmonic $Y_{JM}(\theta, \phi)$, one obtains for integral values of J a relationship between the D-functions and the spherical harmonics. This is easily seen by considering the special rotation (α, β, γ) in which

(a) α is a rotation through an angle ϕ about the z axis
(b) β is a rotation through an angle θ about the y_1 axis
(c) $\gamma = 0$.

Thus the arguments of the transformed wave function are $\theta' = \phi' = 0$ and since

$$Y_{JM'}(0, 0) = \left(\frac{2J+1}{4\pi}\right)^{\frac{1}{2}}\delta_{M'0}$$

it follows that

$$D^J_{M0}(\phi, \theta, 0) = \left(\frac{4\pi}{2J+1}\right)^{\frac{1}{2}} Y_{JM}(\theta, \phi). \tag{A7.11}$$

1. Coupling rules for *D*-functions

The coupling rules for the ordinary angular-momentum eigenfunctions

$$\Psi_{JM}(\underline{r}_1\underline{r}_2) = \sum_{M_1M_2} (J_1J_2M_1M_2\,|\,JM)\psi_{J_1M_1}(\underline{r}_1)\psi_{J_2M_2}(\underline{r}_2) \tag{A7.12a}$$

$$\psi_{J_1M_1}(\underline{r}_1)\psi_{J_2M_2}(\underline{r}_2) = \sum_J (J_1J_2M_1M_2\,|\,JM)\Psi_{JM}(\underline{r}_1\underline{r}_2) \tag{A7.12b}$$

can be used to deduce analogous rules for the D-functions. When equation (A7.12b) is expressed in the rotated coordinate system it becomes

$$\sum_{M_1'M_2'} D^{J_1}_{M_1M_1'}(R)D^{J_2}_{M_2M_2'}(R)\psi_{J_1M_1'}(\underline{r}_1')\psi_{J_2M_2'}(\underline{r}_2')$$

$$= \sum_{JM'} (J_1J_2M_1M_2 \mid JM)D^J_{MM'}(R)\Psi_{JM'}(\underline{r}_1'\underline{r}_2')$$

$$= \sum_{JM'M_1''M_2''} (J_1J_2M_1M_2 \mid JM)(J_1J_2M_1''M_2'' \mid JM')\, D^J_{MM'}(R)\psi_{J_1M_1''}(\underline{r}_1')\psi_{J_2M_2''}(\underline{r}_2')$$

where in writing the second line, equation (A7.12a) has been used. When the coefficients of $\psi_{J_1M_2'}(\underline{r}_1')\psi_{J_2M_2'}(\underline{r}_2')$ are equated it follows that

$$D^{J_1}_{M_1M_1'}(R)D^{J_2}_{M_2M_2'}(R) = \sum_J (J_1J_2M_1M_2 \mid JM)(J_1J_2M_1'M_2' \mid JM')D^J_{MM'}(R). \quad \text{(A7.13a)}$$

This equation can be inverted by multiplying by the product of the Clebsch–Gordan coefficients and summing over (M_1M_2) and $(M_1'M_2')$ so that

$$D^J_{MM'} = \sum_{M_1M_2} \sum_{M_1'M_2'} (J_1J_2M_1M_2 \mid JM)(J_1J_2M_1'M_2' \mid JM')D^{J_1}_{M_1M_1'}(R) \times D^{J_2}_{M_2M_2'}(R). \quad \text{(A7.13b)}$$

2. Explicit form for the *D*-functions

From the form of the rotation operator given in equation (A7.4) and the definition of the D-function in equation (A7.6) it follows that

$$D^J_{MM'}(R) = \langle\psi_{JM}(\underline{r})|\, e^{-i\alpha J_z}\, e^{-i\beta J_y}\, e^{i\gamma J_z}|\psi_{J'}(\underline{r})\rangle^* = e^{i\alpha M} d^J_{MM'}(\beta)e^{i\gamma M'} \quad \text{(A7.14)}$$

where

$$d^J_{MM'}(\beta) = \langle\psi_{JM}(\underline{r})|\, e^{-i\beta J_y}\, |\psi_{JM'}(\underline{r})\rangle. \quad \text{(A7.15)}$$

(Since $iJ_y = (J_+ - J_-)/2$, $d^J_{MM'}(\beta)$ is real; consequently the complex conjugate has been omitted in its definition.) For $J = \frac{1}{2}$ it is easy to evaluate this quantity. In this case

$$J_y = \tfrac{1}{2}\sigma_y$$

where σ_y is the Pauli spin matrix

$$\sigma_y = \begin{pmatrix} 0 & -i \\ i & 0 \end{pmatrix}.$$

Since $\sigma_y^2 = 1$ it follows that

$$e^{-i\beta J_y} = 1 - \frac{i\beta}{2}\sigma_y + \frac{1}{2}\left(\frac{i\beta}{2}\right)^2 \sigma_y^2 + \ldots$$

$$= \left\{1 - \frac{1}{2}\left(\frac{\beta}{2}\right)^2 + \ldots\right\} - i\sigma_y\left\{\frac{\beta}{2} - \frac{1}{3!}\left(\frac{\beta}{2}\right)^3 + \ldots\right\}$$

$$= \cos\beta/2 - i\sigma_y \sin\beta/2.$$

Thus

$$\langle \psi_{\frac{1}{2}M}(\underline{r})|\, e^{-i\beta/2)\sigma_y}\, |\psi_{\frac{1}{2}M'}(\underline{r})\rangle = \delta_{MM'} \cos \beta/2 + (-1)^{\frac{1}{2}-M'}(1-\delta_{MM'})\sin \beta/2.$$

Consequently

$$D^{\frac{1}{2}}_{MM'}(R) = \begin{bmatrix} e^{i\alpha/2}(\cos \beta/2)e^{i\gamma/2} & -e^{i\alpha/2}(\sin \beta/2)e^{-i\gamma/2} \\ e^{-i\alpha/2}(\sin \beta/2)e^{i\gamma/2} & e^{-i\alpha/2}(\cos \beta/2)e^{-i\gamma/2} \end{bmatrix}. \qquad \text{(A7.16)}$$

By use of equation (A7.13b) with $J_1 = J_2 = \frac{1}{2}$ it is now possible to construct the explicit form of $D^1_{MM'}(R)$. Once $D^1_{MM'}(R)$ has been obtained $D^{\frac{3}{2}}_{MM'}(R)$ can be found and by a bootstrap procedure the explicit form for any J can be produced. This is, of course, a tedious procedure and fortunately an analytic expression exists for $d^J_{MM'}(\beta)$ (Wigner 1959)

$$d^J_{MM'}(\beta) = \{(J+M')!(J+M)!(J-M')!(J-M)!\}^{\frac{1}{2}}$$
$$\times \sum_k \frac{(-1)^{J-M'-k}(\cos \beta/2)^{2k+M+M'}(\sin \beta/2)^{2J-M-M'-2k}}{k!(J-M-k)!(J-M'-k)!(k+M+M')!}. \qquad \text{(A7.17)}$$

This result can be combined with equation (A7.14) to construct the D-function for any value of J.

3. Integration of *D*-functions

We next consider the integration of products of D-functions over all angles. It is convenient to introduce a normalized integration element

$$\int dR = \frac{1}{8\pi^2} \int_0^{\pi} \sin \beta \, d\beta \int_0^{2\pi} d\alpha \int_0^{2\pi} d\gamma. \qquad \text{(A7.18)}$$

Thus it follows that

$$\int D^J_{MM'}(R)dR = \frac{1}{8\pi^2} \int_0^{\pi} d^J_{MM'}(\beta) \sin \beta \, d\beta \int_0^{2\pi} e^{iM\alpha}\, d\alpha \int_0^{2\pi} e^{iM'\gamma}\, d\gamma$$
$$= \frac{1}{4\pi}\, \delta_{M'0} \int_0^{\pi} \int_0^{2\pi} D^J_{M0}(\alpha, \beta, 0) \sin \beta \, d\beta \, d\alpha$$
$$= \frac{1}{4\pi} \left(\frac{4\pi}{2J+1}\right)^{\frac{1}{2}} \delta_{M'0} \int_0^{\pi} \int_0^{2\pi} Y_{JM}(\beta, \alpha) \sin \beta \, d\beta \, d\alpha$$

where use has been made of equation (A7.11). Since the integral of Y_{JM} vanishes unless $J = M = 0$ and because $Y_{00} = 1/\sqrt{(4\pi)}$

$$\int D^J_{MM'}(R)\, dR = \delta_{J0}\delta_{M0}\delta_{M'0}. \qquad \text{(A7.19)}$$

From the explicit form of $d^J_{MM'}$ in equation (A7.17) it follows that

$$d^J_{-M-M'} = (-1)^{M'-M} d^J_{MM'}.$$

Thus

$$D^{J*}_{MM'}(R)=(-1)^{M'-M}D^{J}_{-M-M'}(R). \tag{A7.20}$$

We may combine this result with the coupling rule in equation (A7.13a) to show that

$$\int D^{J_1^*}_{M_1M_1'}(R)D^{J_2}_{M_2M_2'}(R)\,\mathrm{d}R=(-1)^{M_1'-M_1}\sum_J(J_1J_2-M_1M_2\mid JM)$$

$$\times(J_1J_2-M_1'M_2'\mid JM')\int D^{J}_{MM'}(R)\,\mathrm{d}R$$

$$=(-1)^{M_1'-M_1}\sum_J(J_1J_2-M_1M_2\mid JM)$$

$$\times(J_1J_2-M_1'M_2'\mid JM')\delta_{J0}\delta_{M0}\delta_{M'0}.$$

For $J=0$ the Clebsch–Gordan coefficients have the simple form

$$(J_1J_2-M_1M_2\mid 00)=((-1)^{J_1+M_1}/\sqrt{(2J_1+1)})\delta_{J_1J_2}\delta_{M_1M_2}$$

so that

$$\int D^{J_1*}_{M_1M_1'}(R)D^{J_2}_{M_2M_2'}(R)\,\mathrm{d}R=\frac{\delta_{J_1J_2}\delta_{M_1M_2}\delta_{M_1'M_2'}}{(2J_1+1)}. \tag{A7.21}$$

These results are easily extended to obtain an expression for the integral of three D-functions. By use of equations (A7.13b) and (A7.20) it follows that

$$\int D^{J_1*}_{M_1M_1'}(R)D^{J_2}_{M_2M_2'}(R)D^{J_3}_{M_3M_3'}(R)\,\mathrm{d}R$$

$$=\sum_{JMM'}(J_2J_3M_2M_3\mid JM)(J_2J_3M_2'M_3'\mid JM')\int D^{J_1*}_{M_1M_1'}(R)D^{J}_{MM'}(R)\,\mathrm{d}R$$

$$=\frac{1}{(2J_1+1)}(J_2J_3M_2M_3\mid J_1M_1)(J_2J_3M_2'M_3'\mid J_1M_1'). \tag{A7.22}$$

4. The D-functions as angular-momentum eigenfunctions

So far the D-functions have appeared only as coefficients in the transformation of angular-momentum eigenfunctions. However, these quantities are also eigenfunctions of the symmetric top and as such enter into the wave functions describing heavy deformed nuclei. Because of this we need to know how these functions themselves transform under rotations. To deduce these properties consider two coordinate systems S and S'. As before the wave functions describing a physical state will be denoted by $\psi_{JM}(\underline{r})$ and $\psi_{JM}(\underline{r}')$. If the operator $\mathcal{R}'$ corresponds to a rotation of S', leaving the system S alone, it follows that

$$\mathcal{R}'\psi_{JM}(\underline{r})=0$$

because $\psi_{JM}(\underline{r})$ will be unaffected by rotations of S'. By use of equation (A7.10) it therefore follows that

$$0=\sum_{M'}\mathcal{R}'D^J_{MM'}(R)\psi_{JM'}(\underline{r}')$$

$$=\sum_{M'}[\{\mathcal{R}'D^J_{MM'}(R)\}\psi_{JM'}(\underline{r}')+D^J_{MM'}(R)\mathcal{R}'\psi_{JM'}(\underline{r}')] \tag{A7.23}$$

where $\{\mathcal{R}'D^J_{MM'}(R)\}$ means the result of the operation of $\mathcal{R}'$ on $D^J_{MM'}(R)$ alone.

Let $\underline{\mathrm{J}}$ be the angular-momentum operator for the particles described by ψ_{JM}. Thus as already shown $\underline{\mathrm{J}}$ is the operator associated with rotation of the particles. If one subjects both the particles and the axes of reference to *the same rotation*, the infinitesimal change in the wave function describing the particles brought about by such an operation must be zero. Consequently

$$(\mathcal{R}'+\underline{\mathrm{J}})\psi_{JM'}(\underline{r}')=0. \tag{A7.24}$$

When this condition is combined with equation (A7.23) we arrive at the result

$$\sum_{M'}[\{\mathcal{R}'D^J_{MM'}(R)\}\psi_{JM'}(\underline{r}')-D^J_{MM'}(R)\underline{\mathrm{J}}\psi_{JM'}(\underline{r}')]=0. \tag{A7.25}$$

We now examine the consequences of this equation for the various spherical components of $\mathcal{R}'$ and $\underline{\mathrm{J}}$. First, we consider the components $\mathcal{R}'_0, \mathcal{R}'_{\pm1}$ referred to the S' coordinate system (the body-fixed system). Since $\mathrm{J}'_0=\mathrm{J}'_z$ operating on $\psi_{JM'}(\underline{r}')$ gives $M'\psi_{JM'}(\underline{r}')$, the z component of equation (A7.25) is

$$\sum_{M'}\{\mathcal{R}'_0D^J_{MM'}(R)-M'D^J_{MM'}(R)\}\psi_{JM'}(\underline{r}')=0.$$

The functions $\psi_{JM'}(\underline{r}')$ are linearly independent so that we arrive at the result

$$\mathcal{R}'_0D^J_{MM'}(R)=M'D^J_{MM'}(R). \tag{A7.26a}$$

Since $\mathrm{J}'_1=-\mathrm{J}'_+/\sqrt{2}$ we may use equation (A1.4) to find the effect of $\mathcal{R}'_1$. For this component equation (A7.25) yields

$$0=\sum_{M'}\left[(\mathcal{R}'_1D^J_{MM'}(R))\psi_{JM'}(\underline{r}')+\left\{\frac{(J-M')(J+M'+1)}{2}\right\}^{\frac{1}{2}}D^J_{MM'}(R)\psi_{JM'+1}(\underline{r}')\right]$$

$$=\sum_{M'}\left[(\mathcal{R}'_1D^J_{MM'}(R)+\left\{\frac{(J+M')(J-M'+1)}{2}\right\}^{\frac{1}{2}}D^J_{MM'-1}(R)\right]\psi_{JM'}(\underline{r}').$$

In writing the second line of this equation the dummy index M' in the second term has been replaced by $M'-1$. Because the $\psi_{JM'}(\underline{r}')$ are linearly independent their coefficient must be zero for each value of M'.

A similar discussion can be carried out to find the effect of $\mathcal{R}'_{-1}$ on $D^J_{MM'}(R)$. When this is done we arrive at the result

$$\mathcal{R}'_{\pm1}D^J_{MM'}(R)=\mp\left\{\frac{(J\pm M')(J\mp M'+1)}{2}\right\}^{\frac{1}{2}}D^J_{MM'\mp1}(R). \tag{A7.26b}$$

The components of $\mathcal{R}'$ along the space-fixed axes (i.e. the S axes) will be denoted by $\mathcal{R}_0$ and $\mathcal{R}_{\pm1}$, and we shall now consider the effect of $\mathcal{R}_\mu$ on $D^J_{MM'}(R)$. In order to calculate the effect of J_μ in equation (A7.25) it is convenient to carry out the sum over M' in the last term and write

$$\sum_{M'} \{\mathcal{R}_\mu D^J_{MM'}(R)\}\psi_{JM'}(\underline{r}') - \mathrm{J}_\mu \psi_{JM}(\underline{r}) = 0. \tag{A7.27}$$

Since $\psi_{JM}(\underline{r})$ is now a function of the space-fixed coordinates, the effect of the various spherical components of $\underline{\mathrm{J}}$ referred to this set of axes yields the standard results given by equations (A1.2b). Thus

$$\begin{aligned} \mathrm{J}_0 \psi_{JM}(\underline{r}) &= M\psi_{JM}(\underline{r}) \\ &= M \sum_{M'} D^J_{MM'}(R)\psi_{JM'}(\underline{r}') \end{aligned}$$

so that the z component of equation (A7.27) leads to the result

$$\mathcal{R}_0 D^J_{MM'}(R) = M\, D^J_{MM'}(R). \tag{A7.28a}$$

In a similar manner

$$R_{\pm1} D^J_{MM'}(R) = \mp\left\{\frac{(J \mp M)(J \pm M + 1)}{2}\right\}^{\frac{1}{2}} D^J_{M\pm1M'}(R). \tag{A7.28b}$$

From equations (A7.26a) and (A7.26b) or equations (A7.28a) and (A7.28b) one easily shows that

$$\mathcal{R}^2 D^J_{MM'}(R) = J(J+1) D^J_{MM'}(R). \tag{A7.29}$$

Since the operator that generates infinitesimal rotations is the angular-momentum operator, it follows that $D^J_{MM'}(R)$ is an angular-momentum eigenfunctions with eigenvalue $J(J+1)$. Moreover, it is also an eigenfunction of J_z with z component referred to the space-fixed axes (S) equal to M and when referred to the body-fixed axes (S') equal to M'. In summary, if $\underline{\mathrm{I}}$ is the total angular-momentum operator

$$\begin{aligned} &\mathrm{I}^2 D^J_{MM'}(R) = J(J+1) D^J_{MM'}(R) \\ &\mathrm{I}_\mu D^J_{MM'}(R) = \sqrt{\{J(J+1)\}}(J1M\mu \mid J\bar{M}) D^J_{\bar{M}M'}(R) \\ &\mathrm{I}'_\mu D^J_{MM'}(R) = (-1)^\mu \sqrt{\{J(J+1)\}}(J1M'-\mu \mid J\bar{M}') D^J_{M\bar{M}'}(R) \end{aligned} \tag{A7.30}$$

where I_μ are the spherical components of $\underline{\mathrm{I}}$ referred to the space-fixed axes (S) and I'_μ are the spherical components referred to the body-fixed axes (S').

REFERENCES

Ajzenberg-Selove, F. (1970). *Nucl. Phys.* **A152,** 1.
—— (1972). *Nucl. Phys.* **A190,** 1.
—— (1977). *Nucl. Phys.* **A281,** 1.
—— and Lauritsen, T. (1974). *Nucl. Phys.* **A227,** 1.
Alburger, D. E. and Wilkinson, D. H. (1973). *Phys. Rev.* **C8,** 657.
Altman, A. and MacDonald, W. M. (1962). *Nucl. Phys.* **35,** 593.
Anderson, J. D. and Wong, C. (1961). *Phys. Rev. Lett.* **7,** 250.
Arima, A. and Horie, H. (1954). *Prog. theoret. Phys.* **11,** 509.
—— and Huang-Lin, L. J. (1972). *Phys. Lett.* **41B,** 429.
—— —— (1972a). *Phys. Lett.* **41B,** 435.
—— and Ichimura, M. (1966). *Prog. theoret. Phys.* **36,** 296.
—— and Kawarda, H. (1964). *J. Phys. Soc. Japan* **19,** 1768.
—— and Teresawa, T. (1960). *Prog. theoret. Phys.* **23,** 115.
——, Horie, H. and Sano, M. (1957). *Prog. theoret. Phys.* **17,** 567.
——, Horiuchi, H. and Sebe, T. (1967). *Phys. Lett.* **24B,** 129.
——, Horie, H. and Tanabe, Y. (1954a). *Prog. theoret. Phys.* **11,** 284.
Armstrong, D. D. and Blair, A. G. (1965). *Phys. Rev.* **140,** B1226.
Armstrong, J. C. and Quisenberry, K. S. (1961). *Phys. Rev.* **122,** 150.
Arvieu, R. and Moszkowski, S. A. (1966). *Phys. Rev.* **145,** 830.
Ascuitto, R. J. and Glendenning, N. K. (1970). *Phys. Rev.* **C2,** 1260.
Astner, G., Bergström, L., Blomqvist, J., Fant, B. and Wikström, K. (1972). *Nucl. Phys.* **A182,** 219.
Auble, R. L. (1975). *Nucl. Data Sheets* **16,** 1.
Auerbach, K., Braunsfurth, J., Maier, M., Bodenstedt, E., and Flender, H. W. (1967). *Nucl. Phys.* **A94,** 427.
Austern, N. (1970). *Direct nuclear reaction theories.* John Wiley, New York.
Bacher, R. F. and Goudsmit, S. (1934). *Phys. Rev.* **46,** 948.
Bakri, M. M. (1967). *Nucl. Phys.* **A96,** 115.
Bansal, R. K. and French, J. B. (1964). *Phys. Lett.* **11,** 145.
Baranger, E. U. and Lee, C. W. (1961). *Nucl. Phys.* **22,** 157.
Baranger, M. (1960). *Phys. Rev.* **120,** 957.
—— and Davies, K. T. R. (1966). *Nucl. Phys.* **79,** 403.
Bardeen, J., Cooper, L. N., and Schrieffer, J. R. (1957). *Phys. Rev.* **108,** 1175.
Bardin, T. T., Becker, J. A., and Fisher, T. R. (1973). *Phys. Rev.* **C7,** 190.
Barrett, B. R. (1967). *Phys. Rev.* **154,** 955.
—— (1975). *Effective interactions and operators in nuclei*, Lecture Notes in Physics, **40**. Springer-Verlag, Berlin.

—— and Kirson, M. W. (1973). *Advances in nuclear physics*, Vol. 6 (eds. M. Baranger and E. Vogt). Plenum Press, New York.

Bartlett, J. H., Jr. (1936). *Phys. Rev.* **49,** 102.

Bar-Touv, J. and Kelson, I. (1965). *Phys. Rev.* **138,** B1035.

Bassichis, W. H., Kerman, A. K., and Svenne, J. P. (1967). *Phys. Rev.* **160,** 746.

Bayman, B. F. (1966). *Am. J. Phys.* **34,** 216.

—— and Hintz, N. M. (1968). *Phys. Rev.* **172,** 1113.

—— and Kallio, A. (1967). *Phys. Rev.* **156,** 1121.

—— and Lande, A. (1966). *Nucl. Phys.* **77,** 1.

Becker, J. A. and Warburton, E. K. (1971). *Phys. Rev. Lett.* **26,** 143.

Bell. J. S. (1959). *Nucl. Phys.* **12,** 117.

Belyaev, S. T. (1959). *K. Dan. Vidensk. Selsk. Mat.-fys. Medd.* **31,** No. 11.

Benson, H. G. and Flowers, B. H. (1969). *Nucl. Phys.* **A126,** 305.

Berant, Z., Broude, C., Engler, G., and Start, D. F. H. (1975). *Nucl. Phys.* **A225,** 55.

Bethe, H. A. and Bacher, R. F. (1936). *Rev. mod. Phys.* **8,** 162.

—— and Butler, S. T. (1952). *Phys. Rev.* **85,** 1045.

Betts, R. R., Gaarde, C., Hansen, O., Larsen, J. S., and Van der Werf, S. Y. (1975). *Nucl. Phys.* **A253,** 380.

Bhatt, K. H. (1962). *Nucl. Phys.* **39,** 375.

Biedenharn, L. C., Blatt, J. M., and Rose, M. E. (1952). *Rev. mod. Phys.* **24,** 249.

Bizetti, P. G. (1971). *Topical conference on the structure of $f_{\frac{7}{2}}$ nuclei, Legnaro (Padova)* (ed. R. A. Ricci). Editrice Compositori, Bologna.

Bjerregaard, J. H., Hansen, O., Nathan, O., Chapman, R., Hinds, S., and Middleton, R. (1967). *Nucl. Phys.* **A103,** 33.

Blin Stoyle, R. J. (1953). *Proc. phys. Soc. Lond.* **A66,** 1158.

—— and Nair, S. C. K. (1963). *Phys. Lett.* **7,** 161.

—— and Perks, M. A. (1954). *Proc. phys. Soc. Lond.* **A67,** 885.

Bloom, S. D., McGrory, J. B., and Moszkowski, S. A. (1973). *Nucl. Phys.* **A199,** 369.

Bogoliubov, N. N., (1958), JETP **34,** 58, 73; Nuovo Cimento **7,** 794.

Bohr, A. and Mottelson, B. R. (1953). *Dan. Mat. Fys. Medd.* **27,** No. 16.

—— and Mottelson, B. R. (1975). *Nuclear Structure, Vol. II,* W. A. Benjamin, Inc., New York.

——, Damgaard, J., and Mottelson, B. R. (1967). *Nuclear Structure*, edited by A. Hossain, Harun-ar-Rashid and M. Islam, North Holland Publishing Co., Amsterdam, page 1.

——, Mottelson, B. R., and Pines, D. (1958). *Phys. Rev.* **110,** 936.

Brandolini, F., Rossi Alvarez, C., Vingiari, C. B., and De Poli, M. (1974). *Phys. Lett.* **49B,** 261.

Breit, G., Condon, E. U., and Present, R. D. (1936). *Phys. Rev.* **50,** 825.

Brody, T. A. and Moshinsky, M. (1960). *Tables of transformation brackets.* Monografias del Instituto de Fisica, Mexico.

Broglia, R. A., Hansen, O., and Riedel, C. (1973). *Advances in nuclear physics,* Vol. 6 (eds. M. Baranger and E. Vogt). Plenum Press, New York.

Brown, B. A., Fossan, D. B., McDonald, J. M., and Snover, K. A. (1974). *Phys. Rev.* **C9,** 1033.
Brown, G. E. (1964). *Proc. Int. Congress on Nuclear Physics* (ed. P. Gugenberger). Centre National de la Recherche Scientifique, Paris.
—— and Bolsterli, M. (1959). *Phys. Rev. Lett.* **3,** 472.
Butler, S. T. (1951). *Proc. R. Soc.* **A208,** 559.
Carraz, L. C., Blachot, J., Monnand, E., and Moussa, A. (1970). *Nucl. Phys.* **A158**, 403.
Cassen, B. and Condon, E. U. (1936). *Phys. Rev.* **50,** 846.
Chasman, R. R. and Wahlborn, S. (1967). *Nucl. Phys.* **A90,** 401.
Chemtob, M. (1969). *Nucl. Phys.* **A123**, 449.
Christy, A. and Hausser, O. (1973). *Nucl. Data Tabl.* **A11,** 281.
Clegg, A. B. (1961). *Phil. Mag.* **6,** 1207.
Cohen, B. L. (1968). *Proc. Dubna Symposium on Nuclear Structure.* IAEA, Vienna.
Cohen, S., and Kurath, D. (1965). *Nucl. Phys.* **73,** 1.
——, Lawson, R. D., and Soper, J. M. (1966). *Phys. Lett.* **21,** 306.
——, ——, Macfarlane, M. H., and Soga, M. (1964). *Phys. Lett.* **9,** 180.
——, ——, ——, Pandya, S. P. and Soga, M. (1967). *Phys. Rev.* **160,** 903.
Comfort, J. R., Maher, J. V., Morrison, G. C., and Schiffer, J. P. (1970). *Phys. Rev. Lett.* **25,** 383.
Condon, E. U. and Shortley, G. H. (1951). *The theory of atomic spectra.* Cambridge University Press, Cambridge.
Cosman, E. R. Schramm, D. N., Enge, H. A., Sperduto, A., and Paris, C. H. (1967). *Phys. Rev.* **163,** 1134.
Cox, A. J., Caraça, J. M. G., Schlenk, B., Gill, R. D., and Rose, H. J. (1973). *Nucl. Phys.* **A217,** 400.
Crozier, D. J., Fortune, H. T., Middleton, R., Wiza, J. L., and Wildenthal, B. H. (1972). *Phys. Lett.* **41B,** 291.
Daehnick, W. W. and Sherr, R. (1973). *Phys. Rev.* **C7,** 150.
Davidson, J. P. (1968). *Collective models of the nucleus.* Academic Press, New York.
DeBenedetti, S. (1964). *Nuclear interactions.* John Wiley, New York.
DeJager, J. L. and Boeker, E. (1973). *Nucl. Phys.* **A216,** 349.
DeShalit, A. (1961). *Phys. Rev.* **122,** 1530.
—— and Talmi, I. (1963). *Nuclear shell theory.* Academic Press, New York.
—— and Walecka, J. D. (1966). *Phys. Rev.* **147,** 763.
Dupont, Y., Martin, P., and Chabre, M. (1973). *Phys. Rev.* **C7,** 637.
Eckart, C. (1930). *Rev. mod. Phys.* **2,** 305.
Eisenberg, J. M. and Greiner, W. (1970). *Excitation mechanisms of the nucleus*, Vol. 2. North Holland, Amsterdam.
Elliott, J. P. (1958). *Proc. R. Soc.* **A245,** 128, 562.
—— and Flowers, B. H. (1955). *Proc. R. Soc.* **A229,** 536.
——, ——, (1957). *Proc. R. Soc.* **A242,** 57.
—— and Skyrme, T. H. R. (1955). *Proc. R. Soc.* **A232,** 561.
—— and Wilsdon, C. (1968). *Proc. R. Soc.* **A302,** 509.

——, Jackson, A. D., Mavromatis, H. A., Sanderson, E. A., and Singh, B. (1968). *Nucl. Phys.* **A121,** 241.
Ellis, P. J. and Engeland, T. (1970). *Nucl. Phys.* **A144,** 161.
—— and Osnes, E. (1977). *Rev. mod. Phys.* **49,** 777.
Endt, P. and Van der Leun, C. M. (1973). *Nucl. Phys.* **A214,** 1.
——, —— (1974). *Nucl. Data Tabl.* **13,** 67.
——, —— (1974a). *Nucl. Phys.* **A235,** 27.
Engeland, T. (1965). *Nucl. Phys.* **72,** 68.
—— and Ellis, P. J. (1972). *Nucl. Phys.* **A181,** 368.
Engelbertink, G. A. P. and Olness, J. W. (1972). *Phys. Rev.* **C5**, 431.
——, Warburton, E. K., and Olness, J. W. (1972a). *Phys. Rev.* **C5,** 128.
Erskine, J. R., Marinov, A., and Schiffer, J. P. (1966). *Phys. Rev.* **142,** 633.
Fanger, U., Heck, D., Michaelis, W., Ottmar, H., Schmidt, H., and Gaeta, R. (1970). *Nucl. Phys.* **A146,** 549.
Federman, P. and Zamick, L. (1969). *Phys. Rev.* **177,** 1534.
Feenberg, E. and Knipp, J. K. (1935). *Phys. Rev.* **48,** 906.
Fermi, E. (1934). *Z. Phys.* **88,** 161.
Fermi, E. and Amaldi, E. (1936). *Ric. Sci.* **1,** 1.
Feshbach, H. and Iachello, F. (1974). *Ann. Phys.* (*N.Y.*) **84,** 211.
Fiarman, S. and Meyerhof, W. E. (1973). *Nucl. Phys.* **A206,** 1.
Flowers, B. H. and Skouras, L. D. (1969). *Nucl. Phys.* **A136,** 353.
Fortune, H. T. and Headley, S. C. (1974). *Phys. Lett.* **51B,** 136.
Freedman, M. S., Wagner, E., Jr., Porter, F. T., and Bolotin, H. H. (1966). *Phys. Rev.* **146,** 791.
French, J. B. (1960). *Nucl. Phys.* **15,** 393.
—— and Macfarlane, M. H. (1961). *Nucl. Phys.* **26,** 168.
Fujita, J. I. and Ikeda, K. (1965). *Nucl. Phys.* **67,** 145.
Fuller, G. H. and Cohen, V. W. (1968). *Nucl. Data* **5,** 433.
Gaarde, C., Kemp, K., Petresch, C., and Folkmann, F. (1972). *Nucl. Phys.* **A184,** 241.
Gamow, G. and Teller, E. (1936). *Phys. Rev.* **49,** 895.
Gartenhaus, S. and Schwartz, C. (1957). *Phys. Rev.* **108,** 482.
Giraud, B. (1965). *Nucl. Phys.* **71,** 373.
Glaudemans, P. W. M., Wiechers, G., and Brussaard, P. J. (1964). *Nucl. Phys.* **56,** 529.
——, Brussaard, P. J., and Wildenthal, B. H. (1967). *Nucl. Phys.* **A102,** 593.
Glendenning, N. K. (1965). *Phys. Rev.* **137,** B102.
—— (1968). UCRL Report No. 18268 (unpublished).
Gloeckner, D. H. and Lawson, R. D. (1974). *Phys. Lett.* **53B,** 313.
——, —— (1975). *Phys. Lett.* **56B,** 301.
——, ——, and Serduke, F. J. D. (1973). *Phys. Rev.* **7C,** 1913.
——, Macfarlane, M. H., Lawson, R. D. and Serduke, F. J. D. (1972). *Phys. Lett.* **40B,** 597.
—— and Serduke, F. J. D. (1974). *Nucl. Phys.* **A220,** 477.
Goldstein, S. and Talmi, I. (1956). *Phys. Rev.* **102,** 589.
Gove, N. B. and Martin, M. J. (1971). *Nucl. Data* **10,** 206.

Grayson, W. C., Jr. and Nordheim, L. W. (1956). *Phys. Rev.* **102,** 1084.
Green, I. M. and Moszkowski, S. A. (1965). *Phys. Rev.* **139,** B790.
Gupta, K. K. and Lawson, R. D. (1959). *Phys. Rev.* **114,** 326.
Haefele, J. C. and Woods, R. (1966). *Phys. Lett.* **23,** 579.
Halbert, E. C. and French, J. B. (1957). *Phys. Rev.* **105,** 1565.
——, McGrory, J. B., Wildenthal, B. H., and Pandya, S. P. (1971). *Advances in physics,* Vol. 4 (Eds. M. Baranger and E. Vogt). Plenum Press, New York.
Hamada, T. and Johnston, I. D. (1962). *Nucl. Phys.* **34,** 382.
Hamamoto, I. (1969). *Nucl. Phys.* **A126,** 545.
—— (1969a). *Nucl. Phys.* **A135,** 576.
—— (1974). *Phys. Rep.* **10C,** No. 2.
Hansen, O., Lien, J. R., Nathan, O., Sperduto, A., and Tjøm, P. O. (1975). *Nucl. Phys.* **A243,** 100.
Hardy, J. C., Brunnader, H., and Cerny, J. (1969). *Phys. Rev. Lett.* **22,** 1439.
Hartmann, R., Grawe, H., and Kändler, K. (1973). *Nucl. Phys.* **A203,** 401.
Harvey, M. (1968). *Advances in physics,* Vol. 1, (eds. M. Baranger and E. Vogt). Plenum Press, New York.
—— and Sebe, T. (1968). *Nucl. Phys.* **A117,** 289.
Hecht, K. T. (1967). *Nucl. Phys.* **A102,** 11.
Heisenberg, W. (1932). *Z. Phys.* **77,** 1.
Heitler, W. (1936). *Proc. Camb. phil. Soc.* **32,** 112.
Henley, E. M. (1969). *Isospin in nuclear physics* (ed. D. H. Wilkinson). North Holland, Amsterdam.
Hertel, J. W., Fleming, D. G., Schiffer, J. P., and Gove, H. E. (1969). *Phys. Rev. Lett.* **23,** 488.
Hill, D. L. and Wheller, J. A. (1953). *Phys. Rev.* **89,** 1102.
Hodgson, P. E. (1971). *Nuclear reactions and nuclear structure.* Clarendon Press, Oxford.
Holland, R. E., Lawson, R. D., and Lynch, F. J. (1971). *Ann. Phys. (N.Y.)* **63,** 607.
Horie, H. and Sugimoto, K. (1973). *J. Phys. Soc. Japan,* suppl. **34,** 1.
Horoshko, R. N., Cline, D., and Lesser, P. M. S. (1970). *Nucl. Phys.* **A149,** 562.
Hubbard, L. B. (1971). *Nucl. Data* **A9,** 85.
Ichimura, M. (1968). *Prog. Nucl. Phys.* **10,** 307.
Inglis, D. R. (1936). *Phys. Rev.* **50,** 783.
Inoue, T., Sebe, T., Hagiwara, H., and Arima, A. (1964). *Nucl. Phys.* **59,** 1.
Jahn, H. A. and Hope, J. (1954). *Phys. Rev.* **93,** 280.
Jensen, J. H. D. and Mayer, M. G. (1952). *Phys. Rev.* **85,** 1040.
Johns, M. W., Park, J. Y., Shafroth, S. M., Van Patter, D. M., and Way, K. (1970). *Nucl. Data Tables* **A8,** 373.
Johnstone, I. P. (1968). *Nucl. Phys.* **A110,** 429.
Kennedy, J. M., Sears, B. J., and Sharp, W. T. (1954). *Tables of X-coefficients.* Atomic Energy of Canada, Ltd., Chalk River Report CRT-569, AECL-106.
Kerman, A. K. (1956). *K. Dan. Vidensk. Selsk. Mat.-fys. Medd.* **30,** No. 15.
—— (1961). *Ann. Phys. (N.Y.)* **12,** 300.
——, Lawson, R. D., and Macfarlane, M. H. (1961). *Phys. Rev.* **124,** 162.

Kirson, M. W. (1974). *Ann. Phys. (N.Y.)* **82,** 345.
Kisslinger, L. S. and Sorensen, R. A. (1960). *K. Dan. Vidensk. Selsk. Mat.-fys. Medd.* **32,** No. 9.
——, —— (1963). *Rev. mod. Phys.* **35,** 853.
Klapdor, H. V. (1971). *Phys. Lett.* **35B,** 405.
Kocher, D. C. and Haeberli, W. (1972). *Nucl. Phys.* **A196,** 225.
Konopinski, E. J. (1966). *The theory of beta radioactivity*. Clarendon Press, Oxford.
Kozub, R. L. (1968). *Phys. Rev.* **172,** 1078.
Krohn, V. E. and Ringo, G. R. (1975). *Phys. Lett.* **55B,** 175.
Kumar, K. (1966). *J. math. Phys.* **7,** 671.
Kumar, N. (1973). *Lett. Nuovo Cim.* **6,** 224.
Kuo, T. T. S. (1967). *Nuclei. Phys.* **A103,** 71.
Kuo, T. T. S. (1974). *A. Rev. nucl. Sci.* **24,** 101.
—— and Brown, G. E. (1966). *Nucl. Phys.* **85,** 40.
——, —— (1968). *Nucl. Phys.* **A114,** 241.
——, Baranger, E. U., and Baranger, M. (1966). *Nucl. Phys.* **79,** 513.
Kurath, D. (1956). *Phys. Rev.* **101,** 216.
—— and Lawson, R. D. (1967). *Phys. Rev.* **161,** 915.
——, —— (1972). *Phys. Rev.* **C6,** 901.
—— and Pĭcman, L. (1959). *Nucl. Phys.* **10,** 313.
Lane, A. M. (1962). *Nucl. Phys.* **35,** 676.
—— and Mekjian, A. Z. (1973). *Advances in Physics*, Vol. 7, (eds. M. Baranger and E. Vogt). Plenum Press, New York.
—— and Soper, J. M. (1962). *Nucl. Phys.* **37,** 663
Lanford, W. A. and Wildenthal, B. H. (1973). *Phys. Rev.* **C7,** 668.
Lauritsen, T. and Ajzenberg-Selove, F. (1966). *Nucl. Phys.* **78,** 1.
Lawson, R. D. (1961). *Phys. Rev.* **124,** 1500.
—— (1971). *Nucl. Phys.* **A173,** 17.
—— and Goeppert-Mayer, M. (1960). *Phys. Rev.* **117,** 174.
—— and Macfarlane, M. H. (1965). *Nucl. Phys.* **66,** 80.
—— and Soper, J. M. (1967). *Proc. International Nuclear Physics Conference, Gatlinburg, Tenn.* (eds. R. L. Becker, C. D. Goodman, P. H. Stelson, and A. Zuker). Academic Press, New York.
—— and Uretsky, J. L. (1957). *Phys. Rev.* **108,** 1300.
—— and Zeidman, B. (1962). *Phys. Rev.* **128,** 821.
——, Serduke, F. J. D., and Fortune, H. T. (1976). *Phys. Rev.* **C14,** 1245.
Lewis, M. B. (1970). *Nucl. Data Sheets* **4B,** 237.
—— (1970a). *Nucl. Data Sheets* **4B,** 313.
MacDonald, W. M. (1954). Thesis, Univ. of Princeton, UCRL 2746.
Macfarlane, M. H. (1966). *Lectures in theoretical physics*, Vol. 8C, (eds. P. D. Kunz, D. A. Lind, and W. E. Brittin). University of Colorado Press, Boulder.
—— and French, J. B. (1960). *Rev. mod. Phys.* **32,** 567.
Maier, K. H., Nakai, K., Leigh, J. R., Diamond, R. M., and Stephens, F. S. (1972). *Nucl. Phys.* **A183,** 289.
Majorana, E. (1933). *Z. Phys.* **82,** 137.

Malik, F. B. and Scholz, W. (1966). *Phys. Rev.* **150,** 919.
——, —— (1967). *Phys. Rev.* **153,** 1071.
Maripuu, S. (1969). *Nucl. Phys.* **A123,** 357.
—— (1970). *Phys. Lett.* **31B,** 181.
——, Wildenthal, B. H., and Evwaraye, A. O. (1973). *Phys. Lett.* **43B,** 368.
Markham, R. G. and Fulbright, H. W. (1973). *Nucl. Phys.* **A203,** 244.
Martin, P., Buenerd, M., Dupont, Y., and Chabre, M. (1972). *Nucl. Phys.* **A185,** 465.
Mayer, M. G. and Jensen, J. H. D. (1955). *Elementary theory of nuclear structure.* John Wiley, New York.
McCullen, J. D., Bayman, B. F., and Zamick, L. (1964). *Phys. Rev.* **134,** B515.
McGrory, J. B. and Wildenthal, B. H. (1973). *Phys. Rev.* **7C,** 974.
Melvin, M. A. and Swamy, N. V. V. J. (1957). *Phys. Rev.* **107,** 186.
Meshkov, S. and Ufford, C. W. (1956). *Phys. Rev.* **101,** 734.
Middleton, R. and Pullen, D. J. (1964). *Nucl. Phys.* **51,** 63.
Miyazawa, H. (1951). *Prog. theoret. Phys.* **6,** 801.
Moinester, M., Schiffer, J. P., and Alford, W. P. (1969). *Phys. Rev.* **179,** 984.
Morpurgo, G. (1958). *Phys. Rev.* **110,** 721.
Morse, P. M. and Feshbach, H. (1953). *Methods of theoretical physics.* McGraw-Hill, New York.
Moshinsky, M. (1959). *Nucl. Phys.* **13,** 104.
Moszkowski, S. A. (1953). *Phys. Rev.* **89,** 474.
—— (1966). *Alpha, beta and gamma-ray spectroscopy*, Vol. 2 (ed. K. Siegbahn). North Holland, Amsterdam.
Müller-Arnke, A. (1973). *Nucl. Phys.* **A215,** 205.
Negele, J. W. (1971). *Nucl. Phys.* **A165,** 305.
Nilsson, S. G. (1955). *K. Dan. Vidensk. Selsk. Mat.-fys. Medd.* **29,** No. 16.
Nomura, T., Gil, C., Saito, H., Yamazaki, T., and Ishibara, M. (1970). *Phys. Rev. Lett.* **25,** 1342.
Noya, H., Arima, A., and Horie, H. (1958). *Prog. theoret. Phys. Suppl.* **8,** 33.
Olness, J. W., Warburton, E. K., and Becker, J. A. (1973). *Phys. Rev.* **C7,** 2239.
Pal. M. K., Gambhir, Y. K., and Raz, R. (1967). *Phys. Rev.* **155,** 1144.
Palumbo, F. and Prosperi, D. (1968). *Nucl. Phys.* **A115,** 296.
Pandya, S. P. (1956). *Phys. Rev.* **103,** 956.
—— (1963). *Nucl. Phys.* **43,** 636.
Paul, E. B. (1957). *Phil. Mag.* **2,** 311.
Racah, G. (1942). *Phys. Rev.* **62,** 438.
—— (1943). *Phys. Rev.* **63,** 367.
—— (1949). *Phys. Rev.* **76,** 1352.
—— and Talmi, I. (1952). *Physica* **18,** 1097.
Raman, S. (1967). *Nucl. Data Sheets* **B2,** Vol. 5, p. 41.
—— (1970). *Nucl. Data Sheets* **B4,** 397.
——, Walkiewicz, T. A., and Behrens, H. (1975). *Nucl. Data Sheets* **16,** 451.
Rao, M. N. and Rapaport, J. (1970). *Nucl. Data Sheets* **B3,** Vol. 6, p. 37.
Redlich, M. G. (1958). *Phys. Rev.* **110,** 468.
Reid, R. V. (1968). *Ann. Phys.* (N.Y.) **50,** 411.

Ripka, G. (1966). *Lectures in theoretical physics*, Vol. 8C, (eds. P. D. Kunz, D. A. Lind, and W. E. Brittin). University of Colorado Press, Boulder.
—— (1968). *Advances in physics*, Vol. 1 (eds. M. Baranger and E. Vogt). Plenum Press, New York.
Rose, H. J. and Brink, D. M. (1967). *Rev. mod. Phys.* **39,** 306.
Rose, M. E. (1957). *Elementary theory of angular momentum.* John Wiley, New York.
Ross, A. A., Mark, H. and Lawson, R. D. (1956). *Phys. Rev.* **102,** 1613.
Rotenberg, M., Bivens, R., Metropolis, N. and Wooten, J. K. Jr. (1959). *The 3-j and 6-j symbols.* Technology Press, MIT, Cambridge, Mass.
Sachs, R. G. and Austern, N. (1951). *Phys. Rev.* **81,** 705.
Sack, R. A. (1964). *J. math. Phys.* **5,** 252.
Sandhya-Devi, K. R., Khandkikar, S. B., Parikh, J. C., and Banerjee, B. (1970). *Phys. Lett.* **32B,** 179.
Satchler, G. R. (1964). *Nucl. Phys.* **55,** 1.
—— (1966). *Lectures in theoretical physics.* Vol. 8C, (eds. P. D. Kunz, D. A. Lind, and W. E. Brittin). University of Colorado Press, Boulder.
Sauer, P. U. (1974). *Phys. Rev. Lett.* **32,** 626.
Schiffer, J. P. and True, W. W. (1976). *Rev. mod. Phys.* **48,** 191.
Schlegel, W., Schmitt, D., Santo, R. and Pühlhofer, F. (1970). *Nucl. Phys.* **A153,** 502.
Schmidt, T. (1937). *Z. Phys.* **106,** 358.
Schmorak, M. R. and Auble, R. L. (1971). *Nucl. Data Sheets* **B5,** 205.
Schneid, E. J., Prakash, A., and Cohen, B. L. (1967). *Phys. Rev.* **156,** 1316.
Schüler, H. (1937). *Z. Phys.* **107,** 12.
Schwartz, C. and DeShalit, A. (1954). *Phys. Rev.* **94,** 1257.
Schwartz, J. J. and Watson, B. A. (1969). *Phys. Lett.* **29B,** 567.
Segel, R. E., Wedberg, G. H., Beard, G. B., Puttaswamy, N. G., and Williams, N. (1970). *Phys. Rev. Lett.* **25,** 1352.
Sekiguchi, M., Shida, Y., Soga, F., Hirao, Y., and Sakai, M. (1977). *Nucl. Phys.* **A278,** 231.
Sen, S., Hollas, C. L., Bjork, C. W., and Riley, P. J. (1972). *Phys. Rev.* **C5,** 1278.
Serduke, F. J. D., Lawson, R. D., and Gloeckner, D. H. (1976). *Nucl. Phys.* **A256,** 45.
Seth, K. K., Saha, A., and Greenwood, L. (1973). *Phys. Rev. Lett.* **31,** 552.
Sharma, S. K. and Bhatt, K. H. (1972). *Nucl. Phys.* **A192,** 625.
Sherr, R. and Talmi, I. (1975). *Phys. Lett.* **56B,** 212.
Sherr, R., Kouzes, R., and Del Vecchio, R. (1974). *Phys. Lett.* **52B,** 401.
Shimizu, K., Ichimura, M., and Arima, A. (1974). *Nucl. Phys.* **A226,** 282.
Shlomo, S. (1972). *Nucl. Phys.* **A184,** 545.
Silverberg, L. and Winther, A. (1963). *Phys. Lett.* **3,** 158.
Slater, J. C. (1929). *Phys. Rev.* **34,** 1293.
Smirnov, Yu. F. (1961). *Nucl. Phys.* **27,** 177.
Smith, K. and Stephenson, J. W. (1957). *A table of the Wigner 9-j coefficients.* Argonne National Laboratory Report, ANL 5776.
Soloviev, V. G. (1958). *Nucl. Phys.* **9,** 655.

Soper, J. M. (1969). *Isospin in nuclear physics* (ed. D. H. Wilkinson). North Holland, Amsterdam.
—— (1970). *Theory of nuclear structure.* Trieste Lectures, IAEA, Vienna.
Stephenson, G. J., Jr and Marion, J. B. (1966). *Isobaric spin in nuclear physics* (eds. J. D. Fox and D. Robson), p. 766. Academic Press, New York.
Strominger, D. and Rasmussen, J. O. (1957). *Nucl. Phys.* **3,** 197.
Swamy, N. V. V. J. and Green, A. E. S. (1958). *Phys. Rev.* **112,** 1719.
Talman, J. D. and Lande, A. (1971). *Nucl. Phys.* **A163,** 249.
Talmi, I. (1952). *Helv. Phys. Acta* **25,** 185.
—— (1962). *Rev. mod. Phys.* **34,** 704.
—— and Unna, I. (1960). *Ann. Rev. Nucl. Sci.* **10,** 353.
Thankappan, V. K. and True, W. W. (1965). *Phys. Rev.* **137,** B793.
Thomas, L. H. (1926). *Nature* (*Lond.*) **117,** 514.
Towner, I. S. and Hardy, J. C. (1969). *Nucl. Data* **A6,** 153. Erratum in *Nucl. Data* **A10,** 319 (1971).
——, —— (1969a). *Adv. Phys.* **18,** No. 74, 401.
True, W. W., Ma, C. W., and Pinkston, W. T. (1971). *Phys. Rev.* **C3,** 2421.
Unna, I. and Talmi, I. (1958). *Phys. Rev.* **112,** 452.
Valatin, J. G. (1958). *Nuovo Cimento* **7,** 843.
Verheul, H. (1971). *Nucl. Data Sheets* **B5,** 457.
—— (1974). *Nucl. Data Sheets* **13,** 443.
—— and Ewbank, W. B. (1972). *Nucl. Data Sheets* **B8,** 477.
Vervier, J. (1967). *Nucl. Data Sheets* **B2,** Vol. 5, p. 1.
Vignon, B., Bruadet, J. F., Longequeue, N., and Towner, I. S. (1971). *Nucl. Phys.* **A162,** 82.
Wahlborn, S. and Blomqvist, J. (1969). *Nucl. Phys.* **A133,** 50.
Wapstra, A. H. and Gove, N. B. (1971). *Nuclear Data Tables*, Vol. 9, July 1971.
Warburton, E. K. (1958). *Phys. Rev. Lett.* **1,** 68.
—— and Weneser, J. (1969). *Isospin in nuclear physics* (ed. D. H. Wilkinson). North Holland, Amsterdam.
Wedberg. G. H. and Segel, R. E. (1973). *Phys. Rev.* **C7,** 1956.
Weiffenbach, C. V. and Tickle, R. (1971). *Phys. Rev.* **C3,** 1668.
Weigt, P., Herzog, P., Hubel, H., and Bodenstedt, E. (1968). *Nucl. Phys.* **A106,** 570.
Weisskopf, V. F. (1951). *Phys. Rev.* **83,** 1073.
Wigner, E. P. (1933). *Phys. Rev.* **43,** 252.
—— (1951). On the matrices which reduce the Kronecker products of representations of simply reducible groups (hectographed paper, unpublished). Princeton University.
—— (1959). *Group theory.* Academic Press, New York.
Wildenthal, B. H. and Larson, D. (1971). *Phys. Lett.* **37B**, 266.
Wilkinson, D. H. (1960). *Nuclear spectroscopy, Part B.* (ed. F. Ajzenberg-Selove). Academic Press, New York.
—— (1973). *Nucl. Phys.* **A209,** 470.
—— and Macefield, B. C. E. (1974). *Nucl. Phys.* **A232,** 58.
Wiza, J. L. Middleton, R., and Hewka, P. V. (1966). *Phys. Rev.* **141,** 975.

Woods, R. D. and Saxon, D. S. (1954). *Phys. Rev.* **95,** 577.
Wu, C. S. and Moszkowski, S. A. (1966). *Beta decay*, Interscience Publishers, New York.
Yntema, J. L. (1962). *Phys. Rev.* **127,** 1659.
—— (1964). *Phys. Lett.* **11,** 140.
—— and Satchler, G. R. (1964). *Phys. Rev.* **134B,** 984.
Yoshida, S. (1961). *Phys. Rev.* **123,** 2122.
—— (1962). *Nucl. Phys.* **38,** 380.
—— and Zamick, L. (1972). *Ann. Rev. Nucl. Sci.* **22,** 121.
Zamick, L. (1965). *Phys. Lett.* **19,** 580.
—— (1973). *Ann. Phys.* (N.Y.) **77,** 230.

AUTHOR INDEX

Only primary references to data compilations and numerical tables are included.

Ajzenberg-Selove, F., 6, 8, 289, 393, 403
Alburger, D. E., 419
Alford, W. P., 174
Altman, A., 89
Amaldi, E., 89
Anderson, J. D., 99
Arima, A., 35, 127, 277, 296, 300, 315, 316, 442, 472, 496
Armstrong, D. D., 155
Armstrong, J. C., 258
Arvieu, R., 438
Ascuitto, R. J., 245
Astner, G., 319
Auble, R. L., 85, 348
Auerbach, K., 303
Austern, N., 150, 269

Bacher, R. F., 49, 98
Bakri, M. M., 206, 496
Banerjee, B., 397
Bansal, R. K., 137
Baranger, E. U., 225, 244, 356
Baranger, M., 353, 356, 496
Bardeen, J., 339
Bardin, T. T., 306, 327
Barrett, B. R., 2, 232, 287, 319
Bartlett, J. H. Jr., 97
Bar-Touv, J., 396
Bassichis, W. H., 395
Bayman, B. F., 40, 95, 115, 255, 261, 263, 398, 480
Beard, G. B., 305
Becker, J. A., 286, 306, 327
Behrens, H., 105, 373
Bell, J. S., 176
Belyaev, S. T., 339
Benson, H. G., 128
Berant, Z., 285
Bergström, L., 319
Bethe, H. A., 98, 151
Betts, R. R., 165
Bhatt, K. H., 397, 406
Biedenharn, L. C., 468
Bivins, R., 33
Bizetti, P. G., 190
Bjerregaard, J. H., 72, 262
Bjork, C. W., 117
Blachot, J., 325
Blair, A. G., 155
Blatt, J. M., 468
Blin-Stoyle, R. J., 91, 277, 300
Bloom, S. D., 104, 315
Blomqvist, J., 296, 319
Bodenstedt, E., 303
Boeker, E., 133
Bogoliubov, N. N., 355
Bohr, A., 103, 291, 319, 339, 388, 409, 502
Bolotin, H. H., 46, 191
Bolsterli, M., 336
Brandolini, F., 301
Braunsfurth, J., 303
Breit, G., 89
Brink, D. M., 267
Brody, T. A., 206
Broglia, R. A., 245
Broude, C., 285
Brown, B. A., 285, 317
Brown, G. E., 2, 35, 148, 218, 287, 332, 336, 397, 413
Bruadet, J. F., 111
Brunnader, H., 264
Brussaard, P. J., 109, 201
Buenerd, M., 153
Butler, S. T., 150, 151

Caraça J. M. G., 88
Carraz, L. C., 325
Cassen, B., 92

Cerny, J., 264
Chabre, M., 153, 186
Chapman, R., 72, 262
Chasman, R. R., 496
Chemtob, M., 296
Christy, A., 322
Clegg, A. B., 406
Cline, D., 44
Cohen, B. L., 164, 368
Cohen, S., 2, 84, 135, 148, 154, 212, 261, 380, 381, 403
Cohen, V. W., 297
Comfort, J. R., 176
Condon, E. U., 5, 89, 92, 421
Cooper, L. N., 339
Cosman, E. R., 86
Cox, A. J., 88
Crozier, D. J., 67

Daehnick, W. W., 186
Damgaard, J., 103
Davidson, J. P., 409
Davies, K. T. R., 496
DeBenedetti, S., 267
DeJager, J. L., 133
Del Vecchio, R., 144, 166
De Poli, M., 301
DeShalit, A., 1, 73, 125, 154, 232, 485
Diamond, R. M., 296
Dupont, Y., 153, 186

Eckart, C., 29, 431
Eisenberg, J. M., 267
Elliott, J. P., 2, 35, 238, 240, 319, 389, 399
Ellis, P. J., 2, 287, 319, 412
Endt, P., 13, 105
Enge, H. A., 86
Engeland, T., 35, 148, 332, 412, 413
Engelbertink, G. A. P., 117, 305
Engler, G., 285
Erskine, J. R., 314
Evwaraye, A. O., 306
Ewbank, W. B., 135

Fanger, U., 22, 133
Fant, B., 319
Federman, P., 318
Feenberg, E., 89
Fermi, E., 89, 372
Feshbach, H., 203, 267, 332
Fiarman, S., 231
Fisher, T. R., 306, 327
Fleming, D. G., 131
Flender, H. W., 303
Flowers, B. H., 35, 128, 238, 413
Folkmann, F., 314
Fortune, H. T., 37, 67, 257, 413
Fossan, D. B., 285, 317
Freedman, M. S., 46, 191
French, J. B., 1, 16, 49, 73, 106, 137, 162, 228
Fujita, J. I., 380
Fulbright, H. W., 133
Fuller, G. H., 297

Gaarde, C., 165, 314
Gaeta, R., 22, 133
Gambhir, Y. K., 356
Gamow, G., 372
Gartenhaus, S., 239
Gil, C., 43
Gill, R. D., 88
Glaudemans, P. W. M., 109, 201
Glendenning, N. K., 245, 250, 253
Gloeckner, D. H., 74, 90, 105, 118, 136, 193, 229, 326, 336, 361
Goldstein, S., 172
Goudsmit, S., 49
Gove, H. E., 131
Gove, N. B., 6, 373
Grawe, H., 88
Grayson, W. C. Jr., 154, 414, 485
Green, A. E. S., 98
Green, I. M., 438
Greenwood, L., 153
Greiner, W., 267
Gupta, K. K., 363

Haeberli, W., 158
Hafele, J. C., 130
Hagiwara, H., 35
Halbert, E. C., 228, 286, 319
Hamada, T., 2, 218
Hamamoto, I., 131
Hansen, O., 72, 152, 165, 245, 262
Hardy, J. C., 41, 245, 264
Hartmann, R., 88
Harvey, M., 123, 387, 388
Hausser, O., 322
Headley, S. C., 257
Hecht, K. T., 442

Heck, D., 22, 133
Heisenberg, W., 92, 96
Heitler, W., 267
Henley, E. M., 87, 89
Hertel, J. W., 131
Herzog, P., 303
Hewka, P. V., 258
Hill, D. L., 399
Hinds, S., 72, 262
Hintz, N. M., 263
Hirao, Y., 167
Hodgson, P. E., 150
Holland, R. E., 326
Hollas, C. L., 117
Hope, J., 472
Horie, H., 277, 295, 300, 315, 472
Horiuchi, H., 127
Horoshko, R. N., 44
Huang-Lin, L. J., 296, 316
Hubbard, L. B., 106,
Hubel, H., 303

Iachello, F., 332
Ichimura, M. 296, 442
Ikeda, K., 380
Inglis, D. R., 271
Inoue T., 35
Ishibara, M., 43

Jackson, A. D., 2
Jahn, H. A., 472
Jensen, J. H. D., 1, 270
Johns, M. W. 193
Johnston, I. D., 2, 218
Johnstone, I. P., 413

Kallio, A., 255, 261
Kändler, K., 88
Kawarda, H., 442
Kelson, I., 396
Kemp, K., 314
Kennedy, J. M., 213
Kerman, A. K., 76, 347, 395, 406, 440
Khadkikar, S. B., 397
Kirson, M. W., 2, 287, 319
Kisslinger, L. S., 348, 363, 368, 385
Klapdor, H. V., 312
Knipp, J. K., 89
Kocher, D. C., 158
Konopinski, E. J., 105, 371
Kouzes, R., 144, 166
Kozub, R. L., 153
Krohn, V. E., 373
Kumar, K., 496
Kumar, N., 403
Kuo, T. T. S., 2, 218, 287, 356, 397
Kurath, D., 212, 218, 295, 306, 380, 398, 403

Lande, A., 40, 480, 496
Lane, A. M., 103, 105
Lanford, D., 380
Larsen, J. S. 165
Larson, D., 325
Lauritsen, T., 393, 403
Lawson, R. D., 2, 5, 37, 74, 84, 90, 105, 107, 118, 125, 135, 136, 148, 154, 193, 229, 261, 295, 302, 306, 326, 336, 347, 363, 381, 398, 413, 416, 442, 496
Lee, C. W., 225, 244
Leigh, J. R., 296
Lesser, P. M., 44
Lewis, M. B., 414
Lien, J. R., 152
Longequene, N., 111
Lynch, F. J., 326

Ma, C. W., 238
MacDonald, W. M., 89, 99
Macefield, B. C. E., 373
Macfarlane, M. H., 1, 16, 49, 74, 84, 135, 136, 148, 162, 180, 261, 326, 347, 381, 442
Maher, J. V., 176
Maier, K. H., 296
Maier, M., 303
Majorana, E., 97
Malik, F. B., 406
Marinov, A., 314
Marion, J. B., 105
Maripuu, S., 306, 312, 315
Mark, H., 5
Markham, R. G., 133
Martin, M. J., 373
Martin, P., 153, 186
Mavromatis, H. A., 2
Mayer, M. G., 1, 270, 496
McCullen, J. D., 115, 398
McDonald, J. M., 285, 317
McGrory, J. B., 104, 240, 286, 315, 319

Mekjian, A. Z., 105
Melvin, M. A., 427
Meshkov, S., 2
Metropolis, N., 33
Meyerhof, W. E., 231
Michaelis, W., 22, 133
Middleton, R., 67, 72, 258, 261, 262
Miyazawa, H., 296
Moinester, M., 174
Monnand, E., 325
Morpurgo, G., 308
Morrison, G. C., 176
Morse, P. M., 203, 267
Moshinsky, M., 205, 206, 496
Moszkowski, S. A., 104, 105, 267, 273, 315, 371, 438, 439
Mottelson, B. R., 103, 291, 319, 339, 388, 409, 502
Moussa, A., 325
Müller-Arnke, A., 178

Nair, S. C. K., 91
Nakai, K., 296
Nathan, O., 72, 152, 262
Negele, J. W., 87
Nilsson, S. G., 7, 291, 390
Nomura, T., 43
Nordheim, L. W., 154, 414, 485
Noya, H., 277

Olness, J. W., 117, 286, 305
Osnes, E., 2, 287, 319
Ottmer, H., 22, 133

Pal, M. K., 356
Palumbo, F., 229
Pandya, S. P., 63, 84, 135, 171, 286, 319, 381
Parikh, J. C., 397
Paris, C. H., 86
Park, J. Y., 193
Paul, E. B., 406
Perks, M. A., 277
Petresch, C., 314
Pĭcman, L., 398
Pines, D., 339
Pinkston, W. T., 238
Porter, F. T., 46, 191
Prakash, A., 368
Present, R. D., 89
Prosperi, D., 229
Pühlhofer, F., 256
Pullen, D. J., 261
Puttaswamy, N. G., 305

Quisenberry, K. S., 258

Racah, G., 25, 40, 49, 71, 73, 184, 425, 440, 459, 477
Raman, S., 30, 42, 105, 328, 373
Rao, M. N., 303
Rapaport, J., 303
Rasmussen, J. O., 293
Raz, R., 356
Redlich, M. G., 398
Reid, R. V., 2, 218
Riedel, C., 245
Ringo, G. R., 373
Ripley, P. J., 117
Ripka, G., 387
Rose, H. J., 88, 267
Rose, M. E., 468
Ross, A. A., 5
Rossi Alvarez, C., 301
Rotenberg, M., 33

Sachs, R. G., 269
Sack, R. A., 496
Saha, A., 153
Saito, H., 43
Sakai, M., 167
Sanderson, E. A., 2
Sandhya-Devi, K. R., 397
Sano, M., 315
Santo, R., 256
Satchler, G. R., 142, 150, 245, 247
Sauer, P. U., 87
Saxon, D. S., 5
Schafroth, S. M., 193
Schiffer, J. P., 68, 131, 174, 176, 219, 314
Schlegel, W., 256
Schlenk, B., 88
Schmidt, H., 22, 133
Schmidt, T., 298
Schmitt, D., 256
Schmorak, M. R., 348
Schneid, E. J., 368
Scholz, W., 406
Schramm, D. N., 86
Schrieffer, J. R., 339
Schüler, H., 298
Schwartz, C., 73, 239

Schwartz, J. J., 173
Sears, B. J., 213
Sebe, T., 35, 123, 127, 387
Segel, R. E., 305
Sekiguchi, M., 167
Sen, S., 117
Serber, R., 236
Serduke, F. J. D., 37, 74, 90, 118, 136, 193, 326, 361, 413
Seth, K. K., 153
Sharma, S. K., 397
Sharp, W. T., 213
Sherr, R., 89, 144, 166, 186
Shida, Y., 167
Shimizu, K., 296
Shlomo, S., 106
Shortley, G. H., 5, 421
Silverberg, L., 379
Singh, B., 2
Skouras, L. D., 413
Skyrme, T. H. R., 240
Slater, J. C., 10, 203
Smirnov, Yu F., 496
Smith, K., 213
Snover, K. A., 285, 317
Soga, F., 167
Soga, M., 84, 135, 148, 261, 381
Soloviev, V. G., 339
Soper, J. M., 2, 103, 105, 154
Sorensen, R. A., 348, 363, 368, 385
Sperduto, A., 86, 152
Start, D. F. H., 285
Stephens, F. S., 296
Stephenson, G. J. Jr., 105
Stephenson, J. W., 213
Strominger, D., 293
Sugimoto, K., 295, 300
Svenne, J. P., 395
Swamy, N. V. V. J., 98, 427

Talman, J. D., 496
Talmi, I., 1, 2, 49, 73, 80, 89, 154, 172, 205, 228, 485, 493
Tanabe, Y., 472
Teller, E., 372
Teresawa, T., 496
Thankappan, V. K., 132,
Thomas, L. H., 6, 271
Tickle, R., 167
Tjøm, P. O., 152
Towner, I. S., 41, 111, 245
True, W. W., 68, 132, 219, 238

Ufford, C. W., 2
Unna, I., 2, 49, 228
Uretsky, J. L., 125

Valatin, J. G., 355
Van der Leun, C. M., 13, 105
Van der Werf, S. Y., 165
Van Patter, D. M., 193
Verheul, H., 85, 135, 193
Vervier, J., 327
Vignon, B., 111
Vingiari, C. B., 301

Wagner, E. Jr., 46, 191
Wahlborn, S., 296, 496
Walecka, J. D., 232
Walkiewicz, T. A., 105, 373
Wapstra, A. H., 6
Warburton, E. K., 117, 286, 289, 294, 306
Watson, B. A., 173
Way, K., 193
Wedberg, G. H., 305
Weiffenbach, C. V., 167
Weigt, P., 303
Weisskopf, V. F., 273
Weneser, J., 289
Wheeler, J. A., 399
Wiechers, G., 109
Wigner, E. P., 29, 97, 425, 431, 505
Wikström, K., 319
Wildenthal, B. H., 67, 201, 240, 286, 306, 319, 325, 380
Wilkinson, D. H., 277, 373, 380, 419
Williams, N., 305
Wilsdon, C., 319
Winther, A., 379
Wiza, J. L., 67, 258
Wong, C., 99
Woods, R., 130
Woods, R. D., 5
Wooten, J. K. Jr., 33
Wu, C. S., 105, 371

Yamazaki, T., 43
Yntema, J. L., 107, 135, 142
Yoshida, S., 295, 364, 386

Zamick, L., 115, 127, 137, 200, 295, 318, 398
Zeidman, B., 107

SUBJECT INDEX

References to individual nuclei are collected at the end of the Index.

Allowed beta decay (*see* beta decay)
Analogue states, 99, 144
  gamma decay, 309–313
  population in one nucleon transfer, 167
Analogue quenching, 104
Angular momentum (*see also* Clebsch–Gordan, Racah and $9j$ coefficients)
  allowable states p-shell, 212
    scs nuclei j^n, 16–21
      many particles, 21–25
      two particles, 8–16
    neutron-proton j^2, 91
      several particles, 113, 116
  closed shell, 10
  coupling, 9–13
    Clebsch–Gordan coefficients, 14, 424–429
    [×] notation, 15
  diagrams, 49–52
    coupling four, 51–52
      three, 24, 51
      two, 16, 49, 50
    identical nucleon configuration j^n, 22,
    including isospin, 107–109
    parentage coefficients, 42, 50
    reduced matrix element, 30, 50
  operator, 3, 6
    commutation relationship, 8
    many particles, 9
    step down/up, 8, 420–421
  recoupling four, 51, 52
    three, 24, 51
  single particle eigenfunctions, 5, 422–423
Annihilation operator
  single particle, 18, 95
    in terms of quasi-particles, 353
  two particles, 27
    including isospin, 95
  zero-coupled pair, 76
Anti-Analogue State, 310
Anticommutation relationship
  Fermions, 17, 18
    including isospin, 95
  Quasi particles, 351, 352
Antisymmetric wavefunctions, 10
Axial symmetry, 397

Bansal–French formula, 142
barn, 190
Bartlett exchange, 97
BCS Approximation, 339–344
  number projected states, 347, 349
  variational calculation, 341–344
  validity, degenerate model, 345–348
    non degenerate, 348–351
  wave function, 339
    number distribution, 350
    properties, 339, 341
Beta decay, 370–373, 413–419
  ft value, 373
  isospin mixing, 105
  μ-selection rule, 416
  operators for allowed
    Fermi, 372
    Gamow–Teller, 373
  reduced transition probability, 373
    $v = 1$ states of j^n, 414
  use in determining wavefunctions, 376–379, 413–419
Body-fixed system, 507

Central Potential (*see* Two-body potentials)
Centre-of-Gravity Theorem, 124
  effect of exclusion principle, 135–136
  including isospin, 142
Centre of Mass
  many particles, coordinate, 224
    momentum, 223

Centre of Mass (cont'd)
two particles, coordinate, 205
momentum, 205
Charge independence, 89, 94, 98
Symmetry, 87
Chemical Potential, 342
Clebsch–Gordon coefficients, 14, 424–429
diagrammatic representation, 16, 49, 50
formulae involving, 427
relationship to 3–j coefficient, 425
tables of, 428–429
[×] notation, 15
Closed Shell
angular momentum, 10
energy, 122
interaction with, 121
Coefficients of Fractional Parentage (*see* Parentage Coefficients)
Commutation relationships
angular momentum, 8
isospin, 92
quasi spin, 76, 179, 440
Configuration mixing, 31, 192, 277–281, 349, 360
induced by quadrupole deformation, 405, 408
Constructive interference, 286
Coordinates
centre-of-mass, 205, 224
relative, 205
Core-excited states, 35–37, 126, 142–148, 166, 199–202, 411–413
Core polarization, 277–281
effect on beta decay operator, 380
electric quadrupole operator, 316–320
magnetic dipole operator, 296–297
Coriolis coupling, 410
Coulomb interaction, 83, 89, 115
particle-hole, 142, 145
mixing, 98–105
total energy, 98
uniform charge distribution, 99
Coupling (*see* Angular momentum & Clebsch–Gordan coefficients)
Creation operators,
relationship to parentage coefficients, 40, 41, 479, 481, 484–485
single particle, 17, 95
two particle, 27, 95
zero-coupled pair, 70, 338,
Deformed Potential, 291, 390 (*see also* Nilsson Model)
Delta function Potential, 32, 436
Modified Surface, 201
Surface, 32, 439
Destruction operators (*see* annihilation operators)
Destructive interference, 286, 293
Diagrams (*see* Angular Momentum)
Dipole Interaction, 132
Dipole Moment (*see* Magnetic Dipole Moment)
Direct Reactions, 128–131, 148–168, 245–257
Inelastic Scattering, 128–131
angular momentum determination, 130
Single-Nucleon Transfer, 148–168
cross section l-dependence, 150, 151
pickup, 149
seniority selection rules, 72, 107
spectroscopic factor, 151
for identical nucleon configuration j^n, 155
quasi particles, 364–367
relationship to parentage coefficients, 154
stripping, 149
transition strength, 150
extraction from data, 150
accuracy, 151
sum rules, 162–164
use in determining wavefunctions, 72, 107, 135, 152–153, 156, 161, 257–259, 314, 328, 368–370
Two-Nucleon Transfer, 245–257
dependence on single-particle angular momentum transferred, 256–257
modified spectroscopic factor, 253–255
oscillator vs Woods–Saxon, 256
role of Moshinsky coefficients, 250
selection rules
j^2 transfer, 251
parity, 251
seniority, 72
spectroscopic factor, 248
extraction from data, 248
relationship to Parentage coefficients, 248
use in determining wavefunctions, 72, 259–261, 263, 264

Distorted Wave Born Approximation, 150, 245
Double-Barred Matrix element, 29 (*see* reduced matrix element)
Double Parentage Coefficients (*see* Parentage Coefficients)

Effective Charge, 284, 317 (*see also* core polarization and polarization charge)
  model space dependence, 317
  state dependence, 319
Effective Interaction (*see also* Two Body Potential)
  for A = 6 and 7 nuclei, 403
    $(0d_{3/2})$ orbit, 110
    $(0f_{7/2})$ proton configuration, 48
    $(0g_{9/2})$ proton configurations, 75
    $(0d_{3/2}, 0f_{7/2})$ neutron-proton, 110
      average, 145
    $(0p)$ shell $T = 1$, 216
      $T = 0$, 221
  model space dependence, 90
  pairing force, 338
  quadrupole force, 132, 203, 386, 389
Effective operators, 277–281 (*see also* effective charge and electromagnetic decays)
Electric Field Strength, 266
Electric Quadrupole Moment, 135, 190, 321
  number dependence, 448, 449
  particle-hole relationship, 188–189
  quadrupole interaction in terms of, 132–133
Electromagnetic Decays, 266–336 (*see also* identical nucleon configuration j^n)
    half life, 277
    mean lifetime, 272
    radiative width, 277
    reduced transition probability, 271
    use in determining wavefunctions, 191, 286, 303–6, 314, 328, 363–4
  Electric Multipole, 268–274, 289–293, 316–336
      operator, 269
      selection rules, 269
      single particle lifetime
        Moszkowski, 273, 274
        Weisskopf, 273, 274
      single particle reduced transition probability, 273
Electromagnetic Decays (cont'd)
    Dipole transitions, 287–293
      collectivity based on separable potential, 333–336
        fragmentation of $\Delta T = 1$ strength, 336
      deformation effects, 290–293
      effective charge, 288
      histograms, 290
      isospin effects, 288–290
      operator, 287
    Quadrupole transitions, 316–328
      between particle-hole multiplets, 325–326
        weak coupling multiplets, 327–328
      calculation for ^{18}O, 283–287
      collectivity based on separable potential, 331–332
      effective operator, 316–320
      half-filled shell, 324–325, 449
      histogram, 321
      isospin retarded, 323–324
      signature rules, 326–327
  Magnetic Multipole, 268–274, 293–316
      selection rules, 270
      single particle lifetime,
          Moszkowski, 274, 276
          Weisskopf, 274, 276
        matrix element, 275
        operator, 271
        reduced transition probability, 275
    Dipole, 303–316
      analogue to anti-analogue, 309–313
      effective operator, 295–297
      histograms, 309
      isospin effects, 308
      l-forbidden, 315–316
      signature selection rule, 306
      weak coupling states, 304–306
    Quadrupole, 293–295
      deformation effects, 295
      isospin effects, 294
      operator, 294
Elliott Generating Procedure, 399
Energy gap, 355
Euler Angles, 500
Exclusion principle, 10, 89
  including isospin, 93
  $0f_{7/2}$ configuration
    allowable states $(f_{7/2})^4$, 19–21

Exclusion principle (*cont'd*)
  effective interaction, 48
  predicted ground state energies, 81
    spectra, 43–47
  parentage coefficients, 489–491
fermi, 5
Fermi operator, 372
  sea, 317–318, 352
Figure of Merit, 215
Fractional Parentage Coefficients (*see* Parentage Coefficients)
ft, 373

g factor, 295
Gamma decay (*see* Electromagnetic Decays)
Gamow–Teller operator, 372
Geometrical quenching, 104

Half life, 277
Hamada Johnston potential, 218, 222–223
Hamiltonian
  centre-of-mass, 229
  deformed potential, 291
    Nilsson model, 390
  dimensionality, 23, 123, 337, 353, 387
  harmonic oscillator, 6
    including Coulomb, 110
  including electromagnetic field, 267
  independent particle, 4
  interacting shell model, general, 96
    scs nuclei, 25–28
  in terms of quasi particles, 354–356
  matrix elements (*see also* particle-hole interaction and two body potentials)
    counting factors, 53
    dipole and quadrupole potentials, 133
    hole-hole interaction, 185
    identical nucleon configuration j^n, 38–42, 80, 450–458
    neutron-proton, 107–122, 137–142
    Nilsson potential, 391
    phase factors, 54
    scs nuclei, 54–62
    single particle-single hole, 171
    two particles, 30, 31
    using oscillator wavefunctions, 208–212
      centre-of-mass, 230–231
      for $0p$-shell $T=1$, 215
      $T=0$, 220
  rotor plus particles, 409

Harmonic oscillator
  constant, 7
  eigenfunction, 7, 205
    convention on number of nodes, 7
      sign at origin, 7
  Hamiltonian, 6
    including Coulomb, 110
  many particle factorization, 224
  matrix element of r^λ, 7
  step down/up operators, 225
    matrix elements, 239
  two particle wavefunctions, 205
    relative centre-of-mass transformation, 204–205, 492–499
  two body matrix elements, 208–212
Hartree–Fock calculations, 387
Heisenberg exchange operator, 96
Hermitian Adjoint Operator, 28, 29, 431
Hill–Wheller integral, 399
Hole-hole interaction, 185

Identical nucleon configuration j^n
  allowable angular momentum states, 14, 16–21
  diagrammatic representation, 42, 50
  electric multipole moments, 449
    centre-of-shell rules, 191, 449
    quadrupole moment, 322
  electromagnetic decays
    electric multipole, 449
      centre-of-shell rules, 191, 324, 449
    magnetic multipole, 303, 448
  energy, 42, 80, 170
  magnetic dipole operator, 297
    multipole moment, 299, 448
  one particle parentage coefficients, 38–40, 477–480
    n-dependence, 443
    tables of, 489–490
    useful formulae, 487–488
  particle-hole conjugation operator, 182
    centre-of-shell phase factor, 184
  seniority, 71
    conservation for delta-function potential, 452–455
    eigenfunctions, 76–79
    mixing, experimental data, 73–76
      centre-of-shell, 188
    relationship to quasi-spin, 440–442
    selection rules for direct reactions, 72

Identical nucleon configuration j^n (cont'd)
spectroscopic factors for single nucleon transfer, 155
two nucleon transfer, 251
two-body potential matrix elements
dependence on radial shape, 70
n-dependence, 450–458
two-particle parentage coefficients, 41, 480–481
calculation from one particle, 482
n-dependence, 447
table of, 491
useful formulae, 487–488
zero coupled pair creation operator, 70, 338, 440
Inelastic scattering, 128–131
Interacting Shell Model (*see* Hamiltonian)
Internal Degrees of Freedom, 226
Intrinsic State, 398
Intruder States, 411–413 (*see also* core excited states)
Irreducible tensor operators, 28, 430–431
matrix elements of 432–436
relationship between operator and Hermitian adjoint, 432
Wigner–Eckart Theorem, 28, 431
including isospin, 139
Isobaric Analog States (*see* Analog States)
Isospin, 91–96
allowable states, 94, 95
diagram coupling rules, 107–109
eigenfunction, one particle, 92
two particles, 94
exclusion principle, 93
mixing, theoretical, 105
experimental, 105
one particle parentage coefficients, 484
tables of, 111, 120
operator, 92, 94
commutation rules, 92
projection operator, neutron, 93
proton, 93
step down/up operator, 93
two particle parentage coefficients, 485
calculation from one, 485

jj coupling, eigenfunction, 5
jj to LS transformation, 206
for antisymmetric states, 208
for the p-shell, 213
$(0s, 0p)$ states, 233
kR for beta decay, 371
gamma decay, 267
K-forbidden (*see* beta decay, μ-forbidden)

Lane potential, 103
Least Squares Fit, 48
Legendre polynomials, 203
Level sequence (*see* single particle)
l-forbidden transitions, 315–316
linear momentum conservation, 149

Magic Numbers, 4, 25
Magnetic Dipole Moment, 297–303
for configuration $(\pi j)^n(\nu j)^{\pm n}$, 302
identical nucleon configuration j^n, 290, 448
Schmidt value, 298
single particle value, 298
weak coupling multiplet, 301
Magnetic Dipole Operator, 295–297 (*see also* Electromagnetic Decays)
effect of core polarization, 296–297
meson exchange, 296
identical nucleon configuration j^n, 297
particle-hole relationship, 189
Magnetic Field Strength, 266
Majorana exchange operator, 97
Matrix elements (*see* Hamiltonian, Single Particle and Two Body Potentials)
Model space, 30
dependence of effective interaction on, 90
electric quadrupole operator on, 317
Modified Hermitian Adjoint Operator, 28, 29, 431
reduced matrix element, 432
Modified Surface Delta Interaction, 201
Monopole Interaction, 131
Moshinsky coefficients, 205, 492–499
for the $(0p)$ wavefunctions, 213–214
$(0s, 0p)$ wavefunctions, 230
$n_1 = n_2 = 0$, 496
$n = l = 0$, 497
general expression, 497
role in two nucleon transfer, 250, 255–256
symmetry properties, 497–499
Moszkowski estimate, 273, 274, 276
Mean lifetime, 272

μ-forbidden beta transitions, 416

Nilsson Model, 390–395
  eigenfunctions, 394, 395
    relationship c_μ to $c_{-\mu}$, 392
      to Hartree–Fock, 395, 397
  Hamiltonian matrix elements, 391
  level sequence, 393
    degeneracy, 392
  many particle intrinsic states, 398
    angular momentum projection, 398–399
      single intrinsic state, 399–402
        comparison with shell model, 405
        seniority mixing, 413–419
      several intrinsic states, 407, 409
        comparison with shell model, 408
Nine-j coefficient, 52, 469–472
  in terms of Racah coefficients, 471
  j-j to LS transformation, 206, 208
  useful formulae, 473–474
Nuclear Forces (*see* Effective and Two Body Interactions)
Nuclear Magneton, 267
Nuclear Radius, 6
Nucleon Transfer (*see* Direct Reactions)
Number operator, 26
  including isospin, 139

Occupation Number Representation, 17
One particle operators, 29
  allowed beta decay, 372
  counting factors, 53
  effective, 277–281, 295–297, 316–320
  electromagnetic, 269, 271
  irreducible tensor, 28, 431
    modified Hermitian adjoint, 431
  nucleon transfer, 150
  one particle parentage coefficients, 40, 477–479
  particle-hole transformation, 188–189
  phase factors, 54
  reduced matrix element, 29, 273, 275, 432–436
    including isospin, 139
  T_z, 139
One particle parentage coefficients (*see* Parentage Coefficients)
One Particle Transfer (*see* Direct Reactions)
Oscillator (*see* Harmonic Oscillator)

Pair degeneracy, 76, 177, 340, 440
Pairing force, 338
  compared to delta function, 338
  degenerate model, 345–348
    BCS solution, 346
    exact solution, 346
  sign of off diagonal matrix element, 361–362
  strength, 348, 368
Pairing gap, 355
Pandya transformation, 171
Parentage Coefficients, 477–491
  One particle parentage coefficients, 38–40, 477–479
    counting factors, 53
    diagrammatic representation, 42, 50, 109
    formulae involving, 487–488
    generalization to many levels, 125
    including isospin, 109, 484
      $(d_{3/2})^3$, 111
      $(s_{3/2})^6$, 120
    matrix elements
      energy, 41, 54–62, 124–125
      one particle, 281–283
    M-independence, 478
    particle-hole relationship, 483
    phase factors, 54
    recursion relationship, 443
    relationship to reduced matrix element, 40, 479
      spectroscopic factors, 150
    tables of, 489–490
  Two Particle parentage coefficients, 41, 480–482
    counting factors, 53
    diagrammatic representation, 42, 50, 109
    energy matrix elements, 42, 54–62
    for $(\frac{1}{2})^3$ and $(\frac{1}{2})^4$, 251
    formulae involving, 487–488
    generalization to many levels, 248
    including isospin, 109, 485–486
    particle-hole relationship, 484
    recursion relationship, 447
    relationship to reduced matrix element, 41, 446, 447, 481
      single particle, 3 particles, 481, 485
        general, 482, 485
      spectroscopic factors, 248
    tables of, 491

Parity, 26
Particle-core coupling, 122–131
Particle-Hole Conjugates, 176
  Conjugation operator, 176–184
    centre-of-shell phase factor, 184
    explicit form, 182
    properties, 178
  energies, 147, 231
  transformation
    energies, 171, 184–185, 193–196
    including isospin, 197–199
    one particle operators, 188–189
Pauli exclusion principle (*see* exclusion principle)
  isospin matrices, 92
  quenching, 104
  spin matrices, 3
Perturbation Theory, 30–31
  for beta decay lifetimes, 373
  for gamma decay lifetimes, 272
  for one nucleon transfer, 148–154
    two nucleon transfer, 245–257
Phase convention
  for harmonic oscillator wavefunction, 7
  for j-j coupling, 5
Pickup, 149
Picosecond, 248
Polarization charge, 317
  neutron-proton difference, 318–319
  sign of, 320
  state dependence, 319
  theory vs experiment, 319
Projected Wave functions (*see* Nilsson Model)
Projection operators, 96–98, 208
  neutron, 93
  proton, 93
Pseudonium, 2, 154

Quadrupole Deformation, 391
Quadrupole Moment (*see* Electric Quadrupole Moment)
Quadrupole–quadrupole interaction, 132, 203, 386, 389
Quasi Spin, 76, 179, 440
  commutation relationship, 440
  number dependence of matrix elements, 442–458
  relationship to seniority, 441–442
  tensors of rank one, 445
Quasi Spin tensors of rank (cont'd)
    one-half, 442
    zero, 445
Quasi Particles, 351–352
  anti-commutation relationship, 351–352
  beta decay, 373–379
    orbit occupation probability, 378
  electromagnetic decay, 361–364
  Hamiltonian, 354–356
  relationship to seniority excitations, 352
  shell model calculations, 355–360
    elimination of unphysical 0^+, 358
    limitations, 381–386
    projected N-particle wavefunctions, 360
  single nucleon transfer, 364–370
    orbit occupation probability, 370
  weak coupling, 385–386
Quenching of beta decay strength, 379–380, 419
    $M4$ transition strength, 362–363
    Magnetic Dipole Moment, 300

Racah coefficients, 25, 459–468
  diagrammatic representation, 25, 51, 108
  M-independence, 460
  relationship to 6-j coefficient, 468
  summation, three to two, 463–466
    two to one, 462–463
  tables of, 475, 476
  useful formulae, 472–473
Radiative width, 277
Recoupling (*see* Angular momentum)
Reduced matrix elements, 29, 430–436
  diagrammatic representation, 30, 50
  encountered in the shell model, 432–439
  including isospin, 139
  M-independence, 431
Reduced Transition probability
  Beta decay, 373
  Electromagnetic decay, 271
Relative Centre-of-Mass transformation coefficients (*see* Moshinsky coefficients)
Residual Two Body Interaction (*see* Effective Interaction and Two Body Potentials)
  experimental evidence for, 8
Rotation Matrix, 397, 500–508
  angular momentum eigenfunctions, 506–508

Rotation Matrix (cont'd)
coupling rules, 503–504
explicit expression for, 505
integration of, 505–506
relationship to spherical harmonics, 503
Rotation Operator, 502
effect on angular momentum eigenfunctions, 502, 503
Rotational Band, 127
Rotational Model, 409–411
eigenfunctions, 409
energy matrix elements, 410

Schiffer–True Interaction, 219, 223
Schmidt value, 298
Second quantization, 17
Seniority, 21, 71
arrow diagrams, 70–71
beta decay between $v = 1$ states, 414
centre-of-shell phase factor, 184
eigenfunctions, 77–79, 488
energy, 80
many levels, 83
mixing
identical nucleon configuration j^n, 73–76
centre-of-shell, 188
delta function potential, 452–455
many valence orbits, 83–86, 381, 384
neutron-proton, 105–107
relationship to quasi particles, 352
quasi spin, 441–442
selection rules, 72, 448–449
centre-of-shell, 191
Separable Potential, 328
Serber Interaction, 235–236
Single Hole energies, 193
Single Nucleon Transfer (*see* Direct Reactions)
Single Particle (*see also* annihilation, creation and one-particle operators)
eigenfunctions, 5, 422–423
energies, 6
experimental spacings, 6, 26, 348, 368, 394–395
from single nucleon transfer, 157
level sequence, spherical, 4
deformed, 393
potential, 4
harmonic oscillator, 6
Single Particle potential (cont'd)
Woods–Saxon, 5
Six-j symbol, 468
Slater decomposition, 203, 278, 387–389
determinant, 10, 38
Spherical Bessell functions, 203
long wavelength limit, 267
Spherical Harmonics, 5, 422
value when $\theta = 0°$, 430
Spin eigenfunctions, 5, 96, 206
Spin Orbit Interaction,
additional electromagnetic interaction due to, 266, 270
single particle, 4
Thomas form, 6, 271
two-particle, 211–212
Spurious Centre-of-Mass Motion, 223–244
collectivity of yrast $T = 0$, 1^- state, 333
excitation energy in ^{4}He, 238
conditions for non-existence, 225–226, 228
construction of $1\hbar\omega$ states, 227
enumeration for $A = 14$, 228
$2\hbar\omega$ states, 241
elimination of, 228–231
estimate of amount, 240, 244
relationship to electric dipole operator, 288
Step down/up operators,
angular momentum, 8, 420–421
harmonic oscillator, 225
matrix elements, 239
isospin, 93
Stretched configuration, 147
Stripping, 149
Sum Rules, 162–164
Surface Delta Interaction, 32, 439
particle-hole transformation, 200
Symmetric wavefunctions, 91

Talmi binding energy formula, 79–80
integrals, 216
Tamm-Dancoff approximation, 357, 381, 384
Taylors series, 178–180
Three-j coefficient, 425
Triple Bar Matrix Element, 139
Two-Body Potentials
Bartlett, 97
Delta Function, 32, 201, 436
Gaussian, 69, 216–222

Two-Body Potentials (cont'd)
Slater decomposition, 278, 388–389
Hamada–Johnston, 218, 222–223
Heisenberg, 96
Majorana, 97
Matrix Elements
Central using oscillator wavefunctions, 208–209
Slater decomposition, 204
Delta Function, 438
Modified Surface, 202
Particle-hole transformation, 199, 322
Surface, 32, 438–439
sign of off-diagonal, 361–362
spin-orbit potential, 212
symmetry properties, 27
Tensor force, 211
Multipole decomposition, 131–132
Pairing, 338
Quadrupole–quadrupole, 132, 203, 386, 389
Rosenfeld, 396
Schiffer–True, 219, 223
Separable, 328
Serber, 235–236
Singlet even, 98
odd, 98
Triplet even, 98
odd, 98
Wigner, 97
Yukawa, 69, 235
Two Nucleon Transfer (*see* Direct Reactions)
Two-Particle Parentage Coefficients (*see* Parentage Coefficients)

Unphysical 0^+, 358

Vacuum, 17
Variational calculation, 341–344
Vectors, Spherical components, 432–433

Wavefunctions,
BCS, 339
quasi-particle, 355–360
for ^{3}He, *t*, ^{4}He, 249–250
phases, 5, 7, 287, 422–423
projected from deformed orbitals (*see* Nilsson Model)
signature, 301
space-spin antisymmetric, 10, 89
symmetric, 91
Weak Coupling, 122–136 (*see also* Electromagnetic Decays and Magnetic Dipole Moments)
including isospin, 136–148
using quasi-particles, 385–386
Weisskopf estimate, 273, 274–276
Wigner–Eckart Theorem, 29, 431
including isospin, 139
Woods–Saxon Potential, 5
diffuseness, 6
size, 6

[×] convention for angular momentum coupling, 15

Yrast levels definition, 3
transitions, 328–336

Zero Coupled Pair, 70, 338, 440
Range Force, 436 (*see* Two Body Potentials, Delta Function)

NUCLEI

Discussions of properties of Individual Nuclei

${}^{4}_{2}He_2$, spectrum, 237
${}^{7}_{3}Li_4$, spectrum, 404
${}^{14}_{7}N_7$, gamma decay, 294, 308, 323
${}^{17}_{8}O_9$, single particle states, 5, 6
${}^{18}_{8}O_{10}$, spectrum, 34
one and two nucleon transfer, 258–261
gamma decay, 286
${}^{19}_{8}O_{11}$, spectrum, 67
${}^{19}_{9}F_{10}$, negative parity states, 126
gamma decay, 290
${}^{19}_{10}Ne_9$, gamma decay, 290
${}^{20}_{9}F_{11}$, ground state, 265
${}^{20}_{10}Ne_{10}$, spectrum, 126, 265
${}^{25}_{12}Mg_{13}$, gamma decay, 323
${}^{25}_{13}Al_{12}$, gamma decay, 323
${}^{28}_{13}Al_{15}$, excited state, 265
${}^{28}_{14}Si_{14}$, excited state, 265
${}^{32}_{15}P_{17}$, excited state, 265
${}^{32}_{16}S_{16}$, excited state, 265
${}^{35}_{17}Cl_{18}$, spectrum, 112
$T = 3/2$ analog state, 111
gamma decay, 276
${}^{36}_{17}Cl_{19}$, excited states, 265
${}^{36}_{18}Ar_{18}$, excited states, 265
gamma decay, 323
${}^{37}_{17}Cl_{20}$, static moments, 190, 306, 322
${}^{38}_{19}K_{19}$, gamma decay, 323
${}^{38}_{17}Cl_{21}$, spectrum, 13, 172, 187
gamma decay, 305
${}^{39}_{18}Ar_{21}$, spectrum, 117
$T = 5/2$ analog state, 145
${}^{39}_{19}K_{20}$, static moments, 190, 300, 306, 322
gamma decay, 316
${}^{39}_{20}Ca_{19}$, pickup to yrast $7/2^-$, 161
gamma decay, 316
Calcium isotopes, ground state energies, 81
⇔ Scandium beta decays, 414
(t, p) reaction on, 262–264
${}^{40}_{19}K_{21}$, gamma decay, 305
spectrum, 172
${}^{40}_{20}Ca_{20}$, negative parity states, 146, 201
single nucleon pickup to, 166
${}^{41}_{20}Ca_{21}$, (d, p) to, 158
magnetic moment, 301, 305
spin-orbit splitting, 271
${}^{42}_{20}Ca_{22}$, gamma decay, 317
single nucleon transfer to, 152
spectrum, 88
${}^{42}_{21}Sc_{21}$, spectrum, 88, 173
${}^{42}_{22}Ti_{20}$, gamma decay, 318
spectrum, 88
${}^{46}_{19}K_{27}$, spectrum, 187
${}^{46}_{20}Ca_{26}$, excited 0^+, 143
${}^{47}_{21}Sc_{26}$, yrast $3/2^+$, 143
${}^{48}_{21}Sc_{27}$, spectrum, 173
${}^{48}_{22}Ti_{26}$, spectrum and gamma decay, 307, 327
${}^{48}_{23}V_{25}$, magnetic moments, 303
${}^{49}_{21}Sc_{28}$, analog and anti-analogue states, 104
${}^{49}_{22}Ti_{27}$, yrast $7/2^-$ state, 107
$N = 28$ isotones ground state energies, 81
${}^{50}_{22}Ti_{28}$, gamma decay, 190
spectrum, 15, 43, 186
${}^{51}_{23}V_{28}$, gamma decay, 303
magnetic moment, 300
spectrum, 45
${}^{52}_{24}Cr_{28}$, gamma decay, 191
single nucleon transfer, 156
spectrum, 47
${}^{53}_{25}Mn_{28}$, magnetic moment, 300
${}^{54}_{26}Fe_{28}$, gamma decay, 190
magnetic moments, 300
spectrum, 186
${}^{55}_{27}Co_{28}$, magnetic moment, 300
${}^{57}_{28}Ni_{29}$, gamma decay, 316
Nickel isotopes ground state energies, 382
${}^{59}_{27}Co_{32}$, yrast $3/2^-$, 327

$^{61}_{28}Ni_{33}$, low-lying states, 85
$^{62}_{28}Ni_{34}$, low-lying states, 85
$^{63}_{29}Cu_{34}$, spectrum, 134
$^{67}_{30}Zr_{37}$, gamma decay, 316
$^{87}_{37}Rb_{50}$, gamma decay, 316
$^{87}_{38}Sr_{49}$, quadrupole moment, 326
 yrast levels, 193
$^{90}_{41}Nb_{49}$, gamma decay, 325–326
$^{92}_{41}Nb_{51}$, spectrum, 175
$^{96}_{41}Nb_{55}$, spectrum, 175
Tin ⇔ Antimony beta decay, 379
 ⇔ Indium beta decay, 377
Tin (cont'd)
 one quasi-particle states, 371
$^{121}_{50}Sn_{71}$, $T = 23/2$ analog states 167
$^{121}_{51}Sb_{70}$, (d, ^{3}He) to low-lying states, 167
$^{136}_{54}Xe_{82}$, gamma decay, 325
$^{175}_{71}Lu_{104}$, quadrupole moment, 320
$^{203}_{81}Tl_{122}$, gamma decay, 316
$^{207}_{82}Pb_{125}$, single-hole states, 348
Lead isotopes ***M*4** lifetimes, 363
$^{208}_{82}Pb_{126}$, 3^- state, 130
$^{209}_{82}Pb_{127}$, gamma decay, 316
$^{209}_{83}Bi_{126}$, positive parity septuplet, 130